# Plant Innate Immunity 2.0

# Plant Innate Immunity 2.0

Special Issue Editor

**Marcello Iriti**

MDPI • Basel • Beijing • Wuhan • Barcelona • Belgrade

**MDPI**

*Special Issue Editor*
Marcello Iriti
Milan State University
Italy

*Editorial Office*
MDPI
St. Alban-Anlage 66
4052 Basel, Switzerland

This is a reprint of articles from the Special Issue published online in the open access journal *International Journal of Molecular Sciences* (ISSN 1422-0067) from 2017 to 2019 (available at: https://www.mdpi.com/journal/ijms/special_issues/plant_innate_immunity_2)

For citation purposes, cite each article independently as indicated on the article page online and as indicated below:

LastName, A.A.; LastName, B.B.; LastName, C.C. Article Title. *Journal Name* **Year**, *Article Number*, Page Range.

ISBN 978-3-03897-580-9 (Pbk)
ISBN 978-3-03897-581-6 (PDF)

# Contents

# About the Special Issue Editor

**Marcello Iriti**, Professor of Plant Biology and Pathology at the Department of Agricultural and Environmental Sciences, Faculty of Agricultural and Food Sciences, Milan State University. He has been studying bioactive phytochemicals that are relevant to human nutrition and health, including melatonin, polyphenols, carotenoids, sterols, and essential oils, focusing on their functional role in planta, as well as on their in vitro/in vivo and in-human biological activities. Furthermore, he has been investigating the effects of elicitors on plant secondary metabolism, as an approach to obtain phytochemical-enriched plant foods and medicinal herbs. He is the author of more than 150 publications (H-index: 33) and a member of the Asian Council of Science Editors and Society of African Journal Editors. He is also a founding member of the Italian Society of Environmental Medicine and a member of the Working Group 'Pharmacognosy and Phytotherapy' of the Italian Pharmacological Society. Main Patent: 'Compositions Comprising Rutin Useful for the Treatment of Tumors Resistant to Chemotherapy' (WO2015036875A1; US20160213698; US9757405B2; EP3043821).

# Preface to "Plant Innate Immunity 2.0"

Even if immunology has been purely regarded as a medical science, immunity represents a trait common to many living organisms. Classically, the mammalian immune system was viewed as consisting of innate and adaptive mechanisms that protect the host from pathogens; in particular, innate mechanisms function independently of previous exposure of the host to the infectious agent. Animal and plant innate immunities share some similarities, and plants, as well as animals, are capable of recognizing and distinguishing between self and non-self. Recognition by the innate immune system is based on germline-encoded receptors (also called pattern recognition receptors, PRRs) expressed on host cells, which sense conserved structural components and metabolism products of fungi, bacteria, and viruses (known as pathogen-associated molecular patterns, PAMPs), including lipids, polysaccharides, proteins, and nucleic acids. During their evolutionary history, plants have developed various defense strategies in order to face pathogens. Although lacking immunoglobulin molecules, circulating cells, and phagocytic processes, the effectors of mammalian immunity, plants possess a rather complex and efficient innate immune system. Therefore, disease is a rare outcome in the spectrum of plant-microbe interactions because plants have (co)evolved a complex set of defense mechanisms to hinder pathogen challenging and, in most cases, prevent infection. The battery of defense reactions includes physical and chemical barriers, both preformed (or constitutive or passive) and inducible (or active), depending on whether they are pre-existing features of the plant or are switched on after challenging. When a pathogen is able to overcome these defenses, disease ceases to be the exception. Three main explanations support this rule: (i) the plant is not a substrate for microbial growth and does not support the lifestyle of the invading pathogen; (ii) constitutive barriers prevent colonization of the plant by pathogens; and (iii) the plant recognizes the pathogen, by its innate immune system, and then activates inducible defenses.

**Marcello Iriti**
*Special Issue Editor*

International Journal of
*Molecular Sciences*

MDPI

*Editorial*

# Evidence-Based Phytoiatry, a New Approach in Crop Protection

**Marcello Iriti [1,*] and Elena Maria Varoni [2]**

[1] Dipartimento di Scienze Agrarie e Ambientali, Università degli Studi di Milano, via G. Celoria 2, 20133 Milan, Italy
[2] Dipartimento di Scienze Biomediche, Chirurgiche e Odontoiatriche, Università degli Studi di Milano, via Beldiletto 1/3, 20142 Milan, Italy; elena.varoni@unimi.it
* Correspondence: marcello.iriti@unimi.it; Tel.: +39-02-50316766

Received: 29 December 2018; Accepted: 3 January 2019; Published: 5 January 2019

**Abstract:** In the past decades, the scientific quality of biomedical studies has been hierarchically depicted in the well-known pyramid of evidence-based medicine (EBM), with higher and higher levels of evidence moving from the base to the top. Such an approach is missing in the modern crop protection and, therefore, we introduce, for the first time, this novel concept of evidence-based phytoiatry in this field. This editorial is not a guideline on plant protection products (PPP) registration, but rather a scientific and technical support for researchers involved in the general area of plant pathology, providing them with evidence-based information useful to design critically new studies.

**Keywords:** plant protection products; agrochemicals; sustainable crop protection; food security

---

The term "evidence-based medicine" (EBM) was first introduced in 1992 in the *Journal of American Medical Association* (*JAMA*), referring to the use and the critical appraisal of the best scientific evidence to support clinical decision-making. This idea purported that all medical activities intended for diagnosis, prognosis and therapy should be based, not primarily on clinical experiences, but on solid scientific evidences coming from clinical research. Since then, EBM approach has continued to grow and spread, taking a pivotal place beside the competence of the clinician and the care preferences of the single patient.

This new way of thinking and acting now accounts the most updated evidence resulting from methodologically rigorous studies, mainly randomized controlled clinical trials (RCTs) and systematic reviews, which correspond to the highest level of scientific quality and represent the top of the EBM pyramid (Figure 1). The base consists of pre-clinical research (in vitro and in vivo experiments), expert opinions and case reports/series, while the middle part is composed of observational studies including case-control and cohort studies, and RCTs and systematic reviews constitute the top. RCTs are intervention studies considered the gold standard of clinical research to evaluate the efficacy/effectiveness of a therapy, because of their ability to minimize the biases over other types of clinical trials [1]. The cornerstone of RCT is randomization, which allows distributing each prognostic factor homogeneously between the test group and the control group. The tip of the pyramid, however, is occupied by systematic reviews, belonging to secondary literature. They are tools aimed at summarizing data coming from primary literature, such as RCTs, cohort and/or case-control studies. They aim at answering a well-focused question about etiology, diagnosis, therapy and prognosis of a certain disease, paying particular attention to the search methodology of studies to be included and providing a critical evaluation of the resulting quality of evidence. Systematic reviews are, whenever possible, associated with meta-analysis, which is a statistical analysis of all data coming from the different studies included; the latter provides a quantitative analysis of findings in support or against a certain treatment or risk factor in the contest of a specific disease. Underlying the importance of systematic review, the Cochrane Collaboration was founded in 1993, an international network with the specific aim of preparing and maintaining continuously updated and publically available systematic reviews on the impact of health interventions, which follow a unique standard of methodology.

**Figure 1.** Pyramid of scientific evidence—Evidence-based medicine (EBM). In biomedical sciences, a pyramid represents the quality of scientific evidence; the base consists of pre-clinical research, with in vitro and in vivo experiments, which provides insights towards potential efficacy of a certain therapy. Clinical research represents a higher level of evidence, referring to in human studies; in this context, randomized controlled trials are methodologically the best current clinical studies. At the apex, updated systematic reviews and meta-analyses are considered the best available knowledge, suitable for clinicians to face decision-making (adapted from Varoni et al., 2015 [1]).

In the EBM, the clinician is trained to identify the best scientific evidence currently available on a medical issue, after constructing an appropriate research question and using the reference database for medical literature (PubMed) for reviewing studies. To construct the clinical question, EBM proposes the PICO strategy, where P stands for Patients (with a particular condition or disease), I for the Intervention of interest (therapeutic, preventive, prognostic, diagnostic), C for Control or Comparison (the former defined as no intervention such as placebo, the latter as the standard intervention to be compared) and O for Outcome (the expected results coming from the intervention).

In the phytoiatric/phytopathological area, an evidence-based approach is still missing. Similar to the medical sciences, the efficacy of plant protection products can be assessed with diverse experimental designs each with different levels of scientific evidence. In general, active substances are initially assayed with in vitro tests evaluating their biostatic/biocidal activity on cultivable plant pathogens. To the next level, in planta experiments in controlled environments, i.e., phytotron cabinets, growth chambers, greenhouses and screenhouses, can be carried out to test the efficacy of both active substances and agrochemicals on model pathosystems as well as obligate pathogens. In these experimental conditions, phytotoxicity and adverse effects on non-target plant species can also be evaluated. Overall, in vitro and in planta experiments (pre-open field studies) can be compared to in vitro/in vivo preclinical research of biomedical area, at the base of the evidence-based phytoiatry pyramid (Figure 2). Following this parallelism, multi-year open field trials, with positive (untreated) and negative (reference standard) controls, performed in a completely randomized design are at the top of the pyramid, providing the highest level of evidence on efficacy of plant protection products including fungicides, elicitors, plant activators, insecticides and herbicides.

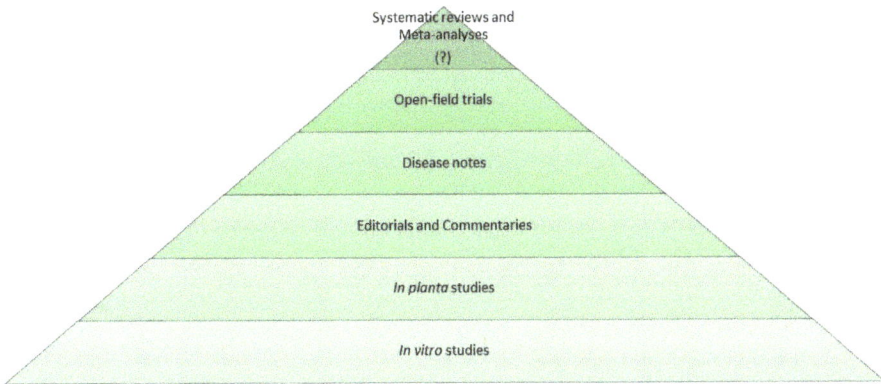

**Figure 2.** Pyramid of scientific evidence—Evidence-based phytoiatry (EBP). Suggested pyramid of the quality of scientific evidence in phytoiatry. In vitro and in planta experiments represent the level of the pre-open field research. Systematic reviews and meta-analyses need to be introduced in the phytoiatric/phytopathological area ('?' on the pyramid top).

Of course, to design and perform an open-field trial is complex, and a number of general principles should be followed. First, the concept of Good Plant Protection Practice (GPP) was defined by the European and Mediterranean Plant Protection Organization (EPPO), a specific set of recommendations including guidelines on decision-making for choice of active substances and formulations, dosage (and appropriate volume), number of applications, timing or frequency of applications, equipment and method of application [2]. In addition, EPPO provides a number of standards covering two application fields: phytosanitary measures and plant protection products. Main topics include analysis of efficacy evaluation trials, efficacy evaluation of plant protection products (PPPs), number of efficacy trials, minimum effective dose, principles of acceptable efficacy and dose expression for PPPs, to name a few [3]. In addition, we have to take into account that legislations on PPPs are demanding, lengthy and costly, especially in the EU where issues related to their assessment and evaluation trials are controlled by the regulation EC 1107/2009 placed on the market of PPPs. During the approval process, the steps concerning the registration of new PPPs are declined because they require multi-year field trials to prove the efficacy of a given PPP against a target pest (sensu lato) under different pedo-climatic situations ruling out any negative effect on non-target organisms.

Not last, systematic reviews (and meta-analyses) are not used in phytoiatry/plant pathology research as a powerful tool to reach the highest level of evidence in a specific phytosanitary issue. Therefore, we believe that the plant protection sector can neither be apart from considering that different levels of scientific evidence actually exist nor disregard a modern evidence-based approach.

## References

1. Varoni, E.M.; Lodi, G.; Iriti, M. Efficacy behind activity–phytotherapeutics are not different from pharmaceuticals. *Pharm. Biol.* **2015**, *53*, 404–406. [CrossRef] [PubMed]
2. European and Mediterranean Plant Protection Organization (EPPO). Principles of good plant protection practice. Principes de bonne pratique phytosanitaire. *Bull. OEPP/EPPO Bull.* **2003**, *33*, 91–97. [CrossRef]
3. EPPO Standards. Available online: https://www.eppo.int/RESOURCES/eppo_standards (accessed on 4 January 2019).

International Journal of
*Molecular Sciences*

MDPI

*Review*

# Signals of Systemic Immunity in Plants: Progress and Open Questions

**Attila L. Ádám** [1,*], **Zoltán Á. Nagy** [2], **György Kátay** [1], **Emese Mergenthaler** [1] and **Orsolya Viczián** [1]

[1]   Plant Protection Institute, Centre for Agricultural Research, Hungarian Academy of Sciences,
      15 Herman Ottó út, H-1022 Budapest, Hungary; katay.gyorgy@agrar.mta.hu (G.K.);
      mergenthaler.emese@agrar.mta.hu (E.M.); viczian.orsolya@agrar.mta.hu (O.V.)
[2]   Phytophthora Research Centre, Department of Forest Protection and Wildlife Management,
      Faculty of Forestry and Wood Technology, Mendel University in Brno, Zemědělská 3,
      613 00 Brno, Czech Republic; zoltan.nagy@mendelu.cz
*    Correspondence: adam.attila@agrar.mta.hu; Tel.: +36-1-487-7575

Received: 15 February 2018; Accepted: 31 March 2018; Published: 10 April 2018

**Abstract:** Systemic acquired resistance (SAR) is a defence mechanism that induces protection against a wide range of pathogens in distant, pathogen-free parts of plants after a primary inoculation. Multiple mobile compounds were identified as putative SAR signals or important factors for influencing movement of SAR signalling elements in *Arabidopsis* and tobacco. These include compounds with very different chemical structures like lipid transfer protein DIR1 (DEFECTIVE IN INDUCED RESISTANCE1), methyl salicylate (MeSA), dehydroabietinal (DA), azelaic acid (AzA), glycerol-3-phosphate dependent factor (G3P) and the lysine catabolite pipecolic acid (Pip). Genetic studies with different SAR-deficient mutants and silenced lines support the idea that some of these compounds (MeSA, DIR1 and G3P) are activated only when SAR is induced in darkness. In addition, although AzA doubled in phloem exudate of *tobacco mosaic virus* (TMV) infected tobacco leaves, external AzA treatment could not induce resistance neither to viral nor bacterial pathogens, independent of light conditions. Besides light intensity and timing of light exposition after primary inoculation, spectral distribution of light could also influence the SAR induction capacity. Recent data indicated that TMV and CMV (*cucumber mosaic virus*) infection in tobacco, like bacteria in *Arabidopsis,* caused massive accumulation of Pip. Treatment of tobacco leaves with Pip in the light, caused a drastic and significant local and systemic decrease in lesion size of TMV infection. Moreover, two very recent papers, added in proof, demonstrated the role of FMO1 (FLAVIN-DEPENDENT-MONOOXYGENASE1) in conversion of Pip to *N*-hydroxypipecolic acid (NHP). NHP systemically accumulates after microbial attack and acts as a potent inducer of plant immunity to bacterial and oomycete pathogens in *Arabidopsis*. These results argue for the pivotal role of Pip and NHP as an important signal compound of SAR response in different plants against different pathogens.

**Keywords:** *Arabidopsis*; azelaic acid; glycerol-3-phosphate; light dependent signalling; methyl salicylate; *N*-hydroxypipecolic acid; pipecolic acid; salicylic acid; SAR signalling; spectral distribution of light; tobacco

## 1. Introduction

   Systemic acquired resistance (SAR) is an inducible defence mechanism that provides protection in distant, pathogen-free parts of plants against a broad range of pathogens. In practice, SAR has been recognised as a strategy to control plant pathogens because of its stability [1], long-lasting effectiveness [2], and transgenerational effect [3]. In the latter case, defence mechanisms are induced faster in the progeny after pathogen infection. The plant "immune" memory involved in the inheritance of SAR has probably an epigenetic character that effects the patterns of DNA methylation at the promoters of defence-related genes [3]. However, it is important to note that the mechanism via which DNA methylation regulates SAR within a single generation may differ from transgenerational SAR responses [4]. Tissues exhibiting SAR in the distant, pathogen-free parts of plants, display a "prepared" state (defence priming) associated with faster and stronger defence mechanisms [5–7]. This memory is based on posttranslational modification of histone and results in changes of the structure of chromatin [8]. This regulation is an integrated part of biotic and abiotic stress responses and is called the "histone code" [9].

The SAR response can be divided into four steps in which signal generation in primary infected leaves and signal movement (Figure 1) into distant organs (leaves) are both important steps [10]. To date, several mobile compounds have been identified as putative SAR signals or important factors for movement of long-distance SAR signals in *Arabidopsis* and tobacco. These include methyl salicylate (MeSA) [11–13] the lipid transfer protein DIR1 (DEFECTIVE IN INDUCED RESISTANCE1) [14,15], dehydroabietinal (DA) [16], azelaic acid (AzA) [17,18], glycerol-3-phosphate dependent factor (G3P) [19,20] and the lysine catabolite amino acid, pipecolic acid (Pip) [21,22]. Chemical structures of signal compounds of SAR are shown in Figure 2. As a last step, after signal perception, the manifestation of SAR (defence priming) in distant leaves (Figure 1) is associated with a massive transcriptional and metabolic reprogramming [7,20,23]. In *Arabidopsis*, this requires several factors in distant, pathogen free leaves: pipecolic acid (Pip) and salicylic acid (SA) accumulation, expression of *ALD1* (*AGD2-LIKE DEFENSE RESPONSE PROTEIN1*), *FMO1* (*FLAVIN-DEPENDENT-MONOOXYGENASE1*), *PAD4* (*PHYTOALEXIN DEFICIENT4*), *ICS1* (*ISOCHORISMATE SYNTHASE1*), *FLD* (*FLOWERING LOCUS D1*), *EDS1* (*ENHANCED DISEASE SUSCEPTIBILITY1*), and *SnRK2.8* (*SNF1-RELATED PROTEIN KINASE2.8*) genes, most of them are components of the SA-amplification loop. SA-dependent activation of a transcriptional regulator, *NPR1* (*NON-EXPRESSOR OF PR GENES1*) is also necessary for defence priming [16,21,23–27]. However, it is important to note that defence priming and signal amplification are dependent on systemic Pip accumulation and Pip orchestrates SA-dependent and SA-independent priming of plant responses in a *FMO1*-dependent manner [21,23].

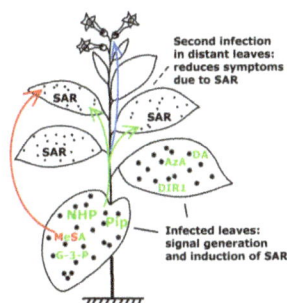

**Figure 1.** Development of systemic acquired resistance (SAR). Putative SAR signal molecules: methyl salicylate (MeSA), lipid transfer protein DIR1 (DEFECTIVE IN INDUCED RESISTANCE1), dehydroabietinal (DA), glycerol-3-phosphate (G3P) or G3P-dependent factor, azelaic acid (AzA), pipecolic acid (Pip) and N-hydroxypipecolic acid (NHP) move from the infected organ (leaves) to pathogen-free distant parts of the plant where they induce SAR against biotrophic and hemibiotrophic pathogens (note the limited symptom expression in the distant, systemic leaves indicated by small spots). Pip and NHP are highlighted. Green and red arrows indicate the movement of signal molecules via phloem transport or the air (putative volatile compounds for inplant or interplant airborne signals), respectively. Blue arrow indicates transgenerational SAR where the epigenetic information is inherited and present in the next generation (re-drawn and modified from [28]). For chemical structures of signal compounds see Figure 2.

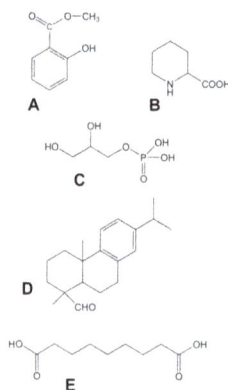

**Figure 2.** Chemical structures of signal compounds participating in SAR induction. (**A**) Phenolic compound, methyl-salicylate; (**B**) Non-protein amino acid, pipecolic acid; (**C**) Glycerol-3-phosphate; (**D**) Geranylgeranyl diphosphate derived dehydroabietinal; (**E**) Dicarboxylic acid, azelaic acid.

Former studies also suggested that the SAR-inducing activity in phloem sap of *Arabidopsis* and cucumber induced by different pathogens was effective in other plants [16,20,29], thus indicating that the mobile SAR signal(s) is neither plant- nor pathogen-specific. However, considering recent data about multiple signal mechanisms with diverse chemical substances described mainly in *Arabidopsis*-bacteria plant-pathogen systems it is reasonable to test whether this multiple signalling is due to environmental or some other circumstantial factors. It is especially true for tobacco where less experimental data is available and a virus-inducible SAR response could be also studied by using a computer-assisted method for symptoms evaluation and statistical analysis [28,30–32].

Plants possess a sophisticated and structured immune response against invading pathogens. The first layer in plant defence is pattern triggered immunity (PTI) which is induced by specific microbe-associated molecular patterns (MAMPs) which are recognized by pattern recognition receptors (PRRs) at the plasma membrane. In addition, endogenous molecules, like extracellular ATP, so-called danger-associated molecular pattern (DAMP) can also induce plant defence response [33,34]. The specific role of PTI in local antiviral defence has been recently seen in *Arabidopsis*. PRR coreceptor kinase mutant, *bak1/serk3* (brassinosteroid insensitive1 (bri1)-associated receptor kinase1 or somatic emryogenesis receptor-like kinase3, *serk3*) and other *SERK* gene family mutants [35] have increased susceptibility to different RNA viruses (note that mutant genotypes are indicated by small letters in italics of the corresponding wild type gene all over the paper). SERKs are a family of closely related leucine-rich repeat receptor kinases involved in different plant signalling routes where they act in a partially redundant manner as coreceptors for the respective PRRs [36]. *BAK1* is also required for PTI responses to bacterial MAMPs such as flagellin-derived flg-22 and translation elongation factor-derived peptide, elf18. In addition, *BAK1* and *SERK4* play a role in peptide elicitor signalling and cell death control [37,38]. Connecting PTI signalling with SAR, Mishina and Zeier [39] found that SAR could be induced by bacterial MAMPs (flagellin and LPS) against bacteria in *Arabidopsis* and the extent of tissue necrosis at the site of inoculation by virulent or avirulent bacteria is dispensable for the SAR response. However, to date, the chemical nature of elicitors of viral origin or virus inducible elicitors of PTI are unknown.

Therefore, we checked the specificity of TMV-induced SAR response in tobacco of which we have no or limited information (Figure 3) [40]. TMV infection resulted in SAR response not only against TMV [2] (Figure 3A) but also against bacterial (*P. syringae* pv. *tabaci*) (Figure 3C) and other viral (necrotic strain of cucumber mosaic virus, CMV Ns) [41] (Figure 3A) pathogens. SAR response induced by TMV was as much as effective against TMV as CMV (58.0% and 54.7% inhibition of mean lesion size as compared to control plants, respectively) (Figure 3A,B). Note on Figure 3B that the difference between the effects of SAR against TMV and CMV infection did not show significant difference. This feature indirectly excludes the activation of a virus species-specific mechanism during SAR induction. Furthermore, effectiveness of TMV-induced SAR against bacteria (Figure 3C) was comparable to bacterial induced SAR response in tobacco and *Arabidopsis* [21,42].

Viral replication is generally associated with the presence of double stranded RNAs (dsRNAs) [43,44]. Synthetic analogues and viral dsRNA itself act as MAMPs and can induce SERK-1 dependent specific antiviral resistance signalling in *Arabidopsis* but this mechanism is independent of RNA silencing [45]. The significance of this fact in SAR induction is unknown. Furthermore, PTI responses were not impaired in plants lacking the two main antiviral dicer-like (DCL) proteins, DCL2 and DCL3 [46] and the short interfering RNA (siRNA)-producing DCL3 [47]. These results also suggest no role of silencing mechanisms during PTI. Moreover, bacterial RNA can also induce PTI-mediated plant defence [48].

A further example of the non-specific nature of SAR response is the accumulation of PR proteins and especially PR1. PR1 is upregulated in SA-dependent manner after microbial (fungal, bacterial or viral) infections and used as marker gene for SA-mediated disease resistance and SAR. Several groups have demonstrated that overexpression of *PR1* in transgenic plants result in increased resistance to oomycetes, fungi, bacteria but not to viruses (for review see [49]). Recently Gamir et al. [50] presented a link between anti-oomycete activity and sterol (especially fungal ergosterol) binding capacity of PR1a from tobacco and P14c from tomato via CAP (sterol binding) domain of PR1 proteins. Moreover, the addition of fungicide that blocks sterol biosynthesis in sterol prototrophic fungi, such as *Aspergillus niger* and *Botrytis cinerea*, made these fungi susceptible to P14c. This model suggests that PR1 proteins have antimicrobial function by sequestering sterols from the membranes of microbes and that they are more effective against sterol auxotrophic organisms (for example oomycetes) which are dependent on sterol supply from the environment. Besides this feature, PR1 proteins harbour an 11-amino acid peptide (CAP-derived CAPE1) at the C terminus with conserved consensus motif PxGNxxxxxPY. CAPE1 peptide can induce an antibacterial effect [51]. Interestingly, CAPE1 did not stimulate the expression of WRKY transcription factor 53 (WRKY53) which is highly upregulated by bacterial MAMPs. Further studies are required to test the putative role of this PR1-derived peptide in the antibacterial property of SAR induction. Furthermore, the lack of antiviral effects of PR1 could be replaced by direct antiviral action of SA on virus replication [52].

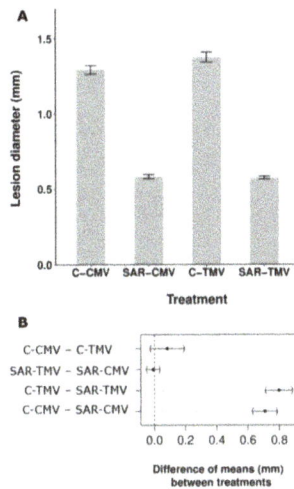

**Figure 3.** *Cont.*

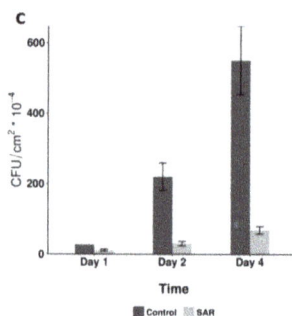

**Figure 3.** Aspecific feature of SAR induction in tobacco (*N. tabacum* cv. Xanthi nc). TMV induced SAR response was effective not only against TMV and another viral (cucumber mosaic virus, CMV necrotic strain) (**A,B**) but bacterial (*Pseudomonas syringae* pv. *tabaci*) (**C**) pathogens as well. SAR was induced by TMV infection as described earlier [31] and 10 days later the 5th and 6th leaves were challenged by CMV (SAR-CMV) and TMV (SAR-TMV) as compared to controls (C-CMV and C-TMV) (half leaves of each case) (**A**) or by bacterial suspension ($10^5$ CFU mL$^{-1}$) (**C**) in triplicates per treatments. (**B**) Multiple comparison of means of selected treatments in (**A**). Dots represent the difference of the estimated means between treatments, brackets flank the 95% confidence intervals. The difference is considered significant if the confidence interval does not contain the 0, represented by a vertical dashed line [31]. (**C**) The SAR induction was also checked by TMV challenge inoculation (showing 69.6% significant inhibition of TMV lesion size). The TMV induced SAR caused time-dependent and significant (generalized linear model) decrease of bacterial multiplication

Taken together, all of these examples argue for the specificity of pathogen signal recognition/perception and aspecificity of the induction of SAR response in plants.

This review will focus on signal transduction mechanisms between pathogen perception and SAR response, especially on the environmental factors that can influence, qualitatively or quantitatively, the signalling processes.

## 2. Multiple Signalling by Chemically Diverse Compounds during SAR Induction

### 2.1. DIR1 (DEFECTIVE IN INDUCED RESISTANCE1)

One of the SAR signal transduction related compounds, DIR1 protein, is involved in the signalling function of other signal molecules (AzA, DA, G3P, MeSA) as a lipid transfer protein (LTP). Therefore, DIR1 will be covered here, before the other chapters. The *dir1* locus was identified in classical mutational analysis as SAR-deficient phenotype [14].

Later on, besides DIR1, DIR1 homolog DIR1-like proteins were also identified in *Arabidopsis* [53,54]. Additional LTPs, like LTP1 and LTP2 and even AZELAIC ACID INDUCED1 (AZI1) and its homologs have a role in SAR signal development [54,55]. Moreover, DIR1 is conserved in other SAR-competent species like cucumber [56] and tobacco (*Nicotiana tabacum*) [57]. The lipid/acyl-CoA binding protein, ACBP6 is also required for SAR induction [58] and present in phloem exudates of SAR induced *Arabidopsis* plants [59].

The small 7 kDa DIR1 protein is present in the petiolar exudate of infected leaves as a high molecular mass oligomeric form or in a complex with other proteins [16]. On the other hand, SAR-inducing capacity of petiolar exudate of infected leaves was proteinase sensitive [16,20]. The overexpression of two *Arabidopsis* proteins, localized in the plasmodesmata, results in the inhibition of the movement of DIR1 protein and SAR induction suggesting a symplastic transport [15]. The effect of external AzA application was DIR1-dependent [17]. Similar results were reported for DA [16] and G3P dependent factor [20]. DIR1 may also interact with MeSA signalling in distant leaves of tobacco [57].

## 2.2. Salicylic Acid (SA) and Methyl Salicylate (MeSA)

Salicylic acid (SA) and its volatile derivative, methyl salicylate (MeSA) were the first two candidates for signalling of SAR (Figures 1 and 2A). In the latter case MeSA was suggested either as a volatile, intra-plant airborne signal or a long-distance inside signal transported via the phloem system.

The first reports in different plants (tobacco and cucumber) strongly suggested that SA could be required either for SAR signalling or development of SAR in the distant leaves [60–62]. These proofs were as follows: (i) SA is accumulated not only in the infected leaves but in distant pathogen-free leaves; (ii) SA is accumulated both in infected leaves and their phloem (petiolar) exudates; (iii) in transgenic plants expressing constitutively a salicylate hydroxylase (*NahG*) gene the SAR response was absent; (iv) externally applied SA was translocated and resulted in SAR-like response [63]. In addition to the effect of *NahG* gene (where there is no SA accumulation), some other biochemical or genetic approaches led to the same conclusion. For example, the inhibition of SA biosynthesis also affected the SAR response. Two *Arabidopsis* mutants, *sard1* (systemic acquired resistance deficient1) and *cbp50g* (calmodulin binding protein 50g) also have SAR-deficient phenotypes. The corresponding proteins are involved in the regulation of SA-biosynthesis via *ICS1* (*ISOCHORISMATE SYNTHASE1*). These mutant proteins were unable to bind to the promoter region of *ICS1* gene (its protein product is a key enzyme of SA biosynthesis localized in the chloroplasts) after infection [64,65]. Plants with *sid1* and *sid2* (salicylic acid induction deficient) mutant loci (the latter is responsible for coding of ICS1 in *Arabidopsis*) were also unable for SA-biosynthesis and induction of SAR [66–68]. Somewhat similar to the effect of *NahG*, *OsBSMT1* (*BENSOIC ACID/SALICYLIC ACID CARBOXYL-METHYLTRANSFERASE1* in rice) gene is also involved in the regulation of free SA accumulation. Transgenic overexpression of this gene resulted in the inhibition of local SA-accumulation (via formation of MeSA from SA) and caused susceptibility to bacterial and fungal pathogens [69]. In tobacco (*N. tabacum*) and cucumber, however, SA accumulation was blocked by biochemical inhibition of PAL (PHENYLAMMONIA LYASE) enzyme and inhibition of *PAL* gene expression [70,71] indicating a putatively different route for SA-biosynthesis. In *N. benthamiana*, however, *ICS1* gene was responsible for SA-accumulation both after biotic and abiotic stress [72].

All of these results, whatever kind of the biochemical pathway for SA biosynthesis is, undoubtedly proved the important role of SA in local resistance and in the induction of SAR, but shad no or little light on the putative signalling function of SA. The first question mark in this context was related to the report on the activation of de novo SA biosynthesis in systemic, pathogen-free tissues [73]. Thus, the only source of systemic SA accumulation could not be the long-distance SA (or MeSA) transport from infected leaves. However, this result does not yet exclude the role of SA in signal transduction of SAR. Later reciprocal grafting studies in tobacco between transgenic *NahG* (deficient in SA-accumulation) or *PAL*-silenced plants (partially deficient in SA-synthesis) and WT plants, indicated a more direct conclusion: (i) the rise in free SA content in primary, pathogen-inoculated leaves is not critical to the induction of SAR in distant leaves; (ii) the long-distance signal is not identical to SA, in spite of the accumulation of SA in petiolar exudate of primary inoculated leaves; (iii) the accumulation/presence of SA in distant, pathogen-free leaves is required for the induction of SAR; (iv) different factors are responsible, at least in part, for local and systemic resistance responses [71,74]. Thus, it is important to note that SA plays differential role in local and systemic resistance responses.

As indicated above, systemic SA accumulation is required for the induction of SAR. In *Arabidopsis*, upregulation of *ICS1* gene expression and de novo SA-biosynthesis in distant leaves is also required for SAR induction [75]. Further experiments with *Arabidopsis sid1* and *sid2* and the respective double mutants, however, suggest a modest SAR induction, indicating an SA-independent route, in addition to a dominant SA-dependent SAR activation pathway [23]. Further support for de novo SA-biosynthesis in the distant, pathogen-free leaves comes from the characterisation of two SAR-deficient mutants, *fld* and *fmo1* [76,77]. Surprisingly, SA content increased in primary pathogen inoculated leaves and the SAR signal was systemically transported in *fld* mutants. On the contrary, distant leaves of *fld* mutants did not accumulate SA. Since FLD affects histone modifications, it is likely that FLD-dependent changes in histone modifications at genes involved in SA accumulation/metabolism are associated with SA-dependent SAR induction mechanisms [76]. Systemic accumulation of SA and SAR induction were also attenuated in the *fmo1* mutant *Arabidopsis* plants and *FMO1* was suggested to be a part of SA amplification loop including Pip [21,77].

Besides free SA, several other compounds are involved in SA metabolism. For example glycosylated SA compounds, like SA-glycosyd (SAG) are also accumulated after infection [75,78]. The methylation of free carboxyl group of SA (2-hydroxy benzoic acid) results in the formation of a volatile compound, methyl-salicylate (MeSA) [79].

The role of MeSA, as a volatile and airborne signal in SAR induction was first suggested by Shulaev et al. [11]. In a closed artificial system infected plants (used as MeSA donor plants) could produce enough amounts of MeSA to induce about 30% decrease in TMV lesion size of acceptor plants.

Later, reciprocal grafting experiments between WT and *SAMT1* (*SALICYLIC ACID METHYLTRANSFERASE1*) silenced tobaccos, however, suggested that MeSA can serve as a signal for SAR induction inside the plants via transport [12]: (i) functional *SAMT1* gene was necessary for SAR induction in primary inoculated leaves; (ii) MeSA accumulated in phloem exudate of infected leaves and thus MeSA could be the long-distance signal for SAR; (iii) competitive, pharmacological inhibitor of MeSA demethylase, 2,2,2,2-tetrafluoro-acetofenon could attenuate SAR development in the distant, pathogen free leaves and finally (iv) further grafting experiments between WT and *SABP2* (*SALICYLIC ACID BINDING PROTEIN2*) silenced plants also suggested the importance of its demethylase function in distant leaves in SA production and SAR induction [12,80]. This scenario was later expanded to other plants, *Arabidopsis* [81,82] and potato [13]. This model suggests that SA is converted to MeSA in primary inoculated leaves and then MeSA is translocated to distant leaves, inside the plants and finally converted to SA and SA accumulation in systemic leaves results in SAR induction [83].

However, there is a controversial step during either the airborne or inside signal function of MeSA [28]. If SA accumulation is inhibited in *NahG* rootstock plants but grafting with WT plants results in SAR induction [74], this conclusion may exclude the possibility that MeSA could be produced from SA in the primary inoculated leaves and subsequently serves as a signal for SAR induction (considering that only one pathway is responsible for MeSA production and this pathway is identical to the production of MeSA from SA). This controversy, however, will be addressed in chapter 3, in connection with the effect of the length of light exposition after primary inoculation on SAR signalling [84].

In fact, Attaran et al. [75] found that *bsmt1* mutants in *Arabidopsis* could induce SAR response, similar to WT plants, in spite of the fact that these plants could not accumulate and evaporate MeSA in/from the primary inoculated leaves. These results suggested that MeSA has role neither in SAR signalling nor in systemic SA accumulation. Four more points were listed against the role of MeSA signalling in SAR induction and generally against volatile signals including MeSA: (i) blockage of phloem transport attenuates SAR induction [85,86]; (ii) the volatile concentration of MeSA (10–1000 $\mu g\,L^{-1}$) required for the induction of resistance in closed systems is two magnitude of orders higher than the concentration that could be present under normal conditions in phytotron chambers; (iii) MeSA was present in very low amounts in petiolar exudates of infected leaves and contrary to SA, MeSA did not accumulate in pathogen-free systemic leaves and finally (iv) ICS gene expression and de novo SA biosynthesis are required for systemic SA accumulation both in WT and *bsmt1* mutant plants suggesting no role of airborne MeSA and/or MeSA transported via the phloem system [74]. Recently, the role of other volatile compounds, monoterpenes ($\alpha$-pinene and $\beta$-pinene) was reported in SAR via induction of reactive oxygen species (ROS) and *AZI1* gene [87].

This controversy of results may indicate that MeSA signalling is influenced by unknown environmental factors and in this case other signal molecules could be responsible for SAR signal transduction.

### 2.3. Lipid-Derived Signalling

#### 2.3.1. Glycerol-3-Phosphate Dependent Factor (G3P-Dependent Factor)

G3P is an important intermediate of lipid biosynthesis and is present both in the cytosol and chloroplast (Figure 2C). Mutational analysis revealed that functional *SFD1* (*SUPPRESSOR OF FATTY ACID DESATURASE DEFICIENCY1* in Nössen genotype) locus was necessary for SAR induction [19]. The same locus is also known as *GLY1* from another *Arabidopsis* genotype [88]. *SFD1* gene codes for a dihydroxy acetone phosphate (DHAP) reductase enzyme, which is responsible for the synthesis of G3P from DHAP [19,89]. In *sfd1* mutant *Arabidopsis* plants, the SA accumulation and *PR1* expression were reduced in distant, pathogen-free leaves but the local response to pathogen infection was not modified. The *sfd1* mutation influences fatty acid composition of galactolipids in chloroplasts, especially the ratio of C16:3 in monogalactosyldiacylglycerol (MGDG) and digalactosyldiacylglycerol (DGDG) is decreased [89]. However, the mode of action of SFD1 is unclear as this protein has no desaturase activity.

The chloroplastic localisation (encoded by an N-terminal signal sequence) and the enzymatic DHAP-reductase activity of SFD1 were necessary for SAR induction [89]. Petiolar exudate from WT plants induced resistance in *sfd1* mutant plants but the opposite experiment gave negative results suggesting a role for G3P and/or a G3P dependent factor in signalling of SAR [90].

To clarify the real function of G3P in SAR signal transduction, other lipid biosynthesis related mutants were also tested. Fatty acid desaturase (*fad7* and *sfd2*) and *mgd1* (monogalactosyl synthase 1, responsible for galactose incorporation into diacylglycerol) mutants had also SAR-deficient phenotype. Therefore, it was concluded that rather the intactness of chloroplastic glycerolipid biosynthesis is required for SAR but not the production of C3 carbon skeleton [90].

Interestingly, another gene mutation, *gli1/nho1* (glycerol insensitive 1/nonhost 1) involved in G3P biosynthesis in the cytosol, also has a SAR-deficient phenotype. The enzymatic glycerol kinase activity of GLI1/NHO1 protein produces G3P from glycerol [20]. Therefore, the effects of these two mutations (*gli1/nho1* localized in the cytosol and *gly1* in the chloroplast) on fatty acid composition of lipids were compared. Surprisingly, *gli1* mutation has no effect on lipid composition, contrary to chloroplastic *gly1* (see earlier results with *sfd1*, which is allelic to *gly1*) [20]. Moreover, the SAR was also inducible in the *act1* (G3P acyltransferase1) mutant, in which the incorporation of C18:1 fatty acid into glycerolipids and other aspects of lipid biosynthesis are inhibited. However, there are three more genes in *Arabidopsis* with DHAR reductase activity. Independently of their subcellular localization, these genes had no effect on lipid composition, but one cytoplasmic and one chloroplastic isoform affected SAR development. Therefore, contrary to former results [90], it was concluded that intactness of glycerolipid biosynthesis is dispensable for SAR induction and the regulation of G3P level is a crucial factor.

Taken together, *sfd1/gly1* and *gli1* mutations in *Arabidopsis* indicated that the effects of these two mutations on glycerolipid biosynthesis are different, but the SAR-deficient phenotypes in both cases could be related to their contribution to G3P synthesis and production of a G3P dependent factor. Supporting this evidence, after infiltration of $^{14}$C-G3P into primary inoculated leaves, this compound was not transported to distant leaves, but was present as an unidentified labelled compound in systemic leaves suggesting a signalling role of a G3P-dependent factor [20]. Importantly, localized application of G3P induces transcriptional changes in distant tissue indicating a role for G3P in systemic transcriptional reprogramming [20].

As indicated above, galactolipids may play an important role in lipid-derived signalling of SAR. The role of two galactolipid mutants *mgd1* and *dgd1* (digalactosyl synthase 1), responsible for MGDG and DGDG synthesis, respectively, in SAR induction was also studied in detail by Gao et al. [90]. Both mutants were compromised in SAR induction. Although the two mutants had differential and multiple impacts on SAR signalling capacity (*mgd1* plants are impaired in pathogen-induced ROS, AzA and G3P accumulation, but *dgd1* mutants in nitric oxide (NO) and SA synthesis), surprisingly these mutants were able to produce biologically active petiole exudate to induce SAR in WT plants. To explain these results one possibility is that these mutants can produce other active signalling compounds but are unable to respond to them. Thus, MGDG and DGDG lipids are rather involved in signal perception than in signal generation. Transgenic expression of a bacterial glucosyltransferase is unable to restore SAR in *dgd1* mutant plants even though it can rescue their morphological and fatty acid phenotypes [91].

### 2.3.2. Azelaic Acid (AzA)

Azelaic acid is a nine-carbon dicarboxylic acid (1,9-nonanedionic acid) (Figure 2E). In human medical and cosmetic practice, AzA is used for treatment of different types of hyperpigmentation, acne and other skin disorders [92]. In plants, AzA is an end-product of lipid peroxidation (LP) under biotic (especially during hypersensitive reaction (HR) caused by incompatible plant-pathogen interactions) and abiotic stress conditions and is produced via different enzymatic and non-enzymatic mechanisms [93].

Considering AzA as a signalling molecule in SAR induction in *Arabidopsis* Jung et al. [17] found that although AzA accumulated at elevated levels locally and in phloem exudates (6–7 times) during bacterial induced SAR, external application of AzA per se did not promote SA accumulation and had a minimal effect on gene expression but could induce local and systemic resistance response. AzA can act as a priming molecule and produces elevated systemic induction of SA accumulation upon bacterial inoculation (*P. syringae* pv. *maculicola* strain *Pma*DG3) of distant leaves with enhanced resistance against the pathogen. Moreover, AzA-induced SAR was dependent on *FMO1* and *ALD1* [17] both of which are involved in the amplification loop of systemic SA and Pip accumulation [21] (see later). Mutation of the *AZI1* gene, which is inducible by azelaic acid, results in the specific loss of systemic immunity triggered by pathogen or AzA [17]. Yu et al. [18] studied the role of fatty acids in SAR induction. Local treatment of *Arabidopsis* leaves with C18 unsaturated fatty acids and its release after bacterial inoculation can serve as precursors of AzA production and trigger SAR induction. AzA in turn can increase G3P accumulation and expression of lipid transfer proteins, DIR1 and AZI1. Wittek et al. [94] also reported that EDS1-dependent SAR is mediated in an AzA- and (its precursor) 9-oxo nonanoic acid-dependent manner in *Arabidopsis*. Metabolomic studies in tobacco cell cultures indicated that AzA treatment can induce the accumulation of the early products of the phenylpropanoid pathway [95]. The putative consequence of this finding on resistance was not tested. On the contrary, Návarová et al. [21] found no increase in accumulation of AzA in petiolar exudate of bacteria (*P. syringae* pv. *maculicola*, *Pma*)-infiltrated *Arabidopsis* leaves. Later studies suggested that AzA can locally induce SAR signal(s) emission in primary infected leaves via AZI1 and its paralog, EARLI1 (EARLY ARABIDOPSIS ALUMINIUM INDUCED1) accumulation at membrane-membrane contact sites (MCS) required for intracellular transport of apolar lipidic signals including AzA [55]. It is important to emphasize that inducible expression of *AZI1* or *EARLI1* only in local tissue of *az1* mutant plants is sufficient to restore SAR [55]. Moreover, an interesting finding is that AZI1:GFP protein was mainly detected in epidermal cells [55]. Importantly, the plastids of these cells are specialized organelles in which fatty acids and cuticle components for the epidermal cell surface are synthesized. The fact an intact cuticle is needed for SAR [58] supports the idea that epidermal cells may be important for long-distance defence signalling.

Therefore, more recently, our own studies have re-examined the role of AzA in SAR induction, in another plant-pathogen system, tobacco mosaic virus (TMV)-tobacco (*N. tabacum* cv. Xanthi nc) [31]. Former results indicated that signal transduction from inducing leaves into distant ones is fully completed within 4 days after primary TMV inoculation [28,30]. Therefore, phloem sap was collected in this time window (2 or 3 days after inoculation) for 24 h from TMV-infected and control leaves. Interestingly, HPLC-MS assays detected (besides C9 AzA) low amounts of two other dicarboxylic acids, suberic acid (1,8-octadienoic acid), and sebacic acid (1,10-decadienoic acid) in petroleum ether extracted petiolar exudates of both TMV-infected and control leaves 2–3 days after TMV inoculation. According to Jung et al. [17] these two other dicarboxylic acids had no biological activity in SAR induction. AzA content was doubled and significantly higher in exudates of TMV-infected leaves as compared to that in control exudates [31]. In our experiments, we focused on the effects of external AzA application on symptom expression. Local and systemic effects of AzA pretreatments on the distribution of TMV lesion size were measured by a semi-automated, computer-assisted method and significance of the results were analysed by a multiple comparison test (R package) [96]. The local application of AzA (0.2–1.0 mM) showed no or limited influence (either increase or decrease) on lesion size of TMV-inoculated leaves. In addition, AzA pretreatment did not modify the multiplication of TMV detected by semiquantitative RT-PCR of coat protein gene. No significant systemic effect of AzA on lesion size of TMV was detectable in distant leaves. Moreover, AzA treatment had no considerable local and systemic effect on symptom expression and multiplication of incompatible (*P. syringae* pv. *tomato* DC3000, *Pst*) and compatible (*P. syringae* pv. *tabaci*) bacteria [31].

In accordance with our results, Zoeller et al. [93] found that in spite of the bacterially inducible AzA accumulation in infected leaves, external AzA treatment does not inhibit the growth of *Pst* (strain DC3000) in *Arabidopsis* leaves. Vicente et al. [97] also found that AzA pretreatment caused a barely detectable inhibition of symptoms and growth of *Pst* DC3000 bacteria in both treated and distant *Arabidopsis* leaves.

Some other aspects of the mode of action of AzA in SAR induction have also been discussed previously. Unlike the priming effect of AzA for pathogen-responsive SA biosynthesis in distant *Arabidopsis* leaves [17], other studies [18] did not find any priming of SA accumulation in systemic leaves. The biosynthesis of AzA is a complicated question. Although the 9-lipoxygenase (9-LOX) pathway is involved in many plant defence responses against bacterial, fungal and viral infections [97] and produces mostly nine-carbon products, Zoeller et al. [93] reported that bacterium-induced AzA is synthesized non-enzymatically from chloroplastic galactolipids. It is more likely that AzA was synthesized via ROS-mediated pathway from chloroplastic galactolipids (MGDG and DGDG) [93,98]. In addition, in *lox1 lox5* double mutant *Arabidopsis*, pathogen-induced AzA production was not compromised suggesting that *9-LOX* genes are not required for AzA biosynthesis. More importantly, *lox1* (9-LOX) mutant *Arabidopsis* plants were not able to develop fully active SAR indicating a putative role of 9-LOX-mediated lipid signal generation in primary inoculated leaves and/or signal manifestation in distant leaves [97].

Taken together, our and some other [21,93,97] results suggest the lack of or limited accumulation of AzA after infection and external AzA treatment does not induce considerable local or systemic effects on viral and bacterial infections. Therefore, its formerly reported role in signal transduction and/or signal generation during induction of SAR in *Arabidopsis* could not be confirmed under our experimental conditions in tobacco [31]. However, it is possible that the role of lipid-derived signalling could be relatively less pivotal under certain conditions as SAR induction is considered independent of tissue necrotisation [39] that is associated with LP.

## 2.4. Dehydroabietinal (DA)

One of the most potent SAR inducers is dehydroabietinal (Figure 2D). This tricyclic diterpenoid compound can induce SAR in picomolar ranges when applied to leaves in *Arabidopsis*, tomato and tobacco [16,99]. In conifers, these abiatene diterpenoids are synthesised from geranylgeranyl diphosphate by cyclization. *Arabidopsis* contains a homolog of a cytochrome monooxigenase P450 class enzyme of *Picea sitchensis* that is capable of synthesizing DA and related compounds from dehydroabiatadiene [100].

SAR induction by DA has common and specific features as compared to other signalling compounds. Like pathogen-induced SAR and the mode of action of Pip (see later), DA requires functional *FMO1* and *ICS1* genes for systemic SA accumulation and SAR induction in distant leaves of *Arabidopsis* [16]. Despite that pure labelled $^2$H-DA is rapidly translocated from treated leaves to the foliage and induced SAR, DA levels did not increase in leaves and petiolar exudate after infection of leaves with a SAR-inducing pathogen. However, due to the infection DA is enriched from a biologically inactive low molecular weight fraction into a trypsin-sensitive high molecular weight signalling form (DA*, 100 kDa) that is capable of SAR induction [99]. It is, however, unclear whether the pure application of DA to leaves also leads to this activation and if so, which factor is responsible for this activation process under pathogen-free conditions. Anyway, DIR1 which could be systemically translocated was associated with the high molecular weight DA* fraction [16,99]. Indeed, DIR1 was required for the full activation of SAR by DA, confirming an important function of DIR1 in DA-induced SAR.

The application of DA also promotes flowering in *Arabidopsis*. Shortly, during vegetative growth, FLC (FLOWERING LOCUS C), a flowering repressor protein suppresses the expression of flowering signal *FLT (FLOWERING LOCUS T)* which is considered to be the phloem-mobile florigen and released from leaves and transported to the shoot meristem to induce transition from vegetative to generative state. DA and bacterial inoculation both promote expression of *FLD (FLOWERING LOCUS D)* which involved in histone modifications. FLD promotes flowering by suppressing the expression of flowering repressor *FLC* gene [101]. FLD function is also required for systemic SA accumulation and priming of *PR1, WRKY6* and *WRKY29* expression in distant leaves [76,102].

### 2.5. Pipecolic Acid (Pip) and N-hydroxypipecolic Acid (NHS)

L-pipecolic acid is an enigmatic heterocyclic non-protein amino acid (Figure 2B) and a catabolite of L-lysine (Lys). In humans it serves as a diagnostic marker of pyridoxine-dependent epilepsy [103] and accumulates in patients with hyperpipecolic acidemia (hyperpipecolatemia), a rare, recessive metabolic disorder related to peroxisomal malfunction [104]. This derivative was also present among others in non-protein amino acids of an extraterrestrial meteorite [105].

In plants, especially in angiosperms, the level of Pip is elevated in a response to different stresses including pathogen infection [106]. Although former genetic studies with an *ald1* mutant indicated the key role of an aminotransferase, ALD1 in local and systemic defence responses [24,107], the function of Pip was discovered only later on [21]. In fact, detailed studies indicated that (i) *ALD1* gene product shows in vitro substrate preference to lysine, a putative precursor of Pip biosynthesis in plants and animals [97,107]; (ii) the biosynthesis of Pip in *Arabidopsis* is dependent on functional *ALD1* locus [21] and (iii) ALD1 enzyme acts as a first step during Lys catabolism and directly transfers the α-amino group of L-Lys to an oxoacid, preferentially pyruvate to form ε-amino-α-ketocaproic acid (KAC) and alanine [22,108]. Next steps from KAC (cyclization, isomerization) via 1,2-dehydropipecolic acid and its in planta detectable, enaminic form, 2,3-dehydropipecolic acid are leading to the formation of Pip [22]. Furthermore, *ALD1* transcript accumulates in the pathogen-inoculated and distant pathogen-free leaves [107]. The local and systemic immune defects of *ald1* mutant *Arabidopsis* after bacterial inoculation could be rescued by external application of Pip. From the point of view of signal transduction during SAR response, it is important to note that Návarová et al. [21] found strong Pip accumulation in petiolar exudate of SAR-inducing *P. syringae* infected leaves. However, whether Pip has a direct role in long-distance SAR signalling remains to be elucidated in the future. Recent results show that transcription factors TGA1 and TGA4 (TGAGG-BINDING FACTOR) also regulate Pip and SA synthesis by modulating the expression of *SARD1* and *CBP60g* genes [109].

To prove further the conserved role of Pip in plant immune responses and to analyse further virus-induced SAR signalling in tobacco we measured by HPLC-MS the level of different amino acids in virus infected tissue (Figure 4) [40]. Both in TMV and CMV infections the level of Pip and tryptophan (Trp) accumulated to high amounts. Chromatograms on Figure 4 indicate about tenfold increase in

Pip level after TMV (Figure 4A,B) and CMV (Figure 4C,D) infections. We also analysed the local and systemic effects of external Pip infiltration (2–10 mM D-L-Pip) into tobacco leaves (Figure 5) [40]. Interestingly, Pip had not only local but systemic effect. Comparison of TMV lesion size distribution in locally treated leaf 6 and systemic leaf 7 to corresponding controls indicated significant reduction in lesion size. This effect, however, was less pronounced (especially after local treatment with 2 mM Pip) in systemic leaves, but even this effect was comparable to the effect of SAR induction by TMV infection (see Figure 3A). Návarová et al. [21] found that exogenous application of Pip via root system is sufficient to induce SAR-like response and primed state in wild type *Arabidopsis* against bacteria. Consequently, Pip can be an important player in SAR induction in different plants against different pathogens.

In two very recent publications, added in proof, a new SAR signalling compound, FMO1-dependent N-oxygenation product of Pip, N-hydroxypipecolic acid (NHS) was described (Figure 1) [110,111].

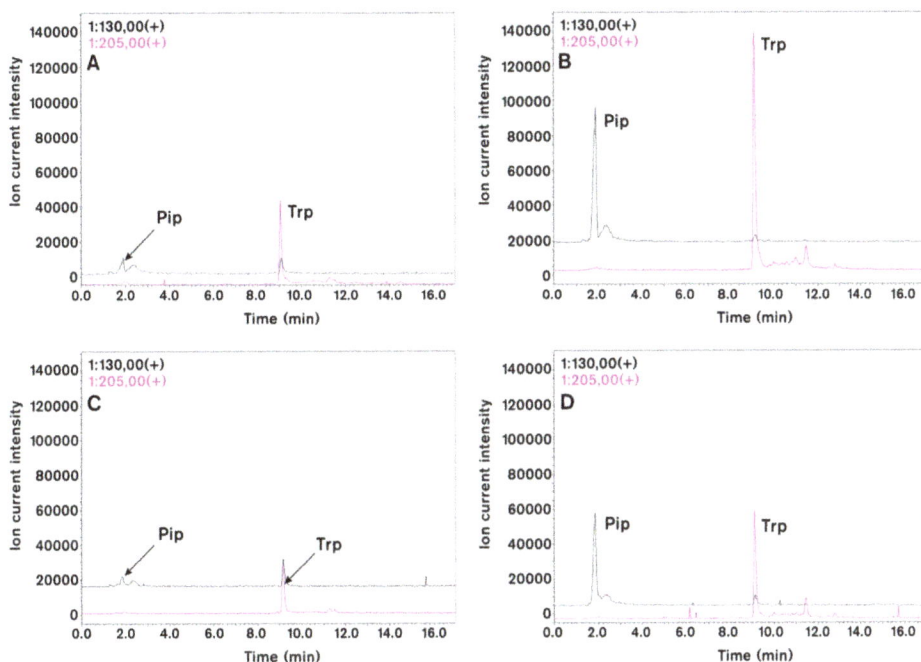

**Figure 4.** Effect of TMV (**A,B**) and CMV (**C,D**) inoculations on selected amino acids of tobacco leaves (4 days after viral inoculation). HPLC-MS chromatograms of pipecolic acid (Pip, $m/z = 130$) and tryptophan (Trp, $m/z = 205$) in virus infected leaves (**B,D**) as compared to control plants (**A,C**). Leaf samples (about 200 mg) were extracted with acetonitrile:water (6:4 v/v in 1.5 mL). This condition supports amino acid extraction but limits protein solubilization. HPLC-ESI-MS analyses were carried out with a Shimadzu LCMS-2020 (Shimadzu Co., Kyoto, Japan) analytical system. Chromatographic separations were performed on a SunShell C18 packed column (2.6 μm, 3.0 mm × 100 mm) by using gradient elution (solvent A, 0.1% formic acid solution; solvent B, 0.1% formic acid in acetonitrile) under non-derivatised conditions. Ions for MS detection were obtained by positive mode of electrospray ionization (ESI). Identity of peaks were also checked by co-chromatography of samples with authentic standards (Sigma-Aldrich Co. St. Louis, MI, USA).

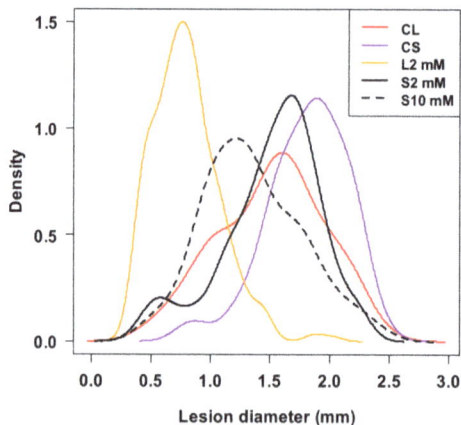

**Figure 5.** Effect of local (CL and L2) and systemic (CS, S2 and S10) pipecolic acid (D-L-Pip) treatments on TMV lesion size distribution by using Kernel density estimation [31,32] (2 days after Pip infiltration). L2, S2 and S10: local (L, leaf 6) and systemic (S, leaf 7) effects of 2 and 10 mM D-L-Pip treatments as compared to corresponding local control (CL) and systemic control (CS) leaves (triplicate per treatments). In a multiple comparison test [31], both concentration of Pip treatment resulted in significant decrease of TMV lesion size. The highest difference was caused by local treatment (L2). The effect of systemic treatment with 10 mM Pip was more pronounced (and significantly different) than with 2 mM Pip

## 3. Role of Light in SAR Induction: Light Intensity, Timing of Exposition and Spectral Distribution

Former results have already indicated the important role of light in plant defence responses and especially in SAR induction. New insights into this question show that not only effectiveness of SAR induction is regulated by light dependent factors, but the quality of signalling compounds. There are several light dependent factors that can influence SAR signalling events. Besides light intensity and timing of illumination after primary inoculation, a third factor, the spectral distribution of light will be also taken into consideration.

Under high light intensity (over 500 $\mu$E m$^{-2}$ s$^{-1}$) SAR can develop without SA accumulation and PR1 expression in distant leaves [112]. This pathway, however, depends on the expression of *FMO1* required also for many other signalling routes during SAR induction (see earlier). The systemic expression of *FMO1* gene depends on phytochromes [76,113]. Earlier results suggested that hypersensitive necrotisation and local resistance response in *Arabidopsis–Pst* AvrRpt2 interaction depended on phytochrome A and B [114]. Griebel and Zeier [113] reported that rather the systemic response depends on phytochromes in *Arabidopsis–Psm* AvrRpm1 interaction. The systemic *FMO1* and *PR1* expression and SA level were not inducible as well as SAR could not develop in the double mutant *phyA-phyB* plants. It is possible that the phytochrome mediated effect is related to the effect of spectral distribution of light on SAR [32].

The interconnection between chloroplastic photoreceptor function and pathogen signalling is suggested by the similarity of photooxidative stress response (ROS generation, PR1 expression and programmed cell death) that can induce resistance against compatible bacteria in local and systemic tissues [115]. In accordance with this finding Fodor et al. [116] found that SAR induction by TMV infection in tobacco was associated with elevated antioxidant capacity of distant resistant leaves.

Local resistance responses (hypersensitive reaction, PR1 expression and SA accumulation) could be inhibited if plants were kept in darkness after infection for a longer period [110,117]. The development of SAR was also inhibited in *Arabidopsis* after exposition to darkness all over the experiment [112]. If plants (*Arabidopsis* and tobacco) were kept in darkness only for overnight period after primary inoculation, the SAR response was present but became weaker as compared to the SAR response of the plants that were exposed to light (at least 3.5 h) after inoculation [84]. The longer light exposition after primary inoculation correlated with the level of SA accumulation, PR1 expression and the strength of SAR response in *Arabidopsis–Psm* AvrRpm1 plant-pathogen system [113].

The discrepancy of the results in signal transduction of MeSA during SAR induction was also explained by the effect of light exposition after primary inoculation [84]. Later studies indicated that several hours (at least 3.5 h or more) of light exposition after primary inoculation with bacteria or TMV restore the SAR-deficient phenotype of *bsmt1* (responsible for MeSA synthesis) mutants [84]. Moreover, two other signal transduction mutants of SAR, *dir1-1* and *sfd1/gly1* mutants were also complemented under this light condition. In other words, the timing of the dark period relative to the primary inoculation severely influences the importance of a certain signal transduction component in SAR induction [57,84].

However, in our experiments with AzA, plants were exposed to light for at least 10 h after treatments and were illuminated with a relatively high daily photon flux. Therefore, we performed experiments with tobacco plants kept in darkness subsequent to AzA treatment to test whether this condition can activate AzA-mediated local and/or systemic response in tobacco. The local AzA treatment did not show significant difference on TMV lesion size of local or systemic leaves after incubation in darkness as compared to control plants. Experiments with the multiplication of a compatible bacterium, *P. syringae* pv. *tabaci*, in local and systemic leaves after AzA treatment in darkness also showed no significant decrease. These data clearly suggest that AzA-mediated signalling does not rely on factors activated in darkness, at least in tobacco plants [31].

The third light-dependent factor that can cause differences in resistance to TMV infection is the spectral distribution of light [32]. Our results clearly indicated that spectral distribution of light sources influences (i) plant growth and development; (ii) local resistance response to TMV infection and (iii) SAR inducing capacity of tobacco plants. Certain light sources with unbalanced light spectrum had negative impact on plant growth and development, local resistance response and SAR induction capacity of tobacco plants. Halogen lamp (HL) and fluorescent tube (FT) light sources showed very different spectral distribution, relative abundance or shortage in red/far red light, respectively. The more similar was the spectrum of the artificial light source to sunshine (greenhouse conditions), the stronger was the inducible SAR response. From a practical point of view, under artificial conditions, metal halide lamp or a mixture of HL and FT light sources can be suggested as optimal test conditions. Consequently, the optimization of the effect of artificial light sources is an important factor in experimental design studying signal transduction and biochemistry of SAR [32].

## 4. Concluding Remarks

At least three separate signalling pathways are present in plants to induce SAR. To date, these pathways could be discriminated by different light conditions indicating the importance of this environmental factor and putatively the functional role of chloroplasts in signal generation. The SAR signalling routes and their main characteristics are as follows: (i) Under high light intensity conditions, SAR is induced without systemic SA accumulation but requires functional *FMO1* gene. The signal is unknown but could be related to ROS-mediated processes; (ii) If after the primary inoculation plants are exposed to a longer (overnight) dark period, SAR induction and signalling may depend on *BSMT1* (MeSA), *DIR1* and *SFD1/GLY1* genes and systemic SA accumulation; (iii) If the SAR induction is dependent on a light period after primary inoculation, it requires functional *ICS1*, *FMO1* and *ALD1* genes, Pip and SA accumulation in the systemic leaves. Pip and its FMO1-dependent N-oxygenation product, *N*-hydroxypipecolic acid can play a critical role in the induction of SAR either after bacterial or viral infection in different host plants. Finally, (iv) DA-mediated signalling is *DIR1-*, *ICS1-*, *FMO1-FLD-* and SA-dependent in systemic leaves, but its relation to light was not yet determined. Optimal SAR induction also requires balanced spectral distribution of light, probably due to phytochrome regulation.

Parallel operation and control of different signals probably can contribute to the plasticity of the SAR response. However, further detailed analysis of the interaction of these overlapping factors is required for the practical application of this resistance mechanism for protecting field crops.

**Acknowledgments:** This work was supported by National Research Development and Innovation Office (Hungary), grant NKFIH-K112146 (Attila L. Ádám). The authors thank Katalin Salánki and Ágnes Móricz (Centre for Agricultural Research, Hungarian Academy of Sciences, Plant Protection Institute) for CMV strain and the help in preparation of Figure 2 (ChemDraw Ultra 9.0 program), respectively. We also thank Gábor Gullner (Centre for Agricultural Research, Hungarian Academy of Sciences, Plant Protection Institute) for critical reading of the manuscript.

**Author Contributions:** Attila L. Ádám and György Kátay designed the experiments; György Kátay, Emese Mergenthaler, Orsolya Viczián and Attila L. Ádám performed the experiments; Zoltán Á. Nagy, Attila L. Ádám, György Kátay and Emese Mergenthaler analysed the data.

**Conflicts of Interest:** The authors declare no conflict of interest. The founding sponsors had no role in the design of the study; in the collection, analyses, or interpretation of data; in the writing of the manuscript, and in the decision to publish the results.

## References

1. Winter, P.S.; Bowman, C.E.; Villani, P.J.; Dolan, T.E.; Hauck, N.R. Systemic acquired resistance in moss: Further evidence for conserved defence mechanisms in plants. *PLoS ONE* **2014**, *9*, e101880. [CrossRef] [PubMed]
2. Ross, A.F. Systemic acquired resistance induced by localized virus infections in plants. *Virology* **1961**, *14*, 340–358. [CrossRef]
3. Luna, E.; Bruce, T.J.; Roberts, M.R.; Flors, V.; Ton, J. Next-generation systemic acquired resistance. *Plant Physiol.* **2012**, *158*, 844–853. [CrossRef] [PubMed]
4. Sánchez, A.L.; Stassen, J.H.; Furci, L.; Smith, L.M.; Ton, J. The role of DNA (de)methylation in immune responsiveness of *Arabidopsis*. *Plant J.* **2016**, *88*, 361–374. [CrossRef] [PubMed]
5. Conrath, U. Molecular aspects of defence priming. *Trends Plant Sci.* **2011**, *16*, 524–531. [CrossRef] [PubMed]
6. Spoel, S.H.; Dong, X. How do plants achieve immunity? Defence without specialized immune cells. *Nat. Rev. Immunol.* **2012**, *12*, 89–100. [CrossRef] [PubMed]
7. Shah, J.; Zeier, J. Long-distance communication and signal amplification in systemic acquired resistance. *Front. Plant Sci.* **2013**, *4*, 30. [CrossRef] [PubMed]
8. Jaskiewicz, M.; Conrath, U.; Peterhänsel, C. Chromatin modification acts as a memory for systemic acquired resistance in the plant stress response. *EMBO Rep.* **2011**, *12*, 50–55. [CrossRef] [PubMed]
9. Strahl, B.D.; Allis, C.D. The language of covalent histone modifications. *Nature* **2000**, *403*, 41–45. [CrossRef] [PubMed]
10. Cameron, R.K.; Paiva, N.T.; Lamb, C.J.; Dixon, R.A. Accumulation of salicylic acid and PR-1 gene transcripts in relation to the systemic acquired resistance (SAR) response induced by *Pseudomonas syringae* pv. *tomato* in *Arabidopsis*. *Physiol. Mol. Plant Pathol.* **1999**, *55*, 121–130. [CrossRef]
11. Shulaev, V.; Silverman, P.; Raskin, I. Airborne signalling by methyl salicylate in plant pathogen resistance. *Nature* **1997**, *385*, 718–721. [CrossRef]
12. Park, S.W.; Kaimoyo, E.; Kumar, D.; Mosher, S.; Klessig, D.F. Methyl salicylate is a critical mobile signal for plant systemic acquired resistance. *Science* **2007**, *318*, 113–116. [CrossRef] [PubMed]
13. Manosalva, P.M.; Park, S.W.; Forouhar, F.; Tong, L.; Fry, W.E.; Klessig, D.F. Methyl esterase 1 (stmes1) is required for systemic acquired resistance in potato. *Mol. Plant Microbe Interact.* **2010**, *23*, 1151–1163. [CrossRef] [PubMed]
14. Maldonado, A.M.; Doerner, P.; Dixon, R.A.; Lamb, C.J.; Cameron, R.K. A putative lipid transfer protein involved in systemic resistance signalling in *Arabidopsis*. *Nature* **2002**, *419*, 399–403. [CrossRef] [PubMed]
15. Carella, P.; Isaacs, M.; Cameron, R.K. Plasmodesmata-located protein overexpression negatively impacts the manifestation of systemic acquired resistance and the long-distance movement of Defective in Induced Resistance1 in *Arabidopsis*. *Plant Biol.* **2015**, *17*, 395–401. [CrossRef] [PubMed]

16. Chaturvedi, R.; Venables, B.; Petros, R.A.; Nalam, V.; Li, M.; Wang, X.; Takemoto, L.J.; Shah, J. An abietane diterpenoid is a potent activator of systemic acquired resistance. *Plant J.* **2012**, *71*, 161–172. [CrossRef] [PubMed]

17. Jung, H.W.; Tschaplinski, T.J.; Wang, L.; Glazebrook, J.; Greenberg, J.T. Priming in systemic plant immunity. *Science* **2009**, *324*, 89–91. [CrossRef] [PubMed]

18. Yu, K.; Soares, J.M.; Mandal, M.K.; Wang, C.; Chanda, B.; Gifford, A.N.; Fowler, J.S.; Navarre, D.; Kachroo, A.; Kachroo, P. A feedback regulatory loop between G3P and lipid transfer proteins DIR1 and AZI1 mediates azelaic-acid-induced systemic immunity. *Cell Rep.* **2013**, *3*, 1266–1278. [CrossRef] [PubMed]

19. Nandi, A.; Welti, R.; Shah, J. The *Arabidopsis thaliana* dihydroxyacetone phosphate reductase gene *SUPRESSOR OF FATTY ACID DESATURASE DEFICIENCY1* is required for glycerolipid metabolism and for the activation of systemic acquired resistance. *Plant Cell* **2004**, *16*, 465–477. [CrossRef] [PubMed]

20. Chanda, B.; Xia, Y.; Mandal, M.K.; Yu, K.; Sekine, K.T.; Gao, Q.M.; Selote, D.; Hu, Y.; Stromberg, A.; Navarre, D. Glycerol-3-phosphate is a critical mobile inducer of systemic immunity in plants. *Nat. Genet.* **2011**, *43*, 421–427. [CrossRef] [PubMed]

21. Návarová, H.; Bernsdorff, F.; Doring, A.C.; Zeier, J. Pipecolic acid, an endogenous mediator of defense amplification and priming, is a critical regulator of inducible plant immunity. *Plant Cell* **2012**, *24*, 5123–5141. [CrossRef] [PubMed]

22. Hartmann, M.; Kim, D.; Bernsdorff, F.; Ajami-Rashidi, Z.; Scholten, N.; Schreiber, S.; Zeier, T.; Schuck, S.; Reichel-Deland, V.; Zeier, J. Biochemical principles and functional aspects of pipecolic acid biosynthesis in plant immunity. *Plant Physiol.* **2017**, *174*, 124–153. [CrossRef] [PubMed]

23. Bernsdorff, F.; Doring, A.C.; Gruner, K.; Schuck, S.; Brautigam, A.; Zeier, J. Pipecolic acid orchestrates plant systemic acquired resistance and defense priming via salicylic acid-dependent and -independent pathways. *Plant Cell* **2016**, *28*, 102–129. [CrossRef] [PubMed]

24. Song, J.T.; Lu, H.; McDowell, J.M.; Greenberg, J.T. A key role for *ALD1* in activation of local and systemic defenses in *Arabidopsis*. *Plant J.* **2004**, *40*, 200–212. [CrossRef] [PubMed]

25. Breitenbach, H.H.; Wenig, M.; Wittek, F.; Jordá, L.; Maldonado-Alconada, A.M.; Satioglu, H.; Colby, T.; Knappe, C.; Bichlmeier, M.; Pabst, E.; et al. Contracting roles of the apoplastic ASPARTYL PROTEASE APOPLASTIC, ENHANCED DISEASE SUSCEPTIBILITY1-DEPENDENT and LEGUME LECTIN-LIKE PROTEIN in *Arabidopsis* systemic acquired resistance. *Plant Physiol.* **2014**, *165*, 791–809. [CrossRef] [PubMed]

26. Lee, H.H.; Park, Y.-J.; Seo, P.J.; Kim, J.-H.; Sim, H.-J.; Kim, S.-K.; Park, C.-M. Systemic immunity requires SnRK2.8-mediated nuclear import of NPR1 in *Arabidopsis*. *Plant Cell* **2015**, *37*, 3425–3438. [CrossRef] [PubMed]

27. Fu, Z.Q.; Yan, S.; Saleh, A.; Wang, W.; Ruble, J.; Oka, N.; Mohan, R.; Spoel, S.H.; Tada, Y.; Zheng, N.; et al. NPR3 and NPR4 are receptors for the immune signal salicylic acid in plants. *Nature* **2012**, *486*, 228–232. [CrossRef] [PubMed]

28. Ádám, A.L.; Nagy, Z.Á. A szisztemikus szerzett rezisztencia szignálátvitele: Eredmények és kihívások. Signal transduction of systemic acquired resistance: Results and new challenges. *Növényvédelem Plant Prot.* **2016**, *77*, 435–461.

29. Jenns, A.; Kuć, J. Graft transmission of systemic resistance of cucumber to anthracnose induced by *Colletotrichum lagenarium* and tobacco necrosis virus. *Phytopathology* **1979**, *69*, 753–756. [CrossRef]

30. Nagy, Z.Á.; Kátay, G.; Gullner, G.; Ádám, A.L. Evaluation of TMV lesion formation and timing of signal transduction during induction of systemic acquired resistance (SAR) in tobacco with a computer-assisted method. In *Biotic and Abiotic Stress—Recent Advances and Future Perspectives*; Shanker, A.K., Shanker, C., Eds.; InTech: London, UK, 2016; pp. 363–372.

31. Nagy, Z.Á.; Kátay, G.; Gullner, G.; Király, L.; Ádám, A.L. Azelaic acid accumulates in phloem exudates of TMV-infected tobacco leaves, but its application does not induce local or systemic resistance against selected viral and bacterial pathogens. *Acta Physiol. Plant.* **2017**, *39*, 9. [CrossRef]

32. Nagy, Z.Á.; Jung, A.; Varga, Z.; Kátay, Gy.; Ádám, A. Effect of artificial light conditions on local and systemic resistance response of tobacco to TMV infection. *Not. Bot. Horti. Agrobot. Cluj-Napoca* **2017**, *45*, 270–275. [CrossRef]

33. Choi, J.; Tanaka, K.; Cao, Y.; Qi, Y.; Qiu, J.; Liang, Y.; Lee, S.Y.; Stacey, G. Identification of a plant receptor for extracellular ATP. *Science* **2014**, *343*, 290–294. [CrossRef] [PubMed]

34. Tanaka, K.; Choi, J.; Cao, Y.; Stacey, G. Extracellular ATP acts as a damage-associated molecular pattern (DAMP) signal in plants. *Front. Plant Sci.* **2014**, *5*, 446. [CrossRef] [PubMed]

35. Kørner, C.J.; Klauser, D.; Niehl, A.; Domínguez-Ferreras, A.; Chinchilla, D.; Boller, T.; Heinlein, M.; Hann, D.R. The immunity regulator BAK1 contributes to resistance against diverse RNA viruses. *Mol. Plant Microbe Interact.* **2013**, *26*, 1271–1280. [CrossRef] [PubMed]

36. Aan den Toorn, M.; Albrecht, C.; de Vries, S. On the origin of SERKs: Bioinformatics analysis of the somatic embryogenesis receptor kinases. *Mol. Plant* **2015**, *8*, 762–782. [CrossRef] [PubMed]

37. He, K.; Gou, X.; Yuan, T.; Lin, H.; Asami, T.; Yoshida, S.; Russell, S.D.; Li, J. BAK1 and BKK1 regulate brassinosteroid-dependent growth and brassinosteroid-independent cell-death pathways. *Curr. Biol.* **2007**, *17*, 1109–1115. [CrossRef] [PubMed]

38. Roux, M.; Schwessinger, B.; Albrecht, C.; Chinchilla, D.; Jones, A.; Holton, N.; Malinovsky, F.G.; Tör, M.; de Vries, S.; Zipfel, C. The *Arabidopsis* leucine-rich repeat receptor-like kinases BAK1/SERK3 and BKK1/SERK4 are required for innate immunity to hemibiotrophic and biotrophic pathogens. *Plant Cell* **2011**, *23*, 2440–2455. [CrossRef] [PubMed]

39. Mishina, T.E.; Zeier, J. Pathogen-associated molecular pattern recognition rather than development of tissue necrosis contributes to bacterial induction of systemic acquired resistance in *Arabidopsis*. *Plant J.* **2007**, *50*, 500–513. [CrossRef] [PubMed]

40. Kátay, Gy.; Mergenthaler, E.; Viczián, O.; Nagy, Z.Á.; Ádám, A.L. Centre for Agricultural Research, Hungarian Academy of Sciences, Plant Protection Institute, Budapest, Hungary. Different aspects of systemic immunity in tobacco. Unpublished work. 2018.

41. Divéki, Z.; Salánki, K.; Balázs, E. The necrotic pathotype of the *Cucumber mosaic virus* (CMV) ns strain is solely determined by amino acid 461 of the 1a protein. *Mol. Plant Microbe Interact.* **2004**, *17*, 837–845. [CrossRef] [PubMed]

42. Vogel-Adghough, D.; Stahl, E.; Navarova, H.; Zeier, J. Pipecolic acid enhances resistance to bacterial infection and primes salicylic acid and nicotine accumulation in tobacco. *Plant Signal. Behav.* **2013**, *8*, e26366. [CrossRef] [PubMed]

43. Kovalev, N.; Pogany, J.; Nagy, P.D. Template role of double-stranded RNA in tombusvirus replication. *J. Virol.* **2014**, *88*, 5638–5651. [CrossRef] [PubMed]

44. Son, K.N.; Liang, Z.; Lipton, H.L. Double-stranded RNA is detected by immunofluorescence analysis in RNA and DNA virus infections, including those by negative-stranded RNA viruses. *J. Virol.* **2015**, *89*, 9383–9392. [CrossRef] [PubMed]

45. Niehl, A.; Wyrsch, I.; Boller, T.; Heinlein, M. Double-stranded RNAs induce a pattern-triggered immune signaling pathway in plants. *New Phytol.* **2016**, *211*, 1008–1019. [CrossRef] [PubMed]

46. Ziebell, H.; Carr, J.P. Effects of dicer-like endoribonucleases 2 and 4 on infection of *Arabidopsis thaliana* by Cucumber mosaic virus and a mutant virus lacking the 2b counter-defence protein gene. *J. Gen. Virol.* **2009**, *90*, 2288–2292. [CrossRef] [PubMed]

47. Henderson, I.R.; Zhang, X.; Lu, C.; Johnson, L.; Meyers, B.C.; Green, P.J.; Jacobsen, S.E. Dissecting *Arabidopsis thaliana* DICER function in small RNA processing, gene silencing and DNA methylation patterning. *Nat. Genet.* **2006**, *38*, 721–725. [CrossRef] [PubMed]

48. Lee, B.; Park, Y.S.; Lee, S.; Song, G.C.; Ryu, C.M. Bacterial RNAs activate innate immunity in *Arabidopsis*. *New Phytol.* **2015**, *209*, 785–797. [CrossRef] [PubMed]

49. Breen, S.; Williams, S.J.; Outram, M.; Kobe, B.; Solomon, P.S. Emerging insights into the functions of PATHOGENESIS-RELATED PROTEIN1. *Trends Plant Sci.* **2017**, *10*, 871–879. [CrossRef] [PubMed]

50. Gamir, J.; Darwiche, R.; Van't Hof, P.; Choudhary, V.; Stumpe, M.; Schneiter, R.; Mauch, F. The sterol-binding activity of PATHOGENESIS-RELATED PROTEIN1 reveals the mode of action of an antimicrobial protein. *Plant J.* **2016**, *89*, 502–509. [CrossRef] [PubMed]

51. Chen, Y.L.; Lee, C.Y.; Cheng, K.T.; Chang, W.H.; Huang, R.N.; Nam, H.G.; Chen, Y.R. Quantitative peptidomics study reveals that a wound-induced peptide from PR-1 regulates immune signalling in tomato. *Plant Cell* **2014**, *26*, 4135–4148. [CrossRef] [PubMed]

52. Klessig, D.F.; Tian, M.; Choi, H.W. Multiple targets of salicylic acid and its derivatives in plants and animals. *Front. Immunol.* **2016**, *7*, 206. [CrossRef] [PubMed]

53. Champigny, M.J.; Isaacs, M.; Carella, P.; Faubert, J.; Fobert, P.; Cameron, R.K. Long distance movement of DIR1 and investigation of the role of DIR1-like during systemic acquired resistance in *Arabidopsis*. *Front. Plant Sci.* **2013**, *4*, 230. [CrossRef] [PubMed]

54. Carella, P.; Kempthorne, C.J.; Wilson, D.C.; Isaacs, M.; Cameron, R.K. Exploring the role of DIR1, DIR1-like and other lipid transfer proteins during systemic immunity in *Arabidopsis*. *Physiol. Mol. Plant Pathol.* **2017**, *9*, 49–57. [CrossRef]

55. Cecchini, N.M.; Steffes, K.; Schlappi, M.R.; Gifford, A.N.; Greenberg, J.T. *Arabidopsis* AZI1 family proteins mediate signal mobilization for systemic defence priming. *Nat. Commun.* **2015**, *6*, 7658. [CrossRef] [PubMed]

56. Isaacs, M.; Carella, P.; Faubert, J.; Rose, J.K.C.; Cameron, R.K. Orthology analysis and in vivo complementation studies to elucidate the role of DIR1 during systemic acquired resistance in *Arabidopsis thaliana* and *Cucumis sativus*. *Front. Plant Sci.* **2016**, *7*, 566. [CrossRef] [PubMed]

57. Liu, P.P.; von Dahl, C.C.; Park, S.W.; Klessig, D.F. Interconnection between methyl salicylate and lipid-based long-distance signaling during the development of systemic acquired resistance in *Arabidopsis* and tobacco. *Plant Physiol.* **2011**, *155*, 1762–1768. [CrossRef] [PubMed]

58. Xia, Y.; Yu, K.; Gao, Q.-M.; Wilson, R.V.; Navarre, D.; Kachroo, P.; Kachroo, A. Acyl CoA binding proteins are required for cuticle formation and plant responses to microbes. *Front. Plant Sci.* **2012**, *3*, 224. [CrossRef] [PubMed]

59. Carella, P.; Merl-Pham, J.; Wilson, D.C.; Dey, S.; Hauck, S.M.; Vlot, A.C.; Cameron, R.K. Comparative proteomics analysis of phloem exudates collected during the induction of systemic acquired resistance. *Plant Physiol.* **2016**, *171*, 1495–1510. [CrossRef] [PubMed]

60. Gaffney, T.; Friedrich, L.; Vernooij, B.; Negrotto, D.; Nye, G.; Uknes, S.; Ward, E.; Kessmann, H.; Ryals, J. Requirement of salicylic acid for the induction of systemic acquired resistance. *Science* **1993**, *261*, 754–756. [CrossRef] [PubMed]

61. Malamy, J.; Carr, J.P.; Klessig, D.F.; Raskin, I. Salicylic acid: A likely endogenous signal in the resistance response of tobacco to viral infection. *Science* **1990**, *250*, 1002–1004. [CrossRef] [PubMed]

62. Métraux, J.P.; Signer, H.; Ryals, J.; Ward, E.; Wyss-Benz, M.; Gaudin, J.; Raschdorf, K.; Schmid, E.; Blum, W.; Inverardi, B. Increase in salicylic acid at the onset of systemic acquired resistance in cucumber. *Science* **1990**, *250*, 1004–1006. [CrossRef] [PubMed]

63. Delaney, T.P.; Uknes, S.; Vernooij, B.; Friedrich, L.; Weymann, K.; Negrotto, D.; Gaffney, T.; Gut-Rella, M.; Kessmann, H.; Ward, E.; et al. A central role of salicylic acid in plant disease resistance. *Science* **1994**, *266*, 1247–1250. [CrossRef] [PubMed]

64. Zhang, Y.; Xu, S.; Ding, P.; Wang, D.; Cheng, Y.T.; He, J.; Gao, M.; Xu, F.; Li, Y.; Zhu, Z.; et al. Control of salicylic acid synthesis and systemic acquired resistance by two members of a plant-specific family of transcription factors. *Proc. Natl. Acad. Sci. USA* **2010**, *107*, 18220–18225. [CrossRef] [PubMed]

65. Wang, L.; Tsuda, K.; Truman, W.; Sato, M.; Nguyen le, V.; Katagiri, F.; Glazebrook, J. CBP60g and SARD1 play partially redundant critical roles in salicylic acid signaling. *Plant J.* **2011**, *67*, 1029–1041. [CrossRef] [PubMed]

66. Nawrath, C.; Métraux, J.P. Salicylic acid induction-deficient mutants of *Arabidopsis* express PR-2 and PR-5 and accumulate high levels of camalexin after pathogen inoculation. *Plant Cell* **1999**, *11*, 1393–1404. [CrossRef] [PubMed]

67. Wildermuth, M.C.; Dewdney, J.; Wu, G.; Ausubel, F.M. Isochorismate synthase is required to synthesize salicylic acid for plant defence. *Nature* **2001**, *414*, 562–565. [CrossRef] [PubMed]

68. Nawrath, C.; Heck, S.; Parinthawong, N.; Métraux, J.P. EDS5, an essential component of salicylic acid-dependent signaling for disease resistance in *Arabidopsis*, is a member of the MATE transporter family. *Plant Cell* **2002**, *14*, 275–286. [CrossRef] [PubMed]

69. Koo, Y.J.; Kim, M.A.; Kim, E.H.; Song, J.T.; Jung, C.; Moon, J.K.; Kim, J.H.; Seo, H.S.; Song, S.I.; Kim, J.K.; et al. Overexpression of salicylic acid carboxyl methyltransferase reduces salicylic acid-mediated pathogen resistance in *Arabidopsis thaliana*. *Plant Mol. Biol.* **2007**, *64*, 1–15. [CrossRef] [PubMed]

70. Meuwly, P.; Molders, W.; Buchala, A.; Metraux, J.P. Local and systemic biosynthesis of salicylic acid in infected cucumber plants. *Plant Physiol.* **1995**, *109*, 1107–1114. [CrossRef] [PubMed]

71. Pallas, J.A.; Paiva, N.L.; Lamb, C.; Dixon, R.A. Tobacco plants epigenetically suppressed in phenylalanine ammonia-lyase expression do not develop systemic acquired resistance in response to infection by tobacco mosaic virus. *Plant J.* **1996**, *10*, 281–293. [CrossRef]

72. Catinot, J.; Buchala, A.; Abou-Mansour, E.; Metraux, J.P. Salicylic acid production in response to biotic and abiotic stress depends on isochorismate in *Nicotiana benthamiana*. *FEBS Lett.* **2008**, *582*, 473–478. [CrossRef] [PubMed]

73. Rasmussen, J.B.; Hammerschmidt, R.; Zook, M.N. Systemic induction of salicylic acid accumulation in cucumber after inoculation with *Pseudomonas syringae* pv. *syringae*. *Plant Physiol.* **1991**, *97*, 1342–1347. [CrossRef] [PubMed]

74. Vernooij, B.; Friedrich, L.; Morse, A.; Reist, R.; Kolditz-Jawhar, R.; Ward, E.; Uknes, S.; Kessmann, H.; Ryals, J. Salicylic acid is not the translocated signal responsible for inducing systemic acquired resistance. *Plant Cell* **1994**, *6*, 959–965. [CrossRef] [PubMed]

75. Attaran, E.; Zeier, T.E.; Griebel, T.; Zeier, J. Methyl salicylate production and jasmonate signaling are not essential for systemic acquired resistance in *Arabidopsis*. *Plant Cell* **2009**, *21*, 954–971. [CrossRef] [PubMed]

76. Singh, V.; Roy, S.; Giri, M.K.; Chaturvedi, R.; Chowdhury, Z.; Shah, J.; Nandi, A.K. *Arabidopsis thaliana* *FLOWERING LOCUS D* is required for systemic acquired resistance. *Mol. Plant Microbe Interact.* **2013**, *26*, 1079–1088. [CrossRef] [PubMed]

77. Mishina, T.E.; Zeier, J. The *Arabidopsis* flavin-dependent monooxygenase FMO1 is an essential component of biologically induced systemic acquired resistance. *Plant Physiol.* **2006**, *141*, 1666–1675. [CrossRef] [PubMed]

78. Malamy, J.; Hennig, J.; Klessig, D.F. Temperature-dependent induction of salicylic acid and its conjugates during the resistance response to tobacco mosaic virus infection. *Plant Cell* **1992**, *4*, 359–366. [CrossRef] [PubMed]

79. Lee, H.I.; Raskin, I. Purification, cloning, and expression of a pathogen inducible UDP-glucose: Salicylic acid glucosyltransferase from tobacco. *J. Biol. Chem.* **1999**, *274*, 36637–36642. [CrossRef] [PubMed]

80. Park, S.W.; Liu, P.P.; Forouhar, F.; Vlot, A.C.; Tong, L.; Tietjen, K.; Klessig, D.F. Use of a synthetic salicylic acid analog to investigate the roles of methyl salicylate and its esterases in plant disease resistance. *J. Biol. Chem.* **2009**, *284*, 7307–7317. [CrossRef] [PubMed]

81. Vlot, A.C.; Klessig, D.F.; Park, S.W. Systemic acquired resistance: The elusive signal(s). *Curr. Opin. Plant Biol.* **2008**, *11*, 436–442. [CrossRef] [PubMed]

82. Vlot, A.C.; Liu, P.P.; Cameron, R.K.; Park, S.W.; Yang, Y.; Kumar, D.; Zhou, F.; Padukkavidana, T.; Gustafsson, C.; Pichersky, E.; et al. Identification of likely orthologs of tobacco salicylic acid-binding protein 2 and their role in systemic acquired resistance in *Arabidopsis thaliana*. *Plant J.* **2008**, *56*, 445–456. [CrossRef] [PubMed]

83. Dempsey, D.A.; Klessig, D.F. SOS—Too many signals for systemic acquired resistance? *Trends Plant Sci.* **2012**, *17*, 538–545. [CrossRef] [PubMed]

84. Liu, P.P.; von Dahl, C.C.; Klessig, D.F. The extent to which methyl salicylate is required for signaling systemic acquired resistance is dependent on exposure to light after infection. *Plant Physiol.* **2011**, *157*, 2216–2226. [CrossRef] [PubMed]

85. Guedes, M.E.M.; Richmond, S.; Kuć, J. Induced systemic resistance to anthracnose in cucumber as influenced by the location of the inducer inoculation with *Colletotrichum lagenarium* and the onset of flowering and fruiting. *Physiol. Plant Pathol.* **1980**, *17*, 229–233. [CrossRef]

86. Van Bel, A.J.E.; Gaupels, F. Pathogen-induced resistance and alarm signals in the phloem. *Mol. Plant Pathol.* **2004**, *5*, 495–504. [CrossRef] [PubMed]

87. Riedlmeier, M.; Ghirardo, A.; Wenig, M.; Knappe, C.; Koch, K.; Georgii, R.; Dey, S.; Parker, J.E.; Schnitzler, J.P.; Vlot, A.C. Monoterpenes support systemic acquired resistance within and between plants. *Plant Cell* **2017**, *29*, 1440–1459. [CrossRef] [PubMed]

88. Miquel, M.; Cassagne, C.; Browse, J. A new class of *Arabidopsis* mutants with reduced hexadecatrienoic acid fatty acid levels. *Plant Physiol.* **1998**, *117*, 923–930. [CrossRef] [PubMed]

89. Lorenc-Kukula, K.; Chaturvedi, R.; Roth, M.; Welti, R.; Shah, J. Biochemical and molecular-genetic characterization of SFD1's involvement in lipid metabolism and defense signaling. *Front. Plant Sci.* **2012**, *3*, 26. [CrossRef] [PubMed]

90. Chaturvedi, R.; Krothapalli, K.; Makandar, R.; Nandi, A.; Sparks, A.A.; Roth, M.R.; Welti, R.; Shah, J. Plastid omega3-fatty acid desaturase-dependent accumulation of a systemic acquired resistance inducing activity in petiole exudates of *Arabidopsis thaliana* is independent of jasmonic acid. *Plant J.* **2008**, *54*, 106–117. [CrossRef] [PubMed]

91. Gao, Q.M.; Yu, K.; Xia, Y.; Shine, M.B.; Wang, C.; Navarre, D.; Kachroo, A.; Kachroo, P. Mono- and digalactosyldiacylglycerol lipids function nonredundantly to regulate systemic acquired resistance in plants. *Cell Rep.* **2014**, *9*, 1681–1691. [CrossRef] [PubMed]

92. Draelos, Z. Skin lightening preparations and the hydroquinone controversy. *Dermatol. Ther.* **2007**, *20*, 308–313. [CrossRef] [PubMed]

93. Zoeller, M.; Stingl, N.; Krischke, M.; Fekete, A.; Waller, F.; Berger, S.; Mueller, M.J. Lipid profiling of the *Arabidopsis* hypersensitive response reveals specific lipid peroxidation and fragmentation processes: Biogenesis of pimelic and azelaic acid. *Plant Physiol.* **2012**, *160*, 365–378. [CrossRef] [PubMed]

94. Wittek, F.; Hoffmann, T.; Kanawati, B.; Bichlmeier, M.; Knappe, C.; Wenig, M.; Schmitt-Kopplin, P.; Parker, J.E.; Schwab, W.; Vlot, A.C. *Arabidopsis* ENHANCED DISEASE SUSCEPTIBILITY1 promotes systemic acquired resistence via azelaic acid and its precursor 9-oxo nonanoic acid. *J. Exp. Bot.* **2014**, *65*, 5919–5931. [CrossRef] [PubMed]

95. Djami-Tchatchou, A.T.; Ncube, E.N.; Steenkamp, P.A.; Dubery, I.A. Similar, but different: Structurally related azelaic acid and hexanoic acid trigger differential metabolomic and transcriptomic responses in tobacco cells. *BMC Plant Biol.* **2017**, *17*, 227. [CrossRef] [PubMed]

96. R Core Team. *R: A Language and Environment for Statistical Computing*; R Foundation for Statistical Computing: Vienna, Austria, 2015.

97. Vicente, J.; Cascon, T.; Vicedo, B.; Garcia-Agustin, P.; Hamberg, M.; Castresana, C. Role of 9-lipoxygenase and α-dioxygenase oxylipin pathways as modulators of local and systemic defense. *Mol. Plant* **2012**, *5*, 914–928. [CrossRef] [PubMed]

98. Wang, C.; El-Shetehy, M.; Shine, M.B.; Yu, K.; Navarre, D.; Wendehenne, D.; Kachroo, A.; Kachroo, P. Free radicals mediate systemic acquired resistance. *Cell Rep.* **2014**, *7*, 348–355. [CrossRef] [PubMed]

99. Shah, J.; Chaturvedi, R.; Chowdhury, Z.; Venables, B.; Petros, R.A. Signaling by small metabolites in systemic acquired resistance. *Plant J.* **2014**, *79*, 645–658. [CrossRef] [PubMed]

100. Hamberger, B.; Ohnishi, T.; Hamberger, B.; Séguin, A.; Bohlmann, J. Evolution of diterpene metabolism: Sitka spruce CYP720B4 catalyzes multiple oxidations in resin acid biosynthesis of conifer defense against insects. *Plant Physiol.* **2011**, *157*, 1677–1695. [CrossRef] [PubMed]

101. Kim, D.H.; Sung, S. Environmentally coordinated epigenetic silencing of FLC by protein and long noncoding RNA components. *Curr. Opin. Plant Biol.* **2012**, *15*, 51–56. [CrossRef] [PubMed]

102. Singh, V.; Roy, S.; Singh, D.; Nandi, A.K. *Arabidopsis* FLOWERING LOCUS D influences systemic-acquired-resistance- induced expression and histone modifications of WRKY genes. *J. Biosci.* **2014**, *39*, 119–126. [CrossRef] [PubMed]

103. Plecko, B.; Hikel, C.; Korenke, G.C. Pipecolic acid as a diagnostic marker of pyridoxine-dependent epilepsy. *Neuropediatrics* **2005**, *36*, 200–205. [CrossRef] [PubMed]

104. Tranchant, C.; Aubourg, P.; Mohr, M.; Rocchiccioli, F.; Zaenker, C.; Warter, J.M. A new peroxisomal disease with impaired phytanic and pipecolic acid oxidation. *Neurology* **1993**, *43*, 2044–2048. [CrossRef] [PubMed]

105. Kvenholden, K.A.; Lawless, J.G.; Ponnamperuma, C. Nonprotein amino acids in the murchison meteorite. *Proc. Natl. Acad. Sci. USA.* **1971**, *68*, 486–490. [CrossRef]

106. Pálfi, G.; Dézsi, L. Pipecolic acid as an indicator of abnormal protein metabolism in diseased plants. *Plant Soil* **1968**, *29*, 285–291. [CrossRef]

107. Song, J.T.; Lu, H.; Greenberg, J.T. Divergent roles in *Arabidopsis thaliana* development and defense of two homologous genes, *ABERRANT GROWTH AND DEATH2* and *AGD2-LIKE DEFENSE RESPONSE PROTRIN1*, encoding novel aminotransferases. *Plant Cell* **2004**, *16*, 353–366. [CrossRef] [PubMed]

108. Ding, P.; Rekhter, D.; Ding, Y.; Feussner, K.; Busta, L.; Haroth, S.; Xu, S.; Li, X.; Jetter, R.; Feussner, I.; et al. Characterization of a pipecolic acid biosynthesis pathway required for systemic acquired resistance. *Plant Cell* **2016**, *28*, 2603–2615. [CrossRef] [PubMed]

109. Sun, T.; Busta, L.; Zhang, Q.; Ding, P.; Jetter, R.; Zhang, Y. TGAGG-BINDING FACTOR1 (TGA1) and TGA4 regulate salicylic acid and pipecolic acid biosynthesis by modulating the expression of *SYSTEMIC ACQUIRED RESISTANCE DEFICIENT1 (SARD1)* and *CALMODULIN-BINDING PROTEIN 60g (CBP60g)*. *New Phytol.* **2018**, *217*, 344–354. [CrossRef] [PubMed]

110. Hartmann, M.; Zeier, T.; Bernsdorff, F.; Reichel-Deland, V.; Kim, D.; Hohmann, M.; Scholten, N.; Schuck, S.; Bräutigam, A.; Hölzel, T.; et al. Flavin monooxygenase-generated N-hydroxypipecolic acid is a critical element of plant systemic immunity. *Cell* **2018**, *17*, 456–469. [CrossRef] [PubMed]

111. Chen, Y.C.; Holmes, E.C.; Rajniak, J.; Kim, J.-G.; Tang, S.; Fischer, C.R.; Mudgett, M.B.; Sattely, E.S.; Stanford University, Stanford, CA, USA. *N*-hydroxy-pipecolic acid is a mobile signal that induces systemic disease resistance in *Arabidopsis*. 2018. Preprint. Available online: https://www.biorxiv.org/content/early/2018/03/25/288449 (accessed on 9 April 2018).
112. Zeier, J.; Pink, B.; Mueller, M.J.; Berger, S. Light conditions influence specific defence responses in incompatible plant-pathogen interactions: Uncoupling systemic resistance from salicylic acid and PR-1 accumulation. *Planta* **2004**, *219*, 673–683. [CrossRef] [PubMed]
113. Griebel, T.; Zeier, J. Light regulation and daytime dependency of inducible plant defenses in *Arabidopsis*: Phytochrome signaling controls systemic acquired resistance rather than local defense. *Plant Physiol.* **2008**, *147*, 790–801. [CrossRef] [PubMed]
114. Genoud, T.; Buchala, A.J.; Chua, N.H.; Metraux, J.P. Phytochrome signalling modulates the SA-perceptive pathway in *Arabidopsis*. *Plant J.* **2002**, *31*, 87–95. [CrossRef] [PubMed]
115. Muhlenbock, P.; Szechynska-Hebda, M.; Plaszczyca, M.; Baudo, M.; Mateo, A.; Mullineaux, P.M.; Parker, J.E.; Karpinska, B.; Karpinski, S. Chloroplast signaling and lesion simulating disease1 regulate crosstalk between light acclimation and immunity in *Arabidopsis*. *Plant Cell* **2008**, *20*, 2339–2356. [CrossRef] [PubMed]
116. Fodor, J.; Gullner, G.; Ádám, A.L.; Barna, B.; Kőmíves, T.; Király, Z. Local and systemic responses of antioxidants to tobacco mosaic virus infection and to salicylic acid in tobacco (Role in systemic acquired resistance). *Plant Physiol.* **1997**, *114*, 1443–1451. [CrossRef] [PubMed]
117. Chandra-Shekara, A.C.; Gupte, M.; Navarre, D.; Raina, S.; Raina, R.; Klessig, D.; Kachroo, P. Light-dependent hypersensitive response and resistance signaling against *Turnip Crinkle Virus* in *Arabidopsis*. *Plant J.* **2006**, *45*, 320–334. [CrossRef] [PubMed]

International Journal of
*Molecular Sciences*

MDPI

*Review*

# Calcium Signalling in Plant Biotic Interactions

**Didier Aldon, Malick Mbengue, Christian Mazars and Jean-Philippe Galaud \***

Laboratoire de Recherche en Sciences Vegetales, Universite de Toulouse, CNRS, UPS, 24, Chemin de Borde-Rouge, Auzeville, BP 42617, 31326 Castanet-Tolosan, France; aldon@lrsv.ups-tlse.fr (D.A.); malick.mbengue@lrsv.ups-tlse.fr (M.M.); mazars@lrsv.ups-tlse.fr (C.M.)
\* Correspondence: galaud@lrsv.ups-tlse.fr; Tel.: +33-534-323-828; Fax: +33-534-323-802

Received: 26 January 2018; Accepted: 22 February 2018; Published: 27 February 2018

**Abstract:** Calcium ($Ca^{2+}$) is a universal second messenger involved in various cellular processes, leading to plant development and to biotic and abiotic stress responses. Intracellular variation in free $Ca^{2+}$ concentration is among the earliest events following the plant perception of environmental change. These $Ca^{2+}$ variations differ in their spatio-temporal properties according to the nature, strength and duration of the stimulus. However, their conversion into biological responses requires $Ca^{2+}$ sensors for decoding and relaying. The occurrence in plants of calmodulin (CaM) but also of other sets of plant-specific $Ca^{2+}$ sensors such as calmodulin-like proteins (CMLs), $Ca^{2+}$-dependent protein kinases (CDPKs) and calcineurin B-like proteins (CBLs) indicate that plants possess specific tools and machineries to convert $Ca^{2+}$ signals into appropriate responses. Here, we focus on recent progress made in monitoring the generation of $Ca^{2+}$ signals at the whole plant or cell level and their long distance propagation during biotic interactions. The contribution of CaM/CMLs and CDPKs in plant immune responses mounted against bacteria, fungi, viruses and insects are also presented.

**Keywords:** biotic stress responses; calcium; calcium signature; calmodulin; CMLs; CDPKs; plant immunity; symbiosis

## 1. Introduction

Like all living organisms, plants face environmental challenges that can be either of a biotic nature such as interactions with pathogens (e.g., bacteria, fungi, oomycetes, viruses, insects) or of an abiotic nature such as drought, soil salinity, air pollution, extreme temperatures and mechanical injury [1]. These adverse conditions often limit growth and productivity of crops worldwide. The expected global temperature elevation in the coming years and associated climate modifications are creating ever-greater challenges for agriculture [2,3]. To adapt to adverse growth conditions, plants must be able to detect the nature and strength of environmental stimuli, interpret them and activate appropriate physiological responses [3]. Among signalling elements that are involved in plant stress responses and particularly during immune responses to pathogens, reactive oxygen species (ROS) and $Ca^{2+}$ ions are among the earliest actors that coordinate plant adaptive responses [4–7]. The oxidative burst was first described in 1983 following Potato infection by the oomycete *Phytophtora infestans* [8], whereas the importance of $Ca^{2+}$ signalling in plant immunity was reported in tobacco following *Pseudomonas syringae* inoculation in 1990 [9]. A close connection was then established between ROS and $Ca^{2+}$ signalling pathways in plant immunity [10].

In this review, we will focus on the importance of $Ca^{2+}$, a ubiquitous and versatile second messenger [11], in plant biotic interactions. To become informative, the $Ca^{2+}$ message needs to be decoded and relayed in order to activate the appropriate cell response and this is carried out by $Ca^{2+}$-binding proteins termed $Ca^{2+}$ sensors [12]. The complex spatiotemporal patterns of $Ca^{2+}$ changes at cellular and tissue levels (frequency, amplitude, and distribution within the cell) are proposed to carry information and are denoted as the $Ca^{2+}$ signature [13,14] (Figure 1). The $Ca^{2+}$ signature encodes a first layer of specificity and will be considered first, with a particular emphasis on how methods to monitor $Ca^{2+}$ signatures have evolved and brought new information. $Ca^{2+}$-binding proteins and their downstream targets provide a second layer of specificity. Most $Ca^{2+}$-binding proteins are characterized by the presence in their sequence of the canonical $Ca^{2+}$-binding motif called the EF-hand [15]. For example, the plant model *Arabidopsis thaliana* encodes at least 250 EF-hand-containing proteins [16]. This number is much higher than in mammals and notably, the majority of plant $Ca^{2+}$ sensors do not have homologs in others organisms [17,18]. However, only about half of them have been considered as $Ca^{2+}$ sensors [17]. These plant $Ca^{2+}$ sensors are classified into four major groups: the calcineurin B-like (CBL), the $Ca^{2+}$-dependent protein kinases (CDPK), the calmodulin (CaM) group and its closely related group, the Calmodulin-like protein (CML) family [17,19]. Calmodulin (CaM) is one of the most studied eukaryotic proteins and has been shown to interact with and modulate the activity of numerous target proteins [20]. Plants also possess a remarkable repertoire of CaM-related proteins termed CMLs (7 CaM and 50 CMLs genes in *Arabidopsis*) that are not present in animals, as is also the case for CBLs and CDPKs (ten and 34 genes in Arabidopsis, respectively) [21,22]. To date, the roles of most of these $Ca^{2+}$ sensors remain unknown but recent studies have pointed out the roles for some of them in physiological processes associated with development, abiotic plant stress responses and plant immunity [23–25]. Here, we present recent and relevant data about CaM, CMLs and CDPKs and their involvement in plant responses to various biotic stresses.

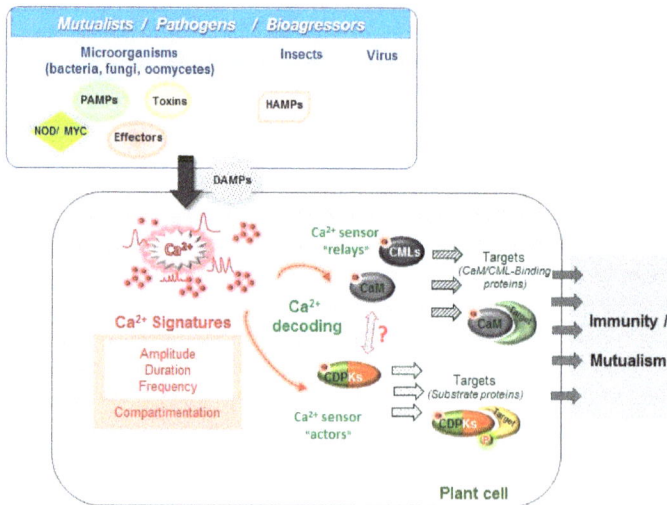

**Figure 1.** Key steps in $Ca^{2+}$ signaling pathways during plant biotic interactions. Plants are exposed to diverse microorganisms, pests or other aggressors leading to beneficial or detrimental interactions. Plant cells possess a large repertoire of sensors that allow to perceive, discriminate and transduce different signals during plant immunity (Pathogen-Associated Molecular Patterns (PAMPs) , effectors, toxins, Damage-Associated Molecular Patterns (DAMPs) or Herbivory-Associated Molecular Patterns (HAMPs)) or during the interaction with mutualistic organisms (Nod and Myc Factors). In response to different stimuli, the earliest steps rely on specific cytosolic $Ca^{2+}$ rises termed calcium signatures occurring in the cytosol and in organelles, including nucleus (Section 2.1). These calcium signatures differ by their spatio-temporal properties and encode a first layer of specificity. A second layer of specificity, relies on the decoding of these calcium transients (Section 3). $Ca^{2+}$ binds to a plethora of sensors such as calmodulin (CaM), CaM-like proteins (CML), calcium-dependent protein kinases (CDPK) that activate target proteins either by direct binding or through phosphorylation (P).

## 2. Ca²⁺ Signatures as the Earliest Signalling Events in Plant-Organism Interactions

*2.1. Discovering the Importance of Ca²⁺ Signalling in Biotic Interactions*

Free $Ca^{2+}$ is a universal second messenger, and increases in cytosolic $Ca^{2+}$ concentration are among the earliest signalling events occurring in plants challenged with mutualistic or pathogenic partners. If we consider plant-pathogen interactions, the plant immune system is schematically organized as a two-tiered system composed of Pathogen-Associated Molecular Patterns (PAMP)-Triggered Immunity (PTI) and Effector-Triggered Immunity (ETI) [26]. Activation of PTI enhances overall plant defence and protects plants from subsequent pathogen attack [27] while activation of ETI often culminates in a localized programmed cell death (PCD) also referred to as the Hypersensitive Response (HR), which blocks pathogen invasion. These immune layers also differ in their $Ca^{2+}$ signature. For instance, PTI activation involves $Ca^{2+}$ transients that returns to basal levels within minutes [28] whereas ETI is associated with a prolonged cytosolic $Ca^{2+}$ increase that can last for hours [29]. The generic $Ca^{2+}$ channel blocker lanthanum strongly impairs the immune responses associated with both types of plant immunity [29,30], thereby placing $Ca^{2+}$ signals and their decoders at the centre of immune signalling pathways.

$Ca^{2+}$ elevations during immune signalling are critical for the control of gene reprogramming which is required to mount the adequate responses (e.g., symbiosis or defence). Such widely acknowledged importance of these $Ca^{2+}$ variations in plant signalling has been made possible through the emergence in the early 1990s of tools allowing the monitoring of the dynamics of $Ca^{2+}$ changes in plant cells. The pioneering work of Trewavas's group and his collaborators demonstrated the occurrence of $Ca^{2+}$ variations in plant cells using the fluorescent Fluo3 $Ca^{2+}$ indicator and caged $Ca^{2+}$ or IP3 as triggers [31]. His group then opened a new research avenue mainly through the application, for the first time in plants, of the non-invasive, organelle-addressable and highly dynamic $Ca^{2+}$ probe aequorin [32]. The use of aequorin allowed tremendous progress in our current understanding of how $Ca^{2+}$ controls plant adaptive responses. In the seminal paper reporting the use of aequorin in plants, Knight et al. established for the first time that biotic stimuli consisting of various yeast preparations were able, like abiotic stimuli, to trigger $Ca^{2+}$ signals in tobacco seedlings [32]. This result suggested that the $Ca^{2+}$ cation should be taken into consideration as an important second messenger in studies dealing with plant-microbe or plant-pest interactions. Due to the numerous studies available in the literature, we will focus in this section, more specifically on several well-documented examples illustrating how exploiting the diversity of $Ca^{2+}$ probes has contributed to increasing our knowledge on the role of $Ca^{2+}$ during plant immunity, symbiosis and response to herbivory attacks.

Chandra and Low pointed out the advantages of the aequorin probe versus other available $Ca^{2+}$ sensors [33]. Using a suspension of aequorin-transformed tobacco cells, they reported that oligogalacturonic acid, a pectin-derived component, was able to induce plant defence and to generate a $Ca^{2+}$ response followed by a delayed oxidative burst [34,35]. Using a pharmacological approach, they demonstrated that the oxidative burst was fully dependent on the $Ca^{2+}$ response. This result was among the first to suggest a possible role of $Ca^{2+}$ transients in the ROS-mediated defence signalling pathway [33]. At the same time, Mithoefer et al. established a similar link between cytosolic $Ca^{2+}$ increases and defence responses using chitotetraose, a fungal cell wall component and β-glucans. A sustained high $Ca^{2+}$ concentration response was observed in the case of β-glucans while this was not the case with the chitotetraose [36]. The sustained $Ca^{2+}$ increase generated after β-glucan application was linked to the increase of a major soybean phytolalexin glyceollin [36]. One year later, Blume et al. demonstrated that a *Phytophtora sojae*-derived peptide (Pep 13) elicited a biphasic $Ca^{2+}$ variation due to an influx of extracellular $Ca^{2+}$ in parsley cells stably expressing aequorin. Only the sustained $Ca^{2+}$ variation reaching a concentration around 250–300 nM could be associated with ROS and phytolalexin production [37]. Interestingly, stimuli that triggered only the first transient peak were unable to stimulate phytolalexin production [37]. Similarly, the use of *Nicotiana plumbaginifolia* cells stably expressing cytosolic aequorin challenged with the *Phytophtora cryptogea* elicitin "cryptogein" allowed us to demonstrate that oligogalacturonic acids and cryptogein each elicit a very specific cytosolic $Ca^{2+}$ response [38]. After a lag phase of 1–2 min, the cryptogein-induced $Ca^{2+}$ response appeared as a biphasic $Ca^{2+}$ increase with a first transient peak lasting 10 min followed by a sustained increase. Suppression of this sustained $Ca^{2+}$ phase using the $Ca^{2+}$ channel blocker lanthanum chloride just after the first $Ca^{2+}$ transient, resulted in the suppression of MAPK activation, accumulation of *PAL* and *hsr203j* immune-related gene transcripts and induced cell death [38].

Most of the above studies were performed with elicitors of various nature and origin coming from fungal or bacterial pathogens such as Microbe Associated Molecular Patterns (MAMPs) or compounds derived from the activity of pathogens on plant cell walls, the Damage-Associated Molecular Patterns (DAMPs) [39,40]. Similarly, elicitors can originate from herbivores and are named Herbivore Associated-Molecular Patterns (HAMPS). They are found in oral secretions or oviposition fluids of insects (for a review see [41]). HAMPs have also been shown to induce cytosolic $Ca^{2+}$ variations, for example in lima bean leaves incubated with the fluorescent $Ca^{2+}$ probe Fluo3-AM or in soybean cell cultures expressing cytosolic aequorin. *Spodoptera littoralis* larvae bites elicited a $Ca^{2+}$ response on the margin of the leaves within a delimited region of 30 to 200 μm from the bite zone whereas several components from the larvae regurgitate such as linolenoyl-L-Glutamine or volicitin were able to elicit $Ca^{2+}$ transients in soybean cells [42]. These $Ca^{2+}$ transients are correlated with early membrane depolarization/hyperpolarization events that are known to be linked to plant defence responses through systemin synthesis, ROS and jasmonate signalling (for a review see [43]).

These few examples showing that different elicitors trigger different $Ca^{2+}$ variations reinforced the concept of the $Ca^{2+}$ signature [13,14]. This concept postulates that a specific $Ca^{2+}$ variation is defined by its form, amplitude, frequency, duration, spatial localization and the $Ca^{2+}$ pool involved. All these parameters are tightly linked to the nature and strength of the stimulus perceived by the cell. Subsequent studies performed on guard cells largely corroborated this concept. Artificial modulation of the frequency, transient number of spikes, duration and amplitude of $Ca^{2+}$ oscillations were found to control the degree of long-term steady-state stomata closure [44]. Other findings supporting the $Ca^{2+}$ signature concept come from work performed with *Medicago truncatula* plants expressing the $Ca^{2+}$ probe cameleon YC2.1. Using a pharmacological approach modulating $Ca^{2+}$ homeostasis, the authors were able to demonstrate a link between the bacterial Nod factor-induced $Ca^{2+}$ oscillations and the activation of some selected nodulation markers genes. For example, the induction of the *ENOD11* *nodulin* gene required about 30 consecutive $Ca^{2+}$ spikes [45].

## 2.2. Monitoring $Ca^{2+}$ Transients at the Whole Tissue or Organism Level

Most of the initial studies on $Ca^{2+}$ variations in plant immunity were performed with elicitors and cultured cells. Although convenient, this model cannot inform us about the downstream events activated during genuine plant-microorganism or plant-pest interactions, such as cell-to-cell communication leading to long distance and systemic signalling. Pathogen elicitors only mimic the recognition step when surface motifs of pathogens are recognized by pattern recognition receptors (PRRs) of the host [7]. In addition to this recognition of surface motifs, pathogens also inject effectors into the plant cell to modulate plant defence responses [46]. Therefore, if we consider pathogenic bacteria, both compatible and incompatible strains possess the same PAMPs but in the first case the interaction leads to plant susceptibility while in the second case the interaction may lead to plant resistance through the execution of the HR due to effector recognition by plant resistance (R) genes [47,48]. Aequorin allows long kinetic studies due to its stability in the cell [49]. It has enabled the analysis of $Ca^{2+}$ signalling in whole plants challenged with a microorganism (pathogen, symbiont) or insect, in a non-destructive manner. One of the first examples of such studies, was the monitoring of $Ca^{2+}$ response in Arabidopsis leaves stably expressing aequorin under control of the 35S promoter challenged either with the compatible strain of *Pseudomonas syringae* pv. *tomato* DC3000 or with incompatible strains carrying the avirulence genes *avrRPM1* or *avrB*, both detected by the resistance Arabidopsis RPM1 protein [29]. In this study, the authors showed that the compatible strain elicited a transient $Ca^{2+}$ peak lasting about 10 min and peaking between 8–12 min whereas the incompatible strains avrRPM1 and avrB induced an additional delayed peak giving a maximal increase at around $105 \pm 10$ and $137 \pm 7$ min, respectively [29]. This second peak was dependent on a functional type III secretion apparatus because DC3000 mutants in this secretion system did not induce this second $Ca^{2+}$ response. In addition, Arabidopsis *rpm1* mutant failed to generate the second $Ca^{2+}$ peak and the HR. Thus, the aequorin technology allowed, in this specific case, to make a clear correlation between the delayed $Ca^{2+}$ peak and the gene for gene interactions responsible for the HR response.

In symbiotic interaction studies, the use of Oregon-dextran dye injected in root hair cells by iontophoresis demonstrated that two important genes in the rhizobia symbiosis were involved upstream of the $Ca^{2+}$ spikes [50]. Using the same fluorescent $Ca^{2+}$ probe, this group demonstrated two years later that spatiotemporal $Ca^{2+}$ spiking studies were helpful in understanding the regulation of nod gene expression. Indeed, they showed that strains differing by their ability to produce Nod factor could be differentiated by the $Ca^{2+}$ oscillations they induce. Thus, two strains derived from the same parent displayed large differences in the kinetics of $Ca^{2+}$ spiking; *Rm1021* triggered a robust $Ca^{2+}$ spiking after a lag phase of 10–15 min in more than 50% of the root hair cells whereas *Rm2011* did not [51]. The new generation of available genetically encoded $Ca^{2+}$ probes [52] facilitated the measurement of $Ca^{2+}$ responses at the organ and cell level during symbiotic relationships with rhizobia and mycorrhizal fungi. Kosuta et al. [53] were able to compare and discriminate the two types of symbiosis through their induced $Ca^{2+}$ oscillations. They demonstrated that the two symbiotic pathways require both DMI1 and DMI2 for both the establishment of functional symbiosis and also for the generation of $Ca^{2+}$ oscillations. This result indicates that $Ca^{2+}$ oscillations are also required for plant colonization by rhizobia or mycorrhizal fungi. Interestingly, the symbiotic fungus induced a $Ca^{2+}$ oscillation that differs from the Nod-factor-induced $Ca^{2+}$ oscillations by its amplitude and its periodicity [53]. In the bacterial symbiotic model, the use of nucleoplasmin-tagged cameleon (NupYC2.1) $Ca^{2+}$ probe demonstrated a regular nuclear $Ca^{2+}$ spiking in *Medicago* root hairs challenged with Nod Factors [54].

### 2.3. Systemic $Ca^{2+}$ Signalling during Plant Defence

The whole plant approach allowed researchers to detect long-range $Ca^{2+}$ waves, but most of these studies were performed in response to abiotic stimuli and were exploiting either cameleon derivatives such as the YCNano-65 FRET sensor [55,56] or Bioluminescence Resonance Energy Transfer (BRET) [57]. Using the BRET system, $Ca^{2+}$ waves propagating from the root to the shoot were reported in response to salt [57]. A similar study reported comparable $Ca^{2+}$ waves and that the Two-Pore Channel (TPC1) vacuolar plant $Ca^{2+}$ channel appeared to be a main player in this spreading [55]. To our knowledge, in response to biotic stimuli, the most significant data displaying systemic $Ca^{2+}$ responses have come from plant–insect interactions [58]. Wounding or caterpillar feeding on a specific Arabidopsis leaf was found to elicit $Ca^{2+}$ signals on the neighbouring leaves. Interestingly, TPC1 appeared to be involved in the systemic propagation of the $Ca^{2+}$ wave but not to be involved in the generation of the primary $Ca^{2+}$ transient observed at the feeding point [58]. The recent emergence of intensity-based $Ca^{2+}$ sensors has made it possible to evaluate the spatio-temporal specificity of immune responses as well as the direction in which the PAMP-induced $Ca^{2+}$ waves propagate [59]. Using R-GECO as an intensity-based $Ca^{2+}$ sensor, Keinath et al. studied $Ca^{2+}$ responses induced by Pathogen-Associated Molecular Patterns (PAMPS) such as flagellin and chitin [60]. They showed that flagellin, known to induce monophasic $Ca^{2+}$ variations in cultured cells when $Ca^{2+}$ changes are monitored with aequorin, was also able to induce $Ca^{2+}$ variations in roots. They localized these variations to the elongation zone where callose deposition takes place and immune-responsive genes are upregulated supporting a clear correlation between $Ca^{2+}$ changes and defence responses [60].

Overall, these few examples highlight the complexity of $Ca^{2+}$ signalling and illustrate how monitoring $Ca^{2+}$ signals at the cell tissue or whole plant level and the use of alternative $Ca^{2+}$ probes from the available panel can be very helpful to decipher both immune and symbiotic responses.

## 3. Ca²⁺ Decoding Processes and Plant Immunity

The information encrypted by Ca$^{2+}$ signals previously described (Section 2.1) needs to be decoded and relayed by Ca$^{2+}$ binding proteins in order to be converted into biological responses (Figure 1). Following Ca$^{2+}$ binding, sensors undergo conformational changes leading either to the regulation of their own catalytic activity or to their interaction with target proteins [61]. In plants, Ca$^{2+}$ sensors are classified into three sub-groups including calmodulin (CaM) and calmodulin-like proteins (CMLs), Ca$^{2+}$-dependent protein kinases (CDPKs) and calcineurin B-Like proteins (CBLs) (for reviews see [62–65]). In this review, we will focus and present information on CaM, CMLs and CDPKs, due to increasing evidence for their involvement in plant immunity. The importance of Ca$^{2+}$ signalling in symbiosis has been recently reviewed, as well as the function of CBLs, which are mainly related to the regulation of membrane proteins involved in plant development, nutrition and abiotic stress [65].

### 3.1. CaM and CaM Binding Proteins in Plant Immune Responses

CaM is the prototype of a Ca$^{2+}$-binding protein and is found in all eukaryotic cells. CaM acts as part of a Ca$^{2+}$ signalling pathway by modifying its interactions with various CaM-binding proteins (CaMBP). Genome analysis of the model plant *A. thaliana* revealed the presence of 3 distinct CaM isoforms (encoded by seven genes) [17,66]. Until now, most of our knowledge about CaM mainly relies on the identification of the repertoire of CaMBPs [67]. During the last decade, the development of large scale yeast two-hybrid protein array screens, as well as proteomic analyses, increased our knowledge of CaM targets [67–69]. We can conclude from these analyses that CaM exerts a regulatory role on a wide variety of cellular processes by modulating the activity of various proteins such as channels, enzymes and transcriptional regulators [5,68,70,71]. Although many CaM-binding transcription factors (TFs) have been identified in diverse families of DNA binding proteins including plant-specific TFs [5,72], in most cases, the biological significance of the interaction with CaM remains to be elucidated.

CaMBPs with predicted or demonstrated functions during plant immunity have been reported [5,24]. For example, several pathogen-induced CaM-binding TFs have been associated with plant defence responses by acting on homeostasis regulation by salicylic acid (SA), a defence-associated hormone in plants [63,73]. The production of SA in *A. thaliana* infected cells is enhanced by up-regulation of the expression of *ICS1* (*Isochorismate synthase 1*) and *EDS1* (*Enhanced Disease Susceptibility 1*) genes and the expression of *ICS1* and *EDS1* is positively and negatively controlled by CBP60g and CAMTA3/AtSR1, respectively, two CaM-binding TFs [73–75].

CBP60g belongs to a plant-specific DNA-binding protein family comprising eight members in Arabidopsis [74,76,77]. The *cbp60g* knock-out mutant exhibits defects in pathogen-induced SA accumulation and shows enhanced susceptibility to the *Pseudomonas syringae* phytopathogenic bacteria [74]. A variant of CBP60g unable to bind CaM, does not restore SA production and defence and failed to complement the *cbp60g* mutant. This shows the importance of CaM in regulating CBP60g function and its contribution to plant immunity [74]. The role of other CBP60g-related proteins have been explored during plant–microbe interaction and to date, only CBP60a seems to contribute to plant immunity [77]. If CaM-binding is also crucial for the biological function of CBP60a, *cbp60a* mutations reduced *P. syringae* growth in planta, indicating that CBP60a acts as a negative regulator of immunity whereas CPB60g acts as a positive regulator [77]. Moreover, it was recently demonstrated that CML46 and CML47 negatively control SA accumulation in Arabidopsis and that this effect is genetically linked and additive to that of CBP60a [78]. Altogether, these data indicate a complex regulation of SA-dependent processes involving related TFs possessing (or not) CaM-binding activity and highlight the importance of the Ca$^{2+}$-CaM/CML complex in the activation of immune responses.

The CBP60 transcriptional regulator family is not the only one regulated by Ca$^{2+}$ and CaM. Indeed, the CaM-binding Transcription Activator (CAMTA) family certainly constitutes one of the most important Ca$^{2+}$/CaM-regulated TF families in plants [79,80]. CAMTAs are key players in various plant biological processes including disease resistance and abiotic stress tolerance [81]. One of the pioneer studies concerned the functional analysis of CAMTA3, also known as AtSR1 (*Arabidopsis thaliana* Signal Responsive 1), in plant defence responses against pathogens [82] and its contribution as a negative regulator of the SA pathway [75]. Arabidopsis *camta3/Atsr1* Knock-Out mutants display elevated SA levels, high expression of *EDS1* and constitutive defence responses. CaM binding is crucial for the function of CAMTA3 in the control of SA production and defence [75]. Surprisingly, *camta3* mutants are more resistant to *P. syringae*, *Botrytis cinerea* and *Sclerotinia sclerotiorum* [82,83] but are more susceptible to insect attack (i.e., *Trichophusia*) [83]. These data suggest that CAMTA3 negatively regulates plant defence against biotrophic and necrotrophic pathogens by controlling endogenous levels of SA [75,82]. In addition, CAMTA3 was proposed to control plant resistance to herbivory insects through the regulation of glucosinolates metabolism [84,85].

Other studies revealed additional roles for CAMTA3/AtSR1 in defence. For instance, AtCAMTA3 negatively regulates plant immunity following PAMP recognition as well as non-host resistance to *Xanthomonas oryzae* [80]. CAMTA3 may negatively regulate PTI (PAMP-triggered Immunity) by targeting BAK1 [83]. More recently, a transcriptome comparative analysis using CAMTA modified transgenic plants has demonstrated activation of defence genes involved in both PTI and ETI (Effector-Triggered Immunity), suggesting that CAMTA could define an early convergence point in these two signalling pathways [86]. To date, it has not yet been elucidated how CaM can balance the activation of CBP60g and/or CAMTA3 upon pathogen attack with how the $Ca^{2+}$ signatures are integrated in this transcriptional pathway. Efforts are being made in this perspective and dynamic mathematical models incorporating several parameters are being developed with the aim of predicting regulation networks according to the generated $Ca^{2+}$ signatures [87,88].

Although the CaM contribution in plant immunity was mainly revealed by the identification of CaMBP, it is now clear how CaM plays a pivotal role in the fine tuning of immune responses by acting either as a positive or a negative regulator of defence responses. Identification of the whole set of CaMBPs is certainly not complete and other CaM-binding TFs including several members of TGA, WRKY, MYB and NAC families also contribute to plant immunity either positively or negatively [63,88,89]. For example, TGA3 and several WRKY transcription factors such as WRKY7 and WRKY53 can interact with CaM in a $Ca^{2+}$-dependent manner [68] but the effects of this interaction on the physiological and/or biochemical function of these TFs remain unknown.

## 3.2. CMLs: Emerging Plant $Ca^{2+}$ Sensors in Immunity

In addition to the presence of the typical CaMs, plants also possess a broad range of divergent forms of CaM called CaM-like proteins (CMLs, with 50 members in *A. thaliana*) [17,19]. Like CaMs, CMLs contain several EF-hand motifs, are predicted to bind $Ca^{2+}$ ions and do not possess other known functional domains [17,19]. Whereas CaM encoding genes are uniformly and highly expressed, the expression patterns of *CMLs* vary according to plant developmental stages, tissues and environmental stimuli, indicating that each CML may have a specific role in plants [17]. Indeed, data on the physiological relevance of CMLs during plant physiology and more specifically in plant immunity have emerged during the last decade (reviewed in [6,90–92]).

Although we cannot rule out functional redundancy between members of the CaM/CML family, accumulating evidence indicate that deregulation of individual *CaM/CML* gene expression or loss of a CML function in mutated plants affect plant defence responses to various pathogens. The first report involving CMLs in plant defence came from gain-of-function experiments by overexpressing CMLs. The over-expression of soybean CMLs (i.e., SCaM-4/-5) in tobacco confers enhanced resistance to a wide spectrum of virulent and avirulent pathogens, including bacteria, fungi and viruses [93]. When constitutively expressed in *Arabidopsis*, SCaM-5 confers enhanced resistance to *Pseudomonas syringae* infection [94] whereas over-expression of SCaM-1 (a typical CaM) does not, which suggests that SCAM-4/-5 are specifically recruited in response to pathogens [94]. Although obtained in heterologous plant systems, these results suggest that

CaM/CML isoforms are components of signal transduction pathways leading to disease resistance. Interestingly, it was later shown that the overexpression of SCaM-4 in soybean stimulated resistance to the oomycete *Phytophtora sojae* and to two necrotrophic fungal pathogens (*Alternaria tenuissima* and *Phomopsis longicolla*) supporting the idea that CaM/CMLs do take part in plant immune responses [95].

Loss-of- function genetic approaches also demonstrated roles in plant immunity for several CMLs. For instance, silencing *APR134* in tomato suppresses the hypersensitive response (HR), whereas overexpression of the APR134 orthologue from *Arabidopsis* CML43 stimulates the HR in response to an avirulent strain of *P. syringae* [96]. Similarly, *cml24* knockout in *Arabidopsis* impairs the HR response and reduces nitric oxide production following PAMP recognition [97]. Gain- and loss-of-function strategies were also developed on Arabidopsis CML8 and CML9 to evaluate their physiological function in plant stress responses. These two CMLs are positive regulators of plant defence against different strains of *P. syringae* [98–100]. CML9 was first described to be involved in plant responses to abiotic stress [101] and later shown to also contribute to plant immune responses [97]. The *cml9* mutants and *CML9* overexpressing lines exhibit enhanced or reduced susceptibility to virulent strains of *P. syringae*, respectively [98]. These phenotypes can be explained by alterations of flagellin-induced responses, including deposition of callose papillae and modifications of defence-related genes expression [98]. CML8 also takes part in plant immune responses against *P. syringae* but compared to CML9, the enhanced resistance observed in CML8 overexpressing lines relies mainly on SA-dependent responses [99]. Emerging data indicate that the regulation of plasmodesmata by plant cells is critical for the establishment of plant defence signalling [102]. Indeed, plasmodesmata are plasma membrane pores that establish cytoplasmic and membrane continuity between cells [102]. It was recently identified that closure of plasmodesmata in response to bacterial flagellin is mediated by the plasmodesmatal-localized CML41 [103]. CML41 is transcriptionally upregulated by PAMPs and facilitates callose deposition at plasmodesmata following flagellin treatment. Using *amiRNA* and *CML41* overexpressing lines, Xu et al. reported that CML41 acts as a positive regulator of defence against *Pseudomonas syringae* [103].

Ca$^{2+}$ signalling is not only required for defence mechanisms upon microbial pathogen attack but also in response to herbivores [58,104–106]. Data from Mithoefer's group indicate that the loss of function of Arabidopsis CML42 enhances resistance to *Spodoptera littoralis* which is correlated with the up-regulation of jasmonic acid-responsive genes and to an accumulation of aliphatic glucosinolates [107]. In contrast to CML42, CML37 acts positively on defence against *S. littoralis* [108]. Indeed, a *cml37* Knock-Out mutant exhibits an enhanced susceptibility to herbivory which is correlated to a lower level of the bioactive form of jasmonate (i.e., JA-Ile) known to be crucial in plant defence coordination against insects [108]. These data suggest opposite roles for CML37 and CML42 in insect herbivory resistance. We cannot exclude that other CMLs such as CML9, CML11, CML12, CML16, CML17 and CML23 could participate in insect defence responses since the corresponding genes are significantly up-regulated in plants treated with oral secretion of the lepidopteran herbivore [109].

Among the range of processes regulated by CaM/CMLs, CMLs are also involved in the suppression of post-transcriptional gene silencing (PTGS), a regulatory mechanism targeting mRNA content [110]. Plant viruses can act both as inducers and as targets of PTGS and this led to the idea that PTGS evolved as a defence mechanism against viruses in plants [110]. Interestingly, a tobacco CML termed rgs-CaM (for regulator of gene silencing) has been reported to act as a suppressor of PTGS [110] and to play a role in antiviral defence by modulating virus-induced RNA silencing [111,112]. The rgs-CaM exerts its antiviral activity by binding and controlling the degradation of viral RNA silencing suppressors. This constitutes the first example of an interaction between a CML and a pathogen protein [111]. Recently, the role of this rgs-CaM in systemic acquired resistance against cucumber mosaic virus has been described in tobacco plants [113]. It was proposed that rgs-CaM functions as an immune receptor that induces salicylic acid signalling by simultaneously perceiving both viral RNA silencing suppressors and Ca$^{2+}$ influx [113].

To date, most of the data describing the involvement of CMLs in different cellular processes associated with plant physiology remain descriptive and new research is now required to decipher the molecular mechanisms controlled by these CMLs. The identification of CML-interacting partners will be crucial to clarify how these CMLs exert their action at the molecular level in plant immunity. For example, CML9, previously described to positively regulate plant immunity, was reported to interact with transcription factors such as the WRKYs and the TGAs, two classes of transcription factors known to play key roles in the regulation of defence processes [69]. PRR2 (*PSEUDO-RESPONSE REGULATOR 2*), a plant specific transcription factor was also described to interact in planta with CML9 but not with the typical CaM [114]. Using a reverse genetic strategy in *A. thaliana*, PRR2 was shown to act as a positive regulator of plant immunity through SA-dependent responses [115].

In many cases, the regulation of these TFs by CML9 or by other CMLs remain presumptive. The next challenge will be to elucidate if all CMLs really act as $Ca^{2+}$ sensors/$Ca^{2+}$ relay proteins, are able to interact with CML-binding proteins and to modulate their activity, as demonstrated for the CaMs. Several questions remain unanswered: Do CaM and CMLs share the same targets? What could be the consequences of such interactions? Interestingly, Yoo et al. showed that the typical CaM (SCaM-1) and the CML (SCaM-4) physically interact with MYB2, a TF that regulates the expression of salt- and dehydration-responsive genes in Arabidopsis [116]. However, SCaM-4 enhances the DNA binding properties of MYB2 whereas SCaM-1 inhibits MYB2 DNA binding [116]. Although these data need validation, different modes of regulation are suggested depending on the interaction with either a CaM or a CML. CaM/CMLs have been reported to interact with different nuclear proteins [68] but do these interactions help to recruit other actors into transcriptional complexes and regulate their activity? What is the contribution of $Ca^{2+}$ in these interactions? Efforts are now made to answer these questions and to better understand the contribution of these sensors in $Ca^{2+}$ signalling with a particular interest on plant specific CMLs.

### 3.3. CDPKs: Positive Regulators of Plant Immune Responses

Beside CaMs and CMLs, plants also possess another class of $Ca^{2+}$ sensors referred to as $Ca^{2+}$ dependent protein kinases or CDPKs. Found only in green algae, land plants and unicellular protists, CDPKs are important $Ca^{2+}$ decoders and relays in plant defence signalling against various types of pathogens (for review: [4]). Their number expanded during evolution of land plants to reach ~30 members in angiosperms (e.g., 34 in *A. thaliana*—29 in *S. lycopersicum*) [117]. The family architecture is nonetheless conserved from mosses to angiosperms, and formed of four distinct groups [22,118]. In contrast to CaMs and CMLs, CDPKs are unique in the repertoire of $Ca^{2+}$ decoders since they combine into a single module $Ca^{2+}$ sensing and downstream signal propagation capabilities. They are composed of a variable N-terminal part, a kinase domain and an activation domain. The activation domain contains an auto-inhibitory pseudo-substrate linked to a CaM-like domain typically containing four EF-hands [119,120]. Upon $Ca^{2+}$ binding by the CaM-like domain, a conformational change displaces the pseudo-substrate from the kinase to allow for downstream phosphorylation events. The dynamic range of $Ca^{2+}$ concentrations that activate a given CDPK therefore depends on the CaM-like domain affinity for $Ca^{2+}$. This likely accounts in part for the specific decoding of different $Ca^{2+}$ variations. Of note, some CDPKs do not show strict $Ca^{2+}$-dependent kinase activity. For example, *Arabidopsis thaliana* CPK7, 8, 13, and CPK30 showed more than half-maximal in vitro syntide-2 phosphorylation activity at very low $Ca^{2+}$ concentrations (1 to 10 nM), with little to no additional kinase activity at higher $Ca^{2+}$ concentrations [121,122]. Whether this reflects true $Ca^{2+}$-independence in vivo or specific $Ca^{2+}$ requirements depending on substrates remains to be clarified [121].

Functional analyses in planta revealed that CDPKs are important components of the plant immune system. Acting alongside and in synergy with the Mitogen Activated Protein Kinases (MAPKs)-dependent signalling cascade, *Arabidopsis* CPK4, 5, 6 and 11 positively regulate defence gene expression upon bacterial flagellin perception [30]. CPK4, 5, 6 and 11 collectively and redundantly contribute to PTI-induced resistance against *Pseudomonas syringae* [30]. CPK5, but also CPK4, 6 and 11, are able to phosphorylate AtRBOHD, the main ROS producing enzyme acting in immunity [122,123]. A model was proposed in which CPK5 and RBOHD sustain a ROS-mediated cell-to-cell communication to reach distal sites from the initial PAMP perception area [124]. Through phosphorylation of the HsfB2a transcription factor, Arabidopsis CPK3 and CPK13 are important regulators of the *PDF1.2* defence gene induction after wounding by the caterpillar *Sodoptera littoralis* [122]. These positive immune regulatory functions of CDPKs prompted their biotechnological use in crop protection against pathogens. For example, overexpression of full length OsCPK4 in rice leads to enhanced disease resistance against *Magnaporthe oryzae* [125]. This enhanced disease resistance in transformed plants is likely due to their higher basal levels of salicylic acid and the resulting potentiation of defence gene induction [125].

In addition to the control of defence gene induction, another important aspect of CDPK function is the control of plant cell death in response to pathogens. Early work on CDPKs in *Nicotiana* sp. showed that CDPK2 and CDPK3 are required for the PCD response triggered after perception of the *Cladosporium fulvum* race-specific Avr4 or Avr9 elicitors [126]. This pioneering work was further expanded in *Arabidopsis* with the demonstration that CPK1 and 2 control the onset of HR upon challenge with avirulent *Pseudomonas syringae* and subsequent NLR-mediated effector recognition [127]. In line with this, several constitutively active CDPKs harbour cell death inducing activity. Auto-active CDPKs (CDPK-VKs) are devoid of their activation domains and therefore do not need $Ca^{2+}$ inputs to be in an active state. Expression of CPK5-VK in Arabidopsis leaf protoplasts leads to cell death and this requires kinase activity [123]. Similar results were obtained in transgenic potato plants stably expressing StCPK5-VK under a pathogen inducible promoter [128]. This latter approach led to increased plant resistance against the hemi-biotrophic oomycete *Phytophthora infestans*. As a trade-off however, plants became more susceptible to the necrotrophic fungus *Alternaria solani* [128]. This contrasted outcome in plant defence toward pathogens with different lifestyles might prevent wider use of these genetic tools in crop protection. PCD induction can also be observed in heterologous expression systems, as demonstrated by the cell death inducing activity of transiently expressed barley CDPK4-VK or Arabidopsis CPK5-VK in tobacco leaves [123,129]. Activation of any given CDPK does not seem sufficient *per se* to provoke PCD since not all CDPKs tested under their auto-active configuration induce PCD [123,129]. The kinase and auto-inhibitory domains of CDPKs being under strong purifying selection [22], it is therefore likely that the specific properties of CDPKs lie, in part, in other features such as their hypervariable termini ends. For example, swapping the N-terminal variable domains of tomato SlCDPK2 and SlCDPK5 exchanged their respective subcellular localizations and their ability to phosphorylate StRBOHB [130]. The existence of cross-species PCD-inducing properties supports nonetheless some degree of functional conservation in CDPKs mode(s) of action through the evolution of land plants.

As already, mentioned, functional redundancy can be problematic when studying particular gene functions within a large family. Although gain-of-function auto-active CDPKs proved of immense value in this context, *cdpk* mutants remain necessary tools and high-order *cdpk* mutants are often required to observe a phenotype of interest such as pathogen susceptibility [30,127]. In contrast to that, a suppressor screen of the *exo70B1* autoimmune phenotype uncovered several *cpk5* loss-of-function alleles [131]. The *exo70B1*-mediated cell death and enhanced resistance to *Golovinomyces cichoracearum* phenotypes were lost in the absence of a functional CPK5, but not of CPK4, 6, or 11 [131]. This advocates for a specific function of CPK5 that cannot be fulfilled here by its otherwise functionally redundant close homologues. In addition to CPK5, the *exo70B1*-mediated autoimmunity also requires the atypical truncated NLR resistance gene TN2 [132]. A tripartite interaction between Exo70B1, CPK5 and TN2 therefore controls cell death. CPK5 kinase activity and its membrane association are both required although the exact mechanism is not completely understood [131]. Whether other CDPKs are also involved in similar interactions, and whether this mechanism represents an innovation present only in *Arabidopsis*, remain open questions.

### 3.4. CDPKs: Also Negative Regulators of Plant Immune Responses

In addition to these positive roles, CDPKs also perform negative functions in plant defence. For example, overexpression of OsCPK12 increased rice susceptibility to avirulent and virulent *Magnaporthe grisae* strains [133] while in barley, overexpressing an auto-active variant of HvCDPK3 increased susceptibility to powdery mildew (*B. graminis* f. sp. *hordei*) [129]. In *Arabidopsis*, perception of several PAMPs by their cognate receptors requires the activity of Somatic Embryogenesis Receptor Kinases 3 (SERK3), a plasma membrane localized receptor and co-regulator also referred to as BAK1 (BRI1-Associated Kinase 1) [134]. In a forward genetic screen set to discover modifiers of the *bak1–5* loss-of-function allele [135], Monaghan and colleagues recovered *cpk28* mutants. The loss of CPK28 reverted the impaired ROS production phenotype observed in *bak1–5* after PAMP treatments [136]. CPK28 buffers immune responses by modulating the proteasome-mediated turnover of the BIK1 receptor-like cytoplasmic kinase, a shared component of signalling pathways controlled by several PRRs [136]. The search for E3 ligases targeting BIK1 led to the identification of PUB25 and 26, two related E3 ligases of the U-box family whose enzymatic activities towards BIK1 are activated through CPK28-dependent phosphorylation [136]. It is tempting to speculate that the $Ca^{2+}$ influx resulting from PRR activation first participates in ROS production through CPK5 activity, before CPK28 dampens immune outputs. In line with this hypothesis, in vitro assays determined that CPK28 is indeed $Ca^{2+}$ responsive [137]. Interestingly, CPK28 is also able to bind $Ca^{2+}$/CaM, however, this interaction negatively affects its kinase activity in vitro [138]. $Ca^{2+}$ therefore seems to have a dual role in regulating CPK28 activity, thereby, adding another layer of complexity that has yet to be resolved.

Globally, plants have a wide repertoire of $Ca^{2+}$-binding proteins whose functions in plant physiology start to be unravelled. Far from being exhaustive, we have summarized here some examples that describe the importance of certain CaM/CMLs and CDPKs in plant immunity [17,18]. The picture remains nevertheless complex as duplications events and the expected neo-functionalization is sometimes confronted with functional redundancy observed in higher plants. Plants belonging to a sister clade of angiosperms such as liverworts or hornworts could be of interest in discovering yet hidden functions of some calcium sensors. The pursuit of research efforts on these plant specific sensors and particularly the identification of novel CaM/CML-binding proteins and CDPK substrates will be important milestones to get a better view of the many $Ca^{2+}$-regulated events.

### 4. Concluding Remarks and Outlooks

Increasing data support that $Ca^{2+}$, $Ca^{2+}$ sensors and their targets are positioned among the first actors setting up the plant response to biotic interactions. One of the challenges of the coming years will be to understand and expect plant behaviour in a changing environment, with the long-term goal to breed stress-tolerant crops for agriculture. However, although $Ca^{2+}$ signals are the very first step in the signaling pathway, it appears that their manipulation to improve plant resistance could be very challenging. Indeed, calcium signals result from the intricate interaction between pathways and components involved in both their generation and their dissipation such as calcium channels, exchangers or calcium pumps [139]. Nevertheless, some reverse genetic approaches have demonstrated that plant immunity can be modulated through the control of calcium signals by interfering with their generation [140,141] or their dissipation through the activity modulation of calcium pumps [142,143]. Such examples in the literature are rather scarce. Envisioning strategies based on calcium variation modulation to improve plant resistance could be hazardous due to a lack of knowledge concerning actors involved in calcium homeostasis associated to specific plant-microorganism interactions.

An alternative and more amenable approach would be rather to interfere with the function of the $Ca^{2+}$ sensors or their targets that decode $Ca^{2+}$ messages. To date, studies about CaM/CML and CDPKs focus mainly on their interaction with plant target proteins. However, recent and promising data indicate that $Ca^{2+}$ signalling components can be themselves the direct targets of pathogen effectors. In the context of plant immunity, studies on these specifically targeted plant calcium sensors by pathogen effectors might provide a new way for breeding plant resistance. Indeed, it was recently demonstrated that plant CaM is required by the HopE1 effector from *Pseudomonas syringae* to target MAP65, a microtubule-associated protein, to reduce plant immune responses [144]. In a similar way, the *Xanthomonas* AvrBsT effector, able to acetylate ACIP1, a microtubule-associated protein in Arabidopsis and required for both PTI and ETI, possesses a CaM-binding region. AvrBsT is the first member of the YopJ family known to suppress effector-triggered plant immunity [145]. It was reported that AvrBsT is able to interact with CaM in a $Ca^{2+}$-dependent manner and that a mutation of the AvrBsT CaM-binding domain alters or delays the hypersensitive response phenotype (MB Mudgett, personal communication and reports from MPMI meeting 2016 [146]). Whereas CaM has long been known to be a co-factor for mammalian pathogen toxins [147]; these new data obtained with plants support new roles for CaM in plant immunity.

In this review, we focused on $Ca^{2+}$ signalling and biotic stress responses. We point out that CMLs and CDPKs can exhibit a dual function as either positive and/or negative regulators in plant biotic stresses. Moreover, many other data highlight the contribution of $Ca^{2+}$ and $Ca^{2+}$ decoding processes following abiotic stress perception [6,23,63,148]. Studies deciphering plant responses to simultaneously applied abiotic and biotic stress remain sparse. More importantly, these different types of stress are often interrelated in the field and several forms of abiotic stress significantly affect the resistance of plants to bacteria, fungi, viruses and insects [149]. Therefore, is it really attractive to hypothesize that $Ca^{2+}$, $Ca^{2+}$ sensors and their respective targets can be at the crossroads of various signalling pathways and that they could be good candidates to act as central integrators involved in the fine tuning of plant physiological responses to pathogens under fluctuating environmental conditions. As stated in the introduction, it is obvious that $Ca^{2+}$ signalling is not acting as a unique contributor in plant stress responses and we cannot rule out crosstalk between $Ca^{2+}$ and ROS, as well as other second messengers. The next mid and long-term challenges should better characterize the interplay between these signalling pathways.

**Acknowledgments:** We would like to thank Julie Cullimore for proofreading and corrections and Julie Mazard for her comments on this manuscript. This work was supported by the Université Paul Sabatier (Toulouse, France), the CNRS (France), the French Laboratory of Excellence project "TULIP" (ANR-10-LABX-41; ANR-11-IDEX-0002-02) and by the Agence Nationale de la Recherche (ANR) (ANR-17-CE20-0017-01) thanks to the CaPPTure project.

**Conflicts of Interest:** The authors declare no conflicts of interest.

## References

1. Van Loon, L.C. The intelligent behavior of plants. *Trends Plant Sci.* **2016**, *21*, 286–294. [CrossRef] [PubMed]
2. Kissoudis, C.; van de Wiel, C.; Visser, R.G.; van der Linden, G. Future-proof crops: Challenges and strategies for climate resilience improvement. *Curr. Opin. Plant Biol.* **2016**, *30*, 47–56. [CrossRef] [PubMed]
3. Kissoudis, C.; van de Wiel, C.; Visser, R.G.; van der Linden, G. Enhancing crop resilience to combined abiotic and biotic stress through the dissection of physiological and molecular crosstalk. *Front. Plant Sci.* **2014**, *5*, 207. [CrossRef] [PubMed]
4. Boudsocq, M.; Sheen, J. CDPKs in immune and stress signaling. *Trends Plant Sci.* **2013**, *18*, 30–40. [CrossRef] [PubMed]
5. Poovaiah, B.W.; Du, L.; Wang, H.; Yang, T. Recent advances in calcium/calmodulin-mediated signaling with an emphasis on plant-microbe interactions. *Plant Physiol.* **2013**, *163*, 531–542. [CrossRef] [PubMed]
6. Ranty, B.; Aldon, D.; Cotelle, V.; Galaud, J.P.; Thuleau, P.; Mazars, C. Calcium sensors as key hubs in plant responses to biotic and abiotic stresses. *Front. Plant Sci.* **2016**, *7*, 327. [CrossRef] [PubMed]
7. Zipfel, C.; Oldroyd, G.E. Plant signalling in symbiosis and immunity. *Nature* **2017**, *543*, 328–336. [CrossRef] [PubMed]
8. Doke, N. Involvement of superoxide anion generation in the hypersensitive response of potato tuber tissues to infection with an incompatible race of phytophthora infestans and to the hyphal wall components. *Physiol. Plant Pathol.* **1983**, *23*, 345–357. [CrossRef]

9.    Atkinson, M.M.; Keppler, L.D.; Orlandi, E.W.; Baker, C.J.; Mischke, C.F. Involvement of plasma membrane calcium influx in bacterial induction of the K$^+$/H$^+$ and hypersensitive responses in tobacco. *Plant Physiol.* **1990**, *92*, 215–221. [CrossRef] [PubMed]

10.   Stael, S.; Kmiecik, P.; Willems, P.; Van Der Kelen, K.; Coll, N.S.; Teige, M.; Van Breusegem, F. Plant innate immunity—Sunny side up? *Trends Plant Sci.* **2015**, *20*, 3–11. [CrossRef] [PubMed]

11.   Berridge, M.J.; Lipp, P.; Bootman, M.D. The versatility and universality of calcium signalling. *Nat. Rev. Mol. Cell Biol.* **2000**, *1*, 11–21. [CrossRef] [PubMed]

12.   Steinhorst, L.; Kudla, J. Signaling in cells and organisms—calcium holds the line. *Curr. Opin. Plant Biol.* **2014**, *22*, 14–21. [CrossRef] [PubMed]

13.   McAinsh, M.R.; Brownlee, C.; Hetherington, A.M. Visualizing changes in cytosolic-free Ca$^{2+}$ during the response of stomatal guard cells to abscisic acid. *Plant Cell* **1992**, *4*, 1113–1122. [CrossRef] [PubMed]

14.   McAinsh, M.R.; Hetherington, A.M. Encoding specificity in Ca$^{2+}$ signalling systems. *Trends Plant Sci.* **1998**, *3*, 32–36. [CrossRef]

15.   Yap, K.L.; Ames, J.B.; Swindells, M.B.; Ikura, M. Diversity of conformational states and changes within the ef-hand protein superfamily. *Proteins* **1999**, *37*, 499–507. [CrossRef]

16.   Day, I.S.; Reddy, V.S.; Shad Ali, G.; Reddy, A.S. Analysis of EF-hand-containing proteins in Arabidopsis. *Genome Biol.* **2002**, *3*. [CrossRef]

17.   Zhu, X.; Dunand, C.; Snedden, W.; Galaud, J.P. Cam and cml emergence in the green lineage. *Trends Plant Sci.* **2015**, *20*, 483–489. [CrossRef] [PubMed]

18.   Edel, K.H.; Marchadier, E.; Brownlee, C.; Kudla, J.; Hetherington, A.M. The evolution of calcium-based signalling in plants. *Curr. Biol.* **2017**, *27*, R667–R679. [CrossRef] [PubMed]

19.   McCormack, E.; Tsai, Y.C.; Braam, J. Handling calcium signaling: Arabidopsis CaMs and CMLs. *Trends Plant Sci.* **2005**, *10*, 383–389. [CrossRef] [PubMed]

20.   Snedden, W.A.; Fromm, H. Calmodulin as a versatile calcium signal transducer in plants. *New Phytol.* **2001**, *151*, 35–66. [CrossRef]

21.   Batistic, O.; Kudla, J. Plant calcineurin B-like proteins and their interacting protein kinases. *Biochim. Biophys. Acta* **2009**, *1793*, 985–992. [CrossRef] [PubMed]

22.   Valmonte, G.R.; Arthur, K.; Higgins, C.M.; MacDiarmid, R.M. Calcium-dependent protein kinases in plants: Evolution, expression and function. *Plant Cell Physiol.* **2014**, *55*, 551–569. [CrossRef] [PubMed]

23.   Zeng, H.; Xu, L.; Singh, A.; Wang, H.; Du, L.; Poovaiah, B.W. Involvement of calmodulin and calmodulin-like proteins in plant responses to abiotic stresses. *Front. Plant Sci.* **2015**, *6*, 600. [CrossRef] [PubMed]

24.   Yuan, P.; Jauregui, E.; Du, L.; Tanaka, K.; Poovaiah, B.W. Calcium signatures and signaling events orchestrate plant-microbe interactions. *Curr. Opin. Plant Biol.* **2017**, *38*, 173–183. [CrossRef] [PubMed]

25.   Zhang, L.; Du, L.; Poovaiah, B.W. Calcium signalling and biotic defense responses in plants. *Plant Signal Behav.* **2014**, *9*, 11. [CrossRef] [PubMed]

26.   Jones, J.D.; Dangl, J.L. The plant immune system. *Nature* **2006**, *444*, 323–329. [CrossRef] [PubMed]

27.   Zipfel, C.; Robatzek, S.; Navarro, L.; Oakeley, E.J.; Jones, J.D.; Felix, G.; Boller, T. Bacterial disease resistance in arabidopsis through flagellin perception. *Nature* **2004**, *428*, 764–767. [CrossRef] [PubMed]

28.   Lecourieux, D.; Lamotte, O.; Bourque, S.; Wendehenne, D.; Mazars, C.; Ranjeva, R.; Pugin, A. Proteinaceous and oligosaccharidic elicitors induce different calcium signatures in the nucleus of tobacco cells. *Cell Calcium* **2005**, *38*, 527–538. [CrossRef] [PubMed]

29.   Grant, M.; Brown, I.; Adams, S.; Knight, M.; Ainslie, A.; Mansfield, J. The RPM1 plant disease resistance gene facilitates a rapid and sustained increase in cytosolic calcium that is necessary for the oxidative burst and hypersensitive cell death. *Plant J.* **2000**, *23*, 441–450. [CrossRef] [PubMed]

30.   Boudsocq, M.; Willmann, M.R.; McCormack, M.; Lee, H.; Shan, L.; He, P.; Bush, J.; Cheng, S.H.; Sheen, J. Differential innate immune signalling via Ca$^{2+}$ sensor protein kinases. *Nature* **2010**, *464*, 418–422. [CrossRef] [PubMed]

31.   Gilroy, S.; Read, N.D.; Trewavas, A.J. Elevation of cytoplasmic calcium by caged calcium or caged inositol triphosphate initiates stomatal closure. *Nature* **1990**, *346*, 769–771. [CrossRef] [PubMed]

32.   Knight, M.R.; Campbell, A.K.; Smith, S.M.; Trewavas, A.J. Transgenic plant aequorin reports the effects of touch and cold-shock and elicitors on cytoplasmic calcium. *Nature* **1991**, *352*, 524–526. [CrossRef] [PubMed]

33.   Chandra, S.; Stennis, M.; Low, P.S. Measurement of Ca$^{2+}$ fluxes during elicitation of the oxidative burst in aequorin-transformed tobacco cells. *J. Biol. Chem.* **1997**, *272*, 28274–28280. [CrossRef] [PubMed]

34. Hahn, M.G. Microbial elicitors and their receptors in plants. *Annu. Rev. Phytopathol.* **1996**, *34*, 387–412. [CrossRef] [PubMed]

35. Hahn, M.G.; Darvill, A.G.; Albersheim, P. Host-pathogen interactions: XIX. The endogenous elicitor, a fragment of a plant cell wall polysaccharide that elicits phytoalexin accumulation in soybeans. *Plant Physiol.* **1981**, *68*, 1161–1169. [CrossRef] [PubMed]

36. Mithofer, A.; Fliegmann, J.; Ebel, J. Isolation of a french bean (*phaseolus vulgaris* L.) homolog to the beta-glucan elicitor-binding protein of soybean (*glycine max* L.). *Biochim. Biophys. Acta* **1999**, *1418*, 127–132. [CrossRef]

37. Blume, B.; Nurnberger, T.; Nass, N.; Scheel, D. Receptor-mediated increase in cytoplasmic free calcium required for activation of pathogen defense in parsley. *Plant Cell* **2000**, *12*, 1425–1440. [CrossRef] [PubMed]

38. Lecourieux, D.; Mazars, C.; Pauly, N.; Ranjeva, R.; Pugin, A. Analysis and effects of cytosolic free calcium increases in response to elicitors in nicotiana plumbaginifolia cells. *Plant Cell* **2002**, *14*, 2627–2641. [CrossRef] [PubMed]

39. Choi, W.G.; Hilleary, R.; Swanson, S.J.; Kim, S.H.; Gilroy, S. Rapid, long-distance electrical and calcium signaling in plants. *Annu. Rev. Plant Biol.* **2016**, *67*, 287–307. [CrossRef] [PubMed]

40. Newman, M.A.; Sundelin, T.; Nielsen, J.T.; Erbs, G. Mamp (microbe-associated molecular pattern) triggered immunity in plants. *Front. Plant Sci.* **2013**, *4*, 139. [CrossRef] [PubMed]

41. Mithofer, A.; Boland, W. Recognition of herbivory-associated molecular patterns. *Plant Physiol.* **2008**, *146*, 825–831. [CrossRef] [PubMed]

42. Maffei, M.; Bossi, S.; Spiteller, D.; Mithofer, A.; Boland, W. Effects of feeding spodoptera littoralis on lima bean leaves. I. Membrane potentials, intracellular calcium variations, oral secretions, and regurgitate components. *Plant Physiol.* **2004**, *134*, 1752–1762. [CrossRef] [PubMed]

43. Arimura, G.; Ozawa, R.; Shimoda, T.; Nishioka, T.; Boland, W.; Takabayashi, J. Herbivory-induced volatiles elicit defence genes in lima bean leaves. *Nature* **2000**, *406*, 512–515. [PubMed]

44. Allen, G.J.; Chu, S.P.; Harrington, C.L.; Schumacher, K.; Hoffmann, T.; Tang, Y.Y.; Grill, E.; Schroeder, J.I. A defined range of guard cell calcium oscillation parameters encodes stomatal movements. *Nature* **2001**, *411*, 1053–1057. [CrossRef] [PubMed]

45. Miwa, H.; Sun, J.; Oldroyd, G.E.; Downie, J.A. Analysis of calcium spiking using a cameleon calcium sensor reveals that nodulation gene expression is regulated by calcium spike number and the developmental status of the cell. *Plant J.* **2006**, *48*, 883–894. [CrossRef] [PubMed]

46. Collmer, A.; Badel, J.L.; Charkowski, A.O.; Deng, W.L.; Fouts, D.E.; Ramos, A.R.; Rehm, A.H.; Anderson, D.M.; Schneewind, O.; van Dijk, K.; et al. *Pseudomonas Syringae* Hrp type III secretion system and effector proteins. *Proc. Natl. Acad. Sci. USA* **2000**, *97*, 8770–8777. [CrossRef] [PubMed]

47. Heath, M.C. Hypersensitive response-related death. *Plant Mol. Biol.* **2000**, *44*, 321–334. [CrossRef] [PubMed]

48. Keen, N.T. Gene-for-gene complementarity in plant-pathogen interactions. *Annu. Rev. Genet.* **1990**, *24*, 447–463. [CrossRef] [PubMed]

49. Badminton, M.N.; Sala-Newby, G.B.; Kendall, J.M.; Campbell, A.K. Differences in stability of recombinant apoaequorin within subcellular compartments. *Biochem. Biophys. Res. Commun.* **1995**, *217*, 950–957. [CrossRef] [PubMed]

50. Wais, R.J.; Galera, C.; Oldroyd, G.; Catoira, R.; Penmetsa, R.V.; Cook, D.; Gough, C.; Denarie, J.; Long, S.R. Genetic analysis of calcium spiking responses in nodulation mutants of *medicago truncatula*. *Proc. Natl. Acad. Sci. USA* **2000**, *97*, 13407–13412. [CrossRef] [PubMed]

51. Wais, R.J.; Keating, D.H.; Long, S.R. Structure-function analysis of nod factor-induced root hair calcium spiking in rhizobium-legume symbiosis. *Plant Physiol.* **2002**, *129*, 211–224. [CrossRef] [PubMed]

52. Miyawaki, A.; Llopis, J.; Heim, R.; McCaffery, J.M.; Adams, J.A.; Ikura, M.; Tsien, R.Y. Fluorescent indicators for Ca²⁺ based on green fluorescent proteins and calmodulin. *Nature* **1997**, *388*, 882–887. [CrossRef] [PubMed]

53. Kosuta, S.; Hazledine, S.; Sun, J.; Miwa, H.; Morris, R.J.; Downie, J.A.; Oldroyd, G.E. Differential and chaotic calcium signatures in the symbiosis signaling pathway of legumes. *Proc. Natl. Acad. Sci. USA* **2008**, *105*, 9823–9828. [CrossRef] [PubMed]

54. Sieberer, B.J.; Chabaud, M.; Timmers, A.C.; Monin, A.; Fournier, J.; Barker, D.G. A nuclear-targeted cameleon demonstrates intranuclear Ca²⁺ spiking in *medicago truncatula* root hairs in response to rhizobial nodulation factors. *Plant Physiol.* **2009**, *151*, 1197–1206. [CrossRef] [PubMed]

55. Choi, W.G.; Gilroy, S. Plant biologists fret over stress. *eLife* **2014**, *3*, e02763. [CrossRef] [PubMed]

56. Horikawa, K.; Yamada, Y.; Matsuda, T.; Kobayashi, K.; Hashimoto, M.; Matsu-ura, T.; Miyawaki, A.; Michikawa, T.; Mikoshiba, K.; Nagai, T. Spontaneous network activity visualized by ultrasensitive $Ca^{2+}$ indicators, yellow cameleon-nano. *Nat. Methods* **2010**, *7*, 729–732. [CrossRef] [PubMed]

57. Xiong, T.C.; Ronzier, E.; Sanchez, F.; Corratge-Faillie, C.; Mazars, C.; Thibaud, J.B. Imaging long distance propagating calcium signals in intact plant leaves with the BRET-based GFP-aequorin reporter. *Front. Plant Sci.* **2014**, *5*, 43. [CrossRef] [PubMed]

58. Kiep, V.; Vadassery, J.; Lattke, J.; Maass, J.P.; Boland, W.; Peiter, E.; Mithofer, A. Systemic cytosolic $Ca^{2+}$ elevation is activated upon wounding and herbivory in arabidopsis. *New Phytol.* **2015**, *207*, 996–1004. [CrossRef] [PubMed]

59. Akerboom, J.; Carreras Calderon, N.; Tian, L.; Wabnig, S.; Prigge, M.; Tolo, J.; Gordus, A.; Orger, M.B.; Severi, K.E.; Macklin, J.J.; et al. Genetically encoded calcium indicators for multi-color neural activity imaging and combination with optogenetics. *Front. Mol. Neurosci.* **2013**, *6*, 2. [CrossRef] [PubMed]

60. Keinath, N.F.; Waadt, R.; Brugman, R.; Schroeder, J.I.; Grossmann, G.; Schumacher, K.; Krebs, M. Live cell imaging with R-GECO1 sheds light on flg22- and chitin-induced transient $[Ca^{2+}]_{cyt}$ patterns in *arabidopsis*. *Mol. Plant* **2015**, *8*, 1188–1200. [CrossRef] [PubMed]

61. Hashimoto, K.; Kudla, J. Calcium decoding mechanisms in plants. *Biochimie* **2011**, *93*, 2054–2059. [CrossRef] [PubMed]

62. Defalco, T.A.; Bender, K.W.; Snedden, W.A. Breaking the code: $Ca^{2+}$ sensors in plant signalling. *Biochem. J.* **2010**, *425*, 27–40. [CrossRef] [PubMed]

63. Reddy, A.S.; Ali, G.S.; Celesnik, H.; Day, I.S. Coping with stresses: Roles of calcium- and calcium/calmodulin-regulated gene expression. *Plant Cell* **2011**, *23*, 2010–2032. [CrossRef] [PubMed]

64. Kudla, J.; Batistic, O.; Hashimoto, K. Calcium signals: The lead currency of plant information processing. *Plant Cell* **2010**, *22*, 541–563. [CrossRef] [PubMed]

65. Kudla, J.; Becker, D.; Grill, E.; Hedrich, R.; Hippler, M.; Kummer, U.; Parniske, M.; Romeis, T.; Schumacher, K. Advances and current challenges in calcium signaling. *New Phytol.* **2018**. [CrossRef] [PubMed]

66. McCormack, E.; Braam, J. Calmodulins and related potential calcium sensors of arabidopsis. *New Phytol.* **2003**, *159*, 585–598. [CrossRef]

67. Bouche, N.; Yellin, A.; Snedden, W.A.; Fromm, H. Plant-specific calmodulin-binding proteins. *Annu. Rev. Plant Biol.* **2005**, *56*, 435–466. [CrossRef] [PubMed]

68. Popescu, S.C.; Popescu, G.V.; Bachan, S.; Zhang, Z.; Seay, M.; Gerstein, M.; Snyder, M.; Dinesh-Kumar, S.P. Differential binding of calmodulin-related proteins to their targets revealed through high-density arabidopsis protein microarrays. *Proc. Natl. Acad. Sci. USA* **2007**, *104*, 4730–4735. [CrossRef] [PubMed]

69. Popescu, S.C.; Snyder, M.; Dinesh-Kumar, S. Arabidopsis protein microarrays for the high-throughput identification of protein-protein interactions. *Plant Signal Behav.* **2007**, *2*, 416–420. [CrossRef] [PubMed]

70. Hua, B.-G.; Mercier, R.W.; Zielinski, R.E.; Berkowitz, G.A. Functional interaction of calmodulin with a plant cyclic nucleotide gated cation channel. *Plant Physiol. Biochem.* **2003**, *41*, 945–954. [CrossRef]

71. DeFalco, T.A.; Marshall, C.B.; Munro, K.; Kang, H.G.; Moeder, W.; Ikura, M.; Snedden, W.A.; Yoshioka, K. Multiple calmodulin-binding sites positively and negatively regulate Arabidopsis cyclic nucleotide-gated channel12. *Plant Cell* **2016**, *28*, 1738–1751. [CrossRef] [PubMed]

72. Galon, Y.; Finkler, A.; Fromm, H. Calcium-regulated transcription in plants. *Mol. Plant* **2010**, *3*, 653–669. [CrossRef] [PubMed]

73. Pieterse, C.M.; Van der Does, D.; Zamioudis, C.; Leon-Reyes, A.; Van Wees, S.C. Hormonal modulation of plant immunity. *Annu. Rev. Cell Dev. Biol.* **2012**, *28*, 489–521. [CrossRef] [PubMed]

74. Wang, L.; Tsuda, K.; Sato, M.; Cohen, J.D.; Katagiri, F.; Glazebrook, J. Arabidopsis cam binding protein CBP60g contributes to MAMP-induced SA accumulation and is involved in disease resistance against *pseudomonas syringae*. *PLoS Pathog.* **2009**, *5*, e1000301. [CrossRef] [PubMed]

75. Du, L.; Ali, G.S.; Simons, K.A.; Hou, J.; Yang, T.; Reddy, A.S.; Poovaiah, B.W. $Ca^{2+}$/calmodulin regulates salicylic-acid-mediated plant immunity. *Nature* **2009**, *457*, 1154–1158. [CrossRef] [PubMed]

76. Wang, L.; Tsuda, K.; Truman, W.; Sato, M.; Nguyen le, V.; Katagiri, F.; Glazebrook, J. CBP60g and SARD1 play partially redundant critical roles in salicylic acid signaling. *Plant J.* **2011**, *67*, 1029–1041. [CrossRef] [PubMed]

77. Truman, W.; Sreekanta, S.; Lu, Y.; Bethke, G.; Tsuda, K.; Katagiri, F.; Glazebrook, J. The calmodulin-binding protein60 family includes both negative and positive regulators of plant immunity. *Plant Physiol.* **2013**, *163*, 1741–1751. [CrossRef] [PubMed]

78. Lu, Y.; Truman, W.; Liu, X.; Bethke, G.; Zhou, M.; Myers, C.L.; Katagiri, F.; Glazebrook, J. Different modes of negative regulation of plant immunity by calmodulin-related genes. *Plant Physiol.* **2018**. [CrossRef] [PubMed]

79. Finkler, A.; Ashery-Padan, R.; Fromm, H. Camtas: Calmodulin-binding transcription activators from plants to human. *FEBS Lett.* **2007**, *581*, 3893–3898. [CrossRef] [PubMed]

80. Rahman, H.; Yang, J.; Xu, Y.P.; Munyampundu, J.P.; Cai, X.Z. Phylogeny of plant camtas and role of atcamtas in nonhost resistance to xanthomonas oryzae pv. Oryzae. *Front. Plant Sci.* **2016**, *7*, 177. [CrossRef] [PubMed]

81. Shen, C.; Yang, Y.; Du, L.; Wang, H. Calmodulin-binding transcription activators and perspectives for applications in biotechnology. *Appl. Microbiol. Biotechnol.* **2015**, *99*, 10379–10385. [CrossRef] [PubMed]

82. Galon, Y.; Nave, R.; Boyce, J.M.; Nachmias, D.; Knight, M.R.; Fromm, H. Calmodulin-binding transcription activator (CAMTA) 3 mediates biotic defense responses in arabidopsis. *FEBS Lett.* **2008**, *582*, 943–948. [CrossRef] [PubMed]

83. Rahman, H.; Xu, Y.P.; Zhang, X.R.; Cai, X.Z. Brassica napus genome possesses extraordinary high number of *CAMTA* genes and *CAMTA3* contributes to pamp triggered immunity and resistance to *sclerotinia sclerotiorum*. *Front. Plant Sci.* **2016**, *7*, 581. [CrossRef] [PubMed]

84. Laluk, K.; Prasad, K.V.; Savchenko, T.; Celesnik, H.; Dehesh, K.; Levy, M.; Mitchell-Olds, T.; Reddy, A.S. The calmodulin-binding transcription factor signal responsive1 is a novel regulator of glucosinolate metabolism and herbivory tolerance in arabidopsis. *Plant Cell Physiol.* **2012**, *53*, 2008–2015. [CrossRef] [PubMed]

85. Qiu, Y.; Xi, J.; Du, L.; Suttle, J.C.; Poovaiah, B.W. Coupling calcium/calmodulin-mediated signaling and herbivore-induced plant response through calmodulin-binding transcription factor atsr1/camta3. *Plant Mol. Biol.* **2012**, *79*, 89–99. [CrossRef] [PubMed]

86. Jacob, F.; Kracher, B.; Mine, A.; Seyfferth, C.; Blanvillain-Baufume, S.; Parker, J.E.; Tsuda, K.; Schulze-Lefert, P.; Maekawa, T. A dominant-interfering CAMTA3 mutation compromises primary transcriptional outputs mediated by both cell surface and intracellular immune receptors in *arabidopsis thaliana*. *New Phytol.* **2018**, *217*, 1667–1680. [CrossRef] [PubMed]

87. Liu, J.; Whalley, H.J.; Knight, M.R. Combining modelling and experimental approaches to explain how calcium signatures are decoded by calmodulin-binding transcription activators (CAMTAs) to produce specific gene expression responses. *New Phytol.* **2015**, *208*, 174–187. [CrossRef] [PubMed]

88. Lenzoni, G.; Liu, J.; Knight, M.R. Predicting plant immunity gene expression by identifying the decoding mechanism of calcium signatures. *New Phytol.* **2018**, *217*, 1598–1609. [CrossRef] [PubMed]

89. Park, C.Y.; Lee, J.H.; Yoo, J.H.; Moon, B.C.; Choi, M.S.; Kang, Y.H.; Lee, S.M.; Kim, H.S.; Kang, K.Y.; Chung, W.S.; et al. WRKY group IId transcription factors interact with calmodulin. *FEBS Lett.* **2005**, *579*, 1545–1550. [CrossRef] [PubMed]

90. Moore, J.W.; Loake, G.J.; Spoel, S.H. Transcription dynamics in plant immunity. *Plant Cell* **2011**, *23*, 2809–2820. [CrossRef] [PubMed]

91. Bender, K.W.; Snedden, W.A. Calmodulin-related proteins step out from the shadow of their namesake. *Plant Physiol.* **2013**, *163*, 486–495. [CrossRef] [PubMed]

92. Cheval, C.; Aldon, D.; Galaud, J.P.; Ranty, B. Calcium/calmodulin-mediated regulation of plant immunity. *Biochim. Biophys. Acta* **2013**, *1833*, 1766–1771. [CrossRef] [PubMed]

93. Heo, W.D.; Lee, S.H.; Kim, M.C.; Kim, J.C.; Chung, W.S.; Chun, H.J.; Lee, K.J.; Park, C.Y.; Park, H.C.; Choi, J.Y.; et al. Involvement of specific calmodulin isoforms in salicylic acid-independent activation of plant disease resistance responses. *Proc. Natl. Acad. Sci. USA* **1999**, *96*, 766–771. [CrossRef] [PubMed]

94. Park, C.Y.; Heo, W.D.; Yoo, J.H.; Lee, J.H.; Kim, M.C.; Chun, H.J.; Moon, B.C.; Kim, I.H.; Park, H.C.; Choi, M.S.; et al. Pathogenesis-related gene expression by specific calmodulin isoforms is dependent on nim1, a key regulator of systemic acquired resistance. *Mol. Cells* **2004**, *18*, 207–213. [PubMed]

95. Rao, S.S.; El-Habbak, M.H.; Havens, W.M.; Singh, A.; Zheng, D.; Vaughn, L.; Haudenshield, J.S.; Hartman, G.L.; Korban, S.S.; Ghabrial, S.A. Overexpression of GmCaM4 in soybean enhances resistance to pathogens and tolerance to salt stress. *Mol. Plant Pathol.* **2014**, *15*, 145–160. [CrossRef] [PubMed]

96. Chiasson, D.; Ekengren, S.K.; Martin, G.B.; Dobney, S.L.; Snedden, W.A. Calmodulin-like proteins from arabidopsis and tomato are involved in host defense against pseudomonas syringae pv. Tomato. *Plant Mol. Biol.* **2005**, *58*, 887–897. [CrossRef] [PubMed]

97. Ma, W.; Smigel, A.; Tsai, Y.C.; Braam, J.; Berkowitz, G.A. Innate immunity signaling: Cytosolic $Ca^{2+}$ elevation is linked to downstream nitric oxide generation through the action of calmodulin or a calmodulin-like protein. *Plant Physiol.* **2008**, *148*, 818–828. [CrossRef] [PubMed]

98. Leba, L.J.; Cheval, C.; Ortiz-Martin, I.; Ranty, B.; Beuzon, C.R.; Galaud, J.P.; Aldon, D. Cml9, an arabidopsis calmodulin-like protein, contributes to plant innate immunity through a flagellin-dependent signalling pathway. *Plant J.* **2012**, *71*, 976–989. [CrossRef] [PubMed]

99. Zhu, X.; Robe, E.; Jomat, L.; Aldon, D.; Mazars, C.; Galaud, J.P. Cml8, an arabidopsis calmodulin-like protein, plays a role in pseudomonas syringae plant immunity. *Plant Cell Physiol.* **2017**, *58*, 307–319. [PubMed]

100. Zhu, X.; Perez, M.; Aldon, D.; Galaud, J.P. Respective contribution of CML8 and CML9, two arabidopsis calmodulin-like proteins, to plant stress responses. *Plant Signal Behav.* **2017**, *12*, e1322246. [CrossRef] [PubMed]

101. Magnan, F.; Ranty, B.; Charpenteau, M.; Sotta, B.; Galaud, J.P.; Aldon, D. Mutations in atcml9, a calmodulin-like protein from *arabidopsis thaliana*, alter plant responses to abiotic stress and abscisic acid. *Plant J.* **2008**, *56*, 575–589. [CrossRef] [PubMed]

102. Cheval, C.; Faulkner, C. Plasmodesmal regulation during plant-pathogen interactions. *New Phytol.* **2018**, *217*, 62–67. [CrossRef] [PubMed]

103. Xu, B.; Cheval, C.; Laohavisit, A.; Hocking, B.; Chiasson, D.; Olsson, T.S.G.; Shirasu, K.; Faulkner, C.; Gilliham, M. A calmodulin-like protein regulates plasmodesmal closure during bacterial immune responses. *New Phytol.* **2017**, *215*, 77–84. [CrossRef] [PubMed]

104. Arimura, G.; Ozawa, R.; Maffei, M.E. Recent advances in plant early signaling in response to herbivory. *Int. J. Mol. Sci.* **2011**, *12*, 3723–3739. [CrossRef] [PubMed]

105. Arimura, G.; Maffei, M.E. Calcium and secondary cpk signaling in plants in response to herbivore attack. *Biochem. Biophys. Res. Commun.* **2010**, *400*, 455–460. [CrossRef] [PubMed]

106. Verrillo, F.; Occhipinti, A.; Kanchiswamy, C.N.; Maffei, M.E. Quantitative analysis of herbivore-induced cytosolic calcium by using a cameleon (YC 3.6) calcium sensor in *arabidopsis thaliana*. *J. Plant Physiol.* **2014**, *171*, 136–139. [CrossRef] [PubMed]

107. Vadassery, J.; Reichelt, M.; Hause, B.; Gershenzon, J.; Boland, W.; Mithofer, A. Cml42-mediated calcium signaling coordinates responses to spodoptera herbivory and abiotic stresses in arabidopsis. *Plant Physiol.* **2012**, *159*, 1159–1175. [CrossRef] [PubMed]

108. Scholz, S.S.; Vadassery, J.; Heyer, M.; Reichelt, M.; Bender, K.W.; Snedden, W.A.; Boland, W.; Mithofer, A. Mutation of the arabidopsis calmodulin-like protein CML37 deregulates the jasmonate pathway and enhances susceptibility to herbivory. *Mol. Plant* **2014**, *7*, 1712–1726. [CrossRef] [PubMed]

109. Vadassery, J.; Scholz, S.S.; Mithofer, A. Multiple calmodulin-like proteins in Arabidopsis are induced by insect-derived (*spodoptera littoralis*) oral secretion. *Plant Signal Behav.* **2012**, *7*, 1277–1280. [CrossRef] [PubMed]

110. Anandalakshmi, R.; Marathe, R.; Ge, X.; Herr, J.M., Jr.; Mau, C.; Mallory, A.; Pruss, G.; Bowman, L.; Vance, V.B. A calmodulin-related protein that suppresses posttranscriptional gene silencing in plants. *Science* **2000**, *290*, 142–144. [CrossRef] [PubMed]

111. Nakahara, K.S.; Masuta, C.; Yamada, S.; Shimura, H.; Kashihara, Y.; Wada, T.S.; Meguro, A.; Goto, K.; Tadamura, K.; Sueda, K.; et al. Tobacco calmodulin-like protein provides secondary defense by binding to and directing degradation of virus RNA silencing suppressors. *Proc. Natl. Acad. Sci. USA* **2012**, *109*, 10113–10118. [CrossRef] [PubMed]

112. Li, F.; Huang, C.; Li, Z.; Zhou, X. Suppression of RNA silencing by a plant DNA virus satellite requires a host calmodulin-like protein to repress RDR6 expression. *PLoS Pathog.* **2014**, *10*, e1003921. [CrossRef] [PubMed]

113. Jeon, E.J.; Tadamura, K.; Murakami, T.; Inaba, J.I.; Kim, B.M.; Sato, M.; Atsumi, G.; Kuchitsu, K.; Masuta, C.; Nakahara, K.S. Rgs-cam detects and counteracts viral RNA silencing suppressors in plant immune priming. *J. Virol.* **2017**, *91*, e00761-17. [CrossRef] [PubMed]

114. Perochon, A.; Dieterle, S.; Pouzet, C.; Aldon, D.; Galaud, J.P.; Ranty, B. Interaction of a plant pseudo-response regulator with a calmodulin-like protein. *Biochem. Biophys. Res. Commun.* **2010**, *398*, 747–751. [CrossRef] [PubMed]

115. Cheval, C.; Perez, M.; Leba, L.J.; Ranty, B.; Perochon, A.; Reichelt, M.; Mithofer, A.; Robe, E.; Mazars, C.; Galaud, J.P.; et al. Prr2, a pseudo-response regulator, promotes salicylic acid and camalexin accumulation during plant immunity. *Sci. Rep.* **2017**, *7*, 6979. [CrossRef] [PubMed]

116. Yoo, J.H.; Park, C.Y.; Kim, J.C.; Heo, W.D.; Cheong, M.S.; Park, H.C.; Kim, M.C.; Moon, B.C.; Choi, M.S.; Kang, Y.H.; et al. Direct interaction of a divergent CaM isoform and the transcription factor, MYB2, enhances salt tolerance in *Arabidopsis*. *J. Biol. Chem.* **2005**, *280*, 3697–3706. [CrossRef] [PubMed]

117. Hu, Z.; Lv, X.; Xia, X.; Zhou, J.; Shi, K.; Yu, J.; Zhou, Y. Genome-wide identification and expression analysis of calcium-dependent protein kinase in tomato. *Front. Plant Sci.* **2016**, *7*, 469. [CrossRef] [PubMed]

118. Hamel, L.P.; Sheen, J.; Seguin, A. Ancient signals: Comparative genomics of green plant CDPKs. *Trends Plant Sci.* **2014**, *19*, 79–89. [CrossRef] [PubMed]

119. Harper, J.F.; Sussman, M.R.; Schaller, G.E.; Putnam-Evans, C.; Charbonneau, H.; Harmon, A.C. A calcium-dependent protein kinase with a regulatory domain similar to calmodulin. *Science* **1991**, *252*, 951–954. [CrossRef] [PubMed]

120. Satterlee, J.S.; Sussman, M.R. Unusual membrane-associated protein kinases in higher plants. *J. Membr. Biol.* **1998**, *164*, 205–213. [CrossRef] [PubMed]

121. Boudsocq, M.; Droillard, M.J.; Regad, L.; Lauriere, C. Characterization of arabidopsis calcium-dependent protein kinases: Activated or not by calcium? *Biochem. J.* **2012**, *447*, 291–299. [CrossRef] [PubMed]

122. Kanchiswamy, C.N.; Takahashi, H.; Quadro, S.; Maffei, M.E.; Bossi, S.; Bertea, C.; Zebelo, S.A.; Muroi, A.; Ishihama, N.; Yoshioka, H.; et al. Regulation of arabidopsis defense responses against spodoptera littoralis by cpk-mediated calcium signaling. *BMC Plant Biol.* **2010**, *10*, 97. [CrossRef] [PubMed]

123. Dubiella, U.; Seybold, H.; Durian, G.; Komander, E.; Lassig, R.; Witte, C.P.; Schulze, W.X.; Romeis, T. Calcium-dependent protein kinase/nadph oxidase activation circuit is required for rapid defense signal propagation. *Proc. Natl. Acad. Sci. USA* **2013**, *110*, 8744–8749. [CrossRef] [PubMed]

124. Kadota, Y.; Sklenar, J.; Derbyshire, P.; Stransfeld, L.; Asai, S.; Ntoukakis, V.; Jones, J.D.; Shirasu, K.; Menke, F.; Jones, A.; et al. Direct regulation of the NADPH oxidase RBOHD by the PRR-associated kinase BIK1 during plant immunity. *Mol. Cell* **2014**, *54*, 43–55. [CrossRef] [PubMed]

125. Bundo, M.; Coca, M. Enhancing blast disease resistance by overexpression of the calcium-dependent protein kinase oscpk4 in rice. *Plant Biotechnol. J.* **2016**, *14*, 1357–1367. [CrossRef] [PubMed]

126. Romeis, T.; Ludwig, A.A.; Martin, R.; Jones, J.D. Calcium-dependent protein kinases play an essential role in a plant defence response. *EMBO J.* **2001**, *20*, 5556–5567. [CrossRef] [PubMed]

127. Gao, X.; Chen, X.; Lin, W.; Chen, S.; Lu, D.; Niu, Y.; Li, L.; Cheng, C.; McCormack, M.; Sheen, J.; et al. Bifurcation of arabidopsis nlr immune signaling via $Ca^{2+}$-dependent protein kinases. *PLoS Pathog.* **2013**, *9*, e1003127. [CrossRef] [PubMed]

128. Kobayashi, M.; Yoshioka, M.; Asai, S.; Nomura, H.; Kuchimura, K.; Mori, H.; Doke, N.; Yoshioka, H. Stcdpk5 confers resistance to late blight pathogen but increases susceptibility to early blight pathogen in potato via reactive oxygen species burst. *New Phytol.* **2012**, *196*, 223–237. [CrossRef] [PubMed]

129. Freymark, G.; Diehl, T.; Miklis, M.; Romeis, T.; Panstruga, R. Antagonistic control of powdery mildew host cell entry by barley calcium-dependent protein kinases (CDPKs). *Mol. Plant Microbe Interact.* **2007**, *20*, 1213–1221. [CrossRef] [PubMed]

130. Asai, S.; Ichikawa, T.; Nomura, H.; Kobayashi, M.; Kamiyoshihara, Y.; Mori, H.; Kadota, Y.; Zipfel, C.; Jones, J.D.; Yoshioka, H. The variable domain of a plant calcium-dependent protein kinase (CDPK) confers subcellular localization and substrate recognition for NADPH oxidase. *J. Biol. Chem.* **2013**, *288*, 14332–14340. [CrossRef] [PubMed]

131. Liu, N.; Hake, K.; Wang, W.; Zhao, T.; Romeis, T.; Tang, D. Calcium-dependent protein kinase5 associates with the truncated NLR protein TIR-nbs2 to contribute to exo70b1-mediated immunity. *Plant Cell* **2017**, *29*, 746–759. [CrossRef] [PubMed]

132. Zhao, T.; Rui, L.; Li, J.; Nishimura, M.T.; Vogel, J.P.; Liu, N.; Liu, S.; Zhao, Y.; Dangl, J.L.; Tang, D. A truncated nlr protein, tir-nbs2, is required for activated defense responses in the exo70b1 mutant. *PLoS Genet.* **2015**, *11*, e1004945. [CrossRef] [PubMed]

133. Asano, T.; Hayashi, N.; Kobayashi, M.; Aoki, N.; Miyao, A.; Mitsuhara, I.; Ichikawa, H.; Komatsu, S.; Hirochika, H.; Kikuchi, S.; et al. A rice calcium-dependent protein kinase OSCPK12 oppositely modulates salt-stress tolerance and blast disease resistance. *Plant J.* **2012**, *69*, 26–36. [CrossRef] [PubMed]

134. Roux, M.; Schwessinger, B.; Albrecht, C.; Chinchilla, D.; Jones, A.; Holton, N.; Malinovsky, F.G.; Tor, M.; de Vries, S.; Zipfel, C. The arabidopsis leucine-rich repeat receptor-like kinases $BAK_1/SERK_3$ and $BKK_1/SERK_4$ are required for innate immunity to hemibiotrophic and biotrophic pathogens. *Plant Cell* **2011**, *23*, 2440–2455. [CrossRef] [PubMed]

135. Schwessinger, B.; Roux, M.; Kadota, Y.; Ntoukakis, V.; Sklenar, J.; Jones, A.; Zipfel, C. Phosphorylation-dependent differential regulation of plant growth, cell death, and innate immunity by the regulatory receptor-like kinase bak1. *PLoS Genet.* **2011**, *7*, e1002046. [CrossRef] [PubMed]

136. Monaghan, J.; Matschi, S.; Shorinola, O.; Rovenich, H.; Matei, A.; Segonzac, C.; Malinovsky, F.G.; Rathjen, J.P.; MacLean, D.; Romeis, T.; et al. The calcium-dependent protein kinase CPK28 buffers plant immunity and regulates bik1 turnover. *Cell Host Microbe* **2014**, *16*, 605–615. [CrossRef] [PubMed]

137. Wang, J.; Grubb, L.E.; Wang, J.; Liang, X.; Li, L.; Gao, C.; Ma, M.; Feng, F.; Li, M.; Li, L.; et al. A regulatory module controlling homeostasis of a plant immune kinase. *Mol. Cell* **2018**, *69*. [CrossRef] [PubMed]

138. Bender, K.W.; Blackburn, R.K.; Monaghan, J.; Derbyshire, P.; Menke, F.L.; Zipfel, C.; Goshe, M.B.; Zielinski, R.E.; Huber, S.C. Autophosphorylation-based calcium ($Ca^{2+}$) sensitivity priming and $Ca^{2+}$/calmodulin inhibition of *arabidopsis thaliana* $Ca^{2+}$-dependent protein kinase 28 (cpk28). *J. Biol. Chem.* **2017**, *292*, 3988–4002. [CrossRef] [PubMed]

139. McAinsh, M.R.; Pittman, J.K. Shaping the calcium signature. *New Phytol.* **2009**, *181*, 275–294. [CrossRef] [PubMed]

140. Ma, W.; Yoshioda, K.; Berkowitz, G.A. Cyclic nucleotide gated channel and $Ca^{2+}$-mediated signal transduction during plant innate immune responses to pathogen. *Plant Signal Behav.* **2007**, *2*, 548–550. [CrossRef] [PubMed]

141. Ali, R.; Ma, W.; Lemtiri-Chlieh, F.; Tsaltas, D.; Leng, Q.; von Bodman, S.; Berkowitz, G.A. Death don't have no mercy and neither does calcium: Arabidopsis CYCLIC NUCLEOTIDE GATED CHANNEL2 and innate immunity. *Plant Cell* **2007**, *19*, 1081–1095. [CrossRef] [PubMed]

142. Bonza, M.C.; Morandini, P.; Luoni, L.; Geisler, M.; Palmgreen, M.G.; De Michelis, M.I. At-ACA8 encodes a plasma membrane-localized calcium ATPase of Arabidopsis with a calmodulin-binding domain at the N-terminus. *Plant Physiol.* **2000**, *123*, 1495–1506. [CrossRef] [PubMed]

143. Frei dit Frey, N.; Mbengue, M.; Kwaaitaal, M.; Nitsch, L.; Attenbach, D.; Haweker, H.; Lozano-Duran, R.; Njo, M.F.; Beeckman, T.; Huettel, B.; et al. Plasma membrane calcium ATPases are important components of receptor-mediated signalling in plant immune responses and development. *Plant Physiol.* **2012**, *159*, 789–809. [CrossRef] [PubMed]

144. Guo, M.; Kim, P.; Li, G.; Elowsky, C.G.; Alfano, J.R. A bacterial effector co-opts calmodulin to target the plant microtubule network. *Cell Host Microbe* **2016**, *19*, 67–78. [CrossRef] [PubMed]

145. Buttner, D.; Bonas, U. Regulation and secretion of xanthomonas virulence factors. *FEMS Microbiol. Rev.* **2010**, *34*, 107–133. [CrossRef] [PubMed]

146. Ameen, G.; Mechan, M. XVI congress on molecular plant-microbe interactions meeting report. In Proceedings of the Travel Awardees for the 2016 is-MPMI XVII Congress, Portland, OR, USA, 16–21 July 2016.

147. Wolff, J.; Cook, G.H.; Goldhammer, A.R.; Berkowitz, S.A. Calmodulin activates prokaryotic adenylate cyclase. *Proc. Natl. Acad. Sci. USA* **1980**, *77*, 3841–3844. [CrossRef] [PubMed]

148. Knight, H. Calcium signaling during abiotic stress in plants. *Int. Rev. Cytol.* **2000**, *195*, 269–324. [PubMed]

149. Cheng, C.; Gao, X.; Feng, B.; Sheen, J.; Shan, L.; He, P. Plant immune response to pathogens differs with changing temperatures. *Nat. Commun.* **2013**, *4*, 2530. [CrossRef] [PubMed]

International Journal of
*Molecular Sciences*

MDPI

*Review*

# Metabolomics in Plant Priming Research: The Way Forward?

**Fidele Tugizimana, Msizi I. Mhlongo, Lizelle A. Piater and Ian A. Dubery \***

Department of Biochemistry, Research Centre for Plant Metabolomics, University of Johannesburg, Auckland Park 2006, South Africa; Fideletu@gmail.com (F.T.); msizi.mhlongo17@gmail.com (M.I.M.); lpiater@uj.ac.za (L.A.P.)
\* Correspondence: idubery@uj.ac.za; Tel.: +27-011-559-2401

Received: 7 May 2018; Accepted: 4 June 2018; Published: 13 June 2018

**Abstract:** A new era of plant biochemistry at the systems level is emerging, providing detailed descriptions of biochemical phenomena at the cellular and organismal level. This new era is marked by the advent of metabolomics—the qualitative and quantitative investigation of the entire metabolome (in a dynamic equilibrium) of a biological system. This field has developed as an indispensable methodological approach to study cellular biochemistry at a global level. For protection and survival in a constantly-changing environment, plants rely on a complex and multi-layered innate immune system. This involves surveillance of 'self' and 'non-self,' molecule-based systemic signalling and metabolic adaptations involving primary and secondary metabolites as well as epigenetic modulation mechanisms. Establishment of a pre-conditioned or primed state can sensitise or enhance aspects of innate immunity for faster and stronger responses. Comprehensive elucidation of the molecular and biochemical processes associated with the phenotypic defence state is vital for a better understanding of the molecular mechanisms that define the metabolism of plant–pathogen interactions. Such insights are essential for translational research and applications. Thus, this review highlights the prospects of metabolomics and addresses current challenges that hinder the realisation of the full potential of the field. Such limitations include partial coverage of the metabolome and maximising the value of metabolomics data (extraction of information and interpretation). Furthermore, the review points out key features that characterise both the plant innate immune system and enhancement of the latter, thus underlining insights from metabolomic studies in plant priming. Future perspectives in this inspiring area are included, with the aim of stimulating further studies leading to a better understanding of plant immunity at the metabolome level.

**Keywords:** metabolomics; plant defence; plant–microbe interactions; priming; pre-conditioning

---

## 1. Introduction: Multi-Layered Molecular and Cellular Networks Ensure Effective Adaptation to Changing Environments

Evolution dictates that living systems constantly adapt to ever-changing environments in a context-dependent manner. Such adaptation and/or response to environmental or genetic alterations implies complex and dynamic cellular reprogramming [1–3]. These biological responses—which can be phenomenologically described by understanding the cellular or organismal physiological state—are kinetic and highly dynamic events that span the whole cellular biological information network [4–7]. Reflecting on the plant kingdom, one of the epitomes of such adaptation is the constant fine-tuning of physiologies and cellular-scale morphologies, and the dynamic (and complex) biosynthesis of an array of structurally and functionally diverse chemistries [8–10].

Plants are seemingly as adept as animals in responding to environmental conditions. Of necessity, considering their sessile nature, plants have developed dynamic, multi-layered molecular and cellular networks for effective adaptation to unpredictably changing environments [10–12]. The latter is a natural habitat of the continuously-evolving pathogenic microorganisms that represent a biological threat to food security [13–15]. Furthermore, the sustainable production of food plants, considering the exponentially growing world population, is currently one of the challenges facing humanity. Crop losses due to plant pathogens can be quite substantial, with far-reaching effects. Moreover, most of the classical methods for crop protection against pathogenic microorganisms have become less effective and are environmentally unfriendly. Hence, the need for new strategies has led to procedures/metabolites that aid plants in adapting and defending against stress [16,17].

Learning from nature, plant biologists have observed that the interaction of plants with necrotising pathogens, beneficial microbes or agrochemicals can cause a sensitisation of the plant immune system, resulting in a faster and stronger induction of resistance mechanisms upon subsequent infections [18–21]. Memory of a past event may determine the response to future environmental stimuli, thus resulting in phenotypic and stimulus-dependent plasticity of response traits [22]. This unique physiological state in which plants are rendered capable (pre-conditioned) to better or more effectively mount defence responses to biotic or abiotic stresses is termed 'priming' [20,23,24], and differs from adaptation and acclimatisation phenomena in response to environmental stimuli [22]. In view of the competition of plant resource allocation to defence vs. growth, priming demands a small fitness cost that is, however, considerably surpassed by the benefits of an enhanced defensive capability to ward off attacks by potential pathogens [22]. Furthermore, a primed defence state can be inherited epigenetically from defence-expressing plants [19,25–28].

This immune stimulation of plants could be an alternative strategy that holds promise for increasing the capacity of plants to cope with biotic and abiotic stresses. Most of the efforts in characterising and defining the biochemical changes related to priming processes have been driven by targeted approaches. Although these methodologies have played a vital role in elucidating the main elements of priming that includes enhanced perception systems, dormant signal transduction enzymes, chromatin modification and transcription factors [19,23,29–31], there are still gaps in holistically understanding the dynamism and complexity of molecular mechanisms involved in the entire priming event, considering the complexity of multi-layered biological information networks. The phenomenological description of the physiological responses defining the 'prime-ome' is thus rendered possible by systems biology approaches (-omics layers—genomics, transcriptomics, proteomics and metabolomics) [22,24,32,33].

## 2. Metabolomics, a Systems Biology Approach: Prospects and Challenges

A recent resurgence of interest in metabolism and increasing awareness about the physiological insights that can be obtained by measuring the total small-molecule complement of a biological system have made metabolomics a central pillar in systems biology approaches [2,34,35]. Metabolomics can thus be understood as a quantitative measurement of the multi-parametric metabolic responses of living systems to genetic or environmental perturbations [36–38]. Such a description implies that metabolomics can be regarded as the best trade-off for 'functionally' investigating metabolism, offering the finest-grained details: a molecular-level convolution of all upstream biological information (genomic, transcriptomic and proteomic) layers [38–41].

These small molecules (namely metabolites, with molecular masses ≤1500 Da) can be described as the end products of gene expression and define the phenotype of a cell or tissue under defined physiological conditions at a biochemical level. Metabolite profile patterns can thus provide a holistic signature of the physiological state under study as well as deeper knowledge of specific biochemical processes [42–45]. Furthermore, systems biology approaches imply an appreciation of the full complexity and the multi-dimensionality of biochemical networks operating in a biological system to produce physiological and phenotypic coherence (Figure 1). Hence, given that the biochemical actions of metabolites are far-reaching, including regulation of epigenetic mechanisms and gene expression, involvement in signal transduction, post-translational modifications of proteins, protein transport and active roles in defence mechanisms; metabolomics can thus be seen as a powerful tool to investigate cellular biochemistry at the systems level [46–49].

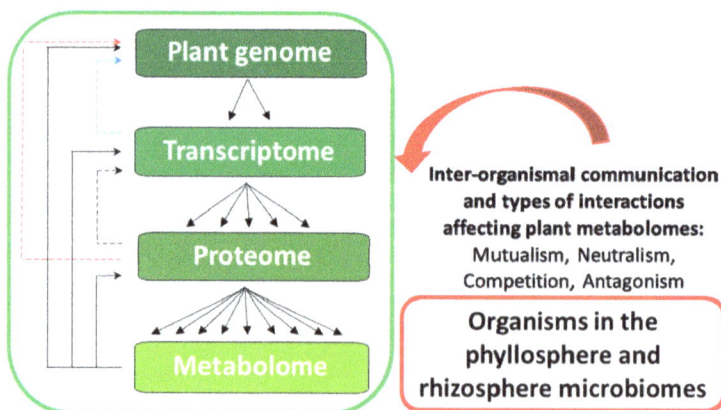

**Figure 1.** Metabolomics in the context of biological information flow, illustrating the complexity of multi-layered biological information networks and mutual interdependence. In biological systems, large numbers of structurally and functionally diverse genes, proteins and metabolites are involved in dynamic, linear and/or non-linear interactions. These interactions may involve a range of time scales and intensities. Some of the types of reciprocal interactions include post-transcriptional control of gene expression (dotted lines). Others include effects of downstream metabolites on transcription through binding to regulatory proteins and feedback inhibition/activation of enzymes (solid lines). Adaptive gene expression in response to environmental influences is ultimately reflected in changes in the pattern and/or concentration of metabolites.

In this review, the term "metabolomics" refers to an untargeted methodology [34,45]. Thus, metabolomics and untargeted or non-targeted metabolomics may be used interchangeably to mean the global metabolic profiling of the entire (measurable) metabolome of the biological system under consideration. This methodology differs from targeted analytical methods in various fundamental aspects such as being a data-driven approach with predictive power that aims to assess (qualitatively and quantitatively) all measurable metabolites without any pre-conception or pre-selection [41,45,50,51].

Being at the interface between biology, chemistry, chemometrics, statistics and computer science; metabolomics is methodologically a multi-disciplinary skillset research field [40,45]. With the innovative developments in analytical technologies, advancement in chemometric and statistical methods, and the integration of orthogonal biological approaches, metabolomic studies have provided remarkable insights into the biochemical mechanisms that underpin various physiological conditions [51–53]. Furthermore, owing to the inherent sensitivity of the metabolome to genetic and environmental perturbations, subtle alterations in biological pathways can be measured [41,50,54,55].

To attain this goal—holistic analysis of the metabolome, considering the complexity and chemo-diversity thereof—a wide range of chemistries, chemometrics methods, novel computational approaches and advanced analytical instrumentation that provide high degrees of sensitivity and reproducibility, are required and employed in metabolomics [56–59]. In contrast to other -omics methodologies, metabolomics faces several unique challenges that make the field particularly demanding. These arise from the inherent characteristics of the metabolome: highly dynamic (continuously changing at different rates), chemically diverse (dramatically different physicochemical properties and biological functions of metabolites, as well as highly diverse and dynamic stereochemistries), a wide range of metabolite levels and the inherent bio-complexity of living systems (biological cycles, organismal and cellular compartmentalisation) [40,45,60,61]. These challenges point to the bottlenecks that have limited (untargeted) metabolomics so far (Figure 2), making the holistic coverage of the whole metabolome currently unrealisable and subsequently impacting the biological insights generated.

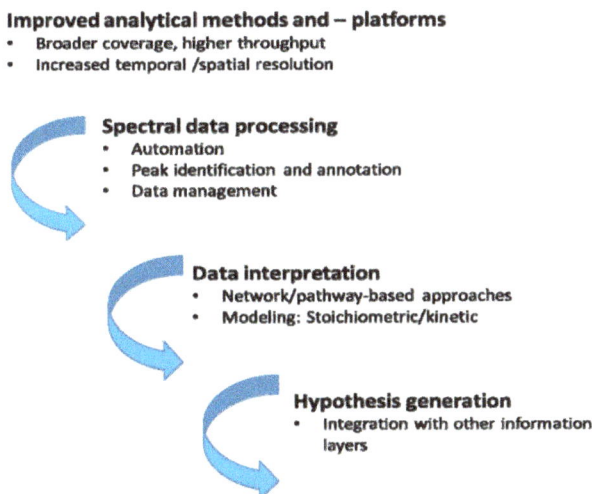

**Improved analytical methods and – platforms**
- Broader coverage, higher throughput
- Increased temporal /spatial resolution

**Spectral data processing**
- Automation
- Peak identification and annotation
- Data management

**Data interpretation**
- Network/pathway-based approaches
- Modeling: Stoichiometric/kinetic

**Hypothesis generation**
- Integration with other information layers

**Figure 2.** Bottlenecks in metabolomics workflows that limit biological insights. Despite the maturation of metabolomics, driven by massive improvements in analytical technologies and impressive advancements in computational and chemometric methods, the realisation of the goal of metabolomics is still a challenge at different levels: metabolome coverage; information extraction from acquired data; and systematic interpretation of complex metabolic changes and derived hypotheses about underlying functional mechanisms.

These holdups can be summarised into three main aspects related to different steps of the metabolomics workflow pipeline. Firstly, the analytical limitations, be it at the metabolite extraction [62–64] or analytical platform level [65–68], hinder the metabolome coverage. The two leading and successful analytical platforms in metabolomics are mass spectrometry (MS) and nuclear magnetic resonance (NMR) spectroscopy. MS platforms provide high sensitivity and detection specificity, thus enabling large-scale coverage of the metabolome. NMR, on the other hand, offers a window into profiling all the most abundant metabolites in sample extracts; and the strengths of this analytical platform include detection of poorly ionisable compounds, identification of compounds with identical masses, and determining structures of unknown metabolites [45,69]. Evidently, MS (often coupled to chromatographic separation) and NMR approaches offer different advantages, which are being explored synergistically [70,71]. However, despite the current analytical advancements, and considering the inherent complexity of metabolomes, the realisation of holistic coverage of the metabolome in toto, in a given biosystem, is not yet possible [69,72].

Secondly, extracting information from acquired data is still a major challenge, thus limiting the maximisation of the value of metabolomics data. This can be at the level of data processing steps: from pre-processing to chemometric and statistical analyses [34,56,58,73,74]; and at the systematic identification of metabolites, with accuracy and high confidence levels [72,75–77]. Thirdly, data interpretation and hypothesis generation are still a bottleneck, as comprehensive strategies (computational and chemometrics) are still limited, and the integration of other biological information layers is still in the developmental phase and has limitations [3,54,78–80]. A detailed description of these three constraints can be found in the cited literature herein.

Despite these challenges, the momentum and maturation of metabolomics have visibly revolutionised life sciences. Innovative and collaborative efforts are continuously providing suggestions to address these limitations: technological advancement, data mining strategies and tools, systematic data interpretation and integration of orthogonal biological information [72,73,76,81,82].

The application of metabolomics spans a wide spectrum of life sciences (fundamental and translational) research [40,44,45]. In the plant sciences, metabolomic approaches are increasingly being used for investigating linkages between genotype and biochemical phenotype [42,83,84], metabolic pathway studies [85–87], silent phenotypes of mutations [6,88], plant–pathogen interactions [89–93] and, as emphasised here, plant priming [24,49,94]. As indicated in the above sections, an overview and reflections on plant–pathogen interactions and priming are articulated, highlighting the metabolomics inputs in decoding the plant priming molecular events. Thus, the following section gives a succinct overview of plant defence mechanisms, briefly highlighting key components and current models that describe plant defence responses.

## 3. Plant Defence Mechanisms—Core Concepts, Key Molecular Components and Current Models

Due to co-evolution between plant hosts and pathogens, plants have developed sophisticated abilities to recognise pathogens and translate this perception into effective immune response strategies for survival. Concurrently, pathogens have evolved their own strategies to evade the host's immune system. The plant–pathogen interactions are undeniably a never-ending evolutionary arms race, and involve key elements for the survival of the host or the pathogen [95–98]. Understanding these interactions at the biochemical level is necessary to develop strategies to aid plants to adapt and defend against continuously evolving pathogens. It may suffice here to briefly point out some classical and current components that define the plant defence mechanisms.

Plant cells are generally protected by an array of structural barriers (waxes, suberin, lignin, etc.) that deny access to a wide range of microbes (Figure 3). This passive protective system can also involve preformed antimicrobial chemicals (metabolites known as phytoanticipins) that form a chemical barrier, in so doing preventing or attenuating invasion by potential attackers [99–102]. In addition to these non-specific defence mechanisms, active immune responses can be activated by the perception of highly conserved molecular features of different classes of bacterial and fungal pathogens, referred to as microbe/pathogen-associated molecular patterns (M/PAMPs) [103,104]. These non-self M/PAMPs are recognised by cell surface-localised pattern recognition receptors (PRRs), which are activated and, in turn, initiate downstream signalling events that ultimately result in the activation of a defence response referred to as M/PAMP-triggered immunity (M/PTI) [105–107]. These responses include newly synthesised antimicrobial metabolites (phytoalexins), antimicrobial hydrolytic enzymes (e.g., ß-glucanases, chitinases) and small-molecule precursors of cell wall-strengthening polymers. In this evolutionary arms race pathogens, on the other hand, can suppress M/PTI by transporting effector molecules into the host cell to target response regulatory components of the immune system. Furthermore, the invading pathogens can be protected by surface polysaccharides, and can produce antioxidants and enzymes to scavenge or detoxify the M/PTI-related toxic reactive oxygen species (ROS) [108–112].

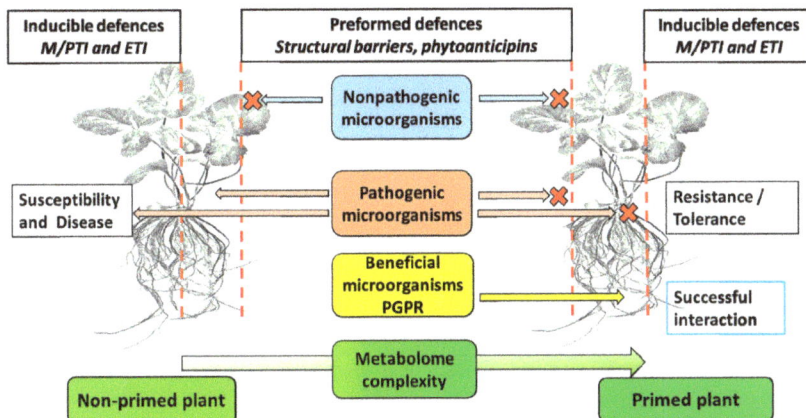

**Figure 3.** Priming, plant–microbe interactions and innate immunity. Physical barriers (waxes, suberin, callose, lignin) and innate immunity defences (MTI, MAMP-triggered immunity, and ETI, effector-triggered immunity, indicated by vertical red lines) may affect priming by biotic inducers. Interactions can either be disease-related due to biotrophic or necrotrophic pathogenic microorganisms, or beneficial due to plant growth-promoting rhizo microorganisms (PGPR) interactions with plant roots. The red crosses indicate the inability of the interacting microbe to overcome the line of defence.

To counter this infection strategy, plants have evolved specialised immune receptors encoded by resistance (*R*) genes that recognise these pathogen-specific effectors, thereby leading to an amplified secondary immune response known as effector-triggered immunity (ETI) [113–115]. ETI is mostly characterised by the induction of localised programmed cell death (referred to as the hypersensitive response or HR) in order to limit the spread of the infection, activation of defence gene expression, induction of local induced/acquired resistance (LAR) to contain the invader at the infection site, and systemic acquired resistance (SAR), which induces defences in distal, non-infected parts of the plant [116–118]. M/PTI, as the first facet of active plant defence, is the primary driving force of plant–pathogen interactions, and has been shown to confer resistance to a wide spectrum of pathogens [102,103,119]. Furthermore, experimental evidences indicate that, in some instances, input and output responses of both M/PTI and ETI converge, pointing to an interplay between M/PTI and ETI to coordinate plant immunity [120–122].

The active plant resistance system is highly complex and involves the coordination of a myriad of highly regulated mechanisms, with phytohormone crosstalk networks as central signalling systems [46,123–125]. These cellular (and biochemical) processes are spatially organised and highly controlled in intracellular compartments, and temporally complex due to the highly dynamic nature [126–128]. The molecular mechanisms of these prompted immune responses are not yet fully understood. Current models indicate that plant defence mechanisms involve cellular and organismal reprogramming expressed at the interconnected systems layers that collectively define the defensive metabolism (the 'defensome') and subsequent physiological state [24,129–131]. Thus, the ability of the pathogen to suppress the immune system of the host, and the capacity of the plant to recognise the pathogen and activate effective defences, define the outcome of the plant defence reaction.

Although plant defence immunity has been extensively studied and some aspects have been thoroughly explored, providing a wealth of detailed biochemical insights that have shaped our understanding [33,103,132–134], the plant–pathogen interactions field still has grey areas and remains an active field of research. Thus, elucidating host receptors and regulatory mechanisms that determine certain responses, unravelling biochemical processes in specific phytopathosystems, elucidating the environmental influences on diverse phyllosphere and rhizosphere interactions with microorganisms, and illuminating epigenetic regulatory mechanisms that result in passing on defensive traits to the progeny are some of the topics that still need to be fully explored.

## 4. Plant Priming: What Drives the Pre-Conditioned State

Lacking specialised mobile immune cells, every plant (and cell) is theoretically capable of establishing an active immune response upon an attack. Of necessity, such a plant immune system is characterised by self-surveillance, systemic signalling and genetic changes as mechanisms to provide successful protection and transgenerational survival [28,135]. Thus, one mechanism by which plants can enhance their resistance capacity is by potentiating the responsiveness of the immune system upon recognition of some danger-related signals from the environment. This phenomenon is known as "priming" or "pre-conditioning", and can be described as an induced state whereby a plant is pre-exposed to an inducing agent, thus rendering it more resistant to secondary stresses, i.e., the "primed" plant responds more rapidly and/or more efficiently to a subsequent stress [23,24,30,33]. Priming agents thus act as response modifiers that can lead to a more intense defence response, a faster response, an earlier response or a more sensitive response compared to the non-primed response to the same stress condition [22].

Priming can occur as a result of interactions between the host plants and beneficial microorganisms (rhizobacteria, mycorrhizal fungi) or virulent/avirulent pathogens, or by natural or synthetic compounds such as certain agrochemicals. Following such interactions, plants are cellularly and organismally reprogrammed in a long-lasting manner, and "remember" such events at a molecular level. Depending on the initial stimulus and the target of priming, primed plants can deploy a diverse set of defence mechanisms that are more rapid and stronger compared to non-primed plants [24,136–139]. Spatially, the priming events involve multiple cellular compartments, and this induced state temporally consists of three stages (Figure 4) namely (i) the priming phase (perception of stimulus), (ii) the post-challenge primed state (challenges by secondary stimulus) and (iii) the transgenerational primed state (primed state inherited from primed parents) [19,24]. These different physiological states (naïve, primed and primed and pathogen-triggered) are reflected in changes to the metabolomes and can best be investigated through untargeted metabolomics approaches [45].

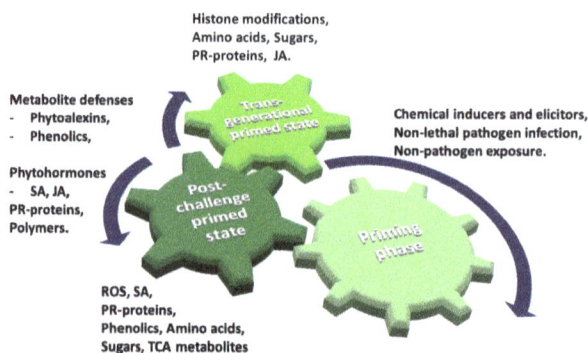

**Figure 4.** Phases in priming events. Priming generally requires sequential environmental stimuli. The priming phase is initiated by a triggering stimulus to last until the plant is exposed to a challenging stress. During this phase, slight alterations in the levels of primary—and secondary metabolites, (e.g., phytohormones, SA and JA) place the plant in a standby state of alertness. When challenged with a secondary stress, primed plants move on to the post-challenge primed state, associated with the induction and rapid deployment of defence reactions. This involves de novo biosynthesis of antimicrobial compounds. Primed plant can revert to the naïve state, but a transgenerational primed state may occur in plants when inherited from primed parental plants [12,22,24,33].

Priming can involve various layers of induced defence mechanisms that are active during different levels of plant–pathogen interactions. Such mechanisms result in a broad span of effectiveness: ranging from early responses controlled by changes in hormone-dependent signalling pathways to longer lasting mechanisms involving chromatin modification and DNA methylation. Despite the grey areas in the mechanistic understanding of priming events; some characteristic key features of priming processes have been elucidated and include 'memory', low fitness costs and more robust defences. These are associated with enhanced levels of PRRs, proteomic and metabolic reprogramming and histone modifications and resulting chromatin changes [22,28,33]. As highlighted in Figure 4, studies have indicated that the priming status of a plant can be inherently stored and passed on to its offspring, lasting for few generations and conferring improved defence responses and resistance to biotic and abiotic stresses [140–142]. This transgenerational primed state implies mechanisms ranging from epigenetic marks to the accumulation of (dormant) defence-related molecules [22,140–142].

Furthermore, as an induced state of resistance in plants, priming can be a result of SAR or induced systemic resistance (ISR) or other forms of induced resistance mechanisms [30,33,143]. In simplified terms, SAR is currently understood as the salicylic acid (SA)-dependent process, involving the transduction protein NPR1 to establish a defensive state in uninfected systemic plant parts [116]. On the other hand, ISR is induced by beneficial microbes (such as growth-promoting rhizobacteria and fungi) and orchestrates a defensive state that depends on other hormones such as jasmonate (JA) and ethylene (ET). However, studies have demonstrated that the induced resistance (IR) state involves interconnected mechanisms (more than just SAR and ISR), creating a network of defences that define the plant immune system in toto. The biochemical and molecular details of these mechanisms are still not yet fully formulated; however, with the advent and progress in systems biology methodologies, in-depth insights are gradually being achieved [138,144–146].

These IR mechanisms imply reprogramming of cellular metabolism and the reciprocal crosstalk with cellular regulatory machinery. Profiling metabolic changes is providing a wealth of descriptive information that advances our understanding of priming mechanisms. These emerging endeavours have pointed to the reprogramming of primary metabolism and differential biosynthesis of secondary metabolites as characteristic processes involved in priming events [12,20,26,49]. Experimental evidences have shown that priming of *Arabidopsis thaliana* plants by β-aminobutyric acid (BABA) involves alterations in tricarboxylic acids fluxes involving malate, oxoglutarate and fumarate, and the intensification of phenylpropanoid biosynthesis and the octadecanoic pathway [146].

Comparative metabolomic analyses of chemically-induced priming by BABA in *Arabidopsis* plants also demonstrated that the resultant defence priming significantly affected sugar metabolism, cell-wall remodelling and levels of shikimic acid derivatives. This metabolic reprogramming resulted in specific changes in amino acid profiles and accumulation of camalexin, indole-acetic acid and indole-3-carboxaldehyde [147]. Similarly, an untargeted metabolomics analysis of *A. thaliana* treated with a pathogen-derived priming agent, bacterial lipopolysaccharides, detailed the importance of tryptophan-derived indolic metabolites that included camalexin, indoleglucosinolates, indole-3-carboxylic acid and indole-acetic acid [92]. Furthermore, treatment of cultured tobacco cells with different chemical—(acibenzolar-S-methyl, azelaic acid and riboflavin) and pathogen-derived (chitosan, lipopolysaccharides and flagellin) inducers resulted in differential metabolic changes involving early phenylpropanoid pathway intermediates and products. The activation hereof is shown to be an important aspect of priming events, as well as the alterations in secondary metabolism pathways involving conjugation of hydroxycinnamic acid derivatives to quinic acid, tyramine, polyamines or glucose [49]. On the other hand, metabolic analysis of resistant progeny has shown that transgenerational priming is associated with enhanced levels of primary metabolites such as amino acids and sugars [20,22,148,149]. However, the altered pathways are highly dependent on the pathogen characteristic, i.e., biotrophic stimuli seems to mainly impact primary metabolism and involves SA signalling, while insects and necrotrophic fungi trigger secondary metabolism via JA/ET-dependent pathways [150]. A recent study has demonstrated that the symbiotic relationship between *Microbacterium* sp. 3J1 and pepper plants confers to the latter protection against drought through a metabolic reprogramming that involves production of osmo-protectants and antioxidants. These metabolic alterations spanned changes in C and N metabolism, resulting in increased levels of sugars and amino acids, phenolics and lignin precursors [151]. Another recent work, involving the interaction of poplar roots with the ectomycorrhizal fungus *Laccaria bicolor*, revealed the systemic adjustment of defence mechanisms in leaves, comprising transcriptional and metabolic reprogramming: enhancement of chitinases, volatiles, nitrogen-bearing compounds and decreased levels of phenolics. This mycorrhiza-primed state influences aboveground plant–insect interactions, conferring protection to the plant [152]. Table 1 gives a summary of available metabolic information on the priming phase, challenged phase and trans-generational priming induced by different stimuli.

These metabolomic studies (and other cited literature) indicate that multiple metabolic pathways are involved in the priming phenomenon. The interconnectedness of metabolic pathways that initially might seem distinct will increasingly be shown to have feedback loops that allow for quick activation of cellular defences to potential attackers present in the external environment, attempted ingress and resultant cellular damage. Furthermore, the combination of a multitude of biotic and abiotic stresses that plants face, and adaptability of priming events, have made the elucidation of the underlying molecular mechanisms a challenging endeavour. Thus, despite overlaps and similarities, priming mechanisms can vary and the same phenotypic traits might be the result of unrelated underlying events [24,138–152]. Unfortunately, most of the reported studies often assess just a few defensive traits related to priming events and overlook the overall multi-layered mechanisms of naïve versus primed plants.

Hence, elucidation of molecular mechanisms in priming events remains an active area of research. Knowledge regarding intracellular metabolic networks that define the dynamic metabolism of priming processes in a biosystem, pinpointing common and unique specific biochemical traits characterising the primed state across species, is crucial for translational applications from model plants to food and industrial crop plants.

**Table 1.** An overview of different stimuli and examples of metabolic changes involved in plant priming.

| Priming Agent | Plant | Phase | Classes of Induced Compounds | References |
|---|---|---|---|---|
| Beta aminobutyric acid (BABA) | *Arabidopsis thaliana* | Priming | TCA metabolites, amino acids, phytohormones, purines, cinnamic acid derivatives and fatty acids. Amino acids, indole compounds, polyamines, SA, ABA *-Plectosphaerella cucumerina* | [20] |
| | | Secondary stimulus | Enhanced levels of amino acids, indolic compounds and polyamines. SA downregulation, enhanced levels of JA and JA-Ile. | [147] |
| Hexanoic acid | *Solanum lycopersicum* | Priming | Fatty acids, oxylipins, phospholipids, chlorophyll metabolism (pheophorbide A), purines (adenosine 2′-monophosphate), sugars. Downregulation of TCA intermediate (citrate) and some amino acids. *-Botrytis cinerea* | [153] |
| | | Secondary stimulus | Glycolytic intermediated and sugars, fatty acids, ascorbate metabolism. Proline downregulation. *-Pseudomonas syringae* | [153] |
| | | Secondary stimulus | Serine upregulation. Downregulation of fatty acids, phytohormones (abscisic acid), signalling molecules (pipecolic acid) and amino acids (valine and threonine). | [153] |
| Lipopolysaccharide (LPS) | *Arabidopsis thaliana* leaves and cells | Priming | Phytohormones (SA and JA) and their methyl esters and sugar conjugates, glucosinolates, indolic compounds, cinnamic acids derivative and other phenylpropanoids. | [92] |
| LPS, chitosan and flagellin flg22 | *Nicotiana tabacum* cells | Priming | Hydroxycinnamic acid conjugates of quinic acid, shikimic acid, tyramine, polyamines or glucose. | [49] |
| Acibenzolar-S-methyl, azelaic acid, riboflavin | *Nicotiana tabacum* cells | Priming | Cinnamic acid derivatives conjugated through ester and amide bonds. | [49] |

**Table 1.** *Cont.*

| Priming Agent | Plant | Phase | Classes of Induced Compounds | References |
|---|---|---|---|---|
| Phenylacetic acid produced by *Bacillus fortis* IAGS162 | Tomato | Priming | Amino acids and sugars. | [154] |
| | | Secondary stimulus | -*Fusarium* wilt SA, sugars, amino acids, hexanoic acid, cinnamic acids (caffeic acid), shikimic acid, quinic acid, TCA metabolites, amino chlorocoumarin and methylquercetin. | [154] |
| *Pseudomonas fluorescens* SS101 | *Arabidopsis thaliana* | Priming | Indolic compounds and glucosinolates. | [155] |
| *Rhizophagus irregularis* | Tomato roots | Priming | Upregulation of cinnamic acid derivatives (ferulic acid, coniferyl alcohol and *p*-coumaroyl alcohol), lignin, yatein and oxylipins, (Z)-jasmone, tuberonic acid, tuberonic acid-12-β-glycoside, methyl-tuberonic acid-12-β-glycoside. Downregulation of phenolic amino acids, some cinnamic acids derivatives (*p*-courmaric acid and *p*-coumaraldehyde) and α-linolenic acid. | [156] |
| *Funneliformis mosseae* | Tomato roots | Priming | Upregulation of cinnamic acid derivatives (ferulic acid, coniferyl alcohol and *p*-coumaryl alcohol), lignin, yatein, oxylipins, (Z)-jasmone, methyljasmonic acid, jasmonoyl-isoluecine, 13-hydroperoxy-9,11,15-octadecatrienoic acid (HPOT), tuberonic acid, tuberonic acid-12-β-glyc, methyl-tuberonic acid-12-β-glyc. Downregulation of phenolic amino acids, some cinnamic acid derivatives (*p*-coumaric acid and *p*-coumaraldehyde), α-linolenic acid. | [156] |
| *Microbacterium* sp 3J1 | Pepper | Priming | Glutamine and α-ketoglutarate, osmoprotectants, antioxidants, sugars, amino acids, phenolics, lignin precursors. | [151] |
| Tobacco mosaic virus (TMV) | *Nicotiana tabacum* | Trans-generational state | Sugars and amino acids | [149] |

## 5. Concluding Remarks and Perspectives

In this review, we give a brief overview of the current mechanistic understanding of the plant innate immune system related to priming. Furthermore, the contribution of metabolomics in plant priming studies is highlighted, pointing out where metabolomics can contribute to new insights and deeper knowledge. Through increased technological advances, scientists are now better equipped to study the detailed metabolomic changes associated with the underlying biochemical mechanisms that support priming. In addition, the inherent constraints of this -omics methodology that represent challenges to be addressed, are discussed.

As emphasised herein, defence priming is a complex natural phenomenon that pre-conditions plants for enhanced defence against a wide range of pathogens. As such, it represents a sustainable alternative or complementary strategy that can provide avenues for plant protection against disease. However, a comprehensive functional and mechanistic understanding of the various layers of priming events is still limited and hence offers opportunities for future research. Even though such studies are still few in number, metabolic profiling of primed and naïve plants interacting with pathogens have certainly provided highly informative insights.

The metabolomic studies thus far performed on priming-related scenarios indicate that this strategy might involve multiple pathways and that the induced resistance state is often broadly specific, and may vary from species to species and in different stressor–plant systems, thereby leading to different outcomes. Depending on the initial stimulus and the target of priming, primed plants can deploy a diverse set of defence mechanisms. This adaptability—'stimulus-dependent plasticity of response traits' [22]—of priming events makes it difficult to exactly define underlying mechanisms. Conversely, despite possible overlaps or similarities, priming mechanisms can and do differ, and the same phenotypic traits might be the result of unrelated causal events. It is thus apparent that priming can involve various 'layers' of induced defence mechanisms that are active during different 'levels' of plant–pathogen interactions.

Major questions regarding priming remain, and here metabolomics techniques and approaches can assist. These include (i) how the molecular dialogue between plant and priming agent (particularly plant-beneficial microbes) drives enhanced stress resistance, yet still benefits plant development; (ii) the switching from normal growth to defence activation and subsequent deactivation of the triggered defensive state; (iii) the dynamic traits of the defensive metabolism that describe the transportation of induced resistance signals (to distal parts of the plant and neighbouring plants) where the interactions between different metabolic networks in a spatial and temporal context need to be dissected; and (iv) transgenerational priming knowledge still needs to be exploited. Furthermore, environmental influences can affect how plant genetic programmes are realised and managed, thus controlling the metabolomic phenotype. Meta-metabolomics, targeted at the phytobiome should therefore be a future approach, i.e., inter-kingdom metabolomics aimed at unravelling the complexity of chemical communication in the rhizosphere or phyllosphere.

Metabolomics, as applied in the plant sciences, is progressing beyond biomarkers towards mechanisms. Here, chemometrics and network analysis can identify participating pathways, and stimulation of pathways can be detected by comparison of metabolite profiles with subsequent quantification of discriminatory biomarkers. Moreover, comparative studies of conserved and unique metabolic pathways from different phyla (phylametabolomics) will help in the annotation of metabolites as well as pointing to important new targets of investigation in plant priming studies.

A recent and future development is that of genome-scale models of metabolism to simulate and comprehensively analyse the metabolism of cells. To attain this, algorithms that use inputs from various-omics data types are used to construct cell-line and tissue-specific metabolic models from genome-scale models. However, it may be more challenging to accurately simulate metabolism in higher plants due to some enzymes being only active in specific cell or tissue types. In addition, it is still unclear how algorithm and parameter selection (e.g., gene expression thresholds, metabolic constraints) would affect model content and predictive accuracy. Further new developments are geared towards a framework for the de novo prediction of metabolic capabilities of a cell or tissue, based on its gene expression and metabolomic profiles. These insights will guide and promote development of tissue- and cell type-specific models, and enable researchers to predict a cell's phenotype from the genotype.

**Author Contributions:** All authors (F.T., M.I.M., L.A.P., I.A.D.) made substantial contributions. Design of the manuscript was by F.T. and I.A.D., F.T. and M.I.M. were responsible for the initial draft and L.A.P. and I.A.D. for revising it critically and for important intellectual content. All authors approved the final version and agree to be accountable for all aspects of the work.

**Funding:** This research was funded by the South African National Research Foundation, grant number [95818] to ID.

**Acknowledgments:** Fidele Tugizimana and Msizi I. Mhlongo acknowledge fellowship support from the NRF and University of Johannesburg (UJ).

**Conflicts of Interest:** The authors declare no financial or non-financial conflicts.

## References

1. Peng, B.; Li, H.; Peng, X.-X. Functional metabolomics: From biomarker discovery to metabolome reprogramming. *Protein Cell* **2015**, *6*, 628–637. [CrossRef] [PubMed]
2. Ray, L.B. Metabolism is not boring. *Science* **2010**, *330*, 1337–1337. [CrossRef] [PubMed]
3. Sévin, D.C.; Kuehne, A.; Zamboni, N.; Sauer, U. Biological insights through nontargeted metabolomics. *Curr. Opin. Biotechnol.* **2015**, *34*, 1–8. [CrossRef] [PubMed]
4. Strange, K. The end of "naive reductionism": Rise of systems biology or renaissance of physiology? *AJP Cell Physiol.* **2005**, *288*, C968–C974. [CrossRef] [PubMed]
5. Carvunis, A.; Gomez, E.; Thierry-mieg, N.; Trilling, L.; Vidal, M. Biologie systémique. *Medecine/Sciences* **2009**, *25*, 578–584. [CrossRef] [PubMed]
6. Sweetlove, L.J.; Obata, T.; Fernie, A.R. Systems analysis of metabolic phenotypes: What have we learnt? *Trends Plant Sci.* **2014**, *19*, 222–230. [CrossRef] [PubMed]
7. Zak, D.E.; Tam, V.C.; Aderem, A. Systems-level analysis of innate immunity. *Annu. Rev. Immunol.* **2014**, *32*, 547–577. [CrossRef] [PubMed]
8. Weng, J. The evolutionary paths towards complexity: A metabolic perspective. *New Phytol.* **2014**, *201*, 1141–1149. [CrossRef] [PubMed]

9. Nissen, K.S.; Willats, W.G.T.; Malinovsky, F.G. Understanding CrRLK1L Function: Cell walls and growth control. *Trends Plant Sci.* **2015**, *21*, 516–527. [CrossRef] [PubMed]

10. Van Loon, L.C. The intelligent behavior of plants. *Trends Plant Sci.* **2015**, *21*, 286–294. [CrossRef] [PubMed]

11. Mhlongo, M.I.; Piater, L.A.; Madala, N.E.; Labuschagne, N.; Dubery, I.A. The chemistry of plant–microbe interactions in the rhizosphere and the potential for metabolomics to reveal signaling related to defense priming and induced systemic resistance. *Front. Plant Sci.* **2018**, *9*, 112. [CrossRef] [PubMed]

12. Tenenboim, H.; Brotman, Y. Omic relief for the biotically stressed: Metabolomics of plant biotic interactions. *Trends Plant Sci.* **2016**, *21*, 781–791. [CrossRef] [PubMed]

13. Mengiste, T. Plant immunity to necrotrophs. *Annu. Rev. Phytopathol.* **2012**, *50*, 267–294. [CrossRef] [PubMed]

14. Pennisi, E. Armed and dangerous. *Science* **2010**, *327*, 804–805. [CrossRef] [PubMed]

15. Gómez-Gómez, L. Plant perception systems for pathogen recognition and defence. *Mol. Immunol.* **2004**, *41*, 1055–1062. [CrossRef] [PubMed]

16. Ahuja, I.; de Vos, R.C.H.; Bones, A.M.; Hall, R.D. Plant molecular stress responses face climate change. *Trends Plant Sci.* **2010**, *15*, 664–674. [CrossRef] [PubMed]

17. Gust, A.A.; Brunner, F.; Nürnberger, T. Biotechnological concepts for improving plant innate immunity. *Curr. Opin. Biotechnol.* **2010**, *21*, 204–210. [CrossRef] [PubMed]

18. Conrath, U.; Beckers, G.J.M.; Flors, V.; García-Agustín, P.; Jakab, G.; Mauch, F.; Newman, M.-A.; Pieterse, C.M.J.; Poinssot, B.; Pozo, M.J.; et al. Priming: Getting ready for battle. *Mol. Plant Microbe Interact.* **2006**, *19*, 1062–1071. [CrossRef] [PubMed]

19. Pastor, V.; Luna, E.; Mauch-Mani, B.; Ton, J.; Flors, V. Primed plants do not forget. *Environ. Exp. Bot.* **2013**, *94*, 45–56. [CrossRef]

20. Pastor, V.; Balmer, A.; Gamir, J.; Flors, V.; Mauch-Mani, B. Preparing to fight back: Generation and storage of priming compounds. *Front. Plant Sci.* **2014**, *5*, 295. [CrossRef] [PubMed]

21. Balmer, A.; De Paoli, E.; Si-Ammour, A.; Mauch-Mani, B.; Balmer, D. Signs of silence: Small RNAs and antifungal responses in *Arabidopsis thaliana* and *Zea mays*. In *Plant Engineering*; Intechopen: London, UK, 2017; pp. 9–31. [CrossRef]

22. Hilker, M.; Schwachtje, J.; Baier, M.; Balazadeh, S.; Bäurle, I.; Geiselhardt, S.; Hincha, D.K.; Kunze, R.; Mueller-Roeber, B.; Rillig, M.C.; et al. Priming and memory of stress responses in organisms lacking a nervous system. *Biol. Rev.* **2016**, *91*, 1118–1133. [CrossRef] [PubMed]

23. Conrath, U.; Pieterse, C.M.J.; Mauch-Mani, B. Priming in plant-pathogen interactions. *Trends Plant Sci.* **2002**, *7*, 210–216. [CrossRef]

24. Balmer, A.; Pastor, V.; Gamir, J.; Flors, V.; Mauch-Mani, B. The "prime-ome": Towards a holistic approach to priming. *Trends Plant Sci.* **2015**, *20*, 443–452. [CrossRef] [PubMed]

25. Chinnusamy, V.; Zhu, J.-K. Epigenetic regulation of stress responses in plants. *Curr. Opin. Plant Biol.* **2009**, *12*, 133–139. [CrossRef] [PubMed]

26. Conrath, U. Molecular aspects of defence priming. *Trends Plant Sci.* **2011**, *16*, 524–531. [CrossRef] [PubMed]

27. Jaskiewicz, M.; Conrath, U.; Peterhänsel, C. Chromatin modification acts as a memory for systemic acquired resistance in the plant stress response. *EMBO Rep.* **2011**, *12*, 50–5. [CrossRef] [PubMed]

28. Holeski, L.M.; Jander, G.; Agrawal, A.A. Transgenerational defense induction and epigenetic inheritance in plants. *Trends Ecol. Evol.* **2012**, *27*, 618–626. [CrossRef] [PubMed]

29. Conrath, U.; Thulke, O.; Katz, V.; Schwindling, S.; Kohler, A. Priming as a mechanism in induced systemic resistance of plants. *Eur. J. Plant Pathol.* **2001**, *107*, 113–119. [CrossRef]

30. Conrath, U.; Beckers, G.J.M.; Langenbach, C.J.G.; Jaskiewicz, M.R. Priming for enhanced defense. *Annu. Rev. Phytopathol.* **2015**, *53*, 97–119. [CrossRef] [PubMed]

31. Návarová, H.; Bernsdorff, F.; Döring, A.-C.; Zeier, J. Pipecolic acid, an endogenous mediator of defense amplification and priming, is a critical regulator of inducible plant immunity. *Plant Cell* **2012**, *24*, 5123–5141. [CrossRef] [PubMed]

32. Windram, O.; Penfold, C.A.; Denby, K.J. Network modeling to understand plant immunity. *Annu. Rev. Phytopathol.* **2014**, *52*, 93–111. [CrossRef] [PubMed]

33. Martinez-Medina, A.; Flors, V.; Heil, M.; Mauch-Mani, B.; Pieterse, C.M.J.; Pozo, M.J.; Ton, J.; Van Dam, N.M.; Conrath, U. Recognizing plant defense priming. *Trends Plant Sci.* **2016**, *21*, 818–822. [CrossRef] [PubMed]

34. Goeddel, L.C.; Patti, G.J. Maximizing the value of metabolomic data. *Bioanalysis* **2012**, *4*, 2199–2201. [CrossRef] [PubMed]

35. Lindon, J.C.; Nicholson, J.K. Analytical technologies for metabonomics and metabolomics, and multi-omic information recovery. *TrAC Trends Anal. Chem.* **2008**, *27*, 194–204. [CrossRef]

36. Nicholson, J.K.; Connelly, J.; Lindon, J.C.; Holmes, E. Metabonomics: A platform for studying drug toxicity and gene function. *Nat. Rev. Drug Discov.* **2002**, *1*, 153–161. [CrossRef] [PubMed]

37. Nicholson, J.K.; Lindon, J.C.; Holmes, E. "Metabonomics": Understanding the metabolic responses of living systems to pathophysiological stimuli via multivariate statistical analysis of biological NMR spectroscopic data. *Xenobiotica* **1999**, *29*, 1181–1189. [CrossRef] [PubMed]

38. Worley, B.; Powers, R. Multivariate Analysis in Metabolomics. *Curr. Metabolomics* **2013**, *1*, 92–107. [PubMed]

39. McKnight, S.L. On Getting There from Here. *Science* **2010**, *330*, 1338–1339. [CrossRef] [PubMed]

40. Beisken, S.; Eiden, M.; Salek, R.M. Getting the right answers: Understanding metabolomics challenges. *Expert Rev. Mol. Diagn.* **2015**, *15*, 97–109. [CrossRef] [PubMed]

41. Kell, D.B.; Brown, M.; Davey, H.M.; Dunn, W.B.; Spasic, I.; Oliver, S.G. Metabolic footprinting and systems biology: The medium is the message. *Nat. Rev. Microbiol.* **2005**, *3*, 557–565. [CrossRef] [PubMed]

42. Fiehn, O.; Kopka, J.; Dörmann, P.; Altmann, T.; Trethewey, R.N.; Willmitzer, L. Metabolite profiling for plant functional genomics. *Nat. Biotechnol.* **2000**, *18*, 1157–1161. [CrossRef] [PubMed]

43. Sumner, L.W.; Mendes, P.; Dixon, R.A. Plant metabolomics: Large-scale phytochemistry in the functional genomics era. *Phytochemistry* **2003**, *62*, 817–836. [CrossRef]

44. Kell, D.B.; Oliver, S.G. The metabolome 18 years on: A concept comes of age. *Metabolomics* **2016**, *12*, 148. [CrossRef] [PubMed]

45. Tugizimana, F.; Piater, L.A.; Dubery, I.A. Plant metabolomics: A new frontier in phytochemical analysis. *S. Afr. J. Sci.* **2013**, *109*, 18–20. [CrossRef]

46. Erb, M.; Glauser, G. Family business: Multiple members of major phytohormone classes orchestrate plant stress responses. *Chem. A Eur. J.* **2010**, *16*, 10280–10289. [CrossRef] [PubMed]

47. Likić, V.A.; McConville, M.J.; Lithgow, T.; Bacic, A. Systems biology: The next frontier for bioinformatics. *Adv. Bioinform.* **2010**, *2010*, 1–10. [CrossRef] [PubMed]

48. Mazid, M.; Khan, T.; Mohammad, F. Role of secondary metabolites in defense mechanisms of plants. *Biol. Med.* **2011**, *3*, 232–249.

49. Mhlongo, M.I.; Steenkamp, P.A.; Piater, L.A.; Madala, N.E.; Dubery, I.A. Profiling of altered metabolomic states in *Nicotiana tabacum* cells induced by priming agents. *Front. Plant Sci.* **2016**, *7*, 1527. [CrossRef] [PubMed]

50. Saito, K.; Matsuda, F. Metabolomics for functional genomics, systems biology, and biotechnology. *Annu. Rev. Plant Biol.* **2010**, *61*, 463–489. [CrossRef] [PubMed]

51. Richards, S.E.; Dumas, M.-E.; Fonville, J.M.; Ebbels, T.M.D.; Holmes, E.; Nicholson, J.K. Intra- and inter-omic fusion of metabolic profiling data in a systems biology framework. *Chemom. Intell. Lab. Syst.* **2010**, *104*, 121–131. [CrossRef]

52. Dikicioglu, D.; Dunn, W.B.; Kell, D.B.; Kirdar, B.; Oliver, S.G. Short- and long-term dynamic responses of the metabolic network and gene expression in yeast to a transient change in the nutrient environment. *Mol. Biosyst.* **2012**, *8*, 1760–1774. [CrossRef] [PubMed]

53. Lazar, A.G.; Romanciuc, F.; Socaciu, M.A.; Socaciu, C. Bioinformatics tools for metabolomic data processing and analysis using untargeted liquid chromatography coupled with mass spectrometry. *Bull. UASVM Anim. Sci. Biotechnol.* **2015**, *72*, 103–115. [CrossRef]

54. Johnson, C.H.; Ivanisevic, J.; Siuzdak, G. Metabolomics: Beyond biomarkers and towards mechanisms. *Nat. Rev. Mol. Cell Biol.* **2016**, *17*, 451–459. [CrossRef] [PubMed]

55. Nägele, T. Linking metabolomics data to underlying metabolic regulation. *Front. Mol. Biosci.* **2014**, *1*, 22. [CrossRef] [PubMed]

56. Goodacre, R.; Vaidyanathan, S.; Dunn, W.B.; Harrigan, G.G.; Kell, D.B. Metabolomics by numbers: Acquiring and understanding global metabolite data. *Trends Biotechnol.* **2004**, *22*, 245–252. [CrossRef] [PubMed]

57. Verpoorte, R.; Choi, Y.H.; Mustafa, N.R.; Kim, H.K. Metabolomics: Back to basics. *Phytochem. Rev.* **2008**, *7*, 525–537. [CrossRef]

58. Boccard, J.; Rudaz, S. Harnessing the complexity of metabolomic data with chemometrics. *J. Chemom.* **2014**, *28*, 1–9. [CrossRef]

59. Lai, Z.; Tsugawa, H.; Wohlgemuth, G.; Mehta, S.; Mueller, M.; Zheng, Y.; Ogiwara, A.; Meissen, J.; Showalter, M.; Takeuchi, K.; et al. Identifying metabolites by integrating metabolome databases with mass spectrometry cheminformatics. *Nat. Methods* **2017**, *15*. [CrossRef] [PubMed]

60. Hall, R.D. Plant metabolomics in a nutshell: Potential and future challenges. In *Annual Plant Reviews*; Hall, R.D., Ed.; Wiley-Blackwell: Chichester, UK, 2011; pp. 1–24. ISBN 9781444339956.

61. Heinig, U.; Gutensohn, M.; Dudareva, N.; Aharoni, A. The challenges of cellular compartmentalization in plant metabolic engineering. *Curr. Opin. Biotechnol.* **2013**, *24*, 239–246. [CrossRef] [PubMed]

62. Yanes, O.; Tautenhahn, R.; Patti, G.J.; Siuzdak, G. Expanding coverage of the metabolome for global metabolite profiling. *Anal. Chem.* **2011**, *83*, 2152–2161. [CrossRef] [PubMed]

63. Choi, Y.H.; Verpoorte, R. Metabolomics: What you see is what you extract. *Phytochem. Anal.* **2014**, *25*, 289–290. [CrossRef] [PubMed]

64. Khoza, B.S.; Dubery, I.A.; Byth-Illing, H.A.; Steenkamp, P.A.; Chimuka, L.; Madala, N.E. Optimization of pressurized hot water extraction of flavonoids from *Momordica foetida* using UHPLC-qTOF-MS and multivariate chemometric approaches. *Food Anal. Methods* **2016**, *9*, 1480–1489. [CrossRef]

65. Dunn, W.B.; Ellis, D.I. Metabolomics: Current analytical platforms and methodologies. *Trends Anal. Chem.* **2005**, *24*, 285–294. [CrossRef]

66. Theodoridis, G.; Gika, H.G.; Wilson, I.D. Mass spectrometry-based holistic analytical approaches for metabolite profiling in systems biology studies. *Mass Spectrom. Rev.* **2011**, *30*, 884–906. [CrossRef] [PubMed]

67. Alonso, A.; Marsal, S.; Juliá, A. Analytical methods in untargeted metabolomics: State of the art in 2015. *Front. Bioeng. Biotechnol.* **2015**, *3*, 23. [CrossRef] [PubMed]

68. Tugizimana, F.; Steenkamp, P.A.; Piater, L.A.; Dubery, I.A. Mass spectrometry in untargeted liquid chromatography/mass spectrometry metabolomics: Electrospray ionisation parameters and global coverage of the metabolome. *Rapid Commun. Mass Spectrom.* **2018**, *32*, 121–132. [CrossRef] [PubMed]

69. Markley, J.L.; Brüschweiler, R.; Edison, A.S.; Eghbalnia, H.R.; Powers, R.; Raftery, D.; Wishart, D.S. The future of NMR-based metabolomics. *Curr. Opin. Biotechnol.* **2017**, *43*, 34–40. [CrossRef] [PubMed]

70. Safer, S.; Cicek, S.S.; Pieri, V.; Schwaiger, S.; Schneider, P.; Wissemann, V.; Stuppner, H. Metabolic fingerprinting of *Leontopodium* species (Asteraceae) by means of $^1$H NMR and HPLC-ESI-MS. *Phytochemistry* **2011**, *72*, 1379–1389. [CrossRef] [PubMed]

71. Marshall, D.D.; Powers, R. Beyond the paradigm: Combining mass spectrometry and nuclear magnetic resonance for metabolomics. *Prog. Nucl. Magn. Reson. Spectrosc.* **2017**, *100*, 1–16. [CrossRef] [PubMed]

72. Da Silva, R.R.; Dorrestein, P.C.; Quinn, R.A. Illuminating the dark matter in metabolomics. *Proc. Natl. Acad. Sci. USA* **2015**, *112*, 12549–12550. [CrossRef] [PubMed]

73. Tugizimana, F.; Steenkamp, P.; Piater, L.; Dubery, I. A conversation on data mining strategies in LC-MS untargeted metabolomics: Pre-processing and pre-treatment Steps. *Metabolites* **2016**, *6*, 40. [CrossRef] [PubMed]

74. Godzien, J.; Ciborowski, M.; Angulo, S.; Barbas, C. From numbers to a biological sense: How the strategy chosen for metabolomics data treatment may affect final results. A practical example based on urine fingerprints obtained by LC-MS. *Electrophoresis* **2013**, *34*, 2812–2826. [CrossRef] [PubMed]

75. Kind, T.; Fiehn, O. Advances in structure elucidation of small molecules using mass spectrometry. *Bioanal. Rev.* **2010**, *2*, 23–60. [CrossRef] [PubMed]

76. Creek, D.J.; Dunn, W.B.; Fiehn, O.; Griffin, J.L.; Hall, R.D.; Lei, Z.; Mistrik, R.; Neumann, S.; Schymanski, E.L.; Sumner, L.W.; et al. Metabolite identification: Are you sure? And how do your peers gauge your confidence? *Metabolomics* **2014**, *10*, 350–353. [CrossRef]

77. Van der Hooft, J.J.J.; Padmanabhan, S.; Burgess, K.E.V.; Barrett, M.P. Urinary antihypertensive drug metabolite screening using molecular networking coupled to high-resolution mass spectrometry fragmentation. *Metabolomics* **2016**, *12*, 125. [CrossRef] [PubMed]

78. Karp, P.; Billington, R.; Holland, T.; Kothari, A.; Krummenacker, M.; Weaver, D.; Latendresse, M.; Paley, S. Computational metabolomics aperations at BioCyc.org. *Metabolites* **2015**, *5*, 291–310. [CrossRef] [PubMed]

79. Gerstl, M.P.; Ruckerbauer, D.E.; Mattanovich, D.; Jungreuthmayer, C.; Zanghellini, J. Metabolomics integrated elementary flux mode analysis in large metabolic networks. *Sci. Rep.* **2015**, *5*, 8930. [CrossRef] [PubMed]

80. Dhondt, S.; Wuyts, N.; Inzé, D. Cell to whole-plant phenotyping: The best is yet to come. *Trends Plant Sci.* **2013**, *18*, 428–439. [CrossRef] [PubMed]

81. Rocca-Serra, P.; Salek, R.M.; Arita, M.; Correa, E.; Dayalan, S.; Gonzalez-Beltran, A.; Ebbels, T.; Goodacre, R.; Hastings, J.; Haug, K.; et al. Data standards can boost metabolomics research, and if there is a will, there is a way. *Metabolomics* **2016**, *12*, 14. [CrossRef] [PubMed]

82. Van der Hooft, J.J.J.; Wandy, J.; Barrett, M.P.; Burgess, K.E.V.; Rogers, S. Topic modeling for untargeted substructure exploration in metabolomics. *Proc. Natl. Acad. Sci. USA* **2016**, *113*, 13738–13743. [CrossRef] [PubMed]

83. Macel, M.; van dam, N.M.; Keurentjes, J.J.B. Metabolomics: The chemistry between ecology and genetics. *Mol. Ecol. Resour.* **2010**, *10*, 583–593. [CrossRef] [PubMed]

84. Fiehn, O. Metabolomics-the link between genotypes and phenotypes. *Plant Mol. Biol.* **2002**, *48*, 155–171. [CrossRef] [PubMed]

85. Wang, H.; Correa, E.; Dunn, W.B.; Winder, C.L.; Goodacre, R.; Lloyd, J.R. Metabolomic analyses show that electron donor and acceptor ratios control anaerobic electron transfer pathways in *Shewanella oneidensis*. *Metabolomics* **2012**, *9*, 642–656. [CrossRef]

86. Croes, D.; Couche, F.; Wodak, S.J.; van Helden, J. Inferring meaningful pathways in weighted metabolic networks. *J. Mol. Biol.* **2006**, *356*, 222–236. [CrossRef] [PubMed]

87. Bernard, T.; Bridge, A.; Morgat, A.; Moretti, S.; Xenarios, I.; Pagni, M. Reconciliation of metabolites and biochemical reactions for metabolic networks. *Brief. Bioinform.* **2014**, *15*, 123–135. [CrossRef] [PubMed]

88. Weckwerth, W.; Loureiro, M.E.; Wenzel, K.; Fiehn, O. Differential metabolic networks unravel the effects of silent plant phenotypes. *Proc. Natl. Acad. Sci. USA* **2004**, *101*, 7809–7814. [CrossRef] [PubMed]

89. Tugizimana, F.; Steenkamp, P.A.; Piater, L.A.; Dubery, I.A. Multi-platform metabolomic analyses of ergosterol-induced dynamic changes in *Nicotiana tabacum* cells. *PLoS ONE* **2014**, *9*, e87846. [CrossRef] [PubMed]

90. Allwood, J.W.; Ellis, D.I.; Goodacre, R. Metabolomic technologies and their application to the study of plants and plant–host interactions. *Physiol. Plant.* **2008**, *132*, 117–135. [CrossRef] [PubMed]

91. López-Gresa, M.P.; Maltese, F.; Bellés, J.M.; Conejero, V.; Kim, H.K.; Choi, Y.H.; Verpoorte, R. Metabolic response of tomato leaves upon different plant-pathogen interactions. *Phytochem. Anal.* **2010**, *21*, 89–94. [CrossRef] [PubMed]

92. Finnegan, T.; Steenkamp, P.A.; Piater, L.A.; Dubery, I.A. The lipopolysaccharide-induced metabolome signature in *Arabidopsis thaliana* reveals dynamic reprogramming of phytoalexin and phytoanticipin pathways. *PLoS ONE* **2016**, *11*, e0163572. [CrossRef] [PubMed]

93. Balmer, D.; Flors, V.; Glauser, G.; Mauch-Mani, B. Metabolomics of cereals under biotic stress: Current knowledge and techniques. *Front. Plant Sci.* **2013**, *4*, 82. [CrossRef] [PubMed]

94. Mhlongo, M.I.; Piater, L.A.; Steenkamp, P.A.; Madala, N.E.; Dubery, I.A. Metabolomic fingerprinting of primed tobacco cells provide the first evidence for the biological origin of *cis*-chlorogenic acid. *Biotechnol. Lett.* **2015**, *37*, 205–209. [CrossRef] [PubMed]

95. Duan, L.; Liu, H.; Li, X.; Xiao, J.; Wang, S. Multiple phytohormones and phytoalexins are involved in disease resistance to *Magnaporthe oryzae* invaded from roots in rice. *Physiol. Plant.* **2014**, *152*, 486–500. [CrossRef] [PubMed]

96. Weiberg, A.; Wang, M.; Bellinger, M.; Jin, H. Small RNAs: A new paradigm in plant-microbe interactions. *Annu. Rev. Phytopathol.* **2014**, *52*, 495–516. [CrossRef] [PubMed]

97. Ahmad, S.; Gordon-Weeks, R.; Pickett, J.; Ton, J. Natural variation in priming of basal resistance: From evolutionary origin to agricultural exploitation. *Mol. Plant Pathol.* **2010**, *11*, 817–827. [CrossRef] [PubMed]

98. Zeier, J. New insights into the regulation of plant immunity by amino acid metabolic pathways. *Plant. Cell Environ.* **2013**, *36*, 2085–2103. [CrossRef] [PubMed]

99. Łaźniewska, J.; Macioszek, V.K.; Kononowicz, A.K. Plant-fungus interface: The role of surface structures in plant resistance and susceptibility to pathogenic fungi. *Physiol. Mol. Plant Pathol.* **2012**, *78*, 24–30. [CrossRef]

100. Aguzzi, A.; Altmeyer, M. Phase separation: Linking cellular compartmentalization to disease. *Trends Cell Biol.* **2016**, *26*, 547–558. [CrossRef] [PubMed]

101. Feussner, I.; Polle, A.; Paszkowski, U.; Scott, D.B. What the transcriptome does not tell—Proteomics and metabolomics are closer to the plants' patho-phenotype. *Curr. Opin. Plant Biol.* **2015**, *26*, 26–31. [CrossRef] [PubMed]

102. Zipfel, C. Pattern-recognition receptors in plant innate immunity. *Curr. Opin. Immunol.* **2008**, *20*, 10–16. [CrossRef] [PubMed]

103. Zipfel, C. Early molecular events in PAMP-triggered immunity. *Curr. Opin. Plant Biol.* **2009**, *12*, 414–420. [CrossRef] [PubMed]

104. Henry, G.; Thonart, P.; Ongena, M. PAMPs, MAMPs, DAMPs and others: An update on the diversity of plant immunity elicitors. *Biotechnol. Agron. Soc. Environ.* **2012**, *16*, 257–268.

105. Jones, J.D.; Dangl, J.L. The plant immune system. *Nature* **2006**, *444*, 323–329. [CrossRef] [PubMed]

106. Sanabria, N.M.; Huang, J.; Dubery, I.A. Self/non-self perception in plants in innate immunity and defense. *Self/Nonself Immune Recognit. Signal.* **2009**, *1*, 1–15. [CrossRef] [PubMed]

107. Zipfel, C. Plant pattern-recognition receptors. *Trends Immunol.* **2014**, *35*, 345–351. [CrossRef] [PubMed]

108. Mendgen, K.; Hahn, M. Plant infection and the establishment of fungal biotrophy. *Trends Plant Sci.* **2002**, *7*, 352–356. [CrossRef]

109. Felix, G.; Boller, T. Molecular sensing of bacteria in plants: The highly conserved RNA-binding motif RNP-1 of bacterial cold shock proteins is recognized as an elicitor signal in tobacco. *J. Biol. Chem.* **2003**, *278*, 6201–6208. [CrossRef] [PubMed]

110. D'Haeze, W.; Holsters, M. Surface polysaccharides enable bacteria to evade plant immunity. *Trends Microbiol.* **2004**, *12*, 555–561. [CrossRef] [PubMed]

111. McDowell, J.M.; Simon, S.A. Molecular diversity at the plant-pathogen interface. *Dev. Comp. Immunol.* **2008**, *32*, 736–744. [CrossRef] [PubMed]

112. Horbach, R.; Navarro-Quesada, A.R.; Knogge, W.; Deising, H.B. When and how to kill a plant cell: Infection strategies of plant pathogenic fungi. *J. Plant Physiol.* **2011**, *168*, 51–62. [CrossRef] [PubMed]

113. Tiffin, P.; Moeller, D.A. Molecular evolution of plant immune system genes. *Trends Genet.* **2006**, *22*, 662–670. [CrossRef] [PubMed]

114. Craig, A.; Ewan, R.; Mesmar, J.; Gudipati, V.; Sadanandom, A. E3 ubiquitin ligases and plant innate immunity. *J. Exp. Bot.* **2009**, *60*, 1123–1132. [CrossRef] [PubMed]

115. Pieterse, C.M.J.; Leon-Reyes, A.; Van der Ent, S.; Van Wees, S.C.M. Networking by small-molecule hormones in plant immunity. *Nat. Chem. Biol.* **2009**, *5*, 308–316. [CrossRef] [PubMed]

116. Fu, Z.Q.; Dong, X. Systemic acquired resistance: Turning local infection into global defense. *Annu. Rev. Plant Biol.* **2013**, *64*, 839–863. [CrossRef] [PubMed]

117. Kachroo, A.; Robin, G.P. Systemic signaling during plant defense. *Curr. Opin. Plant Biol.* **2013**, *16*, 527–533. [CrossRef] [PubMed]

118. Gao, Q.-M.; Zhu, S.; Kachroo, P.; Kachroo, A. Signal regulators of systemic acquired resistance. *Front. Plant Sci.* **2015**, *6*, 228. [CrossRef] [PubMed]

119. Zhang, J.; Zhou, J.-M. Plant immunity triggered by microbial molecular signatures. *Mol. Plant* **2010**, *3*, 783–793. [CrossRef] [PubMed]

120. Tsuda, K.; Katagiri, F. Comparing signaling mechanisms engaged in pattern-triggered and effector-triggered immunity. *Curr. Opin. Plant Biol.* **2010**, *13*, 459–465. [CrossRef] [PubMed]

121. Qi, Y.; Tsuda, K.; Glazebrook, J.; Katagiri, F. Physical association of pattern-triggered immunity (PTI) and effector-triggered immunity (ETI) immune receptors in *Arabidopsis*. *Mol. Plant Pathol.* **2011**, *12*, 702–708. [CrossRef] [PubMed]

122. Thomma, B.P.H. J.; Nürnberger, T.; Joosten, M.H.A. J. Of PAMPs and effectors: The blurred PTI-ETI dichotomy. *Plant Cell* **2011**, *23*, 4–15. [CrossRef] [PubMed]

123. Nafisi, M.; Fimognari, L.; Sakuragi, Y. Interplays between the cell wall and phytohormones in interaction between plants and necrotrophic pathogens. *Phytochemistry* **2014**, *112*, 63–71. [CrossRef] [PubMed]

124. Hou, X.; Ding, L.; Yu, H. Crosstalk between GA and JA signaling mediates plant growth and defense. *Plant Cell Rep.* **2013**, *32*, 1067–1074. [CrossRef] [PubMed]

125. Kohli, A.; Sreenivasulu, N.; Lakshmanan, P.; Kumar, P.P. The phytohormone crosstalk paradigm takes center stage in understanding how plants respond to abiotic stresses. *Plant Cell Rep.* **2013**, *32*, 945–957. [CrossRef] [PubMed]

126. Somssich, I.E.; Hahlbrock, K. Pathogen defence in plants: A paradigm of biological complexity. *Trends Plant Sci.* **1998**, *3*, 86–90. [CrossRef]

127. Caplan, J.L.; Zhu, X.; Mamillapalli, P.; Marathe, R.; Anandalakshmi, R.; Dinesh-Kumar, S.P. Induced ER chaperones regulate a receptor-like kinase to mediate antiviral innate immune response in plants. *Cell Host Microbe* **2009**, *6*, 457–469. [CrossRef] [PubMed]

128. Trujillo, M.; Shirasu, K. Ubiquitination in plant immunity. *Curr. Opin. Plant Biol.* **2010**, *13*, 402–408. [CrossRef] [PubMed]

129. Etalo, D.W.; Stulemeijer, I.J.E.; van Esse, H.P.; de Vos, R.C.H.; Bouwmeester, H.J.; Joosten, M.H.A.J. System-wide hypersensitive response-associated transcriptome and metabolome reprogramming in tomato. *Plant Physiol.* **2013**, *162*, 1599–1617. [CrossRef] [PubMed]

130. Klemptner, R.L.; Sherwood, J.S.; Tugizimana, F.; Dubery, I.A.; Piater, L.A. Ergosterol, an orphan fungal microbe-associated molecular pattern (MAMP). *Mol. Plant Pathol.* **2014**, *15*, 747–761. [CrossRef] [PubMed]

131. Djami-Tchatchou, A.T.; Dubery, I.A. Lipopolysaccharide perception leads to dynamic alterations in the microtranscriptome of *Arabidopsis thaliana* cells and leaf tissues. *BMC Plant Biol.* **2015**, *15*, 79. [CrossRef] [PubMed]

132. Berger, S.; Sinha, A.K.; Roitsch, T. Plant physiology meets phytopathology: Plant primary metabolism and plant-pathogen interactions. *J. Exp. Bot.* **2007**, *58*, 4019–4026. [CrossRef] [PubMed]

133. Xin, X.-F.; Nomura, K.; Aung, K.; Velásquez, A.C.; Yao, J.; Boutrot, F.; Chang, J.H.; Zipfel, C.; He, S.Y. Bacteria establish an aqueous living space in plants crucial for virulence. *Nature* **2016**, *539*, 524–529. [CrossRef] [PubMed]

134. Segonzac, C.; Zipfel, C. Activation of plant pattern-recognition receptors by bacteria. *Curr. Opin. Microbiol.* **2011**, *14*, 54–61. [CrossRef] [PubMed]

135. Spoel, S.H.; Dong, X. How do plants achieve immunity? Defence without specialized immune cells. *Nat. Rev. Immunol.* **2012**, *12*, 89–100. [CrossRef] [PubMed]

136. Goellner, K.; Conrath, U. Priming: It's all the world to induced disease resistance. *Eur. J. Plant Pathol.* **2008**, *121*, 233–242. [CrossRef]

137. Zamioudis, C.; Pieterse, C.M.J. Modulation of host immunity by beneficial microbes. *Mol. Plant. Microbe. Interact.* **2012**, *25*, 139–150. [CrossRef] [PubMed]

138. Venturi, V.; Keel, C. Signaling in the rhizosphere. *Trends Plant Sci.* **2016**, *21*, 187–198. [CrossRef] [PubMed]

139. Cameron, D.D.; Neal, A.L.; van Wees, S.C.M.; Ton, J. Mycorrhiza-induced resistance: More than the sum of its parts? *Trends Plant Sci.* **2013**, *18*, 539–545. [CrossRef] [PubMed]

140. Slaughter, A.; Daniel, X.; Flors, V.; Luna, E.; Hohn, B.; Mauch-Mani, B. Descendants of primed Arabidopsis plants exhibit resistance to biotic stress. *Plant Physiol.* **2012**, *158*, 835–843. [CrossRef] [PubMed]

141. Ramírez-Carrasco, G.; Martínez-Aguilar, K.; Alvarez-Venegas, R. Transgenerational defense priming for crop protection against plant pathogens: A hypothesis. *Front. Plant Sci.* **2017**, *8*, 696. [CrossRef] [PubMed]

142. Tabassum, T.; Farooq, M.; Ahmad, R.; Zohaib, A.; Wahid, A. Seed priming and transgenerational drought memory improves tolerance against salt stress in bread wheat. *Plant Physiol. Biochem.* **2017**, *118*, 362–369. [CrossRef] [PubMed]

143. Pieterse, C.M.J.; Zamioudis, C.; Berendsen, R.L.; Weller, D.M.; Van Wees, S.C.M.; Bakker, P.A.H.M. Induced systemic resistance by beneficial microbes. *Annu. Rev. Phytopathol.* **2014**, *52*, 347–375. [CrossRef] [PubMed]

144. Gozzo, F.; Faoro, F. Systemic acquired resistance (50 years after discovery): Moving from the lab to the field. *J. Agric. Food Chem.* **2013**, *61*, 12473–12491. [CrossRef] [PubMed]

145. Shah, J.; Zeier, J. Long-distance communication and signal amplification in systemic acquired resistance. *Front. Plant Sci.* **2013**, *4*, 30. [CrossRef] [PubMed]

146. Pastor, V.; Luna, E.; Ton, J.; Cerezo, M.; García-agustín, P.; Flors, V. Fine tuning of reactive oxygen species homeostasis regulates primed immune responses in *Arabidopsis*. *Mol. Plant-Microbe Interact.* **2013**, *26*, 1334–1344. [CrossRef] [PubMed]

147. Gamir, J.; Pastor, V.; Kaever, A.; Cerezo, M.; Flors, V. Targeting novel chemical and constitutive primed metabolites against Plectosphaerella cucumerina. *Plant J.* **2014**, *78*, 227–240. [CrossRef] [PubMed]

148. Luna, E.; Bruce, T.J.A.; Roberts, M.R.; Flors, V.; Ton, J. Next-generation systemic acquired resistance. *Plant Physiol.* **2012**, *158*, 844–853. [CrossRef] [PubMed]

149. Mandal, R.; Kathiria, P.; Psychogios, N.; Bouatra, S.; Krishnamurthy, R.; Wishart, D.; Kovalchuk, I. Progeny of tobacco mosaic virus-infected *Nicotiana tabacum* plants exhibit trans-generational changes in metabolic profiles. *Biocatal. Agric. Biotechnol.* **2012**, *1*, 115–123. [CrossRef]

150. Gamir, J.; Sánchez-Bel, P.; Flors, V. Molecular and physiological stages of priming: How plants prepare for environmental challenges. *Plant Cell Rep.* **2014**, *33*, 1935–1949. [CrossRef] [PubMed]

151. Vílchez, J.I.; Niehaus, K.; Dowling, D.N.; González-López, J.; Manzanera, M. Protection of pepper plants from drought by *Microbacterium* sp. 3J1 by modulation of the plant's glutamine and α-ketoglutarate content: A comparative metabolomics approach. *Front. Microbiol.* **2018**, *9*, 284. [CrossRef] [PubMed]

152. Kaling, M.; Schmidt, A.; Moritz, F.; Rosenkranz, M.; Witting, M.; Kasper, K.; Janz, D.; Schmitt-Kopplin, P.; Schnitzler, J.-P.; Polle, A. Mycorrhiza-triggered transcriptomic and metabolomic networks impinge on herbivore fitness. *Plant Physiol.* **2018**, *176*, 2639–2656. [CrossRef] [PubMed]

153. Camañes, G.; Scalschi, L.; Vicedo, B.; González-Bosch, C.; García-Agustín, P. An untargeted global metabolomic analysis reveals the biochemical changes underlying basal resistance and priming in *Solanum lycopersicum*, and identifies 1-methyltryptophan as a metabolite involved in plant responses to *Botrytis cinerea* and *Pseudomonas syringae*. *Plant J.* **2015**, *84*, 125–139. [CrossRef] [PubMed]

154. Akram, W. Phenylacetic acid is ISR determinant produced by *Bacillus fortis* IAGS162, which involves extensive re-modulation in metabolomics of tomato to protect against *Fusarium*. *Front. Plant Sci.* **2016**, *7*, 498. [CrossRef] [PubMed]

155. Van de Mortel, J.E.; De Vos, R.C.H.; Dekkers, E.; Pineda, A.; Guillod, L.; Bouwmeester, K.; Van Loon, J.J.A.; Dicke, M.; Raaijmakers, J.M.; Phytopathology, J.E.; et al. Metabolic and transcriptomic changes induced in *Arabidopsis* by the rhizobacterium *Pseudomonas*. *Plant Physiol.* **2012**, *160*, 2173–2188. [CrossRef] [PubMed]

156. Rivero, J.; Gamir, J.; Aroca, R.; Pozo, M.J.; Flors, V. Metabolic transition in mycorrhizal tomato roots. *Front. Microbiol.* **2015**, *6*, 598. [CrossRef] [PubMed]

International Journal of
*Molecular Sciences*

MDPI

*Review*

# Indispensable Role of Proteases in Plant Innate Immunity

**Anastasia V. Balakireva [1] and Andrey A. Zamyatnin Jr. [1,2,*]**

[1] Institute of Molecular Medicine, Sechenov First Moscow State Medical University, 8, Trubetskaya Str., Moscow 119991, Russia; balakireva.anastacia@gmail.com

[2] Belozersky Institute of Physico-Chemical Biology, Lomonosov Moscow State University, Moscow 119992, Russia

* Correspondence: zamyat@belozersky.msu.ru; Tel.: +7-495-609-1400 (ext. 3028); Fax: +7-495-622-9808

Received: 5 February 2018; Accepted: 19 February 2018; Published: 23 February 2018

**Abstract:** Plant defense is achieved mainly through the induction of microbe-associated molecular patterns (MAMP)-triggered immunity (MTI), effector-triggered immunity (ETI), systemic acquired resistance (SAR), induced systemic resistance (ISR), and RNA silencing. Plant immunity is a highly complex phenomenon with its own unique features that have emerged as a result of the arms race between plants and pathogens. However, the regulation of these processes is the same for all living organisms, including plants, and is controlled by proteases. Different families of plant proteases are involved in every type of immunity: some of the proteases that are covered in this review participate in MTI, affecting stomatal closure and callose deposition. A large number of proteases act in the apoplast, contributing to ETI by managing extracellular defense. A vast majority of the endogenous proteases discussed in this review are associated with the programmed cell death (PCD) of the infected cells and exhibit caspase-like activities. The synthesis of signal molecules, such as salicylic acid, jasmonic acid, and ethylene, and their signaling pathways, are regulated by endogenous proteases that affect the induction of pathogenesis-related genes and SAR or ISR establishment. A number of proteases are associated with herbivore defense. In this review, we summarize the data concerning identified plant endogenous proteases, their effect on plant-pathogen interactions, their subcellular localization, and their functional properties, if available, and we attribute a role in the different types and stages of innate immunity for each of the proteases covered.

**Keywords:** plant proteases; plant immunity; MTI; ETI; SAR; ISR; RNA silencing

## 1. Introduction

Plants are continuously attacked by phytopathogens and have developed various strategies to counter them [1]. Today, the most studied type of immunity in living organisms is the immune system of animals. This is highly complex system and has its own distinct features, such as a highly exquisite adaptive immune structure with an infinite number of antigen-binding receptors that circulate throughout the whole organism and are generated in lymphocytes when a pathogen is encountered [2]. Through the generation of long-lived memory cells, this immune system remembers all antigens that have ever been encountered and multiplies the number of lymphocytes that express such specific antigen-binding receptors, thus allowing the secondary immune response to be faster and more effective [3]. The plant immune system is also very sophisticated, even though compared to the animal system it contains crucial elements, such as high specificity, low self-reactivity, and long-lasting memory, which use unique strategies that seem to have originated independently from animal strategies [4]. Plants have no circulatory system but they do possess cell walls, which are particularly significant in determining the unique features of plant defense.

Plant defense is mainly achieved through the induction of microbe-associated molecular patterns (MAMP)-triggered immunity (MTI), effector-triggered immunity (ETI), systemic acquired resistance (SAR), induced systemic resistance (ISR), and RNA silencing [5]. The development of plant immunity strategies was inextricably intertwined with pathogen strategies and directly depended on them [1]: coevolution and the arms race between plants and pathogens resulted in sophisticated immune responses that need to be tightly regulated. Endogenous plant proteases play an important role in the orchestration of immune processes [6]. Plant genomes encode vast numbers of proteases: the degradome of *Arabidopsis thaliana* L. contains more than 800 proteases from 60 families and the degradome of rice (*Oryza sativa* L.) contains more than 600 proteases [7]. The main function of proteases is proteolysis. Proteases degrade misfolded, damaged and harmful proteins and supply cells with amino acids. Proteases carry out both limited and digestive proteolysis, implying a gain or a switch of function and a loss of function of the proteins, respectively [8]. This makes proteases the major players in the maintenance of cell homeostasis. In addition, proteases also play a regulatory role in a variety of processes that are essential for growth, development, reproduction, immune response, embryogenesis, photosynthesis, programmed cell death (PCD), etc. [7]. Proteases are commonly synthesized as zymogens that determine the folding and function of mature protease. According to the MEROPS database [9], plant proteases are divided into seven classes: serine, cysteine, aspartic, asparagine, threonine, glutamate, and metalloproteases. Serine proteases are the most abundant proteases in plants: they comprise 14 families and nine clans [9]. S8 family contains subtilases that are to date the best-described serine proteases. Serine proteases participate in numerous crucial for vital activity processes such as immunity, symbiosis, PCD, cell differentiation, etc. [10] Cysteine proteases are divided into 15 families of five clans [9]: CA and CE clans contain papain-like folded proteases, while CD clan contains caspase-like folded proteases [7]. Cysteine proteases play an important role in PCD and in responses to biotic and abiotic stresses [11], flowering [12], embryogenesis [7], etc. Metalloproteases are involved in nodulation, plastid degradation, tolerance to stress temperatures, regulation of meristem growth, and meiosis [7,13]. The function of aspartic proteases is not yet well elucidated; however, it is assumed that they are implicated in aging processes [14], plant reproduction [15], and response to stress [16]. *A. thaliana* threonine protease from T1 family subunit beta type (PBA1) is a β-subunit of 26S proteasome and participates in ubiquitin-dependent protein degradation [17]. Glutamate and asparagine proteases have not been studied enough to determine their role in plants. It is noteworthy, that plant proteases are widely used in biotechnology and biomedicine due to their unique features such as wide range of working temperatures and pH values. Among biomedical applications of plant proteases there are antitumor therapy, blood coagulation, wound and burn healing, oral healthcare and treatment of digestive disorders [18].

In terms of immunity, proteases from different families have been shown to participate in almost every stage of immunity establishment, beginning with the pathogen encounter in the apoplast and finishing with their involvement in SAR and transgenerational immune memory [6]. Thus, the aim of the present review is to address the role of plant proteases and their contribution at different stages of distinct types of immunity.

## 2. The Role of Endogenous Plant Proteases in Different Types of Immunity

### 2.1. Plant Immunity Overview

The first event in plant immunity activation occurs when a cell encounters a phytopathogen. Phytopathogens produce a limited number of microbe- (pathogen-) associated molecular patterns (MAMPs [PAMPs]), such as chitin oligomers, flagellin, lipopolysaccharides and peptidoglycans, that are recognized by pattern-recognition receptors (PRRs) [19]. These receptors contain leucine-rich repeats (LRR) and are receptor-like kinases or proteins (RLKs or RLPs). Such plant PRRs are very similar to the Toll-like receptors (TLRs) found in animal analogs and both plant and animal LRR RLKs are able to recognize different epitopes from the same protein, e.g., flagellin [20,21].

These, in particular, support the convergent evolution hypothesis of the receptors in plants and animals and the independent emergence of pattern-triggered immunity in different domains [22]. PRRs bind to MAMPs through the LRR domain and induce the basal defense or MTI that leads to the activation of a mitogen-activated protein kinase (MAPK) cascade, resulting in the expression of defense genes (Figure 1). These also provide biochemical changes, such as reactive oxygen species (ROS) and reactive nitrogen intermediates (RNI) generation, an increase in $Ca^{2+}$ concentration in the extracellular compartments, etc., and structural improvements, such as stomata closure and callose deposition at the plasmodesmata [23] that hardens the cell wall and hinders pathogen intervention at the site of the pathogen attack (Figure 1) [24].

However, MTI can be overcome by pathogens. First, only a very small number of virulent molecules trigger it and, second, pathogens have developed a way to avoid the MTI system through the emission of effector molecules that directly enter the cell. These effector molecules are "avirulent" signals for PRRs and affect the immunity-associated proteins inside the cell. Thus, well-known effectors, such as AvrRpm1, AvrB, AvrRpt2 in *Pseudomonas syringae*, target the RIN4 protein of *A. thaliana* [5]. Perturbations in the RIN4 protein state are monitored by host-resistance proteins (R proteins) that then recognize the effectors that induce such perturbations [25]. This phenomenon is referred to as the Guard Model, where R proteins are the guardees of targeted by pathogen effector proteins, or the Decoy Model, in which the guarded target acts as a co-receptor to the guardee when it is present and is inactive when the guardee is absent [26]. Large amounts of R proteins exist in plants: pathogen effectors are polymorphic in different organisms and R proteins are specific to them [27]. R proteins contain variable coil-coiled N-terminus, nucleotide-binding site (NBS), and the C-terminal LRR domain, which makes R proteins similar to animal nucleotide-oligomerization domain (NOD)-like receptor (NLR) proteins [28]. The ligand binding activity of the R proteins is followed by conformational changes that are mediated by chaperones HSP90 [29]. Active R protein induces ETI—another type of plant immunity that emerged as a result of the arms race between phytopathogens and plants.

It is worth mentioning that the first contact with a pathogen occurs in the apoplast (extracellular space) of the cell surface that is filled with systemically or locally expressed proteases [30]. Although almost all of the apoplastic proteases belong to the papain-like cysteine protease (PLCP) family, C1A, different proteases represent different pathways that could be attributed to either MTI or ETI and the apoplast battleground is crucial for initial pathogen recognition and for further signal transduction for the establishment of different types or stages of immunity.

The ETI transmits the signal further: it activates the biosynthesis of jasmonic acid/ethylene (JA/ET) and salicylic acid (SA) in chloroplasts that trigger the gene expression of the pathogenesis-related (PR) proteins, and SAR-related proteins (Figure 1). ETI often leads to a hypersensitive response (HR)—a local resistance in the infected site. The HR results in the synthesis of PR antimicrobial molecules, such as chitinase and β-1,3-glucanase, and, if such molecules do not succeed, in the programmed cell death (PCD) of the infected cells [31]. It is worth mentioning that phytopathogens use two different strategies when infecting a plant cell: feeding on the living plant cell or destroying the plant cell to feed on its contents. The PCD of the cell infected by the biotrophic pathogens that feed on living cells, such as viruses, bacteria, fungi, nematodes, and oomycetes, is an effective defense strategy. It is well known that caspases are absent in plants, thus, PCD regulation is attributed to proteolytic enzymes from different families, such as the β-subunit of 26S proteasome PBA1 [17], metacaspases [32], PLCPs [33], vacuolar processing enzymes (VPEs) [34], subtilases [35], etc. [11]. Corresponding proteolytic enzymes and, in particular, their involvement in ETI and PCD will be discussed further.

Induced resistance of plants is presented in two forms: SAR and ISR. SAR is primarily associated with SA-dependent signaling, whereas ISR with JA/ET signaling independent of SA [36]. Both types of induced resistance make uninfected plant parts more resistant towards a broad spectrum of plant pathogens. Despite the lack of a plant circulatory system, mobile immune signals may be transmitted from the ETI-associated local resistance infected site to distal healthy areas, inducing SAR in order to protect the whole plant organism from the invader. In plants infected with the Tobacco mosaic virus (TMV), *Nicotiana benthamiana* L., SAR lasted for 20 days against TMV and other pathogens [37]. Immune signals travel through the apoplast and the phloem and include methylsalicylic acid (MeSA), lipid-derived compounds azelaic acid and JA, and glycerol-3-phosphate [22]. The lipid-transfer proteins, Defective in Induced Resistance 1 (DIR1) and Azelaic Acid Induced 1 (AZI1), transport mobile immune molecules [38]. As a result, MeSA is converted back into SA and the accumulation of SA in the phloem occurs, thus establishing SAR. Subsequently, SA accumulation leads to the induction of immune-related genes (PR proteins) in distal tissues: the synthesis of β-1,3-glucanase, chitinase, defensins, proteases, etc., and transgenerational immune memory (Figure 1) [22].

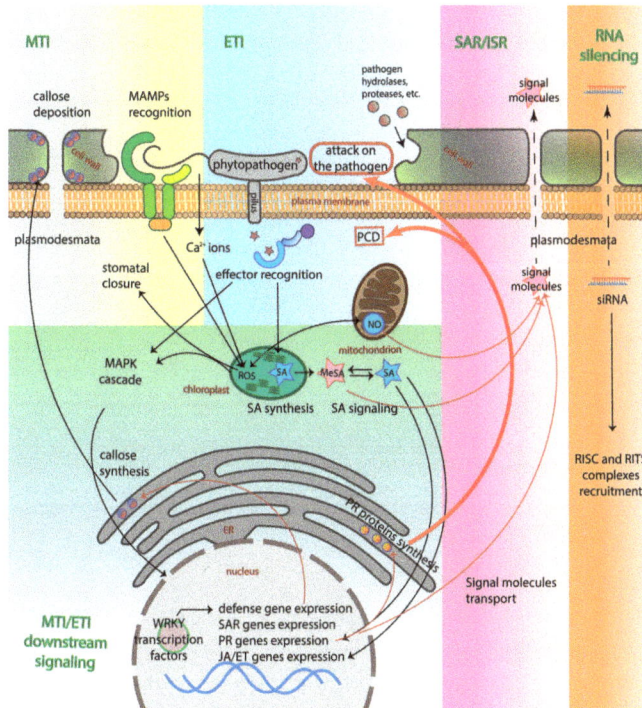

**Figure 1.** Plant defense mechanisms. MTI is triggered by MAMPs, leading to the elevation of cytosolic calcium ions, ROS and RNI generation, callose deposition at plasmodesmata and stomatal closure [39]. Effectors trigger ETI through binding to R proteins (NB-LRRs) that induce the signaling of SA and the subsequent induction of PR, JA/ET-dependent ISR-related genes and SAR-related genes. PR proteins, such as chitinases, β-1,3-glucanases, proteases, etc., either directly attack the pathogen or induce the PCD of the infected cell. SA is converted into MeSA that is transported into distal parts of the plant, as well as other signal molecules, establishing SAR or ISR. siRNA are also transported into distal parts of the plant through plasmodesmata. Names of immune processes are colored green. The names of cellular compartments are colored red. Black arrows indicate the directions of the activated plant immunity signaling pathways; red arrows indicate the results of genes expression after immunity activation; red bold arrows point to the cell fate in response to the pathogen; dotted arrows indicate the transport of signaling molecules through plasmodesmata.

SAR induces total cell reprogramming: expression of 14 classes of PR proteins, including proteases, mainly regulated by Nonexpresser of PR genes 1 (NPR1) [40] and conferring long-lasting immunity through cell priming—a sensitized state that enables a faster and more effective response to a secondary pathogen attack. Cell priming is thought to be connected to MAPK accumulation, MPK3 and MPK6 [41].

In addition, *R* genes are known to form clusters, thus it was proposed that the transgenerational immune memory is achieved through the duplication events of these genes and the hypomethylation of the chromatin, where these clusters are located, after infection [42].

ISR as well as JA/ET signaling is associated with defense against herbivores [43] and rhizobacteria [36]. ISR does not imply synthesis of PR proteins and SA and could be induced by PAMPs. In *A. thaliana*, SAR is most effective against biotrophic pathogens, downy and powdery mildews, as well as viruses that are sensitive to SA-dependent defenses; SAR inhibits plant growth whereas ISR is more active against nectrotrophic pathogens and promotes plant growth [36]. RNA silencing is a defense pathway of plant immunity that deals with viruses and fungi [44]. RNA silencing implies the use of small regulating RNAs (siRNAs) to specifically inactivate the nucleic acids of the pathogen [45]. This is divided into two directions: transcriptional gene silencing (TGS) and post-transcriptional gene silencing (PTGS) [46]. PTGS implies the elimination of viral mRNA by RNA-induced silencing complex (RISC) through the formation of the siRNA-mRNA complex. siRNAs are generated by Dicer through the cleavage of dsRNA synthesized from the viral genome of RNA-dependent RNA polymerase (RNA viruses) or RNA polymerase II (DNA viruses) [46]. DNA cytosine methylation occurs during TGS: a prime epigenetic event in the defense response to viruses [47]. Heterochromatic ssRNAs are produced by RNA polymerase IV, which is converted into dsRNA, diced by RNA-dependent RNA polymerase 2 (RDR2) and incorporated into the RNA-induced transcriptional silencing complex (RITS complex) that acts as a guiding strand for viral heterochromatin formation and methylation [48]. It is interesting to note that RNA silencing signal factors (ssRNAs and microRNAs) can travel from cell to cell through plasmodesmata and to distal parts of the plant through the phloem, just like MeSA molecules establishing SAR (Figure 1) [49].

## 2.2. Involvement of Endogenous Plant Proteases in Different Types of Plant Defense

### 2.2.1. Plant Proteases Functionality in Immunity Establishment

Proteases perform different functions in respect of plant defense. First, proteases activate different signaling processes by carrying out PRRs and NB-LRRs controlled proteolysis, also known as ectodomain shedding [50]. Although the precise mechanisms of protease action and their substrates are yet not well elucidated, the only known example of ectodomain shedding is the cleavage of the chitin receptor of *A. thaliana*, Chitin Elicitor Receptor Kinase 1 (CERK1), by an obscure protease [51].

Second, proteases are able to release signaling peptides that are perceived as damage-associated molecular patterns (DAMPs) by PRRs and induce immunity. For example, the aspartic protease, Constitutive Disease Resistance 1 (CDR1), generates PAMP or hormone systemin-like peptides that activate basal immunity [52]. Moreover, the recombinant prodomain sequences of C1A proteases from barley, can, alone, control phytophagous arthropods (coleopteran and acari) and reduce leaf damage [53].

Third, proteases orchestrate and regulate a large amount of the signaling pathways of MTI, ETI, SAR, ISR, and the RNA silencing of plant defenses. We focus on this aspect of protease action in the review. Despite the fact that a number of proteases have already been identified as being in some way involved in plant immunity, the exact mechanisms of the majority of proteases action, their substrates, and the signaling pathways they involve remain a mystery. The authors collected all the available data on endogenous plant proteases from the different families that implement different catalytic mechanisms, their functions during plant defense establishment, and their domain architectures (Figure 2) and attempted to attribute the proteases to the different signaling pathways of immunity (Tables 1 and 2). However, it is worth mentioning that in the case of MTI and ETI, the signals received from both pathways through PRRs or NB-LRRs line up in the same MAPK cascade and synthesis of mobile signaling molecules (Figure 1). Thus, one endogenous protease may participate in different types of defense or in shared parts of the pathways. Nevertheless, plant immunity is still a highly complex and obscure field of research.

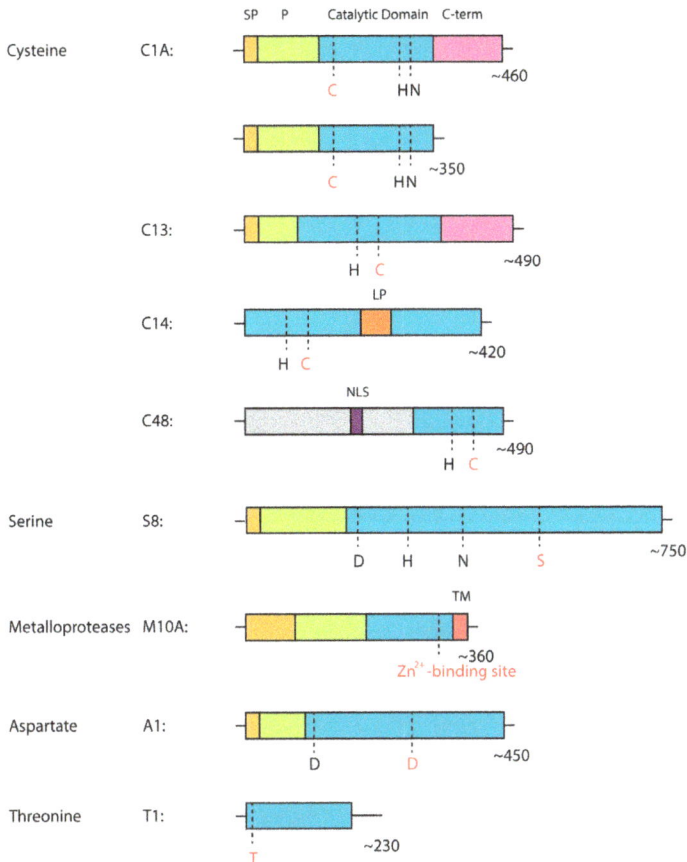

**Figure 2.** The domain architecture of proteases covered in the review from different protease families. Red letters represent catalytic amino acid residues. SP—signal peptide, P—prodomain, C-term—C-terminal domain or granulin domain (for family C1A proteases), LP—linker peptide, NLS—nuclear localization signal, TM—transmembrane region, regions in grey—low complexity regions.

### 2.2.2. Proteases Involved in MTI

As described above, endogenous proteases can produce MAMP- (PAMP-), DAMP-like peptides that are perceived by PRRs. The *Glycine max* L. cryptic signal, 12-aa peptide GmSubPep, was shown to be part of a unique subtilase; the peptide acts as PAMP and causes a pH increase, inducing immunity response when supplied to soybean cultures by induction of the expression of known defense-related genes, such as Cyp93A1, Chib-1b, PDR12, and *achs* [54]. Another fascinating example of the peptide cleavage of a protease is a major hub of the flg22 (flagellin epitope, MAMP)-regulated transcriptional network propeptide of Rapid ALkalinization Factor 23 (proRALF23), confirmed to be a substrate of subtilase SBT6.1 (S1P) in *A. thaliana* [55]. The proRALF23 cleavage of S1P results in the dose-dependent inhibition of elf18 (EF-Tu epitope, MAMP)-induced ROS production and the scaffolding function of EF-Tu receptor (EFR) (the receptor of elf18 that increases the susceptibility to *Pseudomonas syringae* pv. *tomato* DC30000 (*Pst*DC3000)), which makes S1P a negative regulator of plant immunity. The aspartic protease, CDR1, is also able to generate PAMP- or systemin-like peptides and induces the basal defense response and SAR [52].

PRRs might also be targets of plant proteases and not only in terms of ectodomain shedding. Tomato S8 subtilase P69C, which derives from tomato, is able to process the extracellular LRP protein that is possibly involved in immunity [56]. *A. thaliana* cysteine proteases from the family C19 ubiquitin carboxyl-terminal hydrolase 12 and 13 (AtUBP12 and AtUBP13), appear to be negative regulators of plant immunity: they are able to suppress Cf-9-mediated HR in *N. benthamiana* following *Pst*DC3000 infection through the possible deubiquitinylation of PRRs [57].

A conventional sign of MTI is callose accumulation in the plasmodesmata that blocks symplast transport and isolate the infected cells [23]. This occurs following recognition of the flg22 by the Flagellin-Sensing 2 (FLS2)-Brassinosteroid Insensitive 1-associated receptor Kinase 1 (BAK1)-*Botrytis*-Induced Kinase 1 (BIK1) complex and the ROS generation that promotes regulated callose synthesis, which occurs rather not through the one conserved downstream cascade, but its own unique multiple mechanisms [58]. The overexpression of the aspartate protease 13 from *Vitis quinquangularis* (VqAP13) in *A. thaliana* plants increased resistance to *Pst*DC3000 and accumulated more callose in the infected site than in wild-type plants, suggesting the involvement of VqAP13 in MTI signaling. The expression of VqAP13 was upregulated by SA and ET treatment and downregulated by MeJA and *Botritis cinerea* infection, suggesting the participation of the protease in defense against biotrophic pathogen [59].

Another sign of MTI is stomata closure that occurs in response to the elevation of cytosolic $Ca^{2+}$ concentrations and the subsequent $H_2O_2$ and NO accumulation in the guard cells [60]. VPE cysteine proteases are essential for elicitor-induced stomatal closure in *N. benthamiana*: VPEs are localized in the vacuole and control the fusion of the plasma membrane and the vacuole membrane during virus-induced PCD [61]. VPE mediates elicitor-induced stomatal closure by regulating NO accumulation in the guard cells [62]. One study has recently reported that the plant pathogen, *Pst*DC3000, uses the virulence factor coronatine (COR) to actively open stomata, as well as oxalate, which is produced by many fungi species [63]. This makes VPE-dependent stomatal closure an important defense mechanism.

### 2.2.3. Proteases Involved in ETI, PCD, and RNA Silencing

Effectors have emerged as a result of an arms race between plant MTI and pathogen strategies developed to avoid it. Pathogens emit avirulent molecules—effectors—that are primarily dedicated to the activation of host transcription, acting as transcription factors, in order to affect chromatin and histone packaging and to regulate nutrient release for pathogen survival [5]. In addition, it has been concluded that a single dominant host-resistance gene (*R*) incites a phenotype of disease resistance in response to a pathogen expressing a single dominant avirulence gene (*Avr*) [64]. This model was known as the gene-for-gene hypothesis, implying that each dominant pathogen avirulence (*Avr*) gene product is either directly or indirectly recognized by the product of a corresponding dominant host *Cf* resistance gene. For example, in the *Cladopsorium fulvum*-tomato interaction, five *Avr* genes (*Avr2, Avr4, Avr4E, Avr5,* and *Avr9*) have been characterized from *C. fulvum* and their encoded proteins trigger an HR in host plants carrying the corresponding *Cf-2, Cf-4, Cf-4E, Cf-5,* and *Cf-9* genes, respectively [65].

Plant proteases have been shown to participate in Avr/Cf-induced HR (Table 1). *Phytophthora*-inhibited protease 1, from *N. benthamiana*, NbPip1, is associated with an Avr4/Cf-4-induced HR; in NbPip1 mutant HR was delayed for one to two days [66], whereas Rcr3 from tomato acts as co-receptor to Cf-2 for Avr2 effector recognition [67]. *Cf*-encoded R proteins are NB-LRRs that mostly act in the nucleus for the direct rapid regulation of gene expression through the activation of immune responses [68]. The indirect recognition of the effectors is managed by the RIN4 protein that is targeted by many bacterial effectors (AvrRpt2, AvrRpm1, AvrB, and HopF2) and is monitored by NB-LRRs ribose-phosphate pyrophosphokinase PRS2 and disease resistance protein RPM1 that activate ETI in the case of changes to the RIN4 state [69].

**Table 1.** Proteases involved in ETI responses and PCD.

| Plant Species | Plant Protease | Family | Subcellular Localization after Infection | Pathogen | Identified Substrates | Is Inhibited by Effector | Function/Phenotype | Ref. |
|---|---|---|---|---|---|---|---|---|
| A. thaliana | AtMC1, AtMC2 | Cys, C14B | Cytoplasm, Nucleus | PstDC3000 | ND | - | Suppression of hypersensitive cell death response upon infection with avirulent pathogen, AtMC1 and AtMC2 antagonistically control lsd1 runaway cell death | [32] |
| | AtMC9 | Cys, C14B | Nucleus, Cytosol, Apoplast | - | GRI protein, PEPCK1, AtSerpin1 | - | Effector of PCD activation, xylem cell death, degradation of vessel cell contents after vacuolar rupture | [71–73] |
| | CathB | Cys, C1A: CathB-like | Vacuole, Apoplast | PstDC3000 | ND | - | Required for the HR and disease resistance induced by non-host bacterial pathogens, positive regulatory role in senescence | [33] |
| | RD21 | Cys, C1A: CathL-like | ER vesicles, Vacuole | B. cinerea | ND | - | 'Pro-death' signal activated during elicitation of cell death, targeted by plant AtSerpin1, AtWSCP; processed by PtMC13 and PtMC14 | [74–76] |
| | RD19A | Cys, C1A: CathL-like | Vacuole, Nucleus | Ralstonia solanacearum | ND | - | RRS1-R-mediated resistance, inhibited by effector PopP2 | [77] |
| | VPEs | Cys, C13 | Vesicles, Vacuole | PstDC3000 | Storage proteins (12S globulins and 2S albumins) | - | Activate vacuolar enzymes and disintegrate the vacuolar membrane to release hydrolytic enzymes during PCD, involved in the HR elicited by infection with TMV | [17,63,78] |
| | AtCEP1 | Cys, C1A: CathL-like | ER | Erysiphe cruciferarum | ND | - | Restriction of powdery mildew controlling late stages of compatible interaction including late epidermal PCD | [79] |
| | PBA1 | Thr, T1B | Cytosol, Nucleus | PstDC3000 | ND | - | Caspase-3-like (DEVDase) activity in the vacuolar and plasma membranes proteasome-regulating membrane fusion | [17] |
| | AtSBT5.2 (a and b) | Ser, S8 | Endosomes | PstDC3000 | ND | - | Independent of protease activity attenuation of MYB30-mediated HR | [80] |
| Picea abies H. Karst. | mcII-Pa | Cys, C14 | Cytoplasm, Nucleus | - | TSN | - | Induces autophagy, which triggers PCD mechanisms during the terminal differentiation of embryonic suspensor cells, and participates in further development of PCD | [81,82] |

**Table 1.** *Cont.*

| Plant Species | Plant Protease | Family | Subcellular Localization after Infection | Pathogen | Identified Substrates | Is Inhibited by Effector | Function/Phenotype | Ref. |
|---|---|---|---|---|---|---|---|---|
| Solanum lycopersicum L. | P69B | Ser, S8 | Apoplast | Phytophthora infestans, P. syringae | ND | Kazal-like inhibitors EPI1 and EPI10 | Local apoplast surveillance, substrate of Sl2-, Sl3-MMP, positive regulator of PCD | [83,84] |
| | P69C | Ser, S8 | Apoplast | P. syringae | LRP protein | - | LRP protein processing | [56] |
| | Sl2-,Sl3-MMPs | Metallo, M10A | Apoplast | B. cinerea PstDC3000 | P69B | - | Extracellular cascade of epidermal cell death | [84] |
| | RCR3 | Cys, C1A: CathL-like | Apoplast | C. fulvum, P. infestans, Globodera rostochiensis | ND | Avr2, EPICs, Gr-VAP1 | Extracellular defense; co-receptor to Cf-2 for effector recognition in the case of C. fulvum | [67,85] |
| | PIP1 | Cys, C1A: CathL-like | Plasma membrane, Apoplast | C. fulvum, P. infestans, P. syringae | ND | Avr2, EPICs | Broad-range extracellular defense | [85,86] |
| | CYP1 | Cys, C1A: CathL-like | Apoplast | TYLCV | ND | V2 | Involved in hypersensitive response reactions | [87] |
| | C14 | Cys, C1A: CathL-like | Apoplast | P. infestans | ND | EPICs, AVRblb2 | Defense-related secretion in haustoriated plant cells | [88] |
| | Sl-SBT3 | Ser, S8 | Apoplast | P. infestans | ND | - | Caspase-3-like DEVDase activity, HR-like PCD induction | [89] |
| Populus tremula x tremuloides | PttMC13, PttMC14 | Cys, C14B | Cytoplasmic aggregates | - | RD21, TSN, PASPA3 | - | Type II metacaspases, AtMC9 homologues, involvement of stress granules in the metacaspase-TSN pathway and xylem vessel and fiber cells PCD, processing of RD21, TSN, PASPA3—postmortem autolytic processes | [90] |
| Solanum tuberosum L. | StSBTc-3 | Ser, S8 | Apoplast | P. infestans | ND | - | Caspase-3-like DEVDase activity, HR-like PCD induction | [89] |
| Avena sativa L. | Saspase | Ser, S8 | Apoplast | Cochliobolus victoriae | RuBisCO | - | RuBisCO proteolysis in victorin-induced PCD, IETDase and LEHDase activities | [91] |
| Nicotiana tabacum L. | Phytaspase | Ser, S8 | Cytosol, Apoplast | TMV | VirD2 from Agrobacterium tumefasciens | - | Activated in tobacco mosaic virus (TMV)-induced HR, VirD2 cleavage preventing protein transport to nucleus, VEIDase, IETDase, LEHDase, and VDVADase | [92] |

ND–not defined.

**Table 2.** Proteases involved in MTI/ETI downstream pathways, SAR and ISR.

| Plant Species | Plant Protease | Family | SUBCELLULAR LOCALIZATION | Pathogen | Function/Phenotype | Ref. |
|---|---|---|---|---|---|---|
| S. lycopersicum | Sl3-MMP | Metallo, M10A | Plasma membrane | PstDC3000, B. cinerea | Enhanced resistance to B. cinerea and upregulated expression of defense-related genes | [93] |
| | SBT3-Sl | Ser, S8 | Tomato vasculature | Manduca sexta larvae | Herbivore defense, involved in systemin processing and JA-mediated resistance response | [94] |
| Gossypium babardense L. | GbSBT1 | Ser, S8 | Plasma membrane, cytoplasm | Verticillium dahliae | Associated with JA signaling | [95] |
| V. quinquangularis | AP13 | Asp, A1 | - | PstDC3000, Powdery mildew, B. cinerea | Promotion of the SA dependent signal transduction pathway; but suppression of the JA signal transduction pathway; enhanced callose deposition | [59] |
| A. thaliana | SBT3.3 | Ser, S8 | Apoplast | PstDC3000 | $H_2O_2$-inducible positive regulator of innate immunity operating upstream of the SA pathway, MPK activation, concurrent chromatin remodeling at SA-responsive genes | [35] |
| | OTS1, OTS2 | Cys, C48 | Nucleus | P. syringae | OTS1 and -2 negatively regulate SA biosynthesis restricting biosynthesis gene ICS1 expression and propose that de novo synthesis and SA-promoted degradation of OTS1/2 antagonistically adjust the abundance of this negative regulator depending on the level of pathogen threat | [96] |
| | CDR1 | Asp, A1 | ER and apoplast | P. syringae | Induction of a SA-dependent resistance response; could generate endogenous extracellular peptides that act as mobile signals for SAR | [52] |
| | AED1 | Asp, A1 | Apoplast | P. syringae | Induced locally (EDS1-independent) and systemically (EDS1-dependent) during SAR signaling and locally by SA, homeostatic mechanism to limit SAR, tradeoff between defense and plant growth | [97] |
| | TOP1, TOP2 | Metallo, M3 | TOP1-chloroplasts, TOP2—cytosol | - | Non-competitive SA-binding, mediate SA-dependent signaling and are necessary for the immune response to avirulent pathogens | [98] |
| Maize Black Mexican Sweetcorn (BMS-33) | MIR1 | Cys, C1A | Maize midwhorl | Caterpillars Heliothis virescens, corn leaf aphids | Proteolysis of caterpillar peritrophic matrix, ET-dependent, long-distance transport signal | [99] |

The ETI response subsequently activates the MAPK cascade that induces transcription from the PR genes. This induces the biosynthesis of SA, JA, and ET and antimicrobial enzymes, as described above. Degradation of NPR1 is a molecular switch that depends on its paralogues NPR3 and NPR4, receptors of SA, that bind it with different affinities and regulate degradation of NPR1 in SA-regulated manner [100]. NPR3 and NPR4 are adaptors of the Cullin 3 ubiquitin E3 ligase, and *npr3 npr4* double mutant plants accumulate a high amount of NPR1 protein and are not able to launch ETI, PCD, or establish SAR. Plant PCD is characterized by chromatin condensation, shrinkage of cytoplasm, swelling of mitochondria, vacuolization, and chloroplast disruption [101]. The chloroplast plays a central role in the defense responses and HR of plants. It is a source of defense signaling molecules, such as ROS, RNI, SA, and JA [102]. In addition, pathogen effectors that possess chloroplast localization signals suppress immunity [103].

Metacaspases are associated with PCD and RNA silencing in plants. When infected with a necrotrophic pathogen, infection leads to the necrosis of the cell, whereas when infected with a biotrophic pathogen, the plant organism launches PCD to eliminate the infected cells. However, at the same time, such pathogens suppress PCD to feed on living cells. The main executioners of PCD in animals are caspases—cysteine aspartate-specific proteases that are absent in plants—and PCD in plants is associated with caspase-like activities but not with caspases (Table 1). Metacaspases are very distant homologs of caspases and their biochemical properties also differ: their substrate specificity is arginyl-/lysil-specific, whereas caspases are aspartate-specific proteases. Metacaspases are divided into types I (AtMC1-AtMC3) and II (AtMC4-AtMC9). Type I metacaspases contain an additional N-terminal proline-rich prodomain with a zinc finger motif [11]. *A. thaliana* metacaspases type I, AtMC1 and AtMC2, have been shown to regulate autolytic PCD (involving rapid cytoplasm clearance): they suppress the hypersensitive cell death response upon infection with the avirulent pathogen, *Pst*DC3000. AtMC1 and AtMC2 antagonistically control Lesion Simulating Disease 1 (LSD1) runaway cell death [32]. The *A. thaliana* type II metacaspase, AtMC9, functions extracellularly during the autolysis of xylem elements after their cell death: it provides post-mortem clearance of the cell contents after vacuolar rupture [70]. In extracellular space, AtSerpin1 inhibits AtMC9 [104]. AtMC9 has been shown to cleave 11 amino acids from GRI protein involved in PCD initiation [72]. Another substrate of AtMC9 is phosphoenolpyruvate carboxykinase 1 (PEPCK1)—one of the components of gluconeogenesis in plants. PEPCK1 cleavage by AtMC9 leads to the enhancement of enzyme activity, suggesting the role of AtMC9 in limited proteolysis and its effector properties [71]. The homologs of AtMC9 from hybrid aspen (*P. tremula* x *tremuloides*), PttMC13 and PttMC14, were shown to be involved in a variety of xylem processes: in the metacaspase-Tudor Staphylococcal Nuclease (TSN) pathway, in the regulatory proteolysis of Responsive To Dehydration 21 (RD21) during xylem maturation, and in the cell death of xylem elements (PASPA3 aspartate protease processing) [90]. They are also expressed during the non-autolytic type of PCD, i.e., vacuolar collapse [90]. It is worth mentioning that although caspases and metacaspases possess different substrate specificities and biochemical properties, they are functionally similar: both proteases can hydrolyze a single conserved substrate which is very important for the vital activities of different species. It has been shown that the protein TSN, which is found in plants and animals (including humans) and is essential for transcription, RNA splicing, RNA editing, etc., is a substrate for both caspase-3 and Norway spruce metacaspase type II (mcII-Pa) [81,105]. TSN manages activation of transcription, activation of mRNA splicing, regulation of RNA silencing as a member of RISC complex [105]. TSN was shown to be an anti-apoptotic agent: uncleavable TSN stimulates cell proliferation and protects cells from death. Degradation of TSN by mcII-Pa results in impairment of its ability to activate mRNA splicing, inhibition of its ribonuclease activity, and in PCD of the cell. mcII-Pa was shown to execute PCD of embryo suspensor cells during plant embryogenesis [81]. These data suggest that TSN cleavage by mcII-Pa represents a regulatory mechanism that limits RNA silencing and promotes PCD of the cells infected by virus (Figure 3).

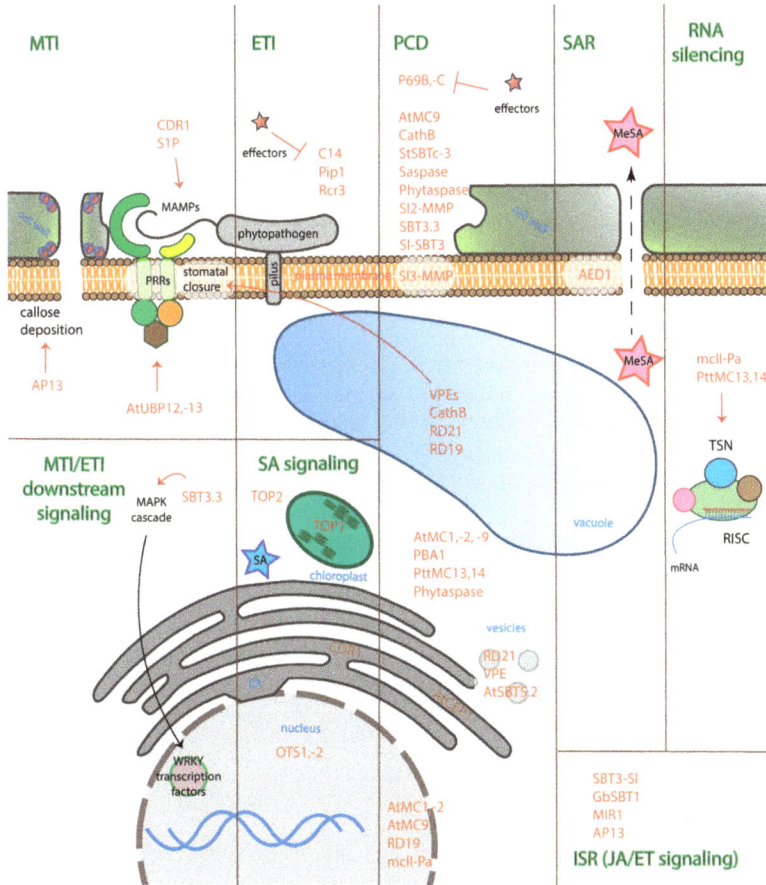

**Figure 3.** Involvement of plant proteases in different immunity pathways and their subcellular localization. Proteases covered in the review are colored orange. Immune processes are colored green. The names of cellular compartments are colored blue. Red arrows point to the object of the protease action (hydrolysis of the substrate or influence on the process); T-like arrows imply inhibition of proteases by effectors; black arrow indicates the direction of the MAPK cascade action; dotted arrow indicates the transport of signaling molecules through plasmodesmata.

Vacuole collapse is a type of PCD that is characterized by the absence of rapid cytoplasm clearance and is held by cysteine proteases—VPE enzymes [34]. VPEs exhibit caspase-1-like activity and substrate specificity toward an asparagine residue; they localize in the vacuolar membrane and mediate virus-induced hypersensitive cell death by regulating the collapse of the vacuole membrane and the release of vacuolar hydrolytic enzymes for attacking the virus in response to infection. VPEs possess YVADase activity to mediate TMV-induced PCD [62]. Another protease—the threonine-dependent PBA1, β-subunit of proteasome—orchestrates a further vacuole-associated defense mechanism: proteasome-regulating membrane fusion of the vacuolar and plasma membranes that provides plants with a mechanism for attacking intercellular bacterial pathogens. Following bacterial infection, proteasome-regulating vacuolar-plasma membrane fusion occurs in the intact cell wall, resulting in the discharge of the vacuolar contents in the apoplast—the extracellular antibacterial fluid that contains PBA1 manifests caspase-3-like DEVDase activity and mediates the consequent PCD [17].

Subtilases (serine proteases, family S8) also exhibit caspase-like activity and contribute to plant PCD. Tomato S8 subtilases, P69B and P69C, were the first identified plant apoplastic proteases. These subtilases are PR proteins that provide basal levels of surveillance; their promoters are induced by SA and they are synthesized as preproenzymes, both locally (in the infected site) and systemically (in distal, non-infected sites), in response to *P. infestans* and *P. syringae* [106]. The P69B protease is targeted and inhibited by the Kazal-like inhibitors EPI1 and EPI10 from *P. infestans* [83]. P69B was recently shown, in vitro, to be a substrate of the tomato matrix metalloproteases Sl2- and Sl3-MMPs. The overexpression of P69B leads to the PCD of the epidermis cells of tomato hypocotyls, suggesting that P69B is a positive regulator and Sl2- and Sl3-MMPs are negative regulators of the PCD of the epidermis [84]. Sl2- and Sl3-MMPs are also secreted into the apoplast, they belong to the M10A family of metalloproteases and are synthesized in the form of preproenzymes.

Saspases from oats possess IETDase and LEHDase activities and cleave RuBisCO molecules during chloroplast rapture in victorin-induced PCD (toxin from *C. victoriae*) [91], whereas phytaspases exhibit VEIDase, IETDase, LEHDase, and VDVADase activities and are activated during TMV-induced PCD [92]. Saspases are constitutively expressed in an inactive form; processed and relocalized into the apoplast following PCD induction. Although the exact role of saspases remains obscure, phytaspases cleave off the nuclear localization signal (NLS) from NLS-containing the *A. tumefasciens* protein, T-DNA border endonuclease VirD2, which is essential for the nuclear uptake of foreign DNA within the plant cell during bacterial infection and plant transformation [107]. Phytaspases are constitutively synthesized and located in an active form in the apoplast in normal conditions and are relocalized into cytoplasm upon PCD induction [92,108,109]. Another subtilase, from *S. tuberosum*, StSBTc-3, was shown to possess caspase-3-like—DEVDase—activity. It is located in the apoplast and accumulated in detached leaves following *P. infestans* infection. This protease is constitutively expressed and, in vitro, may induce PCD and cytoplasm shrinkage [110]. The subtilase AtSBT5.2, from *A. thaliana*, was shown to be a negative regulator of MYB30 activity—a transcription factor that promotes the cell death-associated response activation of genes that participate in lipid biosynthesis [80]. The variant protein, AtSBT5.2(b), retains MYB30 in vesicles that are not allowed to enter a nucleus during infection.

PLCPs have also been shown to regulate plant PCD. The vast majority of studied apoplastic proteases are represented by papain-like cysteine proteases (PLCP) that comprise the C1A subfamily of cysteine proteases. The C1A family was subclassified by Richau [111] into nine classes according to their homology and domain architecture. These classes are combined into four groups in accordance with their similarity to human cathepsins: CathL-like, CathF-like, CathH-like, and CathB-like. All of the apoplastic PLCPs described below are CathL-like proteases. The accumulated data on PLCPs specificities suggest that they have a rather low specificity. However, nonpolar (including Pro) or aromatic amino acid residue was found to be preferential at the P2 position of the substrate [112–115]. PLCP are key players in a variety of processes, including growth, development, responses to stresses and defense. PLCP are upregulated in the *atg* mutants (autophagy mutants) of *A. thaliana*, suggesting either their backup role in nutrient recycling and remobilization or their cell-death promoting role [116].

In plant immunity, in contrast to subtilases, pathogen effectors inhibit apoplastic PLCPs that are strongly associated with the ETI response. For example, tomato and potato RD21-like C14 protease are inhibited by the cystatin-like effectors EPICs [117] and the AVRblb2 effector from *P. infestans* [88]. C14 is synthesized as preproenzyme and contains a unique granulin domain with an unknown function (Figure 2). It is processed through cleavage of the prodomain (iC14 form) and the subsequent granulin domain cleavage (mC14 form). The silencing of C14 leads to susceptibility to *P. infestans* in *N. benthamiana* C14 is a highly conserved protease; it is targeted by pathogen effectors and its homologs are present in different species of the plant kingdom, e.g., in tomato Cysteine Protease 1 (CYP1)—the orthologue of C14 that does not contain granulin domain and is targeted by the V2 protein of the tomato yellow leaf curl virus (TYLCV) [87]. Cysteine proteases CP1A, CP1B, CP2, and XCP2 from maize are also orthologues of C14, they do not contain the granulin domain and are targeted by the Pit2 effector of *Ustilago maydis* [118]. A lack of these enzymes results in increased susceptibility to pathogens.

Fascinating examples of the coevolution of pathogen effectors and plant defenses, driven by the arms race, are tomato paralogous apoplastic Senescence-Associated Gene 12 (SAG12)-like PLCPs RCR3 and *Phytophthora*-inhibited protease 1 (PIP1), which are targeted by the different effectors of different pathogens, suggesting the importance of these proteases for plant defense [85]. Both proteases are expressed constitutively in leaves but their expression increases after inoculation with *C. fulvum* (both virulent and avirulent strains), *P. infestans*, and *P. syringae*. Both proteases are targeted by the effector Avr2 and the cystatin-like effectors, EPICs [86]. RCR3 is also targeted by the nematode effector, Gr-VAP1 [67]. PIP1 expression is 10× higher than RCR3, and this is a major broad-range immune protease against various apoplastic pathogens: the absence of PIP1 leads to hyper-susceptibility to fungal, bacterial and oomycete plant pathogens, whereas RCR3 acts as a co-receptor to the immune receptor Cf-2 for the recognition of Avr2. The absence of Rcr3 causes increased susceptibility only to the *P. infestans*. The functional divergence of paralogous genes provides more effective defense against different pathogens. In addition, a homolog of the tomato PIP1, NbPip1 from *N. benthamiana*, contributes to the Avr4/Cf-4 induced hypersensitive response [66].

Non-apoplastic PLCPs are associated with PCD. Cathepsins in animals are lysosomal proteases that are associated with cell death and survival [119–122]. In *A. thaliana*, AtCathB is likely to be a positive regulator of cell death and basal resistance, mediated through interactions with LSD1. A key marker gene of senescence, SAG12, is downregulated in *atcathb* triple mutants. Thus, cathepsin B is required for HR and for resistance to non-host bacterial pathogens [33].

*A. thaliana* RD21 is known to be a pro-death cysteine protease, which is accumulated in ER bodies in an inactive form. RD21 is a homolog of C14. The proteolytic domain, RD21, also contains a signal peptide, prodomain and granulin domain (Figure 2), which are cleaved during processing: prodomain cleavage results in the intermediate form, iRD21, and granulin domain cleavage, in the mature form, mRD21 [123]. RD21 is physically and reversibly inhibited by the Kunitz-type inhibitor, *A. thaliana* water-soluble chlorophyll protein (AtWSCP), which regulates PCD during flower development [75]. In detached leaves, RD21 is also targeted by the inhibitor, AtSerpin1: AtSerpin1 is localized in cytoplasm, while RD21 is localized in ER bodies, and elicitors of PCD increase the permeability of vesicles membranes, which leads to the colocalization of RD21 and AtSerpin1 in cytoplasm and the irreversible inhibition of the protease. This suggests the pro-survival role of AtSerpin1 [76,124]. In addition, as already mentioned, RD21 is also processed by the type II metacaspases, PttMC13 and PttMC14 [90]. The RD21 mutants are unaffected in interactions with *P. syringae* but are more susceptible to the necrotrophic fungal pathogen, *B. cinerea*, demonstrating that RD21 provides immunity to a necrotrophic pathogen.

The orthologue of RD21, RD19, is also involved in plant defense: it is inhibited by the effector of *R. solanacearum* PopP2 that interacts with the *A. thaliana* Resistant to *R. solanacearum* 1-R (RRS1-R) R protein [77]. RD19 is induced during *R. solanacearum* infection and is required for RRS1-R-mediated resistance: RD19 is normally localized in mobile vacuole-associated ER compartments and, on interaction with the effector PopP2, it is relocalized to the plant nucleus. The nuclear complex RD19-PopP2 is recognized by RRS1-R and is required for the ETI response.

A KDEL-containing cysteine endopeptidase 1 from *A. thaliana* (AtCEP1) is also located in ER and functions during developmental cell death and tissue remodeling. However, it has been shown to be associated with late defense reactions: it restricts the growth of the biotrophic fungus *E. cruciferarum* and participates in late epidermal cell death; it is a possible target for pathogen effectors [79]. AtCEP1 is regulated by hormone molecules rather than by MAMPs, and it accumulates around the haustoria, inducing PCD. AtCEP1 is likely to be delivered from the ER to the vacuole through late endosomes with a subsequent fusion of the vacuole and plasma membranes, resulting in the relocalization of AtCEP1 into the apoplast or extrahaustorial space [17]. KDEL-cysteine endopeptidases, such as AtCEP1, are considered to be late-acting proteases that digest cell wall proteins during the final stages of PCD and tissue remodeling following cellular disintegration [125].

### 2.2.4. Proteases Involved in MTI/ETI Downstream Signaling Pathways, SAR and ISR

Downstream MTI and ETI events partially overlap and include the activation of the MAPK cascade and WRKY transcription factors that contain the WRKY domain that is defined by the conserved amino acid sequence WRKYGQK at its N-terminal end [126]. WRKY transcription factors manage the rapid activation of the PR genes associated with the biosynthesis of the signal molecules, SA, JA, and ET, with lignifications, the production of antimicrobial agents, etc. [127,128]. The subtilase SBT3.3, from *A. thaliana*, activates MAPKs and the Oxidative Signal-Inducible 1 (OXI1) kinase following pathogen attack (Table 2). In knockout SBT3.3 plants, the MAPK cascade is inhibited, as is chromatin remodeling, suggesting its involvement in the regulation of kinases and in epigenetic regulation [35].

Plants produce SA from chorismic acid through two biosynthetic pathways, one catalyzed by the phenylalanine lyase1 to 4 (AtPAL1–4) and the other by isochorismate synthase1 and 2 (AtICS1 and 2) in chloroplasts [129]. The accumulation of SA leads to the establishment of SAR—a long-lasting broad-spectrum disease resistance that implies the activation of defense mechanisms in infected sites and is targeted to protect distal healthy tissues [130]. SAR leads to the total reprogramming of PR genes and cell priming. This is achieved through the production of the mobile signal molecule, MeSA, in the infected site and its transport, through the plasmodesmata and phloem, to distal parts of the plant [131]. Delivered to healthy cells, MeSA is converted into SA and is bound by its NPR3 and NPR4 receptors, resulting in NPR1 degradation and induction of SAR-related genes [100]. SAR-associated proteases are summarized in Table 2.

SA-activated downstream genes are divided into immediate-early genes and late genes, which include the SAR-marker gene, *pr1* [132]. Thimet oligopeptidases (TOPs) 1 and 2 are metalloproteases from *A. thaliana* that interact with SA, resulting in the loss of their proteolytic activity, both in vitro and in plant extracts [98]. The absence of TOPs results in increased susceptibility to *P. syringae* [98]. It has been suggested that TOPs mediate the SA-dependent signaling pathways and act as modulators of chloroplast and cytosolic PCD-related processes that may not be exclusively activated by pathogen infection.

Overly Tolerant to Salt 1 and 2 (OTS1 and 2*)* are cysteine proteases, from *A. thaliana* that cleave off Small Ubiquitin-Related Modifier 1 and 2 (SUMO1 and 2) and manage the SUMO1/2-mediated regulation of SA signaling [96]. Double mutant *ots1/2* displayed enhanced resistance to virulent *P. syringae*, higher levels of SA and upregulated expressions of the *ics1* SA synthetic gene as compared with wild-type plants. SA stimulated OTS1/2 degradation and promoted the accumulation of SUMO1/2 conjugates. The accumulation of SUMO1/2 conjugates coincides with SA-promoted OTS degradation and may play a positive role in SA-mediated signaling.

The apoplastic aspartate protease, Apoplastic EDS1-Dependent protein 1 (AED1) from *A. thaliana*, was shown to be induced locally and systemically by SA and its analog, benzothiadiazole [97]. Enhanced Disease Susceptibility 1 (EDS1) protein is essential, both for SAR signal generation in an infected site and its perception in the systemic tissues. In *eds1* mutant plants, the conditional overaccumulation of AED1-hemagglutinin inhibited SA-induced resistance and SAR (EDS1-dependent), but not local resistance (EDS1-independent). The data suggest that AED1 is part of a homeostatic feedback mechanism that regulates systemic immunity. Another aspartate apoplastic protease is CDR1, from *A. thaliana*, whose hyperactivation induces an SA-dependent disease resistance response [52]. This activates SAR and induces the accumulation of SA and the transcripts of the *pr1* and *pr2* genes, which are markers of SAR. The CDR1 protein is accumulated in the apoplast in response to the inoculation of avirulent bacterial pathogens. CDR1 may generate extracellular peptide elicitors that activate the basal defense response.

JA is a main signal molecule of ISR and is produced through the induction of the octadecanoid pathway. JA is a pleiotropic hormone and JA-dependent genes encode an arsenal of plant defense proteins involved in resistance to insects and necrotrophic pathogens [133]. JA and ET defense signaling pathways are transduced and integrated through the ET-responsive transcription factor superfamily, binding to the GCC box of PR proteins [134]. JA-induced (ISR) and SA-induced (SAR) signaling pathways negatively regulate each other: the SA signaling pathway is involved in the resistance to biotrophic pathogens, while the JA and ET signaling pathways principally mediate resistance to necrotrophic pathogens [135]. However, only a few genes that regulate the interplay between JA and SA signaling networks have been identified (Table 2). The overexpression of the aspartate protease 13, AP13, from *V. quinquangularis*, described above, not only enhances resistance to powdery mildew but also contributes to SA, JA, and ET regulation: its transcript levels decreased following *B. cinerea* (necrotrophic pathogen) infection and MeJA treatment but increased following ET and SA treatments. This suggests that VqAP13 action promotes the SA-dependent signal transduction pathway but suppresses the JA signal transduction pathway. The overexpression of VqAP13 suppresses the biosynthesis of JA, as well as the expression of downstream genes, e.g., lipoxygenase 3 and plant defensin 1.2, and the upregulation of the expression of the genes *ics1* and *pr1*, which are components of the SA biosynthesis pathway and the SA signaling pathways, respectively [59].

The tomato matrix metalloprotease, Sl3-MMP, expression is induced by infection with *B. cinerea*, *Pst*DC3000, and by the SA, JA and ET treatment [93]. Sl2- and Sl3-MMP prefer hydrophobic amino acid residues in the P1 position and proline residue in the P3 position of a substrate [84]. The absence of Sl3-MMP results in increased susceptibility to *B. cinerea*, *Pst*DC3000. In addition, treatment with SA, MeJA, and the precursor of ET 1-amino cyclopropane-1-carboxylic acid (ACC) showed a three to four-fold increase as compared to a control, suggesting the activation of the Sl3-MMP gene in late ETI responses.

The subtilase from *G. barbadense*, GbSBT1, is closely related to the AtSBT5.2 described above and is associated with the defense response to infection by *V. dahliae* [95]. The isochorismatase, VdISC1, is secreted by *V. dahliae* to suppress the SA-mediated innate immunity of host cells. Verticillium wilt resistance is, thus, related to the JA signaling pathway. The GbSBT1 is normally localized in the plasma membrane, whereas after treatment with JA and ET, it is relocalized into the cytoplasm, suggesting the protease to be a sensor of the combined signals of ET and JA treatments and its involvement in ISR. Moreover, the overexpression of GbSBT1 in *A. thaliana* results in the activation of MAPK signaling and JA-responsive genes [136].

Proteases that contribute to ISR and to the defense against insects have also been identified (Table 2). An interesting example of a plant protease that participates in the defense against herbivorous insect attack is cysteine protease MIR1, from maize, which is activated in response to caterpillar feeding [99]. It is ET-dependent and accumulates in maize midwhorl, resulting in a reduction in caterpillar growth and affecting the caterpillar peritrophic matrix. Another protease is subtilase SBT3-Sl, from tomato, which is induced after wounding and insect attack [94]. SBT3-Sl silencing results in decreased resistance to the larvae of the specialist herbivore, *M. sexta*, and attenuates the induction of systemic wound defense genes. SBT3-Sl was found to be stable and active in the insect's digestive system, from where it may act on unidentified proteins of insect or plant origin. In addition, SBT3 is involved in the regulation of pectin methylesterases.

## 3. Conclusions

To date, plant proteases are known to be crucial components of plant immunity. In this review, we have summarized the available data on the proteases of different plant species and participation of the proteases in immune processes. We have also attempted to classify the proteases according to their involvement in different types and at different stages of plant defense. It is worth mentioning that proteases seem to be implemented in every step of immunity establishment: pathogen encounter, generation of DAMPs and MAMPs, effector recognition, regulation of PRRs and NB-LRRs (R proteins) action, signal transduction (including MAPK cascade activation), involvement in the synthesis of signal molecules, the orchestration of PCD, cell priming, the regulation of PR proteins expression, SAR and

ISR establishment, and, finally, RNA silencing. It is noteworthy that proteases are extremely powerful tools in any organism and tight regulation of their action is required. Pathogen effectors that target plant proteases are mainly represented by protease inhibitors [137] but the plant itself synthetizes inhibitors to regulate their activity. For example, an irreversible inhibitor, AtSerpin1, and a reversible Kunitz-type protease inhibitor, AtWSCP, tightly regulate the activity of the *A. thaliana* cysteine protease, RD21, in plant development and defense [124]. In addition, the production of maize CC9 cystatin inhibitor, which is induced by an *U. maydis* infection simultaneous with the production of cysteine proteases, results in the inhibition of the proteases in the apoplast; the pathogen manipulates the transcription of the plant inhibitors to facilitate infection [138].

Moreover, plant proteases possess unique features that could be applied in biotechnology and biomedicine, e.g., papain from *Carica papaya* L. and bromelain from *Ananas comosus* L. are already used in very different fields of industry. However, the question concerning the role of plant proteases in the immune system and in vital processes of plant organisms, in general, is still highly relevant and remains a mystery. Nowadays, more and more complete plant proteomes are becoming available. With the use of predictive bioinformatic tools relying on already existing data, it has become a great opportunity to identify novel enzymes, including proteases. Identified proteases could be studied through the investigation of generated mutant plants carrying loss-of-function (or other) mutations in the corresponding genes. In addition, the knowledge concerning already known proteases could be expanded further: nowadays, a little data on the substrate specificity and functional substrates is available. Modern techniques such as molecular modeling and molecular dynamics methods, based on already identified 3D structures homologous to the studied protease enzymes, could be applied for the prediction of its substrate specificity and potent functional substrates in the proteome. The identification of novel proteases and the elucidation of the physiological functions of already identified plant proteases will certainly contribute to the development of modern science and biotechnology in general [18,139,140].

**Acknowledgments:** This research was funded by the Russian Science Foundation (grant # 16-15-10410).

**Conflicts of Interest:** The authors declare no conflict of interest.

## Abbreviations

| | |
|---|---|
| AED1 | apoplastic EDS1-dependent protein 1 |
| AP13 | aspartate protease 13 |
| AtCEP1 | cysteine endopeptidase 1 from *A. thaliana* |
| AtICS1/2 | isochorismate synthase1/2 from *A. thaliana* |
| AtMC1/2/9 | metacaspase 1/2/9 from *A. thaliana* |
| AtPAL1–4 | phenylalanine lyase1 to 4 from *A. thaliana* |
| AZI1 | azelaic acid induced 1 |
| BAK1 | brassinosteroid insensitive 1-associated receptor kinase 1 |
| BIK1 | *Botrytis*-induced kinase 1 |
| CDPK | calcium-dependent protein kinase |
| CDR1 | constitutive disease resistance 1 |
| CERK1 | chitin elicitor receptor kinase 1 |
| CYP1 | cysteine protease 1 |
| DAMP | damage-associated molecular pattern |
| DIR1 | defective in induced resistance 1 |
| EDS1 | enhanced disease susceptibility 1 |
| EFR | EF-Tu receptor |
| ER | endoplasmic reticulum |
| ET | ethylene |
| ETI | effector-triggered immunity |
| FLS2 | flagellin-sensing 2 |

| HR | hypersensitive response |
| ISR | induced systemic resistance |
| JA | jasmonic acid |
| LRR | leucine-rich repeats |
| LSD1 | lesion simulating disease 1 |
| MAMPs | microbe-associated molecular pattern |
| MAPK | mitogen-activated protein kinase |
| mcII-Pa | metacaspase type II from *P. abies* |
| MeSA | methylsalicylic acid |
| MTI | MAMP-triggered immunity |
| NBS | nucleotide-binding site |
| NLR | NOD-like receptor |
| NLS | nuclear localization signal |
| NPR1/3/4 | nonexpresser of PR genes 1/3/4 |
| OTS1/2 | overly tolerant to salt 1/2 |
| OXI1 | oxidative signal-inducible 1 kinase |
| PAMP | pathogen-associated molecular pattern |
| PCD | programmed cell death |
| PEPCK1 | phosphoenolpyruvate carboxykinase 1 |
| PIP1 | *Phytophthora*-inhibited protease 1 |
| PLCP | papain-like cysteine protease |
| PR | pathogenesis-related |
| PRR | pattern-recognition receptors |
| PstDC3000 | *Pseudomonas syringae* pv. *tomato* DC30000 |
| PTGS | post-transcriptional gene silencing |
| RALF23 | rapid alkalinization factor 23 |
| RD21 | responsive to dehydration 21 |
| RDR2 | RNA-dependent RNA polymerase 2 |
| RISC | RNA-induced silencing complex |
| RITS complex | RNA-induced transcriptional silencing complex |
| RLK | receptor-like kinase |
| RLP | receptor-like protein |
| RNA | ribonucleic acid |
| RNI | reactive nitrogen intermediate |
| ROS | reactive oxygen species |
| RRS1-R | Resistant to *R. solanacearum* 1-R |
| RuBisCO | ribulose-1,5-bisphosphate carboxylase/oxygenase |
| SA | salicylic acid |
| SAG12 | senescence-associated gene 12 |
| SAR | systemic acquired resistance |
| SBT | subtilase |
| Sl2/3-MMPs | matrix metalloprotease 2/3 from *S. lycopersicum* |
| SUMO1/2 | small ubiquitin-related modifier 1/2 |
| TGS | transcriptional gene silencing |
| TMV | tobacco mosaic virus |
| TOPs | thimet oligopeptidases |
| TSN | Tudor staphylococcal nuclease |
| VPE | vacuolar processing enzyme |
| WSCP | water-soluble chlorophyll protein |

## References

1. Anderson, J.P.; Gleason, C.A.; Foley, R.C.; Thrall, P.H.; Burdon, J.B.; Singh, K.B. Plants versus pathogens: An evolutionary arms race. *Funct. Plant Biol.* **2010**, *37*, 499–512. [CrossRef] [PubMed]
2. Santori, F.R. The immune system as a self-centered network of lymphocytes. *Immunol. Lett.* **2015**, *166*, 109–116. [CrossRef] [PubMed]
3. Farber, D.L.; Netea, M.G.; Radbruch, A.; Rajewsky, K.; Zinkernagel, R.M. Immunological memory: Lessons from the past and a look to the future. *Nat. Rev. Immunol.* **2016**, *16*, 124–128. [CrossRef] [PubMed]
4. Ronald, P.C.; Beutler, B. Plant and animal sensors of conserved microbial signatures. *Science* **2010**, *330*, 1061–1064. [CrossRef] [PubMed]
5. Muthamilarasan, M.; Prasad, M. Plant innate immunity: An updated insight into defense mechanism. *J. Biosci.* **2013**, *38*, 433–449. [CrossRef] [PubMed]
6. Misas-Villamil, J.C.; van der Hoorn, R.A.; Doehlemann, G. Papain-like cysteine proteases as hubs in plant immunity. *New Phytol.* **2016**, *212*, 902–907. [CrossRef] [PubMed]
7. Van der Hoorn, R.A. Plant proteases: From phenotypes to molecular mechanisms. *Annu. Rev. Plant Biol.* **2008**, *59*, 191–223. [CrossRef] [PubMed]
8. Minina, E.A.; Moschou, P.N.; Bozhkov, P.V. Limited and digestive proteolysis: Crosstalk between evolutionary conserved pathways. *New Phytol.* **2017**, *215*, 958–964. [CrossRef] [PubMed]
9. Rawlings, N.D.; Waller, M.; Barrett, A.J.; Bateman, A. MEROPS: The database of proteolytic enzymes, their substrates and inhibitors. *Nucleic Acids Res.* **2014**, *42*, D503–D509. [CrossRef] [PubMed]
10. Antao, C.M.; Malcata, F.X. Plant serine proteases: Biochemical, physiological and molecular features. *Plant Physiol. Biochem.* **2005**, *43*, 637–650. [CrossRef] [PubMed]
11. Zamyatnin, A.A., Jr. Plant Proteases Involved in Regulated Cell Death. *Biochemistry* **2015**, *80*, 1701–1715. [CrossRef] [PubMed]
12. Shahri, W.; Tahir, I. Flower senescence: Some molecular aspects. *Planta* **2014**, *239*, 277–297. [CrossRef] [PubMed]
13. Buckley, J.J.; Jessen, J.R. Matrix metalloproteinase function in non-mammalian model organisms. *Front. Biosci. (Schol. Ed.)* **2015**, *7*, 168–183. [PubMed]
14. Chen, H.J.; Huang, Y.H.; Huang, G.J.; Huang, S.S.; Chow, T.J.; Lin, Y.H. Sweet potato SPAP1 is a typical aspartic protease and participates in ethephon-mediated leaf senescence. *J. Plant Physiol.* **2015**, *180*, 1–17. [CrossRef] [PubMed]
15. Niu, N.; Liang, W.; Yang, X.; Jin, W.; Wilson, Z.A.; Hu, J.; Zhang, D. EAT1 promotes tapetal cell death by regulating aspartic proteases during male reproductive development in rice. *Nat. Commun.* **2013**, *4*, 1445. [CrossRef] [PubMed]
16. Guo, R.; Xu, X.; Carole, B.; Li, X.; Gao, M.; Zheng, Y.; Wang, X. Genome-wide identification, evolutionary and expression analysis of the aspartic protease gene superfamily in grape. *BMC Genom.* **2013**, *14*, 554. [CrossRef] [PubMed]
17. Hatsugai, N.; Iwasaki, S.; Tamura, K.; Kondo, M.; Fuji, K.; Ogasawara, K.; Nishimura, M.; Hara-Nishimura, I. A novel membrane fusion-mediated plant immunity against bacterial pathogens. *Genes Dev.* **2009**, *23*, 2496–2506. [CrossRef] [PubMed]
18. Balakireva, A.V.; Kuznetsova, N.V.; Petushkova, A.I.; Savvateeva, L.; Zamyatnin, A.A. Trends and Prospects of Plant Proteases in Therapeutics. *Curr. Med. Chem.* **2018**. [CrossRef] [PubMed]
19. Boller, T.; Felix, G. A renaissance of elicitors: Perception of microbe-associated molecular patterns and danger signals by pattern-recognition receptors. *Annu. Rev. Plant Biol.* **2009**, *60*, 379–406. [CrossRef] [PubMed]
20. Danna, C.H.; Millet, Y.A.; Koller, T.; Han, S.W.; Bent, A.F.; Ronald, P.C.; Ausubel, F.M. The Arabidopsis flagellin receptor FLS2 mediates the perception of Xanthomonas Ax21 secreted peptides. *Proc. Natl. Acad. Sci. USA* **2011**, *108*, 9286–9291. [CrossRef] [PubMed]
21. Smith, K.D.; Andersen-Nissen, E.; Hayashi, F.; Strobe, K.; Bergman, M.A.; Barrett, S.L.; Cookson, B.T.; Aderem, A. Toll-like receptor 5 recognizes a conserved site on flagellin required for protofilament formation and bacterial motility. *Nat. Immunol.* **2003**, *4*, 1247–1253. [CrossRef] [PubMed]
22. Spoel, S.H.; Dong, X. How do plants achieve immunity? Defence without specialized immune cells. *Nat. Rev. Immunol.* **2012**, *12*, 89–100. [PubMed]

23. Nedukha, O.M. Callose: Localization, functions, and synthesis in plant cells. *Cytol. Genet.* **2015**, *49*, 49–57. [CrossRef]

24. Luna, E.; Pastor, V.; Robert, J.; Flors, V.; Mauch-Mani, B.; Ton, J. Callose deposition: A multifaceted plant defense response. *Mol. Plant. Microbe Interact.* **2011**, *24*, 183–193. [CrossRef] [PubMed]

25. Liu, J.; Elmore, J.M.; Lin, Z.J.; Coaker, G. A receptor-like cytoplasmic kinase phosphorylates the host target RIN4, leading to the activation of a plant innate immune receptor. *Cell Host Microbe* **2011**, *9*, 137–146. [CrossRef] [PubMed]

26. Dangl, J.L.; Jones, J.D. Plant pathogens and integrated defence responses to infection. *Nature* **2001**, *411*, 826–833. [CrossRef] [PubMed]

27. Goff, S.A.; Ricke, D.; Lan, T.H.; Presting, G.; Wang, R.; Dunn, M.; Glazebrook, J.; Sessions, A.; Oeller, P.; Varma, H.; et al. A draft sequence of the rice genome (*Oryza sativa* L. ssp. *japonica*). *Science* **2002**, *296*, 92–100. [CrossRef] [PubMed]

28. Tenthorey, J.L.; Haloupek, N.; Lopez-Blanco, J.R.; Grob, P.; Adamson, E.; Hartenian, E.; Lind, N.A.; Bourgeois, N.M.; Chacon, P.; Nogales, E.; et al. The structural basis of flagellin detection by NAIP5: A strategy to limit pathogen immune evasion. *Science* **2017**, *358*, 888–893. [CrossRef] [PubMed]

29. Shirasu, K. The HSP90-SGT1 chaperone complex for NLR immune sensors. *Annu. Rev. Plant Biol.* **2009**, *60*, 139–164. [CrossRef] [PubMed]

30. Misas-Villamil, J.C.; Toenges, G.; Kolodziejek, I.; Sadaghiani, A.M.; Kaschani, F.; Colby, T.; Bogyo, M.; van der Hoorn, R.A. Activity profiling of vacuolar processing enzymes reveals a role for VPE during oomycete infection. *Plant J.* **2013**, *73*, 689–700. [CrossRef] [PubMed]

31. Fomicheva, A.S.; Tuzhikov, A.I.; Beloshistov, R.E.; Trusova, S.V.; Galiullina, R.A.; Mochalova, L.V.; Chichkova, N.V.; Vartapetian, A.B. Programmed cell death in plants. *Biochemistry* **2012**, *77*, 1452–1464. [CrossRef] [PubMed]

32. Coll, N.S.; Vercammen, D.; Smidler, A.; Clover, C.; Van Breusegem, F.; Dangl, J.L.; Epple, P. Arabidopsis type I metacaspases control cell death. *Science* **2010**, *330*, 1393–1397. [CrossRef] [PubMed]

33. McLellan, H.; Gilroy, E.M.; Yun, B.W.; Birch, P.R.; Loake, G.J. Functional redundancy in the Arabidopsis Cathepsin B gene family contributes to basal defence, the hypersensitive response and senescence. *New Phytol.* **2009**, *183*, 408–418. [CrossRef] [PubMed]

34. Hatsugai, N.; Kuroyanagi, M.; Yamada, K.; Meshi, T.; Tsuda, S.; Kondo, M.; Nishimura, M.; Hara-Nishimura, I. A plant vacuolar protease, VPE, mediates virus-induced hypersensitive cell death. *Science* **2004**, *305*, 855–858. [CrossRef] [PubMed]

35. Ramirez, V.; Lopez, A.; Mauch-Mani, B.; Gil, M.J.; Vera, P. An extracellular subtilase switch for immune priming in Arabidopsis. *PLoS Pathog.* **2013**, *9*, e1003445. [CrossRef] [PubMed]

36. Choudhary, D.K.; Prakash, A.; Johri, B.N. Induced systemic resistance (ISR) in plants: Mechanism of action. *Indian J. Microbiol.* **2007**, *47*, 289–297. [CrossRef] [PubMed]

37. Ross, A.F. Systemic acquired resistance induced by localized virus infections in plants. *Virology* **1961**, *14*, 340–358. [CrossRef]

38. Chanda, B.; Xia, Y.; Mandal, M.K.; Yu, K.; Sekine, K.T.; Gao, Q.M.; Selote, D.; Hu, Y.; Stromberg, A.; Navarre, D.; et al. Glycerol-3-phosphate is a critical mobile inducer of systemic immunity in plants. *Nat. Genet.* **2011**, *43*, 421–427. [CrossRef] [PubMed]

39. Li, L.; Li, M.; Yu, L.; Zhou, Z.; Liang, X.; Liu, Z.; Cai, G.; Gao, L.; Zhang, X.; Wang, Y.; et al. The FLS2-associated kinase BIK1 directly phosphorylates the NADPH oxidase RbohD to control plant immunity. *Cell Host Microbe* **2014**, *15*, 329–338. [CrossRef] [PubMed]

40. Tada, Y.; Spoel, S.H.; Pajerowska-Mukhtar, K.; Mou, Z.; Song, J.; Wang, C.; Zuo, J.; Dong, X. Plant immunity requires conformational changes [corrected] of NPR1 via S-nitrosylation and thioredoxins. *Science* **2008**, *321*, 952–956. [CrossRef] [PubMed]

41. Beckers, G.J.; Jaskiewicz, M.; Liu, Y.; Underwood, W.R.; He, S.Y.; Zhang, S.; Conrath, U. Mitogen-activated protein kinases 3 and 6 are required for full priming of stress responses in *Arabidopsis thaliana*. *Plant Cell* **2009**, *21*, 944–953. [CrossRef] [PubMed]

42. Baumgarten, A.; Cannon, S.; Spangler, R.; May, G. Genome-level evolution of resistance genes in *Arabidopsis thaliana*. *Genetics* **2003**, *165*, 309–319. [PubMed]

43. Heil, M.; Bostock, R.M. Induced systemic resistance (ISR) against pathogens in the context of induced plant defences. *Ann. Bot.* **2002**, *89*, 503–512. [CrossRef] [PubMed]

44. Zhang, T.; Jin, Y.; Zhao, J.H.; Gao, F.; Zhou, B.J.; Fang, Y.Y.; Guo, H.S. Host-Induced Gene Silencing of the Target Gene in Fungal Cells Confers Effective Resistance to the Cotton Wilt Disease Pathogen Verticillium dahliae. *Mol. Plant.* **2016**, *9*, 939–942. [CrossRef] [PubMed]

45. Sharma, N.; Sahu, P.P.; Puranik, S.; Prasad, M. Recent advances in plant-virus interaction with emphasis on small interfering RNAs (siRNAs). *Mol. Biotechnol.* **2013**, *55*, 63–77. [CrossRef] [PubMed]

46. Martinez de Alba, A.E.; Elvira-Matelot, E.; Vaucheret, H. Gene silencing in plants: A diversity of pathways. *Biochim. Biophys. Acta* **2013**, *1829*, 1300–1308. [CrossRef] [PubMed]

47. Lister, R.; O'Malley, R.C.; Tonti-Filippini, J.; Gregory, B.D.; Berry, C.C.; Millar, A.H.; Ecker, J.R. Highly integrated single-base resolution maps of the epigenome in Arabidopsis. *Cell* **2008**, *133*, 523–536. [CrossRef] [PubMed]

48. Yadav, R.K.; Chattopadhyay, D. Enhanced viral intergenic region-specific short interfering RNA accumulation and DNA methylation correlates with resistance against a geminivirus. *Mol. Plant Microbe Interact.* **2011**, *24*, 1189–1197. [CrossRef] [PubMed]

49. Ruiz-Medrano, R.; Xoconostle-Cazares, B.; Kragler, F. The plasmodesmatal transport pathway for homeotic proteins, silencing signals and viruses. *Curr. Opin. Plant Biol.* **2004**, *7*, 641–650. [CrossRef] [PubMed]

50. Antolin-Llovera, M.; Ried, M.K.; Binder, A.; Parniske, M. Receptor kinase signaling pathways in plant-microbe interactions. *Annu. Rev. Phytopathol.* **2012**, *50*, 451–473. [CrossRef] [PubMed]

51. Petutschnig, E.K.; Stolze, M.; Lipka, U.; Kopischke, M.; Horlacher, J.; Valerius, O.; Rozhon, W.; Gust, A.A.; Kemmerling, B.; Poppenberger, B.; et al. A novel Arabidopsis CHITIN ELICITOR RECEPTOR KINASE 1 (CERK1) mutant with enhanced pathogen-induced cell death and altered receptor processing. *New Phytol.* **2014**, *204*, 955–967. [CrossRef] [PubMed]

52. Xia, Y.; Suzuki, H.; Borevitz, J.; Blount, J.; Guo, Z.; Patel, K.; Dixon, R.A.; Lamb, C. An extracellular aspartic protease functions in Arabidopsis disease resistance signaling. *EMBO J.* **2004**, *23*, 980–988. [CrossRef] [PubMed]

53. Santamaria, M.E.; Arnaiz, A.; Diaz-Mendoza, M.; Martinez, M.; Diaz, I. Inhibitory properties of cysteine protease pro-peptides from barley confer resistance to spider mite feeding. *PLoS ONE* **2015**, *10*, e0128323. [CrossRef] [PubMed]

54. Pearce, G.; Yamaguchi, Y.; Barona, G.; Ryan, C.A. A subtilisin-like protein from soybean contains an embedded, cryptic signal that activates defense-related genes. *Proc. Natl. Acad. Sci. USA* **2010**, *107*, 14921–14925. [CrossRef] [PubMed]

55. Stegmann, M.; Monaghan, J.; Smakowska-Luzan, E.; Rovenich, H.; Lehner, A.; Holton, N.; Belkhadir, Y.; Zipfel, C. The receptor kinase FER is a RALF-regulated scaffold controlling plant immune signaling. *Science* **2017**, *355*, 287–289. [CrossRef] [PubMed]

56. Tornero, P.; Mayda, E.; Gomez, M.D.; Canas, L.; Conejero, V.; Vera, P. Characterization of LRP, a leucine-rich repeat (LRR) protein from tomato plants that is processed during pathogenesis. *Plant J.* **1996**, *10*, 315–330. [CrossRef] [PubMed]

57. Ewan, R.; Pangestuti, R.; Thornber, S.; Craig, A.; Carr, C.; O'Donnell, L.; Zhang, C.; Sadanandom, A. Deubiquitinating enzymes AtUBP12 and AtUBP13 and their tobacco homologue NtUBP12 are negative regulators of plant immunity. *New Phytol.* **2011**, *191*, 92–106. [CrossRef] [PubMed]

58. Zhang, J.; Shao, F.; Li, Y.; Cui, H.; Chen, L.; Li, H.; Zou, Y.; Long, C.; Lan, L.; Chai, J.; et al. A *Pseudomonas syringae* effector inactivates MAPKs to suppress PAMP-induced immunity in plants. *Cell Host Microbe* **2007**, *1*, 175–185. [CrossRef] [PubMed]

59. Guo, R.; Tu, M.; Wang, X.; Zhao, J.; Wan, R.; Li, Z.; Wang, Y.; Wang, X. Ectopic expression of a grape aspartic protease gene, AP13, in *Arabidopsis thaliana* improves resistance to powdery mildew but increases susceptibility to Botrytis cinerea. *Plant Sci.* **2016**, *248*, 17–27. [CrossRef] [PubMed]

60. Wang, W.H.; Yi, X.Q.; Han, A.D.; Liu, T.W.; Chen, J.; Wu, F.H.; Dong, X.J.; He, J.X.; Pei, Z.M.; Zheng, H.L. Calcium-sensing receptor regulates stomatal closure through hydrogen peroxide and nitric oxide in response to extracellular calcium in Arabidopsis. *J. Exp. Bot.* **2012**, *63*, 177–190. [CrossRef] [PubMed]

61. Zhang, H.; Dong, S.; Wang, M.; Wang, W.; Song, W.; Dou, X.; Zheng, X.; Zhang, Z. The role of vacuolar processing enzyme (VPE) from *Nicotiana benthamiana* in the elicitor-triggered hypersensitive response and stomatal closure. *J. Exp. Bot.* **2010**, *61*, 3799–3812. [CrossRef] [PubMed]

62. Zhang, H.; Zheng, X.; Zhang, Z. The role of vacuolar processing enzymes in plant immunity. *Plant Signal. Behav.* **2010**, *5*, 1565–1567. [CrossRef] [PubMed]

63. Melotto, M.; Underwood, W.; Koczan, J.; Nomura, K.; He, S.Y. Plant stomata function in innate immunity against bacterial invasion. *Cell* **2006**, *126*, 969–980. [CrossRef] [PubMed]

64. Cook, D.E.; Mesarich, C.H.; Thomma, B.P. Understanding plant immunity as a surveillance system to detect invasion. *Annu. Rev. Phytopathol.* **2015**, *53*, 541–563. [CrossRef] [PubMed]

65. Iida, Y.; van't Hof, P.; Beenen, H.; Mesarich, C.; Kubota, M.; Stergiopoulos, I.; Mehrabi, R.; Notsu, A.; Fujiwara, K.; Bahkali, A.; et al. Novel Mutations Detected in Avirulence Genes Overcoming Tomato Cf Resistance Genes in Isolates of a Japanese Population of Cladosporium fulvum. *PLoS ONE* **2015**, *10*, e0123271. [CrossRef] [PubMed]

66. Xu, Q.F.; Cheng, W.S.; Li, S.S.; Li, W.; Zhang, Z.X.; Xu, Y.P.; Zhou, X.P.; Cai, X.Z. Identification of genes required for Cf-dependent hypersensitive cell death by combined proteomic and RNA interfering analyses. *J. Exp. Bot.* **2012**, *63*, 2421–2435. [CrossRef] [PubMed]

67. Lozano-Torres, J.L.; Wilbers, R.H.; Gawronski, P.; Boshoven, J.C.; Finkers-Tomczak, A.; Cordewener, J.H.; America, A.H.; Overmars, H.A.; Van't Klooster, J.W.; Baranowski, L.; et al. Dual disease resistance mediated by the immune receptor Cf-2 in tomato requires a common virulence target of a fungus and a nematode. *Proc. Natl. Acad. Sci. USA* **2012**, *109*, 10119–10124. [CrossRef] [PubMed]

68. Cheng, Y.T.; Germain, H.; Wiermer, M.; Bi, D.; Xu, F.; Garcia, A.V.; Wirthmueller, L.; Despres, C.; Parker, J.E.; Zhang, Y.; et al. Nuclear pore complex component MOS7/Nup88 is required for innate immunity and nuclear accumulation of defense regulators in Arabidopsis. *Plant Cell* **2009**, *21*, 2503–2516. [CrossRef] [PubMed]

69. Wilton, M.; Subramaniam, R.; Elmore, J.; Felsensteiner, C.; Coaker, G.; Desveaux, D. The type III effector HopF2Pto targets Arabidopsis RIN4 protein to promote *Pseudomonas syringae* virulence. *Proc. Natl. Acad. Sci. USA* **2010**, *107*, 2349–2354. [CrossRef] [PubMed]

70. Bollhoner, B.; Zhang, B.; Stael, S.; Denance, N.; Overmyer, K.; Goffner, D.; Van Breusegem, F.; Tuominen, H. Post mortem function of AtMC9 in xylem vessel elements. *New Phytol.* **2013**, *200*, 498–510. [CrossRef] [PubMed]

71. Tsiatsiani, L.; Timmerman, E.; De Bock, P.J.; Vercammen, D.; Stael, S.; van de Cotte, B.; Staes, A.; Goethals, M.; Beunens, T.; Van Damme, P.; et al. The Arabidopsis metacaspase9 degradome. *Plant Cell* **2013**, *25*, 2831–2847. [CrossRef] [PubMed]

72. Wrzaczek, M.; Vainonen, J.P.; Stael, S.; Tsiatsiani, L.; Help-Rinta-Rahko, H.; Gauthier, A.; Kaufholdt, D.; Bollhoner, B.; Lamminmaki, A.; Staes, A.; et al. GRIM REAPER peptide binds to receptor kinase PRK5 to trigger cell death in Arabidopsis. *EMBO J.* **2015**, *34*, 55–66. [CrossRef] [PubMed]

73. Watanabe, N.; Lam, E. Arabidopsis metacaspase 2d is a positive mediator of cell death induced during biotic and abiotic stresses. *Plant J.* **2011**, *66*, 969–982. [CrossRef] [PubMed]

74. Shindo, T.; Misas-Villamil, J.C.; Horger, A.C.; Song, J.; van der Hoorn, R.A. A role in immunity for Arabidopsis cysteine protease RD21, the ortholog of the tomato immune protease C14. *PLoS ONE* **2012**, *7*, e29317. [CrossRef] [PubMed]

75. Boex-Fontvieille, E.; Rustgi, S.; Reinbothe, S.; Reinbothe, C. A Kunitz-type protease inhibitor regulates programmed cell death during flower development in *Arabidopsis thaliana*. *J. Exp. Bot.* **2015**, *66*, 6119–6135. [CrossRef] [PubMed]

76. Lampl, N.; Alkan, N.; Davydov, O.; Fluhr, R. Set-point control of RD21 protease activity by AtSerpin1 controls cell death in Arabidopsis. *Plant J.* **2013**, *74*, 498–510. [CrossRef] [PubMed]

77. Bernoux, M.; Timmers, T.; Jauneau, A.; Briere, C.; de Wit, P.J.; Marco, Y.; Deslandes, L. RD19, an Arabidopsis cysteine protease required for RRS1-R-mediated resistance, is relocalized to the nucleus by the Ralstonia solanacearum PopP2 effector. *Plant Cell* **2008**, *20*, 2252–2264. [CrossRef] [PubMed]

78. Shimada, T.; Yamada, K.; Kataoka, M.; Nakaune, S.; Koumoto, Y.; Kuroyanagi, M.; Tabata, S.; Kato, T.; Shinozaki, K.; Seki, M.; et al. Vacuolar processing enzymes are essential for proper processing of seed storage proteins in *Arabidopsis thaliana*. *J. Biol. Chem.* **2003**, *278*, 32292–32299. [CrossRef] [PubMed]

79. Howing, T.; Huesmann, C.; Hoefle, C.; Nagel, M.K.; Isono, E.; Huckelhoven, R.; Gietl, C. Endoplasmic reticulum KDEL-tailed cysteine endopeptidase 1 of Arabidopsis (AtCEP1) is involved in pathogen defense. *Front. Plant Sci.* **2014**, *5*, 58. [CrossRef] [PubMed]

80. Serrano, I.; Buscaill, P.; Audran, C.; Pouzet, C.; Jauneau, A.; Rivas, S. A non canonical subtilase attenuates the transcriptional activation of defence responses in *Arabidopsis thaliana*. *eLife* **2016**, *5*, e19755. [CrossRef] [PubMed]

81. Bozhkov, P.V.; Suarez, M.F.; Filonova, L.H.; Daniel, G.; Zamyatnin, A.A., Jr.; Rodriguez-Nieto, S.; Zhivotovsky, B.; Smertenko, A. Cysteine protease mcII-Pa executes programmed cell death during plant embryogenesis. *Proc. Natl. Acad. Sci. USA* **2005**, *102*, 14463–14468. [CrossRef] [PubMed]

82. Minina, E.A.; Filonova, L.H.; Fukada, K.; Savenkov, E.I.; Gogvadze, V.; Clapham, D.; Sanchez-Vera, V.; Suarez, M.F.; Zhivotovsky, B.; Daniel, G.; et al. Autophagy and metacaspase determine the mode of cell death in plants. *J. Cell Biol.* **2013**, *203*, 917–927. [CrossRef] [PubMed]

83. Tian, M.; Benedetti, B.; Kamoun, S. A Second Kazal-like protease inhibitor from Phytophthora infestans inhibits and interacts with the apoplastic pathogenesis-related protease P69B of tomato. *Plant Physiol.* **2005**, *138*, 1785–1793. [CrossRef] [PubMed]

84. Zimmermann, D.; Gomez-Barrera, J.A.; Pasule, C.; Brack-Frick, U.B.; Sieferer, E.; Nicholson, T.M.; Pfannstiel, J.; Stintzi, A.; Schaller, A. Cell Death Control by Matrix Metalloproteinases. *Plant Physiol.* **2016**, *171*, 1456–1469. [CrossRef] [PubMed]

85. Ilyas, M.; Horger, A.C.; Bozkurt, T.O.; van den Burg, H.A.; Kaschani, F.; Kaiser, M.; Belhaj, K.; Smoker, M.; Joosten, M.H.; Kamoun, S.; et al. Functional Divergence of Two Secreted Immune Proteases of Tomato. *Curr. Biol.* **2015**, *25*, 2300–2306. [CrossRef] [PubMed]

86. Tian, M.; Win, J.; Song, J.; van der Hoorn, R.; van der Knaap, E.; Kamoun, S. A Phytophthora infestans cystatin-like protein targets a novel tomato papain-like apoplastic protease. *Plant Physiol.* **2007**, *143*, 364–377. [CrossRef] [PubMed]

87. Bar-Ziv, A.; Levy, Y.; Citovsky, V.; Gafni, Y. The Tomato yellow leaf curl virus (TYLCV) V2 protein inhibits enzymatic activity of the host papain-like cysteine protease CYP1. *Biochem. Biophys. Res. Commun.* **2015**, *460*, 525–529. [CrossRef] [PubMed]

88. Bozkurt, T.O.; Schornack, S.; Win, J.; Shindo, T.; Ilyas, M.; Oliva, R.; Cano, L.M.; Jones, A.M.; Huitema, E.; van der Hoorn, R.A.; et al. Phytophthora infestans effector AVRblb2 prevents secretion of a plant immune protease at the haustorial interface. *Proc. Natl. Acad. Sci. USA* **2011**, *108*, 20832–20837. [CrossRef] [PubMed]

89. Cedzich, A.; Huttenlocher, F.; Kuhn, B.M.; Pfannstiel, J.; Gabler, L.; Stintzi, A.; Schaller, A. The protease-associated domain and C-terminal extension are required for zymogen processing, sorting within the secretory pathway, and activity of tomato subtilase 3 (SlSBT3). *J. Biol. Chem.* **2009**, *284*, 14068–14078. [CrossRef] [PubMed]

90. Bollhoner, B.; Jokipii-Lukkari, S.; Bygdell, J.; Stael, S.; Adriasola, M.; Muniz, L.; Van Breusegem, F.; Ezcurra, I.; Wingsle, G.; Tuominen, H. The function of two type II metacaspases in woody tissues of Populus trees. *New Phytol.* **2018**, *217*, 1551–1565. [CrossRef] [PubMed]

91. Coffeen, W.C.; Wolpert, T.J. Purification and characterization of serine proteases that exhibit caspase-like activity and are associated with programmed cell death in Avena sativa. *Plant Cell* **2004**, *16*, 857–873. [CrossRef] [PubMed]

92. Chichkova, N.V.; Shaw, J.; Galiullina, R.A.; Drury, G.E.; Tuzhikov, A.I.; Kim, S.H.; Kalkum, M.; Hong, T.B.; Gorshkova, E.N.; Torrance, L.; et al. Phytaspase, a relocalisable cell death promoting plant protease with caspase specificity. *EMBO J.* **2010**, *29*, 1149–1161. [CrossRef] [PubMed]

93. Li, D.; Zhang, H.; Song, Q.; Wang, L.; Liu, S.; Hong, Y.; Huang, L.; Song, F. Tomato Sl3-MMP, a member of the Matrix metalloproteinase family, is required for disease resistance against Botrytis cinerea and *Pseudomonas syringae* pv. tomato DC3000. *BMC Plant Biol.* **2015**, *15*, 143. [CrossRef] [PubMed]

94. Meyer, M.; Huttenlocher, F.; Cedzich, A.; Procopio, S.; Stroeder, J.; Pau-Roblot, C.; Lequart-Pillon, M.; Pelloux, J.; Stintzi, A.; Schaller, A. The subtilisin-like protease SBT3 contributes to insect resistance in tomato. *J. Exp. Bot.* **2016**, *67*, 4325–4338. [CrossRef] [PubMed]

95. Duan, X.; Zhang, Z.; Wang, J.; Zuo, K. Characterization of a Novel Cotton Subtilase Gene GbSBT1 in Response to Extracellular Stimulations and Its Role in Verticillium Resistance. *PLoS ONE* **2016**, *11*, e0153988. [CrossRef] [PubMed]

96. Bailey, M.; Srivastava, A.; Conti, L.; Nelis, S.; Zhang, C.; Florance, H.; Love, A.; Milner, J.; Napier, R.; Grant, M.; et al. Stability of small ubiquitin-like modifier (SUMO) proteases overly tolerant to salt1 and -2 modulates salicylic acid signalling and SUMO1/2 conjugation in *Arabidopsis thaliana*. *J. Exp. Bot.* **2016**, *67*, 353–363. [CrossRef] [PubMed]

97. Breitenbach, H.H.; Wenig, M.; Wittek, F.; Jorda, L.; Maldonado-Alconada, A.M.; Sarioglu, H.; Colby, T.; Knappe, C.; Bichlmeier, M.; Pabst, E.; et al. Contrasting Roles of the Apoplastic Aspartyl Protease APOPLASTIC, ENHANCED DISEASE SUSCEPTIBILITY1-DEPENDENT1 and LEGUME LECTIN-LIKE PROTEIN1 in Arabidopsis Systemic Acquired Resistance. *Plant Physiol.* **2014**, *165*, 791–809. [CrossRef] [PubMed]

98. Moreau, M.; Westlake, T.; Zampogna, G.; Popescu, G.; Tian, M.; Noutsos, C.; Popescu, S. The Arabidopsis oligopeptidases TOP1 and TOP2 are salicylic acid targets that modulate SA-mediated signaling and the immune response. *Plant J.* **2013**, *76*, 603–614. [CrossRef] [PubMed]

99. Pechan, T.; Cohen, A.; Williams, W.P.; Luthe, D.S. Insect feeding mobilizes a unique plant defense protease that disrupts the peritrophic matrix of caterpillars. *Proc. Natl. Acad. Sci. USA* **2002**, *99*, 13319–13323. [CrossRef] [PubMed]

100. Fu, Z.Q.; Yan, S.; Saleh, A.; Wang, W.; Ruble, J.; Oka, N.; Mohan, R.; Spoel, S.H.; Tada, Y.; Zheng, N.; et al. NPR3 and NPR4 are receptors for the immune signal salicylic acid in plants. *Nature* **2012**, *486*, 228–232. [CrossRef] [PubMed]

101. Hayward, A.P.; Tsao, J.; Dinesh-Kumar, S.P. Autophagy and plant innate immunity: Defense through degradation. *Semin. Cell Dev. Biol.* **2009**, *20*, 1041–1047. [CrossRef] [PubMed]

102. Coll, N.S.; Epple, P.; Dangl, J.L. Programmed cell death in the plant immune system. *Cell Death Differ.* **2011**, *18*, 1247–1256. [CrossRef] [PubMed]

103. Fu, Z.Q.; Guo, M.; Jeong, B.R.; Tian, F.; Elthon, T.E.; Cerny, R.L.; Staiger, D.; Alfano, J.R. A type III effector ADP-ribosylates RNA-binding proteins and quells plant immunity. *Nature* **2007**, *447*, 284–288. [CrossRef] [PubMed]

104. Vercammen, D.; Belenghi, B.; van de Cotte, B.; Beunens, T.; Gavigan, J.A.; De Rycke, R.; Brackenier, A.; Inze, D.; Harris, J.L.; Van Breusegem, F. Serpin1 of *Arabidopsis thaliana* is a suicide inhibitor for metacaspase 9. *J. Mol. Biol.* **2006**, *364*, 625–636. [CrossRef] [PubMed]

105. Sundstrom, J.F.; Vaculova, A.; Smertenko, A.P.; Savenkov, E.I.; Golovko, A.; Minina, E.; Tiwari, B.S.; Rodriguez-Nieto, S.; Zamyatnin, A.A., Jr.; Valineva, T.; et al. Tudor staphylococcal nuclease is an evolutionarily conserved component of the programmed cell death degradome. *Nat. Cell Biol.* **2009**, *11*, 1347–1354. [CrossRef] [PubMed]

106. Jorda, L.; Vera, P. Local and systemic induction of two defense-related subtilisin-like protease promoters in transgenic Arabidopsis plants. Luciferin induction of PR gene expression. *Plant Physiol.* **2000**, *124*, 1049–1058. [CrossRef] [PubMed]

107. Reavy, B.; Bagirova, S.; Chichkova, N.V.; Fedoseeva, S.V.; Kim, S.H.; Vartapetian, A.B.; Taliansky, M.E. Caspase-resistant VirD2 protein provides enhanced gene delivery and expression in plants. *Plant Cell Rep.* **2007**, *26*, 1215–1219. [CrossRef] [PubMed]

108. Chichkova, N.V.; Kim, S.H.; Titova, E.S.; Kalkum, M.; Morozov, V.S.; Rubtsov, Y.P.; Kalinina, N.O.; Taliansky, M.E.; Vartapetian, A.B. A plant caspase-like protease activated during the hypersensitive response. *Plant Cell* **2004**, *16*, 157–171. [CrossRef] [PubMed]

109. Chichkova, N.V.; Tuzhikov, A.I.; Taliansky, M.; Vartapetian, A.B. Plant phytaspases and animal caspases: Structurally unrelated death proteases with a common role and specificity. *Physiol. Plant.* **2012**, *145*, 77–84. [CrossRef] [PubMed]

110. Fernandez, M.B.; Daleo, G.R.; Guevara, M.G. DEVDase activity is induced in potato leaves during Phytophthora infestans infection. *Plant Physiol. Biochem.* **2012**, *61*, 197–203. [CrossRef] [PubMed]

111. Richau, K.H.; Kaschani, F.; Verdoes, M.; Pansuriya, T.C.; Niessen, S.; Stuber, K.; Colby, T.; Overkleeft, H.S.; Bogyo, M.; Van der Hoorn, R.A. Subclassification and biochemical analysis of plant papain-like cysteine proteases displays subfamily-specific characteristics. *Plant Physiol.* **2012**, *158*, 1583–1599. [CrossRef] [PubMed]

112. Savvateeva, L.V.; Gorokhovets, N.V.; Makarov, V.A.; Serebryakova, M.V.; Solovyev, A.G.; Morozov, S.Y.; Reddy, V.P.; Zernii, E.Y.; Zamyatnin, A.A., Jr.; Aliev, G. Glutenase and collagenase activities of wheat cysteine protease Triticain-alpha: Feasibility for enzymatic therapy assays. *Int. J. Biochem. Cell. Biol.* **2015**, *62*, 115–124. [CrossRef] [PubMed]

113. Niemer, M.; Mehofer, U.; Verdianz, M.; Porodko, A.; Schahs, P.; Kracher, D.; Lenarcic, B.; Novinec, M.; Mach, L. *Nicotiana benthamiana* cathepsin B displays distinct enzymatic features which differ from its human relative and aleurain-like protease. *Biochimie* **2016**, *122*, 119–125. [CrossRef] [PubMed]

114. Paireder, M.; Mehofer, U.; Tholen, S.; Porodko, A.; Schahs, P.; Maresch, D.; Biniossek, M.L.; van der Hoorn, R.A.; Lenarcic, B.; Novinec, M.; et al. The death enzyme CP14 is a unique papain-like cysteine proteinase with a pronounced S2 subsite selectivity. *Arch. Biochem. Biophys.* **2016**, *603*, 110–117. [CrossRef] [PubMed]

115. Paireder, M.; Tholen, S.; Porodko, A.; Biniossek, M.L.; Mayer, B.; Novinec, M.; Schilling, O.; Mach, L. The papain-like cysteine proteinases NbCysP6 and NbCysP7 are highly processive enzymes with substrate specificities complementary to *Nicotiana benthamiana* cathepsin B. *Biochim. Biophys. Acta* **2017**, *1865*, 444–452. [CrossRef] [PubMed]

116. Have, M.; Balliau, T.; Cottyn-Boitte, B.; Derond, E.; Cueff, G.; Soulay, F.; Lornac, A.; Reichman, P.; Dissmeyer, N.; Avice, J.C.; et al. Increase of proteasome and papain-like cysteine protease activities in autophagy mutants: Backup compensatory effect or pro cell-death effect? *J. Exp. Bot.* **2017**. [CrossRef]

117. Kaschani, F.; Shabab, M.; Bozkurt, T.; Shindo, T.; Schornack, S.; Gu, C.; Ilyas, M.; Win, J.; Kamoun, S.; van der Hoorn, R.A. An effector-targeted protease contributes to defense against Phytophthora infestans and is under diversifying selection in natural hosts. *Plant Physiol.* **2010**, *154*, 1794–1804. [CrossRef] [PubMed]

118. Mueller, A.N.; Ziemann, S.; Treitschke, S.; Assmann, D.; Doehlemann, G. Compatibility in the Ustilago maydis-maize interaction requires inhibition of host cysteine proteases by the fungal effector Pit2. *PLoS Pathog.* **2013**, *9*, e1003177. [CrossRef] [PubMed]

119. Repnik, U.; Cesen, M.H.; Turk, B. The endolysosomal system in cell death and survival. *Cold Spring Harb Perspect. Biol.* **2013**, *5*, a008755. [CrossRef] [PubMed]

120. Stoka, V.; Turk, V.; Turk, B. Lysosomal cysteine cathepsins: Signaling pathways in apoptosis. *Biol. Chem.* **2007**, *388*, 555–560. [CrossRef] [PubMed]

121. Turk, B.; Stoka, V.; Rozman-Pungercar, J.; Cirman, T.; Droga-Mazovec, G.; Oresic, K.; Turk, V. Apoptotic pathways: Involvement of lysosomal proteases. *Biol. Chem.* **2002**, *383*, 1035–1044. [CrossRef] [PubMed]

122. Turk, V.; Stoka, V.; Vasiljeva, O.; Renko, M.; Sun, T.; Turk, B.; Turk, D. Cysteine cathepsins: From structure, function and regulation to new frontiers. *Biochim. Biophys. Acta* **2012**, *1824*, 68–88. [CrossRef] [PubMed]

123. Yamada, K.; Matsushima, R.; Nishimura, M.; Hara-Nishimura, I. A slow maturation of a cysteine protease with a granulin domain in the vacuoles of senescing Arabidopsis leaves. *Plant Physiol.* **2001**, *127*, 1626–1634. [CrossRef] [PubMed]

124. Rustgi, S.; Boex-Fontvieille, E.; Reinbothe, C.; von Wettstein, D.; Reinbothe, S. Serpin1 and WSCP differentially regulate the activity of the cysteine protease RD21 during plant development in *Arabidopsis thaliana*. *Proc. Natl. Acad. Sci. USA* **2017**, *114*, 2212–2217. [CrossRef] [PubMed]

125. Helm, M.; Schmid, M.; Hierl, G.; Terneus, K.; Tan, L.; Lottspeich, F.; Kieliszewski, M.J.; Gietl, C. KDEL-tailed cysteine endopeptidases involved in programmed cell death, intercalation of new cells, and dismantling of extensin scaffolds. *Am. J. Bot.* **2008**, *95*, 1049–1062. [CrossRef] [PubMed]

126. Eulgem, T.; Rushton, P.J.; Robatzek, S.; Somssich, I.E. The WRKY superfamily of plant transcription factors. *Trends Plant. Sci.* **2000**, *5*, 199–206. [CrossRef]

127. Nomura, H.; Komori, T.; Uemura, S.; Kanda, Y.; Shimotani, K.; Nakai, K.; Furuichi, T.; Takebayashi, K.; Sugimoto, T.; Sano, S.; et al. Chloroplast-mediated activation of plant immune signalling in Arabidopsis. *Nat. Commun.* **2012**, *3*, 926. [CrossRef] [PubMed]

128. Eichmann, R.; Schafer, P. The endoplasmic reticulum in plant immunity and cell death. *Front. Plant Sci.* **2012**, *3*, 200. [CrossRef] [PubMed]

129. Dempsey, D.A.; Vlot, A.C.; Wildermuth, M.C.; Klessig, D.F. Salicylic Acid biosynthesis and metabolism. *Arabidopsis Book* **2011**, *9*, e0156. [CrossRef] [PubMed]

130. Durrant, W.E.; Dong, X. Systemic acquired resistance. *Annu. Rev. Phytopathol.* **2004**, *42*, 185–209. [CrossRef] [PubMed]

131. Kiefer, I.W.; Slusarenko, A.J. The pattern of systemic acquired resistance induction within the Arabidopsis rosette in relation to the pattern of translocation. *Plant Physiol.* **2003**, *132*, 840–847. [CrossRef] [PubMed]

132. Lebel, E.; Heifetz, P.; Thorne, L.; Uknes, S.; Ryals, J.; Ward, E. Functional analysis of regulatory sequences controlling PR-1 gene expression in Arabidopsis. *Plant J.* **1998**, *16*, 223–233. [CrossRef] [PubMed]

133. Bosch, M.; Wright, L.P.; Gershenzon, J.; Wasternack, C.; Hause, B.; Schaller, A.; Stintzi, A. Jasmonic acid and its precursor 12-oxophytodienoic acid control different aspects of constitutive and induced herbivore defenses in tomato. *Plant Physiol.* **2014**, *166*, 396–410. [CrossRef] [PubMed]

*Int. J. Mol. Sci.* **2018**, *19*, 629

134. Huang, P.Y.; Catinot, J.; Zimmerli, L. Ethylene response factors in Arabidopsis immunity. *J. Exp. Bot.* **2016**, *67*, 1231–1241. [CrossRef] [PubMed]

135. Hind, S.R.; Pulliam, S.E.; Veronese, P.; Shantharaj, D.; Nazir, A.; Jacobs, N.S.; Stratmann, J.W. The COP9 signalosome controls jasmonic acid synthesis and plant responses to herbivory and pathogens. *Plant J.* **2011**, *65*, 480–491. [CrossRef] [PubMed]

136. Christians, M.J.; Larsen, P.B. Mutational loss of the prohibitin AtPHB3 results in an extreme constitutive ethylene response phenotype coupled with partial loss of ethylene-inducible gene expression in Arabidopsis seedlings. *J. Exp. Bot.* **2007**, *58*, 2237–2248. [CrossRef] [PubMed]

137. Jashni, M.K.; Mehrabi, R.; Collemare, J.; Mesarich, C.H.; de Wit, P.J. The battle in the apoplast: Further insights into the roles of proteases and their inhibitors in plant-pathogen interactions. *Front. Plant Sci.* **2015**, *6*, 584. [CrossRef] [PubMed]

138. Van der Linde, K.; Hemetsberger, C.; Kastner, C.; Kaschani, F.; van der Hoorn, R.A.; Kumlehn, J.; Doehlemann, G. A maize cystatin suppresses host immunity by inhibiting apoplastic cysteine proteases. *Plant Cell* **2012**, *24*, 1285–1300. [CrossRef] [PubMed]

139. Ben-Nun, Y.; Fichman, G.; Adler-Abramovich, L.; Turk, B.; Gazit, E.; Blum, G. Cathepsin nanofiber substrates as potential agents for targeted drug delivery. *J. Control. Release* **2017**, *257*, 60–67. [CrossRef] [PubMed]

140. Shekhter, A.B.; Balakireva, A.V.; Kuznetsova, N.V.; Vukolova, M.N.; Litvitsky, P.F.; Zamyatnin, A.A. Collagenolytic enzymes and their applications in biomedicine. *Curr. Med. Chem.* **2018**. [CrossRef] [PubMed]

*International Journal of*
**Molecular Sciences**

MDPI

*Article*

# Quantitative Proteomic Analysis Provides Insights into Rice Defense Mechanisms against *Magnaporthe oryzae*

Siyuan Lin [1,2,†], Pingping Nie [1,2,3,†], Shaochen Ding [1,2], Liyu Zheng [1,2], Chen Chen [1,2], Ruiying Feng [1,2], Zhaoyun Wang [1,2], Lin Wang [1,2], Jianan Wang [1,2], Ziwei Fang [1,2], Shaoxia Zhou [1,2], Hongyu Ma [1,2,*] and Hongwei Zhao [1,2,*]

[1] College of Plant Protection, Nanjing Agricultural University, Nanjing 210095, China; 2016102048@njau.edu.cn (S.L.); sdniepingping@163.com (P.N.); 2017102050@njau.edu.cn (S.D.); 2017102052@njau.edu.cn (L.Z.); 2016102047@njau.edu.cn (C.C.) 2016802185@njau.edu.cn (R.F.); 2014202018@njau.edu.cn (Z.W.); 2015202020@njau.edu.cn (L.W.); 12115104@njau.edu.cn (J.W.); 12115106@njau.edu.cn (Z.F.); sxzhou@njau.edu.cn (S.Z.)
[2] Key Laboratory of Integrated Management of Crop Diseases and Pests, Nanjing Agricultural University, Ministry of Education, Nanjing 210095, China
[3] College of Life Sciences, Zaozhuang University, Zaozhuang 277160, China
* Correspondence: mahongyu@njau.edu.cn (H.M.); hzhao@njau.edu.cn (H.Z.); Tel.: +86-25-843-99552 (H.Z.)
† These authors contributed equally to this work.

Received: 9 May 2018; Accepted: 28 June 2018; Published: 3 July 2018

**Abstract:** Blast disease is one of the major rice diseases, and causes nearly 30% annual yield loss worldwide. Resistance genes that have been cloned, however, are effective only against specific strains. In cultivation practice, broad-spectrum resistance to various strains is highly valuable, and requires researchers to investigate the basal defense responses that are effective for diverse types of pathogens. In this study, we took a quantitative proteomic approach and identified 634 rice proteins responsive to infections by both *Magnaporthe oryzae* strains Guy11 and JS153. These two strains have distinct pathogenesis mechanisms. Therefore, the common responding proteins represent conserved basal defense to a broad spectrum of blast pathogens. Gene ontology analysis indicates that the "responding to stimulus" biological process is explicitly enriched, among which the proteins responding to oxidative stress and biotic stress are the most prominent. These analyses led to the discoveries of OsPRX59 and OsPRX62 that are robust callose inducers, and OsHSP81 that is capable of inducing both ROS production and callose deposition. The identified rice proteins and biological processes may represent a conserved rice innate immune machinery that is of great value for breeding broad-spectrum resistant rice in the future.

**Keywords:** innate immunity; basal defense; rice blast; *Magnaporthe oryzae*; proteomics; iTRAQ

## 1. Introduction

Rice is the staple food that feeds about one third of the world population. Rice blast disease is one of the major diseases threatening rice production, which is estimated to cause about 30% annual yield loss [1]. Due to its enormous economic importance, studying the interaction between rice and the causal agent of the rice blast disease, *Magnaporthe oryzae*, is of great scientific and economic significance. Also, as more and more research has been focused on the mechanism governing this mutual interaction, the rice-*M. oryzae* system has become a model system of cereal plants and their fungal pathogens [2]. The recently discovered *M. oryzae* colonization of wheat [3] strengthened its significance.

Plants are capable of defending themselves against various pathogens. The plant innate immune system is composed of multiple components located at both the plasma membrane and inside the cells. The receptors on the membrane sense the pathogen-associated molecular patterns (PAMP) and initiate the PAMP-triggered immunity (PTI). Due to the extreme conserved nature of PAMPs, PTI is constitutively effective against broad-spectrum pathogens, which is characterized by the rapid launch of reactive oxygen species (ROS) and callose deposition around the infection loci [4]. Some pathogens win the combat over PTI, most of which through secreting effectors that specifically interrupt PTI. Plants have evolved machinery that recognizes effectors and activates effector-triggered immunity (ETI). ETI is usually associated with massive gene expression reprogramming, including activation of defense-related genes, alteration of cellular redox status, and activation of phytohormone signaling pathways such as salicylic acid (SA) and jasmonic acid (JA) [5].

To breed blast-resistant rice lines that are effective for various strains in the field, we need understand the common mechanism that rice employs to sense the diverse type of blast pathogens, to identify the signals that are passed downstream, and find out defense responses that are activated and are efficient to restrain the progression of the disease. Previous effort discovered critical immune modules that are important for rice defense responses against the blast disease, such as the enhanced ability of ROS production and callose deposition [4], activation of the mitogen-activated protein kinase (MAPK) signaling cascade [6], and preferential employment of the SA or JA signaling pathways [7]. However, most of our knowledge is from genetic examinations, which lack interpretation from a proteomic view. Moreover, essential genes playing primary defensive roles to a specific strain are over-emphasized, with molecules and biological processes responding to broad-spectrum resistance overlooked.

Proteome is the entire set of proteins expressed by a genome, cell, tissue, or organism at a certain time. By comparing the proteomic profiles between mock- and *M. oryzae*-treated rice seedlings or suspension cultured cells, critical rice immune components against blast disease were identified [8,9]. This was particularly facilitated by the tandem utilization of high-performance liquid chromatography (HPLC) and mass spectrometry (MS) over the past two decades. By virtue of the superb fractionation capability of HPLC and excellent sensitivity of MS, high throughput proteomic profiling has emerged as a powerful tool to investigate the protein machineries involved in blast disease resistance at trace amount-level [10,11].

In this study, we employed an isobaric tag for relative and absolute quantitation (iTRAQ) technique that can compare protein expression levels between different rice samples. We focused on rice proteins responding to both the virulent (Guy11) and the avirulent (JS153) *M. oryzae* strains, but not just to any one of them. We aimed to identify the conserved basal defensive components of broad-spectrum blast disease pathogens. By applying the iTRAQ method, we found that both Guy11 and JS153 typically induce proteins involved in biological processes such as "responses to oxidative stress" (such as OsAPX1, OsPRX59, and OsPRX62) and "response to biotic stress" (such as OsHSP81, OsPBZ1, and OsPR10). We further proved that OsPRX59 and OsPRX62 are robust cell wall synthetic enhancers while OsHSP81, OsPBZ1, and OsPR10 participate in both ROS accumulation and callose deposition. Specific expression variation of several SA signaling components was also observed, suggesting its involvement in responding to either Guy11 or JS153 infection. Our discovery identified the critical innate immune machinery that will facilitate breeding of rice with broad-spectrum blast resistance.

## 2. Results

To identify rice proteins that are potentially involved in defense against rice blast disease, we employed the iTRAQ peptide labeling approach and liquid chromatography–tandem mass spectrometry (LC–MS/MS) that can determine the amount of proteins from different sources in a single experiment [12,13]. Total rice (Nipponbare; three-leaf-stage) proteins from both *M. oryzae*-infected (24 and 72 h post inoculation; hpi) and healthy rice (0 hpi) were examined. Both a relative virulent (Guy11) and an avirulent strain (JS153) were used to explore the interaction between rice and the blast pathogens. Guy11 is a relative virulent strain that causes moderate disease symptoms on Nipponbare, which is weaker than on Kongyu 131 but stronger than on Lijiangxintuanheigu (LTH) [14,15]. JS153 is an avirulent strain that shows no disease symptoms on Nipponbare. Therefore, the similarities between Guy11 and JS153 responses represent rice basal defense against blast disease, while the differences represent defense responses initiated by the effector-triggered immunity (ETI) [4]. We examined two independent biological replicates, from which 3109 and 2990 rice proteins were identified, respectively (Table S1). Between these two repeats, 1618 proteins were found in both assays (Table S2), representing reproducible proteins from our quantitative proteomic measurements. Differentially expressed (DE) rice proteins were identified by comparing the protein expression profiles between the *M. oryzae*-treated and the healthy rice (24/0 and 72/0 hpi, respectively). We only selected for further study the proteins that were highly confident ($p < 0.5$), had more than 10% peptide coverage from mass spectrometry, and were either more than 1.25-fold or less than 0.8-fold in the *M. oryzae*-treated samples than in the healthy samples [16]. According to these criteria, we obtained 634 DE-proteins upon *M. oryzae* infection (Table S3).

We analyzed the 634 DE-proteins according to their expression preferences in different treatments. We found that 390 proteins were explicitly differentially expressed after Guy11 infection, among which 30 proteins expressed only at 24 hpi, and 170 proteins just expressed at 72 hpi, while 194 proteins expressed at both 24 and 72 hpi. After JS153 infection, 561 were specifically differentially expressed (Figure 1A), among which 81 proteins only expressed at 24 hpi, 279 only expressed at 72 hpi, while 201 proteins expressed at both 24 and 72 hpi. These proteins represent distinct rice responses against both the virulent and the avirulent blast strains. Importantly, 317 DE-proteins from both Guy11 and JS153-infected rice, but not from just one of the treatments, were identified, accounting for 50.1% of the total DE-proteins.

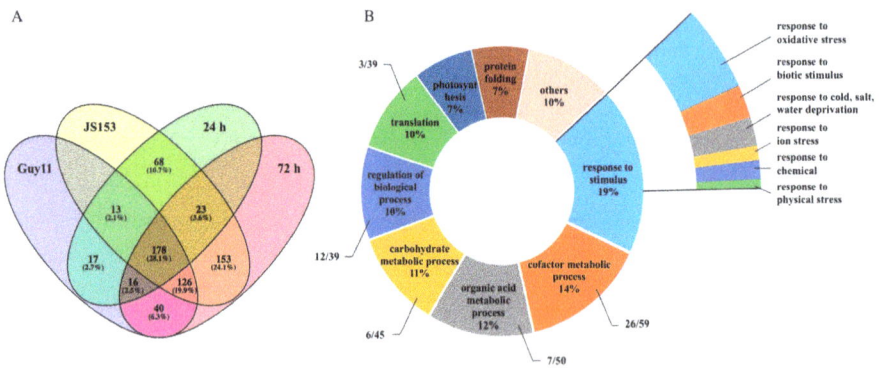

**Figure 1.** Gene ontology (GO) analysis on differentially expressed (DE)-proteins and their distribution map. (**A**) Venn diagram shows the expression preference of the 634 DE-proteins; (**B**) Gene ontology analysis of the 634 DE-proteins. The pie chart indicates the biological process of 634 DE-proteins. The fan-shaped chart shows the subterms of the 40 DE-proteins responding to stimulus.

To find what biological processes are employed for defense to blast disease by rice, we performed a gene ontology (GO) analysis of these 634 DE-proteins. Six biological processes with more than 10% enrichment were identified. These enriched biological processes were "response to stimulus" (19%), "cofactor metabolic process" (14%), "organic acid metabolic" (12%), "carbohydrate metabolic" (11%), "regulation of biological process" (10%), and "translation" (10%). "Response to stimulus" is composed of proteins related to major biological responses to stimuli such as biotic, abiotic, chemical, and wounding stresses. The specific enrichment of proteins responding to stimulus indicates that after the *M. oryzae* infection, rice makes a considerable adjustment by shifting many biological processes toward palliating and preventing cellular damage by pathogen infection (Figure 1B). "Cofactor metabolic process" refers to chemical reactions and pathways involving a cofactor that is required for the activity of an enzyme or other protein. Our results indicate that 26 out of the 59 cofactors are associated with response to stimulus (Table S4). "Organic acid metabolic" includes chemical reactions and pathways involving acidic compounds containing carbon in covalent linkage. Seven out of the 50 proteins in this category are associated with response to stimulus. We further found that six out of the 45 "carbohydrate metabolic" proteins, 12 out of the 39 proteins belong to the "regulation of biological process", and three out of the 39 proteins belonging to "translation" are related to "response to stimulus".

When the DE-proteins responding both to Guy11 and JS153 infections were compared to those just responding to either Guy11 or JS153, we found defense-related biological processes were highly enriched (Figure 2). For example, when biological processes (BP) were aligned according to their preferentially differential expression by both Guy11 and JS153 infection, many defense-related biological processes were more enriched, including "regulation of protein serine/threonine phosphatase activity", "response to oxidative stress", "protein folding", and "glutathione metabolic process". In contrast, other biological processes, such as "mRNA splicing", "cellular amino acid biosynthesis", "photosynthesis", "response to light stimulus", and "cytoplasm translation" are preferentially enriched in responses just to Guy11 or JS153 infection (Figure 2). The preferential enrichment of defense-related biological processes both in Guy11- and JS153-infected rice indicates that a consensus set of defense responses are allocated to defend both virulent and avirulent blast disease pathogens, whereby we may identify the innate immune components essential for a broad spectrum of blast pathogens.

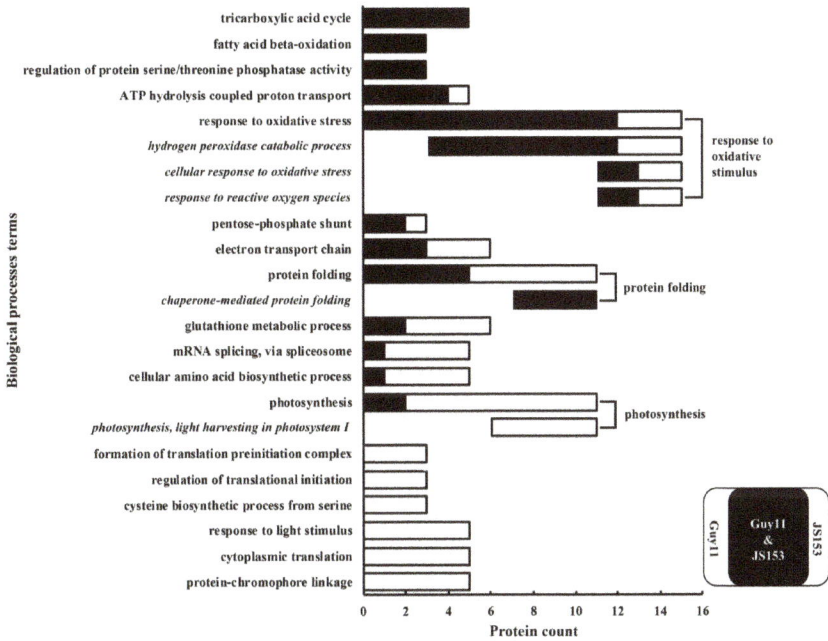

**Figure 2.** Biological processes (BP) analysis on DE-proteins and their expression preference. The 634 DE-proteins were analyzed by using the DAVID Bioinformatics Resources 6.8. Biological processes were sorted according to their enrichment in both Guy11 and JS153 infections. The terms in italic indicate sub-terms that belong to the major terms above.

Based on the fact that defense responses, both to virulent and avirulent infection represent a conserved defense machinery against broad-spectrum blast pathogens, and considering the preferential enrichment of rice proteins related to the "response to stimulus" (Figure 1B), we further narrowed down the DE-proteins that [1] respond to both the virulent and avirulent pathogen infections at either 24 or 72 hpi, and [2] belong to the "response to stimulus" GO category. The 40 proteins satisfying these two criteria (Table 1) were regarded as basal defense-related proteins and were pursued for further study. Among the 40 basal defense-related proteins, 18 proteins have functions related to "responses to oxidative stress", seven proteins are related to "responses to biotic stress", and four proteins are related to "responses to cold", respectively. Proteins involved in "responses to salt stress", "responses to ion stress", "responses to chemical stress", and "responses to water stress" were also identified (Table 1 and Figure 1B). Interestingly, more than half of the defense-related proteins (25/40) are related to two principal GO categories: "responses to oxidative stress" (18/40) and "responses to biotic stress" (7/40). These results indicate that after *M. oryzae* infection, these two biological processes are preferentially activated. Therefore, we concentrated our study on proteins from these two biological processes.

We selected three proteins from each category for further validation. From the "responses to oxidative stress" group, we selected OsAPX1 (B7E6Z4), OsPRX62 (Q7XSU7), and OsPRX59 (Q9ST80). From the "responses to biotic stress" group, we selected OsPBZ1 (Q40707), OsPR10 (Q75T45), and OsHSP81 (I1QJW3). Both Guy11 and JS153 infection induced the expression of OsAPX1 at 72 hpi but only by Guy11 infection at 24 hpi. qRT-PCR validation indicated that the transcription level of OsAPX1 was elevated in both Guy11- and JS153-infected rice at 24 and 72 hpi (Figure 3). Similarly, the protein levels of OsPRX59 and OsPRX62 were all induced both by Guy and JS153 infections at 72 hpi but not in JS153-infected rice at 24 hpi. Their transcriptional expression showed durative elevation at both 24 and 72 hpi, no matter whether infected by Guy11 or by JS153 (Figure 3). Our results indicate that the expression of these three proteins involved in responses to oxidative stress is induced by both virulent and avirulent *M. oryzae* strains. OsHSP81, OsPBZ1, and OsPR10 were induced by both Guy and JS153 at 72 hpi. However, the qRT-PCR results agreed with both OsHSP81 and OSPR1 but not OsPBZ1 by JS153 at 72 hpi. In general, our results showed a good coalition between transcription and translation on these proteins (Figure 3).

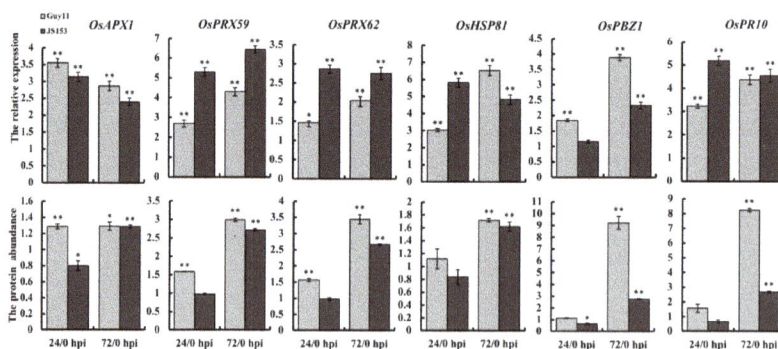

**Figure 3.** Expression coalition between the transcription and the translation levels. Top: expression of the six selected genes examined by qRT-PCR at the indicated time points by Guy11 infection or JS153 infection. Values represent the means ± SD of three independent samples (* $p < 0.05$, ** $p < 0.01$). Similar results were obtained from three biological repeats. Bottom: protein abundance of the six selected genes. Error bars indicated SD. Asterisks indicate significant differences (* $p < 0.05$, ** $p < 0.01$).

We transiently expressed these proteins in *Nicotiana benthamiana* leaves and examined their ability to induce ROS production. OsAPX1 did not induce discernable ROS production, as manifested by comparable ROS signals between OsAPX1 and the empty vector (EV), as well as by the dramatic difference when compared to PcINF1, a *Phytophthora* effector with known ROS-eliciting function (Figure 4A). In contrast, both OsPRX59 and OsPRX62 induced significant ROS production, which was almost five to eight times more than that in the EV inoculation. OsHSP1 induced around five times more ROS than the EV did. Significant ROS production was also observed in *N. Benthamiana* leaves transiently expressing OsPR10 (4 folds) and OsPBZ1 (about 3-fold).

*Int. J. Mol. Sci.* **2018**, *19*, 1950

**Table 1.** Significant rice differentially expressed (DE)-proteins responding to *M. oryzae*.

| Protein ID | Name | Annotation | Guy11 | | JS153 | |
|---|---|---|---|---|---|---|
| | | | 24/0 hpi | 72/0 hpi | 24/0 hpi | 72/0 hpi |
| | | response to oxidative stress | | | | |
| Q652L6 | MDAR3 | cellular oxidant detoxification | 1.20 | 1.33 | 0.85 | 1.35 |
| B7E6Z4 | APX1 | hydrogen peroxide catabolic process | 1.29 | 1.29 | 0.80 | 1.29 |
| I1Q8M5 | TRXh1 | oxidoreductase activity | 1.11 | 1.35 | 0.95 | 1.47 |
| A0A0E0Q2V5 | CATB | hydrogen peroxide catabolic process | 1.17 | 1.81 | 0.88 | 1.86 |
| Q7XSU7 | PRX62 | hydrogen peroxide catabolic process | 1.57 | 3.44 | 0.98 | 2.67 |
| Q9ST80 | PRX59 | hydrogen peroxide catabolic process | 1.59 | 2.99 | 0.98 | 2.72 |
| Q5U1T0 | PRX13 | hydrogen peroxide catabolic process | 1.55 | 2.70 | 0.91 | 3.05 |
| Q5Z4D3 | PRX78 | hydrogen peroxide catabolic process | 1.44 | 1.71 | 0.75 | 1.72 |
| Q5Z7J2 | PRX86 | hydrogen peroxide catabolic process | 1.71 | 2.04 | 1.14 | 1.73 |
| Q654S1 | PRX12 | hydrogen peroxide catabolic process | 1.42 | 1.54 | 0.96 | 1.82 |
| Q6AVZ8 | PRX65 | hydrogen peroxide catabolic process | 1.26 | 1.41 | 0.78 | 1.35 |
| Q6ESJ0 | GPX3 | glutathione peroxidase activity | 1.15 | 1.36 | 0.93 | 1.58 |
| Q6EUS1 | PRX27 | hydrogen peroxide catabolic process | 1.35 | 1.56 | 0.84 | 1.79 |
| Q624E4 | ALDH6B2 | methylmalonate-semialdehyde dehydrogenase activity | 1.27 | 1.66 | 0.87 | 1.82 |
| Q7XHB3 | PRX125 | hydrogen peroxide catabolic process | 1.39 | 1.77 | 1.02 | 1.75 |
| Q9FEV2 | riPHGPX | glutathione peroxidase activity | 1.36 | 1.75 | 0.8 | 1.54 |
| Q8W317 | NADH dehydrogenase | NADH dehydrogenase activity | 1.25 | 1.37 | 0.79 | 1.53 |
| A0A0E0PU51 | alkaline α-galactosidase | catalytic activity | 1.54 | 1.62 | 1.02 | 1.83 |
| | | response to biotic stimulus | | | | |
| Q7XPU1 | Harpin-induced 1 domain containing protein | signal transducer activity | 1.34 | 1.50 | 0.93 | 1.47 |
| Q4O7O7 | PBZ1 | response to biotic stimulus, defense response | 1.14 | 9.22 | 0.63 | 2.77 |
| I1QJW3 | HSP81 | response to stress, ATP binding | 1.12 | 1.72 | 0.84 | 1.62 |
| Q75T45 | RSOsPR10 | pathogenesis-related protein | 1.58 | 8.25 | 0.66 | 2.68 |
| Q75L45 | OsRLCK178 | cell surface receptor signaling pathway | 1.49 | 1.55 | 1.02 | 1.66 |
| A0A0E0PMK8 | OsGDI1 | protein transport | 1.17 | 1.6 | 0.77 | 1.37 |
| Q945E9 | JIOsPR10 | response to biotic stimulus | 1.92 | 2.55 | 1.03 | 2.19 |

Table 1. *Cont.*

| Protein ID | Name | Annotation | Guy11 | | JS153 | |
|---|---|---|---|---|---|---|
| | | | 24/0 hpi | 72/0 hpi | 24/0 hpi | 72/0 hpi |
| | | **response to cold, salt stress, water deprivation** | | | | |
| Q8LHG8 | Os01g0542000 | isomerase activity | 1.35 | 1.53 | 0.87 | 1.56 |
| I1QGF2 | YchF1 | hydrolyzes ATP | 1.22 | 1.31 | 0.85 | 1.37 |
| Q7XXS0 | RMtATPd2 | mitochondrial membrane ATP synthase | 1.29 | 1.46 | 0.92 | 1.41 |
| Q7XUC9 | Histone H4 | transcription regulation, DNA repair, DNA replication | 0.96 | 1.44 | 0.91 | 1.32 |
| I1PYW0 | Os6PGDH1 | phosphogluconate dehydrogenase activity | 1.36 | 1.70 | 0.89 | 1.59 |
| I1PUR5 | UspA | response to stress | 3.26 | 1.18 | 2.17 | 1.38 |
| | | **response to ion stress** | | | | |
| Q5JK10 | Os01g0926300 | response to cadmium ion | 1.15 | 1.59 | 0.98 | 1.52 |
| Q6H734 | Os02g0198600 | ubiquitin binding | 1.35 | 1.53 | 0.87 | 1.50 |
| S4TZU3 | Os02g0621700 | magnesium ion binding | 1.23 | 1.49 | 0.79 | 1.31 |
| | | **response to chemical** | | | | |
| Q5W676 | HXK5 | fructose and glucose phosphorylating enzyme | 1.06 | 1.27 | 0.79 | 1.43 |
| Q2QYK6 | chalcone isomerase | chalcone isomerase activity | 1.15 | 1.4 | 0.94 | 1.59 |
| Q852M0 | GDH1 | glutamate dehydrogenase activity | 1.37 | 1.49 | 0.83 | 1.45 |
| Q8S718 | OsGSTU23 | glutathione transferase activity | 1.28 | 1.73 | 0.78 | 1.68 |
| | | **response to physical stress** | | | | |
| Q10LV7 | LOC_Os03g21560 | cellular response to light intensity | 1.20 | 1.36 | 0.81 | 1.44 |
| Q7XRB6 | Os04g0435700 | response to UV-B, photoreceptor activity | 1.39 | 1.46 | 0.75 | 1.46 |

Callose deposition was also monitored in *N. Benthamiana* leaves when these proteins were transiently expressed. Proteins associated with "responses to oxidative stress" (OsAPX1, OsPRX59, and OsPRX62) induced dramatic callose deposition in all the samples examined, which is comparable to PsNLP, a *Phytophthora* effector eliciting significant callose deposition. The proteins from the "response to biotic stress" category also induced a significant amount of callose deposition when they were transiently expressed in *N. Benthamiana* leaves, but to a lesser degree than OsAPX1, OsPRX59, and OsPRX62 (Figure 4B).

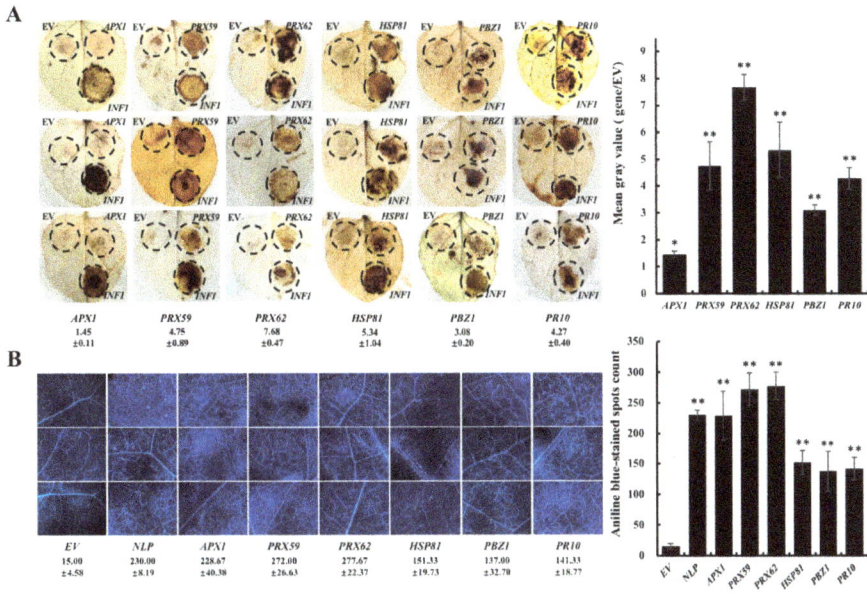

**Figure 4.** The DEP candidates contribute to callose deposition and ROS accumulation. (**A**) Left: 3,3′-diaminobenzidine (DAB) staining for reactive oxygen species (ROS) accumulation in *N. benthamiana* leaves after infiltration with *A. tumefaciens* carrying empty vector, *PcINF1*, and the six selected genes. The reddish-brown at the injection site shows the accumulation of ROS. Numbers are the relative accumulation and standard deviations of ROS using Image J. Right: Histogram represents the relative accumulation of ROS in the images. Error bars indicate SD from three technical replicates. Asterisks indicate significant differences (* $p < 0.05$, ** $p < 0.01$); (**B**) Left: Aniline blue staining for callose deposition in the leaves (40× magnification) expressing EV (empty vector), *PsNLP* and six selected genes. Numbers are the means and standard deviations of three 1 cm$^2$ microscopic fields of view. Right: Histogram represents the means of three 1 cm$^2$ microscopic fields of view. Error bars indicate SD. Asterisks indicate significant differences (** $p < 0.01$).

Guy11 and JS153 are *M. oryzae* strains with distinct pathogenesis mechanisms and pathogenicity. To demonstrate these two strains can be distinguished by rice and can elicit distinct phytohormone responses, we examined several critical phytohormones along the time course of infection. SA, JA, and ET (ethylene) are critical phytohormones associated with rice innate immunity against blast disease [14,17]. We evaluated the responses of these critical phytohormones after both Guy11 and JS153 infection by examining the expression of selected key components [14]. *OsPAD4* and *OsEDS1* are SA synthetic regulators. *OsPAD4* was induced by Guy11 but not JS153 at 24, 72, and 96 hpi, while *OsEDS1* was suppressed by both pathogens at most of the time points except 24 hpi, which was slightly induced (Figure S1). *OsSID2* is an SA synthetic component, which was induced by both Guy11 and JS153 at 72 and 96 hpi. *OsPAL1* is associated with both SA and lignin synthesis, which was induced only by Guy11 but not JS153. *OsPR1* showed similar induction profile as *OsPAL1* but to a much higher degree (Figure S1). We also examined key components involved in JA synthesis and signaling. *OsJAZ1* was slightly induced by JS153 at 72 and 96 hpi but not by Guy11. *OsMYC2* was suppressed by both Guy11 and JS153 at 72 and 96 hpi while it was only temporarily induced by Guy11 at 24 hpi. These results indicate that the SA and JA signaling is perturbed by *M. Oryzae* infection. Aminocyclopropane carboxylic acid synthase (ACS) is the committed and rate-limiting enzyme in the biosynthesis of ethylene. *OsACS2* was suppressed by both Guy11 and JS153 infection. *EIN3* (*ethylene-insensitive 3*) encodes a nuclear transcription factor that initiates downstream transcriptional cascades for ethylene responses. The expression of *OsEIN3* was suppressed by JS153 and by Guy11 except at 72 hpi. Ethylene Response Factors (ERFs) encode transcription factors that regulate the molecular response to pathogen attacks [18]. We examined the expression of *OsERF1/2* and found that the expression level was marginally altered (Figure S1). Taken together, the SA signaling pathway is induced by Guy11 infection while remains unaffected or even slighted suppressed by the JS153 infection. In contrast, the JA and ET signaling pathways are unaltered or suppressed by either Guy11 or JS153 infection (Figure 5).

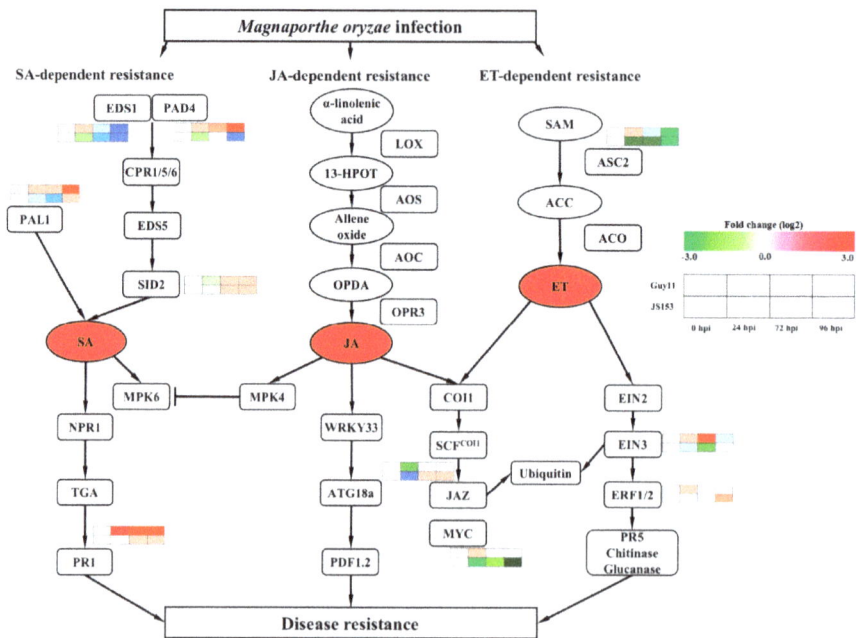

**Figure 5.** The defense signaling pathway in response to *M. oryzae* in rice. Critical components involved in the SA, JA, and ET signaling pathways are analyzed by real-time PCR. Rectangles indicate genes or proteins; ovals indicate chemical compounds, red ovals indicate phytohormones. The chart with different colors indicates the expression of indicated genes at the indicated time points by Guy11 infection or JS153 infection.

## 3. Discussion

Quantitative proteomics greatly prompts biological studies by not only identifying trace amount proteins in the samples but also comparing the relative abundance of proteins among different samples. The 1618 proteins presented in this report were from two independent quantitative proteomic analyses, by which the credibility of our proteomic measurement is guaranteed. DE-proteins were filtered by both confidence (coverage > 10%) and accuracy ($p < 0.5$) so that only a condensed but highly relevant group of proteins were further studied and validated. The agreement between the transcriptional and translational expressions of the six examinees manifests a close coalition between the proteomic data and their in vivo expression levels (Figure 3). In this study, we compared protein expression profiles between healthy rice and rice infected with the virulent (Guy11) and avirulent (JS153) strains. We focused on proteins exhibiting similar responses to both these *M. oryzae* strains, which represent common rice defense responses against both the virulent and avirulent strains. By investigating the general defense machinery to broad-spectrum rice blast disease, we expect to identify common mechanisms that would be useful for developing broad-spectrum blast disease resistance.

Our research strategy is supported by the preferential enrichment of defense-related biological processes shared by rice either infected by Guy11 or by JS153, and the meagerness of such proteins in rice only infected by Guy11 or by JS153 (Figure 2). For example, the "regulation of protein serine/threonine phosphatase activity" biological process is enriched by both Guy11 and JS153 infection. OsPBZ1 and OsPR10 belonging to this biological process have been demonstrated to exhibit fundamental roles in rice basal defense in other genetic studies [19–22]. Involvement of proteins belonging to the "response to oxidative stress" in plant disease resistance has also been demonstrated in many plant species [23,24]. Other biological processes such as "protein folding" and "glutathione metabolic" have also been linked to plant basal defense [25–28]. Therefore, our proteomic investigation of rice proteins responding to both Guy11 and JS153 infections will facilitate our understanding of broad-spectrum resistant mechanism.

Of the DE-proteins 19% were associated with "response to stimulus", an unexpected result, verifying a great enrichment of the defense-related proteins after the blast disease pathogens' infection. When further narrowed down to stimulus-responding proteins that differentially expressed upon both Guy11 and JS153 infections, the list condensed to 40 proteins (Table 1). Outstandingly, 18 out of the 40 proteins belong to "response to oxidative stress", and five proteins belong to the "response to biotic stress" category. These two biological processes are well known for their involvement in defense responses to pathogen infections [29]. Therefore, these results further justify that our analysis strategy targeted authentic immune players in response to blast disease.

Our proteomic discovery was further validated by whether the examinees could contribute to consensus basal defense responses such as inducing ROS production and callose deposition. Leaf discs transiently expressing OsAPX1 caused callose deposition but did not accumulate detectable ROS, which is in line with its ROS-scavenger activity in other plants [30]. In contrast, transient expression of OsPRX59 and OsPRX62 induced a significant amount of ROS. OsPRX59 and OsPRX62 are class III plant peroxidases with multiple functions, such as ROS removal and cell wall biosynthesis [31,32]. Significant ROS accumulation upon OsPRX59 and OsPRX62 expression overrules their potential of ROS removal and makes their involvement in cell wall synthesis more plausible. In fact, when callose deposition was checked, expression of OsPRX59 and OsPRX62 caused most extensive callose deposition signals (Figure 4B), suggesting these two proteins participate in cell wall synthesis or reinforcement.

OsHSP81 belongs to the HSP90 family that, in rice, is involved in resistance to insect, bacterial, viral, and fungal pathogen infections [25,26,33,34]. Our results showed that OsHSP81 boosted both ROS production and callose deposition, suggesting a decisive role in response to blast disease. OsPBZ1 is highly conserved in plant species [20], which induces programmed cell death. OsPR10 has also been reported in response to various pathogens in multiple plant species, including rice. Previous genetic studies revealed the involvement of OsPBZ1 and OsPR10 in response to blast disease [19,21,35,36]. However, our study confirmed at the proteomic level that OsPBZ1 and OsPR10 are induced by both virulent and avirulent infection, and showed that OsPBZ1 and OsPR10 are capable of inducing both ROS accumulation and callose deposition (Figure 4).

Our previous studies showed that Guy11 and JS153 caused obvious different disease symptoms on the Nipponbare cultivar [14,15]. The phytohormone experiments further demonstrated the difference between these two strains (Figure 5). In general, JA and ET showed suppressed profiles after Guy11 or JS153 infection. In contrast, SA showed a positive association with defense responses to the blast disease (Figure S1). This is inconsistent with the antagonistic relation between the JA/ET and the SA signaling pathways. In specifics, *OsEDS1* and *OsSID2* showed similar reactions to both Guy11 and JS153 infection, indicating at least part of the SA synthesis is regulated by a common mechanism to both virulent and avirulent blast pathogen. However, *OsPAD4*, *OsPAL1*, and *OsPR1* demonstrated a more preferential response to Guy11, instead of to JS153, indicating that rice may recognize and respond to Guy11 individually. Therefore, the proteomic components revealed in this study represent common rice defense responses against these representational strains, indicating a general defense machinery to broad-spectrum rice blast disease. Specifically, our results indicate that "responses to oxidative stress" and "responses to biotic stress" are critical biological processes to blast disease. Breeding projects concentrated on modification of several critical components such as OsPRX59 and OsPRX62 should foresee the success of resistant rice lines to a broad spectrum of blast disease pathogens.

## 4. Materials and Methods

### 4.1. Plants and Inoculation

Rice (*Oryza sativa* L. japonica. cv. Nipponbare) and *N. benthamiana* were grown in a growth room maintained at 25 °C and 70% relative humidity with a 12-h/12-h light/dark photoperiod. Three-leaf-stage plants were spray-inoculated with gelatin or indicated *M. oryzae* conidial suspensions ($1 \times 10^5$ spores/mL in 0.2% gelatin) [37]. The inoculated plants were kept in darkness at 80% RH for 24 h before they were transferred to a growth chamber at 25 °C, 80% relative humidity, and a 12-h/12-h light/dark photoperiod.

### 4.2. Protein Extraction

Rice seedlings (0.5 g) were ground to powder in liquid nitrogen and then dissolved in 10% tricarboxylic acid (TCA)/acetone ($w/v$) containing 0.1% DL-dithiothreitol (DTT)at $-20$ °C for 2 h. The supernatant was discarded after centrifugation at 40,000 $g$ for 20 min. The pellet was washed twice with cold acetone, lyophilized, and dissolved in 300 μL of a lysis solution containing 7 M urea, 2 M thiourea, 4% $w/v$ 3-[(3-cholamidopropyl)-dimethylammonio]-1-propane sulfonate (CHAPS), 65 mM DTT, and 1 mM phenylmethanesulfonyl fluoride (PMSF). The proteins were labeled by an iTRAQ 8-plex kit (AB Sciex) and measured by a Triple TOF 5600 mass spectrometer.

### 4.3. Protein Digestion, iTRAQ Labeling, and Strong Cation Exchange

Protein samples (100 μg of each protein) were mixed with dissolution buffer from AB Sciex (Framingham, MA, USA), digested with trypsin at a 20:1 mass ratio at 37 °C for 14 h, then labeled using the iTRAQ Reagents 8-plex kit according to the manufacturer's instructions (AB Sciex).

The labeled samples were then pooled and dried in an Eppendorf vacuum concentrator. Then, the samples were mixed and lyophilized before dissolving in 4 mL of strong cation exchange (SCX) buffer A (25 mM NaH₂PO₄ in 25% acetonitrile, pH 2.7). The peptides fractionated on an Ultremex SCX column (4.6 mm × 250 mm) using an Agilent 1200 HPLC were grouped into ten components. An Exigent Nano LC-Ultra 2D system (AB Sciex) was used for sample separation. A Triple TOF 5600 mass spectrometer and a Nano Spray III Source (AB Sciex) were used to perform mass spectrometer data acquisition.

### 4.4. Database Search and iTRAQ Quantification

ProteinPilot™ software (version 4.2) was used for raw data processing against the database of *Oryza sativa* from UniProt (http://www.uniprot.org). The primary database search parameters were as follows: the instrument was TripleTOF 5600, iTRAQ quantification, cysteine modified with iodoacetamide; and biological modifications were selected as ID and trypsin digestion. Peptides with a global false discovery rate (FDR) <1% were used for further protein annotation.

To minimize the incidence of false positives, a strict cutoff of unused ProtScore >1.3 was applied for protein identification.

*4.5. Gene Ontology Analysis and Biological Processes Analysis*

Differentially expressed proteins were classified according to GO analysis in Protein Information Resource (https://pir.georgetown.edu). DAVID (https://david.ncifcrf.gov) was used to predict biological processes.

*4.6. RT-PCR Analysis of the Small RNAs and Predicted Targets*

Total RNA was extracted using the Trizol method, reverse-transcribed using a reverse transcription kit (Takara, Shiga, Japan), and the expression levels of the genes of interest were detected using a Real-time PCR Kit (Takara). Primers used for real-time PCR amplification are listed in Table S5.

*4.7. Transient Expression Analysis in N. Benthamiana*

Transient co-expression assays in *N. benthamiana* were performed by infiltrating 3-week-old *N. benthamiana* leaves with *Agrobacterium* GV3101 ($OD_{600}$ = 0.8) harboring constructs containing the gene of interest (pEG104), or an empty vector as a control [38]. Leaf tissue was collected 48 hpi, and protein expression was detected by Western blot.

*4.8. DAB Staining*

*N. benthamiana* leaves were placed in 1 mg/ml of 3,3′-diaminobenzidine (DAB) (Sigma-Aldrich, St. Louis, MO, USA) and shaken at 26 °C for 8 h in a dark place. Then the leaves were decolorized in a 94:4 ethanol:acetic acid ($v/v$) solution at 26 °C for 8 h in dark place. Stained leaf was observed under a camera [39].

*4.9. Aniline Blue Staining*

Leaves were fixed in 94:4 ethanol:acetic acid ($v/v$) solution for 8 h and then stained with 1 mg/mL aniline blue in 150 mM sodium phosphate (pH = 7.0) in the dark for more than 1 h at room temperature (25 °C). Stained leaves were mounted using 50% glycerol. Callose was observed under a microscope (Olympus, Tokyo, Japan) [40,41].

**Supplementary Materials:** Supplementary materials can be found at http://www.mdpi.com/1422-0067/19/7/1950/s1.

**Author Contributions:** Conceptualization, H.M. and H.Z.; Funding acquisition, J.W., Z.F. and H.Z.; Investigation, S.L., P.N., S.D., L.Z., C.C., R.F., Z.W., L.W., J.W., Z.F. and S.Z.; Supervision, H.Z.; Validation, S.L., P.N., S.D., L.Z., C.C., R.F., Z.W., L.W., J.W., Z.F. and S.Z.; Writing—original draft, S.L.; Writing—review and editing, H.Z.

**Funding:** This work was supported by the Fundamental Research Funds for the Central Universities (KYT201805), the Fundamental Research Funds for the Central Universities (KYTZ201403), an Innovation Team Program for Jiangsu Universities (2017) to H.Z, a Top-notch Academic Programs Project of Jiangsu Higher Education Institutions (PPZY2015B157) and a National Training Program for Student Innovative Research and Training (201710307021) to J.W. and Z.F.

**Acknowledgments:** We are grateful to Tanzeela Zia for her critical proofreading of this manuscript.

**Conflicts of Interest:** The authors declare no conflict of interest.

## References

1. Nalley, L.; Tsiboe, F.; Durand-Morat, A.; Shew, A.; Thoma, G. Economic and Environmental Impact of Rice Blast Pathogen (*Magnaporthe oryzae*) Alleviation in the United States. *PLoS ONE* **2016**, *11*. [CrossRef] [PubMed]
2. Talbot, N. On the trail of a cereal killer: Exploring the biology of *Magnaporthe grisea*. *Annu. Rev. Microbiol.* **2003**, *57*, 177–202. [CrossRef] [PubMed]

3.  Wang, G.; Valent, B. Durable resistance to rice blast. *Science* **2017**, *355*, 906–907. [CrossRef] [PubMed]
4.  Jones, J.; Dangl, J. The plant immune system. *Nature* **2006**, *444*, 323–329. [CrossRef] [PubMed]
5.  Liu, W.; Liu, J.; Ning, Y.; Ding, B.; Wang, X.; Wang, Z.; Wang, G. Recent Progress in Understanding PAMP- and Effector-Triggered Immunity against the Rice Blast Fungus *Magnaporthe oryzae*. *Mole. Plant* **2013**, *6*, 605–620. [CrossRef] [PubMed]
6.  Wang, C.; Wang, G.; Zhang, C.; Zhu, P.; Dai, H.; Yu, N.; He, Z.; Xu, L.; Wang, E. OsCERK1-Mediated Chitin Perception and Immune Signaling Requires Receptor-like Cytoplasmic Kinase 185 to Activate an MAPK Cascade in Rice. *Mol. Plant* **2017**, *10*, 619–633. [CrossRef] [PubMed]
7.  Xie, X.; Xue, Y.; Zhou, J.; Zhang, B.; Chang, H.; Takano, M. Phytochromes Regulate SA and JA Signaling Pathways in Rice and Are Required for Developmentally Controlled Resistance to *Magnaporthe grisea*. *Mol. Plant* **2011**, *4*, 688–696. [CrossRef] [PubMed]
8.  Kim, S.; Wang, Y.; Lee, K.; Park, Z.; Park, J.; Wu, J.; Kwon, S.; Lee, Y.; Agrawal, G.; Rakwal, R.; et al. In-depth insight into in vivo apoplastic secretome of rice–*Magnaporthe oryzae* interaction. *J. Proteom.* **2013**, *78*, 58–71. [CrossRef] [PubMed]
9.  Kim, S.; Kang, Y.; Wang, Y.; Wu, J.; Park, Z.; Rakwal, R.; KumarAgrawal, G.; Lee, S.; Kang, K. Secretome analysis of differentially induced proteins in rice suspension-cultured cells triggered by rice blast fungus and elicitor. *Proteomics* **2009**, *9*, 1302–1313. [CrossRef] [PubMed]
10. Cao, J.; Yang, C.; Li, L.; Jiang, L.; Wu, Y.; Wu, C.; Bu, Q.; Xia, G.; Liu, X.; Luo, Y.; et al. Rice Plasma Membrane Proteomics Reveals Magnaporthe oryzae Promotes Susceptibility by Sequential Activation of Host Hormone Signaling Pathways. *Mol. Plant Microbe Interact.* **2016**, *29*, 902–913. [CrossRef] [PubMed]
11. Franck, W.; Gokce, E.; Oh, Y.; Muddiman, D.; Dean, R. Temporal Analysis of the *Magnaporthe oryzae* Proteome During Conidial Germination and Cyclic AMP (cAMP)-mediated Appressorium Formation. *Mol. Cell. Proteom.* **2013**, *12*, 2249–2265. [CrossRef] [PubMed]
12. Ross, P.; Huang, Y.; Marchese, J.; Williamson, B.; Parker, K.; Hattan, S.; Khainovski, N.; Pillai, S.; Dey, S.; Daniels, S.; et al. Multiplexed protein quantitation in Saccharomyces cerevisiae using amine-reactive isobaric tagging reagents. *Mol. Cell. Proteom.* **2004**, *3*, 1154–1169. [CrossRef] [PubMed]
13. Zieske, L. A perspective on the use of iTRAQ (TM) reagent technology for protein complex and profiling studies. *J. Exp. Bot.* **2006**, *57*, 1501–1508. [CrossRef] [PubMed]
14. Zhang, X.; Bao, Y.; Shan, D.; Wang, Z.; Song, X.; Wang, Z.; Wang, J.; He, L.; Wu, L.; Zhang, Z.; et al. *Magnaporthe oryzae* Induces the Expression of a MicroRNA to Suppress the Immune Response in Rice. *Plant Physiol.* **2018**, *177*, 352–368. [CrossRef] [PubMed]
15. Wang, Z.; Xia, Y.; Lin, S.; Wang, Y.; Guo, B.; Song, X.; Ding, S.; Zheng, L.; Feng, R.; Chen, S.; et al. Osa-miR164a targets OsNAC60 and negatively regulates rice immunity against the blast fungus *Magnaporthe oryzae*. *Plant J.* **2018**, doi:10.1111/tpj.13972. [CrossRef] [PubMed]
16. Xu, D.; Li, Y.; Li, X.; Wei, L.-L.; Pan, Z.; Jiang, T.-T.; Chen, Z.-L.; Wang, C.; Cao, W.-M.; Zhang, X.; et al. Serum protein S100A9, SOD3, and MMP9 as new diagnostic biomarkers for pulmonary tuberculosis by iTRAQ-coupled two-dimensional LC-MS/MS. *Proteomics* **2015**, *15*, 58–67. [CrossRef] [PubMed]
17. Yang, C.; Li, W.; Cao, J.; Meng, F.; Yu, Y.; Huang, J.; Jiang, L.; Liu, M.; Zhang, Z.; Chen, X.; et al. Activation of ethylene signaling pathways enhances disease resistance by regulating ROS and phytoalexin production in rice. *Plant J.* **2017**, *89*, 338–353. [CrossRef] [PubMed]
18. Mueller, M.; Munne-Bosch, S. Ethylene Response Factors: A Key Regulatory Hub in Hormone and Stress Signaling. *Plant Physiol.* **2015**, *169*, 32–41. [CrossRef] [PubMed]
19. Kim, S.; Kim, S.; Hwang, D.; Kang, S.; Kim, H.; Lee, B.; Lee, J.; Kang, K. Proteomic analysis of pathogen-responsive proteins from rice leaves induced by rice blast fungus, *Magnaporthe grisea*. *Proteomics* **2004**, *4*, 3569–3578. [CrossRef] [PubMed]
20. Liu, J.; Ekramoddoullah, A. The family 10 of plant pathogenesis-related proteins: Their structure, regulation, and function in response to biotic and abiotic stresses. *Physiol. Mol. Plant Pathol.* **2006**, *68*, 3–13. [CrossRef]
21. Takahashi, A.; Kawasaki, T.; Henmi, K.; Shii, K.; Kodama, O.; Satoh, H.; Shimamoto, K. Lesion mimic mutants of rice with alterations in early signaling events of defense. *Plant J.* **1999**, *17*, 535–545. [CrossRef] [PubMed]

22. Wang, D.; Qin, Y.; Han, J.; Zhang, L.; Xu, X.; Liu, X.; Wang, C. Expression analysis of innate immunity related genes in the true/field blast resistance gene-mediated defence response. *Biotechnol. Biotechnol. Equip.* **2014**, *28*, 999–1007. [CrossRef] [PubMed]

23. Delledonne, M.; Xia, Y.J.; Dixon, R.A.; Lamb, C. Nitric oxide functions as a signal in plant disease resistance. *Nature* **1998**, *394*, 585–588. [CrossRef] [PubMed]

24. Yoshimura, K.; Yabuta, Y.; Ishikawa, T.; Shigeoka, S. Expression of spinach ascorbate peroxidase isoenzymes in response to oxidative stresses. *Plant Physiol.* **2000**, *123*, 223–233. [CrossRef] [PubMed]

25. Jiang, S.; Lu, Y.; Li, K.; Lin, L.; Zheng, H.; Yan, F.; Chen, J. Heat shock protein 70 is necessary for Rice stripe virus infection in plants. *Mol. Plant Pathol.* **2014**, *15*, 907–917. [CrossRef] [PubMed]

26. Kim, N.; Hwang, B. Pepper Heat Shock Protein 70a Interacts with the Type III Effector AvrBsT and Triggers Plant Cell Death and Immunity. *Plant Physiol.* **2015**, *167*. [CrossRef] [PubMed]

27. Li, J.; Zhang, H.; Hu, J.; Liu, J.; Liu, K. A Heat Shock Protein Gene, *CsHsp45.9*, Involved in the Response to Diverse Stresses in Cucumber. *Biochem. Genet.* **2012**, *50*, 565–578. [CrossRef] [PubMed]

28. Mou, Z.; Fan, W.H.; Dong, X.N. Inducers of plant systemic acquired resistance regulate NPR1 function through redox changes. *Cell* **2003**, *113*, 935–944. [CrossRef]

29. Jing, M.; Ma, H.; Li, H.; Guo, B.; Zhang, X.; Ye, W.; Wang, H.; Wang, Q.; Wang, Y. Differential regulation of defense-related proteins in soybean during compatible and incompatible interactions between *Phytophthora sojae* and soybean by comparative proteomic analysis. *Plant Cell Rep.* **2015**, *34*, 1263–1280. [CrossRef] [PubMed]

30. Yang, H.; Mu, J.; Chen, L.; Feng, J.; Hu, J.; Li, L.; Zhou, J.; Zuo, J. S-Nitrosylation Positively Regulates Ascorbate Peroxidase Activity during Plant Stress Responses. *Plant Physiol.* **2015**, *167*, 1604–1753. [CrossRef] [PubMed]

31. Choi, H.; Kim, Y.; Lee, S.; Hong, J.; Hwang, B. Hydrogen peroxide generation by the pepper extracellular peroxidase CaPO2 activates local and systemic cell death and defense response to bacterial pathogens. *Plant Physiol.* **2007**, *145*, 890–904. [CrossRef] [PubMed]

32. Bestwick, C.; Brown, I.; Mansfield, J. Localized changes in peroxidase activity accompany hydrogen peroxide generation during the development of a nonhost hypersensitive reaction in lettuce. *Plant Physiol.* **1998**, *118*, 1067–1078. [CrossRef] [PubMed]

33. Diaz, F.; Orobio, R.; Chavarriaga, P.; Toro-Perea, N. Differential expression patterns among heat-shock protein genes and thermal responses in the whitefly *Bemisia tabaci* (MEAM 1). *J. Therm. Biol.* **2015**, *52*, 199–207. [CrossRef] [PubMed]

34. Thao, N.; Chen, L.; Nakashima, A.; Hara, S.; Umemura, K.; Takahashi, A.; Shirasu, K.; Kawasaki, T.; Shimamoto, K. RAR1 and HSP90 form a complex with Rac/Rop GTPase and function in innate-immune responses in rice. *Plant Cell* **2007**, *19*, 4035–4045. [CrossRef] [PubMed]

35. Kim, S.; Kim, S.; Wang, Y.; Yu, S.; Choi, I.; Kim, Y.; Kim, W.; Agrawal, G.; Rakwal, R.; Kang, K. The RNase activity of rice probenazole-induced protein1 (PBZ1) plays a key role in cell death in plants. *Mol. Cells* **2011**, *31*, 25–31. [CrossRef] [PubMed]

36. Wang, Y.; Kwon, S.; Wu, J.; Choi, J.; Lee, Y.; Agrawal, G.; Tamogami, S.; Rakwal, R.; Park, S.; Kim, B.; et al. Transcriptome Analysis of Early Responsive Genes in Rice during *Magnaporthe oryzae* Infection. *Plant Pathol. J.* **2014**, *30*, 343–354. [CrossRef] [PubMed]

37. Qu, S.; Liu, G.; Zhou, B.; Bellizzi, M.; Zeng, L.; Dai, L.; Han, B.; Wang, G. The broad-spectrum blast resistance gene *Pi9* encodes a nucleotide-binding site-leucine-rich repeat protein and is a member of a multigene family in rice. *Genetics* **2006**, *172*, 1901–1914. [CrossRef] [PubMed]

38. Zhang, X.; Zhao, H.; Gao, S.; Wang, W.; Katiyar-Agarwal, S.; Huang, H.; Raikhel, N.; Jin, H. Arabidopsis Argonaute 2 Regulates Innate Immunity via miRNA393*-Mediated Silencing of a Golgi-Localized SNARE Gene, MEMB12. *Mol. Cell* **2011**, *42*, 356–366. [CrossRef] [PubMed]

39. Xiao, S.; Brown, S.; Patrick, E.; Brearley, C.; Turner, J. Enhanced transcription of the arabidopsis disease resistance genes *RPW8.1* and *RPW8.2* via a salicylic acid-dependent amplification circuit is required for hypersensitive cell death. *Plant Cell* **2003**, *15*, 33–45. [CrossRef] [PubMed]

40. Clay, N.; Adio, A.; Denoux, C.; Jander, G.; Ausubel, F. Glucosinolate Metabolites Required for an Arabidopsis Innate Immune Response. *Science* **2009**, *323*, 95–101. [CrossRef] [PubMed]
41. Luna, E.; Pastor, V.; Robert, J.; Flors, V.; Mauch-Mani, B.; Ton, J. Callose Deposition: A Multifaceted Plant Defense Response. *Mol. Plant Microbe Interact.* **2011**, *24*, 183–193. [CrossRef] [PubMed]

International Journal of
*Molecular Sciences*

MDPI

*Article*

# Cloning, Characterization, and Functional Investigation of *VaHAESA* from *Vitis amurensis* Inoculated with *Plasmopara viticola*

Shaoli Liu [1,†], Chi Zhang [1,†], Nan Chao [2], Jiang Lu [3,4] and Yali Zhang [1,*]

[1]   The Viticulture and Enology Program, College of Food Science and Nutritional Engineering,
      China Agricultural University, Beijing 100083, China; shaolil@126.com (S.L.); cauzhangchi@163.com(C.Z.)
[2]   Center for Plant Biology, TSinghua University, Beijing 100084, China; chaonan1989@126.com
[3]   Center for Viticulture and Enology, School of Agriculture and Biology, Shanghai Jiao Tong University,
      Shanghai 200024, China; jiang.lu@sjtu.edu.cn
[4]   Guangxi Crop Genetic Improvement and Biotechnology Laboratory, Guangxi Academy of Agricultural
      Sciences, Nanning 530007, China
*     Correspondence: zhangyali@cau.edu.cn; Tel.: +86-181-0120-6227
†     These authors contributed equally to this work.

Received: 26 March 2018; Accepted: 6 April 2018; Published: 16 April 2018

**Abstract:** Plant pattern recognition receptors (PRRs) are essential for immune responses and establishing symbiosis. Plants detect invaders via the recognition of pathogen-associated molecular patterns (PAMPs) by PRRs. This phenomenon is termed PAMP-triggered immunity (PTI). We investigated disease resistance in *Vitis amurensis* to identify PRRs that are important for resistance against downy mildew, analyzed the PRRs that were upregulated by incompatible *Plasmopara viticola* infection, and cloned the full-length cDNA of the *VaHAESA* gene. We then analyzed the structure, subcellular localization, and relative disease resistance of *VaHAESA*. *VaHAESA* and PRR-receptor-like kinase 5 (RLK5) are highly similar, belonging to the leucine-rich repeat (LRR)-RLK family and localizing to the plasma membrane. The expression of PRR genes changed after the inoculation of *V. amurensis* with compatible and incompatible *P. viticola*; during early disease development, transiently transformed *V. vinifera* plants expressing *VaHAESA* were more resistant to pathogens than those transformed with the empty vector and untransformed controls, potentially due to increased $H_2O_2$, NO, and callose levels in the transformants. Furthermore, transgenic *Arabidopsis thaliana* showed upregulated expression of genes related to the PTI pathway and improved disease resistance. These results show that *VaHAESA* is a positive regulator of resistance against downy mildew in grapevines.

**Keywords:** downy mildew; grapevine; PRRs; PTI; *VaHAESA*

## 1. Introduction

Downy mildew is caused by the oomycete *Plasmopara viticola* and is one of the major diseases affecting grapevines worldwide. However, grapevines possess natural resistance against *P. viticola* as a result of disease resistance synergy. These resistance mechanisms involve physiological, ecological, and morphological changes in the plant [1,2]. An important mode of active defense in plant immunity is the detection of pathogen-associated molecular patterns (PAMPs) by pattern recognition receptors (PRRs) [3], otherwise known as PAMP-triggered immunity (PTI) [4,5]. PTI comprises a wide array of responses, including the rapid generation of reactive oxygen species (ROS), deposition of callose, activation of mitogen-activated protein kinases (MAPKs), and expression of immune-related genes [3,5]. Consequently, PTI plays a major role in preventing the pathogenic invasion of plants.

Plant PRRs are either receptor-like kinases (RLKs) or receptor-like proteins (RFLPs), taking the form of single-pass transmembrane proteins with extracellular domains. While RLKs have an intracellular kinase domain, RFLPs lack this cytosolic signaling domain [6]. PRRs exhibit both high sensitivity to and specialization for plant pathogens, where a certain PRR can recognize PAMPs at nanomolar quantities [4], and conserved functional domains of PAMPs are often recognized by PRRs. For example, FLS2 is one of the best-studied plant PRRs in *Arabidopsis* and recognizes bacterial flagellin via perception of the conserved 22-aminoacid epitope flg22 [7]. However, while FLS2 is highly conserved in plant species, the FLS2 homolog of tomato (*LeFLS2*) recognizes the flg15 polypeptide of *Escherichia coli* but does not recognize *Pseudomonas syringae* flg22 [8]. It has been hypothesized that the ligand-induced endocytosis and degradation of FLS2 may regulate receptor signaling [9]. In addition, the elongation factor Tu (EF-Tu) receptor (EFR) is a PRR in *Arabidopsis* that recognizes EF-Tu from bacteria [10]; however, this kind of PRR exists only in cruciferous species [11]. At present, the complete genome sequences of plants that contain homologous EFR genes are not well characterized [3].

Although the PAMPs of pathogenic microorganisms have been extensively studied, few of these studies have focused on the corresponding PRRs in plants. Furthermore, there are relatively few reports regarding the PRRs of grapevines. With advances in whole-genome sequencing, research focusing on characterization of resistance genes in grapevines is increasing, and recently published transient expression assays have been widely used for the characterization of newly discovered genes, including their functions and metabolic pathways [12–14]. Transcriptomic and proteomic analyses of grapevines infected with *P. viticola* are likely to result in the discovery of novel genes involved in pathways related to resistance against downy mildew, in addition to helping to elucidate the molecular mechanisms involved in the resistance response.

The present study describes the novel *V. amurensis* PRR gene *VaHAESA*. This gene was identified by analyzing the transcriptome of the *V. amurensis* cultivar "Shuanghong" while infected with either the compatible *P. viticola* strain "ZJ-1-1" or the incompatible *P. viticola* strain "JL-7-2" [13]. Here, we demonstrate that *VaHAESA* belongs to the LRR-RLK (leucine-rich repeat receptor-like protein kinase) family of proteins. Transient expression studies indicate that *VaHAESA* can trigger a series of PTI responses, including the accumulation of $H_2O_2$ and NO as well as the deposition of callose. This research provides a better understanding of the characteristics and function of a novel PRR gene in grape.

## 2. Results

### 2.1. PRR Expression in Vitis amurensis 'Shuanghong' Infected with Incompatible and Compatible Strains of Plasmopara viticola

Quantitative RT-PCR revealed several differences in the expression patterns of PRR genes in *V. amurensis* "Shuanghong" (Figures 1 and 2) after inoculation with compatible ("ZJ-1-1") and incompatible *P. viticola* ("JL-7-2") strains. After inoculation with either compatible *P. viticola* "ZJ-1-1" or incompatible *P. viticola* "JL-7-2", clear and consistent trends were observed in the expression (both up- and downregulation) of some genes. These affected genes included GSVIVT01035611001, GSVIVT01014117001, and GSVIVT01014147001. For example, the expression of GSVIVT01014147001 increased 2-fold within 0.5 h after inoculation with the *P. viticola* strains "ZJ-1-1" and "JL-7-2". In contrast, some PRRs were upregulated after inoculation with *P. viticola* "ZJ-1-1" but downregulated after inoculation with incompatible *P. viticola* "JL-7-2" (i.e., GSVIVT01035304001, GSVIVT01026000001, GSVIVT01014110001, and GSVIVT01023113001).

However, among the remaining genes, we observed downregulation after inoculation with "ZJ-1-1" and upregulation after inoculation with 'JL-7-2' (i.e., GSVIVT01036966001, GSVIVT01035315001, GSVIVT01014138001, GSVIVT01015298001, and GSVIVT01023369001). Within 0.5 h of inoculation with JL-7-2, the expression of all PRR genes that we evaluated was initially induced and then decreased. GSVIVT01015298001 was upregulated 4-fold, while GSVIVT01036966001, GSVIVT01035315001, GSVIVT01014138001, and GSVIVT01023369001 were upregulated 2- to 3-fold.

In addition, the expression levels of these five genes indicated inhibition within the first 12 h of "ZJ-1-1" inoculation. According to these results, we hypothesize that GSVIVT01036966001, GSVIVT01035315001, GSVIVT01014138001, GSVIVT01023369001, and, most notably, GSVIVT01015298001 play an active role in pathogen resistance during the early phase of resistance against *P. viticola* infection. Next, based on the results of expression analyses, we chose the PRRs with the potential to confer stronger resistance against downy mildew in 'huanghong' (i.e., upregulated genes showing significant differences after inoculation with incompatible *P. viticola* and downregulated genes showing significant differences after inoculation with the compatible strain). Thus, we chose GSVIVT01015298001 for further functional verification of *P. viticola* resistance in *V. amurensis* "Shuanghong", which was expressed at the earliest post-inoculation time point (0.5 h) and exhibited maximum upregulation.

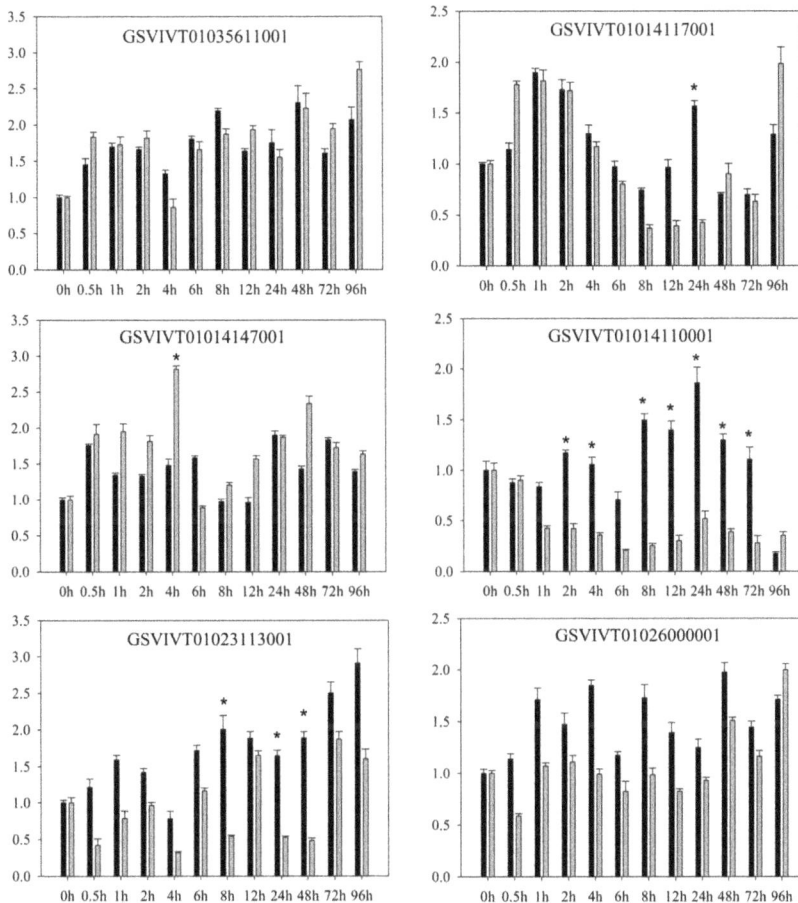

**Figure 1.** Relative expression of the pattern recognition receptors (PRR) genes of *Vitis amurensis* "Shuanghong" after inoculation with *Plasmopara viticola* "ZJ-1-1" (black bars) and *P. viticola* "JL-7-2" (gray bars). The values on the vertical axes indicate the fold-changes in gene expression normalized to the expression level of Vitis elongation factor 1-$\alpha$ (EF1-$\alpha$), SAND, and ubiquitin-conjugating enzyme (UBQ). The x-axes represent the time since inoculation. The error bars represent the standard deviation calculated from three replicates. * indicates significant differences ($p < 0.05$) as determined with Student's *t*-test. Significant differences were identified by comparing the two gene expression levels at each time point.

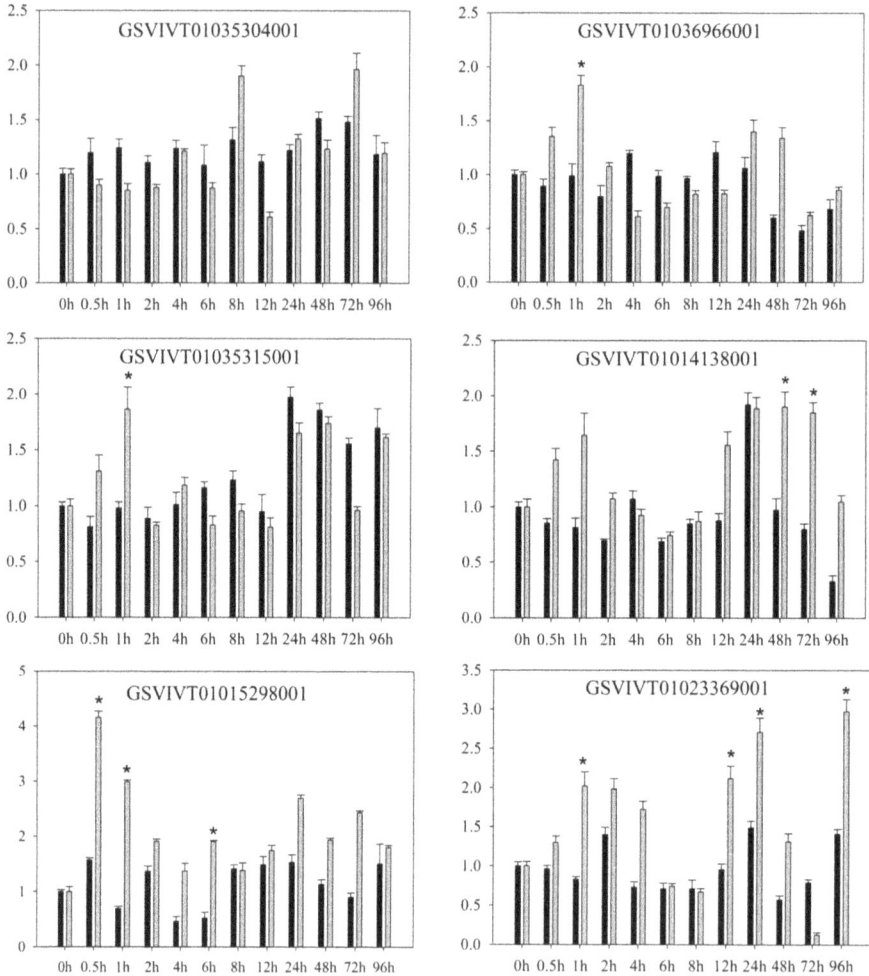

**Figure 2.** Relative expression of the PRR genes of *Vitis amurensis* 'Shuanghong' after inoculation with *Plasmopara viticola* "ZJ-1-1" (black bars) and *P. viticola* "JL-7-2" (gray bars). Values on the vertical axes indicate the fold-change in gene expression, normalized to the expression level of Vitis EF1-$\alpha$, SAND, and UBQ. The *x*-axes represent the time since inoculation. Error bars represent the standard deviation calculated from three replicates. * indicates significant differences ($p < 0.05$) as determined with Student's *t*-test. Significant differences were identified by comparing the two gene expression levels at each time point.

## 2.2. Characterization and Phylogenetic Analysis of VaHAESA

We cloned *VaHAESA* from the "Shuanghong" subspecies of grape based on the aforementioned transcriptome data. The *VaHAESA* gene is 3066 bp long and encodes 1021 amino acids. Further characterization of the *VaHAESA* protein sequence using the Pfam and SUPERFAMILY 2 databases confirmed that *VaHAESA* belongs to the LRR-RLK protein family, due to the presence of a leucine-rich repeat N-terminal domain (LRRNT), two LRR_8 domains, and a protein kinase C-terminal domain. These findings are consistent with a sequence alignment and phylogenetic analysis using putative *VaHAESA* and RLK homologs from a variety of plant species (Figure 3). We detected several LRR core motifs (Figure 3, Box I–IV); these LRR repeats constitute a novel class of α/β folds. The LRR core (LXXLXLXXNXL) motifs, which form an α/β sheet, are thought to form an exposed face involved in protein-protein interactions [15]. In addition, we detected the key motifs constituting the catalytic core of the kinase. These motifs include (i) GXGXXG (Figure 3, box V), which is thought to be an integral part of many nucleotide-binding proteins; (ii) a highly conserved Lys amino acid (Figure 3, box VI); and (iii) a DFG(Asp-Phe-Gly) motif (Figure 3, box VII). Finally, the conserved motif G(T/S)PXYXAPE (Figure3, box VIII) is characteristic of serine/threonine kinases [15].

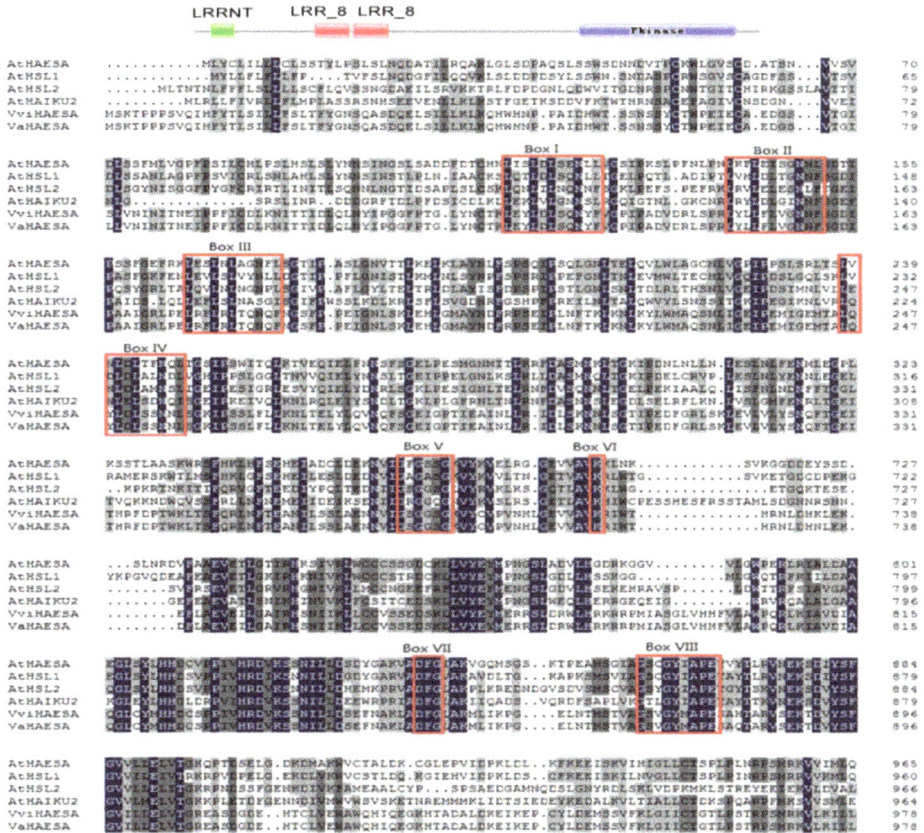

**Figure 3.** Alignment of the putative *Vitis amurensis* HAESA (*VaHAESA*) protein sequence with other HAESA and HAESA-like proteins. Box I–IV: leucine-rich repeat (LRR) core LXXLXLXXNXL motifs; Box V–VII: motifs involved in the kinase catalytic core.

A phylogenetic analysis revealed that the LRR-RLK that we identified in the present study (*VaHAESA*) clustered with HAESA and HAESA homologs from *Arabidopsis* and five other species (Figure 4). These results further confirmed that this gene is likely to be the *HAESA* gene of *V. amurensis* 'Shuanghong'. HAESA, which has also been called RLK5, is a critical component required for floral organ abscission, belonging to LRR XI based on the classification of LRR-RLKs [16]. A homology matrix for these sequences is provided as supplemental material. In addition, we also established a model of *VaHAESA* based on the reported crystal structure for HAESA from *Arabidopsis* (5IXQ) (Figure 5). The crystallization data for PDB 5 IXQ are shown in yellow and correspond to the crystal structure of the *Arabidopsis* receptor kinase HAESA LRR ectodomain. The crystal structure of *VaHAESA* is shown in magenta and was obtained by Swiss-model homology. The comparison of the alignment results shows that the overall structures are very similar. The details are shown in Figure 5.

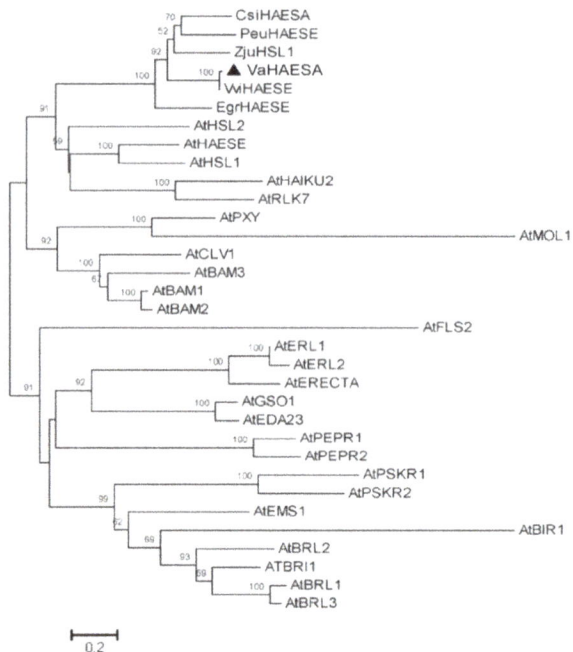

**Figure 4.** Phylogenetic tree of LRR-RLKs. *Vitis amurensis* HAESA (*VaHAESA*) is indicated by the filled triangle.

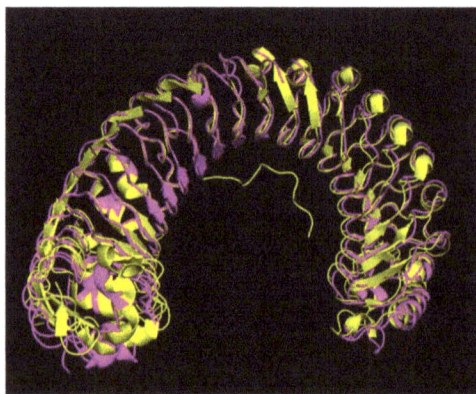

**Figure 5.** Alignment of the 3-D structures of *VaHAESA* and *Arabidopsis thaliana* HAESA (*AtHAESA*) (5IXQ).

## 2.3. Subcellular Location of VaHAESA in Nicotiana benthamiana

Based on the differences in expression observed for the PRRs of *V. amurensis*, we selected *VaHAESA* for further investigation. To study the distribution and cellular localization of *VaHAESA* in the mesophyll cells of *N. benthamiana*, we designed green fluorescent protein-tagged constructs of *VaHAESA* downstream of the signal peptide cleavage site. The fusion constructs were expressed transiently in *N. benthamiana* using Agrobacteria infiltration. The results showed that the *VaHAESA* protein was only observed in the cytoplasm (Figure 6).

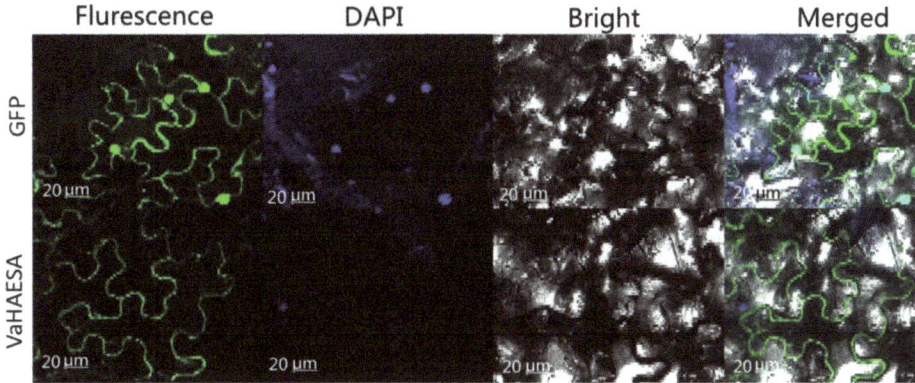

**Figure 6.** Subcellular localization of the *GFP-VaHAESA* protein in transiently transformed *Nicotiana benthamiana*.

## 2.4. Expression of VaHAESA Promoted Resistance against Plasmopara viticola in Grapevine

The transient expression analysis showed that 3 d after inoculation, grape leaves transiently transformed with the *VaHAESA* construct exhibited smaller infected areas than those transformed with the empty vector and untransformed *V. vinifera* "Thompson Seedless" (Figure 7). The number of spores on the *VaHAESA*-expressing plants was $0.18 \times 10^5$, whereas those on the empty vector-transformed plants and untransformed controls were $1.45 \times 10^5$ and $1.38 \times 10^5$, respectively. These results show that disease resistance was improved accordingly in the transient grape leaves. However, five days after inoculation, the differences in the infected areas and spore concentrations on the leaves from all groups decreased (Figure 7).

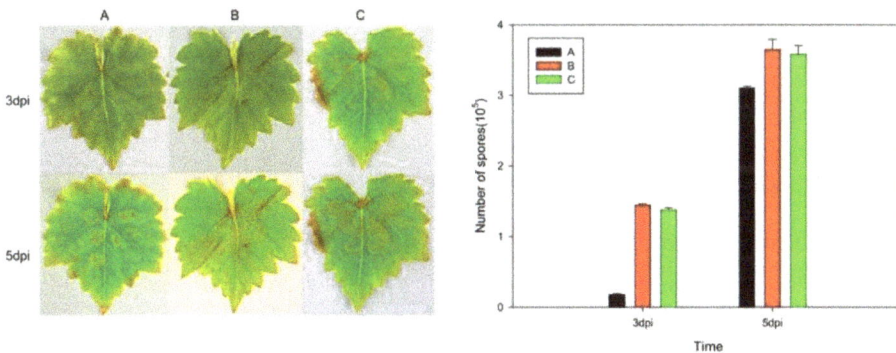

**Figure 7.** Phenotypes of leaves from *Vitis vinifera* after infection with *Plasmopara viticola*. (**A**) *V. vinifera* expressing *VaHAESA*; (**B**) *V. vinifera* transformed with the empty vector (pBI121); (**C**) untransformed *V. vinifera* "Thompson Seedless". The representative images were taken three and five days post-infection (dpi) with *Plasmopara viticola*.

Three days after inoculation, the density of sporophores and spores in grape leaves transiently transformed with *VaHAESA* was lower than that in leaves transformed with the empty vector (Figure 8A1,B1). In addition, we observed necrosis in the guard cells of grape leaves transiently transformed with the *VaHAESA* gene (Figure 8A2), whereas the leaves transformed with the empty vector did not present similar necrosis (Figure 8B2).

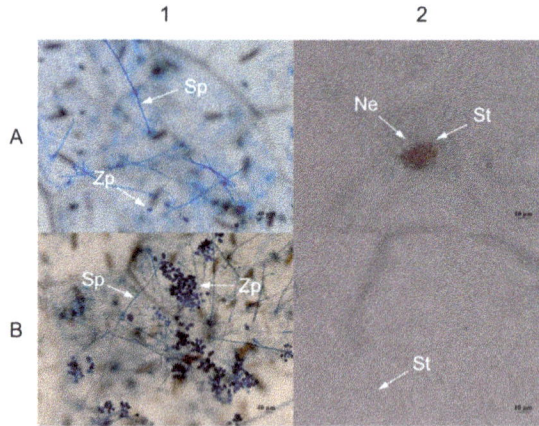

**Figure 8.** Microscopy examination of *Vitis vinifera* after infection with *Plasmopara viticola*. (**A**) *V. vinifera* expressing *VaHAESA*; (**B**) *V. vinifera* transformed with the empty vector (pBI121). Representative images were taken 3 days after inoculation with *Plasmopara viticola*. Sp: Sporophore; Zp: Zoospore; St: Stomata; Ne: Necrosis.

### 2.5. Measurements of $H_2O_2$, NO, and Callose in Vitis vinifera Transiently Expressing VaHAESA

We observed that 24 h after infection in *V. vinifera*, the $H_2O_2$ and NO contents of leaves transformed with *VaHAESA* were higher than those of leaves transformed with the empty vector or the untransformed control (Figure 9).

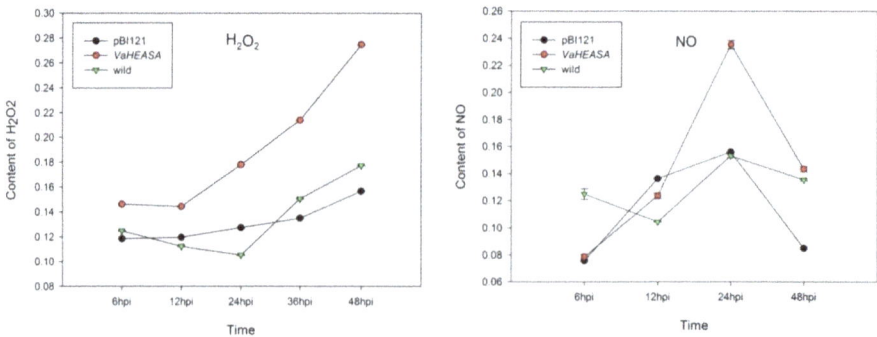

**Figure 9.** Analysis of $H_2O_2$ and NO levels in *Vitis vinifera* after inoculation with *Plasmopara viticola*.

Microscopic observations indicated that after inoculation with *P. viticola*, callose formation at 6 hpi and 12 hpi was much greater in the *VaHAESA*-expressing leaves than in leaves transformed with empty vector and untransformed *V. vinifera* (Figure 10). In addition, we observed few sporophores on leaves transformed with *VaHAESA* at 24 hpi, while a mass of sporophores was observed on the leaves transformed with the empty vector and wild Thompson Seedless (untransformed control).

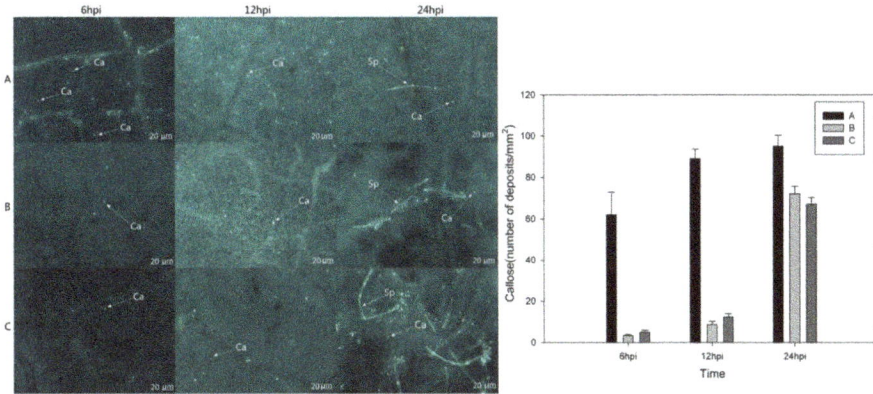

**Figure 10.** Microscopic assessment and statistics of callose deposition in the leaves of *Vitis vinifera* after inoculation with *Plasmopara viticola*. (**A**) *V. vinifera* expressing *VaHAESA*; (**B**) *V. vinifera* transformed with the empty vector (pBI121); (**C**) untransformed *V. vinifera* 'Thompson Seedless'. Representative images were taken 6, 12, 24, and 48 h post-infection (hpi). Ca: Callose, Sp: sporophore.

*2.6. Identification and Analysis of Disease Resistance in Transgenic Arabidopsis thaliana*

To characterize the physiological functions of *VaHAESA* involved in disease resistance, ten transgenic lines of *Arabidopsis* were obtained, and three transgenic lines exhibiting stable expression and improved disease resistance, designated "line2, line3, and line4" (confirmed through PCR and qRT-PCR, Figure 11A,B), were selected for further experiments. The results show that the expression level of *VaHEASA* was up to 40-fold higher than that of the *AtActin*gene in transgenic *Arabidopsis*. To determine the resistance of the transgenic plants, wild-type and transgenic *A. thaliana* plants were inoculated with *Hyaloperonospora arabidopsidis*. After infection, we analyzed the expression patterns of key genes in the PTI pathway. Our results showed that after 5 days of inoculation, *H. arabidopsidis* was abundant on wild-type leaves, while transgenic *A. thaliana* did not harbor the pathogen (Figure 11CI,CII). As key genes in the PTI pathway, the expression of MPK3 (mitogen-activated protein kinase 3), MPK6 (mitogen-activated protein kinase 6), SCD1 (stomatal cytokinesis-defective 1), BAK1 (BRI1 associated receptor kinase 1), and BIK1 (Botrytis-induced kinase 1) in the transgenic plants was substantially higher than in wild-type plants. These results showed that the expression of MAPKs and receptor-like cytoplasmic kinases (RLCKs) was 5-fold higher in transgenic plants than in wild-type plants at 2 h to 8 h after infection. The expression of these genes in wild-type *A. thaliana* was only slightly increased at 8 h after inoculation. These results clearly show that compared to wild-type plants, the transgenic plants exhibited an enhanced ability to defend against the invading pathogen.

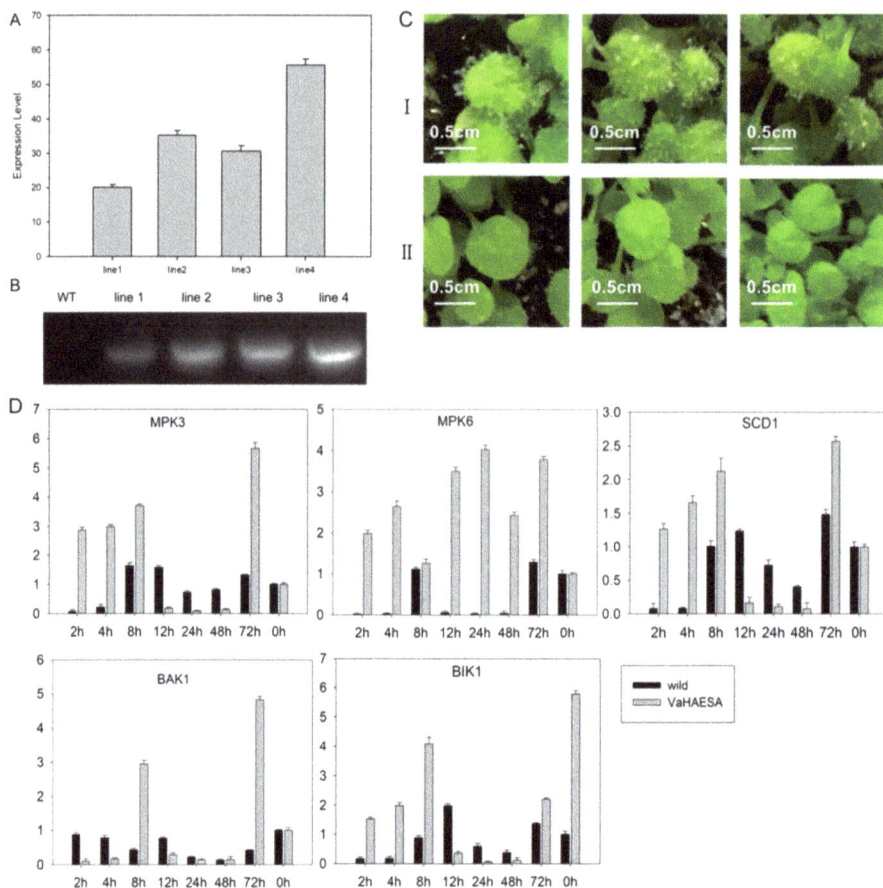

**Figure 11.** Identification of disease resistance in transgenic *Arabidopsis thaliana* and the expression patterns of related genes. (**A**) qRT-PCR analysis of *VaHAESA* expression levels in transgenic *Arabidopsis thaliana*; (**B**) PCR detection of positive transgenic *Arabidopsis thaliana*; (**C**) phenotypic identification of *Arabidopsis thaliana* after inoculation with *H. arabidopsidis*: I. wild type; II, transgenic *Arabidopsis thaliana*; (**D**) the expression patterns of related genes in the pathogen-associated molecular patterns (PAMP)-triggered immunity (PTI) pathway. Values on vertical axes indicate the fold-change in gene expression, normalized to the expression levels of *AtACTIN*, *AtSAND*, and *AtUBQ*. The *x*-axes represent the time since inoculation. Error bars represent the standard deviation calculated from three replicates.

## 3. Discussion

In recent years, it has become apparent that PRRs exist in the plasma membrane within intricate protein complexes resembling supramolecular structures and require numerous regulators to initiate and fine-tune plant immune responses [17–19]. In plants, the RLKs have been implicated in the prevention of self-pollination, pathogen responses, hormone perception, signal transduction, and plant development. These functions are divided into two categories [20]. One category consists of kinases involved in cell growth and development. For instance, some studies have shown that in *Arabidopsis*, LRR-RLK HAESA (HAE) and the peptide hormone IDA (inflorescence deficient in abscission) control floral organ abscission [21–23]. HAESA is a plasma membrane-associated protein with serine/threonine protein kinase activity. The other category includes RLK proteins involved in plant–pathogen interactions and defense responses [7]. However, the functional classification of RLK on the basis of structure may actually be more complex due to cross talk between disease and developmental pathways or due to recognition of multiple ligands by a signal receptor [24,25].

Our analysis of the conserved domain of *VaHAESA*, which contains an intracellular kinase domain, revealed that it belongs to the LRR-RLK family. These PRRs contain extracellular domains that allow MAMP/DAMP(microbe/damage—associated molecular patterns) perception [26]. LRR-type PRRs localize to the cell membrane and bind to proteins or peptides such as bacterial flagellin, EF-Tu, or endogenous PEP peptides [27]. Our results regarding the subcellular location of *VaHAESA* were consistent with the findings of previous studies. Interestingly, FLS2 forms a complex with the regulatory LRR-RLK BAK1 quasi-instantaneously upon flg22 perception, suggesting that FLS2 and BAK1 already exist in proximity to each other in the plasma membrane [28]. The paradigm of signaling activation by receptor kinases implies that ligand binding via the extracellular domain causes activation of the intracellular kinase domain and phosphorylation of substrates that contribute to intracellular signal transduction.

Therefore, *VaHAESA* could receive the signal through the extracellular LRR domain structure and transmit that signal with the help of the intracellular protein kinase domain. Recent transcriptome analysis of abscission zones from wild-type and *hae* mutants indicate that the IDA-HAE signaling module triggers cell wall-degrading and cell wall-remodeling genes that are necessary for the abscission process, in addition to genes commonly related to defense against bacteria and fungi [29]. Finally, the observed oxidative burst could also trigger the activation of pathogen defense genes because shedding exposes a fresh cell surface, which may be highly susceptible to pathogen infection.

Plant receptor kinases ectopically expressed in plant cells can be expected to be processed and localized correctly. RLK involved in signaling mainly recognize exogenous elicitor complexes that activate defense-related pathways. Recognition results in the formation of ROS, changes in ion flux, and rapid phosphorylation of the kinase domains. These effects all invoke the activation of the MAPK cascade and activation of defense-related genes [30]. It is also highly likely that the adaptors and coreceptors required for the activation of downstream signaling are present in heterologous cells and allow a functional signal output upon stimulation with appropriate ligands. There are indications that the signaling pathways involved in defense and development exhibit common features, for example, sharing MAPKs signaling components [31,32]. Many RLCKs serve as the core of PRRs and downstream defense systems [33]. In these RLCKs, BIK1 (Botrytis-induced kinase 1) is an important central component [34]. When plants are induced by PAMPs, the complex of BAK1 and PAMP phosphorylates BIK1, after which BIK1 dissociates from the complex and activates the downstream signal. Subsequently, the downstream reaction is accomplished by the MAPK signaling pathway [35]. Early research on parsley cells showed that the activation of elicitor-responsive MAPK, a homolog of AtMPK3, by a fungal elicitor results in the translocation of MAPK into the nucleus [31]. In *Arabidopsis*, AtMPK6 is activated by the bacterial flagellin peptide or by xylanase from the fungus *Trichoderma viride* [36]. These results suggested that MAPK might phosphorylate transcription factors that are involved in the plant defense response. Numerous studies have shown that PTI occurs in the early phase of plant defense [3]. Most PRRs are present at low concentrations in the plasma membrane, where the expression of these PRRs does not show any obvious changes in response to challenge by pathogenic bacteria. However, some studies have shown that PRRs may be upregulated after pathogenic invasion, thereby enhancing plant resistance [37]. Our qRT-PCR results indicated that *VaHAESA* expression was upregulated in the early stages of invasion after inoculation with a *P. viticola* strain, which is consistent with early immune responses to pathogenic challenge.

Previous research has shown that $H_2O_2$ is one of the earliest measurable indicators of PTI activity, representing a stable measure of the ROS produced in plants in response to pathogenic infection and PAMPs. This production of $H_2O_2$ is typically apoplastic but is subsequently associated with intracellular immunity-related pathways that regulate disease resistance, such as systemic acquired resistance (SAR) and PTI [38,39]. The transient expression of *VaHAESA* in the *V. vinifera* 'Thompson Seedless' variety resulted in a decrease in disease incidence, accumulation of $H_2O_2$, increased NO levels, and deposition of callose. These results indicate that *VaHAESA* enhances resistance to *P. viticola* in grapevines via the induction of resistance signaling and other molecular and cell wall modifications. Moreover, our results indicate that these phenomena, which occurred at the very early stages of infection in the leaves, were caused by the transient expression of *VaHAESA* after inoculation with *P. viticola*. Our observations of increased $H_2O_2$ levels further suggest a close relationship between ROS- and NO-signaling during the responses to pathogenic attacks [40,41]. Concerning the lag phase of NO generation, cryptogein, and chitosan (a deacylated derivative of chitin) induce NO production within two minutes, and specifically in the case of chitosan, NO production was shown to increase constantly until the last measured time point [42]. The ultimate outcome of PTI is the induction of resistance responses that prevent microbial colonization. Callose deposition is typically one of the late defense responses to pathogen invasion, with accumulation beginning approximately 16 h after the initiation of PTI [3]. Callose deposition is an important feature of plant immunity and is thought to reinforce the cell wall at fungal penetration sites to impede further infection [39,43,44]. In the present study, we showed that *VaHAESA* expression increased the deposition of callose in transgenic *V. vinifera* inoculated with *P. viticola*, suggesting that callose plays an important role in the PTI resistance of grapevines to downy mildew.

In conclusion, there is a large genotypic component in the resistance of grapevines to downy mildew, and defense reactions occur with variable timing and intensity. Several studies have indicated that increased production of ROS (superoxide radicals, 4 to 6 hpi) is followed by a hypersensitive response (6 to 8 hpi); subsequent increased activity of peroxidase in cells flanking the infection area and in the vascular tissue (10 to 12 hpi); and finally, increased production, accumulation, or conversion of phenolic compounds (12 to 15 hpi) [45,46]. The present study shows that *VaHAESA* acts as a PRR in grapevines that could initiate responses against pathogenic attacks; however, more detailed characterization of the genes and pathways involved is necessary to improve our understanding of the role of *VaHAESA* in the resistance of grapevines to downy mildew.

## 4. Materials and Methods

### 4.1. Plant Materials, Plasmopara viticola Strains, and Pathogen Infection

One-year old *V. amurensis* grapevines of the "Shuanghong" variety were maintained in a greenhouse under a 16:8 h light:dark cycle at 25 °C, with 85% relative humidity. The *Plasmopara viticola* strains "ZJ-1-1" and "JL-7-2" [13] were subcultured on *V. amurensis* leaf discs every 10 d at 22 °C under a 16:8 h photoperiod. The third to fifth unfolded leaves from the shoot apex of *V. amurensis* were inoculated with a suspension of $10^5$ *P. viticola* sporangia·mL$^{-1}$. Three leaves were pooled to represent one replicate, and three independent replicates were collected from each sample. Infected leaves were collected at 0, 0.5, 1, 2, 4, 6, 8, 12, 24, 48, 72, and 96 h post-infection (hpi). These samples were used for subsequent reverse transcription polymerase chain reaction (RT-PCR) experiments.

### 4.2. qRT-PCR

We measured the expression patterns of PRRs in the leaves of "Shuanghong" at different time points post-inoculation with P. viticola strains "ZJ-1-1" and "JL-7-2". Total RNA was extracted from leaves using a modified CTAB method [47]. One microgram of total RNA was reverse transcribed into first-strand cDNA using a cDNA Synthesis Kit (TaKaRa Biotechnology, Dalian, China). The PRR genes were identified through transcriptome analysis [13]. Primers for qRT-PCR were designed using Beacon Designer ver. 8.10 (Premier Biosoft, Palo Alto, CA, USA). Vitis elongation factor 1-α (EF1-α), ubiquitin-conjugating enzyme (UBQ), and SAND family protein (SAND) were used as internal controls to normalize all data [48,49]. The fold-change in gene expression was estimated based on threshold cycles via the $2^{-\Delta\Delta CT}$ method [50].

*4.3. Cloning, Sequencing, and Phylogenetic Characterization of VaHAESA*

The full open reading frames of *VaHAESA* genes were amplified from cDNA isolated from *V. amurensis* leaves inoculated with *P. viticola* "ZJ-1-1" using gene-specific primers (forward: 5′-ATGTCGAAAACACCCCCACCTTCTG-3′, reverse: 5′-TCACACATTGGACGCAAAATTC-3′). PCR amplification was performed in a final volume of 50 μL under the following conditions: initial denaturation at 98 °C for 3 min; followed by 34 cycles of 98 °C for 30 s, 55 °C for 30 s, and 72 °C for 1 min; with a final extension at 72 °C for 5 min (Phusion High-Fidelity PCR Kit; NEB). The amplification products were cloned into the expression vector pBI121. The putative VaRLK protein sequences were submitted to Pfam (available online: http://pfam.xfam.org) and SUPERFAMILY 2 (available online: http://supfam.org) to further characterize this gene.

Phylogenetic analyses were performed using 27 known LRR-RLKs from *Arabidopsis* and several HAESA or HAESA-like proteins from five different plants. Mega 5.05 software (Koichiro Tamura, et al., 2011) was employed to infer and align protein sequences with the default parameters [51]. A phylogenetic tree was constructed using the maximum-likelihood method, the JTT substitution model, and the "G + I rates among sites" model. Bootstrapping with 500 replicates was performed to assess the reliability of internal branches, and the nodes with bootstrap values greater than 50 were marked.

*4.4. Agrobacterium-Mediated Transient Expression in Plants*

To explore the function of the *VaHAESA* gene in disease resistance in grapevines, we transiently transformed *Vitis vinifera* variety "Thompson Seedless" plants with the expression vector carrying *VaHAESA* and the empty vector and then inoculated the transgenic plants with compatible *P. viticola* to test the effect of *VaHAESA* on resistance. The constructs and empty vectors were transformed into *Agrobacterium tumefaciens* "GV3101" according the manufacturer's instructions (BC304-01, Biomed, Beijing, China). Agrobacterium-mediated transient expression in *Nicotiana benthamiana* was performed as previously described to assess localization [52], and *Agrobacterium*-mediated transformation of *Vitis vinifera* "Thompson Seedless" was performed using vacuum infiltration according to previously published methods [53].

*4.5. Analysis of the Subcellular Localization of VaHAESA and Its Effect on Pathogen Infection*

To verify that the expression of *VaHAESA* improved disease resistance in *V. vinifera*, we performed a microscopic examination of the leaves three days after inoculation. To analyze the subcellular localization of *VaHAESA* in *N. benthamiana*, the leaves were immersed in PBS buffer containing 5 mg·L$^{-1}$ 4,6-diamidino-2-phenylindole (DAPI) for 10 min to stain the nuclei. Subsequently, leaf patches were mounted on microscope slides and observed using a Nikon C1 Si/TE2000E confocal laser-scanning microscope (Nikon, Minato, Tokyo, Japan).

To observe the development of *P. viticola* after leaf inoculation, *V. vinifera* 'Thompson Seedless' leaves from untransformed wild-type plants transiently expressing *VaHAESA* were collected at the corresponding time points. The leaves were stained with lactophenol-trypan blue (10 mL of lactic acid, 10 mL of glycerol, 10 g of phenol, and 10 mg of trypan blue dissolved in 10 mL of distilled water) following Keogh et al. [54]. For the analysis of callose, the leaves were treated according to the KOH-aniline blue fluorescence method [55]. Callose deposits were visualized under a UV filter using a fluorescence microscope and were counted using ImageJ 1.43U software (available online: https://imagej.nih.gov/ij/index.html).

The number of deposits was expressed as the mean of three different leaf areas. Under the microscope at a constant magnification, five fields of view were selected for each leaf, and the number of calli was counted. The average value was calculated for the final statistics.

*4.6. Analysis of $H_2O_2$ and NO Levels in Transgenic Vitis vinifera Expressing VaHAESA*

The detection of $H_2O_2$ was performed using a Hydrogen Peroxide Assay Kit (S0038, Beyotime, Shanghai, China). Each sample was ground to powder with liquid nitrogen, and 100 mg of the sample was then transferred to a 1.5 mL screw-cap tube containing 1.5 mL of lysate solvent. The tubes were shaken at a speed of 12,000 rpm at 4 °C for 5 min, after which the suspension was placed on ice. The lysate was used to dilute $H_2O_2$ to concentrations of 1, 3, 10, 30, and 100 µM, which were employed as standards. Then, 50 µL of the samples or standard products was added to a 96-well plate, and 100 µL of the peroxide detection reagent was added. After a mild shock, the 96-well plate was placed at room temperature for 30 min. The absorbance was detected at A560. The content of $H_2O_2$ in the sample was calculated according to the standard curve. The detection of NO was performed using the Total Nitric Oxide Assay Kit (S0023, Beyotime, Shanghai, China) according to the manufacturer's instructions. First, 1 g of frozen leaf tissue was ground and then added to 50 µL of Griess Reagent I. The mixture was then heated for 5 min in a boiling water bath to denature the proteins, followed by centrifugation for 5 min at 12,000 *g*. The supernatants were subsequently collected, and 5 µL of 20 mM NADPH, 10 µL of FAD and 5 µL of nitrate reductase were added. After mixing, the samples were incubated at 37 °C for 30 min. Next, 10 µL LDH buffer and 10 µL LDH were added to the mixture, followed by incubation at 37 °C for 30 min. Finally, 50 µL of Griess Reagent I and Griess Reagent II were added to the mixture. A540 was determined after incubation for 10 min at room temperature (20–30 °C).

*4.7. Screening and Identification of Transgenic Arabidopsis thaliana*

To identify the function of the *VaHAESA* gene, the transformation vector 35S::*VaHAESA*-pBI121 was transformed into *Agrobacterium tumefaciens* "GV3101". Using the floral dip transformation method [56], these strains were transferred to wild-type *Arabidopsis thaliana*. Transgenic plants were selected using 1/2 MS media containing 50 mg·L$^{-1}$ kanamycin and then transferred to soil (24 °C; 16/8 h light/dark). Screening was performed until stable homozygous T$_3$ lines were obtained, and transgenic plants were subsequently tested for *VaHAESA* expression via PCR and qRT-PCR. Disease resistance in wild-type and transgenic *A. thaliana* was monitored through inoculation with *H. arabidopsidis*, and the detection of related genes was performed via qRT-PCR. *A. thaliana* was inoculated with a suspension of $10^5$ *H. arabidopsidis* sporangia·mL$^{-1}$. Three independent lines were collected from each sample. Infected leaves were collected at 0, 2, 4, 8, 12, 24, 48, and 72 h post-infection (hpi). These samples were used for subsequent qRT-PCR experiments. *AtACTIN*, *AtUBQ*, and *AtSAND* were employed as internal controls to normalize all data. The phenotypic observation of disease resistance was conducted at 5 dpi.

**Acknowledgments:** We are grateful to the China Agriculture Research System (grant No. CARS-29-yc-2) and the Guangxi Bagui Scholar Fund (2013-3).

**Author Contributions:** Shaoli Liu, Chi Zhang, Jiang Lu, and Yali Zhang conceived and designed the experiments; Shaoli Liu and Chi Zhang performed the experiments; Shaoli Liu and Nan Chao analyzed the data; Shaoli Liu wrote the paper.

**Conflicts of Interest:** The authors declare no conflict of interest.

**References**

1. Casagrande, K.; Falginella, L.; Castellarin, S.D.; Testolin, R.; Di Gaspero, G. Defence responses in Rpv3-dependent resistance to grapevine downy mildew. *Planta* **2011**, *234*, 1097–1099. [CrossRef] [PubMed]
2. Milli, A.; Cecconi, D.; Bortesi, L.; Persi, A.; Rinalducci, S.; Zamboni, A.; Zoccatelli, G.; Lovato, A.; Zolla, L.; Polverari, A. Proteomic analysis of the compatible interaction between Vitis vinifera and Plasmopara viticola. *J. Proteom.* **2012**, *75*, 1284–1302. [CrossRef] [PubMed]
3. Boller, T.; Felix, G. A renaissance of elicitors: Perception of microbe-associated molecular patterns and danger signals by pattern-recognition receptors. *Annu. Rev. Plant Biol.* **2009**, *60*, 379–406. [CrossRef] [PubMed]
4. Jones, J.D.; Dangl, J.L. The plant immune system. *Nature* **2006**, *444*, 323. [CrossRef] [PubMed]

5.	Ausubel, F.M. Are innate immune signaling pathways in plants and animals conserved? *Nat. Immunol.* **2005**, *6*, 973. [CrossRef] [PubMed]

6.	Monaghan, J.; Zipfel, C. Plant pattern recognition receptor complexes at the plasma membrane. *Curr. Opin. Plant Biol.* **2012**, *15*, 349–357. [CrossRef] [PubMed]

7.	Gómezgómez, L.; Boller, T. FLS2: An LRR receptor-like kinase involved in the perception of the bacterial elicitor flagellin in Arabidopsis. *Mol. Cell* **2000**, *5*, 1003–1011. [CrossRef]

8.	Robatzek, S.; Bittel, P.; Chinchilla, D.; Köchner, P.; Felix, G.; Shiu, S.H.; Boller, T. Molecular identification and characterization of the tomato flagellin receptor LeFLS2, an orthologue of Arabidopsis FLS2 exhibiting characteristically different perception specificities. *Plant Mol. Biol.* **2007**, *64*, 539–547. [CrossRef] [PubMed]

9.	Smith, J.M.; Salamango, D.J.; Leslie, M.E.; Collins, C.A.; Heese, A. Sensitivity to Flg22 is modulated by ligand-induced degradation and de novo synthesis of the endogenous flagellin-receptor FLAGELLIN-SENSING2. *Plant Physiol.* **2014**, *164*, 440–454. [CrossRef] [PubMed]

10.	Zipfel, C.; Robatzek, S.; Navarro, L.; Oakeley, E.J.; Jones, J.D.; Felix, G.; Boller, T. Bacterial disease resistance in Arabidopsis through flagellin perception. *Nature* **2004**, *428*, 764. [CrossRef] [PubMed]

11.	Zipfel, C.; Kunze, G.; Chinchilla, D.; Caniard, A.; Jones, J.D.; Boller, T.; Felix, G. Perception of the bacterial PAMP EF-Tu by the receptor EFR restricts Agrobacterium-mediated transformation. *Cell* **2006**, *125*, 749–760. [CrossRef] [PubMed]

12.	Xu, W.; Li, R.; Zhang, N.; Ma, F.; Jiao, Y.; Wang, Z. Transcriptome profiling of Vitis amurensis, an extremely cold-tolerant Chinese wild Vitis species, reveals candidate genes and events that potentially connected to cold stress. *Plant Mol. Biol.* **2014**, *86*, 527–541. [CrossRef] [PubMed]

13.	Li, X.; Wu, J.; Yin, L.; Zhang, Y.; Qu, J.; Lu, J. Comparative transcriptome analysis reveals defense-related genes and pathways against downy mildew in Vitis amurensis grapevine. *Plant Physiol. Biochem.* **2015**, *95*, 1–14. [CrossRef] [PubMed]

14.	Han, J.; Fang, J.; Wang, C.; Yin, Y.; Sun, X.; Leng, X.; Song, C. Grapevine microRNAs responsive to exogenous gibberellin. *BMC Genom.* **2014**, *15*, 111. [CrossRef] [PubMed]

15.	Walker, J.C. Receptor-like protein kinase genes of Arabidopsis thaliana. *Plant J. Cell Mol. Biol.* **2015**, *3*, 451–456. [CrossRef]

16.	Cho, S.K.; Larue, C.T.; Chevalier, D.; Wang, H.; Jinn, T.L.; Zhang, S.; Walker, J.C. From the Cover: Regulation of floral organ abscission in Arabidopsis thaliana. *Proc. Natl. Acad. Sci. USA* **2007**, *105*, 15629–15634. [CrossRef] [PubMed]

17.	Roux, M.; Schwessinger, B.; Albrecht, C.; Chinchilla, D.; Jones, A.; Holton, N.; Malinovsky, F.G.; Tör, M.; De Vries, S.; Zipfel, C. The Arabidopsis leucine-rich repeat receptor-like kinases BAK1/SERK3 and BKK1/SERK4 are required for innate immunity to hemibiotrophic and biotrophic pathogens. *Plant Cell* **2011**, *23*, 2440–2455. [CrossRef] [PubMed]

18.	Tintor, N.; Ross, A.; Kanehara, K.; Yamada, K.; Fan, L.; Kemmerling, B.; Nürnberger, T.; Tsuda, K.; Saijo, Y. Layered pattern receptor signaling via ethylene and endogenous elicitor peptides during Arabidopsis immunity to bacterial infection. *Proc. Natl. Acad. Sci. USA* **2013**, *110*, 6211–6216. [CrossRef] [PubMed]

19.	Shi, H.; Shen, Q.; Qi, Y.; Yan, H.; Nie, H.; Chen, Y.; Zhao, T.; Katagiri, F.; Tang, D. BR-SIGNALING KINASE1 physically associates with FLAGELLIN SENSING2 and regulates plant innate immunity in Arabidopsis. *Plant Cell* **2013**, *25*, 1143–1157. [CrossRef] [PubMed]

20.	Shiu, S.H.; Karlowski, W.M.; Pan, R.; Tzeng, Y.H.; Mayer, K.F.; Li, W.H. Comparative analysis of the receptor-like kinase family in Arabidopsis and rice. *Plant Cell* **2004**, *16*, 1220–1234. [CrossRef] [PubMed]

21.	Jinn, T.L.; Stone, J.M.; Walker, J.C. HAESA, an Arabidopsis leucine-rich repeat receptor kinase, controls floral organ abscission. *Gene Dev.* **2000**, *14*, 108–117. [PubMed]

22.	Butenko, M.A.; Wildhagen, M.; Albert, M.; Jehle, A.; Kalbacher, H.; Aalen, R.B.; Felix, G. Tools and Strategies to Match Peptide-Ligand Receptor Pairs. *Plant Cell* **2014**, *26*, 1838–1847. [CrossRef] [PubMed]

23.	Julia, S.; Benjamin, B.; Mari, W.; Ulrich, H.; Hothorn, L.A.; Butenko, M.A.; Michael, H. Mechanistic insight into a peptide hormone signaling complex mediating floral organ abscission. *eLife* **2016**, *5*, e15075.

24.	Wong, F.P.; Burr, H.N.; Wilcox, W.F. Heterothallism in Plasmopara viticola. *Plant Pathol.* **2010**, *50*, 427–432. [CrossRef]

25.	Kemmerling, B.; Zipfel, C.; Chinchilla, D.; Felix, G.; Jones, J.D.G.; Robatzek, S.; Boller, T.; Nürnberger, T. A flagellin-induced complex of the receptor FLS2 and BAK1 initiates plant defence. *Nature* **2007**, *448*, 497.

26. Böhm, H.; Albert, I.; Fan, L.; Reinhard, A.; Nürnberger, T. Immune receptor complexes at the plant cell surface. *Curr. Opin. Plant Biol.* **2014**, *20*, 47–54. [CrossRef] [PubMed]

27. Sun, Y.; Li, L.; Macho, A.P.; Han, Z.; Hu, Z.; Zipfel, C.; Zhou, J.M.; Chai, J. Structural Basis for flg22-Induced Activation of the Arabidopsis FLS2-BAK1 Immune Complex. *Science* **2013**, *342*, 624–628. [CrossRef] [PubMed]

28. Schulze, B.; Mentzel, T.; Jehle, A.K.; Mueller, K.; Beeler, S.; Boller, T.; Felix, G.; Chinchilla, D. Rapid heteromerization and phosphorylation of ligand-activated plant transmembrane receptors and their associated kinase BAK1. *J. Biol. Chem.* **2010**, *285*, 9444–9451. [CrossRef] [PubMed]

29. Niederhuth, C.E.; Rahul, P.O.; Walker, J.C. Transcriptional profiling of the Arabidopsis abscission mutanthae hsl2by RNA-Seq. *BMC Genom.* **2013**, *14*, 37. [CrossRef] [PubMed]

30. Gómez-Gómez, L.; Boller, T. Flagellin perception: A paradigm for innate immunity. *Trends Plant Sci.* **2002**, *7*, 251–256. [CrossRef]

31. Jonak, C.; Okrész, L.; Bögre, L.; Hirt, H. Complexity, cross talk and integration of plant MAP kinase signalling. *Curr. Opin. Plant Biol.* **2002**, *5*, 415–424. [CrossRef]

32. Betsuyaku, S.; Takahashi, F.; Kinoshita, A.; Miwa, H.; Shinozaki, K.; Fukuda, H.; Sawa, S. Mitogen-activated protein kinase regulated by the CLAVATA receptors contributes to shoot apical meristem homeostasis. *Plant Cell Physiol.* **2011**, *52*, 14–29. [CrossRef] [PubMed]

33. Lin, W.; Ma, X.; Shan, L.; He, P. Big Roles of Small Kinases:The Complex Functions of Receptor-Like Cytoplasmic Kinases in Plant Immunity and Development. *Chin. J. Plant Ecol.* **2013**, *55*, 1188–1197.

34. Veronese, P.; Nakagami, H.; Bluhm, B.; Abuqamar, S.; Chen, X.; Salmeron, J.; Dietrich, R.A.; Hirt, H.; Mengiste, T. The membrane-anchored BOTRYTIS-INDUCED KINASE1 plays distinct roles in Arabidopsis resistance to necrotrophic and biotrophic pathogens. *Plant Cell* **2006**, *18*, 257–273. [CrossRef] [PubMed]

35. Meng, X.; Zhang, S. MAPK cascades in plant disease resistance signaling. *Annu. Rev. Phytopathol.* **2013**, *51*, 245–266. [CrossRef] [PubMed]

36. Nühse, T.S.; Peck, S.C.; Hirt, H.; Boller, T. Microbial elicitors induce activation and dual phosphorylation of the Arabidopsis thaliana MAPK 6. *J. Biol. Chem.* **2000**, *275*, 7521–7526. [CrossRef] [PubMed]

37. Sun, W.; Dunning, F.M.; Pfund, C.; Weingarten, R.; Bent, A.F. Within-species flagellin polymorphism in Xanthomonas campestris pv campestris and its impact on elicitation of Arabidopsis FLAGELLIN SENSING2-dependent defenses. *Plant Cell* **2006**, *18*, 764. [CrossRef] [PubMed]

38. Torres, M.A. ROS in biotic interactions. *Physiol. Plant.* **2010**, *138*, 414. [CrossRef] [PubMed]

39. Tian, S.; Wang, X.; Li, P.; Wang, H.; Ji, H.; Xie, J.; Qiu, Q.; Shen, D.; Dong, H. Plant Aquaporin AtPIP1;4 Links Apoplastic H2O2 Induction to Disease Immunity Pathways. *Plant Physiol.* **2016**, *171*. [CrossRef] [PubMed]

40. Groß, F.; Durner, J.; Gaupels, F. Nitric oxide, antioxidants and prooxidants in plant defence responses. *Front. Plant Sci.* **2013**, *4*, 419. [CrossRef] [PubMed]

41. Scheler, C.; Durner, J.; Astier, J. Nitric oxide and reactive oxygen species in plant biotic interactions. *Curr. Opin. Plant Biol.* **2013**, *16*, 534–539. [CrossRef] [PubMed]

42. Raho, N.; Ramirez, L.; Lanteri, M.L.; Gonorazky, G.; Lamattina, L.; Have, A.T.; Laxalt, A.M. Phosphatidic acid production in chitosan-elicited tomato cells, via both phospholipase D and phospholipase C/diacylglycerol kinase, requires nitric oxide. *J. Plant Physiol.* **2011**, *168*, 534–539. [CrossRef] [PubMed]

43. Underwood, W. The Plant Cell Wall: A Dynamic Barrier Against Pathogen Invasion. *Front. Plant Sci.* **2012**, *3*, 85. [CrossRef] [PubMed]

44. Yu, Y.; Zhang, Y.; Yin, L.; Lu, J. The mode of host resistance to Plasmopara viticola infection of grapevines. *Phytopathology* **2012**, *102*, 1094–1101. [CrossRef] [PubMed]

45. Liu, S.L.; Jiao, W.; Pei, Z.; Hasi, G.; Yu, H.; Jiang, L.; Zhang, Y.L. Response of phytohormones and correlation of SAR signal pathway genes to the different resistance levels of grapevine against Plasmopara viticola infection. *Plant Physiol. Biochem.* **2016**, *107*, 56–66. [CrossRef] [PubMed]

46. Jürges, G.; Kassemeyer, H.H.; Dürrenberger, M.; Düggelin, M.; Nick, P. The mode of interaction between Vitis and Plasmopara viticola Berk. & Curt. Ex de Bary depends on the host species. *Plant Biol.* **2009**, *11*, 886–898. [PubMed]

47. Iandolino, A.B.; Silva, F.G.D.; Lim, H.; Choi, H.; Williams, L.E.; Cook, D.R. High-Quality RNA, cDNA, and Derived EST Libraries From Grapevine (*Vitis vinifera* L.). *Plant Mol. Biol. Rep.* **2004**, *22*, 269–278. [CrossRef]

48. Monteiro, F.; Sebastiana, M.; Pais, M.S.; Figueiredo, A. Reference Gene Selection and Validation for the Early Responses to Downy Mildew Infection in Susceptible and Resistant Vitis vinifera Cultivars. *Minerva Stomatol.* **2012**, *56*, 611. [CrossRef] [PubMed]

49. Selim, M.; Legay, S.; Berkelmannlöhnertz, B.; Langen, G.; Kogel, K.H.; Evers, D. Identification of suitable reference genes for real-time RT-PCR normalization in the grapevine-downy mildew pathosystem. *Plant Cell Rep.* **2012**, *31*, 205–216. [CrossRef] [PubMed]

50. Livak, K.J.; Schmittgen, T.D. Analysis of relative gene expression data using real-time quantitative PCR and the 2(-Delta Delta C(T)) Method. *Methods* **2001**, *25*, 402–408. [CrossRef] [PubMed]

51. Tamura, K.; Peterson, D.; Peterson, N.; Stecher, G.; Nei, M.; Kumar, S. MEGA5: Molecular Evolutionary Genetics Analysis Using Maximum Likelihood, Evolutionary Distance, and Maximum Parsimony Methods. *Mol. Biol. Evol.* **2011**, *28*, 2731–2739. [CrossRef] [PubMed]

52. Jiang, X.; Li, X.; Wu, J.; Yin, L.; Zhang, Y.; Lu, J. Studying the Mechanism of Plasmopara viticola RxLR Effectors on Suppressing Plant Immunity. *Front. Microbiol.* **2016**, *7*, 709.

53. Guan, X.; Zhao, H.; Xu, Y.; Wang, Y. Transient expression of glyoxal oxidase from the Chinese wild grape Vitis pseudoreticulata can suppress powdery mildew in a susceptible genotype. *Protoplasma* **2011**, *248*, 415–423. [CrossRef] [PubMed]

54. Keogh, R.C.; Deverall, B.J.; Mcleod, S. Comparison of histological and physiological responses to Phakopsora pachyrhizi resistant and susceptible soybean. *Trans. Br. Mycol. Soc.* **1980**, *74*, 329–333. [CrossRef]

55. Díez-Navajas, A.M.; Greif, C.; Poutaraud, A.; Merdinoglu, D. Two simplified fluorescent staining techniques to observe infection structures of the oomycete Plasmopara viticola in grapevine leaf tissues. *Micron* **2007**, *38*, 680–683. [CrossRef] [PubMed]

56. Clough, S.J.; Bent, A.F. Floral dip: A simplified method for Agrobacterium-mediated transformation of Arabidopsis thaliana. *Plant J.* **1998**, *16*, 735–743. [CrossRef] [PubMed]

International Journal of
*Molecular Sciences*

MDPI

*Article*

# Comparative Analysis of Impatiens Leaf Transcriptomes Reveal Candidate Genes for Resistance to Downy Mildew Caused by *Plasmopara obducens*

Krishna Bhattarai, Weining Wang, Zhe Cao and Zhanao Deng *

Department of Environmental Horticulture, Gulf Coast Research and Education Center, IFAS, University of Florida, Wimauma, FL 33598, USA; krishnabhattarai@ufl.edu (K.B.); billwang@ufl.edu (W.W.); cjun01@gmail.com (Z.C.)
* Correspondence: zdeng@ufl.edu; Tel.: +1-813-419-6605; Fax: +1-813-419-6641

Received: 13 June 2018; Accepted: 9 July 2018; Published: 15 July 2018

**Abstract:** Impatiens downy mildew (IDM) is a devastating disease to garden impatiens. A good understanding of IDM resistance in New Guinea impatiens is essential for improving garden impatiens resistance to this disease. The present study was conducted to sequence, assemble, annotate and compare the leaf transcriptomes of two impatiens cultivars differing in resistance to IDM, reveal sequence polymorphisms and identify candidate genes for IDM resistance. RNA-Seq was performed on cultivars Super Elfin® XP Pink (SEP) and SunPatiens® Compact Royal Magenta (SPR). De novo assembly of obtained sequence reads resulted in 121,497 unigenes with an average length of 1156 nucleotides and N50 length of 1778 nucleotides. Searching the non-redundant protein and non-redundant nucleotide, Swiss-Prot, Kyoto Encyclopedia of Genes and Genomes and Clusters of Orthologous Groups and Gene Ontology databases, resulted in annotation of 57.7% to 73.6% of the unigenes. Fifteen unigenes were highly similar to disease resistance genes and more abundant in the IDM-resistant cultivar than in the susceptible cultivar. A total of 22,484 simple sequence repeats (SSRs) and 245,936 and 120,073 single nucleotide polymorphisms (SNPs) were identified from SPR and SEP respectively. The assembled transcripts and unigenes, identified disease resistance genes and SSRs and SNPs sites will be a valuable resource for improving impatiens and its IDM resistance.

**Keywords:** candidate disease resistance gene; disease resistance; downy mildew; garden impatiens; leaf transcriptome; New Guinea impatiens; RNA-Seq

---

## 1. Introduction

The Genus *Impatiens* belongs to the family Balsaminaceae, which consists of about 850 species of mostly succulent annual or perennial herbs [1]. Wild *Impatiens* plants are primarily found in the mountainous regions of South-East Asia, south China, India and Africa while some species are also found in Japan, Europe, Russia and North America [2]. Garden impatiens (*Impatiens walleriana* Hook.f.) is one of the most popular flowers in the world, widely grown in garden beds, borders and woodland gardens as bedding plants and in containers, window boxes and hanging baskets as house plants [3,4]. Impatiens flowers are available virtually in all colors [3,4].

Impatiens downy mildew (IDM), caused by *Plasmopara obducens* (J. Impatiens noli-tangere), is a devastating disease to impatiens and can cause rapid defoliation and plant death. This disease was first reported in Germany in 1877 on *Impatiens noli-tangere*, a wild species of *Impatiens* native to many temperate countries in the northern hemisphere [5]. During the 1880s, *P. obducens* was identified in North America in native *Impatiens* species including *I. capensis* (synonym *I. biflora*), *I. fulva* and *I. pallida* [5]. The first recent occurrence of IDM on *I. walleriana* was reported in the UK in 2003 [6].

Subsequently it spread to other countries in Europe (including Norway (2010), Serbia (2011) and Hungary (2011)) and other continents [7–9]. IDM was reported in Taiwan and Japan in 2003 and in Australia in 2008 [10–12]. Since then, IDM has been problematic in these countries. In the USA, IDM appeared on *I. walleriana* plants in California in 2004 [13]. In 2011, IDM was reported in eleven states in the USA including Massachusetts, New York and Minnesota [14]. By 2012, this disease spread to 33 states [13]. In 2013, IDM was reported in Hawaii [15]. Outbreak of this disease in the USA in this short period of time has caused the loss of hundreds of millions of dollars, reducing the national annual wholesale value of impatiens from approximately $150 million in 2005 to $65 million in 2015 [16]. While *I. walleriana* is highly susceptible to IDM, New Guinea Impatiens (NGI) (*Impatiens hawkeri*) and its interspecific hybrids have shown resistance to IDM [17]. NGI has been grown as a substitute of garden impatiens currently and occupies a $55-million wholesale market [16]. Although the market of NGI is increasing, it ranked third among the alternatives of garden impatiens [18].

There has been a strong interest in introducing IDM resistance into *I. walleriana* cultivars due to continued strong consumer demand. *Impatiens walleriana* cultivars are available in a wide array of vibrant colors, possess excellent shade tolerance and are well adapted to a wide range of growing conditions in containers and garden beds [3]. Moreover, the ease of culturing, wide and greater availability of the seeds and cutting materials has increased the demand of garden impatiens [3]. Use of traditional breeding approaches to transferring the resistance from NGI to garden impatiens has been impeded by the differences between NGI and garden impatiens in chromosome number ($2n = 2\times = 16$ in *I. walleriana* and $2n = 2\times = 32$ in NGI), size and morphology [19]. Hence, it is necessary to identify and isolate the gene(s) conferring IDM resistance in NGI and transfer the identified gene(s) into garden impatiens.

Disease resistance (*R*) genes are the most important component of the plant defense mechanism to confer resistance to pathogens carrying matching avirulence genes [20]. A great majority of isolated plant *R* genes encode nucleotide-binding site leucine-rich repeats (NB-LRR) proteins [21]. Based on structure, NB-LRR proteins are separated into Toll/interleukin-1 receptor (TIR)-domain-containing (TNL) and coiled coil (CC)-domain-containing (CNL) subfamilies. Plant NB-LRR proteins function in a network through signaling pathways and induce a series of defense responses such as initiation of oxidative burst, calcium and ion fluxes, mitogen-associated protein kinase cascade, induction of pathogenesis-related genes and hypersensitive responses [22–25]. Recent studies have shown that the TNL subfamily R proteins transduce signals through the enhanced disease susceptibility 1 (EDS1) protein and the CNL subfamily R proteins do so through non-race specific disease resistance 1 (NDR1) protein with some interchanges occurring in different plants [26]. There also exists a separate independent signal transduction pathway activated by Arabidopsis CNL proteins RPP8 and RPP13.

Few genomic resources are available in impatiens. Genomic and transcriptomic data are needed in impatiens to facilitate the identification of gene(s) conferring IDM resistance in NGI and the development of simple sequence repeat (SSR) and single nucleotide polymorphism (SNP) markers for impatiens breeding. With the rapid development of next generation sequencing and analytical tools, whole-genome and transcriptome sequencing has become possible in non-model organisms. Transcriptome sequencing and characterization can provide a descriptive and biological insight of the functionality of an organism. Recently, transcriptome sequencing has been used commonly to identify candidate *R* genes and to develop molecular markers for disease resistance breeding [27,28]. There has been very few or no study on genes involved in disease resistance using transcriptome sequencing in impatiens. Here, we report the sequencing, assembly, annotation and characterization of the leaf transcriptomes of two impatiens cultivars (Super Elfin® XP Pink (SEP) and SunPatiens® Compact Royal Magenta (SPR)) that differ in IDM resistance and the use of the transcriptome data to identify differentially expressed transcripts and candidate *R* genes and SSR and SNP sites for development of molecular markers. Results of this study, including the assembled transcriptomes and the identified candidate genes and polymorphic sites, will greatly facilitate the development of molecular markers for dissecting the resistance mechanism for downy mildew resistance in NGI, tagging the responsible gene loci and expediting the improvement of downy mildew resistance in impatiens.

## 2. Results

### 2.1. HiSeq Sequencing and De Novo Assembly

A total of over 126 million raw reads were generated from the leaf transcriptomes of the two impatiens cultivars, resulting in approximately 117 million clean reads for the two cultivars and approximately 58 million clean reads per cultivar. More than 97% of the reads had a Phred score of 20. The GC content of impatiens sequence reads was approximately 45% (Table 1). De novo assembly of the impatiens leaf transcriptomes in Trinity resulted in 121,497 unigenes containing approximately 140 million bases of nucleotides. The average length of these unigenes was 1156 nucleotides and the N50 was 1778 nucleotides. The total number of contigs for SPR and SEP was 122,166 and 104,752, respectively. These contigs were functionally annotated, which produced 87,415 and 69,369 annotated unigenes for SPR and SEP, respectively (Table 2).

**Table 1.** Summary of HiSeq sequencing data for two impatiens leaf transcriptomes.

| Samples | Total Raw Reads | Total Clean Reads | Total Clean Nucleotides (Mb) | Q20 (%) [1] | N (%) [2] | GC (%) |
|---|---|---|---|---|---|---|
| SunPatiens® Compact Royal Magenta (SPR) | 63,769,252 | 58,266,032 | 5826 | 97.59 | 0.01 | 44.79 |
| Super Elfin® XP Pink (SEP) | 62,996,124 | 58,834,368 | 5883 | 97.50 | 0.01 | 45.11 |
| Total | 126,765,376 | 117,100,400 | 11,710 | | | |

[1] Q20 (%) is the proportion of nucleotides with Q20 quality value greater than 20. [2] N (%) is the proportion of unknown nucleotides in clean reads.

**Table 2.** Description of the impatiens transcriptome assembly and quality of unigenes and contigs.

| Types | Sample | Total Number | Total Length (nt) | Mean Length (nt) | N50 | Total Consensus Sequences | Distinct Clusters | Distinct Singletons |
|---|---|---|---|---|---|---|---|---|
| Contig | SunPatiens® Compact Royal Magenta (SPR) | 122,166 | 48,323,299 | 396 | 965 | | | |
| | Super Elfin® XP Pink (SEP) | 104,752 | 44,944,832 | 429 | 1052 | | | |
| Unigene | SunPatiens® Compact Royal Magenta (SPR) | 87,415 | 88,434,291 | 1012 | 1774 | 87,415 | 44,629 | 42,789 |
| | Super Elfin® XP Pink (SEP) | 69,369 | 67,514,711 | 973 | 1726 | 69,369 | 27,516 | 41,853 |
| | All | 121,497 | 140,506,651 | 1156 | 1778 | 121,497 | 78,448 | 43,049 |

There were 42,789 and 41,853 distinct singletons identified in SPR and SEP, respectively (Table 2). There were 12,835 (14.68%) and 9431 (13.60%) unigenes that were of 2000 or more nucleotides in length in SPR and SEP cultivars (Table S1). Similarly, 22,656 (18.55%) and 22,175 (21.17%) of contigs with more than 500 nucleotides length were assembled in SPR and SEP cultivars respectively (Table S2).

### 2.2. Functional Annotation

Unigenes were annotated by searching against protein databases including the non-redundant protein (NR), Swiss-Prot, Kyoto Encyclopedia of Genes and Genomes (KEGG) and Clusters of Orthologous Groups (COG) using the BLASTx with an *E*-value cut-off of $1 \times 10^{-5}$ and to the nucleotide database NT using the BLASTn with *E*-value cutoff of $1 \times 10^{-5}$. Out of the 121,497 unigenes, 91,187 were annotated (Table 3).

**Table 3.** Annotation of impatiens unigenes using the NT (the non-redundant nucleotide database), NR (the non-redundant protein database), SWISS-PROT, COG (Clusters of Orthologous Groups of proteins), KEGG (Kyoto Encyclopedia of Genes and Genomes) and GO (Gene Ontology Consortium) databases.

| Database | Unigene Annotated | Percentage |
|---|---|---|
| NR | 89,490 | 73.66 |
| NT | 71,482 | 58.83 |
| Swiss-Prot | 59,403 | 48.89 |
| KEGG | 54,521 | 44.87 |
| COG | 37,576 | 30.93 |
| GO | 70,190 | 57.77 |
| Total annotated genes | 91,187 | 75.05 |
| Total unigenes | 121,497 | 100.00 |

As shown in Table 3, more than 75% of the impatiens unigenes were annotated with at least one of the databases. At the nucleotide level, significant hits ($E$-value $1 \times 10^{-5}$) were observed for 71,482 (58.83%) unigenes in the non-redundant nucleotide database. Comparing the unigenes in the non-redundant protein database, 89,490 (73.66%) unigenes had significant hits ($E$-value $1 \times 10^{-5}$). The similar distribution of impatiens unigenes to other plant species, whose transcriptome sequences and gene annotations are available, show that the impatiens transcriptomes are closely related with *Vitis vinifera* (35.8%), *Lycopersicum esculentum* (13.5%), *Amygdalus persica* (9.2%), *Ricinus communis* (8.6%), *Populus balsamifera* subsp. *trichocarpa* (7.0%), *Fragaria vesca* subsp. *vesca* (4.6%) and *Glycine max* (4.4%) (Figure 1). Similarly, 59,403 (48.89%), 54,521 (44.87%), 37,576 (30.93%) and 70,190 (57.77%) of the impatiens unigenes had similarity hits in SWISS-Prot, KEGG, COG and GO databases, respectively (Table 3).

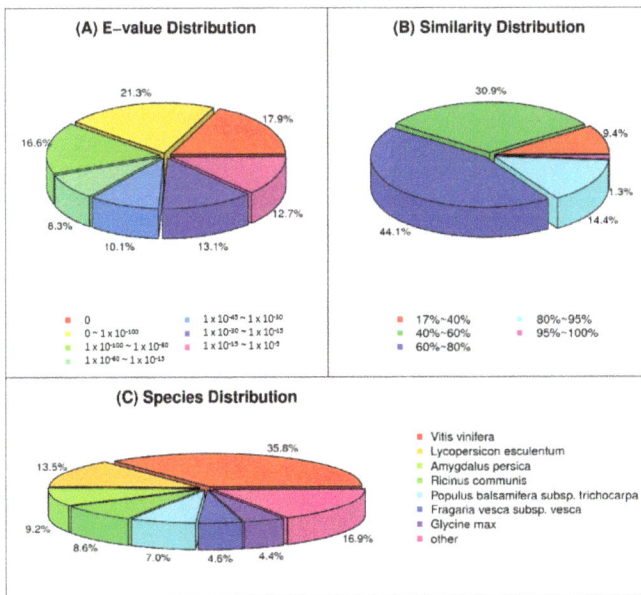

**Figure 1.** Distribution of similarities of impatiens unigenes with other plant species based on searching the non-redundant protein database according to (**A**) $E$-value (**B**) Similarity and (**C**). In relation to other species with unigenes similarity.

## 2.3. COG Classification and KEGG Pathway Mapping

More than 54,521 impatiens unigenes (44%) were annotated in the KEGG database (Table 3). These unigenes were classified into 128 pathways, of which the primary five classes are: cellular processes (3.13%), environmental information processing (2.34%), genetic information processing (16.41%), metabolism (75%) and organismal systems (4%) (Table 4).

**Table 4.** Classification of annotated unigene pathways according to categorical processes based on KEGG Pathway Mapping.

| Pathway Category | Number | Percentage |
|---|---|---|
| Cellular Processes | 4 | 3.13 |
| Environmental Information Processing | 3 | 2.34 |
| Genetic Information Processing | 21 | 16.41 |
| Metabolism | 96 | 75.00 |
| Organismal Systems | 4 | 3.13 |
| Total | 128 | 100 |

The top 25 mapped pathways annotated by the KEGG database are shown in Figure 2, among which metabolic pathways (24.70%) was the predominant one, followed by biosynthesis of secondary metabolites (12.01%), plant-pathogen interaction (7.50%), plant hormone signal transduction (6.93%) and spliceosome (4.30%) (Table 5).

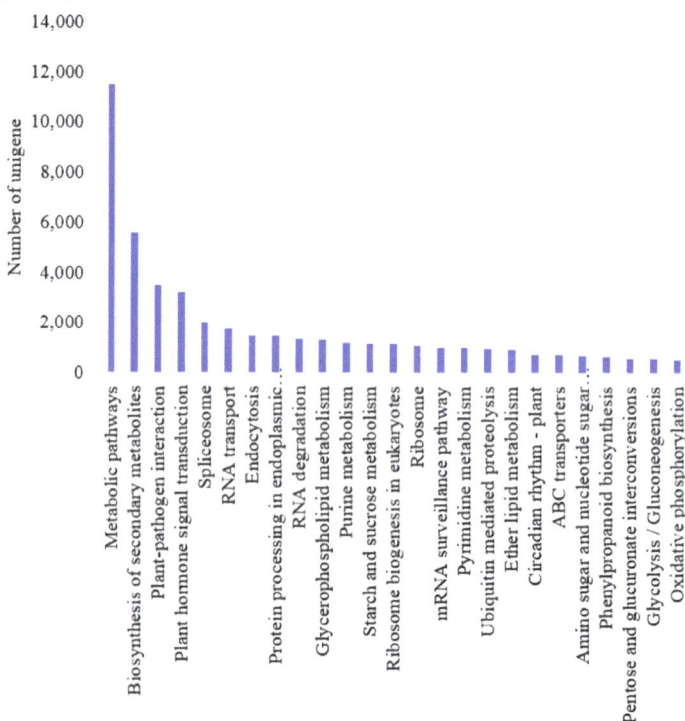

**Figure 2.** Top 25 pathways to which impatiens leaf transcriptomes were mapped using the Kyoto Encyclopedia of Genes and Genome Database (KEGG).

**Table 5.** Number of impatiens unigenes involved in top 25 pathways listed by KEGG database.

| Pathway | Genes with Pathway Annotation | % of Genes |
|---|---|---|
| Metabolic pathways | 11,493 | 24.70 |
| Biosynthesis of secondary metabolites | 5590 | 12.01 |
| Plant-pathogen interaction | 3491 | 7.50 |
| Plant hormone signal transduction | 3224 | 6.93 |
| Spliceosome | 2001 | 4.30 |
| RNA transport | 1760 | 3.78 |
| Endocytosis | 1488 | 3.20 |
| Protein processing in endoplasmic reticulum | 1484 | 3.19 |
| RNA degradation | 1366 | 2.94 |
| Glycerophospholipid metabolism | 1334 | 2.87 |
| Purine metabolism | 1237 | 2.66 |
| Starch and sucrose metabolism | 1186 | 2.55 |
| Ribosome biogenesis in eukaryotes | 1171 | 2.52 |
| Ribosome | 1117 | 2.40 |
| mRNA surveillance pathway | 1033 | 2.22 |
| Pyrimidine metabolism | 1027 | 2.21 |
| Ubiquitin mediated proteolysis | 967 | 2.08 |
| Ether lipid metabolism | 943 | 2.03 |
| Circadian rhythm–plant | 764 | 1.64 |
| ABC transporters | 758 | 1.63 |
| Amino sugar and nucleotide sugar metabolism | 712 | 1.53 |
| Phenylpropanoid biosynthesis | 660 | 1.42 |
| Pentose and glucoronate interconversions | 585 | 1.26 |
| Glycolysis / Gluconeogenesis | 584 | 1.25 |
| Oxidative phosphorylation | 564 | 1.21 |
| Total | 46,539 | 100 |

COG classification distributed the annotated impatiens unigenes into 25 categories of functional class. The General Function Prediction Only class contained 13,290 impatiens unigenes, followed by the Transcription class with 6959 impatiens unigenes. The Nuclear Structure class category had only one impatiens unigene (Figure 3). All the categories are shown in Figure 3.

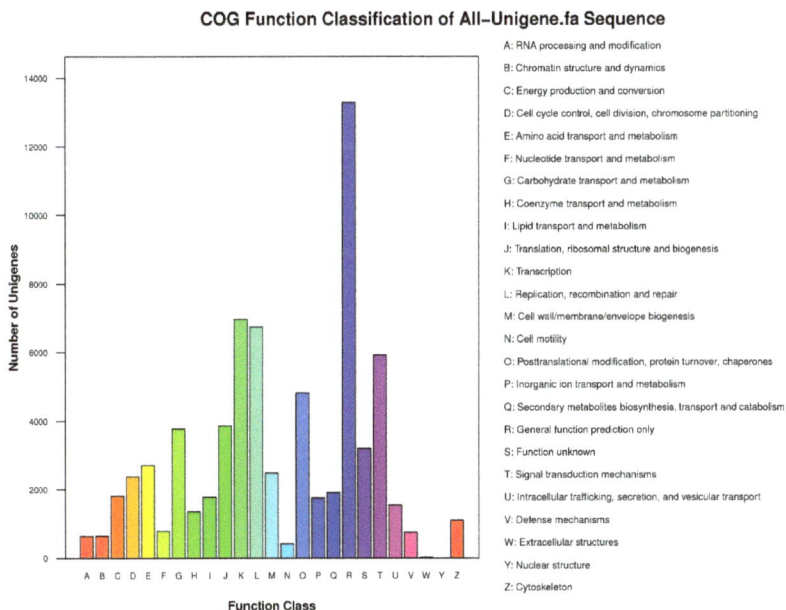

**Figure 3.** Functional classes of unigenes based on the Cluster of Orthologous Group (COG).

### 2.4. Functional Annotation Based on GO Classification

GO classification differentiated 70,190 impatiens unigenes into at least one of the three categories: biological process, cellular component and molecular function and 55 sub-categories. There were 22 sub-categories in the biological process, among which "cellular process," "metabolic process" and "single-organism process" had the three highest number of unigenes, involving 44,175, 42,260 and 31,456 unigenes, respectively (Figure 4).

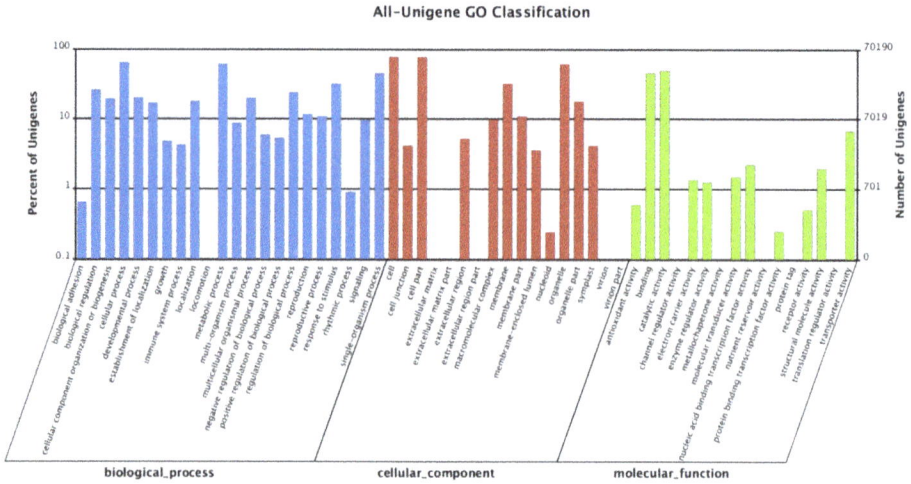

**Figure 4.** Classification of impatiens unigenes based on the Gene Ontology database.

Similarly, in the "cellular component" category, the top three components involving highest number of unigenes were "cell part", "cell" and "organelle" sub-categories which involved 53,480; 53,480 and 42,491 unigenes, respectively (Figure 4). In the third category "molecular function" the three sub-categories involving the highest number of unigenes were "catalytic activity", "binding" and "transporter activity" with 34,479; 31,890 and 4764 unigenes, respectively (Figure 4).

### 2.5. Disease Resistance/Defense Genes

In this study, we were particularly interested in genes that are potentially involved in disease resistance and defense-related mechanisms in impatiens. Hence, the impatiens unigenes that were functionally annotated to disease resistance with different databases were selected for further characterization. A total of 164 impatiens unigenes were identified through functional annotation to be potentially involved in disease resistance (Table S3). Fifteen of these unigenes were more than 2000 bp in length each and were found in a greater abundance in the IDM-resistant cultivar SPR than in the IDM-susceptible cultivar SEP (Table 6).

Two of these unigenes were similar to the NB-LRR class *R* genes and another two unigenes were similar to the LRR class *R* genes. In addition, two unigenes showed similarity to the recognition of *Peronospora parasitica 13* (*RPP13*)-like protein 1 family and four unigenes hit putative disease resistance proteins in *Arabidopsis thaliana*. A number of unigenes were similar to the LETM1-like protein, cation efflux family protein, receptor like protein 35, or DNA-binding storekeeper protein-related transcriptional regulator protein classes. There was an uncharacterized unigene, uncharacterized mitochondrial protein AtMg00860 that was found to be involved in the disease resistance mechanism (Table 6).

Sequence similarity comparison of the impatiens unigenes (Table 6) using their coding region sequences and NCBI Blast (https://blast.ncbi.nlm.nih.gov/Blast.cgi) resulted in the separation of these genes into five groups: RPP13, RPM1, RGA, LRR and disease resistance. Three unigenes CL12505.Contig5, CL8803.Contig4, CL8803.Contig5 exhibited similarity to RPP13-like proteins; three unigenes CL10.Contig1, Unigene2713 and Unigene2747 were similar to LRR-proteins; two unigenes CL14259.Contig1 and Unigene13240 similar to RPM1-like proteins; and two unigenes CL1855.Contig4 and Unigene6688 similar to RGA proteins. Some of the unigenes were regulatory proteins like CL14017.Contig1 was similar to zinc transporter like protein, unigene3612 similar to transcription factor mediator associated-protein and CL2138.Contig3 similar to retrotransposon proteins.

The abundance of these unigenes varied between the IDM-resistant cultivar SPR and susceptible cultivar SEP. CL1855.Contig4, similar to *RPP13*-like protein 1 family, reached a FPKM value of 96.38 in SPR but only 0.02 in SEP. Another *RPP13*-like protein 1 family resembling unigene, CL12505.Contig5, had a FPKM value of 2.40 in SPR but 0.45 in SEP. The unigene CL14259.Contig1, similar to CNL *R* genes, reached a FPKM value of 69.99 in SPR but was absent in SEP. Similarly, unigene6688 was quite abundant in SPR (FPKM = 13.23) but absent in SEP and it was similar to TNL *R* genes. CL8803.Contig4 and CL8803.Contig5, both encoding LRR class *R* proteins, were more abundant in SPR (FPKM = 4.60 and 2.61, respectively) than in SEP (FPKM = 0.02). Unigenes CL10.Contig1, CL3381.Contig3, unigene13240 and unigene2747 had a FPKM value of 47.92, 22.67, 24.54 and 37.32 in SRP, respectively, whereas they were present at very low FPKM values or absent in SEP (0.04, 0.06, 0.05 and 0.0, respectively) (Table 6). All these four unigenes were similar to putative *Impatiens noli-tangere*, in *A. thaliana*.

**Table 6.** List of impatiens unigenes selected based on more than 2-fold higher abundancy in IDM-resistant impatiens (SPR) compared to IDM-susceptible impatiens (SEP) that were annotated to previously known genes for disease resistance or defense using NR (non-redundant protein), NT (non-redundant nucleotide), SWISS-PROT, KEGG (Kyoto Encyclopedia of Genes and Genomes), COG (Clusters of Orthologous Groups of proteins) and GO (Gene Ontology) databases.

| Unigene ID | Length (bp) | Abundance (FPKM[1]) in SPR | Abundance (FPKM[1]) in SEP | p-Value | FDR | Score | Annotation |
|---|---|---|---|---|---|---|---|
| CL10.Contig1 | 5073 | 47.92 | 0.04 | $7.13 \times 10^{-24}$ | $2.66 \times 10^{-23}$ | 164 | Putative disease resistance protein At5g05400 |
| CL12505.Contig5 | 4428 | 2.40 | 0.45 | $4.91 \times 10^{-29}$ | $2.04 \times 10^{-28}$ | 689 | Putative disease resistance RPP13-like protein 1 |
| CL12796.Contig1 | 2489 | 9.11 | 0.38 | $8.20 \times 10^{-108}$ | $7.28 \times 10^{-107}$ | 103 | LETM1-like protein |
| CL14017.Contig1 | 2600 | 18.20 | 0.95 | $2.35 \times 10^{-212}$ | $3.47 \times 10^{-211}$ | 805 | Cation efflux family protein |
| CL14259.Contig1 | 2360 | 69.99 | 0.00 | 0 | 0 | 400 | Disease resistance protein (CC-NB-LRR class) family |
| CL1855.Contig4 | 2257 | 96.38 | 0.02 | $5.25 \times 10^{-09}$ | $1.17 \times 10^{-08}$ | 473 | Putative disease resistance RPP13-like protein 1 |
| CL2138.Contig3 | 2505 | 5.82 | 0.00 | $1.69 \times 10^{-88}$ | $1.32 \times 10^{-87}$ | 75.5 | Uncharacterized mitochondrial protein AtMg00860 |
| CL3381.Contig3 | 4495 | 22.67 | 0.06 | 0 | 0 | 543 | Probable disease resistance protein At5g45510 |
| CL8803.Contig4 | 2187 | 4.60 | 0.02 | $2.46 \times 10^{-59}$ | $1.51 \times 10^{-58}$ | 54.7 | Leucine-rich repeat (LRR) family protein |
| CL8803.Contig5 | 2321 | 2.61 | 0.02 | $1.99 \times 10^{-35}$ | $9.18 \times 10^{-35}$ | 73.2 | Leucine-rich repeat (LRR) family protein |
| Unigene2713 | 2869 | 7.12 | 0.00 | $5.52 \times 10^{-94}$ | $4.49 \times 10^{-93}$ | 254 | Receptor-like protein 35 |
| Unigene13240 | 2592 | 24.54 | 0.05 | $1.00 \times 10^{-123}$ | $9.81 \times 10^{-123}$ | 414 | Putative disease resistance protein At1g59780 |
| Unigene2747 | 2575 | 37.32 | 0.00 | $2.08 \times 10^{-23}$ | $7.68 \times 10^{-23}$ | 308 | Putative disease resistance protein At3g14460 |
| Unigene3612 | 3233 | 2.05 | 0.04 | 0.000496 | 0.000733 | 177 | DNA-binding storekeeper protein-related transcriptional regulator |
| Unigene6688 | 2671 | 13.23 | 0.00 | | | | disease resistance protein (TIR-NB-LRR class) |

[1] FPKM: Fragments per kilobase of transcripts per million mapped reads.

## 2.6. Discovery of SSRs and SNPs

A total of 22,484 SSRs were identified in 19,017 out of the 121,497 impatiens unigene sequences. There were 2986 unigene sequences that each contained more than one SSR and 757 SSRs contained compound repeats. The distribution of SSRs consisted of 2053 mononucleotides, 8195 dinucleotides, 10,501 trinucleotides, 599 tetranucleotides, 488 pentanucleotides and 648 hexanucleotides, among which the trinucleotide type AAG/GTT was the most abundant (3466), followed by the ATC/ATG type (1966) and dinucleotide type AT/AT (1165). The distribution of SSR repeat types among the impatiens unigenes are shown in Figure 5 Primers were developed using the SSRs identified from the unigenes and before and after application of the filtration, there were 44890 and 10389 primers developed (Table S4).

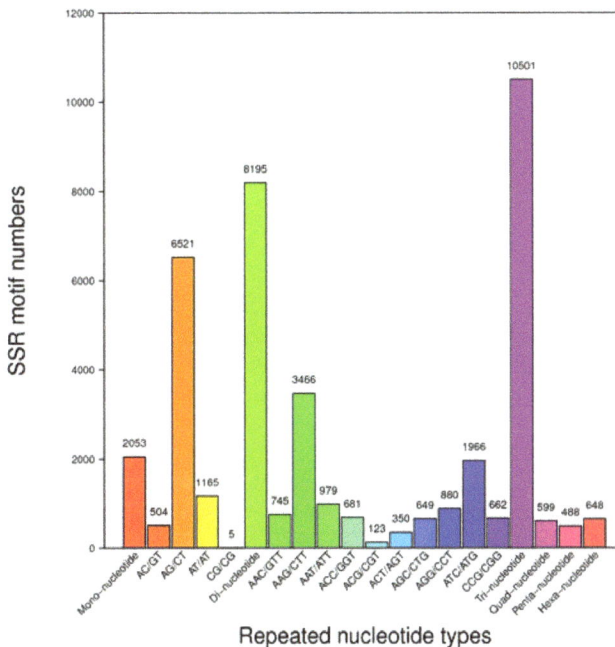

**Figure 5.** Distribution of simple sequence repeats (SSR) repeat types in impatiens unigenes.

The SOAPsnp software was used to identify SNPs in the impatiens transcriptome sequences. The combined assembled impatiens sequence was used as reference and the sequence reads from each cultivar were aligned to the reference individually to call SNPs. SNPs with the number of reads less than seven and the quality score below 20 were discarded. As a result, there were 245,936 and 120,073 SNPs discovered in SPR and SEP cultivars, respectively. The most dominant SNPs type was the transition type, consisting 139,794 (56.84%) and 71,181 (59.28%) for SPR and SEP, respectively, followed by the transversion type consisting of 106,142 (43.16%) and 48,892 (40.72%) for SPR and SEP, respectively. The number of SNPs identified in SPR was higher in all categories and in total in comparison to that of SEP (Figure 6). The list of SNPs in SPR and SEP are supplemented in Tables S5 and S6, respectively. Supplementary Table S7 contain the SNPs, their positions in specific contigs and presence in specific cultivars and supplementary Table S8 contains the SNPs that have been identified in the contigs that are involved in disease resistance and explained in Table 6.

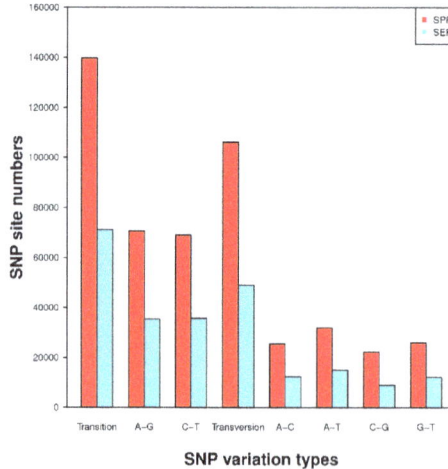

**Figure 6.** Description and classification of single nucleotide polymorphisms (SNPs) identified in SPR and SEP lines.

## 3. Discussion

RNA-Seq has become a very powerful technique for analysis of gene expression, identification of candidate genes involved in the expression of important traits and rapid discovery of large numbers of SSRs and SNPs in non-model plant species [29–31]. In this study, the leaf transcriptomes of impatiens were sequenced, assembled, annotated and compared to identify candidate genes potentially involved in resistance against IDM. We found fifteen unigenes that were more than 2000 nucleotides long and had at least 2-fold more transcripts in the IDM-resistant cultivar SPR than in the IDM-susceptible cultivar SEP. Most of the identified candidate genes showed high levels of similarity to the NB-LRR or LRR gene families. Members of the NB-LRR gene family are known to confer plants resistance to diseases caused by bacteria, fungi and viruses and are localized in the cytoplasm [32–35]. LRR genes have been shown to be involved in fungal resistance. The LRR domain in the NB-LRR and LRR R-proteins plays a critical role in defense response by perceiving signals from effectors released by pathogens or pathogens themselves and transducing the signals to initiate the defense responses [36–39]. The NB-LRR proteins are divided into two groups based on the presence of Toll/interleukin-1 receptor (TIR) or coiled-coil (CC) domain in the amino terminal end. Despite the common function of pathogen recognition, the sequence and signaling mechanism of TNL and CNL proteins are different [40–42]. In this study, there were two genes identified as TNLs or CNL encoding proteins.

RPP13 is a part of a NB-ARC domain known for the presence in APAF-1 (apoptopic protease-activating factor-1), R proteins and CED-4 (*Caenorhabditis elegans* death-4 protein) and belongs to the CC-NBS-LRR family that encodes NBS-LRR type R protein with a putative amino-terminal leucine zipper [43]. RPP13 has been reported to confer resistance to five isolates of *Peronospora parasitica* that caused downy mildew in *A. thaliana* [44]. More than 20 loci for *RPP* genes has been already identified in *A. thaliana* conferring recognition of *P. parasitica* [45]; genes from three loci, *RPP1*, *RPP5* and *RPP8* have been cloned [46–48]. *RPP13* has been placed in the subclass of NB-LRR type R proteins that have putative leucine zipper (LZ) domain located near the N-terminus [49]. Before it was known that these loci confer resistance to downy mildew in *A. thaliana*, *RPP13* was only known to possess resistant activity to *Pseudomonas syringae* pathovars [36–39]. After the understanding of TIR type NB-LRR R genes found at the complex *RPP1* and *RPP5* loci [47–49]. In *Arabidopsis*, *RPP13* has been known to be reminiscent of R genes *RPM1* [42] and *RPS2* [36,38]. A recent study shows functionally diverged alleles as a distinction in *RPP13* at a simple R gene locus [44]. In this study, there were two putative disease resistance RPP13-proteins identified in the IDM-resistant cultivar with higher amount than in the susceptible cultivar. Presence of genes similar to *RPP13* indicate the possibility of *RPP*-like genes conferring the resistance to IDM. Resistance to *P. parasitica* by *RPP13* in *A. thaliana* Niederzenz accession is known to be independent to either Salicylic acid, *NRD1* or *EDS1* [50] triggered by the avirulence gene *ATR13* [51]. *RPP13* encodes protein that is predicted to be located in cytoplasm [43]. In addition to *RPP*-like genes, increased expression of other genes similar to LETM1-like protein, Cation efflux family protein, Receptor like protein 35 and DNA-binding storekeeper protein-related transcriptional regulator indicates the complexity of defense mechanism in the resistant line to IDM. Further research on the mode of resistance in New Guinea Impatiens for IDM could reveal the involvement of these proteins in conjunction with RPP13 to confer resistance or exhibit different mode of action.

Functional annotation of these genes also grouped them into different categories on the basis of similarity. The common method of recognition and signaling mechanisms for resistance exhibited by different classes of gene families differently in unrelated group of pathogens indicates the possibility of functional diversity of *R* genes to defend the host from different kinds of pathogens in different species of host. It has been previously reported that *R* genes exhibit polymorphism regarding the role in defense mechanism. There were other unigenes that were identified to be functionally similar to RPS2-like proteins family in the transcriptome but not all of them were significantly upregulated in the resistant cultivar in this experiment.

Being abundant and randomly distributed in plant genomes, SNPs and SSRs are very useful loci for development of molecular markers for high-density, high-resolution gene or genome mapping and marker-assisted breeding. RNA-Seq and bioinformatics analytical tools have made the discovery of SSR and SNP sites easy and fast. So far, molecular markers have not been reported for downy mildew resistance in impatiens. This study resulted in the discovery of 22,484 SSRs and 245,936 and 120,073 SNPs in SPR and SEP, respectively. These SSR and SNP sites offer a very useful resource for developing molecular markers that can be used by geneticists and breeders for genetic studies and improvement of impatiens.

## 4. Materials and Methods

### 4.1. Plant Materials

IDM-susceptible cultivar Super Elfin® XP Pink (SEP, *I. walleriana*) and IDM-resistant cultivar SunPatiens® (SPR, *I. hawkeri*) were used in this study. Three plants of each cultivar were grown individually in 15-cm plastic containers filled with the commercial soilless potting mixture Faffard® 3B (50% Canadian peat and 50% of the mixture of vermiculite, pine bark and perlite) (Agawam, MA, USA) in a greenhouse facility at the University of Florida's Gulf Coast Research and Education Center (UF/GCREC), Wimauma, FL, USA. Plants were watered as needed and no fungal pesticides were applied. Mature leaves from both resistant and susceptible plants were sampled and they were instantly frozen in liquid nitrogen and stored in a deep freezer at −80 °C until use.

### 4.2. RNA Isolation, cDNA Synthesis and Sequencing

Frozen leaf tissue samples of SEP and SPR were shipped on dry ice to Beijing Genomics Institute (BGI) in Shenzhen, China. Total RNA was extracted from the impatiens leaf tissues using pBiozol Total RNA Extraction Reagent (BioFlux, Hangzhou, China) following manufacturer's instructions. The RNA concentration, RNA integrity number (RIN) and the 28S/18S ratio were determined on an Agilent 2100 Bioanalyzer (Agilent Technologies, Santa Clara, CA, USA). An equal amount of RNA from each of the three biological replicates was combined to form a pool of resistant and susceptible RNA samples for mRNA isolation and cDNA synthesis. The combined RNA samples were treated with DNase I to remove residual genomic DNA and mRNA was isolated using magnetic beads with Oligo (dT). The isolated mRNA molecules were fragmented using the Elute Prime Fragment Mix from Illumina

TruSeq™ RNA sample prep kit v2 (Illumina, San Diego, CA, USA) at 94 °C for 8 min. The fragmented mRNAs were used as templates for cDNA synthesis. First strand cDNA was synthesized by using the First Strand Master Mix and SuperScript II from Invitrogen (Carlsbad, CA, USA) and amplified at 25 °C for 10 min, 42 °C for 50 min and 70 °C for 15 min. Second cDNA strand was synthesized using the Second Strand Master Mix from Invitrogen at 16 °C for 1 h. The resulted cDNA fragments were purified, their ends were repaired using the End Repair Master Mix from Invitrogen at 30 °C for 30 min and the Ampure XP beads (Beckman Coulter, Brea, CA, USA) and a single adenine (A) base were added to each end of the cDNA fragments. End-repaired and A-added short fragments were joined with adapters. The adapter-ligated cDNA was PCR-amplified to generate sequencing libraries. The libraries were analyzed with an Agilent 2100 Bioanalyzer and the ABI StepOnePlus Real-Time PCR System for quantifying and qualifying the sample library. Sequencing of cDNA was performed on an Illumina HiSeq™ 2000, generating 100 bp paired-end reads.

*4.3. Sequence Filtering, De Novo Assembly and Transcriptome Analysis*

Raw sequence reads were cleaned by removing the adapters attached at the ends and reads with unknown nucleotides more than 5% were removed using BGI's internal software (filter_fq). Low quality reads with more than 20% of quality value < 10 were removed, Q20 percentage, proportion of unknown nucleotides in clean reads (N) percentage and GC percentage among total nucleotides were calculated and the remaining clean reads were used in further analysis. De novo transcriptome assembly was done using the Trinity software (http://trinityrnaseq.sourceforge.net/) [52] with the following parameters: minimum contig length of 100, minimum glue 3, group pair distance 250, path reinforcement distance 85 and minimum kmer coverage 3 using the cleaned short reads. The longest, non-redundant, unique transcripts were defined as unigenes. Further processing of the sequences for redundancy removal and splicing was done using the sequence clustering software, TIGR Gene Indices Clustering Tools (TGICL) v2.1 (http://sourceforge.net/projects/tgicl/files/tgicl%20v2.1/) with minimum overlap length 40, base quality cutoff of clipping 10 and maximum length of unmatched overhangs 20. Homologous transcript clustering was done using Phrap release 23.0 (http://www.phrap.org/) with repeat stringency 0.95, minimum match 35 and minimum score 35.

Functional annotation of the unigenes was done by comparing them using the NR (non-redundant protein) and NT (non-redundant nucleotide) databases of National Center for Biotechnology Information (http://blast.ncbi.nlm.nih.gov/Blast.cgi), SWISS-PROT database (European Bioinformatics Institute, ftp://ftp.ebi.ac.uk/pub/databases/swissprot/), KEGG (Kyoto Encyclopedia of Genes and Genomes database), COG (Clusters of Orthologous Groups of proteins database) and GO (Gene Ontology) with BLASTx using *E*-value cutoff of $1 \times 10^{-5}$. The best aligning results were used to determine the sequence direction of unigenes. When the results of databases conflicted each other, a priority order of NR, SWISS-PROT, KEGG and COG was followed to determine the sequence direction. When unigenes were found to be unaligned with any of the databases, ESTScan was used to determine the sequence direction [53]. Blast2Go (http://www.blast2go.com/b2ghome) was used to annotate unigenes generated by NR annotation to get GO annotation [54]. After GO annotation, WEGO software was used for functional classification of the unigenes and to understand the distribution of gene functions from the macro level [55]. KEGG database was used for metabolic pathway analysis of gene products in cell and function of gene products. Further study of unigene pathways was done using Path_finder (http://www.genome.jp/) with default parameters [56].

*4.4. Unigenes Sequence Comparison and Phylogenetic Analysis*

The sequences of selected unigenes from impatiens were compared to the Arabidopsis Information Resource (TAIR) Database (https://www.arabidopsis.org/) [57] to show the functional similarity and the best aligning results were used. BLAST was done using CDS sequences of fifteen unigenes (Table 6) involved in disease resistance to the nr/nt database for sequence similarity.

*4.5. Identification of SSRs and SNPs*

SSRs in the unigene sequences were detected using Microsatellite (MISA) (http://pgrc.ipk-gatersleben.de/misa/misa.html). The minimum cut-off values for the identification of mono-, di-, tri, tetra-, penta- and hexa-nucleotide SSRs were 12, 6, 5, 5, 4 and 4, respectively. Only SSRs with 150-bp flanking sequences on both ends in the unigenes were retained. SSR primers were designed using the Primer3 software (Release 2.3.4, http://www.onlinedown.net/soft/51549.htm) and the default parameters [58]. The SOAPsnp software Release 1.03 (Short Oligonucleotide Analysis Package single nucleotide polymorphism) (http://soap.genomics.org.cn/soapsnp.html) was used to identify SNPs by aligning consensus sequences from each cultivar to the transcriptome assembly [59].

## 5. Conclusions

This study represents the first application of RNA-Seq and sequence analysis to the leaf transcriptomes of garden impatiens, a very important floriculture crop and a very popular garden plant in the world and a representative species of the family Balsaminaceae. The leaf transcriptome sequences will become an invaluable resource for understanding the genetic makeup of this important plant. The transcripts and unigenes assembled can provide valuable template sequences for gene mining and gene expression analysis in impatiens. The unigenes that were abundantly expressed in IDM-resistant impatiens while absent or expressed at very low levels in the IDM-susceptible impatiens and show strong similarity to plant *R* genes can serve as candidate genes for understanding the genetic basis of the IDM resistance in *I. hawkeri* and over-expressing in IDM-susceptible *I. walleriana* for improved IDM resistance. The SSR and SNP sites identified in the impatiens transcriptomes will be very useful for molecular marker development, gene and genome mapping and identification of specific markers for important plant, foliar and flower traits in impatiens.

**Supplementary Materials:** Supplementary materials can be found at http://www.mdpi.com/1422-0067/19/7/2057/s1. Raw sequence reads of the reported impatiens leaf transcriptomes for SPR and SEP have been deposited in the Sequence Read Archive (SRA, http://www.ncbi.nlm.nih.gov/Traces/sra/) under BioSample accession number SAMN09394245 and SAMN09393910 respectively.

**Author Contributions:** K.B. analyzed the data and drafted and revised the manuscript; W.W. prepared plant materials and performed the greenhouse experiments; Z.C. performed initial mining of the transcriptome data; Z.D. conceived the study, reviewed data and results and revised the manuscript. All authors read and consented to the final version of the manuscript.

**Funding:** This study was supported in part by the American Floral Endowment, the Frederic C. Gloeckner Foundation, Inc. and USDA-NIFA hatch projects FLA-GCR-005065 and FLA-GCC-005507.

**Acknowledgments:** Authors would like to thank the Beijing Genomics Institute (BGI) for sequencing and bioinformatics analysis and Joyce Jones and Gail Bowman for caring impatiens plants in the greenhouse.

**Conflicts of Interest:** The authors declare no conflicts of interest.

## References

1. Grey-Wilson, C. *Impatiens of Africa: Morphology, Pollination and Pollinators, Ecology, Phytogeography, Hybridisation, Keys and a Systematic Treatment of all African Species, with a Note on Collecting and Cultivation;* Rotterdam A.A. Balkema: Rotterdam, The Netherlands, 1980; p. 248.
2. Utami, N.; Shimizu, T. Seed morphology and classification of impatiens (Balsaminaceae). *Blumea Biodivers. Evol. Biogeogr. Plants* **2005**, *50*, 447–456. [CrossRef]
3. Uchneat, M.S. Impatiens. In *Flower Breeding and Genetics: Issues, Challenges and Opportunities for the 21st Century;* Anderson, N.O., Ed.; Springer: Dordrecht, The Netherlands, 2006; pp. 277–299.
4. Lim, T.K. Impatiens walleriana. In *Edible Medicinal and Non-Medicinal Plants;* Springer: Berlin/Heidelberg, Germany, 2013; Volume 7, pp. 548–550. Available online: http://link.springer.com/book/10.1007/978-94-007-7395-0 (accessed on 1 March 2016).
5. Jones, D.R.; O'Neill, T. *Impatiens Downy Mildew;* Factsheet 05/04; Horticultural Development Council: East Malling, UK, 2004.

6. Lane, C.R.; Beales, P.A.; O'Neill, T.M.; McPherson, G.M.; Finlay, A.R.; David, J.; Constantinescu, O.; Henricot, B. First report of Impatiens downy mildew (*Plasmopara obducens*) in the UK. *Plant Pathol.* **2005**, *54*, 243. [CrossRef]

7. Toppe, B.; Brurberg, M.B.; Stensvand, A.; Herrero, M.L. First report of *Plasmopara obducens* (downy mildew) on *Impatiens walleriana* in Norway. *Plant Pathol.* **2010**, *59*, 800. [CrossRef]

8. Bulajić, A.; Vučurović, A.; Stanković, I.; Ristić, D.; Jović, J.; Stojković, B.; Krstić, B. First report of *Plasmopara obducens* on *Impatiens walleriana* in Serbia. *Plant Dis.* **2011**, *95*, 491. [CrossRef]

9. Vajna, L. First report of *Plasmopara obducens* on impatiens (*Impatiens walleriana*) in Hungary. *New Dis. Rep.* **2011**, *24*, 13. [CrossRef]

10. Shen, Y.M.; Huang, J.H.; Liu, H.L. First report of downy mildew caused by *Plasmopara obducens* on impatiens in Taiwan. *Plant Dis.* **2013**, *97*, 1512. [CrossRef]

11. Satou, M.; Sugawara, K.; Nagashima, S.; Tsukamoto, T.; Matsushita, Y. Downy mildew of busy lizzie caused by *Plasmopara obducens* in Japan. *J. Gen. Plant Pathol.* **2013**, *79*, 205–208. [CrossRef]

12. Cunnington, J.H.; Aldaoud, R.; Loh, M.; Washington, W.S.; Irvine, G. First record of *Plasmopara obducens* (downy mildew) on impatiens in Australia. *Plant Pathol.* **2008**, *57*, 371. [CrossRef]

13. Wegulo, S.N.; Koike, S.T.; Vilchez, M.; Santos, P. First report of downy mildew caused by *Plasmopara obducens* on impatiens in California. *Plant Dis.* **2004**, *88*, 909. [CrossRef]

14. Impatiens Downy Mildew Alert: URI Plant Protection Clinic. Available online: https://www.ecolandscaping. org/05/pests-pest-management/impatiens-downy-mildew-alert-uri-plant-protection-clinic/ (accessed on 4 March 2018).

15. Crouch, J.A.; Ko, M.P.; McKemy, J.M. First report of impatiens downy mildew outbreaks caused by *Plasmopara obducens* throughout the Hawai'ian Islands. *Plant Dis.* **2014**, *98*, 696. [CrossRef]

16. National Agricultural Statistics Service (NASS); USDA. *Floriculture Crops 2015 Summary (April 2016)*; US Department of Agriculture: Washington, DC, USA, 2016; p. 58. Available online: http://usda.mannlib. cornell.edu/usda/current/FlorCrop/FlorCrop-04-26-2016.pdf (accessed on 5 April 2018).

17. Suarez, S.N.; Palmateer, A.J. *Overview of Impatiens Downy Mildew in Florida*; Conference Paper; American Phytopathological Society: Minneapolis, MI, USA, 2014.

18. Getter, K.L.; Bridget, K.B. Consumer willingness to purchase *Impatiens walleriana* alternatives. *HortScience* **2013**, *48*, 1370–1377.

19. Song, Y.; Yong-Ming, Y.; Philippe, K. Chromosomal evolution in Balsaminaceae, with cytological observations on 45 species from Southeast Asia. *Caryologia* **2003**, *56*, 463–481. [CrossRef]

20. Flor, H.H. Current status of the gene-for-gene concept. *Annu. Rev. Phytopathol.* **1971**, *9*, 275–296. [CrossRef]

21. De Young, B.J.; Roger, W.I. Plant NBS-LRR proteins in pathogen sensing and host defense. *Nat. Immunol.* **2006**, *7*, 1243. [CrossRef] [PubMed]

22. Jones, J.D.; Dangl, J.L. The plant immune system. *Nature* **2006**, *7117*, 323. [CrossRef] [PubMed]

23. Hammond-Kosack, K.E.; Parker, J.E. Deciphering plant-pathogen communication: Fresh perspectives for molecular resistance breeding. *Curr. Opin. Biotechnol.* **2003**, *14*, 177–193. [CrossRef]

24. Nimchuk, Z.; Eulgem, T.; Holt, B.F.; Dangl, J.L. Recognition and response in the plant immune system. *Annu. Rev. Genet.* **2003**, *37*, 579–609. [CrossRef] [PubMed]

25. Pedley, K.F.; Martin, G.B. Role of mitogen-activated protein kinases in plant immunity. *Curr. Opin. Plant Biol.* **2005**, *8*, 541–547. [CrossRef] [PubMed]

26. Glazebrook, J. Genes controlling expression of defense responses in Arabidopsis-2001 status. *Curr. Opin. Plant Biol.* **2001**, *4*, 301–308. [CrossRef]

27. Mosquera, T.; Alvarez, M.F.; Jiménez-Gómez, J.M.; Muktar, M.S.; Paulo, M.J.; Steinemann, S.; Li, J.; Draffehn, A.; Hofmann, A.; Lübeck, J.; et al. Targeted and untargeted approaches unravel novel candidate genes and diagnostic SNPs for quantitative resistance of the potato (*Solanum tuberosum* L.) to Phytophthora infestans causing the late blight disease. *PLoS ONE* **2016**, *11*, e0156254. [CrossRef] [PubMed]

28. Ashrafi, H.; Hill, T.; Stoffel, K.; Kozik, A.; Yao, J.; Chin-Wo, S.R.; Van Deynze, A. De novo assembly of the pepper transcriptome (Capsicum annuum): A benchmark for in silico discovery of SNPs, SSRs and candidate genes. *BMC Genom.* **2012**, *13*, 571. [CrossRef] [PubMed]

29. Wang, W.; Wang, Y.; Zhang, Q.; Qi, Y.; Guo, D. Global characterization of Artemisia annua glandular trichome transcriptome using 454 pyrosequencing. *BMC Genom.* **2009**, *10*, 465. [CrossRef] [PubMed]

30. Wall, P.K.; Leebens-Mack, J.; Chanderbali, A.S.; Barakat, A.; Wolcott, E. Comparison of next generation sequencing technologies for transcriptome characterization. *BMC Genom.* **2009**, *10*, 347. [CrossRef] [PubMed]

31. Collins, L.J.; Biggs, P.J.; Voelckel, C.; Joly, S. An approach to transcriptome analysis of non-model organisms using short-read sequences. *Genome Inf.* **2008**, *21*, 3–14. [CrossRef]

32. Jones, D.A.; Jones, J.D.G. The role of leucine-rich repeat proteins in plant defenses. *Adv. Bot. Res.* **1997**, *24*, 90–167.

33. Ellis, J.; Dodds, P.; Pryor, T. Structure, function and evolution of plant disease resistance genes. *Curr. Opin. Plant Biol.* **2000**, *3*, 278–284. [CrossRef]

34. Belkhadir, Y.; Subramaniam, R.; Dangl, J.L. Plant disease resistance protein signaling: NBS-LRR proteins and their partners. *Curr. Opin. Plant Biol.* **2004**, *7*, 391–399. [CrossRef] [PubMed]

35. McHale, L.; Tan, X.; Koehl, P.; Michelmore, R.W. Plant NBS-LRR proteins: Adaptable guards. *Genome Biol.* **2006**, *7*, 212. [CrossRef] [PubMed]

36. Bent, A.F.; Kunkel, B.N.; Dahlbeck, D.; Brown, K.L.; Schmidt, R.; Giraudat, J.; Leung, J.; Staskawicz, B.J. *RPS2* of *Arabidopsis thaliana*: A leucine-rich repeat class of plant disease resistance genes. *Science* **1994**, *265*, 1856–1860. [CrossRef] [PubMed]

37. Grant, M.R.; Godiard, L.; Straube, E.; Ashfield, T.; Lewald, J.; Sattler, A.; Innes, R.W.; Dangl, J.L. Structure of the *Arabidopsis RPM1* gene enabling dual specificity disease resistance. *Science* **1995**, *269*, 843–846. [CrossRef] [PubMed]

38. Mindrinos, M.; Katagiri, F.; Yu, G.-L.; Ausubel, F.M. The *A. thaliana* disease resistance gene *RPS2* encodes a protein containing a nucleotide-binding site and leucine-rich repeats. *Cell* **1994**, *78*, 1089–1099. [CrossRef]

39. Warren, R.F.; Henk, A.; Mowery, P.; Holub, E.B.; Innes, R.W. A mutation within the leucine-rich repeat domain of the *Arabidopsis* disease resistance gene *RPS5* partially suppresses multiple bacterial and downy mildew resistance genes. *Plant Cell* **1998**, *10*, 1439–1452. [CrossRef] [PubMed]

40. Mondragon-Palomino, M.; Gaut, B.S. Gene conversion and the evolution of three leucine-rich repeat gene families in *Arabidopsis thaliana*. *Mol. Biol. Evol.* **2005**, *22*, 2444–2456. [CrossRef] [PubMed]

41. Meyers, B.C.; Dickerman, A.W.; Michelmore, R.W.; Sivaramakrishnan, S.; Sobral, B.W.; Young, N.D. Plant disease resistance genes encode members of an ancient and diverse protein family within the nucleotide-binding superfamily. *Plant J.* **1999**, *20*, 317–332. [CrossRef] [PubMed]

42. Pan, Q.; Wendel, J.; Fluhr, R. Divergent evolution of plant NBS-LRR resistance gene homologues in dicot and cereal genomes. *J. Mol. Evol.* **2000**, *50*, 203–213. [CrossRef] [PubMed]

43. Rose, L.E.; Bittner-Eddy, P.D.; Langley, C.H.; Holub, E.B.; Michelmore, R.W.; Beynon, J.L. The maintenance of extreme amino acid diversity at the disease resistance gene, *RPP13*, in *Arabidopsis thaliana*. *Genetics* **2004**, *166*, 1517–1527. [CrossRef] [PubMed]

44. Bittner-Eddy, P.D.; Crute, I.R.; Holub, E.B.; Beynon, J.L. *RPP13* is a simple locus in *Arabidopsis thaliana* for alleles that specify downy mildew resistance to different avirulence determinants in *Peronospora parasitica*. *Plant J.* **2000**, *21*, 177–188. [CrossRef] [PubMed]

45. Holub, E.B.; Beynon, J. Symbiology of mouse-ear cress (*Arabidopsis thaliana*) and oomycetes. *Adv. Bot. Res.* **1997**, *24*, 228–273.

46. Milligan, S.B.; Bodeau, J.; Yaghoobi, J.; Kaloshian, I.; Zabel, P.; Williamson, V.M. The root knot nematode resistance gene *Mi* from tomato is a member of the leucine-rich repeat family of plant genes. *Plant Cell* **1998**, *10*, 1307–1319. [CrossRef] [PubMed]

47. Botella, M.A.; Parker, J.E.; Frost, L.N.; Bittner-Eddy, P.D.; Beynon, J.L.; Daniels, M.J.; Holub, E.B.; Jones, J.D.G. Three genes of the *Arabidopsis RPP.* 1 complex resistance locus recognise distinct *Peronospora parasitica* avirulence determinants. *Plant Cell* **1998**, *10*, 1847–1860. [CrossRef] [PubMed]

48. Parker, J.E.; Coleman, M.J.; Szabò, V.; Frost, L.N.; Schmidt, R.; Van Der Biezen, E.A.; Moores, T.; Dean, C.; Daniels, M.J.; Jones, J.D.G. The *Arabidopsis* downy mildew resistance gene *RPP. 5* shares similarity to the Toll and Interleukin-1 receptors with N and L6. *Plant Cell* **1997**, *9*, 879–894. [CrossRef] [PubMed]

49. Hammond-Kosack, K.E.; Jones, J.D. Resistance gene-dependent plant defense responses. *Plant Cell* **1996**, *8*, 1773. [CrossRef] [PubMed]

50. Bittner-Eddy, P.D.; Beynon, J.L. The *Arabidopsis* downy mildew resistance gene, *RPP13-Nd*, functions independently of *NDR1* and *EDS1* and does not require the accumulation of salicylic acid. *Mol. Plant-Microbe Interact.* **2001**, *14*, 416–421. [CrossRef] [PubMed]

51. Allen, R.L.; Bittner-Eddy, P.D.; Grenville-Briggs, L.J.; Meitz, J.C.; Rehmany, A.P.; Rose, L.E.; Beynon, J.L. Host-parasite coevolutionary conflict between *Arabidopsis* and downy mildew. *Science* **2004**, *306*, 1957–1960. [CrossRef] [PubMed]

52. Grabherr, M.G.; Haas, B.J. Full-length transcriptome assembly from RNA-Seq data without a reference genome. *Nat. Biotechnol.* **2011**, *29*, 644–652. [CrossRef] [PubMed]

53. Iseli, C.; Jongeneel, C.V. ESTScan: A program for detecting, evaluating and reconstructing potential coding regions in EST sequences. *Proc. Int. Conf. Intell. Syst. Mol. Biol.* **1999**, *99*, 138–148.

54. Conesa, A.; Götz, S. Blast2GO: A universal tool for annotation, visualization and analysis in functional genomics research. *Bioinformatics* **2005**, *21*, 3674–3676. [CrossRef] [PubMed]

55. Ye, J.; Fang, L. WEGO: A web tool for plotting GO annotations. *Nucleic Acids Res.* **2006**, *34*, W293–W297. [CrossRef] [PubMed]

56. Kanehisa, M.; Araki, M. KEGG for linking genomes to life and the environment. *Nucleic Acids Res.* **2008**, *36*, D480–D484. [CrossRef] [PubMed]

57. Arabidopsis Information Resource Database. Available online: https://www.arabidopsis.org/ (accessed on 3 April 2018).

58. Untergasser, A.; Cutcutache, I.; Koressaar, T.; Ye, J.; Faircloth, B.C. Primer3-new capabilities and interfaces. *Nucleic Acids Res.* **2012**, *40*, e115. [CrossRef] [PubMed]

59. Li, R.; Yu, C. SOAP2: An improved ultrafast tool for short read alignment. *Bioinformatics* **2009**, *25*, 1966–1967. [CrossRef] [PubMed]

International Journal of
*Molecular Sciences*

MDPI

*Article*

# CaWRKY22 Acts as a Positive Regulator in Pepper Response to *Ralstonia Solanacearum* by Constituting Networks with CaWRKY6, CaWRKY27, CaWRKY40, and CaWRKY58

Ansar Hussain [1,2,3,†], Xia Li [1,2,3,†], Yahong Weng [1,2,3], Zhiqin Liu [1,2,3], Muhammad Furqan Ashraf [1,2,3], Ali Noman [1,2,3,4], Sheng Yang [1,2,3], Muhammad Ifnan [1,2,3], Shanshan Qiu [1,2,3], Yingjie Yang [1,2,3], Deyi Guan [1,2,3] and Shuilin He [1,2,3,*]

[1] Ministry of Education Key Laboratory of Plant Genetic Improvement and Comprehensive Utilization, Fujian Agriculture and Forestry University, Fuzhou 350002, China; ahtraggar@yahoo.com (A.H.); lixiatrista@163.com (X.L.); wwwyh0915@163.com (Y.W.); lzqfujian@126.com (Z.L.); furqanashraf@hotmail.com (M.F.A.); alinoman@gcuf.edu.pk (A.N.); yangsheng2061@163.com (S.Y.); mifnan@yahoo.com (M.I.); B.miracle@hotmail.com (S.Q.); hnyingjieyang@163.com (Y.Y.); gdyfujian@126.com (D.G.)
[2] College of Crop Science, Fujian Agriculture and Forestry University, Fuzhou 350002, China
[3] Key Laboratory of Applied Genetics of Universities in Fujian Province, Fujian Agriculture and Forestry University, Fuzhou 350002, China
[4] Department of Botany, Government College University, Faisalabad 38040, Pakistan
* Correspondence: shlhe201304@aliyun.com
† These authors contributed equally to the paper.

Received: 26 March 2018; Accepted: 1 May 2018; Published: 10 May 2018

**Abstract:** The WRKY web, which is comprised of a subset of WRKY transcription factors (TFs), plays a crucial role in the regulation of plant immunity, however, the mode of organization and operation of this network remains obscure, especially in non-model plants such as pepper (*Capsicum annuum*). Herein, *CaWRKY22*, a member of a subgroup of IIe WRKY proteins from pepper, was functionally characterized in pepper immunity against *Ralstonia Solanacearum*. CaWRKY22 was found to target the nuclei, and its transcript level was significantly upregulated by *Ralstonia Solanacearum* inoculation (RSI) and exogenously applied salicylic acid (SA), Methyl jasmonate (MeJA), or ethephon (ETH). Loss-of-function *CaWRKY22*, caused by virus-induced gene silencing (VIGS), enhanced pepper's susceptibility to RSI. In addition, the silencing of *CaWRKY22* perturbed the hypersensitive response (HR)-like cell death elicited by RSI and downregulated defense-related genes including *CaPO2*, *CaPR4*, *CaACC*, *CaBPR1*, *CaDEF1*, *CaHIR1*, and *CaWRKY40*. CaWRKY22 was found to directly bind to the promoters of *CaPR1*, *CaDEF1*, and *CaWRKY40* by chromatin immuno-precipitation (ChIP) analysis. Contrastingly, transient overexpression of *CaWRKY22* in pepper leaves triggered significant HR-like cell death and upregulated the tested immunity associated maker genes. Moreover, the transient overexpression of *CaWRKY22* upregulated the expression of *CaWRKY6* and *CaWRKY27* while it downregulated of the expression of *CaWRKY58*. Conversely, the transient overexpression of *CaWRKY6*, *CaWRKY27*, and *CaWRKY40* upregulated the expression of *CaWRKY22*, while transient overexpression of *CaWRKY58* downregulated the transcript levels of *CaWRKY22*. These data collectively recommend the role of CaWRKY22 as a positive regulator of pepper immunity against *R. Solanacearum*, which is regulated by signaling synergistically mediated by SA, jasmonic acid (JA), and ethylene (ET), integrating into WRKY networks with WRKY TFs including CaWRKY6, CaWRKY27, CaWRKY40, and CaWRKY58.

**Keywords:** *Capsicum annuum*; CaWRKY22; immunity; *Ralstonia Solanacearum*; WRKY networks

## 1. Introduction

Being sessile, plants frequently encounter various biotic and abiotic stresses individually, and in some cases, collectively [1]. To protect against stresses, plants have evolved a sophisticated defense system developed under frequent selection pressures from the main environmental constraints. This system is largely regulated at the transcriptional level by the action of different transcription factors (TFs) interconnected to make a complicated transcriptional networks [2]. Defense responses to different stresses need to be appropriately coordinated and strictly controlled since they are costly to the plants in terms of energy expenditure and development [3]. The defense system might differ in different plant species due to diverse ecological conditions affecting their evolution in the regions/habitats [4]. Therefore, defense mechanisms found in model plants cannot be totally suggested for other non-model plants directly. Plant defense mechanisms have been intensively studied in the past decades, but the majority of these studies have focused on model plants such as *Arabidopsis* and rice. However, the organization of transcriptional networks and their functional coordination to regulate plant responses to different stresses, especially in non-model plants, remains poorly understood.

WRKY proteins constitute one of the largest TF families in plants. WRKY TFs are characterized by one or two conserved WRKY domains and the almost invariant WRKYGQK sequence at the N-terminus followed by a C2H2 or C2HC zinc-finger motif [5]. Based on the number of WRKY domains and the structure of zinc-finger motif, WRKY proteins are phylogenetically classified into three major groups (groups I–III). Group II is further divided into five subgroups (IIa, IIb, IIc, IId, and IIe) [6,7]. By forming a unique wedge shape that inserts perpendicularly into the major groove of the DNA [8], WRKY TFs primarily bind W-boxes [TTGAC(C/T)] present in the promoter regions of target genes through the WRKYGQK motif on the second b-strand, and thereby, transcriptionally modulate the expression of these target genes. By activating or repressing the transcription of their target genes, WRKY TFs have been implicated in plant biological processes like senescence, seed development, dormancy, and germination as well as biotic and abiotic stress responses [9]. It has been found that a subset of WRKY genes was transcriptionally modified by a single stress, or a single WRKY TF participates in multiple stresses. The W-boxes are enriched within the promoter regions [10,11]. These results indicate the existence of WRKY networks involved in plant responses to a specific stress or combined stresses. Although the roles of WRKY proteins in plant responses to biotic and abiotic stresses and their underlying mechanisms have been intensively studied over the years, the majority of these studies have mainly focused on a single gene in the response of plants to a single stress in model plants such as *Arabidopsis* and rice. A significant functional divergence among close structural homologs of WRKY proteins from different plant species was recently suggested [12]. The roles of WRKY TFs and their networks in the response of non-model plants to different single stresses, or to closely related stress combinations, remain poorly understood.

Pepper (*Capsicum annuum*) is a vegetable of great economic importance and is a Solanaceae distributed or planted in uplands during warm seasons, where it is confronted with various soil-borne pathogens, such as *Phytophthora capsici* and *Ralstonia Solanacearum*, the causal agents for pepper blight and bacterial wilt diseases, respectively [13,14]. The coexistence of these soil-borne pathogens causes severe destruction, once the pepper plants experience the combined stress of high temperature and high humidity (HTHH) which attenuates R protein mediated immunity and accelerates the development of pathogens. On the other hand, the combination of pathogen attack and HTHH constitutes most natural selection pressures on pepper that have historically affected its evolution [15]. However, this pressure might not act in the evolution of rice and *Arabidopsis*, which grow in paddy fields during warm seasons or dry lands during cool seasons with fewer soil-borne pathogens. Thus, pepper seems more suitable as a plant for investigation on the coordination of resistance or tolerance against abiotic or biotic stresses, for example, *Ralstonia Solanacearum* inoculation (RSI) and HTHH. Some native pepper varieties from subtropical regions specifically show augmented disease resistance even under HTHH [15]. HSE, a high temperature responsive *cis*-element, was ubiquitously found to co-occur with Salicylic acid (SA)-, Jasmonic acid (JA)-, Ethylene (ET)-, or pathogen-responsive elements in the promoters of the majority of CDPKs and MAPKs, which have been frequently involved in plant immunity [10,11]. This suggests the existence of cross-talk mechanisms between immunity, high temperature and humidity.

Since the genome of pepper is about 27 and 7.5 times larger than that of *Arabidopsis* and rice, respectively, a total of 73 WRKY genes were found in the genome of pepper [16], which is much less than what we expected when compared to the 72 WRKY genes in *Arabidopsis* and 122 in rice [3,6]. Our previous studies indicated that *CaWRKY6* [17], *CaWRKY27* [18], *CaWRKY40* [19], and *CaWRKY58* [20] have been implicated in the pepper response to RSI. Of these, CaWRKY6, CaWRKY27, and CaWRKY40 act as positive regulators, while CaWRKY58 acts as a negative regulator. A subset of W-boxes and HSE elements were found in the promoters of these genes as well as that of other WRKY promoters, implying WRKY networks are involved in pepper's response to RSI. In addition, CaWRKY40 was also found to be regulated directly by CaWRKY6 [17] and CabZIP63 [21] and indirectly by CaCDPK15 [22]. However, the majority of pepper WRKY TFs have not been characterized in terms of pepper's response to pathogen infection. In the present study, we report that CaWRKY22, a new IIe WRKY TF of pepper, acts as a positive regulator in pepper's response to *Ralstonia Solanacearum* inoculation by directly targeting *CaWRKY40* and incorporating a WRKY network including *CaWRKY6*, *CaWRKY27*, *CaWRKY40*, and *CaWRKY58*.

## 2. Results

### 2.1. Cloning and Sequence Analysis of CaWRKY22 cDNA

By *cis*-element scanning within the promoters of WRKY genes in the genome sequence of *Capsicum annuum* (http://peppergenome.snu.ac.kr), *CaWRKY22* was selected for further functional characterization. The presence of HSE and immunity associated *cis*-elements such as the TCA, TGACG-motif, and W-box in the *CaWRKY22* promoter region imply its potential role in pepper immunity (Figure S1). By using gene specific primers (Table S1), we cloned a cDNA fragment of *CaWRKY22* (CA08g07730) of 1500 bp in length that contained a 1122 bp open-reading frame (ORF). Its deduced amino acid sequence was 373 amino acid residues in length, containing one conserved WRKY-domain and was classified into subgroup IIe [23] (Figure 1). The size and theoretical pI of the predicted protein were 41.29 kDa and 5.87, respectively. CaWRKY22 shares 91%, 91%, 88%, and 55% of amino acid identities with SpWRKY22, StWRKY22, NsWRKY22, and GrWRKY22, respectively (Figure S2).

### 2.2. The Transcriptional Expression of CaWRKY22 Is Upregulated by R. Solanacearum Infection and Exogenous Applied Phytohormones Including SA, MeJA, and ETH

The presence of a subset of putative immunity responsive *cis*-elements in the promoter of *CaWRKY22* implies its inducible expression upon pathogen attack. To test this possibility, qRT-PCR was performed to examine the expression pattern of *CaWRKY22* in response to inoculation of *R. Solanacearum*. The results showed that the transcriptional levels of *CaWRKY22* were upregulated in pepper leaves inoculated with *R. Solanacearum*, compared to that in the mock treated leaves (Figure 2A). The increased *CaWRKY22* transcriptional levels were maintained between 6 and 24 hpi (hours post inoculation) and exhibited maximal levels at 6 hpi, implying the involvement of *CaWRKY22* in the response of pepper toward *R. Solanacearum* (Figure 2A).

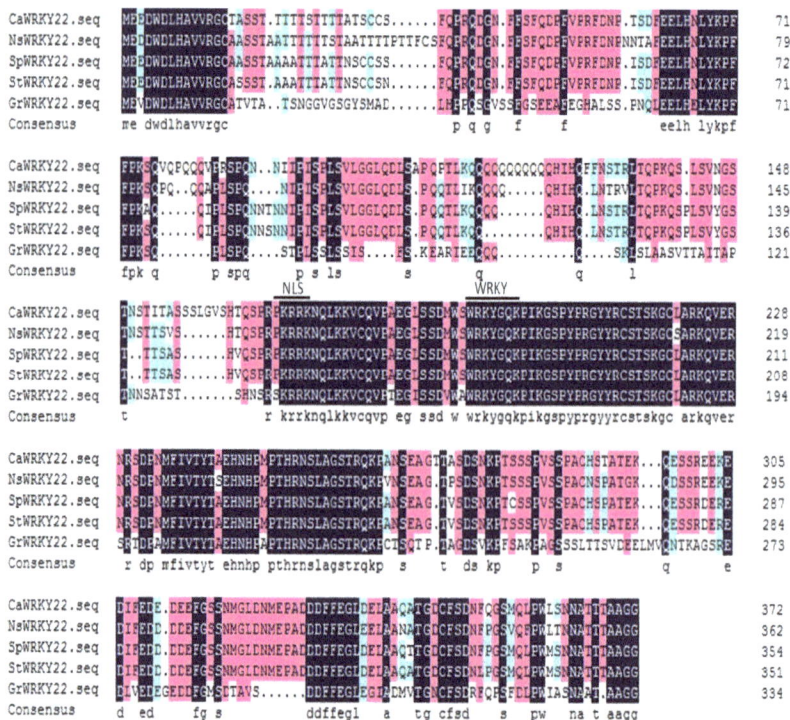

**Figure 1.** Multiple sequence alignment analysis of proteins related to CaWRKY22. Comparison of the deduced amino acid sequence of CaWRKY22 with that of representative related proteins from *Nicotiana sylvestris* NsWRKY22 (XP009768958.1), *Solanum pennellii* SpWRKY22 (XP015061786.1), *Solanum tuberosum* StWRKY58 (XP006339631.1), and *Gossypium raimondii* GrWRKY22 (XP012490022.1). Green shading, 50–75% identity; red shading, 75–100% identity; black shading, 100% identity. Alignment was carried out by DNAMAN5.

Signaling pathways which are mediated by phytohormones such as SA, JA, ET, or ABA (abscisic acid) are involved in the regulation of plant responses to biotic or abiotic stresses. To confirm the data that *CaWRKY22* is involved in the pepper response to RSI, and to test if it is regulated by signaling pathways mediated by these hormones, the relative abundance of *CaWRKY22* against exogenous application of SA, MeJA, ETH, or ABA was measured by qRT-PCR. The results showed that the relative abundance of *CaWRKY22* was enhanced after treatment with 1 mM SA from 1 to 24 hpt (hours post treatment) and exhibited highest levels at 1 hpt (Figure 2B). The exogenous application of ABA resulted in a significant decrease in *CaWRKY22* expression from 1 to 24 hpt (Figure 2C). Similar to the application of SA, treatment of the pepper plant with 100 μM MeJA or 100 μM ETH (ethephon) upregulated the transcription of *CaWRY22* from 1 to 24 hpt compared to the mock treatment (Figure 2D,E).

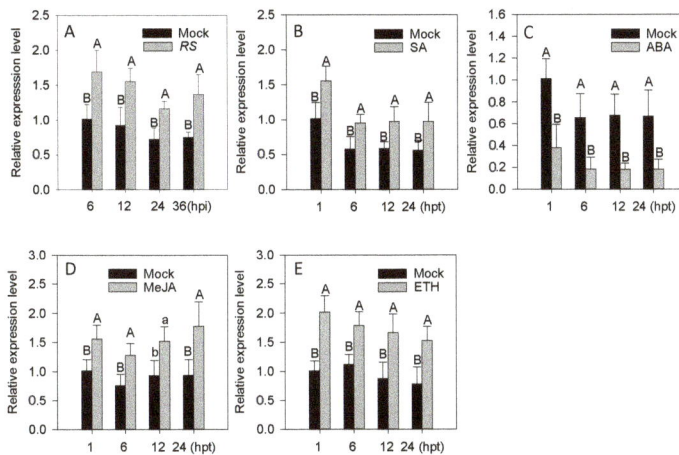

**Figure 2.** QRT-PCR analysis of relative *CaWRKY22* transcriptional levels in pepper plant leaves exposed to *Ralstonia Solanacearum* inoculation (RSI) and different phytohormones. QRT-PCR was performed to detect the expression levels of *CaWRKY22* in pepper leaves under several treatments including *R. Solanacearum* inoculation, (**A**) application of 1 mM SA; (**B**) application of 100 μm ABA; (**C**) application of 100 μm MeJA; (**D**) and application of 100 μm ETH; (**E**) at different time-points. (**A**) The transcriptional levels in RSI-treated pepper leaves were compared with those in MgCl₂-treated control plants (mock), whose relative expression level was set to "1". (**B–E**) The transcriptional levels in hormone-treated pepper leaves were compared with those in ddH₂O-treated plants (mock), whose expression level was set to "1". Error bars indicated the standard error. Different letters above the bar show a significant difference between the means of the three biological replicates based on the Fisher's protected LSD test: uppercase letters, $p < 0.01$; lower case letters, $p < 0.05$.

## 2.3. CaWRKY22 Is Located in the Nuclei

Sequence analysis using WoLFPSORT ("http://www.genscript.com/psort/wolf_psort.html\T1\textquotedblrightolf_psort.html) showed that the predicted CaWRKY22 amino acid sequence contains a putative nuclear localization signal (Figure 1), indicating its potential nucleus targeting. To confirm this speculation, we constructed a *CaWRKY22*–GFP fusion construct driven by the constitutive promoter of CaMV35S, and the generated vector was transformed into *Agrobacterium tumefaciens strain* GV3101. *CaWRKY22*–GFP was transiently overexpressed in *N. benthamiana* leaves by *Agrobacterium* infiltration, and the GFP signals were observed using a confocal fluorescence microscope. The result showed that the GFP signal of the *CaWRKY22*–GFP was exclusively found in the nuclei, whereas the GFP control was found in multiple subcellular compartments, including the cytoplasm and nuclei (Figure 3A). Several studies have mentioned the binding of WRKY proteins to the W-box manner [TTGAC(C/T)] present in the promoter region of the defense associated target genes, which frequently serve as pathogen responsive regulatory elements. To check whether this also relates to *CaWRKY22*, we conducted a transient coexpression experiment with an effector vector carrying the full length *CaWRKY22* cDNA controlled by a *CaMV35S* promoter (*35S:CaWRK22-HA*), and a reporter vector having *GUS*-control driven by the *CaMV35S* core promoter (−46 to +8 bp), with two copies of W-box (*2xW-p35Score:GUS*) or muted W-box (*2xW-m-p35Score:GUS*) in its proximal upstream region (Figure 3B). Reporter vectors were either transformed individually or co-transformed with the effector construct into *N. benthamiana* leaves by infiltration, and the co-infected leaves were sampled for GUS activity measurement. The GUS quantification result revealed that *N. benthamiana* leaves coinfected with *35S:CaWRK22-HA* and *2xW-p35Score:GUS* exhibited strong GUS activity, compared to the mock leaves, suggesting that CaWRKY22 is capable of active transcription expression of the downstream target gene (Figure 3C).

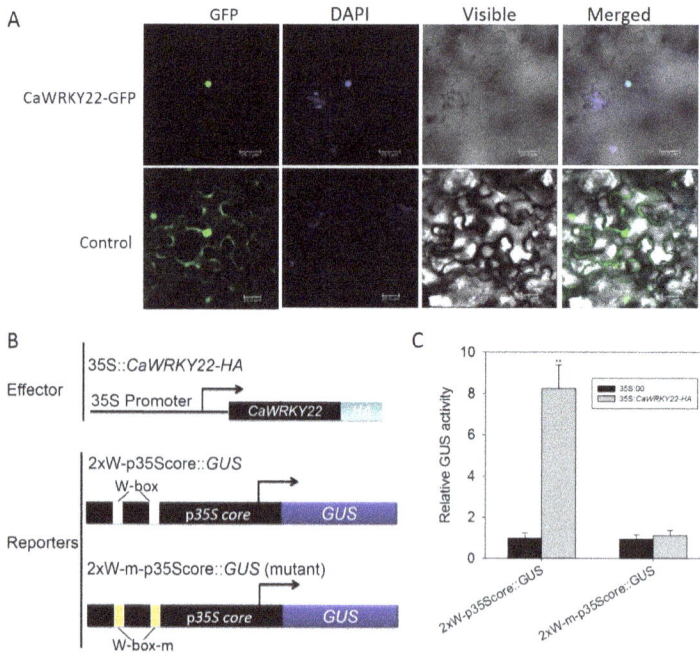

**Figure 3.** Subcellular localization of CaWRKY22 and its transactivation confirmation experiment. (**A**) CaWRKY22 was exclusively found in the nucleus of *N. benthamiana* leaves, the transiently expressed 35S:*CaWRKY22*–GFP. Green color shows GFP. Blue color shows DAPI staining of nucleus. Cyan clor shows merger of green GFP and DAPI stained nucleus. GFP signal (Green) for the control *N. benthamiana* leaves was found throughout the cell. Images were taken by confocal microscopy at 48 hpi (hours post inoculation). Bars = 25 μm; (**B**) schematic diagram of the effector and reporter constructs used for transient coexpression; (**C**) leaves of pepper were cotransfected with the reporter and effector plasmids and the infiltrated leaves were harvested for GUS activity measurement. The GUS activity of pepper leaves transient coexpressing 2xW-p35Score:GUS and the empty effector vector was set to "1". Error bars indicated the standard error. The data represents the means ± SD from three biological replicates. The asterisk indicates significant differences, as determined by Fisher's protected LSD test (* $p < 0.05$, ** $p < 0.01$).

## 2.4. Effect of CaWRKY22 Loss-of-Function by VIGS on Response of Pepper to R. Solanacearum Inoculation

The loss-of-function experiment was performed in pepper seedlings by Virus Induced Gene Silencing (VIGS) to investigate the role of *CaWRKY22* in immunity. 50 plants of TRV:00 and 50 plants of TRV:*CaWRKY22* were acquired. Six plants from the TRV:*CaWRKY22* plants were randomly selected to assess their gene silencing efficiency by root inoculation with cells of the virulent *R. Solanacearum* strain FJC100301. The result showed that in *R. Solanacearum* challenged TRV:*CaWRKY22* pepper plants, transcriptional levels of *CaWRKY22* were reduced to ~30% of that in TRV:00 plants, showing the successful silencing of *CaWRKY22* (Figure 4A). After *R. Solanacearum* inoculation, TRV:*CaWRKY22* pepper plants exhibited a significantly enhanced susceptibility to the pathogen compared to TRV:00 (control). This susceptibility was coupled with an increase in the growth of *R. Solanacearum* in *CaWRKY22*-silenced pepper plants, manifested by higher cfu values compared with those in the control plants at 3 dpi (days post inoculation) (Figure 4B). Histochemical staining was performed to assess cell death and $H_2O_2$ production in *R. Solanacearum*-infected, *CaWRKY22*-silenced, and control pepper leaves, an intensive DAB (dark brown color) staining (indicator of $H_2O_2$ accumulation) and hypersensitive reaction (HR) mimic cell death, manifested by darker trypan blue staining, were detected in the control leaves at 48 hpi, whereas the intensities of DAB and trypan blue staining were distinctly reduced in *CaWRKY22*-silenced leaves (Figure 4C). Ion leakage was estimated to analyze the severity of cell death and plasma membrane damage after inoculation by *R. Solanacearum*, the results showed that the unsilenced pepper plants exhibited higher ion leakage compared to *CaWRKY22*-silenced pepper plants at 24, 48, and 72 hpi (Figure 4D). A fluorescent modulation meter was used to take the photos of leaf cell death in TRV:*CaWRKY22* and TRV:00 after infection with *R. Solanacearum* FJC100301. The cell death detected in the unsilenced pepper leaves was very noticeable and strong, while very weak and negligible cell death was detected in *CaWRKY22*-silenced plants leaves (Figure 4E). 10 pepper plants of TRV:*CaWRKY22* and TRV:00 were randomly selected and root inoculated with *R. Solanacearum*. At 7 dpi, definite wilting symptoms were observed in *CaWRKY22*-silenced pepper plants, whereas unsilenced plants exhibited only faint wilting symptoms (Figure 4F). QRT-PCR was used to check the transcriptional expression levels of known defense-related genes, the results showed that transcriptional levels of the defense-related pepper genes, including *CaPO2*, *CaPR4*, *CaACC*, *CaBPR1*, *CaDEF1*, and *CaHIR1*, were lessened in *CaWRKY22*-silenced pepper plant leaves compared to that in control pepper plants at 24 hpi (Figure 4G).

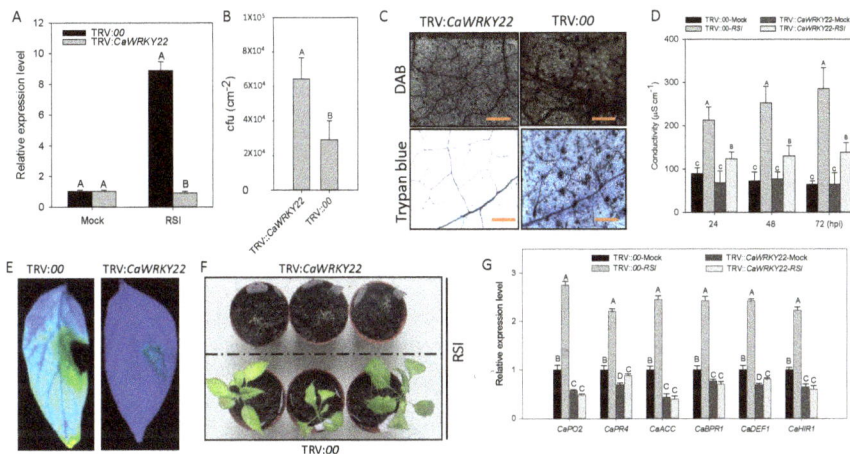

**Figure 4.** Distinctive responses of *CaWRKY22*-knockout attenuates the pepper's resistance to RSI. (**A**) QRT-PCR analysis of *CaWRKY22* expression in *R. Solanacearum*-inoculated, mock (inoculated with MgCl$_2$ solution) *CaWRKY22*-silenced pepper plants (TRV:*CaWRKY22*), and control plants (TRV:00); (**B**) difference in *R. Solanacearum* growth between *CaWRKY22*-silenced and control pepper plants inoculated with *R. Solanacearum* at 3 dpi (days post inoculation); (**C**) DAB and trypan blue staining in *R. Solanacearum*-inoculated *CaWRKY22*-silenced (TRV:*CaWRKY22*) and control (TRV:00) pepper leaves at 48 hpi. Scale bar = 50 μm; (**D**) electrolyte leakage as ion conductivity to assess the cell death responses in the leaf discs of *CaWRKY22*-silenced (TRV:*CaWRKY22*) and control (TRV:00) pepper after 24, 48, and 72 h of inoculation with and without *R. Solanacearum*; (**E**) cell death in *R. Solanacearum* inoculated *CaWRKY22*-silenced (TRV:*CaWRKY22*) and control (TRV:00) pepper leaves under fluorescent modulation meter; (**F**) phenotypic effect of *R. Solanacearum* treatment on *CaWRKY22*-silenced (TRV:*CaWRKY22*) and control (TRV:00) pepper plants at 7 dpi; (**G**) qRT-PCR analysis of transcriptional levels of defense-related marker genes in *CaWRKY22*-silenced (TRV:*CaWRKY22*) and control (TRV:00) pepper plants 24 h post inoculation with *R. Solanacearum*. The relative expression level of mock treated unsilenced plants was set to "1". Error bars indicated the standard error. Data represents the means ± SD from four biological replicates. Different letters indicate significant differences, as determined by Fisher's protected LSD test: uppercase letters, $p < 0.01$; lower case letters, $p < 0.05$.

### 2.5. Transient Overexpression of CaWRKY22-Triggered HR-Like Cell Death and Accumulation of $H_2O_2$ in the Leaves of Pepper Plants

The results of the *CaWRKY22* silencing experiment indicate that *CaWRKY22* acts as a positive regulator in pepper's response to RSI. To further confirm this speculation, a transient overexpression assay was performed in pepper leaves to investigate the effect of *CaWRKY22* transient overexpression on the induction of plant HR cell death. A Western blotting assay confirmed that CaWRKY22 was successfully expressed in pepper plants (Figure 5A). An intensive cell death was manifested by darker trypan blue staining, and a higher level of accumulation of $H_2O_2$ displayed with darker DAB staining was detected in pepper leaves infiltrated with *Agrobacterium* cells carrying 35S:*CaWRKY22* compared to the control plant leaves infiltrated with *Agrobacterium* cells carrying 35S:00 (Figure 5B). Consistently, the pepper leaves transiently overexpressing *CaWRKY22* exhibited significantly higher ion leakage at 24, 48, and 72 h post inoculation compared to the leaves expressing the empty vector (Figure 5C). QRT-PCR was performed to examine the relative transcriptional levels of defense-related genes including the SA-responsive *CaPR4* and *CaBPR1*, JA-responsive *CaDEF1*, ET biosynthesis-associated *CaACC*, HR marker *CaHIR1*, and ROS detoxification-associated *CaPO2*. The results showed that *CaWRKY22* transient overexpression in pepper leaves significantly increased the transcriptional levels of *CaBPR1*, *CaPR4*, *CaDEF1*, *CaACC*, *CaHIR1*, and *CaPO2* as compared to that in pepper leaves transiently overexpressing empty vector (Figure 5D). All these data demonstrate that CaWRKY22 acts as a positive regulator of plant cell death.

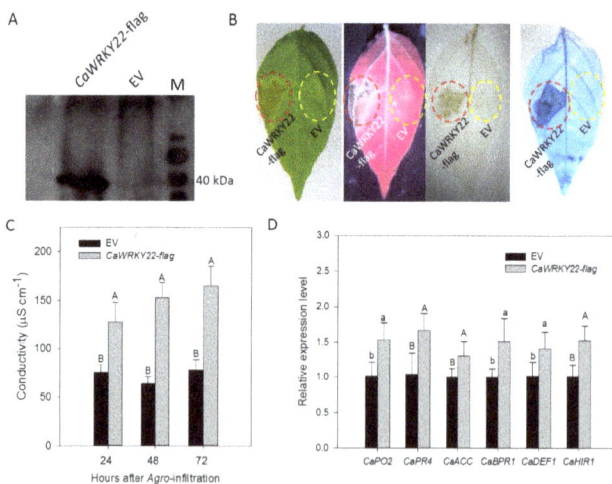

**Figure 5.** Transient over expression of *CaWRKY22* in pepper leaves triggered intense HR-like cell death and ROS accumulation. (**A**) Western blotting was performed to confirm the successful overexpression of CaWRKY22-Flag; (**B**) HR caused by transient overexpression of 35S:*CaWRKY22*, confirmed by phenotype detection, UV light exposure, and DAB and Trypan Blue staining at 4 dpi, respectively; (**C**) measurement of electrolyte leakage (ion conductivity) to evaluate the cell death response in leaf discs at 24, 48, and 72 h post agro-infiltration, respectively; (**D**) qRT-PCR analysis of the expression of immunity-related marker genes, including *CaPO2*, *CaPR4*, *CaACC*, *CaBPR1*, *CaDEF1*, and *CaHIR1*, in 35S:*CaWRKY22* expressed pepper leaves at 24 hpi, respectively. The relative expression level of marker genes in pepper leaves transiently expressing the empty vector were set to "1". Data represent the means ± SD from four biological replicates. Error bars indicated the standard error. Different letters above the bars shows significant differences between the means, as analyzed by Fisher's protected LSD test: uppercase letters, $p < 0.01$; lower case letters, $p < 0.05$.

*2.6. CaWRKY22 Binds to the W-Box Containing Promoter Fragment but Not the Fragments without W-Box*

It has been generally reported that the majority of members of the WRKY family fulfill their functions by targeting and binding the conserved W-box [TTGAC(C/T)] present in the promoters of their target genes [23]. To test if CaWRKY22 can activate the transcription of its target genes in a W-box dependent manner, a chromatin immuno-precipitation (ChIP) assay was performed on chromatins isolated from *CaWRKY22*-HA which were transiently overexpressed in pepper leaves. The chromatin was sheared into fragments from 300 to 500 bps in length, and immuno-precipitated (IPed) with antibody of HA. The IPed DNA fragments were collected as a template for PCR with specific primer pairs of the conserved W-box containing promoter fragments of *CaPR1*, *CaDEF1*, and *CaWRKY40*. The results showed that CaWRKY22 could directly bind to the 250 bp W-box-containing fragments within the promoters of tested immunity associated genes, whereas no binding signal was detected in the promoter fragments within the promoter regions of the tested genes without W-box (Figure 6). These results indicate that CaWRKY22 binds to the W-box containing promoter regions, but does not bind to the promoter regions without W-box. In particular, CaWRKY40, a positive regulator in the pepper response to RSI, high temperature, and high humidity stresses, as identified in our previous study [9] , is also found to be a target of CaWRKY22.

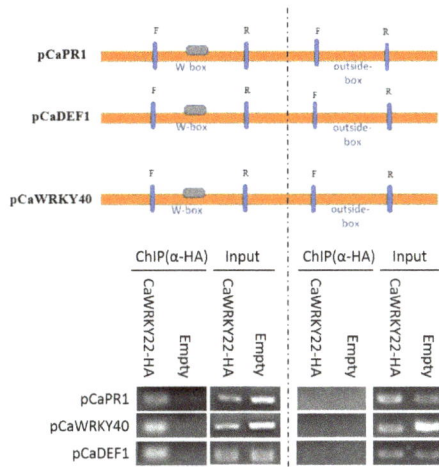

**Figure 6.** CaWRKY22 binds to the W-boxes of different marker genes as detected by chromatin immunoprecipitation (ChIP) analysis. CaWRKY22-HA was transiently overexpressed in pepper leaves, the chromatin was isolated, the DNA–protein complex was immunoprecipitated using anti-HA antibodies and adjusted to the same concentration. PCR was performed using primer pairs based on the sequence flanking the W-box within the promoters of *CaPR1*, *CaDEF1*, and *CaWRKY40*. "F" stands for forward while "R" represent reverse.

*2.7. The Inter-Relationship between CaWRKY22 and CaWRKY40*

As the promoter of *CaWRKY40* was bound by CaWRKY22, we speculated that *CaWRKY40* was transcriptionally regulated by *CaWRKY22*. To test this possibility, and to further assay the possible feedback regulation of *CaWRKY40* by *CaWRKY22*, QRT-PCR was employed to check the possible modulation of *CaWRKY22* by transient overexpression of *CaWRKY40*, or its silencing, as well as the possible effect of *CaWRKY22* transient overexpression or silencing on the expression of *CaWRKY40*. The results showed that transcription of *CaWRKY22* in *CaWRKY40*-transiently-expressing pepper leaves increased at 24 and 48 hpi compared to the control (Figure 7A). On the other hand, the transcriptional abundance of *CaWRKY40* was also increased in *CaWRKY22*-overexpressing leaves at 24 and 48 hpi (Figure 7B). By contrast, it was found that transcriptional levels of *CaWRKY40* were significantly downregulated in *R. Solanacearum* inoculated *CaWRKY22*-silenced pepper plants compared to that in the control plants (Figure 7C). In addition, *CaWRKY22*-silencing fully or partially suppressed the upregulation of the tested defense associated marker genes including *CaBPR1*, *CaPO2*, *CaPR4*, *CaDEF1*, and *CaHIR1*; triggered by *CaWRKY40* transient overexpression (Figure 7D). This suggests direct transcriptional regulation of *CaWRKY40* by *CaWRKY22* and that there exists a positive regulatory loop between *CaWRKY22* and *CaWRKY40*.

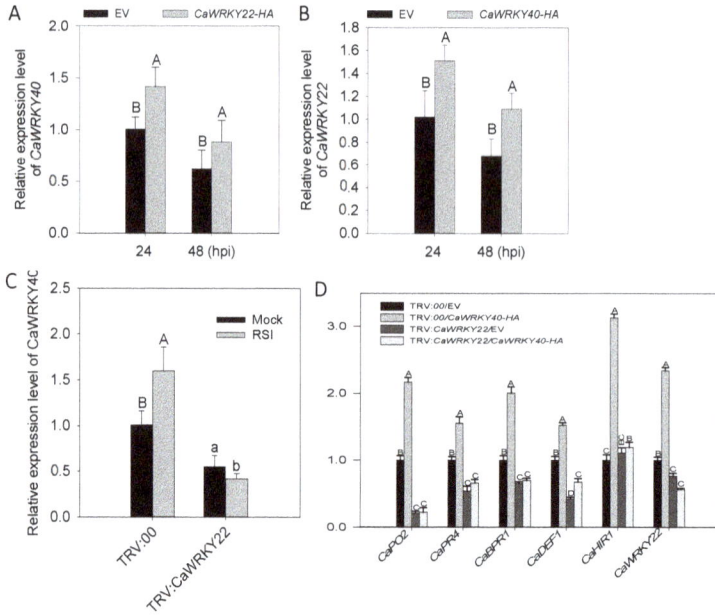

**Figure 7.** The inter-relationship between *CaWRKY22* and *CaWRKY40*. (**A**) Transcriptional expression of *CaWRKY40* in pepper leaves transiently overexpressing *CaWRKY22* at 24 and 48 hpi. The expression level of pepper leaves transiently overexpressing the empty vector was set to "1"; (**B**) transcriptional expression of *CaWRKY22* in pepper leaves transiently overexpressing *CaWRKY40* at 24 and 48 hpi. The expression level of pepper leaves transiently overexpressing the empty vector was set to "1"; (**C**) qRT-PCR analysis of *CaWRKY40* expression levels in *CaWRKY22*-silenced and control pepper plants. The expression level of mock treated unsilenced pepper leaves was set to "1"; (**D**) qRT-PCR analysis of the transcriptional levels of defense-related marker genes in *CaWRKY22*-silenced and control pepper plants transiently overexpressing 35S:*CaWRKY40-HA* and 35S:00. (**A–D**) Data represents the means ± SD from four biological replicates. The relative expression level of mock treated unsilenced plants was set to "1". Error bars indicated the standard error. Different letters indicate significant differences between means, as determined by Fisher's protected LSD test: uppercase letters, $p < 0.01$; lower case letters, $p < 0.05$.

### 2.8. The Inter-Relationship between CaWRKY22 and CaWRKY6, CaWRKY27 and CaWRKY58

As *CaWRKY40* was previously found to be expressionally and functionally related to other WRKYs, including *CaWRKY6*, the close relationship between *CaWRKY22* and *CaWRKY40* implies that *CaWRKY22* might also be associated with other WRKY TFs in pepper immunity against RSI. To test this speculation, the relationship between *CaWRKY22* and *CaWRKY6*, *CaWRKY27* and *CaWRKY58*, which have been implicated in pepper immunity against RSI by our previous studies, were assayed by qRT-PCR, the results showed that the transcriptional levels of *CaWRKY22* were increased in *CaWRKY6* and *CaWRKY27* transiently overexpressing leaves but were decreased in *CaWRKY58*-expressing leaves at 24 hpi (Figure 8A). On the other hand, the transcriptional levels of *CaWRKY6* and *CaWRKY27* increased while that of *CaWRKY58* decreased in *CaWRKY22*-expressing leaves at 24 hpi (Figure 8B). These results suggest that *CaWRKY22* and *CaWRKY6*, *CaWRKY27*, and *CaWRKY58* are expressionally and functionally interrelated.

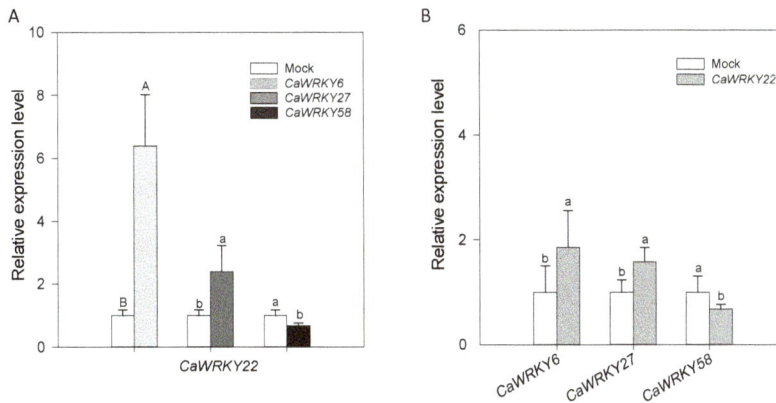

**Figure 8.** The inter-relationship between *CaWRKY22* and *CaWRKY6*, *CaWRKY27* and *CaWRKY58*. (**A**) Transcriptional expression of *CaWRKY22* in pepper leaves transiently overexpressing *CaWRKY6*, *CaWRKY27*, and *CaWRKY58* at 24 hpi. The expression level of *CaWRKY22* in pepper leaves transiently overexpressing the mock (empty vector) was set to "1"; (**B**) transcriptional expression of *CaWRKY6*, *CaWRKY27*, and *CaWRKY58* in pepper leaves transiently overexpressing *CaWRKY22* at 24 hpi. The relative expression level of target genes in mock treated plants (empty vector) was set to "1". Data represents the means ± SD from four biological replicates. Error bars indicated the standard error. Different letters on the bars indicate significant differences between means, as determined by Fisher's protected LSD test: uppercase letters, $p < 0.01$; lower case letters, $p < 0.05$.

## 3. Discussion

WRKY proteins constitute one of the largest TF families in plants. A subset of members in this family, which have been found in *Arabidopsis* and rice, participate and play important roles in the regulation of plant immunity. Significant functional divergence among close structural homologs of WRKY proteins from different plant species has been already described [12], and the roles of WRKY TFs in plant immunity in non-model plants such as pepper remain to be elucidated. Our present study functionally characterized CaWRKY22, an IIe WRKY TF in pepper, and our data indicates that CaWRKY22 is a positive regulator in the response of pepper to RSI and a component of a WRKY transcriptional network including *CaWKRY6*, *CaWRKY27*, *CaWRKY40*, and *CaWRKY58*.

The involvement of *CaWRKY22* in pepper immunity is implied by the presence of pathogen responsive *cis*-elements such as TCA, TGACG-motif, and W-box in the promoter of *CaWRKY22*, and was further supported by the data that *CaWRKY22* transcriptional levels were significantly upregulated after RSI. Since the genes upregulated by a given stress have been frequently found to play roles in response to that stress [24], we postulated that *CaWRKY22* might act as a positive regulator in pepper resistance to RSI. This hypothesis was confirmed by data from loss-of-function of CaWRKY22 by VIGS and gain-function analyses by transient overexpression, respectively. The silencing of *CaWRKY22* by VIGS significantly increased the susceptibility of pepper plants to RSI. This impaired immunity was consistent with the enhanced growth of the inoculated *R. Solanacearum* and downregulated HR-associated *CaHIR1* [25–27], ROS scavenging related *CaPO2* [28], SA-dependent *CaBPR1* [29], *CaPR4* [30,31], JA-associated *CaDEF1* [32–35], and ethylene-dependent *CaACC* [36].

In contrast, the transient overexpression of *CaWRKY22* significantly triggered HR-like cell death and $H_2O_2$ accumulation, accompanied with upregulation of *CaHIR1, CaPO2, CaBPR1, CaPR4, CaDEF1*, and *CaACC*. These results strongly suggest a role of CaWRKY22 as a positive regulator in pepper cell death and immunity. Similarly, it has been previously found that *AtWRKY22*, a homologue of CaWRKY22 in Arabidopsis, functions as a positive regulator in pattern-triggered immunity (PTI) triggered by flg22, chitin, or submergence modulating by the MAP kinase cascade (MEKK1, MKK4/MKK5, and MPK3/MPK6) [37–40]. It can be inferred that, upon attack by *R. Solanacearum, CaWRKY22* is upregulated and therefore decreases the susceptibility of pepper plants to RSI.

Hormones such as SA, JA, and ET are involved in plant immune signaling networks. SA activates resistance against biotrophic pathogens, while JA and ET are generally important for immunity to necrotrophic pathogens [41]. Frequently the production of SA, JA, and ET is coupled with ETI or PTI. Depending on their concentrations, these phytohormones can act either synergistically or antagonistically during defense signaling [42,43]. Synergistic relationships among the three signaling components have been found in PTI. Compensatory relationships among the sectors have been found in ETI [44,45]. *CaWRKY22* was consistently found to be induced by exogenous application of SA, MeJA, or ETH. The tested SA-, JA-, and ET-dependent immunity associated marker genes, SA-dependent *CaBPR1* [29] and *CaPR4* [30,31], JA-associated *CaDEF1* [32–35] and ethylene-dependent *CaACC* [37], were all downregulated by silencing of *CaWRKY22*, but upregulated by the transient overexpression of *CaWRKY22* in pepper plants, indicating that *CaWRKY22* participates in defense signaling which is synergistically mediated by SA, JA, and ET and, therefore, leads to PTI.

Genome-wide analyses indicated the participation of multiple WRKY TFs in plant immunity [46–50]. By functional genomic studies, *WRKY11, -17* [49], *-18* [50], *-25* [51], *-28* [52], *-33* [53], *-38* [54], *-45* [55], *-46* [56], *-53* [56], *-62* [54], *-70* [56], and *-75* [52] have been functionally characterized in *Arabidopsis* immunity, acting as either positive or negative regulators. These TFs have been suggested to integrate into a transcriptional network composed of positive and negative feedback loops and feed-forward modules [57–59]. However, the composition of these networks in different plant species remains poorly understood. Our previous studies found that CaWKRY6, CaWRKY27, and CaWRKY40 act as positive regulators in pepper's response towards RSI [9,18,19], and that CaWRKY58 acts as a negative regulator in the immune process [20]. Our present study suggests that CaWRKY22 acts as a positive regulator in plant cell death and the response of pepper to RSI, its transcription was upregulated by transient overexpression of *CaWRKY6 -27* and *-40*, while down regulated by the transient overexpression of *CaWRKY58*. On the other hand, the transient overexpression of *CaWRKY22* also upregulated the expressions of *CaWKRY6, CaWRKY27*, and *CaWRKY40*, while downregulated the expression of *CaWRKY58*, suggesting the existence of WRKY networks and positive feedback loops between *CaWRKY22* and *CaWRKY6, CaWRKY27* or *CaWRKY40*. Similar positive feedback loops are believed to be present in plant immunity [60]. Alike positive feedback loops have been found between CaWRKY40 and CaCDPK15, CaWRKY40 and CabZIP63, CaWRKY40 and CaWRKY6. By chromatin immuno-precipitation, CaWRKY22 was found to bind the promoter of *CaWRKY40*. Similarly, CaWRKY6 [9] and CabZIP63 [21] have also been found to directly and transcriptionally regulate the expression of *CaWRKY40* during pepper response to RSI. In light of these evidences, it can be inferred that CaWRKY40 might be orchestrated by multiple TFs. However, unlike *CaWRKY6* and *CabZIP63*, which were upregulated by exogenous application of ABA and phenocopy *CaWRKY40* in response to RSI or HTHH, *CaWRKY22* was downregulated by exogenously applied ABA and its silencing or transient overexpression in pepper plants did not exhibit any phenotypic effect on thermotolerance or the expression of thermotolerance-associated marker genes in the present study. ABA is implicated in the plant response to heat stress and drought stress. As a crucial signaling molecule [61–64], ABA affects plant immunity antagonistically [65–68]. In conclusion, the expression of *CaWRKY40* is coregulated at the transcriptional level by CaWRKY6, CabZIP63, and CaWRKY22 upon the challenge of RSI, and is regulated by CaWRKY6 and CabZIP63 but not CaWRKY22 when pepper plants are exposed to heat stress.

A functional model of CaWRKY22 in the pepper immune response to *R. Solanacearum* was proposed, based on the data in the present study (Figure 9). When pepper plants are challenged by *Ralstonia Solanacearum*, SA, JA, and ET mediated signaling is activated, while ABA signaling is depressed. These signaling pathways might be transmitted to the nuclei where they are integrated in some way by various TFs including *CaWRKY6, -27, -40, or -58*. These proteins modulate directly or indirectly the transcription of *CaWRKY22*. CaWRKY22 in turn activates the transcription of *CaWRKY6, CaWRKY27*, and *CaWRKY40*, while depressing the transcription of *CaWRKY58*, therefore, CaWRKY22 activates pepper immunity against *R. Solanacearum*, forming WRKY networks with CaWRKY6, CaWRKY27, CaWRKY40, and CaWRKY58.

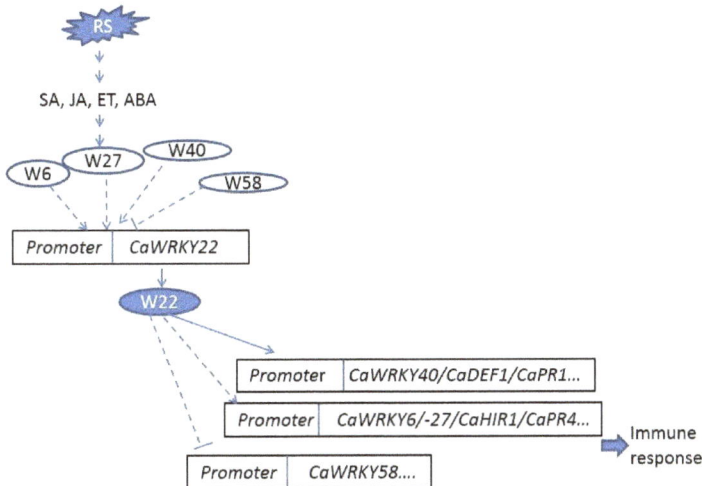

**Figure 9.** The proposed model of CaWRKY22 functions in pepper immunity against *Ralstonia Solanacearum*. RS: *Ralstonia Solanacearum*; W6: CaWRKY6; W22: CaWRKY22; W27: CaWRKY27; W40: CaWRKY40; W58: CaWRKY58. The dashed lines: indirect or direct regulation; the full line: direct regulation. ⟶ . Positive regulation, ⁻⁻⁻⊣ . Negative regulation

## 4. Experimental Procedures

### 4.1. Plant Materials and Growth Conditions

The seeds of Pepper (*Capsicum annuum*) cultivar GZ03 and *Nicotiana benthamiana* were procured from the pepper breeding group at the Fujian Agriculture and Forestry University (www.fafu.edu.cn). The seeds were sown in a soil mix (peat moss: perlite; 2/1, *v*/*v*) in plastic pots and placed in a greenhouse, and grew in a growth room at 25 °C, 60–70 μmol photons m$^{-2}$·s$^{-1}$, a relative humidity of 70%, and under a 16-h light/8-h dark photoperiod.

### 4.2. Vectors Construction

The vectors were constructed using the gateway technique. The full length ORF of *CaWRKY22* (with or without termination codon) was cloned into the entry vector pDONR207 by BP reaction to generate satellite vectors, and then transferred into destination vectors pMDC83, CD3688 (Flag-tag), pK7WG2, and CD3687 (HA-tag) to construct vectors for overexpression, subcellular localization, and ChIP assay by LR reaction, respectively. To prepare vectors for VIGS, a 308 bps fragment in the 3′-untranslated region (UTR) of *CaWRKY22* was selected, and the specificity was confirmed by a BLAST against the genome sequence in database of CM334 (http://peppergenome.snu.ac.kr/) and Zunla-1 (http://peppersequence.genomics.cn/page/species/blast.jsp). The specific fragment was cloned into the entry vector pDONR207 by Briggs–Rauscher (BR) reaction and then into the PYL279 vector by LR reaction.

### 4.3. Pathogens and Inoculation Procedures

*R. Solanacearum* strain FJC100301 was isolated from disease affected pepper plants from Fujian province (China) [19]. By using the tetrazolium chloride method stem exudates were purified [68]. *R. Solanacearum* was cultured in SPA medium (200 g potato, 20 g sucrose, 3 g beef extract, 5 g tryptone, and 1 L of double-distilled $H_2O$) overnight at 28 °C and 200 rpm. The cultured *R. Solanacearum* was centrifuged and the pellet was suspended in distilled sterilized 10 mM $MgCl_2$. Bacterial cell density was adjusted to $10^8$ cfu·mL$^{-1}$ (OD600 nm = 0.8). To assay the effect of *R. Solanacearum* inoculation on transcription of *CaWRKY22* and the resistance of pepper to RSI, pepper plants were inoculated with 10 μL *R. Solanacearum* into the top third leaf with the help of a syringe without a needle. The respective leaves were harvested at the indicated time points for RNA extraction and histochemical staining, including DAB and trypan blue staining. To study the *CaWRKY22*-silenced phenotype in pepper plants under *R. Solanacearum* inoculation, the roots were injured by giving little cuts with a glass rod and then inoculated with the *R. Solanacearum*. After inoculation, the plants were grown in a growth chamber at $28 \pm 2$ °C, 60–70 μmol photons m$^{-2}$·s$^{-1}$, relative humidity of 70%, and under a 16-h light/8-h dark photoperiod.

### 4.4. Plant Treatments with Exogenously Applied Phytohormones

For the phytohormone treatments, the healthy pepper plants at the four-leaf stage were sprayed with 1 mM SA, 100 μM methyl jasmonate (MeJA), 100 μM ABA, and 100 μM ETH, respectively. Mock plants were sprayed with sterile ddH$_2$O. The treated samples above were harvested for RNA extraction at the indicated time points.

### 4.5. Analysis of CaWRKY22 Subcellular Localization

*Agrobacterium* containing *35S:CaWRKY22-GFP* or *35S:GFP* (used as control) were cultured overnight in a LB medium containing the corresponding antibiotics. Bacteria were centrifuged and the pellet was suspended in induction medium (10 mM 2-Morpholinoethanesulfonic acid [MES], 10 mM $MgCl_2$ pH = 5.4, and 200 μM acetosyringone), and adjusted to $OD_{600}$ = 0.8. *Agrobacterium* cells harboring the *35S:CaWRKY22-GFP* and *35S:GFP* were infiltrated into *N. benthamiana* leaves with a needleless syringe. 4′,6-Diamidino-2-phenylindole (DAPI) staining was performed as described previously [20] to specifically stain the nuclei. GFP and DAPI fluorescence signals were detected and images were taken using a Leica fluorescence light microscope (Tokyo, Japan) with an excitation wavelength of 488 nm, a band-pass emission filter, and an excitation wavelength of 405 nm with a 435–480 nm band-pass emission filter.

### 4.6. Histochemical Staining

Trypan blue and 3,3′-diaminobenzidine (DAB) staining was performed according to the previously published method of [21,22,69]. For trypan blue staining, pepper leaves were boiled in trypan blue solution (10 mL lactic acid, 10 mL glycerol, 10 mL phenol, 40 mL ethanol, 10 mL ddH$_2$O, and 1 mL trypan blue) for 20 min, kept at room temperature for 8 hours, then placed into a chloral hydrate solution (2.5 g of chloral hydrated dissolved in 1 mL of distilled water) and boiled for 25 min for destaining; this process was repeated in triplicate. Finally, samples were kept in 70% glycerol. For DAB staining, the leaves were immersed in 1 mg/mL of DAB solution and maintained at room temperature overnight. Lactic acid:glycerol:absolute ethanol [1:1:3 (v/v/v)] solution was used to destain the DAB-stained pepper leaves and kept in 95% absolute ethanol [70]. A camera and light microscope (Leica, Wetzlar, Germany) were used to take the images of trypan blue and DAB staining.

*4.7. Virus-Induced Gene Silencing (VIGS) of CaWRKY22 in Pepper Plants*

The Tobacco Rattle Virus (TRV)-based virus induced gene silencing (VIGS) system was employed for *CaWRKY22* silencing in pepper plants, following the method of our previous studies [21,22,71,72]. *Agrobacterium* cells containing TRV1 and TRV2, TRV2-*CaWRKY22*, or TRV2-*PDS* ($OD_{600}$ = 0.8) constructs were mixed in a 1:1 ratio, respectively. This mixture was infiltrated into cotyledons of 2-week-old pepper plants using a syringe without a needle. The *Agrobacterium*-inoculated pepper plants were grown in a growth chamber at 16 °C in the dark for 56 h with 45% relative humidity, and then transferred into a growth room at $25 \pm 2$ °C, 60–70 µmol photons $m^{-2} \cdot s^{-1}$ and a relative humidity of 70%, under a 16-h light/8-h dark cycle.

*4.8. Transient CaWRKY22 Expression Assay*

*Agrobacterium* cells containing the 35S:*CaWRKY22-flag* vector were cultured to OD600 = 1.0 in LB medium containing the corresponding antibiotics overnight. These cells were then centrifuged at 7800 rpm, 28 °C for 10 min. The pellet was suspended into the induction medium (10 mM MES, 10 mM $MgCl_2$, pH 5.4, and 200 µM acetosyringone) and adjusted to $OD_{600}$ = 0.8. The suspension was infiltrated into pepper leaves using a needleless syringe. Later, the infiltrated plants were regularly observed for HR cell death, and harvested for trypan blue or DAB staining. RNA extraction used for detection of the expression of immunity associated genes.

*4.9. Total RNA Isolation and qRT-PCR*

Total RNA was extracted from pepper leaves samples and wild type seedlings, using TRIzol reagent (Invitrogen, Carlsbad, CA, USA), were reverse transcribed by using the PrimeScript RT-PCR kit (TaKaRa, Dalian, China). To determine the relative transcription level of targeted genes, qRT-PCR and the corresponding data processing were performed, according to the method used in our previous studies [17,21,22,72], with specific primers (see Table S2) using the Bio-Rad Real-time PCR system (Bio-Rad, Foster City, CA, USA) and SYBR premix Ex Taq II system (TaKaRa Perfect Real Time).

*4.10. Measurement of Ion Conductivity*

Ion leakage was measured as previously described with slight modifications [9]. The six leaf discs (4 mm in diameter) were cut with a hole-punch, washed with sterilized $ddH_2O$ thrice, and immediately incubated in 20 mL of double distilled water. Discs were kept in a shaker with gentle shaking (60 rpm) for 1 h at room temperature. Ion conductivity was recorded using a conductivity meter (Mettler Toledo 326 Mettler, Zurich, Switzerland).

*4.11. Chromatin Immuno-Precipitation Analysis (ChIP)*

The ChIP assay was carried out according to a previously described protocol [9]. The *Agrobacterium* cells carrying 35S:*CaWRKY22*-HA or 35S:*CaWRKY40*-HA were infiltrated into the pepper leaves at the eight leaf stage. The infiltrated leaves were harvested at the required time point and ~2 g leaves of pepper were fixed with 1.0% formaldehyde for 5 min. The chromatin was sheared by sonication to an average length of 300–500 bp and immuno-precipitated with antibody against hemagglutinin (anti-HA; Santa Cruz Biotechnology, Dallas, TX, USA). The immuno-precipitated DNA was analyzed for enrichment of CaWRKY22 at the promoter region of targeted genes by common ChIP-PCR. The primers used for ChIP-PCR analysis are listed in Table S3.

## 5. Conclusions

The data in the present study indicates that CaWRKY22 targets to nuclei, is up-regulated by RSI and confer enhanced immunity against *Ralstonia solanccearum* by directly targeting *CaWRKY40* and incorporating into a WRKY network including *CaWRKY6*, *CaWRKY27*, *CaWRKY40*, and *CaWRKY58*.

**Supplementary Materials:** Supplementary materials can be found at http://www.mdpi.com/1422-0067/19/5/1426/s1..

**Author Contributions:** H.S., A.H., and Z.L. conceived the idea and provided resources for experimentation. A.H., X.L., Y.W., and Z.L. designed the study and performed the experiments. A.H., M.F.A., A.N., M.I., S.Y., and S.Q. helped in experimentation and writing. A.H., Y.Y., Y.W., X.L., and D.G. analyzed the data.

**Funding:** This work was supported by grants from the National Natural Science Foundation of China (31501767, 31572136), the Natural Science Foundation of Fujian Province, China (2017J01436), and the Young and Middle-aged Teachers Education Scientific Research Project of the Fujian Provincial Department of Education (2016). The funders had no role in study design, data collection, data analysis, the decision to publish, or the preparation of the manuscript.

**Acknowledgments:** We also want to thank Mark D. Curtis (University of Zurich) for kindly providing Gateway-destination vectors and S.P. Dinesh-Kumar (Yale University) for providing pTRV1 and pTRV2.

**Conflicts of Interest:** The Authors declare no conflict of interest.

## References

1. Noman, A.; Ali, Q.; Maqsood, J.; Iqbal, N.; Javed, M.T.; Rasool, N.; Naseem, J. Deciphering physio-biochemical, yield, and nutritional quality attributes of water-stressed radish (*Raphanus sativus* L.) plants grown from Zn-Lys primed seeds. *Chemosphere* **2018**, *195*, 175–189. [CrossRef] [PubMed]

2. Jones, J.D.; Dangl, J.L. The plant immune system. *Nature* **2006**, *444*, 323. [CrossRef] [PubMed]

3. Eulgem, T.; Somssich, I.E. Networks of WRKY transcription factors in defense signaling. *Curr. Opin. Plant Biol.* **2007**, *10*, 366–371. [CrossRef] [PubMed]

4. Phukan, U.J.; Mishra, S.; Timbre, K.; Luqman, S.; Shukla, R.K. Mentha arvensis exhibit better adaptive characters in contrast to mentha piperita when subjugated to sustained waterlogging stress. *Protoplasma* **2014**, *251*, 603–614. [CrossRef] [PubMed]

5. Ülker, B.; Somssich, I.E. Wrky transcription factors: From DNA binding towards biological function. *Curr. Opin. Plant Biol.* **2004**, *7*, 491–498. [CrossRef] [PubMed]

6. Zhang, Y.; Wang, L. The WRKY transcription factor superfamily: Its origin in eukaryotes and expansion in plants. *BMC Evolut. Biol.* **2005**, *5*, 1.

7. Rushton, P.J.; Somssich, I.E.; Ringler, P.; Shen, Q.J. WRKY transcription factors. *Trends Plant Sci.* **2010**, *15*, 247–258. [CrossRef] [PubMed]

8. Yamasaki, K.; Kigawa, T.; Watanabe, S.; Inoue, M.; Yamasaki, T.; Seki, M.; Shinozaki, K.; Yokoyama, S. Structural basis for sequence-specific DNA recognition by an arabidopsis WRKY transcription factor. *J. Biol. Chem.* **2012**, *287*, 7683–7691. [CrossRef] [PubMed]

9. Cai, H.; Yang, S.; Yan, Y.; Xiao, Z.; Cheng, J.; Wu, J.; Qiu, A.; Lai, Y.; Mou, S.; Guan, D. CaWRKY6 transcriptionally activates CaWRKY40, regulates ralstonia *Solanacearum* resistance, and confers high-temperature and high-humidity tolerance in pepper. *J. Exp. Bot.* **2015**, *66*, 3163–3174. [CrossRef] [PubMed]

10. Chi, Y.; Yang, Y.; Zhou, Y.; Zhou, J.; Fan, B.; Yu, J.-Q.; Chen, Z. Protein–protein interactions in the regulation of WRKY transcription factors. *Mol. Plant* **2013**, *6*, 287–300. [CrossRef] [PubMed]

11. Shan, W.; Chen, J.Y.; Kuang, J.F.; Lu, W.J. Banana fruit nac transcription factor manac5 cooperates with MAWRKYs to enhance the expression of pathogenesis-related genes against colletotrichum musae. *Mol. Plant Pathol.* **2016**, *17*, 330–338. [CrossRef] [PubMed]

12. Yang, Y.; Chi, Y.; Wang, Z.; Zhou, Y.; Fan, B.; Chen, Z. Functional analysis of structurally related soybean GMWRKY58 and GMWRKY76 in plant growth and development. *J. Exp. Bot.* **2016**, *67*, 4727–4742. [CrossRef] [PubMed]

13. Hayward, A.; Elphinstone, J.; Caffier, D.; Janse, J.; Stefani, E.; French, E.; Wright, A. *Round Table on Bacterial Wilt (Brown Rot) of Potato*; Springer: Berlin, Germany, 1998; p. 420.

14. Ristaino, J.B.; Johnston, S.A. Ecologically based approaches to management of phytophthora blight on bell pepper. *Plant Dis.* **1999**, *83*, 1080–1089. [CrossRef]

15. Rizhsky, L.; Liang, H.; Mittler, R. The combined effect of drought stress and heat shock on gene expression in tobacco. *Plant Physiol.* **2002**, *130*, 1143–1151. [CrossRef] [PubMed]

16. Kim, S.; Park, M.; Yeom, S.-I.; Kim, Y.-M.; Lee, J.M.; Lee, H.-A.; Seo, E.; Choi, J.; Cheong, K.; Kim, K.-T. Genome sequence of the hot pepper provides insights into the evolution of pungency in capsicum species. *Nat. Genet.* **2014**, *46*, 270. [CrossRef] [PubMed]

17. Noman, A.; Sheng, Y.; Lei, S.; Hussain, A.; Ashraf, M.F.; Khan, M.I.; He, S.L. Expression and functional evaluation of CaZNF830 during pepper response to Ralstonia solanacearum or high temperature and humidity. *Microb. Pathog.* **2018**, *118*, 336–346. [CrossRef] [PubMed]

18. Dang, F.; Wang, Y.; She, J.; Lei, Y.; Liu, Z.; Eulgem, T.; Lai, Y.; Lin, J.; Yu, L.; Lei, D.; et al. Overexpression of CaWRKY27, a subgroup iie WRKY transcription factor of *Capsicum annuum*, positively regulates tobacco resistance to ralstonia *Solanacearum* infection. *Physiol. Plant* **2014**, *150*, 397–411. [CrossRef] [PubMed]

19. Dang, F.F.; Wang, Y.N.; Yu, L.; Eulgem, T.; Lai, Y.; Liu, Z.Q.; Wang, X.; Qiu, A.L.; Zhang, T.X.; Lin, J.; et al. CaWRKY40, a WRKY protein of pepper, plays an important role in the regulation of tolerance to heat stress and resistance to ralstonia *Solanacearum* infection. *Plant Cell Environ.* **2013**, *36*, 757–774. [CrossRef] [PubMed]

20. Wang, Y.; Dang, F.; Liu, Z.; Wang, X.; Eulgem, T.; Lai, Y.; Yu, L.; She, J.; Shi, Y.; Lin, J.; et al. CaWRKY58, encoding a group I WRKY transcription factor of capsicum annuum, negatively regulates resistance to ralstonia *Solanacearum* infection. *Mol. Plant Pathol.* **2013**, *14*, 131–144. [CrossRef] [PubMed]

21. Shen, L.; Liu, Z.; Yang, S.; Yang, T.; Liang, J.; Wen, J.; Liu, Y.; Li, J.; Shi, L.; Tang, Q.; et al. Pepper cabzip63 acts as a positive regulator during ralstonia *Solanacearum* or high temperature-high humidity challenge in a positive feedback loop with CaWRKY40. *J. Exp. Bot.* **2016**, *67*, 2439–2451. [CrossRef] [PubMed]

22. Shen, L.; Yang, S.; Yang, T.; Liang, J.; Cheng, W.; Wen, J.; Liu, Y.; Li, J.; Shi, L.; Tang, Q.; et al. Cacdpk15 positively regulates pepper responses to ralstonia *Solanacearum* inoculation and forms a positive-feedback loop with CaWRKY40 to amplify defense signaling. *Sci. Rep.* **2016**, *6*, 22439. [CrossRef] [PubMed]

23. Eulgem, T.; Rushton, P.J.; Robatzek, S.; Somssich, I.E. The WRKY superfamily of plant transcription factors. *Trends Plant Sci.* **2000**, *5*, 199–206. [CrossRef]

24. Hsu, F.-C.; Chou, M.-Y.; Chou, S.-J.; Li, Y.-R.; Peng, H.-P.; Shih, M.-C. Submergence confers immunity mediated by the WRKY22 transcription factor in arabidopsis. *Plant Cell* **2013**, *25*, 2699–2713. [CrossRef] [PubMed]

25. Nadimpalli, R.; Yalpani, N.; Johal, G.S.; Simmons, C.R. Prohibitins, stomatins, and plant disease response genes compose a protein superfamily that controls cell proliferation, ion channel regulation, and death. *J. Biol. Chem.* **2000**, *275*, 29579–29586. [CrossRef] [PubMed]

26. Zhou, L.; Cheung, M.Y.; Li, M.W.; Fu, Y.; Sun, Z.; Sun, S.M.; Lam, H.M. Rice hypersensitive induced reaction protein 1 (OSHIR1) associates with plasma membrane and triggers hypersensitive cell death. *BMC Plant Biol.* **2010**, *10*, 290. [CrossRef] [PubMed]

27. Rostoks, N.; Schmierer, D.; Kudrna, D.; Kleinhofs, A. Barley putative hypersensitive induced reaction genes: Genetic mapping, sequence analyses and differential expression in disease lesion mimic mutants. *Theor. Appl. Genet.* **2003**, *107*, 1094–1101. [CrossRef] [PubMed]

28. Karthikeyan, M.; Jayakumar, V.; Radhika, K.; Bhaskaran, R.; Velazhahan, R.; Alice, D. Induction of resistance in host against the infection of leaf blight pathogen (*Alternaria palandui*) in onion (*allium cepa* var aggregatum). *Indian J. Biochem. Biophys.* **2005**, *42*, 371–377. [PubMed]

29. Stevens, C.; Titarenko, E.; Hargreaves, J.A.; Gurr, S.J. Defence-related gene activation during an incompatible interaction between stagonospora (*Septoria*) nodorum and barley (*hordeum vulgare* L.) coleoptile cells. *Plant Mol. Biol.* **1996**, *31*, 741–749. [CrossRef] [PubMed]

30. Gu, Y.Q.; Wildermuth, M.C.; Chakravarthy, S.; Loh, Y.T.; Yang, C.; He, X.; Han, Y.; Martin, G.B. Tomato transcription factors PTI4, PTI5, and PTI6 activate defense responses when expressed in arabidopsis. *Plant Cell* **2002**, *14*, 817–831. [CrossRef] [PubMed]

31. Li, X.; Xia, B.; Jiang, Y.; Wu, Q.; Wang, C.; He, L.; Peng, F.; Wang, R. A new pathogenesis-related protein, lrpr4, from lycoris radiata, and its antifungal activity against magnaporthe grisea. *Mol. Biol. Rep.* **2010**, *37*, 995–1001. [CrossRef] [PubMed]

32. Vicedo, B.; Flors, V.; Finiti, I.; Kravchuk, Z.; Real, M.D.; Garcia-Agustin, P.; Gonzalez-Bosch, C. Hexanoic acid-induced resistance against botrytis cinerea in tomato plants. *Mol. Plant Microbe Interact.* **2009**, *22*, 1455–1465. [CrossRef] [PubMed]

33. Egusa, M.; Ozawa, R.; Takabayashi, J.; Otani, H.; Kodama, M. The jasmonate signaling pathway in tomato regulates susceptibility to a toxin-dependent necrotrophic pathogen. *Planta* **2009**, *229*, 965–976. [CrossRef] [PubMed]

34. Bhattarai, K.K.; Xie, Q.G.; Mantelin, S.; Bishnoi, U.; Girke, T.; Navarre, D.A.; Kaloshian, I. Tomato susceptibility to root-knot nematodes requires an intact jasmonic acid signaling pathway. *Mol. Plant Microbe Interact.* **2008**, *21*, 1205–1214. [CrossRef] [PubMed]

35. O'Donnell, P.J.; Schmelz, E.; Block, A.; Miersch, O.; Wasternack, C.; Jones, J.B.; Klee, H.J. Multiple hormones act sequentially to mediate a susceptible tomato pathogen defense response. *Plant Physiol.* **2003**, *133*, 1181–1189. [CrossRef] [PubMed]

36. Zhu, X.; Wang, A.; Zhu, S.; Zhang, L. Expression of *ACO1*, *ERS1* and *ERF1* genes in harvested bananas in relation to heat-induced defense against colletotrichum musae. *J. Plant Physiol.* **2011**, *168*, 1634–1640. [CrossRef] [PubMed]

37. Asai, T.; Tena, G.; Plotnikova, J.; Willmann, M.R.; Chiu, W.L.; Gomez-Gomez, L.; Boller, T.; Ausubel, F.M.; Sheen, J. Map kinase signalling cascade in arabidopsis innate immunity. *Nature* **2002**, *415*, 977–983. [CrossRef] [PubMed]

38. Wan, J.; Zhang, S.; Stacey, G. Activation of a mitogen-activated protein kinase pathway in arabidopsis by chitin. *Mol. Plant Pathol.* **2004**, *5*, 125–135. [CrossRef] [PubMed]

39. Schikora, A.; Schenk, S.T.; Stein, E.; Molitor, A.; Zuccaro, A.; Kogel, K.H. N-acyl-homoserine lactone confers resistance toward biotrophic and hemibiotrophic pathogens via altered activation of atmpk6. *Plant Physiol.* **2011**, *157*, 1407–1418. [CrossRef] [PubMed]

40. Noman, A.; Liu, Z.; Aqeel, M.; Zainab, M.; Khan, M.I.; Hussain, A.; Ashraf, M.F.; Li, X.; Weng, Y.; He, S. Basic leucine zipper domain transcription factors: The vanguards in plant immunity. *Biotechnol. Lett.* **2017**, *39*, 1779–1791. [CrossRef] [PubMed]

41. Kunkel, B.N.; Brooks, D.M. Cross talk between signaling pathways in pathogen defense. *Curr. Opin. Plant Biol.* **2002**, *5*, 325–331. [CrossRef]

42. Mur, L.A.; Kenton, P.; Atzorn, R.; Miersch, O.; Wasternack, C. The outcomes of concentration-specific interactions between salicylate and jasmonate signaling include synergy, antagonism, and oxidative stress leading to cell death. *Plant Physiol.* **2006**, *140*, 249–262. [CrossRef] [PubMed]

43. Tsuda, K.; Sato, M.; Stoddard, T.; Glazebrook, J.; Katagiri, F. Network properties of robust immunity in plants. *PLoS Genet.* **2009**, *5*, e1000772. [CrossRef] [PubMed]

44. Bagnaresi, P.; Biselli, C.; Orru, L.; Urso, S.; Crispino, L.; Abbruscato, P.; Piffanelli, P.; Lupotto, E.; Cattivelli, L.; Vale, G. Comparative transcriptome profiling of the early response to magnaporthe oryzae in durable resistant vs susceptible rice (*Oryza sativa* L.) genotypes. *PLoS ONE* **2012**, *7*, e51609. [CrossRef] [PubMed]

45. Matic, S.; Bagnaresi, P.; Biselli, C.; Orru, L.; Amaral Carneiro, G.; Siciliano, I.; Vale, G.; Gullino, M.L.; Spadaro, D. Comparative transcriptome profiling of resistant and susceptible rice genotypes in response to the seedborne pathogen fusarium fujikuroi. *BMC Genom.* **2016**, *17*, 608. [CrossRef] [PubMed]

46. Cheng, H.; Liu, H.; Deng, Y.; Xiao, J.; Li, X.; Wang, S. The WRKY45-2 WRKY13 WRKY42 transcriptional regulatory cascade is required for rice resistance to fungal pathogen. *Plant Physiol.* **2015**, *167*, 1087–1099. [CrossRef] [PubMed]

47. Wei, T.; Ou, B.; Li, J.; Zhao, Y.; Guo, D.; Zhu, Y.; Chen, Z.; Gu, H.; Li, C.; Qin, G.; et al. Transcriptional profiling of rice early response to magnaporthe oryzae identified OSWRKYs as important regulators in rice blast resistance. *PLoS ONE* **2013**, *8*, e59720. [CrossRef] [PubMed]

48. Ryu, H.S.; Han, M.; Lee, S.K.; Cho, J.I.; Ryoo, N.; Heu, S.; Lee, Y.H.; Bhoo, S.H.; Wang, G.L.; Hahn, T.R.; et al. A comprehensive expression analysis of the WRKY gene superfamily in rice plants during defense response. *Plant Cell Rep.* **2006**, *25*, 836–847. [CrossRef] [PubMed]

49. Journot-Catalino, N.; Somssich, I.E.; Roby, D.; Kroj, T. The transcription factors WRKY11 and WRKY17 act as negative regulators of basal resistance in arabidopsis thaliana. *Plant Cell* **2006**, *18*, 3289–3302. [CrossRef] [PubMed]

50. Chen, C.; Chen, Z. Potentiation of developmentally regulated plant defense response by ATWRKY18, a pathogen-induced arabidopsis transcription factor. *Plant Physiol.* **2002**, *129*, 706–716. [CrossRef] [PubMed]

51. Zheng, Z.; Mosher, S.L.; Fan, B.; Klessig, D.F.; Chen, Z. Functional analysis of arabidopsis WRKY25 transcription factor in plant defense against pseudomonas syringae. *BMC Plant Biol.* **2007**, *7*, 2. [CrossRef] [PubMed]

52. Chen, X.; Liu, J.; Lin, G.; Wang, A.; Wang, Z.; Lu, G. Overexpression of ATWRKY28 and ATWRKY75 in arabidopsis enhances resistance to oxalic acid and *Sclerotinia sclerotiorum*. *Plant Cell Rep.* **2013**, *32*, 1589–1599. [CrossRef] [PubMed]

53. Lippok, B.; Birkenbihl, R.P.; Rivory, G.; Brummer, J.; Schmelzer, E.; Logemann, E.; Somssich, I.E. Expression of ATWRKY33 encoding a pathogen- or pamp-responsive WRKY transcription factor is regulated by a composite DNA motif containing W-box elements. *Mol. Plant Microbe Interact.* **2007**, *20*, 420–429. [CrossRef] [PubMed]

54. Kim, K.C.; Lai, Z.; Fan, B.; Chen, Z. Arabidopsis WRKY38 and WRKY62 transcription factors interact with histone deacetylase 19 in basal defense. *Plant Cell* **2008**, *20*, 2357–2371. [CrossRef] [PubMed]

55. Shimono, M.; Sugano, S.; Nakayama, A.; Jiang, C.J.; Ono, K.; Toki, S.; Takatsuji, H. Rice WRKY45 plays a crucial role in benzothiadiazole-inducible blast resistance. *Plant Cell* **2007**, *19*, 2064–2076. [CrossRef] [PubMed]

56. Hu, Y.; Dong, Q.; Yu, D. Arabidopsis WRKY46 coordinates with WRKY70 and WRKY53 in basal resistance against pathogen pseudomonas syringae. *Plant Sci.* **2012**, *185–186*, 288–297. [CrossRef] [PubMed]

57. Eulgem, T. Dissecting the WRKY web of plant defense regulators. *PLoS Pathog.* **2006**, *2*, e126. [CrossRef] [PubMed]

58. Moore, J.W.; Loake, G.J.; Spoel, S.H. Transcription dynamics in plant immunity. *Plant Cell* **2011**, *23*, 2809–2820. [CrossRef] [PubMed]

59. Wang, Y.; Chang, H.; Hu, S.; Lu, X.; Yuan, C.; Zhang, C.; Wang, P.; Xiao, W.; Xiao, L.; Xue, G.P.; et al. Plastid casein kinase 2 knockout reduces abscisic acid (ABA) sensitivity, thermotolerance, and expression of aba- and heat-stress-responsive nuclear genes. *J. Exp. Bot.* **2014**, *65*, 4159–4175. [CrossRef] [PubMed]

60. Larkindale, J.; Hall, J.D.; Knight, M.R.; Vierling, E. Heat stress phenotypes of arabidopsis mutants implicate multiple signaling pathways in the acquisition of thermotolerance. *Plant Physiol.* **2005**, *138*, 882–897. [CrossRef] [PubMed]

61. Huang, Y.C.; Niu, C.Y.; Yang, C.R.; Jinn, T.L. The heat stress factor HSFA6B connects ABA signaling and aba-mediated heat responses. *Plant Physiol.* **2016**, *172*, 1182–1199. [CrossRef] [PubMed]

62. Hu, X.J.; Chen, D.; Lynne McIntyre, C.; Fernanda Dreccer, M.; Zhang, Z.B.; Drenth, J.; Kalaipandian, S.; Chang, H.; Xue, G.P. Heat shock factor C2A serves as a proactive mechanism for heat protection in developing grains in wheat via an ABA-mediated regulatory pathway. *Plant Cell Environ.* **2018**, *41*, 79–98. [CrossRef] [PubMed]

63. Moeder, W.; Ung, H.; Mosher, S.; Yoshioka, K. Sa-aba antagonism in defense responses. *Plant Signal. Behav.* **2010**, *5*, 1231–1233. [CrossRef] [PubMed]

64. Liu, S.; Ziegler, J.; Zeier, J.; Birkenbihl, R.P.; Somssich, I.E. Botrytis cinerea B05.10 promotes disease development in arabidopsis by suppressing WRKY33-mediated host immunity. *Plant Cell Environ.* **2017**, *40*, 2189–2206. [CrossRef] [PubMed]

65. Kusajima, M.; Okumura, Y.; Fujita, M.; Nakashita, H. Abscisic acid modulates salicylic acid biosynthesis for systemic acquired resistance in tomato. *Biosci. Biotechnol. Biochem.* **2017**, *81*, 1850–1853. [CrossRef] [PubMed]

66. Yasuda, M.; Ishikawa, A.; Jikumaru, Y.; Seki, M.; Umezawa, T.; Asami, T.; Maruyama-Nakashita, A.; Kudo, T.; Shinozaki, K.; Yoshida, S.; et al. Antagonistic interaction between systemic acquired resistance and the abscisic acid-mediated abiotic stress response in arabidopsis. *Plant Cell* **2008**, *20*, 1678–1692. [CrossRef] [PubMed]

67. Anderson, J.P.; Badruzsaufari, E.; Schenk, P.M.; Manners, J.M.; Desmond, O.J.; Ehlert, C.; Maclean, D.J.; Ebert, P.R.; Kazan, K. Antagonistic interaction between abscisic acid and jasmonate-ethylene signaling pathways modulates defense gene expression and disease resistance in arabidopsis. *Plant Cell* **2004**, *16*, 3460–3479. [CrossRef] [PubMed]

68. Kelman, A. The relationship of pathogenicity of pseudomonas *Solanacearum* to colony appearance in a tetrazolium medium. *Phytopathology* **1954**, *44*, 693–695.

69. Liu, Z.; Shi, L.; Yang, S.; Lin, Y.; Weng, Y.; Li, X.; Hussain, A.; Noman, A.; He, S. Functional and promoter analysis of CHIIV3, a chitinase of pepper plant, in response to phytophthora capsici infection. *Int. J. Mol. Sci.* **2017**, *18*, 1661. [CrossRef] [PubMed]

70. Korasick, D.A.; McMichael, C.; Walker, K.A.; Anderson, J.C.; Bednarek, S.Y.; Heese, A. Novel functions of stomatal cytokinesis-defective 1 (SCD1) in innate immune responses against bacteria. *J. Biol. Chem.* **2010**, *285*, 23342–23350. [CrossRef] [PubMed]

71. Liu, Z.Q.; Qiu, A.L.; Shi, L.P.; Cai, J.S.; Huang, X.Y.; Yang, S.; Wang, B.; Shen, L.; Huang, M.K.; Mou, S.L.; et al. SRC2-1 is required in PCINF1-induced pepper immunity by acting as an interacting partner of PCINF1. *J. Exp. Bot.* **2015**, *66*, 3683–3698. [CrossRef] [PubMed]

72. Liu, Z.; Shi, L.; Liu, Y.; Tang, Q.; Shen, L.; Yang, S.; Cai, J.; Yu, H.; Wang, R.; Wen, J.; et al. Genome-wide identification and transcriptional expression analysis of mitogen-activated protein kinase and mitogen-activated protein kinase kinase genes in *Capsicum annuum*. *Front. Plant Sci.* **2015**, *6*, 780. [CrossRef] [PubMed]

International Journal of
*Molecular Sciences*

MDPI

*Article*

# *CaWRKY40b* in Pepper Acts as a Negative Regulator in Response to *Ralstonia solanacearum* by Directly Modulating Defense Genes Including *CaWRKY40*

Muhammad Ifnan Khan [1,2,3,†], Yangwen Zhang [1,2,3,†], Zhiqin Liu [1,2,3,†], Jiong Hu [1,2,3], Cailing Liu [1,2,3], Sheng Yang [1,2,3], Ansar Hussain [1,2,3], Muhammad Furqan Ashraf [1,2,3], Ali Noman [1,2,3,4], Lei Shen [1,2,3], Xiaoqin Xia [1,2,3], Feng Yang [1,2,3], Deyi Guan [1,2,3] and Shuilin He [1,2,3,*]

[1]   National Education Ministry, Key Laboratory of Plant Genetic Improvement and Comprehensive Utilization, Fujian Agriculture and Forestry University, Fuzhou 350002, China; mifnan@yahoo.com (M.I.K.); zhangzhang201711@163.com; (Y.Z.); lzqfujian@126.com (Z.L.); hjfujian@126.com (J.H.); lclfujian@126.com (C.L.); yangsheng2061@163.com (S.Y.); ahtraggar@yahoo.com (A.H.); furqan2210uaf@hotmail.com (M.F.A.); alinoman@gcuf.edu.pk (A.N.); shorttubelycoris07@163.com (L.S.); s1042657738@a163.com (X.X.); m17750292623@163.com (F.Y.); gdyfujian@126.com (D.G.)
[2]   College of Crop Science, Fujian Agriculture and Forestry University, Fuzhou 350002, China
[3]   Key Laboratory of Applied Genetics of Universities in Fujian Province, Fujian Agriculture and Forestry University, Fuzhou 350002, China
[4]   Department of Botany, Government College University, Faisalabad 38040, Pakistan
[*]   Correspondence: shlhe201304@aliyun.com
[†]   These authors contributed equally to this work.

Received: 2 April 2018; Accepted: 24 April 2018; Published: 8 May 2018

**Abstract:** WRKY transcription factors (TFs) have been implicated in plant growth, development, and in response to environmental cues; however, the function of the majority of pepper WRKY TFs remains unclear. In the present study, we functionally characterized *CaWRKY40b*, a homolog of *AtWRKY40*, in pepper immunity. *Ralstonia solanacearum* inoculation (RSI) in pepper plants resulted in downregulation of *CaWRKY40b* transcript, and green fluorescent protein (GFP)-tagged CaWRKY40b was localized to the nuclei when transiently overexpressed in the leaves of *Nicotiana benthamiana*. Virus-induced gene silencing (VIGS) of *CaWRKY40b* significantly decreased pepper' susceptibility to RSI. Consistently, the transient over-expression of *CaWRKY40b-SRDX* (chimeric repressor version of *CaWRKY40b*) triggered cell death, as indicated by darker trypan blue and DAB staining. CaWRKY40b targets a number of immunity-associated genes, including *CaWRKY40 JAR, RLK1, EIN3, FLS2, CNGIC8, CDPK13*, and *heat shock cognate protein 70* (*HSC70*), which were identified by ChIP-seq and confirmed using ChIP-real time PCR. Among these target genes, the negative regulator HSC70 was upregulated by transient overexpression of *CaWRKY40b* and downregulated by silencing of *CaWRKY40b*, whereas other positive regulators as well as two non-target genes, *CaNPR1* and *CaDEF1*, were downregulated by the transient overexpression of *CaWRKY40b* and upregulated by *CaWRKY40b* silencing or transient overexpression of *CaWRKY40b-SRDX*. In addition, CaWRKY40b exhibited a positive feedback regulation at transcriptional level by directly targeting the promoter of itself. In conclusion, the findings of the present study suggest that *CaWRKY40b* acts as a negative regulator in pepper immunity against *R. solanacearum* by transcriptional modulation of a subset of immunity-associated genes; it also represses immunity in the absence of a pathogen, and derepresses immunity upon pathogen challenge.

**Keywords:** *Capsicum annuum*; *Ralstonia solanacearum*; *CaWRKY40b*; immunity; negative regulator; transcriptional modulation

## 1. Introduction

During the course of evolution, plants have developed a sophisticated defense mechanism for counteracting diverse pathogens. In addition to the physical barriers for damage prevention, plants possess two layers of inducible immunities termed as pathogen-associated molecular pattern (PAMP) triggered immunity (PTI) and effector triggered immunity (ETI). The perception of conserved microbe-associated molecular patterns (MAMPs) in plant triggers PTI via pattern recognition receptors (PRRs) on the plant cell surface. However, ETI in plants is triggered by the perception of strain-specific pathogen effectors that are delivered into host cells via intracellular R proteins, which are generally coupled with hypersensitive response (HR) cell death. Interconnected in a zig-zag manner [1–6], PTI and ETI share a highly overlapping signaling network [7–9]. A key step in both PTI and ETI is massive reprogramming that is mediated by various transcription factors (TFs) [10–12], which play vital roles in plant responses to the pathogen attack [13,14]. However, the roles of the majority of TFs in plant immunity and how these translate upstream stress signals into appropriate transcriptional outputs remain to be elucidated.

WRKYs constitute one of the largest plant TF families and have been implicated in plant growth and development as well as responses to various environmental stresses. These are named after their one or two WRKY domains, which comprise a highly conserved amino acid sequence WRKYGQK, together with a zinc-finger-like motif [15]. WRKY TFs have the ability to activate or repress transcription by directly binding to the W-box (TTGACC/T) within the promoters of their target genes [16], thereby playing important roles in the regulation of plant responses to pathogen and herbivore attacks as well as abiotic stresses, including heat stress, drought, and salinity. A subset of WRKYs have been found to be involved in plant immunity [17–21]. For example, among the 72 WRKY TFs in *Arabidopsis*, 49 *AtWRKY* genes are differentially expressed during the infection with *Pseudomonas syringae* or application of exogenous salicylic acid (SA) [22]. Genetic evidence demonstrates that *WRKY8* [23], *WRKY22* [24], *WRKY25* [25,26], *WRKY28* [23], *WRKY33* [25,27], *WRKY38* [28], *WRKY48* [23], *AtWRKY54* [29], *WRKY62* [28], *AtWRKY72* [30], and *WRKY75* [31] act as positive regulators, whereas *WRKY11*, *WRKY17* [32], *WRKY40* [33], and *WRKY60* [34] act as negative regulators in basal defense or immunity. WRKY TFs are part of a WRKY network and play important roles in the regulation of plant immunity [35,36]. However, the exact functions of other pathogen-responsive WRKY TFs and the modes of coordination among multiple WRKY TFs participating in plant immunity in non-model plant families such as the Solanaceae remain unclear.

Pepper (*Capsicum annuum*) is an important vegetable and a typical member of the Solanaceae. It is distributed in uplands during warm seasons and is constantly challenged with various soil-borne pathogens. The frequent occurrence of high temperature and high humidity (HTHH) conditions usually causes severe diseases in pepper plants. Breeding and the application of pepper cultivars that are highly resistant to disease is one of the most efficient approaches to solving the problem caused by diseases in pepper production, and a better understanding of the mechanism underlying pepper immunity against various pathogens may facilitate in the genetic improvement of pepper disease resistance. Although WRKY TFs play important roles in plant immunity, the number of WRKY TFs in pepper that have been characterized in terms of function and expression in relation to plant immunity is limited. They include: *CaWRKY1* [37], *CaWRKY-a* [38,39], *CaWRKY6* [3], *CaWRKYb* [40], *CaWRKYd* [41], *CaWRKY2* [42], *CaWRKY27* [43], *CaWRKY40* [4], and *CaWRKY58* [44]. In addition, some pepper WRKY TFs such as *CaWRKY6* [3] and *CaWRKY40* [4] also act as positive regulators of responses to high temperature stress, probably reflecting the evolution of special immunity under the combined pressure of HTHH and soil-borne pathogens. However, the exact function of the majority of the WRKY TFs in pepper remains elusive. In the present study, we focused on evaluating the function of *CaWRKY40b* in *R. solanacearum*-infected *C. annuum* plants. *CaWRKY40b* was found to act as a negative regulator in pepper immunity by directly targeting immunity-associated genes, including *CaWRKY40*.

## 2. Results

### 2.1. Sequence Analysis of CaWRKY40b

A pepper WRKY gene (*CA03g32070*) with unknown function previously identified by genome-wide analysis (http://passport.pepper.snu.ac.kr) was selected for functional characterization in relation to plant immunity, and a subset of immunity-associated *cis*-elements, including TGACG motif, TCA elements, TC-rich repeats, and W-box, was identified in its promoter region. The deduced amino acid sequence of *CA03g32070* was 360 amino acids in length, harboring one conserved WRKY domain. Among all members of the WRKY family in *Arabidopsis*, it shares highest sequence identity with *AtWRKY40*. We designated it as *CaWRKY40b* to distinguish from *CaWRKY40*, a positive regulator of pepper immunity to *Ralstonia*. [4]. *CaWRKY40b* shares 87%, 82%, 71%, 58%, and 45% amino acid sequence identity with its homologs in *S. lycopersicum* (XP_006341684.1), *S. tuberosum* (XP_006356251.1), *N. sylvestris* (XP_009802478.1), *A. thaliana* (NP_178199.1), and *O. sativa* (Q6IEK5.1), respectively, and 59.46% sequence identity to *CaWRKY40* [4]. According to the conserved WRKY domain and proposed structure of the zinc finger motif [16], *CaWRKY40b* belongs to Group IIa of the WRKY family-like *AtWRKY40* [16] and *CaWRKY40* [4] (Figure S1).

### 2.2. The Expression of CaWRKY40b Is Modulated Transcriptionally by Ralstonia solanacearum Inoculation (RSI)

The presence of the subset of pathogen-responsive CGTGA motif [45] and W-box bound by WRKY TFs [16,17] in the *CaWRKY40b* promoter region implies its possible transcriptional response to pathogens. To test this possibility, the expression of *CaWRKY40b* in pepper plants against *R. solanacearum* was assessed by qRT-PCR. The results revealed a downregulation of *CaWRKY40b* by RSI in the pepper leaves from 6 hours post inoculation (hpi) to 48 hpi (Figure 1), indicating that *CaWRKY40b* might be involved in the response of pepper to RSI.

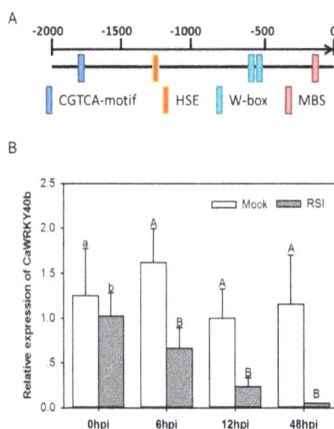

**Figure 1.** qRT-PCR analysis of relative *CaWRKY40b* transcript levels in *Ralstonia solanacearum*-inoculated pepper leaves. (**A**) Defense associated *cis*-element CGTCA-motif and W-boxes were detected in the promoter region of *CaWRKY40b* by plant care (http://bioinformatics.psb.ugent.be/webtools/plantcare/html/). (**B**) *CaWRKY40b* transcript levels measured at different time points in *R. solanacearum*-inoculated leaves after inoculation with virulent *R. solanacearum* strain FJC100301 (OD600nm = 0.8) were compared to that in control plants at relative expression level of "1'. Experiments were performed thrice in triplicate biological repeats. Data are expressed as the mean ± SD of three samples, each containing three plants. Different letters indicate significant differences as determined by Fisher's protected LSD test: uppercase letters, $p < 0.01$; lower case letters, $p < 0.05$.

## 2.3. CaWRKY40b Localizes Exclusively to the Nuclei

To localize *CaWRKY40b* at subcellular level, a fused *CaWRKY40b-GFP* protein was expressed in *N. benthamiana* leaves by infiltrating GV3101 cells containing *35S::CaWRKY40b-GFP* and using *35S::GFP* construct as a control. The GFP signal was observed under a laser scanning confocal microscope (LSCM). In *CaWRKY40b-GFP* transiently overexpressed epidermal cells of *N. benthamiana* leaves, the GFP signal was exclusively observed in the nuclei, whereas the GFP signals in the epidermal cells of *35S::GFP*-infiltrated *N. benthamiana* leaves were found across the entire cell, including plasma membrane, cytosol, and nucleus (Figure 2).

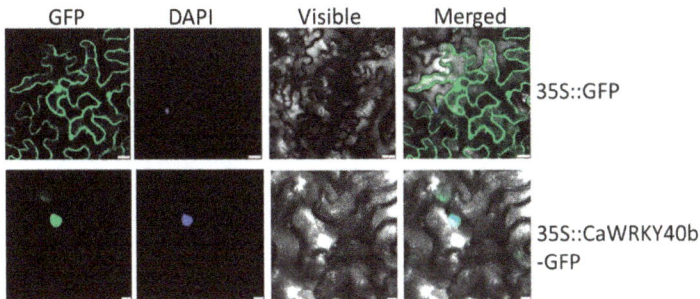

**Figure 2.** Subcellular localization of *CaWRKY40b*. The leaves of 50-day-old *Nicotiana benthamiana* plants were infiltrated with *Agrobacterium* strain GV3101 cells containing the *35S:CaWRKY40b-GFP* and *35S:GFP* construct, respectively. After 48 hpi, GFP fluorescence was imaged under a confocal microscope. The GFP signals in leaves infiltrated with GV3101 cells containing the *35S:CaWRKY40b-GFP* were observed in the nuclei, while that in leaves infiltrated with GV3101 cells containing *35S:GFP* construct were observed throughout the cell. DAPI, 4′,6-diamidino-2-phenylindole; GFP, green fluorescent protein, bar = 25 μm.

## 2.4. Silencing of CaWRKY40b Enhances the Resistance of Pepper to RSI

To investigate the possible role of *CaWRKY40b* in the response of pepper to RSI, loss of function analysis was conducted by VIGS in pepper plants. To silence *CaWRKY40b* in pepper plants, two fragments in the 3′ UTR or ORF of *CaWRKY40b* were employed. The specificity of these two fragments was further confirmed by whole-genome search (http://passport.pepper.snu.ac.kr). GV3101 cells containing TRV::*CaWRKY40b* were infiltrated into the leaves of 20-day-old pepper seedlings. The silencing process was monitored by TRV::*PDS* pepper plants. Using real-time RT-PCR, the transcription of *CaWRKY40b* was assayed and after 20 days post inoculation (dpi), the transcript level of *CaWRKY40b* in TRV::*CaWRKY40b-1* and TRV::*CaWRKKY40b-2* pepper plants was <10% of that in the TRV::00 plants (Figure 3A).

The TRV::*CaWRKY40b-1*, *CaWRKKY40b-2* and TRV::00 pepper plants were inoculated with *R. solanacearum* cells FJC100301 [4] at 3 dpi. A clear wilting phenotype was observed in the majority of FJC100301-inoculated TRV::00 pepper plants, whereas no obvious or only slight wilting was observed in both TRV::*CaWRKY40b-1* and TRV::*CaWRKY40b-2* pepper plants (Figure 3B). To accurately quantify the extent of disease in *R. solanacearum*-inoculated plants, we determined the relative disease index from 1 dpi to 8 dpi in both the TRV::*CaWRKY40b-1* TRV::*CaWRKY40b-2* and TRV::00 pepper plants. To quantify the growth of *R. solanacearum* in the inoculated pepper rootstocks, the cfu of the rootstocks of *R. solanacearum* inoculated pepper plants was measured. TRV2::*CaWRKY40b-1* and TRV2::*CaWRKY40b-2* consistently showed a significant decrease in disease symptoms as well as pathogen growth, respectively, compared to the wild-type plants (Figure 3C,D).

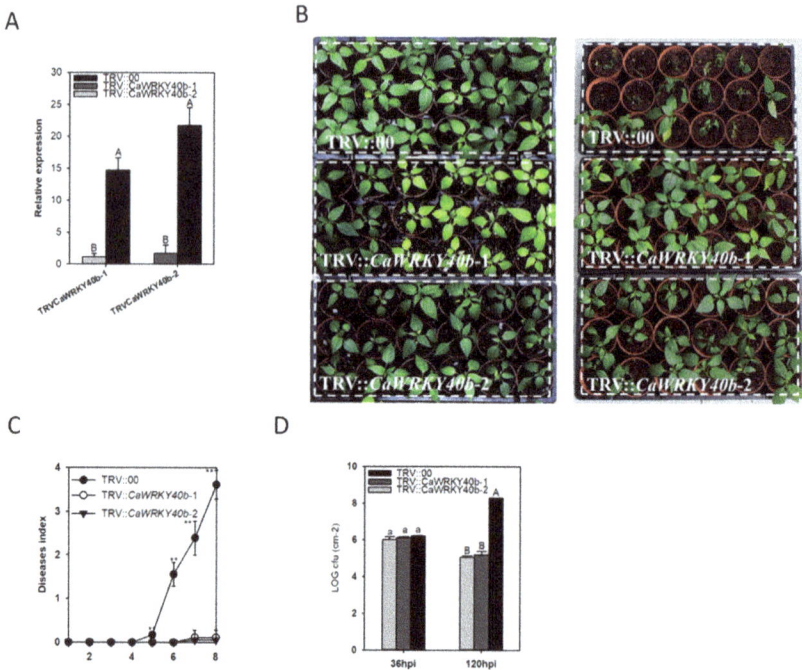

**Figure 3.** Responses of *CaWRKY40b*-silenced pepper plants to *Ralstonia solanacearum* inoculation. (**A**) Real-time RT-PCR analysis of *CaWRKY40b* expression in leaves of *CaWRKY40b*-silenced pepper plants (TRV::*CaWRKY40b1* and TRV::*CaWRKY40b2*) and control (TRV::00) plants. (**B**) *CaWRKY40b*-silenced pepper plants exhibit similar sizes compared to that in the control plants. Phenotypic effect of *R. solanacearum* inoculation on *CaWRKY40b*-silenced (TRV::*CaWRKY40b1* and TRV::*CaWRKY40b2*) and control (TRV::00) plants at 8 dpi. (**C**) Pepper plants inoculated with *R. solanacearum* were scored every 3 dpi using a disease index ranging 0–4: 0 (no wilting), 1 (1–25% wilted), 2 (26–50% wilted), 3 (51–75% wilted), and 4 (76–100% wilted or dead). (**D**). Detection of growth of *R. solanacearum* in rootstocks of the pathogen-inoculated *CaWRKY40b*-silenced (TRV::*CaWRKY40b1* and TRV::*CaWRKY40b2*) or control pepper plants at 24 hpi and 48 hpi, respectively. For *R. solanacearum* inoculation in (**B**,**C**), all the pots containing pepper plants were placed in a tray containing Hoagland's nutrient solution supplement with 10 mL of the FJC100301 suspension (OD600 = 0.8) per liter. In (**A**,**C**,**D**), the mean ± SD was calculated from four independent duplicates, with each duplicate consisting of 10 plants. Asterisks in C indicate statistically significant differences compared to the mock treatment by the LSD test (* $p < 0.05$, ** $p < 0.01$). Different letters in (**A**–**D**) indicate significant differences among means as determined by Fisher's protected LSD test: uppercase letters, $p < 0.01$; lower case letters, $p < 0.05$.

## 2.5. Transient Overexpression of CaWRKY40b-SRDX Triggered Intensive Cell Death in Pepper Plants

The effect of transient overexpression of *CaWRKKY40b* and its chimeric repressor version *CaWRKY40b*-SRDX [46,47] on the hypersensitive response (HR) cell death in leaves of pepper plants was investigated. GV3101 cells containing construct 35S::*CaWRKY40b*, 35S::*CaWRKY40b*-SRDX and 35S::00 were infiltrated into the pepper leaves. Transient overexpression of *CaWRKY40b*-HA in pepper leaves was detected by immunoblotting (IB) against the antibody of HA at 48 hpi (Figure 4A). A clear cell death and darker staining of trypan blue and DAB were consistently noticed around infiltrated sites of *Agrobacterium* cells containing 35S::*CaWRKY40b*-SRDX; however, no obvious cell death or darker staining of trypan blue or DAB was observed in the *CaWRKY40b*-transiently overexpressing and in the mock-treated pepper leaves (Figure 4B). Furthermore, a higher ion leakage was triggered by *35S::CaWRKY40b-SRDX* than that by *35S::CaWRKY40b* (Figure 4C).

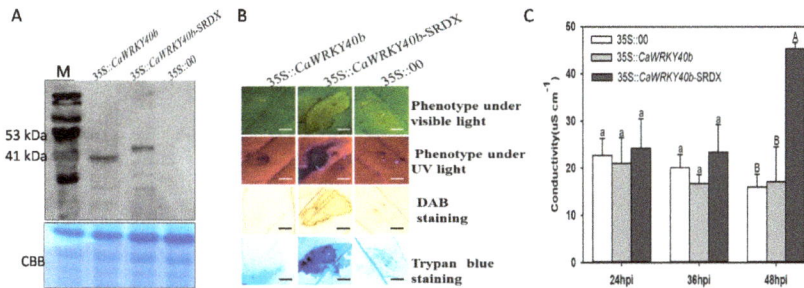

**Figure 4.** The transient overexpression of *CaWRKY40b*-SRDX triggers extensive HR cell death in pepper leaves. (**A**) The transient overexpression of *CaWRKY40b*-HA-SRDX and *CaWRKY40b*-HA in pepper leaves was detected by immune-blotting (IB) against the antibody of HA. CBB, Coomassie brilliant blue. (**B**) Intensive cell death was triggered by transient overexpression of *CaWRKY40b*-SRDX, but not by that of *CaWRKY40b* and displayed by phenotype, DAB, and trypan blue staining at 4 dpi, respectively. Bar = 100 μm. (**C**) Quantification of electrolyte leakage as ion conductivity to assess the cell death response in leaf disks, the means ± SD were calculated from four samples, each containing six disks. Capital and lowercase letters above the bars indicate significant difference at $p < 0.01$ and $p < 0.05$, respectively, as analyzed by Fisher's protected LSD test.

*2.6. Transcriptional Modulation of Marker Genes by Transient Overexpression and Virus-Induced Silencing of CaWRKY40b*

Chromatin Immunoprecipitation sequencing (ChIP-seq) was performed to identify the potential target genes of *CaWRKY40b* (Figure S2). To do this, GV3101 cells harboring *35S::CaWRKY40b-HA* was infiltrated into leaves of six-week-old pepper plants and maintained in the greenhouse. Forty-eight hours post infiltration, the infiltrated leaves were sampled for chromatin isolation, the chromatins were sheared into fragments of 300–500 bps in length and were immunoprecipitated with antibodies of HA, the sequencing of immunoprecipitated DNA was performed on Illumina HiSeq2500 resulted in about one million reads, in which more than 8 hundred binding sites were identified. Gene ontology (GO) was analyzed to determine the biological and functional processes of *CaWRKY40b* target genes (Figure S2 and Table S1). As the role of *CaWRKY40b* in pepper immunity was the focus of the present study, immunity associated genes were selected from the target genes of *CaWRKY40b* to confirm the role of *CaWRKY40b* and to test the possible mode of action, including *CA00g87690* (*CaWRKY40*), *CA05g11520* (*jasmonic acid-amido synthetase JAR1-like isoform X2, JAR1*), *CA06g24540* (*heat shock cognate 70 kDa protein 1, HSC70*), *CA01g13570* (*ETHYLENE INSENSITIVE 3-like, EIN3*), *CA02g12020* (*LRR receptor-like serine/threonine-protein kinase FLS2*), *CA10g01730* (*LRR receptor-like serine/threonine-protein kinase At3g47570 isoform X1, RLK1*), *CA05g11620* (*putative cyclic nucleotide-gated ion channel 8, CNGIC8*), and *CA09g10430* (*calcium-dependent protein kinase 13, CDPK13*). The result from ChIP-PCR showed that DNA within the promoter regions of these genes enriched in the *CaWRKY40b* bound DNA, attesting the validity of the ChIP-seq results (Table S1 and Figure S3).

To test whether *CaWRKY40b* can control the transcriptional expression of all the genes directly targeted by CaWRKY40b, their transcription was comparatively tested in *CaWRKY40b* transiently overexpressed and *CaWRKY40b*-silenced pepper plants by real-time RT-PCR. The results demonstrated that the transcription of *CaWRKY40, JAR1, BPR1, EIN3, FLS2, RLK1, CNGIC8*, and *CDPK13* was significantly enhanced by RSI, and the transcription of these genes was significantly higher in *CaWRKY40b*-silenced plants than in the *CaWRKY40b* un-silenced control plants with or without RSI. In contrast, the transcription of *CaHSC70*, which was downregulated by RSI, was lower in

*CaWRKY40b*-silenced pepper plants than that in the control plants (Figure 5). Consistent to the result in *CaWRKY40b*-silenced pepper plants, transient overexpression of *CaWRKY40b-SRDX* significantly downregulated *CaHSC70*, while elevated the transcript levels of the other tested genes compared to the control (Figure 6).

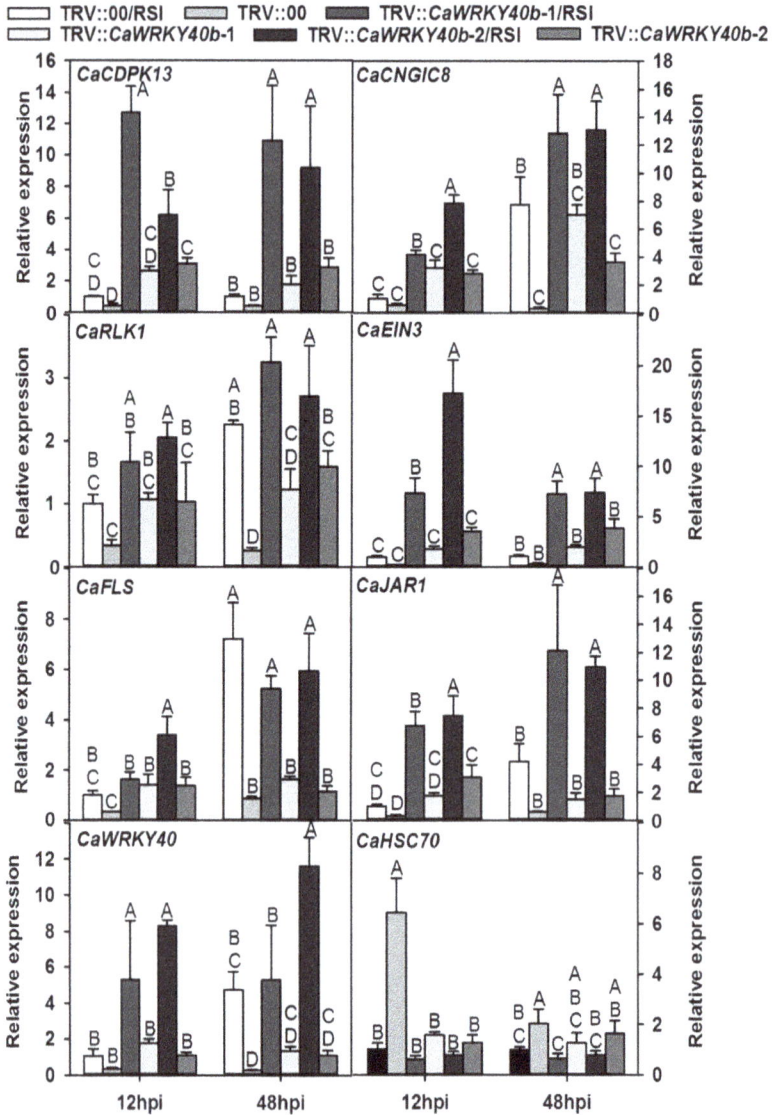

**Figure 5.** q RT-PCR analyses of transcriptional levels of the tested defense-related genes in the leaves of pepper plants. The *CaWRKY40b*-silenced and the control pepper plants were inoculated with 1.0 mL of 108 cfu·mL$^{-1}$ (OD600 = 0.8) virulent *R. solanacearum* strain FJC100301 by root irrigation. Data represent the means ± SD of four independent biological replicates, each containing three plants. Different capital letters indicate significant differences, as determined by Fisher's protected LSD test: uppercase letters, $p < 0.01$; lower case letters, $p < 0.05$.

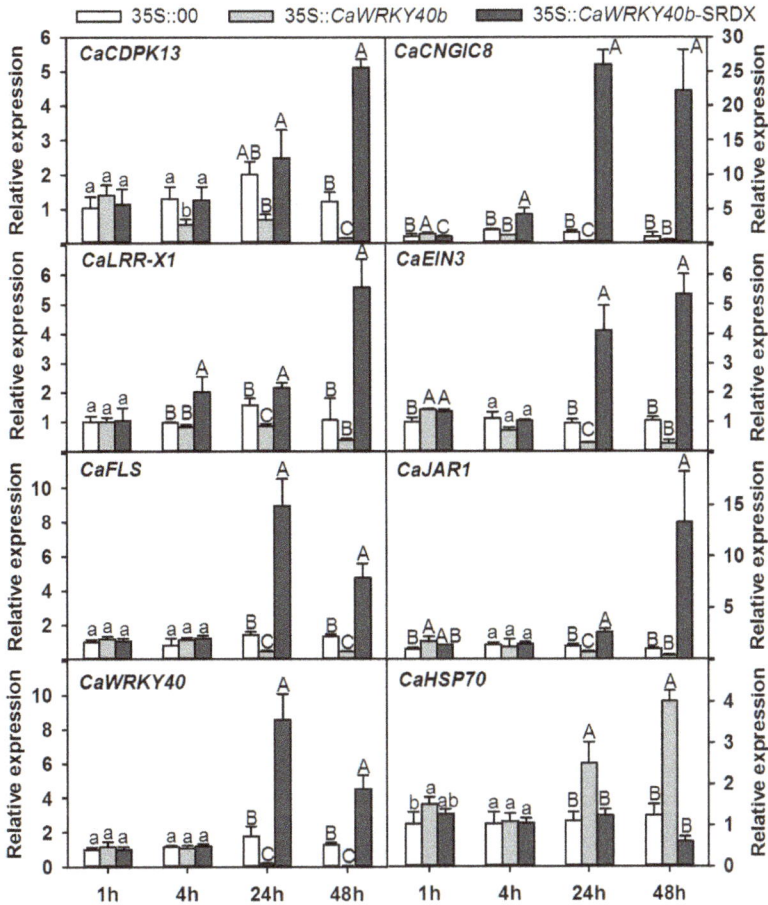

**Figure 6.** qRT-PCR analysis of the expression of immunity-associated marker genes in *35S::00*, *35S::CaWRKY40b* and *35S::CaWRKY40b-SRDX* constructs. Data display the means ± SD of four independent biological replicates, each containing three leaves. Capital and lowercase letters above the bars indicate significantly different means ($p < 0.01$) and significantly different means ($p < 0.01$), respectively, as analyzed by Fisher's protected LSD test.

### 2.7. The Expression of CaWRKY40b Was Directly and Transcriptionally Regulated by CaWRKY40b Itself

As *CaWRKY40b* itself was identified by ChIP-seq as the target gene of *CaWRKY40b*, the possible binding of *CaWRKY40b* to its own promoter was also confirmed by ChIP-seq (Figure 7A). The possible self-regulation of *CaWRKY40b* was also tested comparatively by assaying the effect of transient overexpression of *CaWRKY40b* on the expression of GUS-driven *pCaWRKY40b* and on the transcription of *CaWRKY40b*. The results presented that the transient overexpression of *CaWRKY40b* significantly increased GUS expression (Figure 7B) as well as the transcriptional level of *CaWRKY40b* by real-time RT-PCR using a pair of specific primers based on the sequence of 3'-UTR of *CaWRKY40b* (Figure 7C).

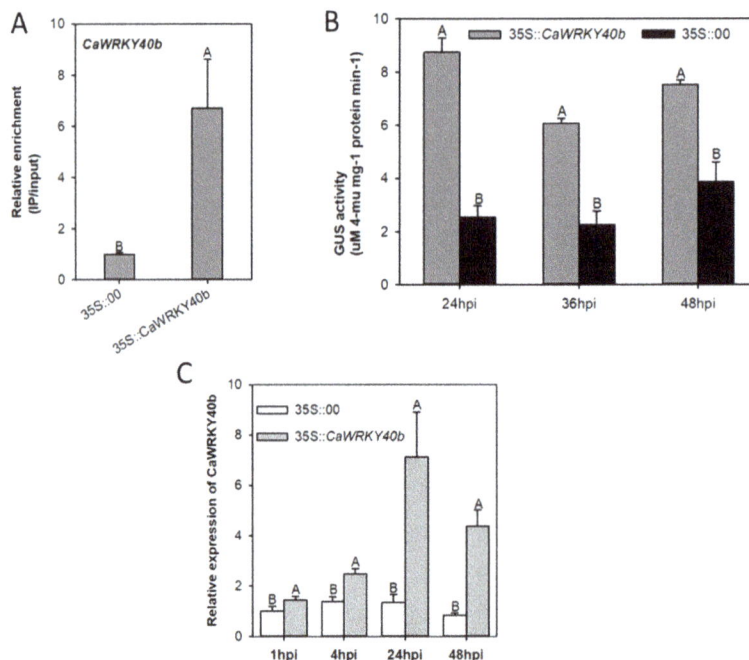

**Figure 7.** The transcriptional self-regulation of *CaWRKY40b*. (**A**) The binding of *CaWRKY40b* to the promoter of *CaWRKY40b* by ChIP and real-time PCR. GV3101 cells containing *35S::CaWRKY40b*-HA infiltrated into pepper leaves, with GV3101 cells containing *35S::00* as mock treatment. The *CaWRKY40b*-HA overexpressing leaves were harvested at 48 hpi for chromatin preparation. Chromatin was isolated from infiltrated pepper leaves crosslinked with 1% formaldehyde, sheared, and immune-precipitated with an anti-HA antibody. Relative enrichment levels of *CaWRKY40b* in the promoter of *CaWRKY40b* were set to 1 after normalization by input. (**B**) *pCaWRKY40b*-driven GUS expression was triggered by transient overexpression of *CaWRKY40b* after infiltration of *Agrobacterium* into the leaves. The pepper leaves were also co-infiltrated with GV3101 possessing *pCaWRKY40b::GUS* or *35S::CaWRKY40b*. (**C**) The transcript level of *CaWRKY40b* against transient overexpression of *CaWRKY40b* in pepper leaves by real time RT-PCR: for transcript level of *CaWRKY40b* in *CaWRKY40b* overexpressing leaves, the primer pair based on the 3′ UTR was used. Data are the means ± SD of four independent biological replicates. Capital and lowercase letters above the bars indicate significantly different means at $p < 0.01$ and $p < 0.05$, respectively, as analyzed by Fisher's protected LSD test.

## 3. Discussion

Data mainly from model plants including *Arabidopsis* and rice suggest that WRKY TFs play important roles in plant immunity, but their role and the underlying mechanism remain poorly understood. Despite functional characterization of several members of this family in pepper [3,4,37,38,40,42,43,48,49], the function of majority of members in this family in pepper remains unknown. The findings of the present study suggest that *CaWRKY40b* acts as a negative regulator in pepper immunity against RSI by directly modulating the transcription of a subset of immunity-associated genes.

WRKY proteins are characterized by their one or two highly conserved *WRKY* domains and bind to the target genes promoters mainly by means of the typical W-box [16,17]. The presence of highly conserved WRKY domains in *CaWRKY40b* and their binding to W-box-containing promoters of their potential target genes as indicated by ChIP assay and real-time PCR suggest that *CaWRKY40b* is a member of the WRKY family of pepper. In addition, our results indicate that *CaWRKY40b* acts as a negative regulator in the response of pepper to *R. solanacearum*, and silencing of *CaWRKY40b* with two independent VIGS vectors significantly and consistently decreased the growth of *R. solanacearum* and the susceptibility of pepper plants to RSI. Similarly, transient overexpression of *CaWRKY40* chimeric repressor version (e.g., *CaWRKY40*-SRDX) [46,47] triggered intensive HR cell death in pepper leaves whereas that of *CaWRKY40b* did not. The role of *CaWRKY40b* as a negative regulator in pepper immunity is similar to that of its homolog, *AtWRKY40*, in *Arabidopsis* [34], but different from its ortholog *CaWRKY40* in pepper [4]. Further investigation into how this functional difference between *CaWRKY40* and *CaWRKY40b* is structurally determined would provide new insights into pepper immunity. Because *CaWRKY40b* was downregulated by RSI, we hypothesize that the constitutive expression of *CaWRKY40b* blocks defense responses to minimize the unnecessary resource cost for defense responses in the absence of pathogens, whereas its downregulation might derepress immunity with RSI challenge. Besides *CaWRKY40b*, *CaWRKY1* [37] and *CaWRKY58* [44] have been characterized in pepper as negative regulators in the immunity [44,50]. Gene silencing of *CaWRKY40b* in this study or *CaWRKY1* and *CaWRKY58* in previous studies [37,44] significantly decreased the susceptibility of pepper to pathogen attack, thereby suggesting their functional specificities. The existence of multiple negative regulators might favor plants to avert inappropriate activation of different defense responses or activate defense responses against pathogens with different lifestyles.

The role of *CaWRKY40b* as a negative regulator in pepper immunity was further supported by its direct modulation of the transcription of a subset of immunity-associated genes. These genes originally identified by ChIP-seq, including *CaWRKY40*, *JAR*, *HSC70*, *RLK1*, *EIN3*, *FLS2*, *CNGIC8*, and *CDPK13*, which act as positive regulators in plant immunity, were downregulated by transient overexpression of *CaWRKY40b*, but upregulated by silencing of *CaWRKY40b*. Although the exact roles of the tested target genes in pepper immunity remain to be identified, *JAR* [51–53] and *EIN3* [54,55] have been implicated in JA and ET signaling pathways, which ubiquitously exist in the immune system of different plant species. *RLK1* [56,57] and *FLS2* [58–61], which are PRRs crucial for perception of conserved MAMPs to activate PTI, are conserved among different plant species [5,6]. Most importantly, our previous study provided concrete evidence that *CaWRKY40* acts as positive regulator in the response of pepper to RSI [3,4]. Therefore, we hypothesize that *CaWRKY40b* functions as a negative regulator in pepper immunity at least partially by repressing these positive regulators. In contrast, *HSC70*, a negative regulator in plant immunity [62] was downregulated by *CaWRKY40b* silencing but enhanced by overexpression of *CaWRKY40b*, indicating that *CaWRKY40b* also acts as a negative regulator by activating negative regulators, and *CaWRKY40b* possesses dual functionality, acting either as a repressor or as an activator in a promoter-context dependent manner, similar to *AtWRKY33* [63]. In addition to these target genes, *CaNPR1* [3,4,64–66] and *CaDEF1* [67–69], which were found to be non-target genes of *CaWRKY40b*, were also consistently upregulated by *CaWRKY40b* silencing and downregulated by transient overexpression of *CaWRKY40b*. All these findings support the role of *CaWRKY40b* as a negative regulator that directly regulates immunity at multiple levels, including signaling regulatory proteins, TFs, and PR proteins.

WRKY genes have been suggested to be functionally connected by forming transcriptional networks [35,36]. Our data in the present study illustrate that *CaWRKY40*, which was previously found to be transcriptionally modulated directly by *CaWRKY6* [3], is directly targeted and transcriptionally regulated by *CaWRKY40b*. Additionally, typical W-boxes were also found to be enriched in the promoters of *CaWRKY6* and *CaWRKY40*, as well as the promoter of *CaWRKY40b* (Figure 1A), indicating the participation of these *WRKYs* in a *WRKY* web during the regulation of pepper responses to *R. solanacearum*. The present study also revealed that *CaWRKY40b* positively modulates the transcription of *CaWRKY40b* by directly binding to the promoter of *CaWRKY40b*. A similar positive feedback regulation has been frequently reported in the *WRKY* network [35,70].

In sum, *CaWRKY40b* acts as a negative regulator of pepper resistance to RSI by transcriptionally regulating a set of immunity-associated genes, including *CaWRKY40*. Its downregulation upon RSI derepresses immunity, but its upregulation in healthy plants impairs pepper immunity.

## 4. Materials and Methods

### 4.1. Plant Materials and Growth Conditions

Seeds of pepper (*C. annuum*) inbred line 68-2, which has an intermediate level of resistance to *R. solanacearum*, and *Nicotiana benthamiana* were obtained from the pepper breeding group at the Fujian Agriculture and Forestry University (Fuzhou, China). Seeds were sown in in plastic pots containing soil with a peat moss:perlite ratio of (2:1, *v*/*v*) at 25 ± 2 °C, 70% relative humidity, 60–70 μmol photons·m$^{-2}$·s$^{-1}$, and 16 h light/8 h dark [71,72].

### 4.2. R. solanacearum Inoculation

A virulent *R. solanacearum* strain FJC100301 was isolated from wilted pepper samples at our laboratory in Fujian Province (China) and amplified as described elsewhere [4]. The *R. solanacearum* strain was cultured in PSA medium (200 g/L of potato, 20 g/L of sucrose, 3 g/L of beef extract, and 5 g/L of tryptone) at 28 ± 2 °C and 200 rpm and then re-suspended in a 10 mM MgCl$_2$ solution. The bacterial cell solution used for inoculation was diluted to 10$^8$ cfu·mL$^{-1}$ (OD600 = 0.8). To inoculate pepper plants with *R. solanacearum* by root irrigation, pepper plants grown in pots were placed in a tray containing Hoagland's nutrient solution supplemented with the FJC100301 suspension. To inoculate pepper leaves with *R. solanacearum*, 10 μL of the *R. solanacearum* suspension (OD600 = 0.8) was infiltrated into the top third leaves of pepper plants at the eight-leaf stage using a needleless syringe, whereas pepper plants inoculated with 10 mM MgCl$_2$ was used as mock control.

### 4.3. Vector Construction

To construct vectors for overexpression, subcellular localization, and ChIP assay, the full-length ORF of *CaWRKY40b* (with or without a termination codon) was cloned into the entry vector pDONR207 by BP reaction. This entry vector was then cloned into destination vectors pMDC83 and 3687-HA by LR reaction using the Gateway® cloning technique (Invitrogen, Carlsbad, CA, USA). In addition, the EAR repression domain (SRDX) was fused to the 3′ terminus of the ORF of *CaWRKY40b*; the resulting *CaWRKY40b*-SRDX was cloned into pEarleyGate201 as previously described. To construct the vector for VIGS, a 229-bp fragment in 3′-untranslated region (UTR) of *CaWRKY40b* and a fragment in its ORF were selected for vector construction, and their specificities were confirmed by BLAST against the genome sequence in the databases of CM334 (http://peppergenome.snu.ac.kr/) and Zunla-1 (http://peppersequence.genomics.cn/page/species/blast.jsp). The specific fragment was cloned into the entry vector pDONR207, and then cloned into the PYL279 vector.

### 4.4. Subcellular Localization

*Agrobacterium tumefaciens* strain GV3101 containing the constructs *35S::CaWRKY40b-GFP* or *35S::GFP* (used as a control) were grown overnight, respectively, and then resuspended in the induction medium. Bacterial suspensions (OD600 = 0.8) were injected into *N. benthamiana* leaves using a syringe without a needle. At 48 hpi, GFP fluorescence was imaged using a Laser Scanning Confocal Microscope (TCS SP8, Leica, Solms, Germany) at an excitation wavelength of 488 nm and a 505–530 nm band-pass emission filter.

### 4.5. VIGS of CaWRKY40b in Pepper Plants

For VIGS of *CaWRKY40b* in 68-2 line pepper plants, *A. tumefaciens* strain GV3101 harboring PYL192, PYL 279-*CaWRKY40b1* and PYL279-*CaWRKY40b2* or PYL279 (resuspended in the induction medium at a 1:1 ratio, OD600 = 0.6) were co-infiltrated into the cotyledons of two-week-old pepper plants. The detailed process was conducted according to our previous studies [3,4].

*4.6. Transient Overexpression of CaWRKY40b in Pepper Leaves*

For transient overexpression analysis, *A. tumefaciens* strain GV3101 harboring the *35S::CaWRKY40b* or *35S::CaWRKY40b-HA*, *35S::CaWRKY40b-SRDX* or *35S::00* vector (the empty vector was used as control) was both shaking-grown overnight. The *agrobacterium* pellets were collected by centrifugation and resuspended in the induction medium (10 mM MES, 10 mM $MgCl_2$, 200 µM acetosyringone, pH 5.6). The adjusted bacterial suspension (OD600 = 0.8) was vacuum-infiltrated into the leaves of 6-week-old pepper plants using a needleless syringe and the injected leaves were maintained in the greenhouse. The samples were collected at the indicated time points for further analysis.

*4.7. Histochemical Staining*

Staining of the leaves with trypan blue and diaminobenzidine (DAB) was used as described elsewhere [73] and in our previous studies [3,74].

*4.8. Quantitative Real-Time RT-PCR*

To determine the relative transcript accumulations of target genes, real-time PCR was performed using specific primers (Table S2) according to manuals of BIO-RAD real-time PCR system (Foster City, CA, USA) and SYBR Premix Ex Taq II system (TaKaRa). Total RNA extraction and real-time RT-PCR were performed as earlier described [3,74]. Four independent biological replicates of each treatment were performed. The Livak method [75] was used to analyze the real-time PCR data. The data were expressed as a normalized relative expression level ($2^{-\Delta\Delta Ct}$) of the respective genes. The transcript accumulation of *CaActin* (GQ339766) and *18S ribosomal RNA* (EF564281) were, respectively, used to normalize the relative transcriptional level of each sample.

*4.9. ChIP Analysis*

ChIP assays were performed by following a previously described protocol with slight modifications [76]. Three to four fully expanded leaves of plants at the eight-leaf stage were inoculated with GV3101 cells containing *35S:CaWKRY40b-HA* or *35S:00* (used as a control). The inoculated leaves were collected at 24 hpi; about 4 g of leaves were treated with 1.0% formaldehyde for 8 min, to which 3 M glycine was added to a final concentration of 0.125 M. The sample was then vacuumed for 5 min to stop cross linking. Nuclear extracts were isolated and were resuspended with the extraction buffer I (0.4 M sucrose, 10 mM Tris-HCl, pH 8.0, 10 mM $MgCl_2$, 5 mM β-mercaptoethanol, 1 U protease Inhibitors), II (0.25 M sucrose, 10 mM Tris-Cl, pH 8.0, 10 mM $MgCl_2$, 1% Triton X-100, 5 mM β-mercaptoethanol, 1 µL protease inhibitors), and III (1.7 M sucrose, 10 mM Tris-HCl, pH 8.0, 2 mM $MgCl_2$, 0.15% Triton X-100, 5 mM β-mercaptoethanol, 1 µL protease Inhibitors) sequentially, and then digested with micrococcal nuclease (Takara, Dalian, China), according to the manufacturer's instructions. Magnetic beads (Invitrogen, Carlsbad, CA, USA) linked to the antibody of HA (anti-HA tag rabbit polyclonal antibody, Sigma, St. Louis, MO, USA) were added to the digested samples, and then eluted. Later, the protein–DNA complex was digested with 2 mL of 10 mg·mL$^{-1}$ proteinase K and incubated at dry bath with 45 °C for 1 h and same volume of Tris-saturated phenol:chloroform:isoamyl alcohol (25:24:1 $v/v$) was used to extract the DNA solution twice. Then, DNA was precipitated by adding 3 mL of 100% ethanol, 1/10 volume of 3 M NaOAc, and 1 mL of 2 M glycogen along with overnight incubation at −20 °C. DNA was pelleted by spinning for 20 min at 16,700× *g*. The DNA pellets were washed with 80% ethanol, dried at room temperature, resuspended in 50 µL TE buffer, and stored at −20 °C until use for ChIP-seq and ChIP-PCR. For ChIP-seq, the immunoprecipitated DNA samples were used to generate sequencing libraries bearing barcodes using a NEBNext ChIP-seq Library PreReagent Set for Illumina kit (New England Biolabs, Ipswich, MA, USA). Sequencing was performed on Illumina HiSeq2500 at Nevogene (Beijing, China).

ChIP-seq data analysis was performed following the method used by Liu et al. [63]. Quantitative real-time PCR was used to analyze the immune-precipitated DNA for enrichment of *CaWRKY40b* at the promoter region of the target genes. Fold increases of immune-precipitated DNA were calculated relative to the input DNA and the internal control *CaACTIN* or *18S rRNA*. ChIP-PCR was performed at least in triplicate.

### 4.10. Fluorometric GUS Enzymatic Assay

A fluorometric GUS enzymatic assay for measuring GUS activity in pepper plant extracts was performed by adopting a previously described protocol [4].

### 4.11. Immunoblotting

Total protein extracts were incubated with anti-HA agarose (Thermo Fisher Scientific, Waltham, MA, USA) overnight at 4 °C. Beads were collected and washed with Tris-buffered saline and Tween-20 (0.05%). Eluted proteins were analyzed by immune-blotting using an anti-HA–peroxidase antibody (Abcam, Cambridge, UK).

### 4.12. Quantification of R. solanacearum Growth in Pepper Plants

The growth of *R. solanacearum* was quantified by measuring the colony-forming units (cfu), and the rootstocks of *R. solanacearum*-inoculated pepper plants were harvested at indicated time points, which were ground into powder in liquid nitrogen. For every 1.0 g of powder, 1.0 mL of 10 mM MgCl$_2$ solution was added and spun down, the supernatant was collected and diluted into 10,000 times, and 1 μL of the supernatant was added to the PSA plate, which was kept at 28 °C. Approximately 48 h later, the cfus were calculated.

**Supplementary Materials:** Supplementary materials can be found at http://www.mdpi.com/1422-0067/19/5/1403/s1.

**Author Contributions:** S.H., M.I.K. and Z.L. conceived the idea and provided resources for experimentation. M.I.K., Y.Z., Z.L., J.H., C.L., S.Y., A.H., M.F.A., A.N., L.S., X.X., F.Y., D.G. designed the study and performed experiments. M.I.K., Y.Z., Z.L. and S.H. helped in experimentation and writing. M.I.K., Y.Z. and Z.L. analyzed the data.

**Acknowledgments:** This work was supported by grants from the National Natural Science Foundation of China (31572136, 31372061, 31501767, 31401890, 31401312, 31260482, and 31060263) and the Natural Science Foundation of Fujian Province, China (2017J01436). We thank Mark D. Curtis for kindly providing the Gateway destination vectors and S. P. Dinesh-Kumar (Yale University) for the pTRV1 and pTRV2 vectors.

**Conflicts of Interest:** The authors declare no conflicts of interest.

## References

1. Fei, Q.; Zhang, Y.; Xia, R.; Meyers, B.C. Small RNAs Add Zing to the Zig-Zag-Zig Model of Plant Defenses. *Mol. Plant Microbe Interact.* **2016**, *29*, 165–169. [CrossRef] [PubMed]
2. Keller, H.; Boyer, L.; Abad, P. Disease susceptibility in the Zig-Zag model of host-microbe interactions: Only a consequence of immune suppression? *Mol. Plant Pathol.* **2016**, *17*, 475–479. [CrossRef] [PubMed]
3. Cai, H.; Yang, S.; Yan, Y.; Xiao, Z.; Cheng, J.; Wu, J.; Qiu, A.; Lai, Y.; Mou, S.; Guan, D.; et al. CaWRKY6 transcriptionally activates CaWRKY40, regulates *Ralstonia solanacearum* Resistance, and confers high-temperature and high-humidity tolerance in pepper. *J. Exp. Bot.* **2015**, *66*, 3163–3174. [CrossRef] [PubMed]
4. Dang, F.F.; Wang, Y.N.; Yu, L.; Eulgem, T.; Lai, Y.; Liu, Z.Q.; Wang, X.; Qiu, A.L.; Zhang, T.X.; Lin, J.; et al. CaWRKY40, a WRKY protein of pepper, plays an important role in the regulation of tolerance to heat stress and resistance to Ralstonia solanacearum infection. *Plant Cell Environ.* **2013**, *36*, 757–774. [CrossRef] [PubMed]
5. Jones, J.D.; Dangl, J.L. The plant immune system. *Nature* **2006**, *444*, 323–329. [CrossRef] [PubMed]
6. Hein, I.; Gilroy, E.M.; Armstrong, M.R.; Birch, P.R. The zig-zag-zig in oomycete-plant interactions. *Mol. Plant Pathol.* **2009**, *10*, 547–562. [CrossRef] [PubMed]

7.  Bozsó, Z.; Ott, P.G.; Kámán-Tóth, E.; Bognár, G.F.; Pogány, M.; Szatmári, Á. Overlapping yet response-specific transcriptome alterations characterize the nature of tobacco-pseudomonas syringae interactions. *Front. Plant Sci.* **2016**, *7*, 251. [CrossRef] [PubMed]

8.  Tsuda, K.; Katagiri, F. Comparing signaling mechanisms engaged in pattern-triggered and effector-triggered immunity. *Curr. Opin. Plant Biol.* **2010**, *13*, 459–465. [CrossRef] [PubMed]

9.  Dong, X.; Jiang, Z.; Peng, Y.L.; Zhang, Z. Revealing shared and distinct gene network organization in *Arabidopsis* immune responses by integrative analysis. *Plant Physiol.* **2015**, *167*, 1186–1203. [CrossRef] [PubMed]

10. Abuqamar, S.; Chen, X.; Dhawan, R.; Bluhm, B.; Salmeron, J.; Lam, S.; Dietrich, R.A.; Mengiste, T. Expression profiling and mutant analysis reveals complex regulatory networks involved in *Arabidopsis* response to botrytis infection. *Plant J.* **2006**, *48*, 28–44. [CrossRef] [PubMed]

11. Eulgem, T. Regulation of the *Arabidopsis* defense transcriptome. *Trends Plant Sci.* **2005**, *10*, 71–78. [CrossRef] [PubMed]

12. Eulgem, T.; Weigman, V.J.; Chang, H.S.; Mcdowell, J.M.; Holub, E.B.; Glazebrook, J.; Zhu, T.; Dangl, J.L. Gene expression signatures from three genetically separable resistance gene signaling pathways for downy mildew resistance. *Plant Physiol.* **2004**, *135*, 1129–1144. [CrossRef] [PubMed]

13. Moore, J.W.; Loake, G.J.; Spoel, S.H. Transcription dynamics in plant immunity. *Plant Cell* **2011**, *23*, 2809–2820. [CrossRef] [PubMed]

14. Bhattacharjee, S.; Garner, C.M.; Gassmann, W. New clues in the nucleus: Transcriptional reprogramming in effector-triggered immunity. *Front. Plant Sci.* **2013**, *4*, 364. [CrossRef] [PubMed]

15. Ulker, B.; Somssich, I.E. WRKY transcription factors: From dna binding towards biological function. *Curr. Opin. Plant Biol.* **2004**, *7*, 491–498. [CrossRef] [PubMed]

16. Eulgem, T.; Rushton, P.J.; Robatzek, S.; Somssich, I.E. The WRKY superfamily of plant transcription factors. *Trends Plant Sci.* **2000**, *5*, 199–206. [CrossRef]

17. Rushton, P.J.; Somssich, I.E.; Ringler, P.; Shen, Q.J. WRKY transcription factors. *Trends Plant Sci.* **2010**, *15*, 247–258. [CrossRef] [PubMed]

18. Agarwal, P.; Reddy, M.P.; Chikara, J. WRKY: Its structure, evolutionary relationship, DNA-binding selectivity, role in stress tolerance and development of plants. *Mol. Biol. Rep.* **2011**, *38*, 3883–3896. [CrossRef] [PubMed]

19. Van Verk, M.C.; Bol, J.F.; Linthorst, H.J. WRKY transcription factors involved in activation of SA biosynthesis genes. *BMC Plant Biol.* **2011**, *11*, 89. [CrossRef] [PubMed]

20. Rushton, D.L.; Tripathi, P.; Rabara, R.C.; Lin, J.; Ringler, P.; Boken, A.K.; Langum, T.J.; Smidt, L.; Boomsma, D.D.; Emme, N.J.; et al. WRKY transcription factors: Key components in abscisic acid signalling. *Plant Biotechnol. J.* **2012**, *10*, 2–11. [CrossRef] [PubMed]

21. Chen, L.; Song, Y.; Li, S.; Zhang, L.; Zou, C.; Yu, D. The role of WRKY transcription factors in plant abiotic stresses. *Biochim. Biophys. Acta* **2012**, *1819*, 120–128. [CrossRef] [PubMed]

22. Dong, J.; Chen, C.; Chen, Z. Expression profiles of the *Arabidopsis* WRKY gene superfamily during plant defense response. *Plant Mol. Biol.* **2003**, *51*, 21–37. [CrossRef] [PubMed]

23. Gao, X.; He, P. Nuclear dynamics of *Arabidopsis* calcium-dependent protein kinases in effector-triggered immunity. *Plant Signal Behav.* **2013**, *8*, E23868. [CrossRef] [PubMed]

24. Hsu, F.C.; Chou, M.Y.; Chou, S.J.; Li, Y.R.; Peng, H.P.; Shih, M.C. Submergence confers immunity mediated by the WRKY22 transcription factor in *Arabidopsis*. *Plant Cell* **2013**, *25*, 2699–2713. [CrossRef] [PubMed]

25. Andreasson, E.; Jenkins, T.; Brodersen, P.; Thorgrimsen, S.; Petersen, N.H.; Zhu, S.; Qiu, J.L.; Micheelsen, P.; Rocher, A.; Petersen, M.; et al. The MAP kinase substrate MKS1 is a regulator of plant defense responses. *EMBO J.* **2005**, *24*, 2579–2589. [CrossRef] [PubMed]

26. Zheng, Z.; Mosher, S.L.; Fan, B.; Klessig, D.F.; Chen, Z. Functional analysis of *Arabidopsis* WRKY25 transcription factor in plant defense against Pseudomonas syringae. *BMC Plant Biol.* **2007**, *7*, 2. [CrossRef] [PubMed]

27. Lippok, B.; Birkenbihl, R.P.; Rivory, G.; Brümmer, J.; Schmelzer, E.; Logemann, E.; Somssich, I.E. Expression of Atwrky33 encoding a pathogen- or PAMP-responsive WRKY transcription factor is regulated by a composite DNA motif containing W box elements. *Mol. Plant Microbe Interact.* **2007**, *20*, 420–429. [CrossRef] [PubMed]

28. Kim, K.C.; Lai, Z.; Fan, B.; Chen, Z. *Arabidopsis* WRKY38 and WRKY62 transcription factors interact with histone deacetylase 19 in basal defense. *Plant Cell* **2008**, *20*, 2357–2371. [CrossRef] [PubMed]

29. Kalde, M.; Barth, M.; Somssich, I.E.; Lippok, B. Members of the *Arabidopsis* WRKY group III transcription factors are part of different plant defense signaling pathways. *Mol. Plant Microbe Interact.* **2003**, *16*, 295–305. [CrossRef] [PubMed]

30. Bhattarai, K.K.; Atamian, H.S.; Kaloshian, I.; Eulgem, T. WRKY72-type transcription factors contribute to basal immunity in tomato and *Arabidopsis* as well as gene-for-gene resistance mediated by the tomato R gene Mi-1. *Plant J.* **2010**, *63*, 229–240. [CrossRef] [PubMed]

31. Encinas-Villarejo, S.; Maldonado, A.M.; Amil-Ruiz, F.; De Los Santos, B.; Romero, F.; Pliego-Alfaro, F.; Muñoz-Blanco, J.; Caballero, J.L. Evidence for a positive regulatory role of strawberry (Fragaria x ananassa) Fa WRKY1 and *Arabidopsis* At WRKY75 proteins in resistance. *J. Exp. Bot.* **2009**, *60*, 3043–3065. [CrossRef] [PubMed]

32. Journot-Catalino, N.; Somssich, I.E.; Roby, D.; Kroj, T. The transcription factors WRKY11 and WRKY17 act as negative regulators of basal resistance in *Arabidopsis thaliana*. *Plant Cell* **2006**, *18*, 3289–3302. [CrossRef] [PubMed]

33. Lozano-Durán, R.; Macho, A.P.; Boutrot, F.; Segonzac, C.; Somssich, I.E.; Zipfel, C. The transcriptional regulator BZR1 Mediates trade-off between plant innate immunity and growth. *Elife* **2013**, *2*, E00983. [CrossRef] [PubMed]

34. Xu, X.; Chen, C.; Fan, B.; Chen, Z. Physical and functional interactions between pathogen-induced *Arabidopsis* WRKY18, WRKY40, and WRKY60 transcription factors. *Plant Cell* **2006**, *18*, 1310–1326. [CrossRef] [PubMed]

35. Eulgem, T.; Somssich, I.E. Networks of WRKY transcription factors in defense signaling. *Curr. Opin. Plant Biol.* **2007**, *10*, 366–371. [CrossRef] [PubMed]

36. Eulgem, T. Dissecting the WRKY web of plant defense regulators. *PLoS Pathog.* **2006**, *2*, E126. [CrossRef] [PubMed]

37. Oh, S.K.; Baek, K.H.; Park, J.M.; Yi, S.Y.; Yu, S.H.; Kamoun, S.; Choi, D. *Capsicum annuum* WRKY protein Cawrky1 is a negative regulator of pathogen defense. *New Phytol.* **2008**, *177*, 977–989. [CrossRef] [PubMed]

38. Park, C.J.; Shin, Y.C.; Lee, B.J.; Kim, K.J.; Kim, J.K.; Paek, K.H. A hot pepper gene encoding WRKY transcription factor is induced during hypersensitive response to Tobacco mosaic virus and Xanthomonas campestris. *Planta* **2006**, *223*, 168–179. [CrossRef] [PubMed]

39. Huh, S.U.; Lee, G.J.; Jung, J.H.; Kim, Y.; Kim, Y.J.; Paek, K.H. Capsicum annuum transcription factor wrkya positively regulates defense response upon TMV infection and is a substrate of CaMK1 and CaMK2. *Sci. Rep.* **2015**, *5*, 7981. [CrossRef] [PubMed]

40. Lim, J.H.; Park, C.J.; Huh, S.U.; Choi, L.M.; Lee, G.J.; Kim, Y.J.; Paek, K.H. Capsicum annuum WRKYb transcription factor that binds to the CaPR-10 promoter functions as a positive regulator in innate immunity upon TMV infection. *Biochem. Biophys. Res. Commun.* **2011**, *411*, 613–619. [CrossRef] [PubMed]

41. Huh, S.U.; Choi, L.M.; Lee, G.J.; Kim, Y.J.; Paek, K.H. Capsicum annuum WRKY transcription factor d (CaWRKYd) regulates hypersensitive response and defense response upon Tobacco mosaic virus infection. *Plant Sci.* **2012**, *197*, 50–58. [CrossRef] [PubMed]

42. Oh, S.K.; Yi, S.Y.; Yu, S.H.; Moon, J.S.; Park, J.M.; Choi, D. CaWRKY2, a chili pepper transcription factor, is rapidly induced by incompatible plant pathogens. *Mol. Cells* **2006**, *22*, 58–64. [PubMed]

43. Dang, F.; Wang, Y.; She, J.; Lei, Y.; Liu, Z.; Eulgem, T.; Lai, Y.; Lin, J.; Yu, L.; Lei, D.; et al. Overexpression of CaWRKY27, a subgroup IIe WRKY transcription factor of Capsicum annuum, positively regulates tobacco resistance to *Ralstonia solanacearum* infection. *Physiol. Plant* **2014**, *150*, 397–411. [CrossRef] [PubMed]

44. Wang, Y.; Dang, F.; Liu, Z.; Wang, X.; Eulgem, T.; Lai, Y.; Yu, L.; She, J.; Shi, Y.; Lin, J.; et al. CaWRKY58, encoding a group I WRKY transcription factor of Capsicum annuum, negatively regulates resistance to *Ralstonia solanacearum* infection. *Mol. Plant Pathol.* **2013**, *14*, 131–144. [CrossRef] [PubMed]

45. Marla, S.S.; Singh, V.K. LOX genes in blast fungus (*Magnaporthe grisea*) resistance in rice. *Funct. Integr. Genom.* **2012**, *12*, 265–275. [CrossRef] [PubMed]

46. Tachikawa, K.; Schröder, O.; Frey, G.; Briggs, S.P.; Sera, T. Regulation of the endogenous VEGF-A Gene by exogenous designed regulatory proteins. *Proc. Natl. Acad. Sci. USA* **2004**, *101*, 15225–15230. [CrossRef] [PubMed]

47. Ishida, T.; Hattori, S.; Sano, R.; Inoue, K.; Shirano, Y.; Hayashi, H.; Shibata, D.; Sato, S.; Kato, T.; Tabata, S.; et al. *Arabidopsis* TRANSPARENT TESTA GLABRA2 is directly regulated by R2R3 MYB transcription factors and is involved in regulation of GLABRA2 transcription in epidermal differentiation. *Plant Cell* **2007**, *19*, 2531–2543. [CrossRef] [PubMed]

48. Zhong, L.; Zhou, W.; Wang, H.; Ding, S.; Lu, Q.; Wen, X.; Peng, L.; Zhang, L.; Lu, C. Chloroplast small heat shock protein HSP21 interacts with plastid nucleoid protein pTAC5 and is essential for chloroplast development in *Arabidopsis* under heat stress. *Plant Cell* **2013**, *25*, 2925–2943. [CrossRef] [PubMed]

49. Diao, W.P.; Snyder, J.C.; Wang, S.B.; Liu, J.B.; Pan, B.G.; Guo, G.J.; Wei, G. Genome-Wide Identification and Expression Analysis of WRKY Gene Family in *Capsicum annuum* L. *Front. Plant Sci.* **2016**, *7*, 211. [CrossRef] [PubMed]

50. Wang, D.; Amornsiripanitch, N.; Dong, X. A genomic approach to identify regulatory nodes in the transcriptional network of systemic acquired resistance in plants. *PLoS Pathog.* **2006**, *2*, E123. [CrossRef] [PubMed]

51. Bohman, S.; Staal, J.; Thomma, B.P.; Wang, M.; Dixelius, C. Characterisation of an *Arabidopsis*-Leptosphaeria maculans pathosystem: Resistance partially requires camalexin biosynthesis and is independent of salicylic acid, ethylene and jasmonic acid signalling. *Plant J.* **2004**, *37*, 9–20. [CrossRef] [PubMed]

52. Fukumoto, K.; Alamgir, K.M.; Yamashita, Y.; Mori, I.C.; Matsuura, H.; Galis, I. Response of rice to insect elicitors and the role of OsJAR1 in wound and herbivory-induced JA-Ile accumulation. *J. Integr. Plant Biol.* **2013**, *55*, 775–784. [CrossRef] [PubMed]

53. Kang, J.H.; Wang, L.; Giri, A.; Baldwin, I.T. Silencing threonine deaminase and JAR4 in Nicotiana attenuata impairs jasmonic acid-isoleucine-mediated defenses against Manduca sexta. *Plant Cell* **2006**, *18*, 3303–3320. [CrossRef] [PubMed]

54. Nie, H.; Zhao, C.; Wu, G.; Wu, Y.; Chen, Y.; Tang, D. SR1, a calmodulin-binding transcription factor, modulates plant defense and ethylene-induced senescence by directly regulating NDR1 and EIN3. *Plant Physiol.* **2012**, *158*, 1847–1859. [CrossRef] [PubMed]

55. Zhao, Y.; Wei, T.; Yin, K.Q.; Chen, Z.; Gu, H.; Qu, L.J.; Qin, G. *Arabidopsis* RAP2.2 plays an important role in plant resistance to Botrytis cinerea and ethylene responses. *New Phytol.* **2012**, *195*, 450–460. [CrossRef] [PubMed]

56. Komjanc, M.; Festi, S.; Rizzotti, L.; Cattivelli, L.; Cervone, F.; de Lorenzo, G. A leucine-rich repeat receptor-like protein kinase (LRPKm1) gene is induced in Malus x domestica by *Venturia inaequalis* infection and salicylic acid treatment. *Plant Mol. Biol.* **1999**, *40*, 945–957. [CrossRef] [PubMed]

57. Song, D.; Li, G.; Song, F.; Zheng, Z. Molecular characterization and expression analysis of OsBISERK1, a gene encoding a leucine-rich repeat receptor-like kinase, during disease resistance responses in rice. *Mol. Biol. Rep.* **2008**, *35*, 275–283. [CrossRef] [PubMed]

58. Wang, S.; Sun, Z.; Wang, H.; Liu, L.; Lu, F.; Yang, J.; Zhang, M.; Zhang, S.; Guo, Z.; Bent, A.F.; et al. Rice OsFLS2-Mediated Perception of Bacterial Flagellins Is Evaded by Xanthomonas oryzae pvs. oryzae and oryzicola. *Mol. Plant* **2015**, *8*, 1024–1037. [CrossRef] [PubMed]

59. Danna, C.H.; Millet, Y.A.; Koller, T.; Han, S.W.; Bent, A.F.; Ronald, P.C.; Ausubel, F.M. The *Arabidopsis* flagellin receptor FLS2 mediates the perception of Xanthomonas Ax21 secreted peptides. *Proc. Natl. Acad. Sci. USA* **2011**, *108*, 9286–9291. [CrossRef] [PubMed]

60. Yeam, I.; Nguyen, H.P.; Martin, G.B. Phosphorylation of the Pseudomonas syringae effector AvrPto is required for FLS2/BAK1-independent virulence activity and recognition by tobacco. *Plant J.* **2010**, *61*, 16–24. [CrossRef] [PubMed]

61. Heese, A.; Hann, D.R.; Gimenez-Ibanez, S.; Jones, A.M.; He, K.; Li, J.; Schroeder, J.I.; Peck, S.C.; Rathjen, J.P. The receptor-like kinase SERK3/BAK1 is a central regulator of innate immunity in plants. *Proc. Natl. Acad. Sci. USA* **2007**, *104*, 12217–12222. [CrossRef] [PubMed]

62. Noël, L.D.; Cagna, G.; Stuttmann, J.; Wirthmüller, L.; Betsuyaku, S.; Witte, C.P.; Bhat, R.; Pochon, N.; Colby, T.; Parker, J.E. Interaction between SGT1 and cytosolic/nuclear HSC70 chaperones regulates *Arabidopsis* immune responses. *Plant Cell* **2007**, *19*, 4061–4076. [CrossRef] [PubMed]

63. Liu, S.; Kracher, B.; Ziegler, J.; Birkenbihl, R.P.; Somssich, I.E. Negative regulation of ABA signaling by WRKY33 is critical for *Arabidopsis* immunity towards Botrytis cinerea 2100. *Elife* **2015**, *4*, E07295. [CrossRef] [PubMed]

64. Kloepper, J.W.; Ryu, C.M.; Zhang, S. Induced systemic resistance and promotion of plant. growth by *Bacillus* spp. *Phytopathology* **2004**, *94*, 1259–1266. [CrossRef] [PubMed]

65. Cao, H.; Li, X.; Dong, X. Generation of broad-spectrum disease resistance by overexpression of an essential regulatory gene in systemic acquired resistance. *Proc. Natl. Acad. Sci. USA* **1998**, *95*, 6531–6536. [CrossRef] [PubMed]

66. Zhang, Y.; Fan, W.; Kinkema, M.; Li, X.; Dong, X. Interaction of NPR1 with basic leucine zipper protein transcription factors that bind sequences required for salicylic acid induction of the PR-1 gene. *Proc. Natl. Acad. Sci. USA* **1999**, *96*, 6523–6528. [CrossRef] [PubMed]

67. Hong, J.K.; Choi, H.W.; Hwang, I.S.; Kim, D.S.; Kim, N.H.; Choi, D.S.; Kim, Y.J.; Hwang, B.K. Function of a novel GDSL-type pepper lipase gene, CaGLIP1, in disease susceptibility and abiotic stress tolerance. *Planta* **2008**, *227*, 539–558. [CrossRef] [PubMed]

68. An, S.H.; Choi, H.W.; Hwang, I.S.; Hong, J.K.; Hwang, B.K. A novel pepper membrane-located receptor-like protein gene CaMRP1 is required for disease susceptibility, methyl jasmonate insensitivity and salt tolerance. *Plant Mol. Biol.* **2008**, *67*, 519–533. [CrossRef] [PubMed]

69. Choi, H.W.; Lee, B.G.; Kim, N.H.; Park, Y.; Lim, C.W.; Song, H.K.; Hwang, B.K. A role for a menthone reductase in resistance against microbial pathogens in plants. *Plant Physiol.* **2008**, *148*, 383–401. [CrossRef] [PubMed]

70. Mao, G.; Meng, X.; Liu, Y.; Zheng, Z.; Chen, Z.; Zhang, S. Phosphorylation of a WRKY transcription factor by two pathogen-responsive MaPKs drives phytoalexin biosynthesis in *Arabidopsis*. *Plant Cell* **2011**, *23*, 1639–1653. [CrossRef] [PubMed]

71. Choi, D.S.; Hwang, I.S.; Hwang, B.K. Requirement of the cytosolic interaction between PATHOGENESIS-RELATED PROTEIN10 and LEUCINE-RICH REPEAT PROTEIN1 for cell death and defense signaling in pepper. *Plant Cell* **2012**, *24*, 1675–1690. [CrossRef] [PubMed]

72. Dang, F.; Wang, Y.; She, J.; Lei, Y.; Liu, Z.; Eulgem, T.; Lai, Y.; Lin, J.; Yu, L.; Lei, D.; et al. Overexpression of CaWRKY27, a subgroup IIe WRKY transcription factor of Capsicum annuum, positively regulates tobacco resistance to *Ralstonia solanacearum* infection. *Physiol. Plant* **2013**, *150*, 397–411. [CrossRef] [PubMed]

73. Zhang, Q.; Kuang, H.; Chen, C.; Yan, J.; Do-Umehara, H.C.; Liu, X.Y.; Dada, L.; Ridge, K.M.; Chandel, N.S.; Liu, J. The Kinase Jnk2 promotes stress-induced mitophagy by targeting the small mitochondrial form of the tumor suppressor ARF for degradation. *Nat. Immunol.* **2015**, *16*, 458. [CrossRef] [PubMed]

74. Liu, Z.Q.; Qiu, A.L.; Shi, L.P.; Cai, J.S.; Huang, X.Y.; Yang, S.; Wang, B.; Shen, L.; Huang, M.K.; Mou, S.L.; et al. SRC2-1 is required in PcINF1-induced pepper immunity by acting as an interacting partner of PcINF1. *J. Exp. Bot.* **2015**, *66*, 3683–3698. [CrossRef] [PubMed]

75. Livak, K.J.; Schmittgen, T.D. Analysis of relative gene expression data using real-time quantitative PCR and the $2^{-\Delta\Delta Ct}$ Method. *Methods* **2001**, *25*, 402–408. [CrossRef] [PubMed]

76. Sun, T.; Zhang, Y.; Li, Y.; Zhang, Q.; Ding, Y.; Zhang, Y. ChIP-seq reveals broad roles of SARD1 and CBP60g in regulating plant immunity. *Nat. Commun.* **2015**, *6*, 10159. [CrossRef] [PubMed]

International Journal of
*Molecular Sciences*

MDPI

*Article*

# *Arabidopsis* RETICULON-LIKE3 (RTNLB3) and RTNLB8 Participate in *Agrobacterium*-Mediated Plant Transformation

Fan-Chen Huang [1,2], Bi-Ju Fu [1], Yin-Tzu Liu [1], Yao-Ren Chang [1], Shin-Fei Chi [1], Pei-Ru Chien [1], Si-Chi Huang [1] and Hau-Hsuan Hwang [1,3,4,*]

[1]   Department of Life Sciences, National Chung Hsing University, Taichung 402, Taiwan;
       seaworld024@hotmail.com (F.-C.H.); seele704@yahoo.com.tw (B.-J.F.); bigbigbig1014@hotmail.com (Y.-T.L.);
       kyopenny@hotmail.com (Y.-R.C.); daicin928@gmail.com (S.-F.C.); so33183.sc@gmail.com (P.R.-C.);
       jwfw28@gmail.com (S.-C.H.)
[2]   Ph.D. Program in Microbial Genomics, National Chung Hsing University and Academia Sinica,
       Taichung 402, Taiwan
[3]   Ph.D. Program in Microbial Genomics, National Chung Hsing University, Taichung 402, Taiwan
[4]   Agricultural Biotechnology Center, National Chung Hsing University, Taichung 402, Taiwan
[*]   Correspondence: hauhsuan@dragon.nchu.edu.tw; Tel.: +886-4-2284-0416-412

Received: 31 January 2018; Accepted: 21 February 2018; Published: 24 February 2018

**Abstract:** *Agrobacterium tumefaciens* can genetically transform various eukaryotic cells because of the presence of a resident tumor-inducing (Ti) plasmid. During infection, a defined region of the Ti plasmid, transfer DNA (T-DNA), is transferred from bacteria into plant cells and causes plant cells to abnormally synthesize auxin and cytokinin, which results in crown gall disease. T-DNA and several virulence (Vir) proteins are secreted through a type IV secretion system (T4SS) composed of T-pilus and a transmembrane protein complex. Three members of *Arabidopsis* reticulon-like B (RTNLB) proteins, RTNLB1, 2, and 4, interact with VirB2, the major component of T-pilus. Here, we have identified that other RTNLB proteins, RTNLB3 and 8, interact with VirB2 in vitro. Root-based *A. tumefaciens* transformation assays with *Arabidopsis rtnlb3*, or *rtnlb5-10* single mutants showed that the *rtnlb8* mutant was resistant to *A. tumefaciens* infection. In addition, *rtnlb3* and *rtnlb8* mutants showed reduced transient transformation efficiency in seedlings. *RTNLB3-* or *8* overexpression transgenic plants showed increased susceptibility to *A. tumefaciens* and *Pseudomonas syringae* infection. *RTNLB1-4* and *8* transcript levels differed in roots, rosette leaves, cauline leaves, inflorescence, flowers, and siliques of wild-type plants. Taken together, *RTNLB3* and *8* may participate in *A. tumefaciens* infection but may have different roles in plants.

**Keywords:** RTNLB; *Agrobacterium*

## 1. Introduction

In nature, the phytopathogenic bacterium *Agrobacterium tumefaciens* of the family Rhizobiaceae infects susceptible plants and causes crown gall tumors. The disease results from the transfer of effector virulence (Vir) proteins and the transfer DNA (T-DNA) derived from a large bacterial tumor-inducing (Ti) plasmid. T-DNA transfer from *A. tumefaciens* into a plant cell requires the expression of several virulence (*vir*) genes that reside on the Ti plasmid [1–4]. The uncontrolled growth of crown gall tumors results from the transfer and expression of oncogenes encoded by the wild-type T-DNA, which directs overproduction of the plant growth hormones cytokinin and auxin [5]. Another set of genes in wild-type T-DNA causes the production of bacterial nutrients, called opines, which are then utilized by *A. tumefaciens* as a carbon and sometimes nitrogen source.

*A. tumefaciens* uses a VirA/VirG two-component regulatory system to sense various environmental signals, including acidity, monosaccharides, and phenolic compounds, and induce *vir* gene expression [6,7]. With the help of VirD1 and VirD2 proteins, the single-stranded T-DNA is processed and then transported into plants via a type IV secretion system (T4SS). The T4SS is used by many pathogens to deliver protein and/or DNA into the cell cytosol and modulate eukaryotic cell functions [8–11]. The process involves the recognition of cognate substrates and delivery of the substrates across membrane barrier(s).

The T4SS consists of two functional components, a transmembrane transporter comprising VirD4 and VirB1-11 proteins, and a filamentous pilus (T-pilus) [12–14]. The T-pilus is a long, semi-rigid, flexuous filament 10 nm in diameter that may play an important role in virulence. The T-pilus contains at least two VirB proteins. The major component, VirB2, is translated as a 12.3-kD pro-pilin protein but is processed to a 7.2-kD pilin protein by removal of a N-terminal signal peptide (1–47 amino acid residues) [15–17]. T-pilin, 74 amino acid residues long, is coupled between the amino terminal residue Gln-48 to Gly-121 at the carboxy terminus in a head-to-tail peptide bond, forming an unusual cyclic peptide [18]. VirB5 co-fractionates as a minor component in T-pilus preparations and contributes to T-pilus assembly [19]. VirB5 is localized at the tips of the cell-bound T-pili and might mediate host cells and bacteria contact via interactions with the host protein during *A. tumefaciens* infection [20]. After T-DNA enters plant cells, T-DNA, along with the attached VirD2 protein, will be transported into the plant nucleus and integrated into the plant chromosome with the assistance of VirE2, VirF, other Vir proteins, and plant proteins. During T-DNA nuclear import, VirE2 may interact with the plant VirE2-interacing plant protein (VIP1) in the cytoplasm to assist in nuclear targeting of T-DNA and to block endogenous VIP1 from activating plant defense responses [4,21,22]. Successful *A. tumefaciens*-mediated plant transformation involves a continuous battle of plant cells activating a defense response to repel bacterial infection and bacteria using Vir proteins and manipulating plant proteins to elude the plant's immunity systems.

A previous study [23] identified plant-encoded proteins that may mediate the initial contact of *A. tumefaciens* T-pilus with the host cell. Yeast two-hybrid and in vitro assays revealed two classes of *Arabidopsis* proteins that interact with VirB2. The first class consists of three related proteins: reticulon-like protein B1 (RTNLB1), 2, and 4. The second class is a RAB8B GTPase. Yeast two-hybrid assay and in vitro interaction studies demonstrated that the three RTNLB proteins interact with themselves, each other, and RAB8B, so these proteins may form a multimeric complex [23]. Pre-incubation of induced *A. tumefaciens* with GST-RTNLB1 protein reduced the *A. tumefaciens* transformation efficiency of *Arabidopsis* suspension cells. The level of RTNLB1 protein transiently increased immediately after *A. tumefaciens* infection. *Arabidopsis rtnlb1* mutant plants were recalcitrant to *Agrobacterium*-mediated transformation, whereas *Arabidopsis* RTNLB1-overexpressing transgenic plants were hypersusceptible to *A. tumefaciens* infection [23]. The three RTNLB proteins all have a carboxyl-terminal 150–201 amino acid reticulon (RTN) homology domain composed of two large hydrophobic regions and a ~66 amino acid loop in between. The RTN1 protein, a membrane-anchored component of the endoplasmic reticulum (ER), is the first identified member of this family and is expressed in the central nervous system and in neuroendocrine cells [24–26]. RTN proteins may interact with themselves or recruit other proteins to form a complex and perform specific functions. In mammalian, yeast, and plant cells, RTN proteins are involved in various endomembrane-related processes, which includes intracellular transport, vesicle formation, and membrane curvature [27–33].

More than 250 reticulon-like (*RTNL*) genes have been identified in divergent eukaryotes, fungi, plants, and animals. *RTNL* genes appear to have evolved from an intron-rich ancestor [27,34]. There are 21 RTNLB proteins in *Arabidopsis thaliana* sharing amino acid sequence similarity to the reticulon domain at the C terminus [30,35]. Consistent with the peripheral location of RTNLB1-GFP [23], RTNLB1 and RTNLB6 were found by proteomic analyses of plasma membrane-enriched preparations [36]. Fluorescent-labeled RTNLB2, 4, [35] and RTNLB13 [31] are localized in ER tubules. RTNLB1-4 and 13 can co-localize and constrict tubular ER membranes, so RTNLB proteins may bend the membrane and form multimeric, arc-like structures to shape the ER tubules [33]. In addition, the C-terminal RHD domain is required for RTNLB1-4 to reside in ER membranes and efficiently constrict ER tubules but is not necessary for their homo- and heterotypic interactions [33].

The RTNLB3 and 6 proteins may participate the formation of the desmotubule, membrane structures derived from the cortical ER that transverse through plasmodesmata (PD) [37]. Many viral movement proteins can help viruses spread via interactions with the PD [37–39]. RTNLB3 and 6 co-localize with the viral movement protein of Tobacco mosaic virus at the primary PD [37]. Potato virus X movement protein is also detected in the desmotubules of *Nicotiana benthamiana* PD [39]. A protein microarray screen identified RTNLB1 and 2 proteins that interact with the *Arabidopsis* FLAGELIN-SENSITIVE2 (FLS2) protein, one of the pattern recognition receptors (PRRs) for the bacterial flagellin [40]. The *rtnlb1,2* double mutant and *RTNLB1* overexpression plants show increased susceptibility to *Pseudomonas syringae* pv. *tomato* DC3000 (*Pst*) infection and decreased FLS2-mediated immunity responses [40]. FLS2 levels at the plasma membrane are lower in the *rtnlb1,2* double mutant and *RTNLB1* overexpression plants, so RTNLB1 and 2 may control the trafficking of the FLS2 protein to the plasma membrane [40]. However, relatively little is known about the function of RTNLB proteins in plant–microbe interactions.

In this study, we further identified two additional RTNLB proteins, RTNLB3 and 8, that interact with the *A. tumefaciens* VirB2 protein. *A. tumefaciens*-mediated transient transformation efficiency was lower in *rtnlb3* and *rtnlb8* mutant than wild-type plants. Furthermore, overexpression of *RTNLB3* or *8* in transgenic *Arabidopsis* plants enhanced both stable and transient *A. tumefaciens* transformation efficiency. Also, *RTNLB3* or *8* overexpression plants were hypersusceptible to *Pst* DC3000 infection. This study further reveals the involvement of RTNLB3 and 8 in plant–microbe interactions.

## 2. Results

### 2.1. Interactions Among RTNLB3 and 8 and Vir Proteins in Yeast and In Vitro

A previous study demonstrated that RTNLB1, 2, and 4 interacted with the C-terminal-processed portion of VirB2 protein in yeast two-hybrid and in vitro assays [23]. From the phylogenetic tree results of the *Arabidopsis* RTNLB family, RTNLB1-8 proteins belong to the Group I proteins containing an N-terminal domain with 43–93 amino acid residues and a short C-terminal domain [27,30]. Therefore, we cloned *RTNLB3* and *RTNLB5-8* from *Arabidopsis* cDNA and examined whether RTNLB3 and RTNLB5-8 could interact with *A. tumefaciens* VirB2 bait protein in yeast two-hybrid assays. The RTNLB8 prey protein but not the RTNLB3 and RTNLB5-7 proteins interacted with the VirB2 bait protein in yeast (Figure 1). RTNLB1, 2, and 4 proteins interacted with the VirB2 protein as well, which was consistent with previous results [23], and were used as positive controls in the yeast two-hybrid assays. As expected, the RTNLB1-8 prey proteins did not interact with the unrelated Lamin C bait protein in yeast and was used as the negative control (Figure 1). We also examined whether RTNLB3 and RTNLB5-8 proteins could interact with other Vir proteins, including VirB5 (the minor component of T-pili), VirB1, ViB1*, VirD2, VirE1, VirE2, and VirF. RTNLB3 and RTNLB5-8 did not interact with other tested Vir proteins, which was similar to the results for RTNLB1, 2, and 4 proteins [23].

Previous studies have demonstrated that RTNLB1, 2, and 4 can interact with each other and with themselves [23,33]. We next tested whether the RTNLB3 and RTNLB5–8 proteins interacted with themselves and/or other RTNLB1-8 proteins in yeast two-hybrid assays. RTNLB2 used as a bait fusion protein interacted with the RTNLB3 or 8 but not RTNLB5, 6, or 7 (Figure 1 and Table S1). RTNLB3 used as the bait fusion protein interacted with RTNLB2, 4 or 8 but not RTNLB3 or RTNLB5-7. Figure 1 results demonstrated that the RTNLB4 protein interacted with RTNLB8 but not RTNLB3 or RTNLB5-7. As well, RTNLB5, 6, 7, or 8 used as bait fusion proteins did not interact with RTNLB1-8 proteins in yeast (Figure 1). Similarly, RTNLB1 did not interact with the RTNLB3 or RTNLB5-8 proteins in yeast two-hybrid assays (Figure 1). Some of the positive yeast two-hybrid interactions were not observed when the tested bait protein was swapped with the prey proteins (Figure 1 and Table S1). For example, the RTNLB2 bait protein interacted with the RTNLB8 prey protein in yeast; whereas the RTNLB8 bait protein did not interact with the RTNLB2 prey protein. These inconsistent findings may result from different conformations of the bait and prey fusion proteins in yeast, as previously reported for interactions between RTNLB and RAB8 [23].

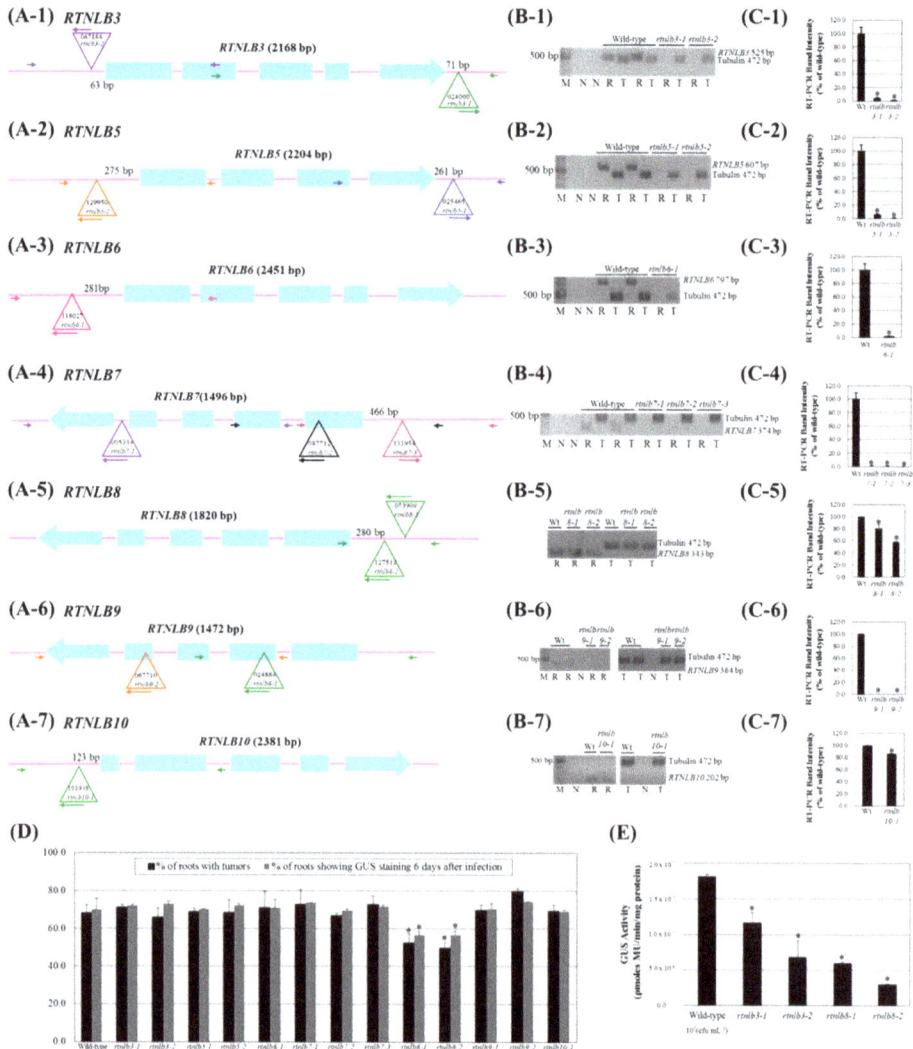

**Figure 1.** RTNLB8, not 3, and 5–7 proteins, interacted with the processed VirB2 in yeast. RTNLB1-8 proteins were tested for interactions with VirB2 or the RTNLB1-8 using a yeast two-hybrid assay. The RTNLB5-7 proteins showed no interactions with RTNLB1-8 proteins in yeast. The unrelated Lamin C bait protein was the negative control.

We next performed β-galactosidase activity assays to quantify the interaction strengths in these yeast strains. The white colony yeast strains on the SD media with X-gal substrates showed zero β-galactosidase activity, so the liquid-based β-galactosidase activity assays showed similar results as the plate-based yeast two-hybrid assays (Table S1). Yeast strains expressing the RTNLB3 bait with RTNLB2, 4, or 8 prey proteins showed relatively lower β-galactosidase activities as compared with yeast strains expressing the RTNLB2 or 4 bait proteins with the same tested prey proteins, which suggests that the interaction strengths might be lower among RTNLB3 interacting with other RTNLB proteins (Table S1).

In vitro glutathione-S-transferase (GST) pull-down assays were used to determine direct protein–protein interactions of RTNLB3 and 8 with VirB2 and other RTNLB proteins. T7-tagged-RTNLB1, 2, 3, and 4, and 8 proteins interacted with the GST-VirB2 fusion protein but not the GST protein in vitro (Figure 2A,B). In addition, GST-RTNLB1, 2, 3, and 4 fusion proteins but not the GST-RTNLB8 fusion protein interacted with the T7-tagged-VirB2 protein. The in vitro protein interactions between RTNLB1-4 and RTNLB8 were examined with GST pull-down assays by using the GST fusions and T7-tagged versions of RTNLB1-4 and 8. GST-RTNLB1 fusion proteins interacted with T7-tagged-RTNLB1, 2, 3, 4, but not T7-tagged-RTNLB8 proteins in vitro, whereas GST-RTNLB2 fusion protein interacted with T7-tagged-RTNLB1, 2, 4, 8, but not T7-tagged-RTNLB3 protein in vitro (Figure 2C,D). Furthermore, GST-RTNLB3 and GST-RTNLB4 fusion proteins interacted with the five tested RTNLB proteins (Figure 2E,F). However, only T7-tagged-RTNLB2 and 3 directly interacted with the GST-RTNLB8 fusion protein (Figure 2G). These interaction results were consistent with previous observations showing that RTNLB1-4 proteins may have homo- and heterotypic interactions [33]. The yeast two-hybrid assay and GST pull-down assay results summarized in Table S1 indicate that RTNLB3 interacted with VirB2, RTNLB1-4 and 8; whereas RTNLB8 interacted with VirB2, and RTNLB2, 3, 4. Interestingly, more positive interaction results of RTNLB3 and 8 with VirB2 and with other RTNLB proteins were obtained with GST pull-down than yeast two-hybrid assays. Previous studies in plants have suggested that RTNLB proteins are membrane proteins and are mainly localized in the plant endomembrane systems [31,33,35,36]. Therefore, the various fusion versions of RTNLB proteins used in our yeast two-hybrid and GST pull-down assays may not form the same conformation as native RTNLB protein in plant cells.

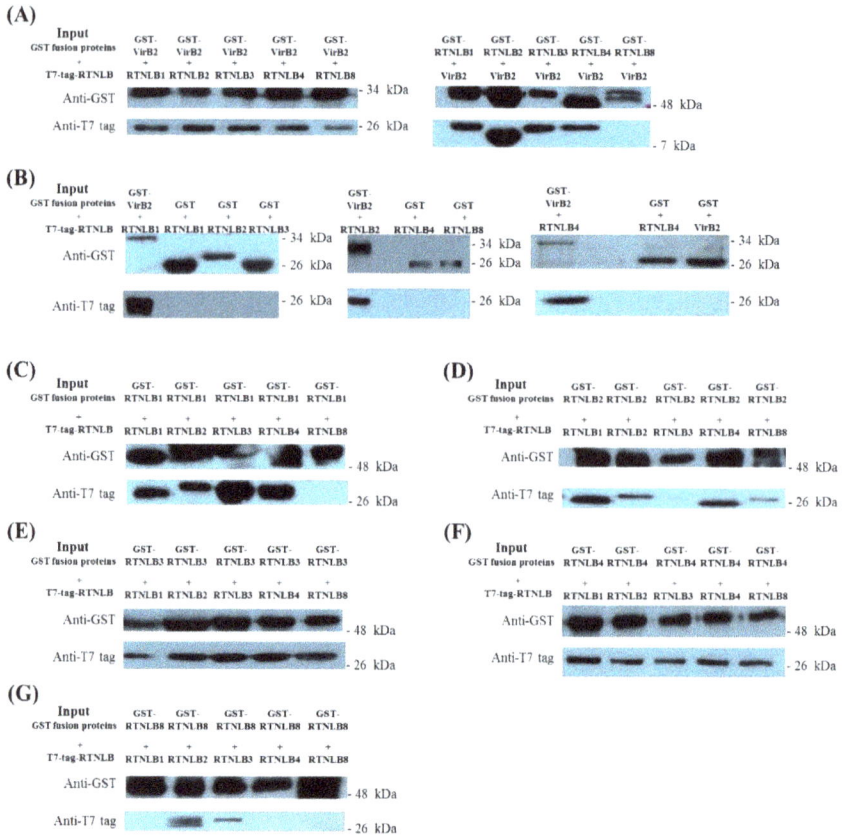

**Figure 2.** The GST-VirB2 fusion protein interacted with RTNLB1-4 and 8 proteins in vitro. The GST-fusion and GST only proteins were linked with glutathione-sepharose beads and incubated with T7-tagged proteins to test their interactions in vitro. Bound proteins were eluted with glutathione and analyzed by protein gel blot using anti-T7 tag and anti-GST antibodies. Panel **A**, interactions between VirB2 and RTNLB1-4 and 8 were determined by using the GST-VirB2 fusion protein and T7-tagged-RTNLB1-4 and 8 or the GST-fusion of RTNLB1-4 and 8 and the T7-tagged-VirB2 protein. Panel **B**, the GST-only protein was used as a negative control in GST pull-down assays. The GST fusions of RTNLB1 (Panel **C**), RTNLB2 (Panel **D**), RTNLB3 (Panel **E**), RTNLB4 (Panel **F**), and RTNLB8 (Panel **G**) were used to investigate their interactions with T7-tagged-RTNLB1-4 and 8 proteins.

## 2.2. Arabidopsis Rtnlb3 and Rtnlb8 Mutants Showed Reduced Levels of A. tumefaciens-Mediated Transformation Efficiency

Because the RTNLB3 and 8 proteins interacted with the *A. tumefaciens* VirB2 protein in vitro, we next examined whether the RTNLB3 and RTNLB5-10 proteins are involved in the *A. tumefaciens* infection process. We obtained several T-DNA insertions *Arabidopsis* mutants of *RTNLB3* or *RTNLB5-10* genes (Table S2) and tested susceptibilities of various *rtnlb* mutant plants to root- and seedling-based *A. tumefaciens* infection assays. At least one T-DNA insertion homozygous mutant was identified for *RTNLB3* and *RTNLB5-10* (Table S2 and Figure 3(A-1–A-7)). In the *rtnlb3, rtnlb5, rtnlb6, rtnlb8,* and *rtnlb10* single mutants, the T-DNA insertion sites were mainly located in the 3′ or 5′ untranslated region (UTR) of *RTNLB* genes (Table S2 and Figure 3(A-1–A-3,A-5,A-7)). The T-DNA insertion sites of the *rtnlb7* and *rtnlb9* single mutants were in the intron or exon (Table S2 and Figure 3(A-4,A-6)).

Semi-quantitative RT-PCR results showed that *RTNLB* target gene transcript levels were reduced to less than 5% of the wild-type level or were not detectable in the *rtnlb3, rtnlb5, rtnlb6, rtnlb7*, and *rtnlb9* single mutants (Figure 3(B-1–B-7)) (Figure 3(C-1–C-4,C-6)), which suggests that T-DNA insertions in these mutants may significantly affect the target *RTNLB* gene transcript stability and accumulation. In the *rtnlb8* and *rtnlb10* single mutants, the target gene transcript levels decreased to 58% to 86% of the wild-type levels (Figure 3(C-5,C-7)). We next examined transformation frequencies of these *rtnlb* mutants with stable and transient *A. tumefaciens*-mediated root transformation. Only *rtnlb8-1* and *rtnlb8-2* mutants showed lower levels of tumor formation and transient transformation efficiency than wild-type plants (Figure 3D), whereas other tested *rtnlb* mutants were as susceptible to transformation by *A. tumefaciens* as wild-type plants. Because transient transformation does not require T-DNA integration into the plant genome [41], these data suggest that the transformation process may be blocked at a step before T-DNA integration in the *rtnlb8* mutants and *RTNLB8* may be involved at the early step(s) in *A. tumefaciens*-mediated root transformation process. Different types of plant tissues may show different susceptibility to *A. tumefaciens* infection [42]. We therefore used *Arabidopsis* seedlings for transient transformation assays [43] with the *rtnlb3* and *rtnlb8* mutants. GUS activities were decreased 36% to 63% in the *rtnlb3-1* and *rtnlb3-2* single mutants and 67% to 84% in the *rtnlb8-1* and *rtnlb8-2* single mutants as compared with wild-type plants (Figure 3E). The decreased levels of GUS activities were greater in the *rtnlb8* than *rtnlb3* mutant (Figure 3E), which suggests that low expression of the *RTNLB8* gene in the mutant plants might affect the *A. tumefaciens*-mediated transient transformation efficiency of seedlings more than the *RTNLB3* gene. The *rtnlb3* mutants showed lower transformation efficiency than wild-type plants with only the seedling-based transformation assay and not the root-based assays (Figure 3D,E), which suggests that seedling tissues might be more sensitive to *A. tumefaciens* infection than root tissues and/or the *RTNLB3* gene might participate in efficient *A. tumefaciens* infection of *Arabidopsis* seedlings. In addition, *RTNLB5, 6, 7, 9*, and *10* might not be directly involved in the *A. tumefaciens* infection process.

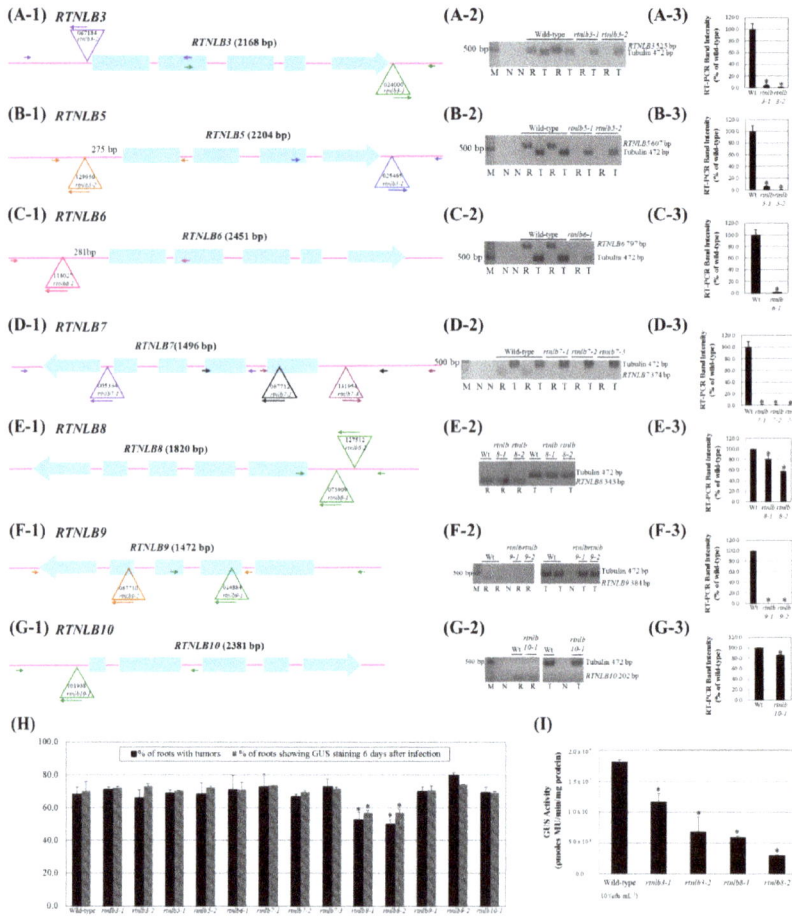

**Figure 3.** The *Arabidopsis rtnlb3* and *rtnlb8* T-DNA insertion mutant seedlings were resistant to *A. tumefaciens* infection. Panel **A**, schematic representations of the T-DNA insertion regions around the *Arabidopsis RTNLB3* (Panel **A-1**), *RTNLB5* (Panel **A-2**), *RTNLB6* (Panel **A-3**), *RTNLB7* (Panel **A-4**), *RTNLB8* (Panel **A-5**), *RTNLB9* (Panel **A-6**), and *RTNLB10* (Panel **A-7**) genes. Blue boxes represented exon regions of each *RTNLB* gene. The large open triangle represents T-DNA insertion sites in each *RTNLB* gene. The long and short arrows indicate the locations of primers used in genomic DNA PCR analysis. Panel B, RT-PCR results of target *RTNLB* transcripts in *rtnlb3* and *rtnlb5-10* single mutants. The α-tubulin was an internal control. Panel **C**, transcript levels of each *RTNLB* gene in *rtnlb* single mutants shown as a relative percentage of wild-type plants. Data are mean ± SE from at least 3 RT-PCR reactions of each mutant. Panel **D**, transformation efficiencies of *rtnlb8-1* and *rtnlb8-2* and wild-type plants. Black bars indicate the percentage of root segments forming tumors 1 month after infection with $10^8$ cfu·mL$^{-1}$ tumorigenic *A. tumefaciens* A208 strain. Grey bars show the percentage of root segments with GUS activity 6 days after infection with $10^8$ cfu·mL$^{-1}$ *A. tumefaciens* At849 strain. Panel **E**, *rtnlb3* and *rtnlb8* mutant seedlings showed decreased susceptibility to transient transformation. Transient transformation efficiency in mutant seedlings infected with $10^7$ cfu·mL$^{-1}$ acetosyringone (AS)-induced *A. tumefaciens* strain for 3 days. Data are mean ± SE. * $p < 0.05$ compared with the wild-type by pairwise Student's *t* test.

### 2.3. RTNLB3 and 8 Overexpression Increased Plant Susceptibility to A. tumefaciens-Mediated Transformation

Because the *rtnlb3* and *rtnlb8* seedling plants were recalcitrant to *A. tumefaciens* infection, we next determined whether overexpression of *RTNLB3* or *8* in plants could enhance the efficiency of *A. tumefaciens* infection. *RTNLB3* and *8* and T7-tagged-*RTNLB3* and *8* genes were individually overexpressed in transgenic plants by using a double CaMV 35S promoter. *RTNLB3* transcript level was increased 1.2- to 1.7-fold in *RTNLB3* and T7-tagged-*RTNLB3* overexpression plants (Figure 4(A-1,B-1)). Similarly, *RTNLB8* transcript level was increased 1.2- to 1.6-fold in *RTNLB8* and T7-tagged-*RTNLB8* overexpression plants (Figure 4(A-2,B-2)). Protein gel blot analysis with anti-T7-tag antibody demonstrated that T7-tagged-RTNLB3 or 8 recombinant proteins were highly accumulated in transgenic plants (Figure 4(C-1,C-2)). Root tissues of the *RTNLB3* and *8* and T7-tagged-*RTNLB3* and *8* overexpression plants were then infected with relatively lower concentrations of *A. tumefaciens*, $10^5$ and $10^6$ cfu·mL$^{-1}$. Overexpression of *RTNLB3* or T7-tagged-*RTNLB3* in transgenic plants increased transient transformation efficiency 1.1- to 1.3-fold and enhanced tumor formation rates 1.2- to 2.1-fold as compared with wild-type plants (Figure 4D). Similarly, both stable and transient transformation rates of *RTNLB8* and T7-tagged-*RTNLB8* overexpression plants were increased 1.3- to 2.0-fold as compared with wild-type plants (Figure 4D). These data demonstrate that overexpression of *RTNLB3* or *RTNLB8* in plants enhanced transgenic plant root tissue susceptibility to *A. tumefaciens* and the presence of the T7 tag sequence in the N-terminal region of RTNLB3 and 8 proteins may not affect the RTNLB protein functions during *A. tumefaciens* infection. With $10^5$ cfu·mL$^{-1}$ *A. tumefaciens* used to infect the *RTNLB3* and *8* overexpression seedlings, *RTNLB3* overexpression plants showed increased GUS activities by 2.4- to 7.2-fold, whereas *RTNLB8* overexpression plants showed increased GUS activities by 2.7- to 10.1-fold as compared with wild-type plants (Figure 4E). Taken together, these data indicated that the *RTNLB3* and *8* may play important roles in plants during *A. tumefaciens* infection.

### 2.4. RTNLB3 and 8 Overexpression Plants were More Susceptible to P. syringae Infection

Because RTNLB1 and 2 may regulate export of FLS2, the PRR for the flagellin of *Pst* DC3000, to the plasma membrane during pathogen infection [40], we next examined whether overexpression of *RTNLB3* or *8* could affect the plant susceptibility of *P. syringae*. Four- to 5-week-old plant leaves were syringe-infiltrated with wild-type *Pst* DC3000 and the *hrcC* mutant, the type III secretion system-defective bacteria, as the negative control. Bacterial growth assays were used to quantify the bacterial proliferation at 0, 1, 3, 5, and 7 days post-infection. Bacterial numbers were significantly increased in plant leaves up to 5 days after infection of wild-type *Pst* DC3000 and decreased at 7 days after infection (Figure 5(A-1,A-2)), which indicates successful infection with *P. syringae* in plants, whereas bacteria numbers only slightly increased after infection of the *hrcC* mutant. *RTNLB3* and *8* overexpression plants had relatively higher viable bacterial numbers than wild-type plants from 1 day after infection with *Pst* DC3000 (Figure 5(A-1)). Overexpression of *RTNLB3* or *8* in transgenic plants formed more severe disease symptoms and more chlorotic haloes than wild-type plants at 5 days after infection with *Pst* DC3000 (Figure 5B). As well, cell death was greater in *RTNLB3* and *8* overexpression than wild-type plants (Figure 5C). On infection with the *hrcC* mutant, wild-type and *RTNLB* overexpression plants showed no difference in bacterial growth (Figure 5(A-2)) and no visible disease symptoms in leaves of both plants types (Figure 5B). These results indicate that increased expression of *RTNLB3* and *8* may lead to enhanced susceptibility to *Pst* DC3000 infection.

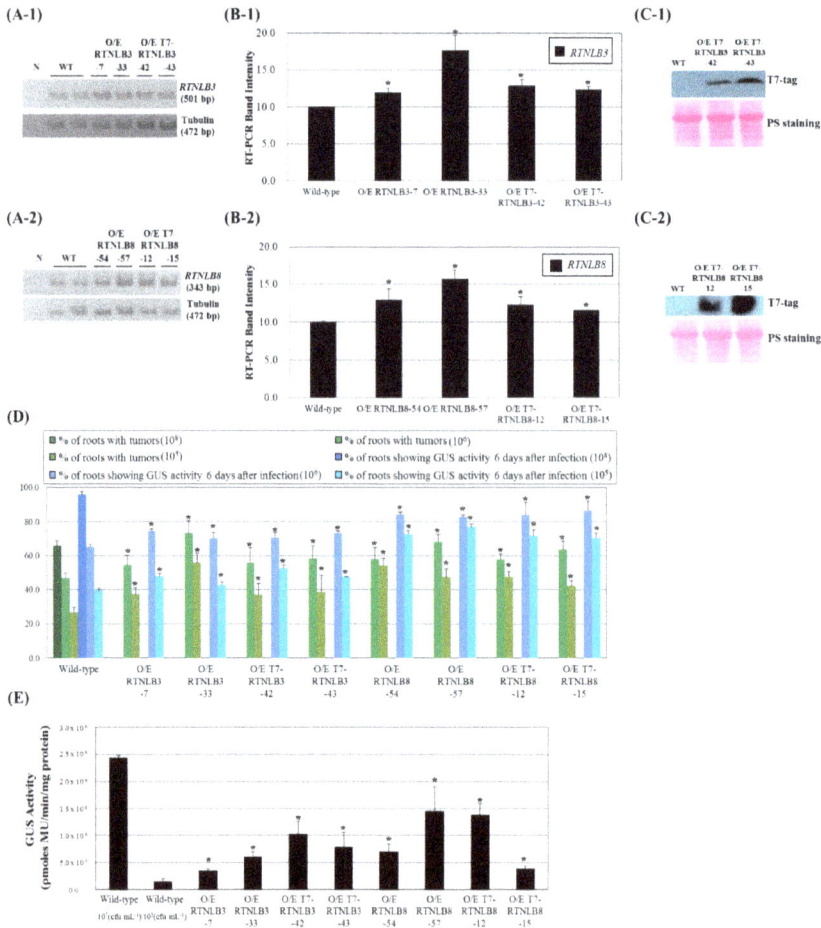

**Figure 4.** *RTNLB3* and *RTNLB8* overexpression (O/E) transgenic plants were hypersusceptible to *A. tumefaciens* infections. Panel **A**, RT-PCR analysis of *RTNLB* transcript levels in *RTNLB3* (Panel **A-1**) and *RTNLB8* (Panel **A-2**) O/E plants and the wild type. The α-tubulin was used as an internal control. Panel **B**, transcript levels of *RTNLB3* (Panel **B-1**) or *RTNLB8* (Panel **B-2**) in O/E plants relative to wild-type expression. Data are mean ± SE from at least 3 RT-PCR reactions of each mutant. Panel **C**, the T7-tagged-*RTNLB3* (Panel **C-1**) and *RTNLB8* (Panel **C-2**) O/E plants accumulated T7-tagged RTNLB proteins. Ponceau S (PS) staining was used to show equivalent loading of total protein in each lane. Panel **D**, Transient transformation efficiency of *RTNLB3* and *8* O/E and wild-type plants. Green bars represent the percentage of root segments developing tumors after infection with $10^8$, $10^6$, or $10^5$ cfu·mL$^{-1}$ of *A. tumefaciens* A208. Blue bars indicate the percentage of root segments with GUS activity after infection with $10^8$, $10^6$, or $10^5$ cfu·mL$^{-1}$ of *A. tumefaciens* At849 strain. The $10^8$ cfu mL$^{-1}$ of *A. tumefaciens* was used to infect wild-type roots as a positive control to indicate successful transformation. Panel **E**, Enhanced transient transformation efficiency in seedlings of *RTNLB3* and *8* O/E plants. Seedlings of O/E plants were infected with $10^5$ cfu·mL$^{-1}$ of AS-induced *A. tumefaciens* strain. Wild-type seedlings were infected with $10^7$ cfu·mL$^{-1}$ of *A. tumefaciens* strain as a positive control. Data are mean ± SE. * $p < 0.05$ compared with the wild-type by pairwise Student's *t* test.

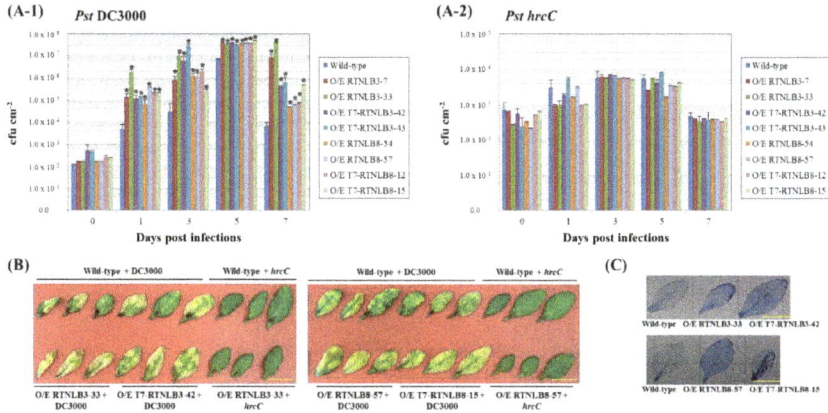

**Figure 5.** *RTNLB3* and *8* O/E plants were more sensitive to *Pseudomonas syringae* pv. *tomato* DC3000 (*Pst* DC3000) infection. Panel **A**, leaves of wild-type and *RTNLB* O/E plants were syringe-infiltrated with *Pst* DC3000 (Panel **A-1**) and *hrcC* mutant (Panel **A-2**). Bacterial numbers in infected leaves were quantified at 0, 1, 3, 5, and 7 days post-infection. Data are mean ± SE. * $p < 0.05$ compared with the wild-type by pairwise Student's *t* test. Panel B, disease symptoms of wild-type and *RTNLB* O/E plant leaves 5 days after infection with *Pst* DC3000 or *hrcC* mutant. Panel **C**, trypan blue staining of infected leaves of wild-type and *RTNLB* O/E plants 5 days after infection with *Pst* DC3000. Yellow bar = 1 cm.

### 2.5. RTNLB1-4 and 8 Gene Levels Differed in Various Plant Tissues

A previous study using the *Arabidopsis* eFP browser suggested that *RTNLB1-4* and *13* genes have tissue-specific expression patterns and levels [33,44]. To understand other functions of *RTNLBs* in plants, we used quantitative real-time RT-PCR of *RTNLB1-4* and *8* expression to determine their transcript levels in root, rosette leaf, cauline leaf, inflorescence, flower, and silique of wild-type *A. thaliana* (ecotype: Columbia). In roots, the transcript level was 4-fold greater for *RTNLB1-3* than *RTNLB4* and *8* (Figure 6). In rosette leaf, *RTNLB1-3* and *8* transcripts accumulated to a similar level, whereas *RTNLB4* level was the lowest among the five examined *RTNLB* genes (Figure 6). In cauline leaf and inflorescence, *RTNLB1* transcript level was the highest, followed by *RTNLB2*, *3*, and *4*, whereas *RTNLB8* level was the lowest (Figure 6). In flower, the transcript level was higher for *RTNLB1* than *RTNLB3* and *8*, whereas *RTNLB2* and *4* levels were lower than those of the other three *RTNLB* genes (Figure 6). In silique, *RTNLB1* transcript level was the highest among the other four *RTNLB* genes (Figure 6). Because we generated the *RTNLB3* and *8* overexpression plants in *A. thaliana* ecotype Wassilewskija (Ws), we also investigated *RTNLB1-4* and *8* transcript levels in Ws plants. The five *RTNLB* transcript levels were lower in Ws plants than in Columbia plants (Figure S1). The transcript levels for *RTNLB1-4* and *8* in various tissues of Ws plants differed from that in Columbia plants (Figure 6 and Figure S1). The RTNLB8 transcript level was the lowest in roots, rosette leaf, cualine leaf, inflorescence, and flower tissues of Ws plants as compared with the other four *RTNLB* genes (Figure S1).

**Wild-type plants (ecotype: Columbia)**

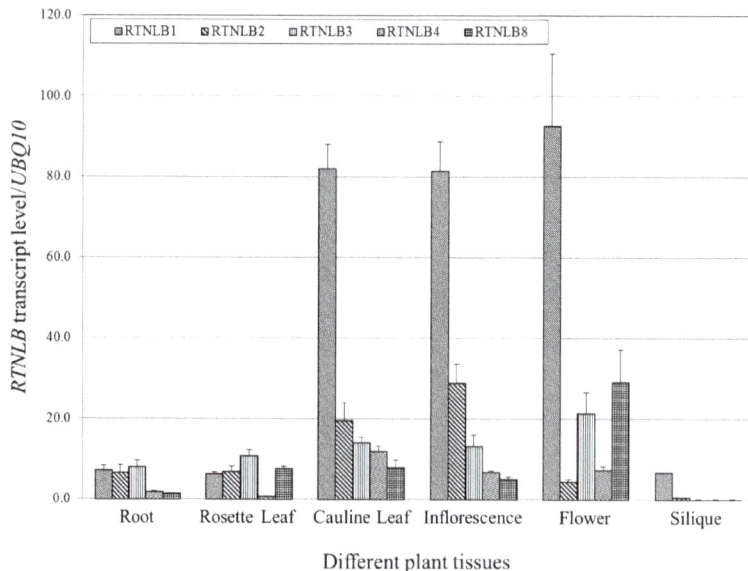

**Figure 6.** Levels of *RTNLB1-4* and *8* in various tissues of wild-type *Arabidopsis* (ecotype: Columbia) plants. RNA from root, rosette leaf, cauline leaf, inflorescence, flower, and silique of wild-type plants were isolated, reverse-transcribed, and used for quantitative real-time PCR. *UBQ10* (polyubiquitin 10) transcript level was an internal control. Data are mean ± SE.

## 3. Discussion

The reticulons (RTNLs) were first identified as endoplasmic reticulum (ER)-localized integral membrane proteins in mammalian neuron cells and have been identified in other eukaryotic cells, including yeast and plant cells [27,35]. So far, only a few members of the plant subfamily of RTNLs, named RTNLB, were demonstrated to localize in the ER and to help shape ER structures [31–33]. In this study, both yeast two-hybrid and GST pull-down assays revealed that RTNLB3 and 8 proteins interacted with the major component of the *A. tumefaciens* T-pilus, the VirB2 protein. Genetic studies have shown that the *rtnlb3* and *rtnlb8* single mutants were recalcitrant on *A. tumefaciens*-mediated transient transformation assays, whereas *RTNLB3* and *8* overexpression plants were hypersusceptible to *A. tumefaciens* and *Pseudomonas syringae* pv. *tomato* DC3000 infection. These data suggest that RTNLB3 and 8 may play important roles in plant–microbe interactions.

Previous studies suggested that RTNLBs and the *ROOT HAIR DEFECTIVE 3* (*RHD3*) may play a significant role in ER tubular structure formation [31–33,45]. RHD3 requires a functional RTNLB13 to work together on ER network alteration [46]. Additionally, in plant cells, RTNLB1-4 and 13 can interact with each other and help with ER tubular structure formation [33]. In our study, RTNLB3 and 8 protein interacted with other RTNLBs in yeast and in vitro, which further supports the possible roles of RTNLB3 and 8 in ER modeling.

Our findings show that RTNLB3 and 8 both interacted with the *A. tumefaciens* VirB2 protein in yeast and in vitro. Subsequently, we examined *rtnlb3* and *rtnlb8* mutants with a seedling-based *A. tumefaciens* transient transformation assay, a more sensitive method than a root-based *A. tumefaciens* transformation assay. Both *rtnlb3* and *rtnlb8* single mutants showed lower GUS activity than did wild-type seedlings. With overexpression of *RTNLB3* or *8*, transgenic plants were hypersensitive on both root- and seedling-based *A. tumefaciens* transformation assays as compared with wild-type plants, which suggests that RTNLB3 and 8 may participate in the *A. tumefaciens* infection process. A previous study indicated that RTNLB1, 2, and 4 may be involved in *A. tumefaciens* transformation [23]. RTNLB3 and/or 8 might affect *A. tumefaciens* infection by interacting with RTNLB1, 2 or 4, or they might have a different role during infection.

Other studies have used a co-immunoprecipitation approach to identify additional RTNLB3-interacting plant proteins and showed that RTNLB3 may also be involved in generation of ER-derived desmotubules [37,46]. Tobacco mosaic virus and Potato virus X may use desmotubules of the plant PD to spread virus particles [37–39] which suggests the possible roles of RTNLB3 during plant virus infection. Furthermore, RTNLB3 interacted with the vesicle-associated protein 27-1 (VAP27-1), which has high homology to the VAP33 family of SNARE-like proteins from animals, possibly involved in vesicular transport of ER [47]. RTNLB3 also interacted with a trafficking protein, synaptotagmin A (SYTA) [47]. SYTA co-localized with VAP27-1 at the ER–plasma membrane (PM) contact site to regulate endocytosis recycling at the ER–PM sites, which is related to Cabbage leaf curl virus and Tobacco mosaic virus movements [48]. A co-immunoprecipitation approach identified other RTNLB3-interacting proteins, RABA1b and RABA2c, which are members of the RAB small GTPase family and are involved in the transport between the *trans*-Golgi network and the PM [47,49,50]. In plant cells, RABA1b participates in the transport of de novo-synthesized FLS2, one of the PRRs for the bacterial flagellin, to the plasma membrane [51]. Most well-studied examples of pathogen-associated molecular pattern (PAMP) are the elongation factor-Tu (EF-Tu) of *A. tumefaciens* and the flagellin of *P. syringae*, which can be recognized by the elongation factor receptor (EFR) and FLS2 in plant cells, respectively [52–55]. PRRs present at the PM and can also be found at specific PM locations, in that FLS2 is enriched at PD as well [56]. A previous study demonstrated that RTNLB1 and 2 regulated ER tubular structure formation and contributed to newly synthesized FLS2 transport to the PM [40]. When the RTNLB1 protein was overexpressed or lost its function, transgenic plants and mutants were more easily infected by *Pst* DC3000 and showed defective FLS2-mediated immunity responses [40]. *RTNLB3* or *8* overexpression plants showed increased susceptibility to *Pst* DC3000 infection. Therefore, RTNLB3 and 8 proteins may contribute to endomembrane trafficking in plant cells and also might participate in plant immune response by affecting plant defense response-related endomembrane trafficking pathways. Overexpression of RTNLB3 or 8 might also perturb secretion of PRRs such as FLS2 to the PM and may therefore affect transgenic plant susceptibility to phytopathogenic bacterial infection.

Although *A. tumefaciens* infection provokes a general defense response during early infection stages, the transfer of T-DNA and several virulence proteins into plant cells at later infection stages could significantly affect gene expression in plants, especially defense-related genes [4,57–59]. During *A. tumefaciens* initial infection, the MPK3 kinase is phosphorylated and activated, which leads to translocation of a defense-related VIP1 transcription factor into the nucleus by interacting with importin alpha and induces defense-related gene expression [21,60,61]. However, the *A. tumefaciens* VirE2 protein could recruit VIP1 to mediate nuclear import of T-DNA [62,63] or to sequester a low-amount of VIP1 into the cytoplasm to dampen the activity of the host defense-related response [22]. *A. tumefaciens* no doubt hijacks several plant proteins to overrule the plant defense response and ensure successful T-DNA transfer and expression. Although endomembrane trafficking of PRRs, including the FLS2 and EFR, is important for plant defense response perception and activation [64,65], relatively little is known about its role during *A. tumefaciens* infection.

In this study, we found that RTNLB3 and 8 proteins interact with VirB2 protein, the essential factor for *A. tumefaciens* infection and the main component of T-pilus. T-pilus is a filament structure protruding from the *A. tumefaciens* surface and might possibly be recognized by the host defense system. RTNLB3 and 8 might possibly participate in endomembrane trafficking of unknown receptors or the EFR protein for *A. tumefaciens*-induced response in plant cells. The direct link between RTNLB3/8 and endomembrane trafficking of PRRs (i.e., EFR and FLS2) awaits further investigations.

We examined *RTNLB1-4* and *8* transcript levels in wild-type *Arabidopsis* plants (ecotype: Columbia) by real-time PCR. As compared with the expression predictions for *RTNLB1-4* by using the *Arabidopsis* eFP browser [33,44], *RTNLB1* expression was relatively higher than *RTNLB2-4* and *8* in several examined tissues, which is similar to our results obtained with real-time PCR (Figure 6). However, expression levels and locations of *RTNLB2-4* and *8* were slightly different from our real-time PCR results. For example, PCR results showed the highest level of *RTNLB8* in flowers but *Arabidopsis* eFP browser results showed *RTNLB8* as most abundant in cauline leaves and rosette leaves [33,44]. Our PCR results showed the level of *RTNLB4* higher in cauline leaves than other tissues, but the *Arabidopsis* eFP browser results revealed higher level of *RTNLB4* in flowers than other plant tissues [33,44]. This discrepancy might be due to two different analysis methods. The *Arabidopsis* eFP browser study used several microarray results from various plant cells and tissues over a long period of growth. In our results, we isolated RNA from various plant tissues of 4- to 5-week-old wild-type plants for PCR analysis. The different plant growth stages might also cause differences in results.

The different gene expression levels of endomembrane trafficking proteins affect plant growth and also have different effects on plant responses to pathogens. A member of a nuclear pore-targeting complex (PTAC), importin α (IMPα), participates in NLS cargo recognition, acts as an adaptor to bring cargo into binding of a PTAC carrier, and further interacts with nucleoporins [66,67]. Although four importin α isoforms interacted with *A. tumefaciens* VirD2 and VirE2 in yeast, in vitro, and in plant cells, only mutation of the importin *IMPa-4* affected *A. tumefaciens* infection efficiency [68]. When overexpressing other importin α in the *impa-4* mutant background, the mutant phenotype can be complemented, which suggests that different expression patterns and/or expression levels might affect different importin α member functions during *A. tumefaciens* infection [68]. Moreover, the histone H2A, *HTA1* gene, is involved in T-DNA integration and T-DNA expression during *A. tumefaciens* infection [69–71]. The histone H2A family contains 13 gene members and accumulates at differing levels in roots and other plant tissues [72]. The *HTA1* mutant, *rat5*, showed resistance to *A. tumefaciens* transformation efficiency. The mutant phenotype could be restored by overexpression of other *HTA* gene members and expression of only *HTA1*, not with other *HTA* genes, under control of its native promoter [72]. In addition, *HTA1* gene expression is induced by wounding and by infection with *A. tumefaciens* in root cells [70,71]. These data suggest that different expression patterns of the *HTA* genes in various plant tissues may affect *A. tumefaciens*-mediated transformation efficiencies. In our results, the *rtnlb8* mutant showed resistance to *A. tumefaciens* infection in both root- and seedling-based assays; whereas the *rtnlb3* mutant showed a resistance phenotype only with the seedling-based assay, which suggests that *RTNLB3* and *8* genes might play different roles in roots and seedlings during *A. tumefaciens* infection.

We also show *RTNLB1-4* and *8* gene transcript levels in various tissues of wild-type *Arabidopsis* (ecotype Wassilewskija, Ws). Five *RTNLB* transcript levels were much lower than the same gene expression in the wild-type ecotype Columbia plants and expression patterns of the five *RTNLB* genes were different as well (Figure S1). Previous studies have shown that expression patterns of various members of a gene family may differ in different ecotype backgrounds [73]. For instance, the *Arabidopsis* sucrose transporter family contains nine members, *AtSUC1-9* [73–75]. The *AtSUC6-9* genes show high sequence homology in coding regions and in introns. From analysis of splice patterns and polymorphic sites of *ATSUC6-9*, *AtSUC7* showed ecotype-specific splice patterns in Ws, C24, Columbia (Col-0), and *Landsberg erecta* (Ler) ecotypes [73]. *AtSUC1* also had ecotype-specific expression and its expression was observed in the funicular epidermis of Ws, C24 and *Landsberg erecta* but not Col-0 [76]. So far, whether expression patterns and levels of different splice variants of *RTNLB* genes in plants differ is unclear. The possible roles and implications of different expression patterns of *RTNLB1-4* and *8* in two *Arabidopsis* ecotypes (Ws and Columbia) still needs further examination.

## 4. Materials and Methods

### 4.1. Bacterial Strains and Culture

*Agrobacterium tumefaciens* and *Escherichia coli* strains used in this study are in Table S3. *A. tumefaciens* strains were grown in 523 medium or on 523 agar supplemented with appropriate antibiotics (rifampicin 50 μg·mL$^{-1}$, gentamycin 50 μg·mL$^{-1}$, and kanamycin 20 μg·mL$^{-1}$; MDBio Inc., Taipei, Taiwan) at 28 °C. *E. coli* strains were grown at 37 °C in 2X YT medium [77] containing appropriate antibiotics (ampicillin 100 μg·mL$^{-1}$, kanamycin 50 μg·mL$^{-1}$).

*4.2. Yeast Two-Hybrid Assays*

Plasmids used for yeast two-hybrid studies are in Table S3. The bait plasmid pSST91 and the prey plasmid pGAD424 [23] were used as vectors for the construction of the various fusions. The bait plasmid pSST91 contains the LexA protein coding sequence under the control of the yeast *ADH1* promoter. The prey plasmid pGAD424 generates a recombinant protein containing the GAL4 activation domain. To generate the bait plasmid expressing the LexA-RTNLB recombinant protein and the prey plasmid expressing the GAL4-RTNLB hybrid protein, *EcoRI-PstI* fragments of the *RTNLB3*, or *RTNLB5-RTNLB8* coding sequences from *Arabidopsis* were obtained from PCR reactions by using *Arabidopsis* cDNA as a template, high-fidelity Phusion DNA polymerase (New England BioLabs Inc., Ipswich, MA, USA), and appropriate primers (Table S4). The PCR products were digested with *EcoRI/PstI* and cloned into the pSST91 or the pGAD424 as in-frame fusion to the LexA or GAL4 coding sequence.

The yeast strain CTY10-5d [23] was used for yeast two-hybrid assays. Yeast transformations involved a lithium acetate method [78]. All yeast strains were cultured at 30 °C in synthetic dropout (SD) medium [78] containing a yeast nitrogen-base, glucose, and all but the selected amino acids. The bait and prey plasmids were transformed into CTY10-5d and colonies were screened for protein interaction by colony color phenotype on SD medium with the chromogenic substrate 5-bromo-4-chloro-3-indolyl-β-D-galactopyranoside (X-Gal) but lacking leucine and tryptophan. To quantify yeast two-hybrid interactions, the yeast liquid cultures with *ortho*-nitrophenyl-β-D-galactopyranosidase (ONPG) as substrates were used for β-galactosidase enzyme activity assays as described [78]. One unit of β-galactosidase enzyme activity was defined as the amount that can hydrolyze 1 μmol of ONPG to *o*-nitrophenol and D-galactose per minute per cell [78].

*4.3. Glutathione-S-Transferase (GST) Protein Affinity Purification Assays*

Plasmids and bacteria used for the GST pull-down assays are in Table S3. The plasmids pGEX4T-1 or pET42a were used to generate recombinant proteins fused in-frame with the GST tag. The plasmids pET23a and pET28a were used to express the T7-tagged fusion proteins in the *E. coli* strain BL21(DE3). The coding sequence of the *RTNLB3* and *8* genes were amplified by PCR with *Arabidopsis* cDNA used as templates, high-fidelity Phusion DNA polymerase, and appropriate primers (Table S4). The PCR fragments were digested with *EcoRI* and *PstI*, which were subsequently cloned into the pBluescript plasmid. The *EcoRI-NotI* fragments containing *RTNLB3* and *8* coding sequences were digested from the plasmids pBluescript-RTNLB3 and pBluescript-RTNLB8, respectively. The *EcoRI-NotI* fragments from pBluescript-RTNLB3 were cloned into pET23a or pGEX4T-1 as an in-frame fusion to the T7 tag or GST coding sequence. Similarly, the *EcoRI-NotI* fragments from pBluescript-RTNLB8 were cloned into pET28a or pET42a to express the T7-tagged or GST fusion proteins in bacteria. Expression and purification of GST fusion proteins and affinity purification of proteins binding to GST fusion proteins were performed as described [23,78]. The isolated protein complexes were analyzed in 12.5% SDS-polyacrylamide gels and immunoblot analysis was performed [78] with a 1:1000 dilution of anti-T7 tag primary antibody (Merck, Danvers, MA, USA) or with a 1:15,000 dilution of anti-GST primary antibody (GE Healthcare, Piscataway, NJ, USA), followed by a 1:20,000 dilution of secondary antibody horseradish peroxidase (HRP)-conjugated goat anti-rabbit IgG (PerkinElmer Life and Analytical Science, Boston, MS, USA) or a 1:15,000 dilution of HRP-conjugated donkey anti-goat IgG- (Santa Cruz Biotechnology, Dallas, TX, USA), to confirm the identities of these fusion proteins. The membranes were developed by chemiluminescent detection and subjected to autoradiography.

*4.4. DNA Isolation from Arabidopsis Plants and Genomic DNA PCR Analysis*

The *Arabidopsis* T-DNA insertion mutants *rtnlb3* and *rtnlb5* to *rtnlb10* (ecotype: Columbia CS60,000) were identified by using the SIGnAL T-DNA Express *Arabidopsis* Gene Mapping Tool (http://signal.salk.edu/) [79]. Seeds of *rtnlb* mutant plants were obtained from the *Arabidopsis* Biological Resource Center (ABRC; Ohio State University). Seedlings from *rtnlb* mutant plants were individually grown in Gamborg's B5 medium and leaves of 3-week-old plants were used to isolate genomic DNA as described [80]. A PCR-based approach similar to that described by Alonso et al. 2003 and the SIGnAL T-DNA Express Gene Mapping Tool (http://signal.salk.edu/) was used to determine the homozygosity of *Arabidopsis rtnlb* mutants. Primers for genomic DNA PCR analysis are in Table S4. The PCR reaction was conducted in a 50 μL reaction volume with 2 units of GenTaq polymerase (GMbiolab Co., Taichung, Taiwan), a 2.5 mM dNTP mixture, 1× Taq polymerase reaction buffer, and 0.25 μm of the PCR primers. The PCR amplification cycle was 95 °C for 1 min (1 cycle); 94 °C for 30 s, 56 °C for 40 s, 72 °C for 1 min (30 cycles) and 72 °C for 5 min (1 cycle).

*4.5. RNA Isolation from Arabidopsis Plants and RT-PCR Analysis*

RNA was extracted from root and above-ground plant tissues from 4- to 5-week-old wild-type plants (ecotypes: Columbia and Wassilewskija [Ws]), the *rtnlb* mutant (ecotype: Columbia), and *RTNLB* overexpression transgenic plants (ecotype: Ws). Plant tissues were ground with a liquid nitrogen-cooled pellet pestle in a 1.5-mL Eppendorf tube. The ground materials were mixed with TRIZOL LS reagents (Total RNA Isolation Reagent for Liquid Samples from Invitrogen, Carlsbad, CA, USA) to isolate RNA according to the manufacturer's instructions. An amount of 1–3 μg RNA was then treated with DNase I (Thermo Fisher Scientific Inc., Waltham, MA, USA) and reactions were stopped with the addition of EDTA and heat inactivation. RT-PCR involved the RevertAid First Strand cDNA Synthesis Kit (Thermo Fisher Scientific Inc., Waltham, MA, USA) and oligo-dT primers were used to generate the first-strand cDNA products. A series of oligonucleotide primers (Table S4) was designed to amplify the sense mRNA strand of *RTNLB3* and *RTNLB5-RTNLB10* genes in the PCR reactions. The level of α-tubulin was an internal control in each RT-PCR reaction. The amplified products were analyzed on an agarose gel and visualized by using a UVP BioImaging System (UVP Inc., Upland, CA, USA) and quantified by using Quantity One software (Bio-Rad Laboratories Inc., Hercules, CA, USA).

An amount of 1 μg RNA samples from various tissues of wild-type plants were reverse transcribed by using the oligo-dT primer to generate cDNA. An amount of 100 ng cDNA was used for quantitative real-time PCR with the IQ$^2$ SYBR Green Fast qPCR System Master Mix (Bio-genesis Technologies Inc., Taipei, Taiwan) in the MS3000P QPCR system (Agilent Technologies, Santa Clara, CA, USA). Another set of oligonucleotide primers (Table S4) was used to determine levels of *RTNLB1-RTNLB4* and *RTNLB8* genes in quantitative real-time PCR reactions. The level of *UBQ10* (polyubiquitin 10) was an internal control in each quantitative real-time PCR reaction. More than 3 independent RT-PCR or real-time PCR reactions were performed with RNA samples isolated from at least 6–12 different *Arabidopsis* plants.

*4.6. Protein Extraction from Arabidopsis Plants and Protein Gel Blot Analysis*

Root tissues from 3- to 4-week-old *RTNLB3* and *8* overexpression transgenic *Arabidopsis* plants, and wild-type plants were used to isolate proteins. Root samples were ground with liquid nitrogen and mixed with CelLytic P (Sigma Chemical Co., St. Louis, MO, USA) containing a protease inhibitor cocktail (1:100 dilution; Sigma) according to the manufacturer's instructions. The final protein concentrations were determined by using a BCA protein assay kit (Pierce, Rockford, IL, USA) and a spectroscopy (PARADIGM Detection Platform, Beckman Coulter Inc., Indianapolis, IN, USA). Equal amounts of plant proteins were analyzed in 12.5% SDS-polyacrylamide gels and immunoblot analysis involved use of a 1:1000 dilution of T7-tag antibody (Abcam, Cambridge, UK), then a 1:20,000 dilution of horseradish peroxidase-conjugated goat anti-rabbit

IgG antibody (PerkinElmer Life and Analytical Sciences, Boston, MA, USA). The membranes were developed by chemiluminescent detection (PerkinElmer Life and Analytical Sciences, Boston, MA, USA) and subjected to autoradiography.

*4.7. Generation of RTNLB3 and 8 Overexpression Arabidopsis Transgenic Plants*

A binary vector, pE1798, containing a double Cauliflower mosaic virus (CaMV) 35S promoter, a nopaline synthase (*Nos*) terminator, and a hygromycin resistance gene (*hptII*) gene as a selectable marker in the T-DNA region [23] were used to overexpress *RTNLB3* or *8* gene in *A. thaliana* transgenic plants. The *Kpn*I-*Sac*I fragments from the pBluescript-RTNLB3 and pBluescript-RTNLB8 were cloned into the same sites of the pE1798 plasmid (Table S3). To overexpress the T7-tagged-RTNLB3 or T7-tagged- RTNLB8 in *Arabidopsis* transgenic plants, the plasmids pET23a-RTNLB3 and pET28a-RTNLB8 were used as templates for PCR with the high-fidelity Phusion DNA polymerase and appropriate primers (Table S4). The PCR fragments were digested with *Kpn*I and *Sac*I, then cloned into the pE1798 plasmid (Table S3). These pE1978 series plasmids were separately transformed into the non-tumorigenic strain *A. tumefaciens* GV3101(pMP90) [81] to generate *Arabidopsis* overexpression plants by a floral dip method [82].

*4.8. Agrobacterium Tumefaciens-Mediated Stable, Transient Root and Seedling Transformation Assays of Rtnlb Mutant Plants and Arabidopsis RTNLB Overexpression Plants*

Seeds from wild-type, *rtnlb* mutants, and *RTNLB* overexpression plants were surface-sterilized and placed on Gamborg's B5 medium (PhytoTechnology Laboratories, Carlsbad, CA) solidified with 0.75% Bactoagar (BD Biosciences, Lenexa, KS, USA) containing appropriate antibiotics (kanamycin $50~\mu g \cdot mL^{-1}$ for *rtnlb* mutant plants and hygromycin $20~\mu g \cdot mL^{-1}$ for overexpression plants). Seedlings were transferred individually to the solidified B5 medium in baby food jars without antibiotics and grown for 3–4 weeks for stable and transient root transformation assays as described [23,83].

All *A. tumefaciens* strains (Table S3) were cultured in 523 medium [84] with appropriate antibiotics (rifampicin $50~\mu g \cdot mL^{-1}$, kanamycin $50~\mu g \cdot mL^{-1}$) at 28 °C. The overnight bacterial culture was inoculated into 25 mL of 523 medium with antibiotics and grown to $10^9$ colony forming units $(cfu) \cdot mL^{-1}$. The bacterial cells were then washed with 0.9% sodium chloride to remove antibiotics and medium. The bacterial cells were resuspended in 0.9% sodium chloride at $10^5$, $10^6$, or $10^8$ cfu·mL$^{-1}$ for root transformation assays.

For stable root transformation assays, root segments were cut from 3- to 4-week-old plants and transferred to solidified Murashige and Skoog medium and co-incubated with the tumorigenic strain *A. tumefaciens* A208 for at 22 to 24 °C for 2 days (Table S3). After co-incubation periods, root segments were separated and transferred to MS medium with antibiotic timentin $(100~\mu g \cdot mL^{-1})$ but lacking hormones for 1 month to score tumor formation efficiencies. For transient root transformation assays, root segments were infected with *A. tumefaciens* At849 containing the pBISN1 binary vector (Table S3). After 2-day co-incubation periods, root segments were placed on the callus induction medium (CIM) including timentin at 22 to 24 °C for 4 additional days. Roots were then stained with 5-bromo-4-chloro-3-indolyl-β-D-glucuronic acid (X-gluc) staining solution for 1 day at 37 °C. Roots were examined with a stereoscopic microscope to obtain transient transformation efficiencies. For root transformation assays, at least 15 different *Arabidopsis* plants were infected with each *A. tumefaciens* strain and more than 60 root segments were examined for each plant for each independent transformation assay.

The transient seedling transformation assays (*Agrobacterium*-mediated enhanced seedling transformation, AGROBAST) were performed as described [43] with minor modifications. The *Arabidopsis* seedlings were first germinated in a 6-well plate containing half-strength MS medium (pH 5.7) and 0.5% sucrose at 22 to 24 °C for 7 days. The *A. tumefaciens* C58C1(pTiB6S3ΔT) strain with a pBISN1 binary vector was first grown in 523 medium with the appropriate antibiotics (rifampicin $50~\mu g \cdot mL^{-1}$, kanamycin $50~\mu g \cdot mL^{-1}$) at 28 °C. The overnight-grown bacteria cells were further cultured for 24 h at 28 °C in acidic AB-MES medium with 200 μM acetosyringone (AS) to induce *vir* gene expression [84]. After AS induction, bacterial cells were washed with sterile water and resuspended in infection solution (half-strength of the MS medium [pH 5.7], one-quarter of the AB-MES medium [pH 5.5, 0.5% sucrose, and 50 μM AS]) at $10^7$ cfu·mL$^{-1}$ for seedling transformation assays. The *Arabidopsis* seedlings were co-incubated with AS-induced bacteria cells at 22 to 24 °C for 3 days. Seedlings were ground with liquid nitrogen and mixed with extraction buffers for fluorescent 4-methylumbelliferyl-β-D-glucuronide (MUG) assays as described [43]. The fluorescence was determined by using a 96 microplate reader (PARADIGM Detection Platform) at 365 nm excitation and 455 nm emission. The protein concentration for each protein sample was determined with a BCA protein assay kit and spectroscopy. The relative GUS activity was the fluorescence signal normalized by an equal amount of proteins. At least 10 different *Arabidopsis* seedlings were infected with the *A. tumefaciens* strain for each independent transformation assay and more than 3 independent transformation assays were performed.

### 4.9. Pseudomonas Syringae Infection Assays of Arabidopsis RTNLB Overexpression Plants

All *P. syringae* strains were grown in King's medium B (KB medium) at 28 °C with the antibiotic rifampicin (20 μg·mL$^{-1}$). After bacterial growth at 28 °C to mid to late log phase, bacterial cells were harvested, washed, and resuspended in 5 mM magnesium chloride solutions at $10^4$ cfu·mL$^{-1}$ for syringe infiltration assays. Leaves of the 4- to 5-week-old pot-grown *Arabidopsis* plants were infected with *P. syringae* strains (Table S3) by syringe infiltrations as described [85,86] with minor modifications. The abaxial side of *Arabidopsis* leaf was infiltrated with bacterial suspensions by using a needless syringe. To determine bacterial populations in plant leaves, leaf discs were excised from infiltrated leaves with use of a 0.6 cm$^2$ cork borer at 0, 1, 3, 5, and 7 days after infiltration. The leaf discs were ground with use of a plastic pestle in a small amount of 5 mM magnesium chloride solutions. The bacterial suspensions were serially diluted with magnesium chloride and cultured on KB agar plates with rifampicin (20 μg·mL$^{-1}$) and cycloheximide (10 μg·mL$^{-1}$) to determine viable cell numbers. For *Pst* DC3000 infection assays, at least 15 different *Arabidopsis* plants were infected with bacteria for each independent infection assay and more than 3 independent infection assays were performed.

**Supplementary Materials:** Supplementary materials can be found at http://www.mdpi.com/1422-0067/19/2/638/s1.

**Acknowledgments:** The authors thank Erh-Min Lai and Wen-Ling Deng for providing *A. tumefaciens* and *Pseudomonas syringae* strains; and the Hwang lab members for discussion and technical assistance. This research was funded by the Ministry of Science and Technology, Taiwan (MOST 105-2313-B-005-008). This research was supported in part by the Ministry of Education, Taiwan, under the ATU plan.

**Author Contributions:** Fan-Chen Huang and Hau-Hsuan Hwang conceived, contributed to experiment design and wrote the manuscript. Fan-Chen Huang, Bi-Ju Fu, Yin-Tzu Liu, Yao-Ren Chang, Shin-Fei Chi, Pei-Ru Chien and Si-Chi Huang conducted experiments and analyzed data. All authors read and approved the manuscript.

**Conflicts of Interest:** The authors declare no conflict of interest.

### References

1. Gelvin, S.B. Plant proteins involved in Agrobacterium-mediated genetic transformation. *Annu. Rev. Phytopathol.* **2010**, *48*, 45–68. [CrossRef] [PubMed]
2. Gelvin, S.B. Traversing the cell: Agrobacterium T-DNA's journey to the host genome. *Front. Plant Sci.* **2012**, *3*, 52. [CrossRef] [PubMed]
3. Lacroix, B.; Citovsky, V. The roles of bacterial and host plant factors in Agrobacterium-mediated genetic transformation. *Int. J. Biochem. Cell Biol.* **2013**, *57*, 467–481. [CrossRef] [PubMed]
4. Pitzschke, A. Infection and plant defense-transformation success hangs by a thread. Front. *Plant Sci.* **2013**, *4*, 519. [CrossRef]
5. Gohlke, J.; Deeken, R. Plant responses to Agrobacterium tumefaciens and crown gall development. *Front. Plant Sci.* **2014**, *5*, 155. [CrossRef] [PubMed]

6. Brencic, A.; Winans, S.C. Detection of and response to signals involved in host–microbe interactions by plant-associated bacteria. *Microbiol. Mol. Biol. Rev.* **2005**, *69*, 155–194. [CrossRef] [PubMed]

7. McCullen, C.A.; Binns, A.N. Agrobacterium tumefaciens and plant cell interactions and activities required for interkingdom macromolecular transfer. *Annu. Rev. Cell Dev. Biol.* **2006**, *22*, 101–127. [CrossRef] [PubMed]

8. Zechner, E.L.; Lang, S.; Schildbach, J.F. Assembly and mechanisms of bacterial type IV secretion machines. Philos. *Trans. R. Soc. B Biol. Sci.* **2012**, *367*, 1073–1087. [CrossRef] [PubMed]

9. Bhatty, M.; Laverde Gomez, J.A.; Christie, P.J. The expanding bacterial type IV secretion lexicon. *Res. Microbiol.* **2013**, *164*, 620–639. [CrossRef] [PubMed]

10. Chandran, V. Type IV secretion machinery: Molecular architecture and function. *Biochem. Soc. Trans.* **2013**, *41*, 17–28. [CrossRef] [PubMed]

11. Christie, P.J.; Whitaker, N.; Gonzalez-Rivera, C. Mechanism and structure of the bacterial type IV secretion systems. *Biochim. Biophys. Acta* **2014**, *1843*, 1578–1591. [CrossRef] [PubMed]

12. Wallden, K.; Rivera-Calzada, A.; Waksman, G. Type IV secretion systems: Versatility and diversity in function. *Cell Microbiol.* **2010**, *12*, 1203–1212. [CrossRef] [PubMed]

13. Waksman, G.; Fronzes, R. Molecular architecture of bacterial type IV secretion systems. *Trends Biochem. Sci.* **2010**, *35*, 691–698. [CrossRef] [PubMed]

14. Waksman, G.; Orlova, E.V. Structural organisation of the type IV secretion systems. *Curr. Opin. Microbiol.* **2014**, *17*, 24–31. [CrossRef] [PubMed]

15. Lai, E.M.; Kado, C.I. Processed VirB2 is the major subunit of the promiscuous pilus of Agrobacterium tumefaciens. *J. Bacteriol.* **1998**, *180*, 2711–2717. [PubMed]

16. Lai, E.M.; Chesnokova, O.; Banta, L.M.; Kado, C.I. Genetic and environmental factors affecting T-pilin export and T-pilus biogenesis in relation to flagellation of Agrobacterium tumefaciens. *J. Bacteriol.* **2000**, *182*, 3705–3716. [CrossRef] [PubMed]

17. Lai, E.M.; Eisenbrandt, R.; Kalkum, M.; Lanka, E.; Kado, C.I. Biogenesis of T pili in Agrobacterium tumefaciens requires precise VirB2 propilin cleavage and cyclization. *J. Bacteriol.* **2002**, *184*, 327–330. [CrossRef] [PubMed]

18. Eisenbrandt, R.; Kalkum, M.; Lai, E.M.; Lurz, R.; Kado, C.I.; Lanka, E. Conjugative pili of IncP plasmids, and the Ti plasmid T pilus are composed of cyclic subunits. *J. Biol. Chem.* **1999**, *274*, 22548–22555. [CrossRef] [PubMed]

19. Schmidt-Eisenlohr, H.; Domke, N.; Angerer, C.; Wanner, G.; Zambryski, P.C.; Baron, C. Vir proteins stabilize VirB5 and mediate its association with the T pilus of Agrobacterium tumefaciens. *J. Bacteriol.* **1999**, *181*, 7485–7492. [PubMed]

20. Aly, K.A.; Baron, C. The VirB5 protein localizes to the T-pilus tips in Agrobacterium tumefaciens. *Microbiology* **2007**, *153*, 3766–3775. [CrossRef] [PubMed]

21. Djamei, A.; Pitzschke, A.; Nakagami, H.; Rajh, I.; Hirt, H. Trojan horse strategy in Agrobacterium transformation: Abusing MAPK defense signaling. *Science* **2007**, *318*, 453–456. [CrossRef] [PubMed]

22. Shi, Y.; Lee, L.Y.; Gelvin, S.B. Is VIP1 important for Agrobacterium-mediated transformation? *Plant J.* **2014**, *79*, 848–860. [CrossRef] [PubMed]

23. Hwang, H.H.; Gelvin, S.B. Plant proteins that interact with VirB2, the Agrobacterium pilin protein, mediate plant transformation. *Plant Cell* **2004**, *16*, 3148–3167. [CrossRef] [PubMed]

24. Roebroek, A.J.; van de Velde, H.J.; Van Bokhoven, A.; Broers, J.L.; Ramaekers, F.C.; Van de Ven, W.J. Cloning and expression of alternative transcripts of a novel neuroendocrine-specific gene and identification of its 135-kDa translational product. *J. Biol. Chem.* **1993**, *268*, 13439–13447. [PubMed]

25. Van de Velde, H.J.; Senden, N.H.; Roskams, T.A.; Broers, J.L.; Ramaekers, F.C.; Roebroek, A.J.; Van de Ven, W.J. NSP-encoded reticulons are neuroendocrine markers of a novel category in human lung cancer diagnosis. *Cancer Res.* **1994**, *54*, 4769–4776. [PubMed]

26. Senden, N.H.; Timmer, E.D.; Boers, J.E.; van de Velde, H.J.; Roebroek, A.J.; Van de Ven, W.J.; Broers, J.L.; Ramaekers, F.C. Neuroendocrine-specific protein C (NSP-C): Subcellular localization and differential expression in relation to NSP-A. *Eur. J. Cell Biol.* **1996**, *69*, 197–213. [PubMed]

27. Oertle, T.; Schwab, M.E. Nogo and its paRTNers. *Trends Cell Biol.* **2003**, *13*, 187–194. [CrossRef]

28. Voeltz, G.K.; Prinz, W.A.; Shibata, Y.; Rist, J.M.; Rapoport, T.A. A class of membrane proteins shaping the tubular endoplasmic reticulum. *Cell* **2006**, *124*, 573–586. [CrossRef] [PubMed]

29. Hu, J.; Shibata, Y.; Voss, C.; Shemesh, T.; Li, Z.; Coughlin, M.; Kozlov, M.M.; Rapoport, T.A.; Prinz, W.A. Membrane proteins of the endoplasmic reticulum induce high-curvature tubules. *Science* **2008**, *319*, 1247–1250. [CrossRef] [PubMed]

30. Nziengui, H.; Schoefs, B. Functions of reticulons in plants: What we can learn from animals and yeasts. *Cell Mol. Life Sci.* **2009**, *66*, 584–595. [CrossRef] [PubMed]

31. Tolley, N.; Sparkes, I.A.; Aunter, P.R.; Craddock, C.P.; Nuttall, J.; Roberts, L.M.; Hawes, C.; Pedrazzini, E.; Frigerio, L. Overexpression of a plant reticulon remodels the lumen of the cortical endoplasmic reticulum but does not perturb protein transport. *Traffic* **2008**, *9*, 94–102. [CrossRef] [PubMed]

32. Tolley, N.; Sparkes, I.; Craddock, C.P.; Eastmond, P.J.; Runions, J.; Hawes, C.; Frigerio, L. Transmembrane domain length is responsible for the ability of a plant reticulon to shape endoplasmic reticulum tubules in vivo. *Plant J.* **2010**, *64*, 411–418. [CrossRef] [PubMed]

33. Sparkes, I.; Tolley, N.; Aller, I.; Svozil, J.; Osterrieder, A.; Botchway, S.; Mueller, C.; Frigerio, L.; Hawes, C. Five Arabidopsis reticulon isoforms share endoplasmic reticulum location, topology, and membrane-shaping properties. *Plant Cell* **2010**, *22*, 1333–1343. [CrossRef] [PubMed]

34. Oertle, T.; Klinger, M.; Stuermer, C.A.; Schwab, M.E. A reticular rhapsody: Phylogenic evolution and nomenclature of the RTN/Nogo gene family. *FASEB J.* **2003**, *17*, 1238–1247. [CrossRef] [PubMed]

35. Nziengui, H.; Bouhidel, K.; Pillon, D.; Der, C.; Marty, F.; Schoefs, B. Reticulon-like proteins in Arabidopsis thaliana: Structure organization and ER localization. *FEBS Lett.* **2007**, *581*, 3356–3362. [CrossRef] [PubMed]

36. Marmagne, A.; Rouet, M.A.; Ferro, M.; Rolland, N.; Alcon, C.; Joyard, J.; Garin, J.; Barbier-Brygoo, H.; Ephritikhine, G. Identification of new intrinsic proteins in Arabidopsis plasma membrane proteome. *Mol. Cell Proteomics* **2004**, *3*, 675–691. [CrossRef] [PubMed]

37. Knox, K.; Wang, P.; Kriechbaumer, V.; Tilsner, J.; Frigerio, L.; Sparkes, I.; Hawes, C.; Oparka, K.J. Putting the squeeze on plasmodesmata: A role for reticulons in primary plasmodesmata formation. *Plant Physiol.* **2015**, *168*, 1563–1572. [CrossRef] [PubMed]

38. Wu, C.H.; Lee, S.C.; Wang, C.W. Viral protein targeting to the cortical endoplasmic reticulum is required for cell-cell spreading in plants. *J. Cell Biol.* **2011**, *193*, 521–535. [CrossRef] [PubMed]

39. Tilsner, J.; Linnik, O.; Louveaux, M.; Roberts, I.M.; Chapman, S.N.; Oparka, K.J. Replication and trafficking of a plant virus are coupled at the entrances of plasmodesmata. *J. Cell Biol.* **2013**, *201*, 981–995. [CrossRef] [PubMed]

40. Lee, H.Y.; Bowen, C.H.; Popescu, G.V.; Kang, H.G.; Kato, N.; Ma, S.; Dinesh-Kumar, S.; Snyder, M.; Popescu, S.C. Arabidopsis RTNLB1 and RTNLB2 Reticulon-like proteins regulate intracellular trafficking and activity of the FLS2 immune receptor. *Plant Cell* **2011**, *23*, 3374–3391. [CrossRef] [PubMed]

41. Mysore, K.S.; Bassuner, B.; Deng, X.B.; Darbinian, N.S.; Motchoulski, A.; Ream, W.; Gelvin, S.B. Role of the Agrobacterium tumefaciens VirD2 protein in T-DNA transfer and integration. *Mol. Plant Microbe Interact.* **1998**, *11*, 668–683. [CrossRef] [PubMed]

42. Mysore, K.S.; Kumar, C.T.; Gelvin, S.B. Arabidopsis ecotypes and mutants that are recalcitrant to Agrobacterium root transformation are susceptible to germ-line transformation. *Plant J.* **2000**, *21*, 9–16. [CrossRef] [PubMed]

43. Wu, H.Y.; Liu, K.H.; Wang, Y.C.; Wu, J.F.; Chiu, W.L.; Chen, C.Y.; Wu, S.H.; Sheen, J.; Lai, E.M. AGROBEST: An efficient Agrobacterium-mediated transient expression method for versatile gene function analyses in Arabidopsis seedlings. *Plant Methods* **2014**, *10*, 19. [CrossRef] [PubMed]

44. Winter, D.; Vinegar, B.; Nahal, H.; Ammar, R.; Wilson, G.V.; Provart, N.J. An 'Electronic fluorescent pictograph' browser for exploring and analyzing large-scale biological data sets. *PLoS ONE* **2007**, *2*, e718. [CrossRef] [PubMed]

45. Chen, J.; Stefano, G.; Brandizzi, F.; Zheng, H. Arabidopsis RHD3 mediates the generation of the tubular ER network and is required for Golgi distribution and motility in plant cells. *J. Cell Sci.* **2011**, *124*, 2241–2252. [CrossRef] [PubMed]

46. Lee, H.; Sparkes, I.; Gattolin, S.; Dzimitrowicz, N.; Roberts, L.M.; Hawes, C.; Frigerio, L. An Arabidopsis reticulon and the atlastin homologue RHD3-like2 act together in shaping the tubular endoplasmic reticulum. *New Phytol.* **2013**, *197*, 481–489. [CrossRef] [PubMed]

47. Kriechbaumer, V.; Botchway, S.W.; Slade, S.E.; Knox, K.; Frigerio, L.; Oparka, K.; Hawes, C. Reticulomics: Protein-protein interaction studies with two plasmodesmata-localized reticulon family proteins identify binding partners enriched at plasmodesmata, endoplasmic reticulum, and the plasma membrane. *Plant Physiol.* **2015**, *169*, 1933–1945. [CrossRef] [PubMed]

48. Lewis, J.D.; Lazarowitz, S.G. Arabidopsis synaptotagmin SYTA regulates endocytosis and virus movement protein cell-to-cell transport. *Proc. Natl. Acad. Sci. USA* **2010**, *107*, 2491–2496. [CrossRef] [PubMed]

49. Feraru, E.; Feraru, M.I.; Asaoka, R.; Paciorek, T.; De Rycke, R.; Tanaka, H.; Nakano, A.; Friml, J. BEX5/RabA1b regulates trans-Golgi network-to-plasma membrane protein trafficking in Arabidopsis. *Plant Cell* **2012**, *24*, 3074–3086. [CrossRef] [PubMed]

50. Asaoka, R.; Uemura, T.; Ito, J.; Fujimoto, M.; Ito, E.; Ueda, T.; Nakano, A. Arabidopsis RABA1 GTPases are involved in transport between the trans-Golgi network and the plasma membrane, and are required for salinity stress tolerance. *Plant J.* **2013**, *73*, 240–249. [CrossRef] [PubMed]

51. Choi, S.W.; Tamaki, T.; Ebine, K.; Uemura, T.; Ueda, T.; Nakano, A. RABA members act in distinct steps of subcellular trafficking of the FLAGELLIN SENSING2 receptor. *Plant Cell* **2013**, *25*, 1174–1187. [CrossRef] [PubMed]

52. Chinchilla, D.; Bauer, Z.; Regenass, M.; Boller, T.; Felix, G. The Arabidopsis receptor kinase FLS2 binds flg22 and determines the specificity of flagellin perception. *Plant Cell* **2006**, *18*, 465–476. [CrossRef] [PubMed]

53. Zipfel, C.; Kunze, G.; Chinchilla, D.; Caniard, A.; Jones, J.D.; Boller, T.; Felix, G. Perception of the bacterial PAMP EF-Tu by the receptor EFR restricts Agrobacterium-mediated transformation. *Cell* **2006**, *125*, 749–760. [CrossRef] [PubMed]

54. Jones, J.D.; Dangl, J.L. The plant immune system. *Nature* **2006**, *444*, 323–329. [CrossRef] [PubMed]

55. Couto, D.; Zipfel, C. Regulation of pattern recognition receptor signaling in plants. *Nat. Rev. Immunol.* **2016**, *16*, 537–552. [CrossRef] [PubMed]

56. Beck, M.; Heard, W.; Mbengue, M.; Robatzek, S. The INs and OUTs of pattern recognition receptors at the cell surface. *Plant Biol.* **2012**, *15*, 367–374. [CrossRef] [PubMed]

57. Ditt, R.F.; Nester, E.; Comai, L. The plant cell defense and Agrobacterium tumefaciens. *FEMS Microbiol. Lett.* **2005**, *247*, 207–213. [CrossRef] [PubMed]

58. Ditt, R.F.; Nester, E.W.; Comai, L. Plant gene expression response to Agrobacterium tumefaciens. *Proc. Natl. Acad. Sci. USA* **2001**, *98*, 10954–10959. [CrossRef] [PubMed]

59. Veena; Doerge, R.W.; Gelvin, S.B. Transfer of T-DNA and Vir proteins to plant cells by Agrobacterium tumefaciens induces expression of host genes involved in mediating transformation and suppresses host defense gene expression. *Plant J.* **2003**, *35*, 219–236. [PubMed]

60. Pitzschke, A.; Djamei, A.; Teige, M.; Hirt, H. VIP1 response elements mediate mitogen-activated protein kinase 3-induced stress gene expression. *Proc. Natl. Acad. Sci. USA* **2009**, *106*, 18414–18419. [CrossRef] [PubMed]

61. Pitzschke, A.; Hirt, H. New insights into an old story: Agrobacterium-induced tumour formation in plants by plant transformation. *EMBO J.* **2010**, *29*, 1021–1032. [CrossRef] [PubMed]

62. Citovsky, V.; Kapelnikov, A.; Oliel, S.; Zakai, N.; Rojas, M.R.; Gilbertson, R.L.; Tzfira, T.; Loyter, A. Protein interactions involved in nuclear import of the Agrobacterium VirE2 protein in vivo and in vitro. *J. Biol. Chem.* **2004**, *279*, 29528–29533. [CrossRef] [PubMed]

63. Tzfira, T.; Vaidya, M.; Citovsky, V. VIP1, an Arabidopsis protein that interacts with Agrobacterium VirE2, is involved in VirE2 nuclear import and Agrobacterium infectivity. *EMBO J.* **2001**, *20*, 3596–3607. [CrossRef] [PubMed]

64. Ben Khaled, S.; Postma, J.; Robatzek, S. A moving view: Subcellular trafficking processes in pattern recognition receptor-triggered plant immunity. *Annu. Rev. Phytopathol.* **2015**, *53*, 379–402. [CrossRef] [PubMed]

65. Gu, Y.; Zavaliev, R.; Dong, X. Membrane trafficking in plant immunity. *Mol. Plant.* **2017**, *10*, 1026–1034. [CrossRef] [PubMed]

66. Gorlich, D.; Kostka, S.; Kraft, R.; Dingwall, C.; Laskey, R.A.; Hartmann, E.; Prehn, S. Two different subunits of importin cooperate to recognize nuclear localization signals and bind them to the nuclear envelope. *Curr. Biol.* **1995**, *5*, 383–392. [CrossRef]

67. Bednenko, J.; Cingolani, G.; Gerace, L. Importin beta contains a COOH-terminal nucleoporin binding region important for nuclear transport. *J. Cell Biol.* **2003**, *162*, 391–401. [CrossRef] [PubMed]

68. Bhattacharjee, S.; Lee, L.Y.; Oltmanns, H.; Cao, H.; Cuperus, J.; Gelvin, S.B. AtImpa-4, an Arabidopsis importin α isoform, is preferentially involved in Agrobacterium-mediated plant transformation. *Plant Cell* **2008**, *20*, 2661–2680. [CrossRef] [PubMed]

69. Mysore, K.S.; Nam, J.; Gelvin, S.B. An Arabidopsis histone H2A mutant is deficient in Agrobacterium T-DNA integration. *Proc. Natl. Acad. Sci. USA* **2000**, *97*, 948–953. [CrossRef] [PubMed]

70. Yi, H.; Mysore, K.S.; Gelvin, S.B. Expression of the Arabidopsis histone H2A-1 gene correlates with susceptibility to Agrobacterium transformation. *Plant J.* **2002**, *32*, 285–298. [CrossRef] [PubMed]

71. Tenea, G.N.; Spantzel, J.; Lee, L.Y.; Zhu, Y.; Lin, K.; Johnson, S.J.; Gelvin, S.B. Overexpression of several Arabidopsis histone genes increases Agrobacterium-mediated transformation and transgene expression in plants. *Plant Cell* **2009**, *21*, 3350–3367. [CrossRef] [PubMed]

72. Yi, H.; Sardesai, N.; Fujinuma, T.; Chan, C.W.; Gelvin, S.B. Constitutive expression exposes functional redundancy between the Arabidopsis histone H2A gene HTA1 and other H2A gene family members. *Plant Cell* **2006**, *18*, 1575–1589. [CrossRef] [PubMed]

73. Sauer, N.; Ludwig, A.; Knoblauch, A.; Rothe, P.; Gahrtz, M.; Klebl, F. AtSUC8 and AtSUC9 encode functional sucrose transporters, but the closely related AtSUC6 and AtSUC7 genes encode aberrant proteins in different Arabidopsis ecotypes. *Plant J.* **2004**, *40*, 120–130. [CrossRef] [PubMed]

74. Lemoine, R. Sucrose transporters in plants: Update on function and structure. *Biochimica et Biophysica Acta* **2000**, *1465*, 246–262. [CrossRef]

75. Williams, L.E.; Lemoine, R.; Sauer, N. Sugar transporters in higher plants—A diversity of roles and complex regulation. *Trends Plant Sci.* **2000**, *5*, 283–290. [CrossRef]

76. Feuerstein, A.; Niedermeier, M.; Bauer, K.; Engelmann, S.; Hoth, S.; Stadler, R.; Sauer, N. Expression of the AtSUC1 gene in the female gametophyte, and ecotype-specific expression differences in male reproductive organs. *Plant Biol.* **2010**, *12*, 105–114. [CrossRef] [PubMed]

77. Sambrook, J.; Russell, D.W. *Molecular Cloning: A Laboratory Manual*, 3rd ed.; Cold Spring Harbor Laboratory Press: Cold Spring Harbor, NY, USA, 2001; ISBN 978–1-936113-42-2.

78. Ausubel, F.M.; Brent, R.; Kingston, R.E.; Moore, D.D.; Seidman, J.G.; Smith, J.A.; Struhl, K. *Current Protocols in Molecular Biology*; John Wiley & Sons Inc.: Hoboken, NJ, USA, 2003; ISBN 978–0-471-50338-5.

79. Alonso, J.M.; Stepanova, A.N.; Leisse, T.J.; Kim, C.J.; Chen, H.; Shinn, P.; Stevenson, D.K.; Zimmerman, J.; Barajas, P.; Cheuk, R.; et al. Genome-wide insertional mutagenesis of Arabidopsis thaliana. *Science* **2003**, *301*, 653–657. [CrossRef] [PubMed]

80. Dellaporta, S.L.; Wood, J.; Hicks, J.B. A plant DNA minipreperation: Version 2. *Plant Mol. Biol. Rep.* **1983**, *1*, 19–22. [CrossRef]

81. Koncz, C.; Schell, J. The promoter of TL-DNA gene 5 controls the tissue-specific expression of chimaeric genes carried by a novel type of Agrobacterium binary vector. *Mol. Gen. Genet.* **1986**, *204*, 383–396. [CrossRef]

82. Clough, S.J.; Bent, A.F. Floral dip: A simplified method for Agrobacterium-mediated transformation of Arabidopsis thaliana. *Plant J.* **1998**, *16*, 735–743. [CrossRef] [PubMed]

83. Zhu, Y.; Nam, J.; Humara, J.M.; Mysore, K.S.; Lee, L.Y.; Cao, H.; Valentine, L.; Li, J.; Kaiser, A.D.; Kopecky, A.L.; et al. Identification of Arabidopsis rat mutants. *Plant Physiol.* **2003**, *132*, 494–505. [CrossRef] [PubMed]

84. Hwang, H.H.; Wang, M.H.; Lee, Y.L.; Tsai, Y.L.; Li, Y.H.; Yang, F.J.; Liao, Y.C.; Lin, S.K.; Lai, E.M. Agrobacterium-produced and exogenous cytokinin-modulated Agrobacterium-mediated plant transformation. *Mol. Plant Pathol.* **2010**, *11*, 677–690. [CrossRef] [PubMed]

85. Deng, W.L.; Preston, G.; Collmer, A.; Chang, C.J.; Huang, H.C. Characterization of the hrpC and hrpRS operons of Pseudomonas syringae pathovars syringae, tomato, and glycinea and analysis of the ability of hrpF, hrpG, hrcC, hrpT, and hrpV mutants to elicit the hypersensitive response and disease in plants. *J. Bacteriol.* **1998**, *180*, 4523–4531. [PubMed]

86. Katagiri, F.; Thilmony, R.; He, S.Y. The Arabidopsis thaliana-Pseudomonas syringae interaction. *Arabidopsis Book* **2002**, *1*, e0039. [CrossRef] [PubMed]

International Journal of
*Molecular Sciences*

MDPI

*Article*

# Studies of Microbiota Dynamics Reveals Association of "*Candidatus* Liberibacter Asiaticus" Infection with Citrus (*Citrus sinensis*) Decline in South of Iran

Alessandro Passera [1], Hamidreza Alizadeh [2], Mehdi Azadvar [3], Fabio Quaglino [1],
Asma Alizadeh [2], Paola Casati [1] and Piero A. Bianco [1,*]

[1]    Department of Agricultural and Environmental Sciences, University of Milan, 20133 Milan, Italy;
       alessandro.passera@unimi.it (A.P.); fabio.quaglino@unimi.it (F.Q.); paola.casati@unimi.it (P.C.)
[2]    Department of Plant Protection, Faculty of Agriculture, University of Jiroft, Jiroft 7867161167, Iran;
       hamid6948@gmail.com (H.A.); Alizadeh.b5273@gmail.com (A.A.)
[3]    Plant Protection Department, South Kerman Agricultural and Natural Resources Research and Education
       Center, Agricultural Research Education and Extension Organization, Jiroft 7867161167, Iran;
       mehdiazadvar@gmail.com
*      Correspondence: piero.bianco@unimi.it; Tel.: +39-02-50316794

Received: 7 May 2018; Accepted: 15 June 2018; Published: 20 June 2018

**Abstract:** Citrus Decline Disease was recently reported to affect several citrus species in Iran when grafted on a local rootstock variety, Bakraee. Preliminary studies found "*Candidatus* Phytoplasma aurantifoliae" and "*Candidatus* Liberibacter asiaticus" as putative etiological agents, but were not ultimately able to determine which one, or if an association of both, were causing the disease. The current study has the aim of characterizing the microbiota of citrus plants that are either asymptomatic, showing early symptoms, or showing late symptoms through amplification of the V1–V3 region of 16S rRNA gene using an Illumina sequencer in order to (i) clarify the etiology of the disease, and (ii) describe the microbiota associated to different symptom stages. Our results suggest that liberibacter may be the main pathogen causing Citrus Decline Disease, but cannot rule out the possibility of phytoplasma being involved as well. The characterization of microbiota shows that the leaves show only two kinds of communities, either symptomatic or asymptomatic, while roots show clear distinction between early and late symptoms. These results could lead to the identification of bacteria that are related to successful plant defense response and, therefore, to immunity to the Citrus Decline Disease.

**Keywords:** citrus decline disease; *Citrus sinensis*; Bakraee; "*Candidatus* Liberibacter"; "*Candidatus* Phytoplasma"; microbiota

## 1. Introduction

Since 2010, a declining condition on trees belonging to different citrus species has widely appeared in various groves of the Southern Kerman region of Iran and has subsequently killed around 10% of cultivated citrus trees. This condition, called Citrus Decline Disease (CDD), causes symptoms that have been classified as (i) early CDD symptoms, which include leaves developing a pale green color, limited production of fresh sprouts, and in general a retardation of growth (Figure S1a), and (ii) late CDD symptoms which manifest as an evident tree decline along with reduction and decay of the root system (Figure S1b) [1]. Plants affected by CDD initially develop the milder, early symptoms and later develop the more severe late symptoms. Affected plants die approximately 5 years after the development of the first symptoms.

These symptoms, similar to those associated with obligate bacterial parasites, brought to the identification of putative etiological agents of the disease in both phytoplasma and liberibacter, as these microorganisms were detected by PCR in infected plants and not in healthy plants [1]. In particular, the putative pathogens were identified as "*Candidatus* Liberibacter asiaticus" and "*Candidatus* Phytoplasma aurantifolia".

The liberibacter-associated citrus disease known as Huanglongbing (HLB) is also present in Iran. The disease was first detected from various locations in Sistan-Baluchistan and Hormozgan provinces of southern Iran on Valencia sweet orange trees (*Citrus Sinensis* (L.) Osbek) in 2009 [2].

Sour orange (*C. aurantium* L.), lime (*C. x aurantifolia* Swigle) and Bakraee (*C. reticulata* Blanco x *C. limettioides* Tan.) are the three most prevalent citrus rootstocks in Iran [3]. Bakraee in particular is being used extensively as a rootstock in the southern part of Iran for its shallow root system and tolerance to high pH [4]. Interestingly, new emerging citrus decline disease can be easily observed on several different citrus species aged six or more years old such as sweet orange, grapefruit (*C. x paradisi* Macfad), and mandarin (*C. reticulata* Blanco), but only when grafted on Bakraee.

Diseases associated to phloem-limited pathogens, such as phytoplasma and liberibacter, are notoriously difficult to manage since there is no direct treatment that can be used against them [5], unlike other pathogens such as fungi that can be controlled through fungicides. For this reason, investigation of the precise etiology of these diseases, as well as of environmental conditions related to their epidemiology, can offer solutions for the management of the disease [6].

Several studies were carried out on HLB, a disease that shares the candidate etiological agent and host with CDD, and it is known that the disease causes a great restructuring of the microbiota of the host and it has been suggested that the dynamics of the bacterial microbiome can affect the concentration of "*Ca*. Liberibacter asiaticus", which are correlated to symptom severity [7,8].

On the basis of these previous studies, it is plausible to hypothesize that a similar phenomenon might occur in the CDD-affected plants, and that the microbiota dynamics might therefore explain how the disease develops.

In the present study, the bacterial microbiota of asymptomatic plants, early symptomatic plants, and late symptomatic plants, determined both from leaf and root, was described and compared with the aims of (i) clearing the etiology of the disease and (ii) identifying the shifts in the microbial community linked to the development of the disease, in order to identify possible taxonomical units associated to a diseased or healthy state.

## 2. Results

### 2.1. Detection of Phytoplasma and Liberibacter in Plant Material

None of the samples from asymptomatic plants showed amplification when phytoplasma or liberibacter PCR assays were performed; on the contrary, samples from the early symptomatic or late symptomatic plants showed amplification for at least one of the putative pathogens. In particular, phytoplasma were detected in one out of two late symptomatic plants (LSP) leaves and one out of three early symptomatic plants (ESP) leaves; phytoplasma were also detected in two out of two LSP roots and one out of three ESP roots. Liberibacter were detected in all five symptomatic roots samples, regardless of symptom stage (Table 1). These amplified fragments were sequenced and the comparison with the database showed that the phytoplasma sequences belonged to the "*Ca*. P. aurantifolia" species while the liberibacter sequences belonged to the "*Ca*. L. asiaticus" species.

**Table 1.** Table reporting the results of detection of phytoplasma and liberibacter in the analyzed samples, both with PCR and with next generation sequencing (NGS).

| Sample | | PCR | | NGS [1] | |
|---|---|---|---|---|---|
| | | Phytoplasma | Liberibacter | Phytoplasma | Liberibacter |
| 1 (ASP) | Leaf | − | − | + | + |
| | Root | − | − | + | + |
| 2 (ASP) | Leaf | − | − | + | + |
| | Root | − | − | + | + |
| 3 (ESP) | Leaf | − | − | − | − |
| | Root | + | + | +++ | +++ |
| 4 (ESP) | Leaf | + | − | + | − |
| | Root | − | + | + | ++ |
| 5 (ESP) | Leaf | − | − | + | + |
| | Root | − | + | + | ++ |
| 6 (LSP) | Leaf | − | − | + | + |
| | Root | + | + | + | +++ |
| 7 (LSP) | Leaf | + | − | + | − |
| | Root | + | + | + | +++ |

[1] NGS data: "−" = 0 operational taxonomic units (OTUs), "+" = 1–100 OTUs, "++" = 101–500 OTUs, "+++" = 501+ OTUs.

## 2.2. Microbial Community in Iranian Citrus Plants

The MiSeq sequencing produced a total of 160,010 reads for asymptomatic plants (ASP) (84,588 leaf, 75,411 root), 283,715 reads of ESP (102,911 leaf, 180,804 root), and 144,744 reads for LSP (72,212 leaf, 72,532 root), allowing the description of 2762 operational taxonomic units (OTUs). After filtering, the remaining reads were 18,153 for ASP (7917 leaf, 10,236 root), 21,859 reads for ESP (7736 leaf, 13,853 root), and 76,385 reads for LSP (1588 leaf, 74,797 root), for a total of 2699 OTUs. For each plant sanitary status and organ, rarefaction curves were described (Figure S2).

Of these OTUs, most were exclusive to the root compartment (1486), followed by those shared among leaf and root (687), and the fewest were identified in leaf only (526) (Figure 1a). Still, these results indicate that relatively few bacterial OTUs are shared among the leaf and root compartments, identifying very diverse microbiota in the different organs. Considering the single sanitary statuses (ASP, ESP, or LSP), the trend is slightly different: While the number of root unique OTUs remains the highest, the leaf unique OTUs are higher than those shared between the two organs (Figure 1b–d). When considering the single plant organ (either leaves or roots), their differences appear clear. In the leaves, the highest amount of different OTUs is registered in ESP, followed by ASP, and lastly LSP; while there is a high amount of shared OTUs between ESP and ASP, ESP and LSP, and the three sanitary statuses, only few OTUs are common among ASP and LSP only (Figure 1e). In the roots the highest amount of different OTUs is found in LSP, followed by ESP, and by ASP; also in roots there is a high number of shared OTUs between the different sanitary statuses: ASP and LSP share the least amount while, the highest is the "core" among all three sanitary statuses (Figure 1f). Only 94 OTUs are shared among all sanitary statuses and plant compartments and could therefore be thought of as the 'core' microbiota of these plants. This core is constituted mostly by Firmicutes (45 OTUs), followed by Proteobacteria (19 OTUs), Bacterioidetes (11 OTUs), and Actinobacteria (9 OTUs) with the remaining 10 OTUs belonging to other phyla. It is interesting to notice that among this core microbiota are both OTUs belonging to "*Ca.* Liberibacter spp." and "*Ca.* Phytoplasma spp".

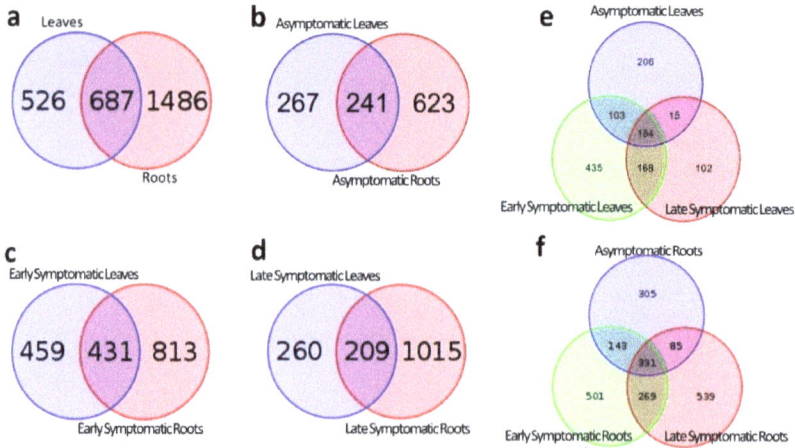

**Figure 1.** Graphs describing the different distribution of OTUs among different organs and/or sanitary statuses of examined plants: (**a**) Between all leaves and roots, regardless of sanitary status; (**b**) between leaves and roots of asymptomatic plants; (**c**) between leaves and roots of early symptomatic plants; (**d**) between leaves and roots of late symptomatic plants; (**e**) between the asymptomatic, early symptomatic, and late symptomatic plants at the leaves; (**f**) between the asymptomatic, early symptomatic, and late symptomatic plants at the roots.

### 2.3. Bacterial Diversity among Different Organs and Sanitary Statuses

Bacterial diversity among the different organs and sanitary statuses was evaluated preliminary by comparing two indices, the Chao-1 and phylogenetic distance (PD) index (Table 2). From these indices it was possible to see that, in general, leaves have a less diverse and complex bacterial community compared to roots and that asymptomatic plants have a less complex bacterial community than symptomatic plants.

**Table 2.** Table reporting the values of Chao-1 and PD indexes, and their standard deviation, for the different categories of simples.

| Sample Category | | Chao-1 | PD |
|---|---|---|---|
| ASP | Leaf | $496 \pm 72$ | $31.6 \pm 3.8$ |
| | Root | $765 \pm 40$ | $42.5 \pm 4.3$ |
| ESP | Leaf | $741 \pm 113$ | $42.5 \pm 6.3$ |
| | Root | $1058 \pm 276$ | $53.2 \pm 11.3$ |
| LSP | Leaf | $572 \pm 60$ | $33.3 \pm 3.1$ |
| | Root | $1187 \pm 351$ | $52.6 \pm 11.4$ |

The first level of diversity analysis, performed at phylum level, was carried out to compare the samples grouped by sanitary status and organ (Figure 2).

While some similarities can be seen among all samples, such as the prevalence of three *phyla* among all others (Actinobacteria, Firmicutes, Proteobacteria), there are some differences that are highlighted already at this taxonomic level. For example, Fibrobacteres and Spirochaetes are relevant groups only in ESP and LSP leaves, and Planctomycetes are relevant only in ESP and LSP roots. In general, it can be said that the leaves are dominated by Firmicutes, while the roots are dominated by Proteobacteria, already depicting very different scenarios that are confirmed by a Principal Component Analysis (Figure 3). From this analysis it is evident that the leaves and roots contain a very different microbiota, which becomes even more different when the plants develop symptoms of CDD. For this reason, the subsequent microbiota analyses carried out in this study were done separately on the leaf and root compartment.

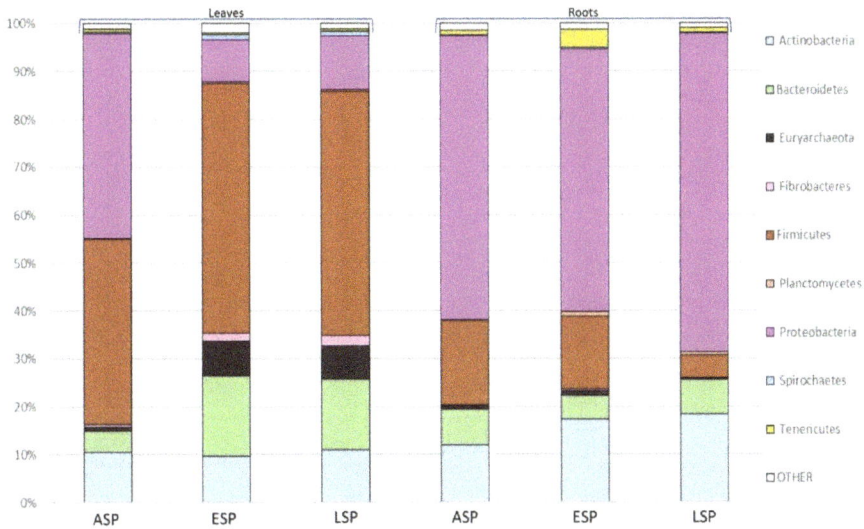

**Figure 2.** Graph describing the taxonomic distribution of OTUs among different organs (leaves or roots, indicated by brackets at the top) and/or sanitary statuses (asymptomatic (ASP), early symptomatic (ESP), or late symptomatic (LSP)) of examined plants at phylum level.

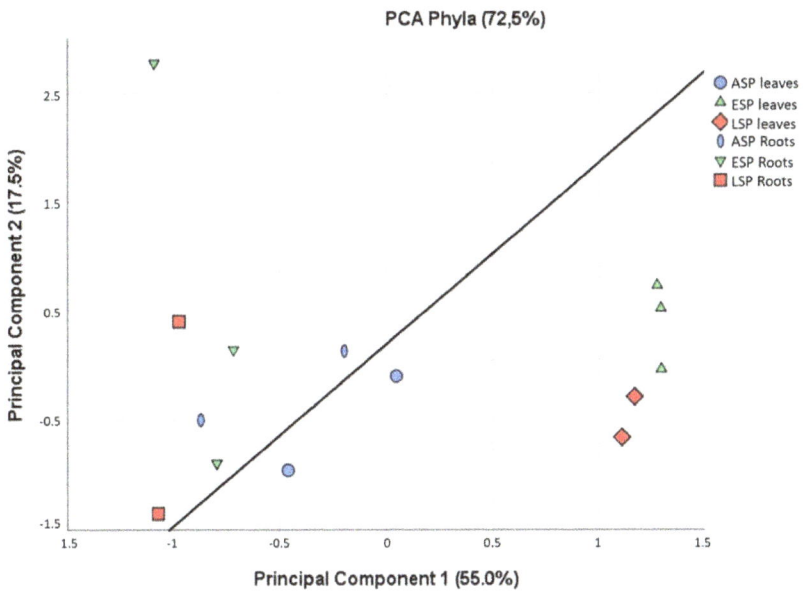

**Figure 3.** Principal Component Analysis calculated among all the samples, based on the relative abundance of OTUs in each phyla. The line was added to highlight the separation between leaves and roots.

## 2.4. Metagenome-Based Diagnosis on Plant Material

The reads obtained from the MiSeq analysis were searched specifically for OTUs assigned to the genera "*Ca.* Phytoplasma" and "*Ca.* Liberibacter", in order to compare the results of this NGS technique to those obtained by regular PCR. OTUs belonging to both pathogens were found in all analyzed samples (Table 1), both symptomatic and asymptomatic, although the asymptomatic plants showed a lower amount of reads belonging to these bacteria (0 to 5 reads for liberibacter, 0 to 72 for phytoplasma), below the 1% threshold used to consider them relevant. On the other hand, both bacteria were present more consistently in symptomatic plants: in particular, while both ESP and LSP showed a high amount of reads belonging to liberibacter (0 to 1286 and 0 to 959 respectively), it is interesting to note that the LSP plants show a presence of phytoplasma comparable to that of ASP (0 to 54) while ESP have a much higher number of reads belonging to phytoplasma (0 to 707) (Figure 4a). It is also of note that both pathogens are found almost exclusively in the roots.

Other than the abundance of these pathogens, this analysis allowed to detect the presence of different OTUs belonging to liberibacter and phytoplasma, 2 and 4 respectively. Each of these genera has one dominant OTU that constitutes the entirety of the detected OTUs in asymptomatic plants and almost the entirety of detected OTUs in symptomatic plants (Figure 4b,c). It is interesting to note that the roots of symptomatic plants show the presence of an OTU for liberibacter and one for phytoplasma that are not present in healthy roots or in leaves. These OTUs represent a small percentage of those belonging to each pathogenic genus (approximately 2.75% for liberibacter and between 0.9 and 2.4 for phytoplasma).

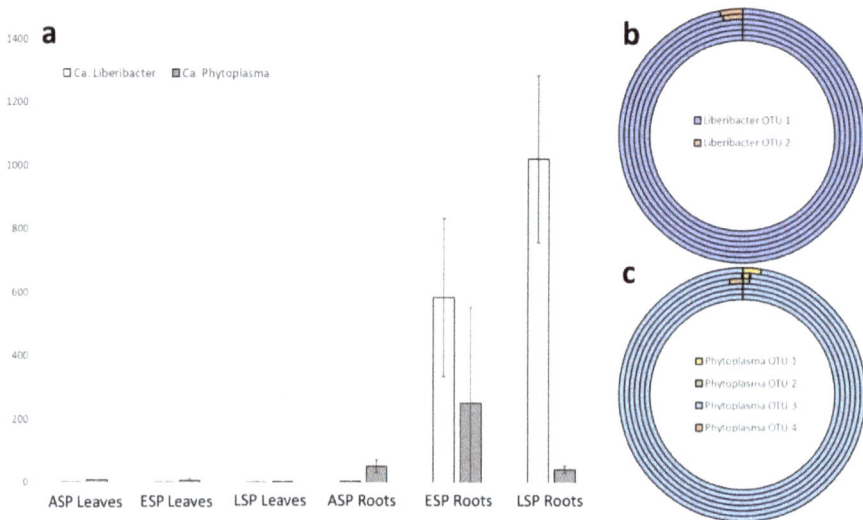

**Figure 4.** Graphs reporting the presence and abundance of pathogenic agents "*Ca.* Phytoplasma" and "*Ca.* Liberibacter" in the analyzed material according to metagenomic analyses: (**a**) Bar graph showing the average absolute abundance (number of OTUs) for each category of sample, with error bars indicating standard deviation among samples of the same category. Light grey bars indicate "*Ca.* Liberibacter" while dark grey bars indicate "*Ca.* Phytoplasma"; (**b**) circular graph indicating the relative abundance of different OTUs associated to "*Ca.* Liberibacter" in the different samples. Each circular bar represents a different category of sample as follow, from innermost circle to outermost: ASP leaf, ESP leaf, LSP leaf, ASP root, ESP root, LSP root; (**c**) circular graph indicating the relative abundance of different OTUs associated to "*Ca.* Phytoplasma" in the different samples. Each circular bar represents a different category of sample as follow, from innermost circle to outermost: ASP leaf, ESP leaf, LSP leaf, ASP root, ESP root, LSP root.

*2.5. Bacterial Diversity in Leaves: Shift of Community in Relation to Sanitary Status*

The composition of the bacterial community associated with the leaves in the three sanitary statuses (ASP, ESP, and LSP) was compared at the family level (Table 3). Significant differences in relative abundance were identified in only 10 out of the 34 considered families. In particular, the Micrococcaeae, Gemellaceae, and Streptococcaeae were significantly more abundant in ASP than in ESP and LSP; while Coriobacteraceae, an unidentified family belonging to the Bacteriodales order, an unidentified family belonging to the Clostridiales order, Lachnospiraceae, Ruminococcaceae, Erysipelotrichaceae, and Desulfovibrionaceae families were significantly more abundant in ESP and LSP than in ASP.

The difference in the number of OTUs for these families, calculated as fold change, showed that the most significant shift in the communities from the asymptomatic to the symptomatic state can be identified in the change in composition in the Firmicutes phylum (Figure 5a): while the percentage of Firmicutes on the total bacteria is almost unchanged between ASP, ESP, and LSP, in ASP plants the Firmicutes are mostly Bacillales and Gemellales, while in ESP and LSP plants most of the Firmicutes belong to the Clostridiales order.

**Table 3.** Taxonomic relations and relative abundance, with standard deviation, of OTUs at family level in leaves. p value is indicated for One-Way ANOVA, followed by Tukey's exact post-hoc test, carried out to determine significant differences between abundances. Significant differences are highlighted by bold text, and * indicates which result is different from the others.

| Phylum | Class | Order | Family | ASP | ESP | LSP | p |
|---|---|---|---|---|---|---|---|
| Actinobacteria | Actinobacteria | Actinomycetales | Corynebacteriaceae | 1.74 ± 0.88 | 0.43 ± 0.06 | 0.35 ± 0.20 | 0.34 |
| | | | **Micrococcaceae** | **1.84 ± 0.31** * | **0.28 ± 0.12** | **0.25 ± 0.19** | **0.02** |
| | | | Propionibacteriaceae | 3.55 ± 1.41 | 3.04 ± 0.87 | 2.69 ± 1.11 | 0.85 |
| | Coriobacteriia | Coriobacteriales | **Coriobacteriaceae** | **1.54 ± 0.71** * | **5.91 ± 0.78** | **5.85 ± 0.59** | **0.05** |
| Bacteroidetes | Bacteroidia | Bacteriodales | **[Paraprevotellaceae]** | **1.77 ± 0.76** * | **10.29 ± 1.10** | **11.39 ± 0.48** | **0.00** |
| | | | Prevotellaceae | 0.48 ± 0.22 | 0.85 ± 0.15 | 0.87 ± 0.24 | 0.47 |
| | | | **Prevotellaceae** | **0.55 ± 0.06** * | **3.09 ± 0.35** | **2.99 ± 0.05** | **0.02** |
| | Flavobacteria | Flavobacteriales | [Weeksellaceae] | 1.12 ± 0.68 | 0.14 ± 0.10 | 0.15 ± 0.12 | 0.37 |
| Euryarchaeota | Methanobacteria | Methanobacteriales | Methanobacteriaceae | 0.64 ± 0.08 | 4.52 ± 0.86 | 4.76 ± 0.84 | 0.05 |
| | Thermoplasmata | E2 | [Methanomassiliicoccaceae] | 0.15 ± 0.00 | 2.56 ± 0.58 | 2.52 ± 0.46 | 0.07 |
| Fibrobacteres | Fibrobacteria | Fibrobacterales | Fibrobacteraceae | 0.54 ± 0.26 | 1.87 ± 0.43 | 1.64 ± 0.53 | 0.12 |
| Firmicutes | Bacilli | Bacillales | Bacillaceae | 1.27 ± 0.68 | 0.16 ± 0.12 | 0.14 ± 0.10 | 0.31 |
| | | | Staphylococcaceae | 1.85 ± 1.01 | 0.61 ± 0.31 | 0.51 ± 0.23 | 0.49 |
| | | Gemellales | **Gemellaceae** | **1.89 ± 0.27** * | **0.04 ± 0.05** | **0.07 ± 0.09** | **0.01** |
| | | Lactobacillales | **Streptococcaceae** | **22.45 ± 6.07** * | **0.63 ± 0.33** | **0.53 ± 0.19** | **0.04** |
| | Clostridia | Clostridiales | **[Mogibacteriaceae]** | **1.77 ± 0.46** * | **9.91 ± 0.51** | **9.85 ± 0.78** | **0.00** |
| | | | Christensenellaceae | 0.92 ± 0.51 | 2.45 ± 0.26 | 2.70 ± 0.19 | 0.07 |
| | | | Clostridiaceae | 0.60 ± 0.32 | 1.13 ± 0.35 | 1.47 ± 0.41 | 0.20 |
| | | | Clostridiaceae | 0.03 ± 0.02 | 1.87 ± 0.62 | 1.59 ± 0.50 | 0.12 |
| | | | **Lachnospiraceae** | **2.33 ± 0.69** * | **15.87 ± 1.99** | **16.03 ± 1.69** | **0.01** |
| | | | **Ruminococcaceae** | **2.94 ± 1.01** * | **15.36 ± 0.81** | **15.92 ± 1.22** | **0.00** |
| | | | Veillonellaceae | 1.01 ± 0.25 | 1.39 ± 0.40 | 1.12 ± 0.36 | 0.45 |
| | Erysipelotrichi | Erysipelotrichales | **Erysipelotrichaceae** | **0.41 ± 0.24** * | **1.48 ± 0.19** | **1.51 ± 0.36** | **0.00** |
| Proteobacteria | Alphaproteobacteria | Rhodobacterales | Rhodobacteraceae | 0.38 ± 0.16 | 1.24 ± 1.25 | 1.81 ± 2.11 | 0.66 |
| | | Sphingomonadales | Sphingomonadaceae | 1.29 ± 0.78 | 0.78 ± 0.25 | 0.95 ± 0.23 | 0.71 |
| | Betaproteobacteria | Burkholderiales | Comamonadaceae | 0.80 ± 0.39 | 0.24 ± 0.06 | 0.23 ± 0.11 | 0.37 |
| | | | Oxalobacteraceae | 0.55 ± 0.35 | 0.59 ± 0.67 | 0.07 ± 0.09 | 0.41 |
| | Deltaproteobacteria | Desulfovibrionales | **Desulfovibrionaceae** | **0.35 ± 0.08** * | **2.74 ± 0.70** | **2.13 ± 0.59** | **0.02** |
| | Gammaproteobacteria | Enterobacteriales | Enterobacteriaceae | 2.53 ± 1.28 | 0.35 ± 0.30 | 0.33 ± 0.32 | 0.29 |
| | | Pasteurellales | Pasteurellaceae | 24.38 ± 15.64 | 0.20 ± 0.12 | 0.24 ± 0.16 | 0.32 |
| | | Pseudomonadales | Moraxellaceae | 4.93 ± 2.64 | 1.49 ± 1.11 | 0.83 ± 0.24 | 0.46 |
| | | | Pseudomonadaceae | 4.99 ± 2.71 | 1.13 ± 0.59 | 0.95 ± 0.93 | 0.39 |

*Int. J. Mol. Sci.* **2018**, *19*, 1817

**Table 3.** *Cont.*

| Phylum | Class | Order | Family | ASP | ESP | LSP | *p* |
|---|---|---|---|---|---|---|---|
| Spirochaetes | Spirochaetes | Spirochaetales | Spirochaetaceae | 0.34 ± 0.18 | 1.01 ± 0.37 | 0.99 ± 0.20 | 0.50 |
| Tenericutes | Mollicutes | Acholeplasmatales | Acholeplasmataceae | 0.67 ± 0.37 | 0.35 ± 0.12 | 0.33 ± 0.22 | 0.72 |
| | | Other | | 7.39 ± 3.15 | 6.05 ± 0.30 | 6.24 ± 0.33 | |

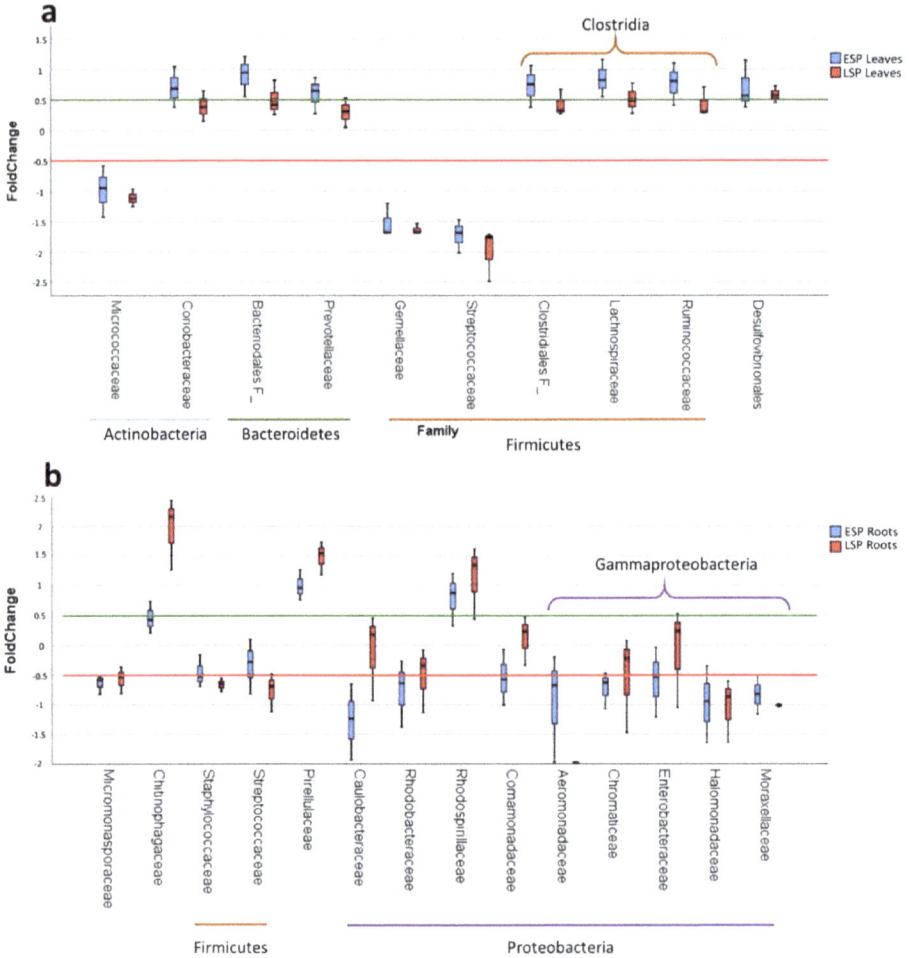

**Figure 5.** Box plots reporting the shift in absolute abundance of relevant taxonomical families in ESP and LSP samples compared to the ASP. Difference in abundance is expressed as fold change in a logarithmic scale (base 10). Median values above 0.5 or above −0.5 were considered relevant differences. (**a**) box plot showing the differences in leaves, highlighting with lines at the bottom families belonging to the same *phylum*, and with brackets on top those belonging to the same order; (**b**) box plot showing the differences in roots, highlighting with lines at the bottom families belonging to the same *phylum*, and with brackets on top those belonging to the same class.

These results are also reflected by the analysis at genus level carried out on the abundance of the genera detected in the three main *phyla* (Actinobactera, Firmicutes, Proteobacteria): the main shift detectable in the Actinobacteria regards the increase of Coriobacteriaceae at the expense of *Propriobacterium* spp. (Figure 6a–c); furthermore, the genus-level analysis confirms the shift that was detected at family-level from *Streptococcus* spp. to various genera of the Clostridiales family (Figure 7a–c); for the Proteobacteria, the shift in bacterial community saw the reduction of several genera in the Gammaproteobacteria and an increase in an unidentified genus in the Oxalobacteraceae family and in the *Desulfovibrio* genus (Figure 8a–c).

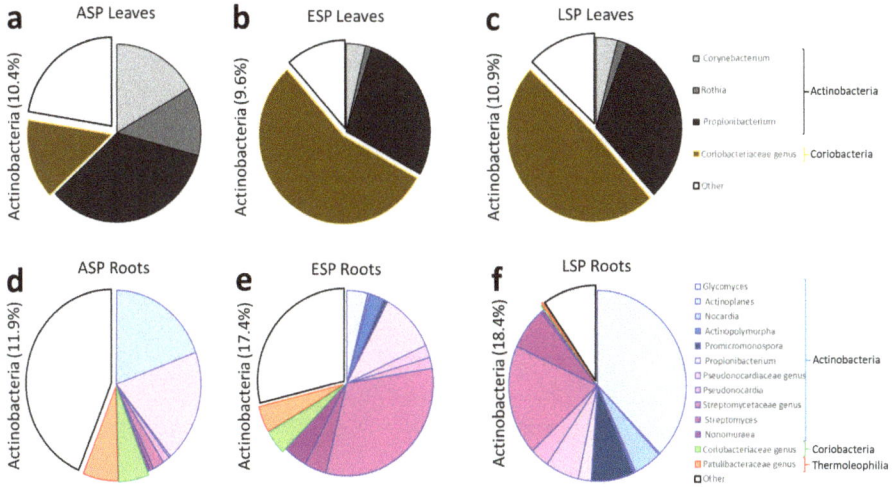

**Figure 6.** Pie graphs describing the relative abundance of genera in the Actinobacteria *phylum*. Between brackets is indicated the total share of the *phylum*, while each genus is represented by the size of the corresponding slice. (**a**) ASP leaves; (**b**) ESP leaves; (**c**) LSP leaves; (**d**) ASP roots; (**e**) ESP roots; (**f**) LSP roots.

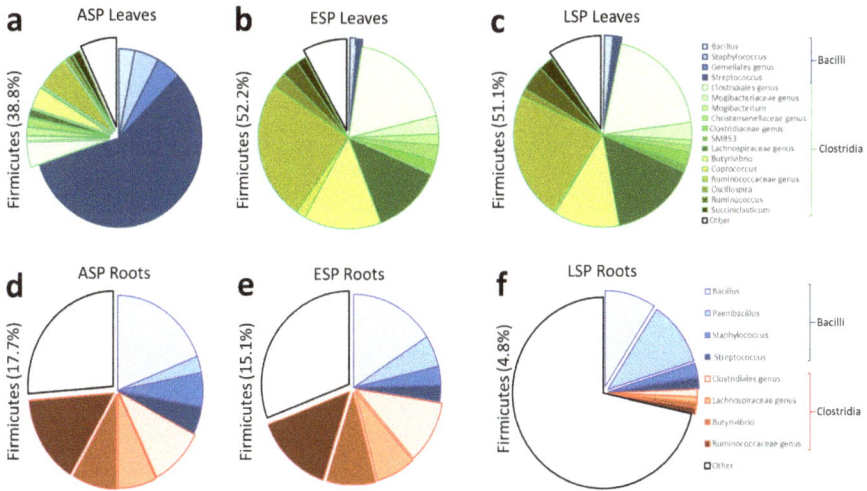

**Figure 7.** Pie graphs describing the relative abundance of genera in the Firmicutes *phylum*. Between brackets is indicated the total share of the *phylum*, while each genus is represented by the size of the corresponding slice. (**a**) ASP leaves; (**b**) ESP leaves; (**c**) LSP leaves; (**d**) ASP roots; (**e**) ESP roots; (**f**) LSP roots.

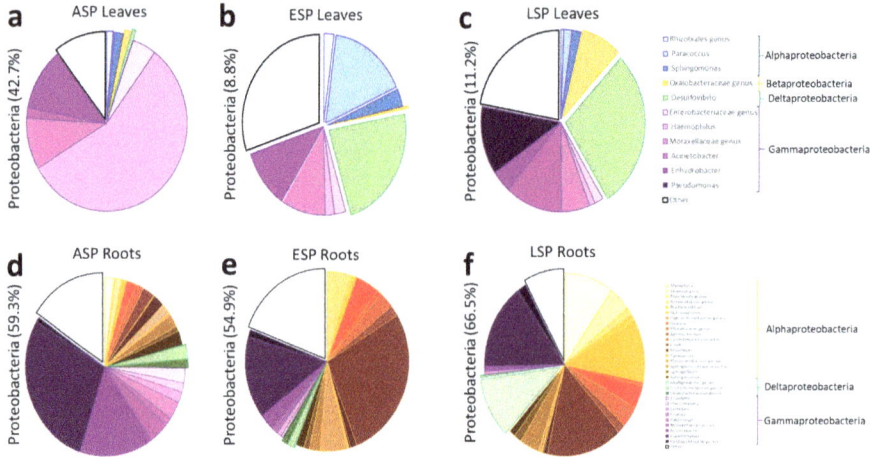

**Figure 8.** Pie graphs describing the relative abundance of genera in the Proteobacteria *phylum*. Between brackets is indicated the total share of the *phylum*, while each genus is represented by the size of the corresponding slice. (**a**) ASP leaves; (**b**) ESP leaves; (**c**) LSP leaves; (**d**) ASP roots; (**e**) ESP roots; (**f**) LSP roots.

From all these analyses, no clear and significant difference could be identified between the bacterial communities in leaves of ESP and LSP, while the leaves of ASP differ from both. This consideration is supported also by the PCA calculated at a family level in leaves only (Figure 9) which clearly shows how the ASP samples are distinct from the ESP and LSP samples, which, instead, are not clearly separated.

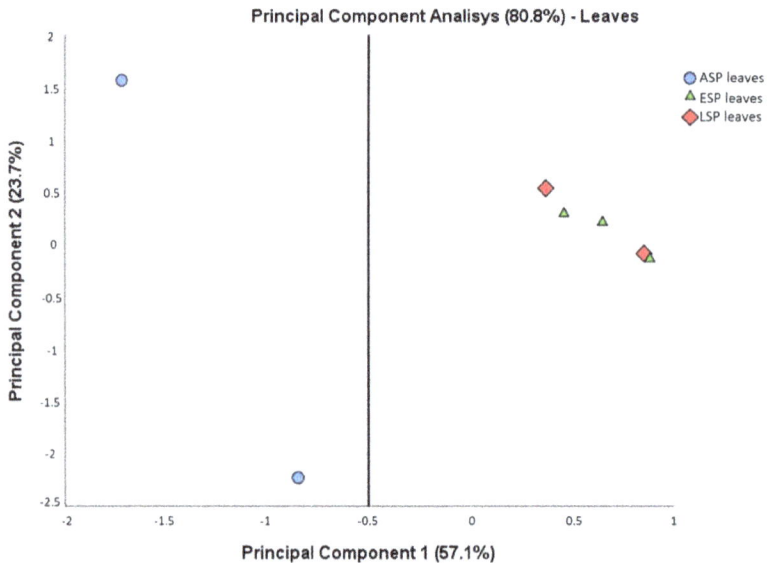

**Figure 9.** Principal Component Analysis calculated among the leaf samples, based on the relative abundance of OTUs in each family. The line was added to highlight the separation between the ASP leaf samples and the ESP/LSP leaf samples.

*2.6. Bacterial Diversity in Roots: Shift of Community in Relation to Sanitary Status*

The composition of the bacterial community associated to the roots in the three sanitary statuses (ASP, ESP, and LSP) was compared at the family level (Table 4). Significant differences in relative abundance were identified in 16 out of the 52 considered families. In particular, the Micrococcaeae, Micromonosporaceae, Staphyloccocaceae, Streptococcaeae, Caulobacteraceae, Rhodobacteraceae, Comamonadaceae, Aeromondaceae, Enterobacteriaceae, Halomonadaceae, and Moraxellaceae were significantly more abundant in ASP; Mycobateriaceae, Pirellulaceae, and Rhodospirillaceae were significantly more abundant in ESP; Chitinophagaceae was the only family over-represented in LSP samples.

The difference in the number of OTUs for these families, calculated as fold change, showed that the most significant shift in the communities from the asymptomatic to the symptomatic state can be identified in the change in the composition in the overall reduction in Firmicutes and Proteobacteria (Figure 5b): With the exception of Rhodospirillaceae, all relevant families of the Proteobacteria are much under-represented in ESP and LSP compared to ASP. In particular, Gammaproteobacteria are all greatly reduced, most of them reaching 10 times less abundance, and the Aeromonadaceae being almost 100 times less represented in LSP compared to ASP.

The analysis at genus level showed how the composition of the microbial community in the three main phyla (Actinobactera, Firmicutes, Proteobacteria) in ESP samples seemed to be intermediate between the ASP and LSP status: in the Actinobacteria community of ASP there is a large presence of lesser genera, and a relevant presence of Cortiobacteria and Thermoleophilia of unidentified genera (Figure 6d). In LSP there is way lesser abundance of these genera, while more than 30% of Actinobacteria belong to the Glycomyces genus (Figure 6f). The ESP plants still show a relevant abundance of Coriobacteria and Thermoleophilia, but also a higher abundance of Actinobacteria, with the appearance of the Glyomyces (Figure 6e). The investigation of the Firmicutes phylum at genus level reveals that the ASP and ESP samples show an almost identical composition of community (Figure 7d,e). The LSP plants instead have a much lower abundance of Firmicutes, furthermore most of them belong to less represented genera, with the the relative abundance of Bacilli remaining similar to ASP and LSP, but losing most of the genera belonging to Clostridia (Figure 7f). Conversely, the ESP and LSP samples are very similar regarding Proteobacteria, while, while the ASP is different. In the ASP plants most of the Proteobacteria belong to the Pseudomonas genus and, in general, Gammaproteobacteria constitute the majority of the community (Figure 8d), while in ESP and LSP the majority of the Proteobacteria belong to the Alphaproteobacteria, with the Rhizobia genus becoming the most represented one instead (Figure 8e,f).

From all these analyses, it can be identified how the ESP have in fact an intermediate community between ASP and LSP, a consideration supported also by the PCA calculated at a family level in roots only (Figure 10), which clearly shows how the ASP, ESP, and LSP samples form their own distinct clusters. On the Component 1, the ESP cluster is found in the middle between the ASP and LSP clusters, while on the Component 2 the ESP cluster occupies the lower end of the scale, whereas the ASP and LSP clusters have similar placements, being closer to the positive values of the component.

**Table 4.** Taxonomic relations and relative abundance, with standard deviation, of OTUs at family level in roots. p value is indicated for One-Way ANOVA, followed by Tukey's exact post-hoc test, carried out to determine significant differences between abundances. Significant differences are highlighted by bold text, and superscript letters ([a,b]) indicate which results are different, while both letters ([a,b]) indicate intermediate results.

| Phylum | Class | Order | Family | ASP | ESP | LSP | p |
|---|---|---|---|---|---|---|---|
| Actinobacteria | Actinobacteria | Actinomycetales | Glycomycetaceae | 0.01 ± 0.01 | 0.67 ± 0.88 | 5.64 ± 5.35 | 0.27 |
| | | | **Micrococcaceae** | **0.85 ± 0.19** [a] | **0.21 ± 0.14** [b] | **0.17 ± 0.10** [b] | **0.02** |
| | | | **Micromonosporaceae** | **2.46 ± 0.28** [a] | **0.73 ± 0.23** [b] | **0.47 ± 0.30** [b] | **0.00** |
| | | | **Mycobacteriaceae** | **0.33 ± 0.11** [b] | **2.04 ± 0.68** [a] | **0.56 ± 0.49** [ab] | **0.05** |
| | | | Nocardiaceae | 0.01 ± 0.01 | 0.09 ± 0.04 | 0.83 ± 0.81 | 0.27 |
| | | | Nocardioidaceae | 0.68 ± 0.04 | 0.87 ± 1.00 | 0.46 ± 0.26 | 0.81 |
| | | | Promicromonosporaceae | 0.00 ± 0.00 | 0.19 ± 0.25 | 1.20 ± 1.13 | 0.29 |
| | | | Propionibacteriaceae | 2.35 ± 0.29 | 1.81 ± 0.65 | 0.72 ± 0.40 | 0.06 |
| | | | Pseudonocardiaceae | 0.24 ± 0.01 | 0.72 ± 0.64 | 1.76 ± 1.52 | 0.39 |
| | | | Streptomycetaceae | 0.34 ± 0.00 | 6.17 ± 2.77 | 5.06 ± 0.54 | 0.05 |
| | | | Streptosporangiaceae | 0.06 ± 0.04 | 0.87 ± 1.01 | 0.27 ± 0.20 | 0.47 |
| | Coriobacteriia | Coriobacteriales | Coriobacteriaceae | 0.84 ± 0.52 | 0.85 ± 0.28 | 0.22 ± 0.21 | 0.31 |
| | Thermoleophilia | Solirubrobacterales | Patulibacteraceae | 0.80 ± 0.08 | 0.79 ± 0.53 | 0.20 ± 0.20 | 0.23 |
| Bacteroidetes | [Saprospirae] | [Saprospirales] | **Chitinophagaceae** | **0.21 ± 0.06** [b] | **0.79 ± 0.26** [b] | **2.96 ± 0.80** [a] | **0.01** |
| | Bacteroidia | Bacteriodales | Prevotellaceae | 1.78 ± 1.18 | 1.86 ± 0.43 | 0.41 ± 0.48 | 0.28 |
| | | | | 0.57 ± 0.34 | 0.56 ± 0.21 | 0.12 ± 0.15 | 0.28 |
| | Flavobacteria | Flavobacteriales | [Weeksellaceae] | 1.49 ± 0.23 | 0.17 ± 0.09 | 1.90 ± 1.77 | 0.34 |
| | | | Flavobacteriaceae | 1.77 ± 0.29 | 0.78 ± 0.55 | 1.12 ± 0.38 | 0.19 |
| Euryarchaeota | Methanobacteria | Methanobacteriales | Methanobacteriaceae | 0.67 ± 0.45 | 0.49 ± 0.08 | 0.13 ± 0.12 | 0.32 |
| Firmicutes | Bacilli | Bacillales | Bacillaceae | 3.71 ± 0.44 | 3.40 ± 1.74 | 1.26 ± 0.72 | 0.17 |
| | | | Paenibacillaceae | 0.57 ± 0.01 | 1.00 ± 0.64 | 0.71 ± 0.37 | 0.66 |
| | | | **Staphylococcaceae** | **1.06 ± 0.15** [a] | **0.50 ± 0.18** [b] | **0.19 ± 0.12** [b] | **0.01** |
| | | Lactobacillales | **Streptococcaceae** | **0.84 ± 0.19** [a] | **0.47 ± 0.18** [ab] | **0.19 ± 0.12** [b] | **0.03** |
| | Clostridia | Clostridiales | | 1.81 ± 1.14 | 1.68 ± 0.43 | 0.39 ± 0.43 | 0.28 |
| | | | Lachnospiraceae | 3.22 ± 2.11 | 2.73 ± 0.65 | 0.63 ± 0.70 | 0.30 |
| | | | Ruminococcaceae | 3.14 ± 2.05 | 2.77 ± 0.70 | 0.61 ± 0.72 | 0.30 |

*Int. J. Mol. Sci.* **2018**, *19*, 1817

**Table 4.** *Cont.*

| Phylum | Class | Order | Family | ASP | ESP | LSP | p |
|---|---|---|---|---|---|---|---|
| Planctomycetes | Planctomycetia | Pirellulales | **Pirellulaceae** | 0.05 ± 0.04 [b] | 0.83 ± 0.20 [a] | 0.56 ± 0.20 [ab] | **0.01** |
| Proteobacteria | | Caulobacterales | **Caulobacteraceae** | 1.75 ± 0.50 [a] | 0.11 ± 0.10 [b] | 0.22 ± 0.09 [b] | **0.01** |
| | | Kiloniellales | Kiloniellaceae | 0.17 ± 0.10 | 0.00 ± 0.00 | 5.25 ± 5.15 | 0.25 |
| | Alphaproteobacteria | Rhizobiales | Bartonellaceae | 0.74 ± 0.01 | 2.55 ± 0.83 | 2.85 ± 1.40 | 0.19 |
| | | | Bradyrhizobiaceae | 0.00 ± 0.00 | 0.16 ± 0.20 | 0.82 ± 0.75 | 0.28 |
| | | | Brucellaceae | 0.38 ± 0.09 | 0.55 ± 0.11 | 0.92 ± 0.54 | 0.37 |
| | | | Hyphomicrobiaceae | 0.31 ± 0.11 | 0.53 ± 0.21 | 6.03 ± 5.74 | 0.27 |
| | | | Phyllobacteriaceae | 1.52 ± 0.09 | 4.56 ± 1.48 | 4.04 ± 1.59 | 0.15 |
| | | | Rhizobiaceae | 0.08 ± 0.02 | 0.41 ± 0.11 | 0.61 ± 0.36 | 0.18 |
| | | | | 4.19 ± 0.02 | 19.23 ± 5.85 | 15.49 ± 6.41 | 0.08 |
| | | Rhodobacterales | **Rhodobacteraceae** | 1.90 ± 0.04 [a] | 0.44 ± 0.21 [b] | 0.17 ± 0.09 [b] | **0.00** |
| | | Rhodospirillales | **Rhodospirillaceae** | 0.38 ± 0.08 [b] | 3.06 ± 0.89 [a] | 1.35 ± 0.57 [ab] | **0.02** |
| | | Sphingomonadales | Sphingomonadaceae | 3.62 ± 0.18 | 2.74 ± 0.44 | 4.40 ± 0.80 | 0.08 |
| | Betaproteobacteria | Burkholderiales | Alcaligenaceae | 0.09 ± 0.04 | 0.13 ± 0.09 | 5.86 ± 5.79 | 0.26 |
| | | | **Comamonadaceae** | 2.61 ± 0.27 [a] | 1.01 ± 0.45 [b] | 0.79 ± 0.14 [b] | **0.00** |
| | | | Oxalobacteraceae | 1.49 ± 0.25 | 1.40 ± 0.77 | 0.44 ± 0.32 | 0.20 |
| | | Aeromonadales | **Aeromonadaceae** | 1.87 ± 0.11 [a] | 0.68 ± 0.45 [b] | 0.14 ± 0.18 [b] | **0.00** |
| | | Alteromonadales | **[Chromatiaceae]** | 1.17 ± 0.22 [a] | 0.25 ± 0.02 [b] | 0.47 ± 0.38 [b] | **0.05** |
| | | | Alteromonadaceae | 1.19 ± 0.26 | 0.53 ± 0.58 | 0.59 ± 0.06 | 0.27 |
| | Gammaproteobacteria | Enterobacteriales | **Enterobacteriaceae** | 2.89 ± 0.27 [a] | 1.08 ± 0.61 [b] | 0.58 ± 0.27 [b] | **0.00** |
| | | Oceanospirillales | **Halomonadaceae** | 0.89 ± 0.26 [a] | 0.19 ± 0.13 [b] | 0.10 ± 0.07 [b] | **0.02** |
| | | Pseudomonadales | **Moraxellaceae** | 11.34 ± 2.37 [a] | 2.21 ± 0.64 [b] | 0.81 ± 0.49 [b] | **0.00** |
| | | | Pseudomonadaceae | 16.96 ± 2.99 | 8.35 ± 3.02 | 9.95 ± 1.61 | 0.06 |
| | | Xanthomonadales | Sinobacteraceae | 0.58 ± 0.20 | 0.87 ± 0.26 | 1.27 ± 0.84 | 0.54 |
| | | | Xanthomonadaceae | 0.55 ± 0.11 | 0.76 ± 0.29 | 0.31 ± 0.15 | 0.20 |
| Tenericutes | Mollicutes | Acholeplasmatales | Acholeplasmataceae | 1.00 ± 0.07 | 3.90 ± 4.45 | 0.99 ± 0.97 | 0.52 |
| | | Other | | 12.45 ± 1.60 | 10.24 ± 0.28 | 6.81 ± 1.63 | |

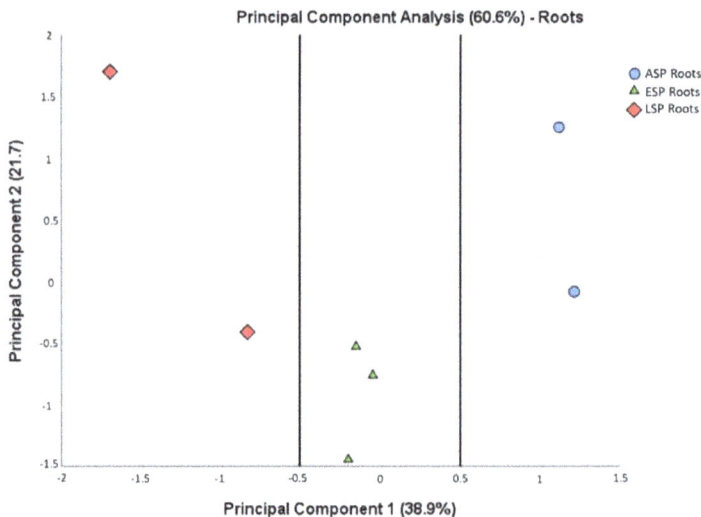

**Figure 10.** Principal Component Analysis calculated among the root samples, based on the relative abundance of OTUs in each family. The lines were added to highlight the separation between the ASP root samples, the ESP root samples, and the LSP root samples.

## 3. Discussion

This study had the principal aim of investigating the bacterial community in Iranian citrus plants in order to describe the different sanitary statuses associated to Citrus Decline Disease and to clear the etiology of this disease.

All the positive detection results for phytoplasma and liberibacter obtained with PCR analyses were confirmed by metagenomics results; however, there are some samples, which are shown to contain pathogens by the metagenomics approach, that were negative to PCR assays, therefore indicating that the genomics technology is more sensitive than PCR [9].

Regarding the etiology of the disease, the results of the study offer some interesting perspectives, but no definitive answer: there is a higher abundance of liberibacter OTUs as the symptoms progress, while there is no particular correlation between symptoms progression and phytoplasma OTUs, as the highest amount is detected in ESP, which suggests that the main, or only, etiological agent of the disease is indeed "*Ca.* Liberibacter asiaticus". Still, the role of phytoplasma cannot be ruled out entirely without further experimentations, since the disease could be caused by a synergistic effect of the two pathogens. In addition, the presence of specific OTUs, for both phytoplasma and liberibacter, in the ESP and LSP samples could suggest that it is not a species, but specific strains that, even at low concentrations, are the etiological agent or agents of the disease.

The presence of the pathogens in the roots, and not in the leaves, seems to suggest that the development of CDD is indeed related to the Bakraee rootstock. It is usually reported that '*Ca.* Liberibacter spp.', despite being easily detected in roots even very early after infection [10], is present mostly in the leaves and not the roots of plants affected by the more common HLB disease [11]. Our results suggest that this is not the case for plants affected by CDD, in which the pathogen seems to replicate only in the rootstock and not in the scion.

The possible role of the rootstock in determining the development of the disease is further supported by some other information: (i) Bakraee variety is already reported as being extremely susceptible to infection by "*Ca.* P. aurantifolia" and its associated disease, Lime Witches' Broom [12,13]; (ii) the higher abundance of liberibacter found in these roots is in accordance to a previous study [14] which found that this pathogen is more present in shallow, horizontally-developed roots than in vertical roots. As Bakraee is characterized by a shallow root system, this could contribute to the spread of liberibacter.

On the other hand, there is also to take into account that the leaves are more exposed to environmental factors, temperature in particular, than the roots. Since high temperatures can dramatically reduce the vitality of bacteria, it is possible that both phytoplasma and liberibacter can be found in leaves during colder months, following a seasonal fluctuation similarly as what already reported for "*Ca.* L. asiaticus" in association to HLB [15].

Since the pathogens, both liberibacter and phytoplasma, are localized in the roots of the symptomatic plants, their mode of transmission from one plant to another needs further investigation: it is possible that they are transmitted to the aerial part of the plant through the most well-known vectors for these kinds of pathogens, such as psyllids and leafhoppers, and then translocated inside the plant to reach the roots, as already reported for HLB [10], but with the pathogen remaining contained to the root compartment afterward; or that they are transmitted directly from root to root by some other, at the moment unknown, soil-borne vector. Another possibility is that the pathogen is transmitted by psyllids directly to the rootstock, as many of the infected plants show the presence of sprouts originating from Bakraee.

To clear the epidemiology of this disease, surveys should be carried out to detect the pathogens in putative insect vectors and reservoir plants in the area, as well as in other asymptomatic citrus plants grafted on different rootstocks from Bakraee.

Similarly as what reported for HLB, the presence of CDD symptoms causes a great restructuring of the microbiota, both in roots where the pathogens are found and in leaves where they are not present [11]. There are theories about how the restructuring of the microbiota happens in HLB, with the replication of liberibacter ousting other bacteria that would occupy the same ecological niche as the pathogen at a leaf level [16], and the lesser amount of nutrients translocated to the roots limiting the growth of endophytic bacteria while allowing the proliferation of opportunistic soil bacteria. The scenario is quite different in CDD. While an interaction with opportunistic soil bacteria can explain the differences at root level, where the pathogens proliferate, the shift at leaf level would be explained only by the change in nutrient content in the pale green canopy. It is unlikely that much of the restructuring detected at leaf level is caused by translocation of bacteria from the roots to the leaves, as the amount of shared OTUs between root and leaves in each sanitary status is quite low compared to that of unique OTUs in each compartment. It is possible that these additional OTUs are always present in leaves, regardless of sanitary status, but become much more abundant in the symptomatic plants, making them more easily detectable.

Since there are no curative treatments that can be used for either liberibacter- or phytoplasma-associated diseases, the perspective of harnessing the native endophytic microbiota of plants, that are key players in determining immunity or tolerance to diseases, to control the spread of diseases is particularly appealing [17]. In addition, the results obtained in this study confirm the general reports that asymptomatic trees are not entirely free from either liberibacter or phytoplasma [11], suggesting that the pathogens are kept below a threshold level needed for pathogenicity [7], or that more pathogenic strains are prevented from proliferating, by other components of the microbiota [18]. This last hypothesis is of particular interest in the light of the obtained results: both liberibacter and phytoplasma had specific OTUs that were found only in symptomatic plants, and in very low concentrations. This could suggest the hypothesis that there are hypervirulent strains that can give rise to the pathogenicity process, while the most common strains found in Iranian citrus, including asymptomatic plants, do not become pathogenic on their own.

The results of this study open possible perspectives for the research of biocontrol agents against this disease: focusing on the differences between the ASP and the symptomatic plants, it is noticeable that the symptomatic plants had lower abundance in bacteria of genera known for including biocontrol agents that are already in use, such as *Pseudomonas* and *Bacillus*. It is therefore possible that, by isolating endophytic bacteria from healthy plants, some cultivable bacteria presenting biocontrol traits that could be employed to protect citrus plants from the CDD disease could be identified. It is interesting to notice that, unlike what has been previously reported for HLB, there seems to be no major involvement of the *Burkholderia* genus, known to be a prevalent genus in biocontrol interaction [19] which is instead often reported as highly abundant in healthy citrus plants and less abundant in diseased plants [20].

Further studies should also be carried out to confirm these results on a higher number of samples, with sampling repeated on more time points, and investigating different citrus varieties, to better characterize the dynamics of the microbiota in relation to CDD in different seasons.

## 4. Materials and Methods

### 4.1. Sample Collection and Processing

The plant samples were collected from the leaves and roots of 8-year-old citrus plants from orchards located in Southern Kerman province during March 2017. Samples were collected from 2 asymptomatic plants (identified as ASP in the rest of the study), 3 plants showing early CDD symptoms (identified as ESP in the rest of the study) that showed the first symptoms in year 2015, and 2 plants showing late CDD symptoms (identified as LSP in the rest of the study) that showed the first symptoms in year 2013. For all these plants, leaf material came from the scion (*Citrus sinensis* L. Osbeck), sampled at one meter from the ground, while root material came from the local rootstock variety, Bakraee. Briefly, the tissues samples were transferred into the sterile biosafety bags, stored in ice boxes and transferred to the laboratory. After washing each sample under running tap water to remove soil particles (5 min), the samples were cut into 3–4 pieces (3–4 cm each) washed with sterile distilled water, and allowed to drain. Tissues were immersed separately in 70% ethanol (3 min), followed by sodium hypochlorite (2%) solution (1 and 3 min for leaves and roots, respectively), and into 70% ethanol (30 s). The samples were rinsed five times in sterile distilled water and were allowed to drain. To verify the accuracy of the surface sterilization procedure, the last rinsing water was inoculated onto nutrient agar sucrose plates and after any bacteria growth in the control agar plates, the sample discarded [21].

DNA was extracted from these samples, following the protocol described in [22].

### 4.2. Detection of Phytoplasma and Liberibacter in Plant Material

To preliminarily assess and identify the possible presence of phytoplasma and liberibacter species, nested PCR amplification was carried out using the universal primers and set of conditions that amplify portions of the 16S rRNA bacterial gene, as described in [1]. The presence of phytoplasmas was assessed using the P1/P7 primer pair in direct PCR, followed by the R16F2n/R16R2 primer pair in nested reaction (Table S1, [23]), while the presence of '*Ca*. L. asiaticus' was assessed using the FD1/RP1 and OI1/OI2c universal primer pairs (Table S1, [24,25]). The amplified fragments were sent to an external service (Eurofins, Hamburg, Germany) for sequencing. Nucleotide sequences were assembled by the Contig Assembling Program in the software BioEdit version 7.2.6 [26]. The obtained nucleotide sequences were compared with related sequences based on the nBlast analysis software available at NCBI GenBank to search for the most similar sequences.

### 4.3. Microbiota Sequencing and OTU Determination

To DNA from sampled citrus plants was sent to an external service (Personal Genomics, Verona (VR), Italy) for sequencing of the hypervariable V1–V3 region of the 16S rRNA (Table S1, [27]) gene using a MiSeq1000 sequencer. The obtained reads (deposited in EMBL-ENA under accession number PRJEB26999; Available online: http://www.ebi.ac.uk/ena/data/view/PRJEB26999) were analyzed using the QIIME pipeline [28] in order to assign them to OTUs and determine the richness of species in the different samples. Reads that mapped on plant-derived sequences (mitochondria, chloroplasts), and reads with low quality, were filtered out. Alpha diversity was calculated for each sample using the Chao-1 and PD indices.

## 4.4. Microbiota Analysis

The sequencing data were analyzed in different ways. A first stage of analysis included the description of rarefaction curves to determine the reliability of the sequencing and the identification of OTUs that were unique to certain sanitary statuses and/or organs, opposed to shared or "core" OTUs. Then, OTUs belonging to the putative pathogens were searched/selected and quantified among samples of different sanitary statuses and/or organs.

The composition of the bacterial community, expressed as relative abundance (1% cutoff threshold), was defined at the *phylum* and family level for all sanitary statuses and organs analyzed. Furthermore, composition of the community was also analyzed at a genus level, only for the bacteria belonging to the three *phyla* that were consistently more represented among the samples (Actinobacteria, Firmicutes, Proteobacteria).

The difference of relative abundance for relevant taxonomical units (either at *phylum*, family, or *genus* level) in different samples (ASP, ESP, LSP) was evaluated through one-way ANOVA performed on the percentage of abundance of these taxonomical units using the SPSS statistical package for Windows v. 23.0 (SPSS Inc., Chicago, IL, USA).

The number of OTUs was compared as fold change (Log10 base) between families that showed significant differences in relative abundance. For this analysis, changes were considered relevant when the fold change was above 0.5 or below −0.5. The relative abundance of OTUs at phylum and family level was also used to calculate the PCA among leaf samples and root samples.

**Supplementary Materials:** Supplementary materials can be found at http://www.mdpi.com/1422-0067/19/6/1817/s1.

**Author Contributions:** Conceptualization, P.A.B. and H.A.; Methodology, P.C.; Data Curation, A.P.; Writing-Original Draft Preparation, A.P. and F.Q.; Writing-Review & Editing, M.A. and A.A. Supervision, P.A.B.

**Funding:** This study was supported by Università degli Studi di Milano Project Linea 2A (project code RV_RIC_AT16FQUAG).

**Acknowledgments:** This work has been supported by the center for international scientific studies and collaboration (CISSC). The authors would like to thank Paola Crepaldi of Università degli Studi di Milano for her valuable suggestions regarding microbiota data analysis.

**Conflicts of Interest:** The authors declare no conflict of interest.

## Abbreviations

| | |
|---|---|
| CDD | Citrus Decline Disease |
| HLB | Huanglongbing |
| ASP | Asymptomatic plants |
| ESP | Early symptomatic plant |
| LSP | Late symptomatic plant |

## References

1. Alizadeh, H.; Quaglino, F.; Azadvar, M.; Kumar, S.; Alizadeh, A.; Bolboli, F.; Casati, P.; Bianco, P.A. First report of a new Citrus Decline Disease (CDD) in association with double and single infection by 'Candidatus Liberibacter asiaticus' and 'Candidatus Phytoplasma aurantifolia' related strains in Iran. *Plant Dis.* **2017**, *101*, 2145. [CrossRef]

2. Faghihi, M.M.; Salehi, M.; Bagheri, A.; Izadpanah, K. First report of citrus huanglongbing disease on orange in Iran. *Plant Pathol.* **2009**, *58*, 793. [CrossRef]

3. Golein, B.; Bigonah, M.; Azadvar, M.; Golmohammadi, M. Analysis of genetic relationship between 'Bakraee' (*Citrus* sp.) and some known Citrus genotypes through SSR and PCR-RFLP markers. *Sci. Hort.* **2012**, *148*, 147–153. [CrossRef]

4. Fotouhi Ghazvini, R.; Fattahi Moghadam, J. *Citrus Growing in Iran*; Guilan University Press: Rasht, Iran, 2010. (In Persian)

5. Bertaccinini, A.; Duduk, B. Phytoplasma and phytoplasma diseases: A review of recent studies. *Phytopathol. Med.* **2009**, *48*, 355–378.

6.   Marcone, C. Molecular biology and pathogenicity of phytoplasmas. *Ann. Appl. Biol.* **2014**, *165*, 199–221. [CrossRef]

7.   Sagaram, U.S.; DeAngelis, K.M.; Trivedi, P.; Andersen, G.L.; Lu, S.E.; Wang, N. Bacterial diversity analysis of Huanglongbing pathogen-infected Citrus, using PhyloChip Arrays and 16S rRNA gene clone library sequencing. *Appl. Environ. Microbiol.* **2009**, *75*, 1566–1574. [CrossRef] [PubMed]

8.   Zhang, M.; Powell, C.A.; Guo, Y.; Benyon, L.; Duan, Y. Characterization of the microbial community structure in 'Candidatus Liberibacter asiaticus'-infected citrus plants treated with antibiotics in the field. *BMC Microbiol.* **2013**, *13*, 112. [CrossRef] [PubMed]

9.   Pereira, R.P.A.; Peplies, J.; Brettar, I.; Höfle, M.G. Development of a genus-specific next generation sequencing approach for sensitiveand quantitative determination of the *Legionella* microbiome in freshwater systems. *BMC Microbiol.* **2017**, *17*, 1–14. [CrossRef] [PubMed]

10.  Johnson, E.G.; Wu, J.; Bright, B.D.; Graham, J.H. Association of 'Candidatus Liberibacter asiaticus' root infection, but not phloem plugging with root loss on huanglongbing-affected trees prior to appearance of foliar symptoms. *Plant Pathol.* **2014**, *63*, 290–298. [CrossRef]

11.  Blaustein, R.A.; Lorca, G.L.; Meyer, J.L.; Gonzalez, C.F.; Teplitski, M. Defining the core Citrus leaf- and root-associated microbiota: Factors associated with community structure and implications of managing Huanglongbing (Citrus Greening) disease. *Appl. Environ. Microbiol.* **2017**, *83*. [CrossRef] [PubMed]

12.  Azadvar, M. Citrus associated phytoplasmas: New findings and challenges. In *Perspectives of Plant Pathology in Genomic Era*; Chowdappa, P., Sharma, P., Singh, D., Misra, A.K., Eds.; Today & Tomorrow's Printers and Publishers: New Delhi, India, 2016; pp. 153–165.

13.  Djavaheri, H.; Rahimian, H. Witches'-broom of bakraee (*Citrus reticulata* hybrid) in Iran. *Plant Dis.* **2004**, *88*, 683. [CrossRef]

14.  Louzada, E.S.; Vazquez, O.E.; Braswell, W.E.; Yanev, G.; Devanaboina, M.; Kunta, M. Distribution of 'Candidatus Liberibacter asiaticus' Above and Below Ground in Texas Citrus. *Phytopathology* **2014**, *106*, 702–709. [CrossRef] [PubMed]

15.  Hu, H.; Roy, A.; Brlansky, R.H. Live population dynamics of 'Candidatus Liberibacter asiaticus', the bacterial agent associated with citrus huanglongbing, in citrus and non-citrus hosts. *Plant Dis.* **2014**, *98*, 876–884. [CrossRef]

16.  Mordecai, E.A. Pathogen impacts on plant communities: Unifying theory, concepts, and empirical work. *Ecol. Monogr.* **2011**, *81*, 429–441. [CrossRef]

17.  Podolich, O.; Ardanov, P.; Zaets, I.; Pirttila, A.M.; Kozyrovska, N. Reviving of the endophytic bacterial community as a putative mechanism of plant resistance. *Plant Soil* **2015**, *388*, 367–377. [CrossRef]

18.  Bové, J.M. Huanglongbing: A destructive, newly-emerging, century-old disease of citrus. *J. Plant Pathol.* **2006**, *88*, 7–37.

19.  Compant, S.; Duffy, B.; Nowak, J.; Clement, C.; Barka, E.A. Use of plant-growth-promoting bacteria for biocontrol of plant diseases: Principles, mechanisms of action, and future prospects. *Appl. Environ. Microbiol.* **2005**, *71*, 4951–4959. [CrossRef] [PubMed]

20.  Trivedi, P.; Duan, Y.; Wang, N. Haunglongbing, a systemic disease, restructures the bacterial community associated with citrus roots. *Appl. Environ. Microbiol.* **2010**, *76*, 3427–3436. [CrossRef] [PubMed]

21.  Araujo, W.L.; Marcon, J.; Maccheroni, W.; van Elsas, J.D.; van Vuurde, J.W.; Azevedo, J.L. Diversity of endophytic bacterial populations and their interaction with *Xylella fastidiosa* in citrus plants. *Appl. Environ. Microbiol.* **2002**, *68*, 4906–4914. [CrossRef] [PubMed]

22.  Angelini, E.; Clair, D.; Borgo, M.; Bertaccini, A.; Boudon-Padieu, E. Flavescence dorée in France and Italy—Occurrence of closely related phytoplasma isolates and their near relationship to Palatinate Grapevine Yellows and an Alder Yellows phytoplasma. *Vitis* **2001**, *40*, 79–86.

23.  Lee, I.M.; Gundersen-Rindal, D.E.; Davis, R.E.; Bartoszyk, I.M. Revised classification scheme of phytoplasmas based on RFLP analyses of 16S rRNA and ribosomal protein gene sequences. *Int. J. Syst. Evol. Microbiol.* **1998**, *48*, 1153–1169. [CrossRef]

24.  Akhtar, N.; Ghauri, M.A.; Iqbal, A.; Anwar, M.A.; Akhtar, K. Biodiversity and phylogenetic analysis of culturable bacteria indigenous to Khewra salt mine of Pakistan and their industrial importance. *Braz. J. Microbiol.* **2008**, *39*, 143–150. [CrossRef] [PubMed]

25.  Jagoueix, S.; Bové, J.M.; Garnier, M. The phloem-limited bacterium of greening disease of citrus is a member of the α subdivision of the Proteobacteria. *Int. J. Syst. Evol. Microbiol.* **1994**, *44*, 379–386. [CrossRef] [PubMed]

26. Hall, T.A. BioEdit: A user-friendly biological sequence alignment editor and analysis program for Windows 95/98/NT. *Nucl. Acids Symp. Ser.* **1999**, *41*, 95–98.
27. Muyzer, G.; Dewaal, E.C.; Uitterlinden, A.G. Profiling of complex microbial-populations by denaturing gradient gel-electrophoresis analysis of polymerase chain reaction-amplified genes-coding for 16S ribosomal-RNA. *Appl. Environ. Microbiol.* **1993**, *59*, 695–700. [PubMed]
28. Caporaso, J.G.; Kuczynski, J.; Stombaugh, J.; Bittinger, K.; Bushman, F.D.; Costello, E.K.; Fierer, N.; Gonalez Pena, A.; Goodrich, J.K.; Gordon, J.I.; et al. QIIME allows analysis of high-throughput community sequencing data. *Nat. Methods* **2010**. [CrossRef] [PubMed]

International Journal of
*Molecular Sciences*

MDPI

*Review*

# Molecular Biology of *Prune Dwarf Virus*—A Lesser Known Member of the *Bromoviridae* but a Vital Component in the Dynamic Virus–Host Cell Interaction Network

Edmund Kozieł [1,*], Józef J. Bujarski [2,3] and Katarzyna Otulak [1,*]

[1] Department of Botany, Faculty of Agriculture and Biology, Warsaw University of Life Sciences—SGGW, Nowoursynowska Street 159, 02-776 Warsaw, Poland
[2] Department of Biological Sciences, Northern Illinois University, DeKalb, IL 60115, USA; jbujarski@niu.edu
[3] Institute of Bioorganic Chemistry, Polish Academy of Sciences, Noskowskiego 12/14, 61-704 Poznań, Poland
* Correspondence: edmund_koziel@sggw.pl (E.K.); katarzyna_otulak@sggw.pl (K.O.)

Received: 17 October 2017; Accepted: 13 December 2017; Published: 16 December 2017

**Abstract:** *Prune dwarf virus* (PDV) is one of the members of *Bromoviridae* family, genus *Ilarvirus*. Host components that participate in the regulation of viral replication or cell-to-cell movement via plasmodesmata are still unknown. In contrast, viral infections caused by some other *Bromoviridae* members are well characterized. *Bromoviridae* can be distinguished based on localization of their replication process in infected cells, cell-to-cell movement mechanisms, and plant-specific response reactions. Depending upon the genus, "genome activation" and viral replication are linked to various membranous structures ranging from endoplasmic reticulum, to tonoplast. In the case of PDV, there is still no evidence of natural resistance sources in the host plants susceptible to virus infection. Apparently, PDV has a great ability to overcome the natural defense responses in a wide spectrum of plant hosts. The first manifestations of PDV infection are specific cell membrane alterations, and the formation of replicase complexes that support PDV RNA replication inside the spherules. During each stage of its life cycle, the virus uses cell components to replicate and to spread in whole plants, within the largely suppressed cellular immunity environment. This work presents the above stages of the PDV life cycle in the context of current knowledge about other *Bromoviridae* members.

**Keywords:** *Bromoviridae*; plant–virus interactions; plant defense response; *Prune dwarf virus*; replication process; systemic and local movement

## 1. Introduction

*Prune dwarf virus* is a member of one of the six genera in the family *Bromoviridae*. Genus *Ilarvirus* includes 19 species. Historically, this genus was divided into four subgroups according to serological characteristics [1] (Table 1), or six groups according to movement protein (MP) and coat protein (CP) coding sequences [2–4]. In 2016, in the genus *Ilarvirus*, new systematics were presented by International Committee on Taxonomy of Viruses (ICTV) , and now this group of viruses includes *American plum line pattern virus* (APLPV), *Apple mosaic virus* (ApMV), *Asparagus virus 2* (AV-2), *Blackberry chlorotic ringspot virus* (BCRV), *Blueberry shock virus* (BlShV), *Citrus leaf rugose virus* (CiLRV), *Citrus variegation virus* (CCV), *Elm mottle virus* (EMoV), *Fragaria chiloensis latent virus* (FCILV), *Humulus japonicus latent virus* (HJLV), *Lilac leaf chlorosis virus* (LLCV), *Lilac ring mottle virus* (LiRMoV), *Parietaria mottle virus* (PMoV), *Prune dwarf virus* (PDV), *Prunus necrotic ringspot virus* (PNRSV), *Spinach latent virus* (SpLV), *Strawberry necrotic shock virus* (SNSV), *Tobacco streak virus* (TSV), *Tulare apple mosaic virus* (TaMV) [5]. Similarly to PDV, viruses in this genus are transferred through seeds, pollen grains, and during vegetative propagation [6]. The generic name refers to the characteristic icosahedral symmetry of virus particles. Icosahedral virions of *Ilavirus* class T = 3 usually consist of 180 molecules of coat protein and encapsidate (+)ssRNA [7,8]. Icosahedral particles are also typical of the remaining genera: *Anulavirus*, *Bromovirus*, and *Cucumovirus*. Certain virus species in the genus *Ilavirus*, including PDV, form two types of particles at the same time: icosahedral and bacilliform [9,10]. The icosahedral (spherical) particles of PDV have a diameter ranging from 26 to 38 nm [6], whereas the bacilliform PDV have the length from 30 to 85 nm and the diameter from 18 to 26 nm. Bacilliform particles are characteristic for genera: *Alfamovirus*, *Ilarvirus*, and *Oleavirus*.

**Table 1.** Taxonomy classification of *Ilarviruses* based on information from Bujarski et al. [1].

| Ilarviruses | |
|---|---|
| Number | Name |
| Subgroup 1 | |
| 1 | Parietaria mottle virus, PMoV |
| 2 | Tobacco streak virus, TSV |
| Subgroup 2 | |
| 3 | Asparagus virus 2, AV-2 |
| 4 | Citrus leaf rugose virus, CiLRV |
| 5 | Citrus variegation virus, CVV |
| 6 | Elm mottle virus, EMoV |
| 7 | Lilac ring mottle virus, LiRMoV |
| 8 | Spinach latent virus, SpLV |
| 9 | Tulare apple mosaic virus, TaMV |
| Subgroup 3 | |
| 10 | Apple mosaic virus, ApMV |
| 11 | Blueberry shock virus, BlShV |
| 12 | Prunus necrotic ringspot virus, PNRSV |
| Subgroup 4 | |
| 13 | Fragaria chiloensis latent virus, FCILV |
| 14 | Prune dwarf virus, PDV |
| No relationships to other existing groups | |
| 15 | American plum line pattern virus, APLPV |
| 16 | Humulus japonicus latent virus, HJLV |

The genomes in all *Bromoviridae* members, including PDV, consist of three single-stranded (+)RNA (ssRNA) components, identified as RNA1, RNA2, and RNA3. Each of the (+)ssRNA particles possesses a cap structure at the 5′-end and a 3′-untranslated region (3′UTR), that forms a tRNA-like structure (TLS) [11]. Furthermore, during virus replication, a subgenomic RNA4 (sgRNA4A) is transcribed, and is responsible for translation of coat protein (CP). Also, in the genus *Cucumovirus*, and in the majority of *Ilarviruses*, an additional open reading frame (ORF2b) has been identified within RNA2, that is expressed from an additional subgenomic sgRNA4A; sgRNA4A encodes a suppressor of RNA interference (RNAi) [12]. Another characteristic of *Bromoviridae* is the occurrence of high concentrations of CP in the infected cells, but rather low amounts of non-structural proteins. The knowledge about replication cycle of PDV is still poorly understood, and limited mostly to the aspects of the structure and RNA genome organization.

## 2. Genome Organization of *Prune Dwarf Virus* (PDV)

The genome of PDV is divided into three single-stranded positive RNA segments RNA1 ($1.3 \times 10^{-6}$ Da), RNA2 ($0.95 \times 10^{-6}$ Da) [13,14], and RNA3 ($0.76 \times 10^{-6}$ Da) [15–17]. Each RNA component is individually packed into viral capsids [10,13].

The monocistronic PDV-RNA1 was first sequenced by Ramptish and Estwell [18] from the Canadian PDV isolate which originated from *Prunus avium* trees of the cultivar Salomo. RNA1 had a length of 3374 nucleotides, and contained single open reading frame (ORF) identified as ORF1, comprising 3168 nucleotides and coding for a non-structural protein P1 (putative replicase component) (Figure 1).

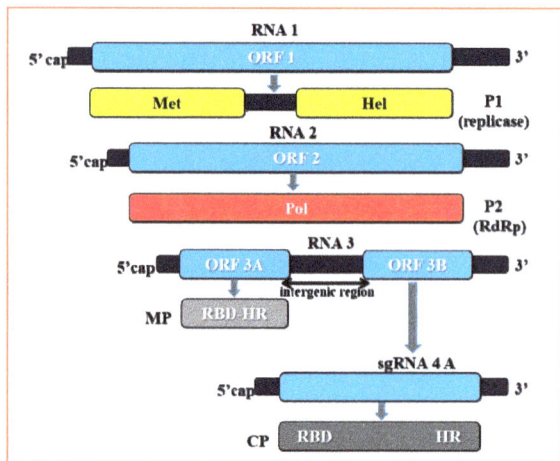

**Figure 1.** Genome structure of *Prune dwarf virus* (PDV). Scheme presenting the individual open reading frames (dark blue) and proteins (different colors) encoded by PDV RNAs. Encoded proteins: P1 with methyltransferase and helicase domains (yellow), P2—polymerase (red), MP—movement protein (grey), CP—coat protein (green). ORF—open reading frame, RBD—RNA binding domain, HR—hydrophobic region. Scheme of genome prepared on the basis of the data in Bujarski et al. [1].

Similarly to RNA1, PDV-RNA2 is a monocistronic molecule carrying ORF 2 (Figure 1). Determined by Scott et al. [19], the full nucleotide sequence of RNA2 amounted to 2593 nucleotides, whereas ORF2 had 2367 nucleotides and coded for a non-structural protein P2, the RNA-dependent RNA polymerase (RdRp) component of the replicase [9]. This feature is in contrast to the structure of RNA2 in other *Ilarviruses*, as most of them carry the additional ORF 2b [7] (in sgRNA4A) well described for *Cucumber mosaic virus* (CMV). Contrary to RNA1 and RNA2, PDV-RNA3 is a dicistronic molecule (Figure 1). Determined by Bachman et al. [20], the full-length sequence of RNA3 of the American PDV-137 isolate consisted of 2129 nucleotides. The two RNA3 ORFs, ORF3a and ORF3b (sometimes identified as MP-ORF and the coat protein (CP)-ORF, respectively) [12] are separated by a 72 nucleotide intergenic region. Little is known about the role and composition of the intergenic region of PDV. The 5′ side ORF3a consists of 882 nucleotides and codes for movement protein MP. The 3′ side ORF 3b consists of 657 nts and codes for CP [21]. RNA3 is a well-studied entity, being most frequently sequenced and phylogenetically compared among different *Bromoviridae*. An analysis of the RNA3 sequences of two Polish isolates (PDV-D1, PDV-D2) and one German isolate (PDV-15/28) demonstrated that the number of nucleotides in RNA3 was 2129, similar to those in PDV-137. The homology of the RNA3 nucleotide sequences and the MP and CP amino acid sequences among PDV-D1, PDV-D2, and PDV-15/28 isolates are 96.9%, 93.8%, and 98.6%, respectively [22]. Also, Vaskova et al. [23] determined the homology of CP-ORFs in 11 PDV isolates from Czech Republic (five isolates from cherry trees, two isolates from peach trees and four from prune trees) to the amount over 88%.

Ulubas-Serce et al. [24] carried out full comparative analysis of the CP nucleotide sequences among PDV isolates obtained from almond trees (nine isolates from Portugal), cherry (20 isolates from Turkey, Germany, Canada, USA, Hungary, and Poland), prune (four isolates from Russia and

Germany), peach (two isolates from Germany), apricot (one isolate from Turkey) and two isolates from unknown sources. The nucleotide sequence homology ranged from 87 to 99%, whereas at the amino acid level (218 amino acids), the homology ranged from 84–99%. Based on these results, the isolates were divided into four groups: cherry 1, cherry 2, almond, and mixed. The homology for mixed group ranged at 92–99%, cherry 1 at 95–99%, cherry 2 at 89–98%, and the almond group at the level of 86–95%.

Another study conducted by Bachman et al. [20], Ramptish and Estewell [18], and Scott et al. [19], indicated that major differences between PDV isolates occurred within the RNA3 sequences, which enabled differentiation of the particular phylogenetic groups. However, the RNA3 sequences were also strongly conservative, indicating key roles of MP and CP at early stages of infection, especially during virus transport and virus interaction with the cell elements [7].

## 3. The Crucial Functions of Proteins Coded by PDV RNA

**P1 protein**—this protein is encoded by PDV-RNA1 (Figure 1), and contains 1055 amino acids (molecular weight 110 kDa) [25]. Sometimes referred to as "putative replicase", it is an enzymatic protein made of two domains and engaged in the viral RNA replication process. The N-terminal domain consists of 340 amino acids, and has a methyltransferase domain (MET). MET is responsible for attaching the 5′-terminal cap that protects the viral RNA against degradation [26]. The C-proximal domain contains 259 amino acids, and it likely has UvrD/REP helicase (HEL) activity, capable of unwinding the RNA strands during replication [18,26,27]. In addition, P1 is engaged in anchoring of the viral RNA, possibly inside a separate vesicle-like membranous structure, within which the replication complex is assembled [7,28,29]. Among the members of *Bromoviridae*, P1 can be linked to different cellular structures. For instance, P1 of BMV locates to ER [25], whereas for *Alfalfa mosaic virus* (AMV) and *Cucumber mosaic virus* (CMV), it targets the tonoplast [30,31]. Kozieł et al. [32] observed PDV-P1 epitope in both palisade and spongy parenchyma cells of tobacco (Figure 2A), as well as in necrotic phloem cells (Figure 2B). The epitope was not spotted in mock inoculated plants (Figure 2C) [32].

Moreover, PDV P1 localized strongly inside tonoplast (Figure 3A,B) and in spherules (Figure 3C,D) of different parenchyma cells or inside sieve tubes (Figure 3E) and vacuoles of companion cells (Figure 3F) [32]. In contrast to infected tobacco, no presence of P1 protein epitopes was observed in control plants (Figure 3G).

**P2 protein**—this protein contains 788 amino acids with molecular weight 89 kDa [21]. P2 is likely an RNA-dependent RNA polymerase (RdRp) (Figure 1) enzyme that, together with P1, forms the RNA replication complex [33]. Translation of P2 from RNA2 is direct, and the protein exhibits strong homology within the C-side throughout the entire *Bromoviridae* family [1].

**MP (movement protein)**—a transport protein with molecular weight of 32 kDa, contains 293 amino acids, and is subjected to direct translation from RNA3 (Figure 1). MP belongs to the superfamily of viral proteins "30K" [34], responsible for local transport of viruses [35]. Proteins of this family are characterized by a strongly conservative RNA-binding domain (RBD). RBD can be found in the genus *Ilarvirus* between residues 56–85, and it possesses a characteristic hydrophobic region, HR. RBD not only enables binding of the viral RNA, but it probably also supports its transport. It has not been determined whether in the genus *Ilarvirus* MP stimulates the formation of tubular structures in infected plant cells. However, it has been found that MP from PNRSV is capable of transporting AMV particles [7]. The immunolabeling analyses conducted with protoplasts from plant cells infected with AMV indicated that MP, similarly to CP, localizes itself in the formed tubular structures. Kasteel et al. [35] demonstrated that the MP proteins can support viral transport through tubular structures. Kozieł et al. [36] showed that the RNA binding domains (RBD) of PDV–MP were most similar to AMV–MP sequences. The similarity reached 34% and 40%, respectively, for the entire MP sequence and RBD (Figure 4A,B).

**CP (*coat protein*)**—coat protein has molecular weight of 23–24 kDa, consists of 218 amino acids, and its translation occurs via subgenomic (sg) RNA4 [21] (Figure 1). CP is a structural protein, but it also functions in genome activation to initiate infection [7,8]. The N-terminal CP fragment in *Ilarvirus* is rich in arginine (R) and/or lysine (K) residues that are responsible for binding to the 3′UTR hairpin structure of viral RNA [37,38]. This binding transforms the hairpin into a pseudoknot structure, similar to that of TLS [39]. The pseudoknot structure is stabilized and maintained with $Mg^{2+}$ [7], which reduces binding of subsequent CP molecules, but stimulates RNA binding to the replication complex. In the case of *Prunus necrotic ring spot virus* (PNRSV), a virus closely related to PDV, a specialized arginine-rich domain of CP displays the affinity to 3′UTR in both RNA3 and sgRNA4A. The domain is located between 25 and 50 N-terminal amino acids, probably in all *Ilaviruses* [40,41]. In two viruses, *Apple mosaic virus* (ApMV) and PNRSV, a zinc finger motif was identified within CP, which probably increases the affinity for RNA binding.

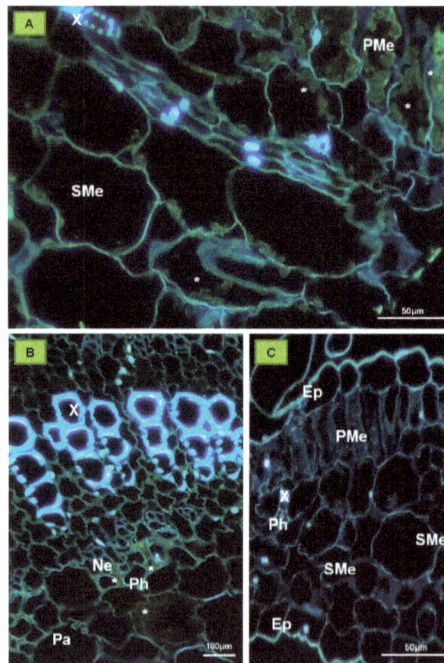

**Figure 2.** Immunofluorescent localization of P1 protein (replicase) in the tobacco leaf tissue of *Samsun* variety. (**A**) Immunofluorescent visualization of the epitopes of P1 protein (green, marked with *) in palisade and spongy mesophyll cells (cross-section of tobacco leaf blade); (**B**) Epitopes of P1 protein (*) visible in parenchyma and necrotic altered phloem; (**C**) Cross section of tobacco leaf 15 days after inoculation with buffer. No locations of P1 protein epitopes. Abbreviations: Ep—epidermis, PMe—palisade mesophyll, SMe—spongy mesophyll, Pa—parenchyma, X—tracheal element, Ph—phloem, Ne—necrosis. Kozieł et al. [32] modified.

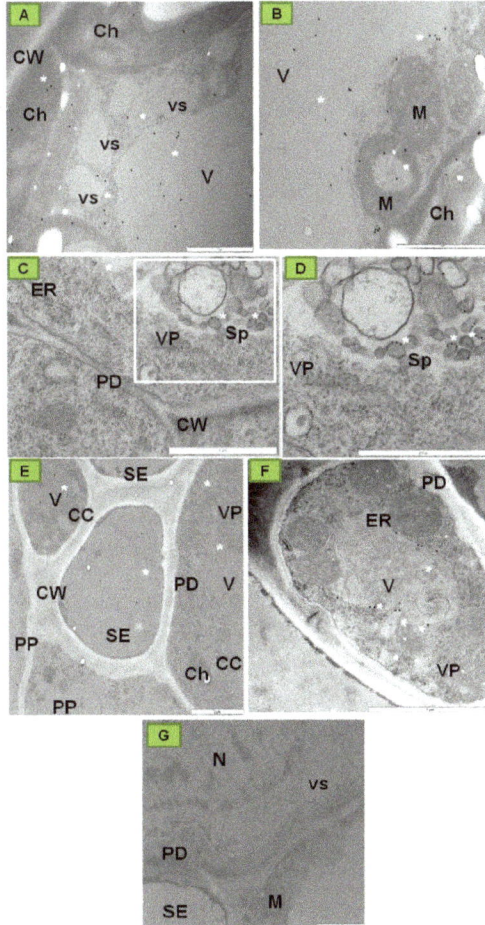

**Figure 3.** Immunogold localization of P1 protein (replicase) in mesophyll and phloem cells of tobacco leaf of *Samsun* variety 15 days after inoculation with PDV. (**A**) Colloidal gold particles associated with P1 epitope (*) in vacuoles, vesicles and chloroplasts of palisade parenchyma cell; Bar 1μm (**B**) Gold particles in palisade parenchyma cell (*) in chloroplast, vacuole, and in vicinity of mitochondria with electron-translucent area; Bar 1μm (**C**) Gold particles (*) in parenchyma cell tonoplast and in membranes of spherules. The white framed area is enlarged in (**D**); Bar 1μm (**D**) Enlarged fragment with spherules in the white frame from (**C**); Bar 0,5 μm (**E**) Epitopes of P1 protein (*) in vacuoles of phloem parenchyma and companion cells, and inside sieve tubes. Viral particles in companion cell; Bar 2μm (**F**) Colloidal gold particles (*) in companion cell vacuoles; Bar 2 μm (**G**) Control tobacco plant (mock-inoculated) phloem without of P1 localization Bar 1μm. Abbreviations: CW—cell wall, Ch—chloroplast, ER—endoplasmic reticulum, V—vacuole, vs—vesicle, M—mitochondrion, Sp—spherule, PD—plasmodesmata, SE—sieve tube, CC—companion cell, PP—phloem parenchyma, VP—viral particles, N—nucleus. Kozieł et al. [32] modified.

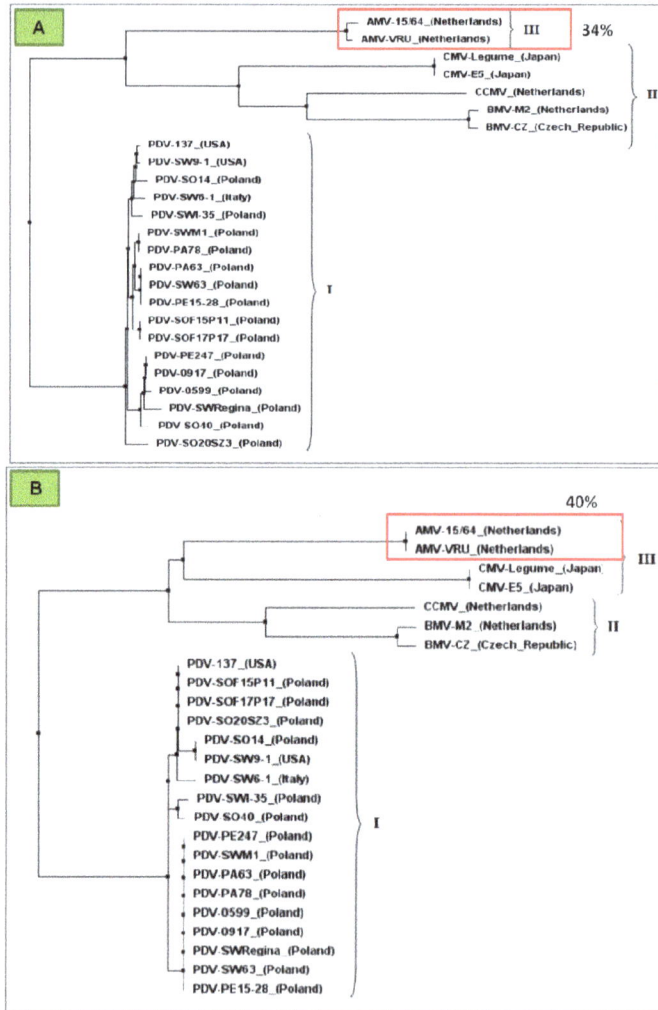

**Figure 4.** Phylogenetic comparison of amino acid sequences of the MP and RNA binding domain of PDV with several members of *Bromoviridae* family: (**A**) MP sequences for the studied groups of virus isolates. The highest similarity (maximum likelihood) between PDV and AMV marked in red frame (34%) (modified [36]); (**B**) RNA binding domain of the MPs. The highest similarity (maximum likelihood) between PDV and AMV marked within the red frame (about level 40%), Kozieł et al. [36] modified. On (**A**)and (**B**): I- group of analyzed PDV isolates sequences of movement protein, II- group of analyzed *Brome mosaic virus* (BMV) and *Cucumber mosaic virus* (CMV) isolates sequences of movement protein, III- group of analyzed AMV and *Cowpea chlorotic mottle virus* (CCMV) isolates sequences of movement protein.

The C-terminal region plays key role in CP dimerization. Bol [28] suggests that CP, apart from genome activation or from virion formation, is also engaged in other processes, including the asymmetric (+)/(−) strand RNA synthesis, translation of viral RNA, and both intercellular and systemic transport of *Ilaviruses*.

Moreover, Neelman and Bol [42] postulated for tobacco protoplasts infected with AMV (member of *Bromoviridae* family like PDV) that collecting and individual packaging of RNA particles to the capsids are influenced by the spatial conformation of CP. Coat protein has always been considered to be responsible for supporting the replication process and later encapsidation, via in *trans* effects on RNA 1 and RNA2 but in *cis*—on RNA3. Rao [43] also developed a model of sequential RNA3 encapsidation process for with particles carrying the subgenomic RNA. In this model, encapsidation consists of three subsequent stages. In stage I, CP subunits recognize a two-part signal consisting of nucleotides at the 3′ TLS end of RNA3, referred to as the nucleating element (NE), and of the ORF coding MP. After the signal recognition, the RNA3 molecule is packed to the capsid. Then, the arginine residues from RBD CP, located on the surface of the virion, bind sgRNA4A (stage II), which is packed together with RNA3 (stage III) [43]. At this moment, however, a similar mechanism of encapsidation in the case of PDV is uncertain. The presence of PDV CP epitopes has been demonstrated in palisade parenchyma (Figure 5A), necrotically altered phloem (Figure 5B,C), and even in xylem treachery elements (Figure 5C) [32]. In the mock-inoculated plants, no CPs were observed (Figure 5D). Similar results were presented by Silva et al. [44] by using the in situ reverse transcription-polymerase chain reaction assay on the almond leaves. These authors [44] have demonstrated the presence of virus (by genome fragment which encoded CP) in both mesophyll and vascular tissues in young leaves. In the case of almond plants, this technique has also enabled the recognition of PDV particles inside the generative organs [44].

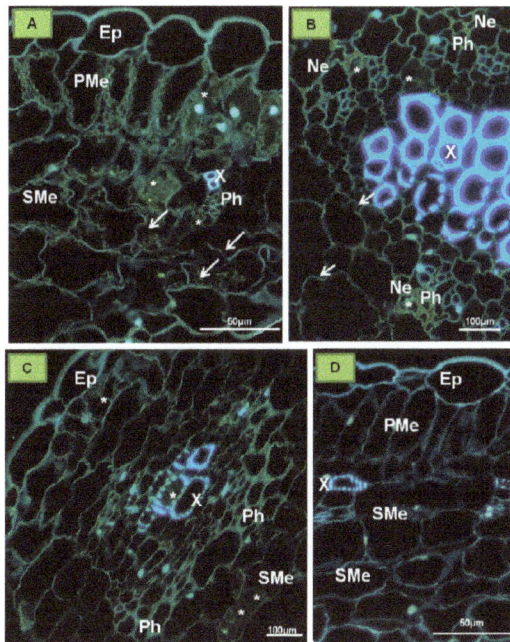

**Figure 5.** Immunofluorescent localization of coat protein in tobacco leaf of *Samsun* variety 15 days after inoculation with PDV. (**A**) The epitopes of coat protein (*) in phloem cells, palisade, and spongy mesophyll. Disintegration of spongy mesophyll cells observed (arrow); (**B**) The epitopes of coat protein (*) in necrotic phloem cells. Visible cell wall invagination of parenchyma cells (arrow); (**C**) The epitopes of coat protein (*) in the spongy mesophyll cell and tracheal element; (**D**) Fragments of leaf blades inoculated only with buffer after 15 days. No CP epitopes observed. Abbreviations: Ep—epidermis, PMe—palisade mesophyll, SMe—spongy mesophyll, X—tracheal element of xylem, Ph—phloem, Ne—necrosis. Kozieł et al. [32] modified.

Moreover, these authors observed PDV CP inside endoplasmic reticulum (ER) (Figure 6A), near plasmodesmata (Figure 6B), in vacuole, and in tonoplasts (Figure 6C) of tobacco cells. The presence of PDV CP in the spherules has also been observed (Figure 6D). In the case of mock inoculated tobacco plants, the PDV CP epitope was absent (Figure 6E).

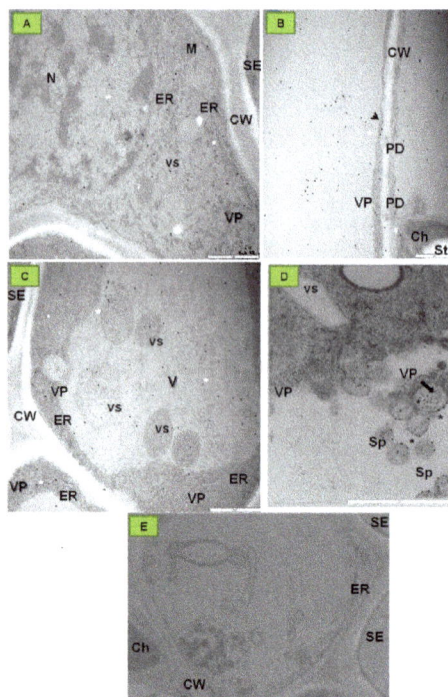

**Figure 6.** Immunogold localization of PDV coat protein (CP) in the tissues of the tobacco leaf blade, *Samsun* variety, 15 days post inoculation with PDV. (**A**) Labeling is observed in nucleus (N) and on the endoplasmic reticulum (ER) surface (*); (**B**) Colloidal gold particles (*) in vacuole of mesophyll cell. Labeling observed in cytoplasm (arrow) and by plasmodesmata; (**C**) Colloidal gold particles (*) in the companion cell protoplast. Labeling is observed both in vesicles and in the vacuole; (**D**) Colloidal gold particles (*) in the spherules' membranes in a parenchyma cell; (**E**) Control tobacco plant phloem without of CP localization. Abbreviations: CW—cell wall, Ch—chloroplast, M—mitochondrion, ER—endoplasmic reticulum, VP—viral particles, vs—vesicle, SE—sieve tube, Sp—spherule, PD–plasmodesmata; St—starch−, Bars 1 µm.Kozieł et al. [32] modified.

Phylogenetic analyses of sequences of PDV CP showed a high level of diversity among the diverse isolates from remote parts of the world. Kalinowska et al. [15] showed a diversity of 86–100% at nucleotide level, and 79–100% at the amino acid level. The ratio of non-synonymous to synonymous polymorphic sites indicated that purifying selection dominated in the case of PDV. However, based on the analysis of six codons, they also showed that the codons were under strong positive selection, including a codon involved in the RNA-binding activity [15]. Among Turkey PDV isolates, Öztürk et al. [45] demonstrated 84–99% to 81–100% nucleotide and amino acid CP sequence identity, respectively, after comparing PDV isolates from Isparta and from other parts of the world. However, despite differences in serological and biological properties among PDV isolates, a molecular characterization of viral coat protein sequences did not confirm a correlation between amino acid composition and host range and/or the origin of viral isolates [23]. It is most likely that the differences between isolates emerged frequently in different parts of the world [15]. Such wide diversity confirms a great ability of PDV to overcome the natural defense responses in a wide spectrum of plant hosts.

## 4. PDV Infection Cycle Based on the *Bromoviridae* Family Model

The initial stage of viral infection is mechanical or biological discontinuity of the plant cell wall, to enable virus penetration through cellular membrane. So far, no specific receptors have been identified for penetration of plant viruses [12,46]. However, the electron microscope (EM) studies carried out on isolated protoplasts have shown that *Bromoviridae* could be absorbed by cellular membrane via pinocytosis [7,47]. In the case of PDV, the virus may be transmitted mechanically among the plants [13]. Greber et al. [48] showed also that PDV under laboratory conditions (but not in the field) may be transmitted to cucumber plants by thrips (*Frankliniella occidentalis*), with the transmission level nearly 20%.

After passing the cellular membrane, certain *Bromoviruses* induce proliferation of endoplasmic reticulum (ER) in the form of spherules and/or the formation of small vacuoles from the external nuclear membrane, which constitute the site of initial viral infection. In the case of PDV, the presence of CP epitopes inside the tobacco spherules has been observed near vacuole rather than ER (Figure 6C,D) [32], suggesting the PDV-induced proliferation in the tonoplast membrane.

Each viral particle encapsidates a single, translation-competent (+)ssRNA, and proteins 1a and 2a are first to be translated. The translation process is stimulated by the 3′TLS region that interacts with the cap at the 5′ side, and secures the circularization of the RNA template [11,49]. Pallas et al. [7] postulate that, similarly to AMV, an efficient translation in *Ilarviruses* (for example PDV) requires the presence of CP molecule at both the 3′ and 5′ sides to maintain a correct RNA conformation. Moreover, the translated replicase proteins are stored in different cellular compartments, depending upon a particular virus. BMV stores them on the ER membrane, AMV, and CMV in the tonoplast or in the vicinity of the vacuole [46,50,51]. Kozieł et al. [32] suggested that PDV P1 protein carries a transmembrane domain between 848 and 869 amino acids (the helicase portion, marked purple on Figure 7A,C) [32]. The rest of this protein comprises a methyltransferase domain (Figure 7A,B).

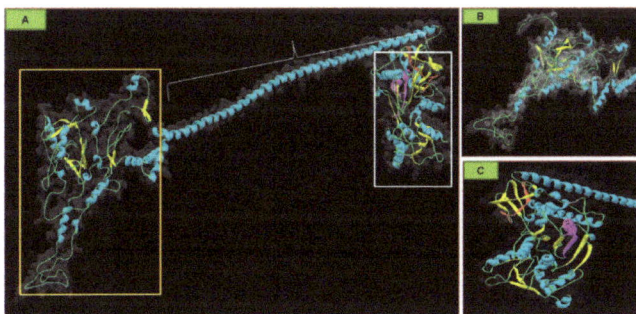

**Figure 7.** Three-dimensional (3D) model structure of the PDV replicase (P1) protein, considering the surface of both N- and C-terminal domain, and the transmembrane domain (modified Kozieł et al., 2017). (**A**) The colors show the particular elements of the secondary structure, as follows. Green indicates the fragments of straight polypeptide chain, blue—α-helical fragments, yellow—β-card fragments; orange—the frame of methyltransferase domain, white—the frame of helicase domain. The central buckle depicts the helical region between both domains. The protein framework of the 3D structure is marked in gray; (**B**) Magnification of the methyltransferase domain from Figure 6a. Gray indicates protein surface; (**C**) Magnification of the helicase domain (from Figure 6A) rotated 90 degrees vis-à-vis the vertical axis, with the region used for immunolocalization of P1 marked in red. Purple indicates the predicted transmembrane domain. The area of the general framework of the helicase C-terminal domain is represented by the gray color. Kozieł et al. [32], modified.

The immunogold labeling of CP and P1 PDV demonstrates that, as for AMV, its replication is primarily associated with the tonoplast [31]. The replication complex also contains the RNA-dependent RNA polymerase (P2 protein) [26]. PDV P1 is firstly attached to the tonoplast membrane (Figure 8A) and then P2 (Figure 8B) is connected (Figure 8C), making a functional replication complex. PDV P1 brings and anchors the viral RNA to the tonoplast membrane [32]. In addition, P1 supports the replication process, whereas P2 is directly responsible for RNA synthesis. The replication complex generates both the (+) strand (the encapsidated strand) and the antisense (−) strand that is used as template for the (+) strand synthesis [8].

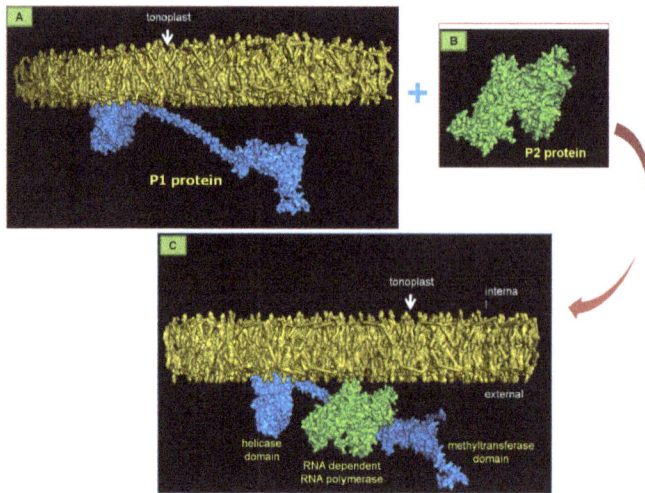

**Figure 8.** (**A**) 3D visualization of a model of the PDV replication complex that assembles during the first steps of viral RNA replication; (**A**) P1 protein (of the replicase complex), marked by blue color, as it anchors to the tonoplast membrane; (**B**) P2 protein (RNA dependent RNA polymerase) 3D surface structure, as marked by green color. (**C**) Fully assembled replication complex, consists of P1 protein (in blue) attached to tonoplast membrane and P2 protein (in green), that is attached to P1 between its both domains. This computer-generated model was generated by using THMM program, in ΔG Prediction Server 1.0, and in AIDA server. To display the results of these calculations, the CELLmicrocosmos Membrane Editor was utilized.

With *Bromoviridae*, the two RNA strands are synthesized asymmetrically; one (−) RNA molecule accounts for approximately 100 (+) RNA molecules [52]. Bol [28] suggested that asymmetrical replication is also characteristic for *Ilarviruses* like PDV. It seems that CP acts as the factor responsible for asymmetrical synthesis of RNA strands [28]. Olsthorn et al. [53] proposed a model of conformational switch affecting the accumulation of sense and antisense strands in AMV. Similarities between AMV and PDV suggest a similar mechanism for both viruses. In this model, CP binds to the 3′UTR, and prevents formation of the pseudoknot structure, and thus prevents RdRp from initiating (−) strand synthesis, promoting the asymmetrical accumulation of (+) strands. The (−) strand promoter includes an AUG loop in 3′UTR TLS, whereas for the plus strands, this is located at the 3′ end of the antisense strand [54]. On the other hand, the 3′UTR of CMV directly interacts with the eukaryotic transcriptional factors eIF4E and eIF4G, and regulates the transcription process [55,56]. The newly-synthesized viral RNA is then interacting with MP, and gets transported to regions where virions are assembled.

## 5. The Specificity of Cell-to-Cell Transport in Different Plant Viruses as Compared to PDV

Plant viruses are specialized pathogens able to move in plants by using cellular structures of the host [43,57,58]. Infection begins in the epidermal cells; the virus first moves from one cell to another to the mesophyll, bundle sheath, and parenchyma, then to phloem via the accompanying cells or xylem [43,57–60]. The cell-to-cell transport enables penetration and spread among the cells, and can be described as a process consisting of three main stages: (i) transfer of newly synthesized genomes/virions from the replication/assembly site to the intracellular transport system [57]; (ii) direct, facilitated transport of genome/virions to reach the plasmodesmata [35]; and (iii) transport to new cells via plasmodesmata [11,61,62].

The long-distance transport of viruses is closely related to the transport that occurs from the phloem parenchyma cells or accompanying cells to the interior parts of the sieve elements in the phloem, where the virus moves rapidly (several centimeters per hour). Then, the virus is actively transported together with assimilates inside the sieve elements [63–66]. Less frequent systemic transport involves the tracheal elements of xylem. In fact, the virus uses the preexisting network of symplastic connections for systemic infection of the plant host [67]. There are two major types of intercellular transport among various plant viruses, and both types can be found in different *Bromoviridae*.

The first group includes two subtypes [68]. One subtype, described well for *Tobacco mosaic virus* (TMV) (*Virgaviridae*, *Tobamovirus*) does not require CP for transport (Figure 8). The virus is transported in a form of its genome (the non-virion form) [69]. Here, the transfer of the viral genome from the cytoplasmic replication site involves the existing intracellular transport systems [70]. The TMV MP forms complexes with the (+) single-stranded viral RNA (vRNA), where one MP monomer accounts for 4–7 nucleotides [71–73]. TMV and other viruses with similar MP proteins of molecular weight around 30 kDa and with a characteristic RNA binding domain, form the "30K" transport protein superfamily [34]. TMV–MP and TMV RNA form a non-undulating structure of the width around 1.5–2.0 nm, that is transported from the membranous replication site to plasmodesmata by using the microtubules and microfilaments of cellular cytoskeleton to direct the transport [73–75] (Figure 9). Studies demonstrated the co-location of MP with microtubules and subunits of actin filaments in protoplasts co-transported with the MP-GFP constructs. TMV–MP has also the potential to interact with tubulin and actin in vitro. Due to facilitated transport, microtubules support direct movement of large complexes, macromolecules, organelles, vesicles, and mRNA [62,76,77]. It is possible that the TMV–MP interaction may constitute an example of molecular mimicry, because MP contains motifs characteristic of tubulin [78]. Viruses with mutation in the tubulin motif did not bind their MPs to tubulin, and were characterized by decreased intercellular spreading rate. Boyko et al. [79,80] demonstrated the necessity of MP interaction with microtubules in order to secure the correct process of TMV transport. So far, two models of the transport of TMV–RNA–MP complex using microtubules have been proposed. The first model indicates the possibility for active transport of the complex, due to participation of kinesin. The second model predicts that the complex is transported thanks to the retraction folding/elongation of microtubules. Studies from recent years lean toward the second model [81,82]. Probably, during the early stages of infection, the TMV–MP co-locates with EB1 protein (end-binding protein 1), that is bound to the plus end of microtubules, which then implies the possibility of MP elongation during transport. Application of herbicides that block microtubule polymerization also blocked the transport of viral complexes [82]. However, in late infection stages, TMV–MP likely remains immobile and bound to microtubules. The complex located near the plasmodesmata penetrates to ER, and via desmotubule, passes the boundary between cells. With the help of MP, TMV can efficiently utilize the cytoskeleton structures, and can protect the pathway of the viral complex to the neighboring cell [73]. The question still remains whether transport of such complexes requires myosin or kinesin, the motor proteins. TMV–MP also has the ability to anchor at ER membrane, enabling the transport from the membrane structures. N and C ends of MP are located on both sides of the cell membrane [83,84]. Thus, viral RNA is most likely transported as a ribonucleoprotein complex (vRNA–MP) (Figure 4). An analogous mechanism to that of TMV was demonstrated by Rao [85] for *Cowpea chlorotic mottle virus* (CCMV), the *Bromoviridae* representative. In this case the transport protein locates itself in plasmodesmata and causes considerable increase of the lower level of size exclusion limit, SEL. As a result, the plasmodesmata widened to the extent that the complex with the viral genome was able to penetrate through to the next cell [85,86].

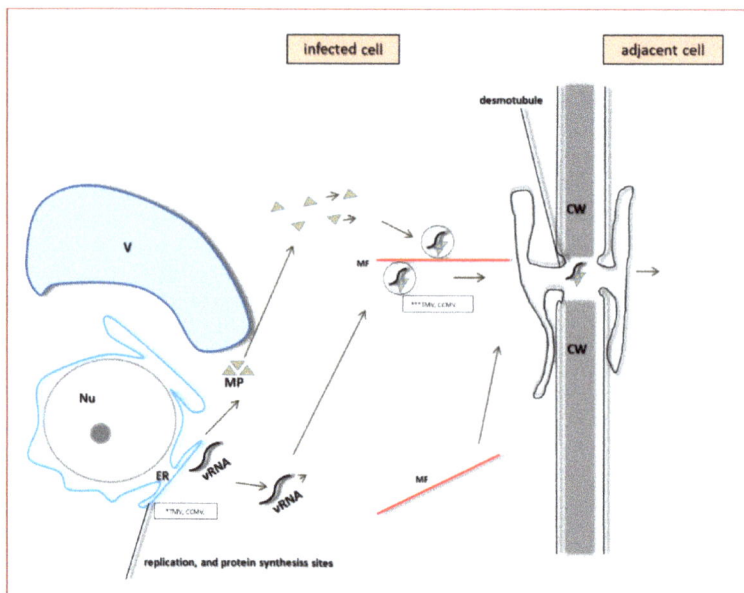

**Figure 9.** Model of the pathogen life cycle, from replication to cell-to-cell transport, by compiling both TMV and CCMV as the examples of viral RNA (vRNA) replicated at the endoplasmic reticulum (ER), and then transported as a complex with movement protein (MP) along actin microfilaments (MF) to the plasmodesmata (PD) region. Simultaneously, MP is also transported alongside microtubules to PD. MP modifies the size exclusion limit (SEL) of PD. Frame with *TMV, CCMV replication sites of vRNA. Frame with ***TMV, CCMV complexes of vRNA and MP, transported with help of MF.

The second subtype depends on both CP and MP (Figure 10). It is characteristic for *Cucumber mosaic virus* (CMV) [87]. Canto and Palukaitis [88] demonstrated that CMV–MP induces formation of tubules in the infected protoplasts. Mutations of C-terminal amino acids of MP blocked tubulin synthesis, without limiting the cell-to-cell transport. Sztuba-Solińska and Bujarski [12] suggested that tubules support transport, but are not indispensable. Su et al. [89] demonstrated that MP binds near the ends of the actin microfilaments, which may indicate the participation of the actin cytoskeleton in the transport process. In the case of CMV the transported form is the ribonucleoprotein complex containing three components: vRNA, MP and CP. Correct CP and MP structures are of paramount importance for the complex formation and their mismatching does block the virus transport. Most likely, MP brings the viral RNA to the ribonucleoprotein complex. The vRNA–CP–MP complex is transported along actin filaments to reach the plasmodesmata (Figure 10), where MP probably stimulates the SEL increase [69].

The second group of movement mechanisms operates e.g., in *Cowpea mosaic virus* (CPMV, *Secoviridae, Comovirus*) or in two *Bromoviridae*: *Alfalfa mosaic virus* (AMV) and *Brome mosaic virus* (BMV) [90–94]. MPs of BMV and AMV induce formation of tubular structures on the surface of infected protoplasts [35,95]. Kaido et al. [96] established that for movement of BMV in tobacco cells, the MP–BMV needs to bind to the cytoplasmic protein *NbNACa1*. This protein has a similar sequence to MP–BMV, and is probably involved in the translocation of newly formed virions.

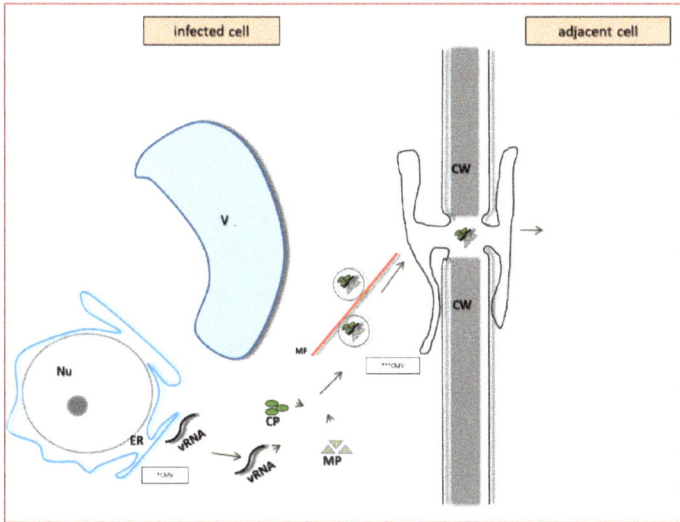

**Figure 10.** Model of a viral pathogen life cycle, from replication to cell-to-cell transport, based upon CMV characteristics. Viral RNA (vRNA) replication occurs at the endoplasmatic reticulum (ER) and the RNA is transported along actin microfilaments (MF) to plasmodesmata as a complex consisting of coat protein (CP) and movement protein (MP). MP modifies the size exclusion limit of plasmodesmata. Frame with *CMV-replication sites of vRNA. Frame with ***CMV vRNA, MP and CP complex transported with help of MF-route (mechanism) of cell-to-cell transport.

Mutations in the gene coding for *Nb*NACa1 limited both the location of MP in plasmodesmata and the BMV movement. Both N- and C-MP termini are responsible for the induction of cytoskeleton protein synthesis, as demonstrated with corresponding MP mutants that blocked the formation of tubules in protoplasts [96]. MP not only induces, but also penetrates to the interior of the tubular structures. Studies on MP from PNRSV, a virus closely related to PDV, demonstrated that the characteristic HR domain has strong affinity to cell membranes, and probably anchors the protein in the membrane [68]. Deletions in HR domain prevents PNRSV translocation through the cell membrane [83]. For this type of mechanism, the transported form is the virion, where CP binds to MP and both cover the interior parts of microtubules that are connected to plasmodesmata (Figure 11).

Kozieł et al. [36] used bioinformatics to analyze the amino acid sequences of MP in PDV and in other *Bromoviridae* members. The authors showed that the sequences of MP RBD (movement protein RNA binding domain) among PDV isolates were most similar to AMV, suggesting a similar mechanism of cell-to-cell transport, likely in the virion form. Van der Vossen et al. [94] demonstrated that the deletion of a considerable C-terminal portion of AMV CP prevented virion formation, but maintained the ability of cell-to-cell transfer. Sanchez-Navarro et al. [97] showed that the normal cell-to-cell transport of AMV, required a 44 amino acid C-terminal MP sequence. Moreover, the same authors have shown that the replacement of this 44 aa sequence with the corresponding region of MP from BMV, PNRSV, or CPMV supported the intercellular movement of AMV. Apparently, these proteins use a very similar mechanism, with a key role for the N-terminal amino acids. MP AMV is often located in ER, and has the capability to move between cells [98].

Apart from the local transport via plasmodesmata, an equally important issue is the fast long-distance movement to secure efficient viral infection within distant plant organs. Phylogenetic comparison of the amino acid sequences of the RBD region (residues 56–85) among PDV strains and with those in the *Bromoviridae* members of known transport mechanisms revealed similarities between AMV and PDV [36], suggesting a similar mechanism of transport. This suggestion supports the presence of PDV particles in cells and CP epitopes near plasmodesmata in infected tobacco [32].

Thus, PDV is most likely transported in a form of viral particles, not only via plasmodesmata but also over long distances.

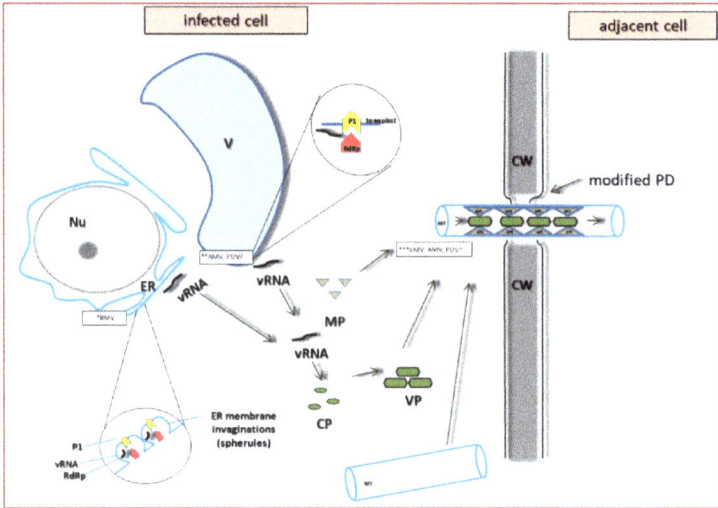

**Figure 11.** A model of PDV life cycle including replication and cell-to-cell transport, based upon research data for BMV and AMV. The BMV replication complex (P1 and P2) is assembled inside the spherules of the endoplasmic reticulum (ER) membranes. Viral RNA of BMV is encapsidated inside virion shells composed of CP molecules. In contrast, in the case of AMV and likely of PDV, the assembly of the replication complex and RNA replication are connected with the tonoplast membrane. As for BMV, the AMV and PDV RNAs are encapsidated with CP molecules. Thereafter, the assembled BMV or AMV (and likely PDV) virions are transported inside the microtubules (MT) that were modified with MP molecules. MP both change the PD size exclusion limit (SEL), but also destroy desmotubule structures inside PD. Frame with *BMV—replication sites of vRNA. Frame with **AMV, PDV replication sites of vRNA. Frame with ***BMV, AMV, PDV viral particles transported with help of MP an MT—in the case of PDV, probably route (mechanism) of cell-to-cell transport.

## 6. Systemic Transport of PDV and Other *Bromoviridae*

Long-distance transport, also referred to as systemic, often requires the virus to move from epidermis or parenchyma cells to penetrate the phloem parenchyma cells, followed by movement to the accompanying cells and/or sieve tubes. Until recently, it was believed that phloem is the only tissue involved in systemic transport. Indeed, Garbaczewska et al. [64] demonstrated that tobacco rattle virus (TRV) utilizes phloem during the transport. In addition, however, TRV moved from external tissues (parenchyma and epidermis) towards xylem parenchyma, and then to the tracheal elements (xylem vessels). The virus particles were observed in the sieve tubes as well as in the vessels. Thus, systemic transport of TRV is linked to both vascular tissues. Regardless of the selected vascular tissue, the next stage always relies on the virus spreading to other plant organs, and then the penetration from the vascular bundles to the neighboring tissues [65,99]. Plasmodesmata link the epidermis, mesophyll cells, and the vascular system, including sieve elements [61].

There are two key points in the pathway to enter and to leave the sieve elements. Plasmodesmata connecting the sieve elements with the accompanying cells display a unique morphology, namely the occurrence of extensive branching at the side of the accompanying cells [62]. At the side of the sieve elements, the plasmodesmata form a pore that does not contain ER, despite the fact that ER cisterns occur inside the sieve elements. Analyzes of the SEL (size exclusion limit) suggest that plasmodesmata between the accompanying cells and sieve elements are gated by volume, differing from those in other plant cells. As a reaction to viral infection, the plant often blocks the sieve elements with callose to reduce viral systemic spread [8]. A study on the accumulation of viruses in the secondary nervation system suggests that selecting a route to the sieve elements always involves the elements of the phloem parenchyma [84]. In all cases, when the accompanying cells became infected, the phloem parenchyma cells were also infected.

Not all viruses can penetrate to the sieve elements through phloem parenchyma; the plasmodesmata that connect directly the phloem parenchyma to the sieve elements can constitute an alternative potential route to phloem [100,101]. Sieve elements are not able to synthesize proteins (lack of ribosomes). Pallas et al. [7] and Hipper et al. [84] indicate that regulation of systemic transport, including virus transport, to and from sieve elements, may require participation of both the viral and host factors. Assembled virions moved efficiently between cells, but were incapable of long-distance transport via phloem [101,102]. CP is definitely required for long-range transport of TMV, suggesting that in phloem, the virus is transported as viral particles. Similarly, *Bromoviridae* require unaltered CP to move systemically [62,84,103]. Fajardo et al. [62] showed that AMV requires an unaltered sequence of 44 amino acids at the C terminus of the MP to move systemically in tobacco. These amino acids probably interact with CP to support systemic transport; similar results were observed for BMV and CMV [104,105]. It is likely that the virions constitute the systemically-transported form of *Bromoviridae* (including PDV), regardless of the differences in the mechanisms of cell-to-cell transport. Virion formation/encapsidation is the last stage in the virus life cycle. CMV is transported between cells as a ribonucleoprotein, and CMV encapsidation takes place within the wall elements of the tube after penetrating to smaller sieve tubes [87]. As demonstrated by Requena et al. [106], CMV is further transported as virion particles. Once in the sieve tubes, CMV particles interact with phloem protein 1 (PP1), and translocate jointly in the tubes. Pallas et al. [103] suggest that additional phloem proteins function in the sieve tubes that likely bind to viral particles and facilitate their transport. Any mutations within the N- or C-terminal sequences of the AMV CP blocked the systemic transport [107]. Pallas et al. [7], Tenllado and Bol [108], and Bol [107] indicate that the tissue associated with the transport of these *Bromoviridae* is most likely the phloem. Until now, there was no reports about the type of vascular tissue responsible for PDV transport. However, Kozieł et al. [32] showed that PDV CP epitope localized in companion cells, sieve tubes, but also in xylem tracheary elements, suggesting both phloem and xylem are responsible for its systemic transport.

## 7. Plant Response to Infections with PDV and Other *Bromoviridae*

Response reactions to PDV infection are mainly undescribed. One example of available information is related to various symptoms induced by different strains in plant hosts [17]. Among PDV strains/isolates, some differences were reported [22]. Nemeth [17] showed that certain PDV isolates could cause different plant diseases (Table 2), as reflected by specific viral names (Table 2). The symptoms can range from chlorotic ringspots, and necrotic changes to even gum leak in apricots.

Regardless of type of disease, PDV infection manifests as a significant decrease in fruit yield. Nemeth [17] noted that crop reduction in PDV-infected cherry cv. *Schattenmorelle* fruits reached 94–96%. Reduction of the number of cherry fruits was accompanied by 9–15% height reduction in infected trees in comparison to healthy plants. In the case of sweet cherry, the range of fruit reduction was 30–90%. Brunt et al. [2] and Kajati [14] observed, respectively, the reduction in diameter (40%) and length (35%) of PDV-infected peach shoots. Moreover, on peach trees, these authors observed 18% less leaves with 73% reduction in leaf surface. Other alternations with PDV infection include flowering disorder, increased low temperature sensitivity of flower buds, flower deformation (formed flowers had no stamens), and premature leaf fall. One of the most important reactions is associated with significant reduction of fruit yield from orchard trees. Nemeth [16] showed that budding effectiveness of different natural PDV hosts ranged from 5% to 99%.

**Table 2.** Different names of diseases on *Prunus* species caused by various strains/isolates of PDV according information from Nemeth [17], modified.

| Infected Species | Strain Name | Disease Name |
| --- | --- | --- |
| *Prunus avium, P. cerasifera, P. cerasus, P. domestica, P. mahaleb* | Cherry chlorotic ringspot of *Prune dwarf virus* | Cherry chlorotic ringspot |
| *Prunus avium, P. cerasifera, P. cerasus, P. domestica, P. mahaleb* | Cherry chlorotic necrotic ringspot of *Prune dwarf virus* | Cherry chlorotic ringspot |
| *Prunus avium, P. cerasus* | Cherry ring mosaic of *Prune dwarf virus* | Cherry chlorotic necrotic ringspot |
| *Prunus avium* | Cherry ring mottleof *Prune dwarf virus* | Cherry ring mosaic |
| *Prunus avium* | Cherry yellow mosaic of *Prune dwarf virus* | Cherry ring mottle |
| *Prunus serrulata cv. Amanogawa, Prunus serrulata cv. Kwanzan, P. avium var. plena, P fontanesiana, P.incisa, P. lannesiana* | Cherry yellow mottle of *Prune dwarf virus* | Cherry yellow mosaic |
| *Prunus domestica* | Type strain of *Prune dwarf virus* | Cherry yellow mottle |
| *Prunus armeniaca, P. avium, P. cerasus* | Apricot gummosis of *Prune dwarf virus* | Chlorotic-necrotic ringspot |

Until now, the mechanisms of plant response to *Bromoviridae* have been only described for CMV. The effects strongly depend upon CMV gene expression at the beginning of infection. *Arabidopsis thaliana* plants (ecotype C24) were resistant to CMV strain Y (*yellow*) [103,109]. The resistance correlated with the expression of *RCY1* 104 kDa protein that belongs to the family of nucleotide-binding leucine-rich repeat proteins (CC-NBS-LRR). These proteins are responsible for signal transduction related to ethylene and salicylic acid pathways, inducing necrotic changes in the infected tissue. Necrosis occurred at the inoculation sites, localizing the virus and limiting its spread to other plant organs [109]. Inaba et al. [110] demonstrated that formation of necrosis in *Arabidopsis thaliana* was caused by the interaction of a CMV suppressor (coded by ORF2b) with a plant catalase. Alike for numerous plant viruses, CMV infection induces the RNA interference (RNAi)-based response, which is suppressed by CMV protein 2b. The interaction of RNAi response machinery with sgRNA2b triggers the silencing of CHLI, the host gene that is involved in chlorophyll biosynthesis, and thus, resulting in the yellowing of infected leaves [110].

## 8. Conclusions

Prune dwarf virus remains an enigmatic pathogen, even in the context of accumulated knowledge about other *Bromoviridae*. The available data mostly concern local transport, missing however, the information about molecular or ultrastructural effects in the infected tissue. Further studies are required for identification of cell components that contribute to PDV infection, and for characterization of pathological changes in the infected plant tissue. It is likely that the accumulating knowledge will reveal new means of resistance against PDV, one of the most dangerous plant viruses debilitating the stone fruit trees.

**Acknowledgments:** The work has been supported by Warsaw University of Life Sciences-SGGW grants number: 505-10-011100-N00208-99 and 505-10-011100-P00122-99).

**Author Contributions:** Edmund Kozieł and Katarzyna Otulak wrote the manuscript, conceived the idea of review and performed some experiments cited in the article (microscopy, bioinformatic, and computer visualizations), Józef J. Bujarski helped to write and edit this article (native speaker), and also helped with his great knowledge about plant viruses, especially about the *Bromoviridae* family.

**Conflicts of Interest:** The authors declare no conflict of interest.

## References

1.  Bujarski, J.J.; Figlerowicz, M.; Gallitelli, D.; Roossinck, M.J.; Scott, S.W. Family Bromoviridae. In *Virus Taxonomy: Classification and Nomenclature of Viruses-Ninth Report of the International Committee on Taxonomy of Viruses*, 1st ed.; King, A.M.Q., Adams, M.J., Carstens, E.B., Lefkowitz, E.J., Eds.; Elsevier Academic Press: San Diego, CA, USA, 2012; pp. 972–976. ISBN 978-0123846846.
2.  Brunt, H.A.; Crabtree, K.; Dallawitz, M.J.; Gibs, A.J.; Watson, L. *Viruses of Plants*, 1st ed.; CAB International UK: Wallingford, UK, 1996; ISBN 978-0851987941.
3.  Codoñer, F.M.; Cuevas, J.M.; Sanchez-Navarro, J.A.; Pallas, V.; Elena, S.F. Molecular evolution of the plant virus family *Bromoviridae* based on RNA3-encoded proteins. *J. Mol. Evol.* **2005**, *61*, 697–705. [CrossRef]
4.  Codoñer, F.M.; Fares, M.A.; Elena, S.F. Adaptive covariation between the coat and movement proteins of *Prunus necrotic ringspot virus*. *J. Virol.* **2006**, *80*, 5833–5840. [CrossRef]
5.  International Committee on Taxonomy of Viruses (ICTV) Official Website. Available online: https://talk.ictvonline.org/ictv-reports/ictv_9th_report/positive-sense-rna-viruses-2011/w/posrna_viruses/251/bromoviridae (accessed on 16 November 2017).
6.  Fulton, R.W. Prune dwarf virus. C.M.I/A.A.B. *Descr. Plant Viruses* **1970**, *1*. Unavailable online.
7.  Pallas, V.; Aparicio, F.; Herranz, M.C.; Amari, K.; Sanchez-Pina, M.A.; Myrta, A.; Sanchez-Navarro, J.A. Ilarviruses of *Prunus* spp.: A continued concern for fruit trees. *Phytopathology* **2012**, *102*, 1108–1120. [CrossRef] [PubMed]
8.  Pallas, V.; Aparicio, F.; Herranz, M.C.; Sanchez-Navarro, J.A.; Scott, S.W. The molecular biology of ilarviruses. *Adv. Virus Res.* **2013**, *87*, 139–183. [CrossRef] [PubMed]
9.  Faquet, C.M.; Mayo, M.A.; Maniloff, J.; Desselberger, U.; Ball, L.A. *Virus Taxonomy-Eight Report of the International Committee on Taxonomy of Viruses*, 1st ed.; Elsevier Academic Press: London, UK, 2005; pp. 200–236. ISBN 978-0122499517.
10. Lee, S.; Shin, Y.G. Development and practical use of RT-PCR for seed-transmitted Prune dwarf virus in quarantine. *Plant Pathol. J.* **2014**, *30*, 178–182. [CrossRef] [PubMed]
11. Gallie, D.R. The cap and poly(A) tail function synergistically to regulate mRNA translational efficiency. *Genes Dev.* **1991**, *5*, 2108–2116. [CrossRef] [PubMed]
12. Sztuba-Solińska, J.; Bujarski, J.J. Insights into the single-cell reproduction cycle of members of the family Bromoviridae: Lessons from the use of protoplast systems. *J. Virol.* **2008**, *82*, 10330–10340. [CrossRef]
13. Fulton, R.W. Ilavirus group. C.M.I/A.A.B. *Descr. Plant Viruses* **1983**, *274*. Unavailable online.
14. Kajati, I. Metody ossledovanija ekomomičeskowo značenija virusnych zabolevanij plodovich kultur. (Methods of investigation on the economic importance of the virus diseses of fruit trees). In Proceedings of the Konferencjia Stran-Tslenov SZEV po Zaščite i Karantenu Rastenij, Budapest, Hungary, 16–19 October 1976; pp. 119–120.
15. Kalinowska, E.; Mroczkowska, K.; Paduch-Cichal, E.; Chodorska, M. Genetic variability among coat protein of *Prune dwarf virus* variants from different countries and different *Prunus* species. *Eur. J. Plant Pathol.* **2014**, *4*, 863–868. [CrossRef]
16. Nemeth, M. Interferencja vizsgǎlatok a csonthějas gyümöcsfǎk gyürüsfoltossǎg (ringspot) virusavial. *Növěnyvědelem* **1972**, *8*, 64–71.
17. Nemeth, M. *Virus, Mycoplasma and Rikettsia Diseases of Fruit Trees*, 1st ed.; Springer: Budapest, Hungary, 1986; pp. 600–650. ISBN 978-90-247-2868-8.
18. Ramptish, C.; Estewell, K.C. The complete nucleotide sequence of prune dwarf ilarvirus RNA-1. *Arch. Virol.* **1997**, *142*, 1911–1918. [CrossRef]
19. Scott, S.W.; Zimmerman, M.T.; Xin, G.; Mackenzie, D.J. The coat proteins and putative movement proteins of isolates of Prunus necrotic ringspot virus from different host species and geographic origins are extensively conserved are extensively conserved. *Eur. J. Plant Pathol.* **1998**, *104*, 155–161. [CrossRef]
20. Bachman, E.J.; Scott, S.W.; Xin, G.; Vance, V.B. The complete nucleotide sequence of prune dwarf ilarvirus RNA 3: Implications for coat protein activation of genome replication in ilarviruses. *Virology* **1994**, *201*, 127–131. [CrossRef] [PubMed]
21. Ramptish, C.; Estewell, K.C.; Hall, J. Setting confidence limits for the detection of prune dwarf virus in Prunus avium with a monoclonal antibody-based triple antibody-sandwich ELISA. *Ann. Appl. Biol.* **1995**, *126*, 485–491. [CrossRef]

22. Paduch-Cichal, E. Characterization of PNRSV and PDV. Associate Professor Thesis, Warsaw University of Life Sciences, Warsaw, Poland, 2000.

23. Vaskova, D.; Petrzik, K.; Spak, J. Molecular variability of the capsid protein of the Prune dwarf virus. *Eur. J. Plant Pathol.* **2000**, *106*, 573–580. [CrossRef]

24. Ulubas-Serce, C.; Ertunc, F.; Özturk, A. Identification and genomic variability of Prune dwarf virus variants infecting stone fruit trees in Turkey. *J. Phytopathol.* **2009**, *157*, 298–305. [CrossRef]

25. Loesch, L.S.; Fulton, R.W. Prunus necrotic ringspot virus as a multicomponent system. *Virology* **1975**, *68*, 71–78. [CrossRef]

26. Rozanov, M.N.; Koonin, E.V.; Gorbalenya, A.E. Conservation of the putative methylotransferase domain: A hallmark of the 'Sindbis like' supergroup of positive-strand RNA viruses. *J. Gen. Virol.* **1992**, *73*, 2129–2134. [CrossRef] [PubMed]

27. Korolev, S.; Hsieh, J.; Gauss, G.H.; Lohman, T.M.; Waksman, G. Major domain swiveling revealed by the crystal structures of complexes of *E. coli* Rep-helicase bound to single stranded DNA and ADP. *Cell* **1997**, *90*, 635–647. [CrossRef]

28. Bol, J.F. Replication of alfamo- and ilarviruses: Role of the coat protein. *Annu. Rev. Phytopathol.* **2005**, *43*, 39–62. [CrossRef] [PubMed]

29. Liu, L.; Westler, W.M.; den Boon, G.; Wang, X.; Diaz, A.; Steinberg, H.A.; Alquist, P. An amphipathic alpha-helix controls multiple roles of Brome mosaic virus protein 1a in RNA replication complex assembly and function. *PLoS Pathog.* **2009**, *5*, 1–18. [CrossRef] [PubMed]

30. Van der Hejden, M.W.; Carette, J.E.; Reinhoud, P.J.; Haegi, A.; Bol, J.F. Alfalfa mosaic virus replicase proteins P1 and P2 interact and colocalize at the vacuolar membrane. *J. Virol.* **2001**, *75*, 1879–1887. [CrossRef] [PubMed]

31. Cillo, F.; Roberts, I.M.; Palukaitis, P. In situ localization and tissue distribution of the replication-associated proteins of cucumber mosaic virus in tobacco and cucumber. *J. Virol.* **2002**, *76*, 10654–10664. [CrossRef] [PubMed]

32. Kozieł, E.; Otulak, K.; Lockhart, B.E.L.; Garbaczewska, G. Subcelullar localization of proteins associated with Prune dwarf virus replication. *Eur. J. Plant Pathol.* **2017**, *149*, 653–668. [CrossRef]

33. Dinant, S.; Janda, M.; Kroner, P.A.; Alquist, P. Bromovirus RNA replication and transcription require compatibility between the polymerase- and helicase-like viral RNA synthesis proteins. *J. Virol.* **1993**, *67*, 7181–7189. [PubMed]

34. Melcher, U. The "30K" superfamily of viral movement proteins. *J. Gen. Virol.* **2000**, *81*, 257–266. [CrossRef] [PubMed]

35. Kasteel, D.T.J.; van der Wel, N.N.; Jansen, K.A.J.; Goldbach, R.W.; van Lent, J.W.M. Tubule-forming capacity of the movement proteins of alfalfa mosaic virus and brome mosaic virus. *J. Gen. Virol.* **1997**, *78*, 2089–2093. [CrossRef] [PubMed]

36. Kozieł, E.; Otulak, K.; Garbaczewska, G. Phylogenetic analysis of PDV movement protein compared to Bromoviridae members as justification of possible intercellular movement. *Acta Biol. Crac. Ser. Bot.* **2015**, *57*, 19–31. [CrossRef]

37. Sanchez-Navarro, J.A.; Pallas, V. Evolutionary relationships in the ilarviruses: Nucleotide sequence of Prunus necrotic ringspot virus RNA 3. *Arch. Virol.* **1997**, *142*, 749–763. [CrossRef] [PubMed]

38. Ansel-McKinney, P.; Scott, S.W.; Swanson, M.; Ge, X.; Gehrke, L. A plant viral coat protein RNA binding consensus sequence contains a crucial arginine. *EMBO J.* **1996**, *15*, 5077–5084. [PubMed]

39. Aparicio, F.; Vilar, M.; Perez-Paya, E.; Pallas, V. The coat protein of Prunus necrotic ringspot virus specifically binds to and regulates the conformation of its genomic RNA. *Virology* **2003**, *313*, 213–223. [CrossRef]

40. Aparicio, F.; Sánchez-Pina, M.A.; Sánchez-Navarro, J.A.; Pallás, V. Location of prunus necrotic ringspot ilarvirus within pollen grains of infected nectarine trees: Evidence from RT-PCR, dot-blot and in situ hybridisation. *Eur. J. Plant Pathol.* **1999**, *105*, 623–627. [CrossRef]

41. Aparicio, F.; Sánchez-Navarro, J.A.; Pallás, V. Implication of the C terminus of the Prunus necrotic ringspot virus movement protein in cell-to-cell transport and in its interaction with the coat protein. *J. Gen. Virol.* **2010**, *91*, 1865–1870. [CrossRef] [PubMed]

42. Neeleman, L.; Bol, J.F. Cis-acting functions of alfalfa mosaic virus proteins involved in replication and encapsidation of viral RNA. *Virology* **1999**, *254*, 324–333. [CrossRef] [PubMed]

43. Rao, A.L.N. Genome packaging by spherical plant RNA viruses. *Annu. Rev. Phytopathol.* **2006**, *44*, 61–87. [CrossRef] [PubMed]

44. Silva, C.; Tereso, S.; Nolasco, G.; Oliveira, M.M. Cellular location of Prune dwarf virus in almond sections by in situ reverse transcription-polymerase chain reaction. *Phytopathology* **2003**, *93*, 278–285. [CrossRef] [PubMed]

45. Öztürk, Y.; Çevik, B. Genetic diversity in the coat protein genes of Prune dwarf virus isolates from sweet cherry in Turkey. *Plant Pathol. J.* **2015**, *31*, 41–49. [CrossRef]

46. Roenhorst, J.W.; van Lent, J.W.; Verduin, B.J. Binding of cowpea chlorotic mottle virus to cowpea protoplasts and relation of binding to virus entry and infection. *Virology* **1988**, *164*, 91–98. [CrossRef]

47. Burgess, J.; Motoyoshi, F.; Fleming, E.N. The mechanism of infection of plant protoplasts by viruses. *Planta* **1973**, *112*, 323–332. [CrossRef] [PubMed]

48. Greber, R.S.; Teakle, D.S.; Mink, G.I. Thrips-facilitated transmission of prune dwarf and prunus necrotic ringspot viruses from cherry pollen to cucumber. *Plant Dis.* **1972**, *76*, 1039–1041. [CrossRef]

49. Gallie, D.R.; Kobayashi, M. The role of the 3′-untranslated region of non-polyadenylated plant viral mRNAs in regulating translational efficiency. *Gene* **1994**, *142*, 152–165. [CrossRef]

50. Ibrahim, A.; Hutchens, H.M.; Berg, R.H.; Loesch-Fries, S. Alfalfa mosaic virus replicase proteins, P1 and P2, localize to the tonoplast in the presence of virus RNA. *Virology* **2012**, *433*, 449–461. [CrossRef] [PubMed]

51. Huang, M.; Jongejan, L.; Zheng, H.; Zhang, L.; Bol, J.F. Intracellular localization and movement phenotypes of Alfalfa mosaic virus movement protein mutants. *Mol. Plant Microbe Interact.* **2001**, *14*, 1063–1074. [CrossRef] [PubMed]

52. Houwing, C.J.; van de Putte, P.; Jaspars, E.M.J. Regulation of single strand RNA synthesis of Alfalfa mosaic virus in non-transgenic cowpea protoplasts by the viral coat protein. *Arch. Virol.* **1998**, *143*, 489–500. [CrossRef] [PubMed]

53. Olsthoorn, R.C.; Mertens, S.; Brederode, F.T.; Bol, J.F. A conformational switch at the 3′ end of a plant virus RNA regulates viral replication. *EMBO J.* **1999**, *18*, 4856–4864. [CrossRef] [PubMed]

54. Sivakumaran, K.; Chen, M.H.; Rossinck, M.J.; Kao, C.C. Core promoter for initiation of cucumber mosaic virus subgenomic RNA4A. *Mol. Plant Pathol.* **2002**, *3*, 43–52. [CrossRef] [PubMed]

55. Schmitz, I.; Rao, A.L.N. Deletions in the conserved amino-terminal basic arm of cucumber mosaic virus coat protein disrupt virion assembly but do not abolish infectivity and cell-to-cell movement. *Virology* **1998**, *248*, 323–331. [CrossRef] [PubMed]

56. Sztuba-Solińska, J.; Dzianott, A.; Bujarski, J.J. Recombination of 5′ subgenomic RNA3a with genomic RNA3 of Brome mosaic bromovirus in vitro and in vivo. *Virology* **2011**, *410*, 129–141. [CrossRef]

57. Schoelz, J.E.; Harries, P.A.; Nelson, R.S. Intracellular transport of plant viruses: Finding the door out of the cell. *Mol. Plant* **2011**, *4*, 813–831. [CrossRef] [PubMed]

58. Niehl, A.; Heinlein, M. Cellular pathways for viral transport through plasmodesmata. *Protoplasma* **2011**, *248*, 75–99. [CrossRef] [PubMed]

59. Otulak, K.; Garbaczewska, G. Ultrastructural events during hypersensitive response of potato cv. Rywal infected with necrotic strains of potato virus Y. *Acta Physiol. Plant.* **2010**, *32*, 635–644. [CrossRef]

60. Otulak, K.; Kozieł, E.; Garbaczewska, G. Seeing is believing. The use of light, fluorescent and transmission electron microscopy in the observation of pathological changes during different plant—Virus interactions. In *Microscopy: Advances in Scientific Research and Education*, 6th ed.; Mendez-Vilas, A., Ed.; Formatex Research Center: Badajoz, Spain, 2014; Volume 1, pp. 367–376. ISBN 978-84-942134-3-4.

61. Carrington, J.C.; Kasschau, K.D.; Mahajan, S.K.; Schaad, M.C. Cell-to-cell and long-distance transport of viruses in plants. *Plant Cell* **1996**, *8*, 1669–1681. [CrossRef] [PubMed]

62. Fajardo, T.V.; Peiro, A.; Pallas, V.; Sanchez-Navarro, J. Systemic transport of alfalfa mosaic virus can be mediated by the movement proteins of several viruses assigned to five genera of the 30K family. *J. Gen. Virol.* **2013**, *94*, 677–681. [CrossRef] [PubMed]

63. Otulak, K.; Garbaczewska, G. Cell-to-cell movement of three genera (+) ssRNA plant viruses. *Acta Physiol. Plant.* **2011**, *33*, 249–260. [CrossRef]

64. Garbaczewska, G.; Chouda, M.; Otulak, K. Ultrastructural studies of plasmodesmatal and vascular translocation of tobacco rattle virus (TRV) in tobacco and potato. *Acta Physiol. Plant.* **2012**, *34*, 1229–1238. [CrossRef]

65. Otulak, K.; Kozieł, E.; Garbaczewska, G. Ultrastructural impact of tobacco rattle virus on tobacco and pepper ovary and anther tissues. *J. Phytopathol.* **2015**, *164*, 217–289. [CrossRef]

66. Leisner, S.; Turgeon, R. Movement of virus and photo-assimilate in the phloem: A comparative analysis. *Bioessays* **1993**, *15*, 741–748. [CrossRef] [PubMed]

67. Lucas, W.J.; Ding, B.; Van der Schoot, C. Plasmodesmata and the supracellular nature of plants. *New Phytol.* **1993**, *125*, 435–476. [CrossRef]

68. Herranz, M.C.; Sanchez-Navarro, J.A.; Sauri, A.; Mingarro, I.; Pallas, V. Mutational analysis of the RNA-binding domain of the prunus necrotic ringspot virus (PNRSV) movement protein reveals its requirement for cell-to-cell movement. *Virology* **2005**, *339*, 31–41. [CrossRef] [PubMed]

69. Tomenius, K.; Clapham, D.; Meshi, T. Localization by immunogold cytochemistry of the virus coded 30K protein in plasmodesmata of leaves infected with tobacco mosaic virus. *Virology* **1987**, *160*, 363–371. [CrossRef]

70. Solovyev, A.G.; Savenkov, E.L. Factors involved in the systemic transport of plant RNA viruses: The emerging role of the nucleus. *J. Exp. Bot.* **2014**, *65*, 1689–1697. [CrossRef] [PubMed]

71. Citovsky, V.; Wong, M.L.; Shaw, A.L.; Prasad, P.V.; Zambryski, P. Visualization and characterization of tobacco mosaic virus movement protein binding to single-stranded nucleic acids. *Plant Cell* **1992**, *4*, 397–411. [CrossRef] [PubMed]

72. Citovsky, V. Probing plasmodesmal transport with plant viruses. *Plant Physiol.* **1993**, *102*, 1071–1076. [CrossRef] [PubMed]

73. Heinlein, M.; Epel, B.L.; Padgett, H.S.; Beachy, R.N. Interaction of Tobamovirus movement proteins with the plant cytoskeleton. *Science* **1995**, *270*, 1983–1985. [CrossRef] [PubMed]

74. McLean, B.G.; Zupan, J.; Zambryski, P.C. Tobacco mosaic virus movement protein associates with the cytoskeleton in tobacco cells. *Plant Cell* **1995**, *7*, 2101–2114. [CrossRef] [PubMed]

75. Hofmann, C.; Niehl, A.; Sambade, A.; Steinmetz, A.; Heinlein, M. Inhibition of tobacco mosaic virus movement by expression of an actin-binding protein. *Plant Physiol.* **2009**, *149*, 1810–1823. [CrossRef] [PubMed]

76. Vale, R.D. Intracellular transport using microtubule-based motors. *Annu. Rev. Cell Biol.* **1987**, *3*, 347–378. [CrossRef] [PubMed]

77. St Johnston, D. The intracellular localization of messenger RNAs. *Cell* **1995**, *81*, 161–170. [CrossRef]

78. Harries, P.A.; Schoelz, J.E.; Nelson, R.S. Intracellular transport of viruses and their components: Utilizing the cytoskeleton and membrane high-ways. *Mol. Plant Microbe Interact.* **2010**, *23*, 1381–1393. [CrossRef] [PubMed]

79. Boyko, V.; Ferralli, J.; Ashby, J.; Schellenbaum, P.; Heinlein, M. Function of microtubules in intercellular transport of plant virus RNA. *Nat. Cell Biol.* **2000**, *2*, 826–832. [CrossRef] [PubMed]

80. Boyko, V.; Ferralli, J.; Heinlein, M. Cell-to-cell movement of TMV RNA is temperature-dependent and corresponds to the association of movement protein with microtubules. *Plant J.* **2000**, *22*, 315–325. [CrossRef] [PubMed]

81. Boyko, V.; Hu, Q.; Seemanpillai, M.; Ashby, J.; Heinlein, M. Validation of microtubule-associated to tobacco mosaic virus RNA movement and involvement of microtubule-aligned particle trafficking. *Plant J.* **2007**, *51*, 589–603. [CrossRef] [PubMed]

82. Sambade, A.; Brandner, K.; Hofmann, C.; Seemanpillai, M.; Mutterer, J.; Heinlein, M. Transport of TMV movement protein particles associated with the targeting of RNA to plasmodesmata. *Traffic* **2008**, *9*, 2073–2088. [CrossRef] [PubMed]

83. Martinez-Gil, L.; Sanchez-Navarro, J.A.; Cruz, A.; Pallas, V.; Perez-Gil, J.; Mingarro, I. Plant virus cell-to-cell movement is not dependent on the transmembrane disposition of its movement protein. *J. Virol.* **2009**, *83*, 5535–5543. [CrossRef] [PubMed]

84. Hipper, C.; Brault, V.; Ziegeler-Graf, V.; Revers, F. Viral and cellular factors involved in phloem transport of plant viruses. *Front. Plant Sci.* **2013**, *4*, 1–24. [CrossRef] [PubMed]

85. Rao, A.L.N. Molecular studies on Bromovirus capsid protein. III. Analysis of cell-to-cell movement competence of coat protein defective variants of cowpea chlorotic mottle virus. *Virology* **1997**, *232*, 385–395. [CrossRef] [PubMed]

86. Kawakami, S.; Watanabe, Y.; Beachy, R.N. Tobacco mosaic virus infection spreads cell to cell as intact replication complexes. *Proc. Natl. Acad. Sci. USA* **2004**, *101*, 6291–6296. [CrossRef] [PubMed]

87. Blackman, L.M.; Boevink, P.; Santa Cruz, S.; Palukaitis, P.; Oparka, K.J. The movement protein of cucumber mosaic virus traffics into sieve elements in minor veins of Nicotiana clevelandii. *Plant Cell* **1998**, *10*, 525–537. [CrossRef] [PubMed]

88. Canto, T.; Palukaitis, P. Are tubules generated by the 3a protein necessary for cucumber mosaic virus movement? *Mol. Plant Microbe Interact.* **1999**, *12*, 985–993. [CrossRef]

89. Su, S.; Liu, Z.; Chen, C.; Zhang, Y.; Wang, X.; Zhu, L.; Miao, L.; Wang, X.C.; Yuan, M. Cucumber mosaic virus movement protein severs actin filaments to increase the plasmodesmal size exclusion limit in tobacco. *Plant Cell* **2010**, *22*, 1373–1387. [CrossRef] [PubMed]

90. Van Lent, J.; Wellink, J.; Goldbach, R.W. Evidence for the involvement of the 58K and 48K proteins in the intracellular movement of cowpea mosaic virus. *J. Gen. Virol.* **1990**, *71*, 219–223. [CrossRef]

91. Van Lent, J.; Storms, M.; Van der Meer, F.; Wellink, J.; Goldbach, R. Tubular structures involved in movement of cowpea mosaic virus are also formed in infected cowpea protoplasts. *J. Gen. Virol.* **1991**, *72*, 2615–2623. [CrossRef] [PubMed]

92. Flasinski, S.; Dzianott, A.; Pratt, S.; Bujarski, J. Mutational analysis of coat protein gene of brome mosaic virus: Effects on replication and movement protein in barley and on Chenopodium hybridum. *Mol. Plant Microbe Interact.* **1995**, *8*, 23–31. [CrossRef] [PubMed]

93. Rao, A.L.; Grantham, G.L. Biological significance of the seven amino-terminal basic residues of brome mosaic virus coat protein. *Virology* **1995**, *211*, 42–52. [CrossRef] [PubMed]

94. Van der Vossen, E.A.; Neeleman, L.; Bol, J.F. Early and late functions of alfalfa mosaic virus coat protein can be mutated separately. *Virology* **1994**, *202*, 891–903. [CrossRef] [PubMed]

95. Van der Wel, N.N.; Goldbach, R.W.; van Lent, J. The movement protein and coat protein of alfalfa mosaic virus accumulate in structurally modified plasmodesmata. *Virology* **1998**, *244*, 322–329. [CrossRef] [PubMed]

96. Kaido, M.; Inoue, Y.; Takeda, Y.; Sugiyama, K.; Takeda, A.; Mori, M.; Tamai, A.; Meshi, T.; Okuno, T.; Mise, K. Downregulation of the *NbNACa1* gene encoding a movement-protein-interacting protein reduces cell-to-cell movement of brome mosaic virus in Nicotiana benthamiana. *Mol. Plant Microbe Interact.* **2007**, *20*, 671–681. [CrossRef] [PubMed]

97. Sanchez-Navarro, J.A.; Herranz, M.C.; Pallas, V. Cell-to-cell movement of alfalfa mosaic virus can be mediated by the movement proteins of Ilar-, bromo-, cucumo-, tobamo- and comoviruses and does not require virion formation. *Virology* **2006**, *346*, 66–73. [CrossRef] [PubMed]

98. Huang, M.; Zhang, L. Association of the movement protein of alfalfa mosaic virus with the endoplasmic reticulum and its trafficking in epidermal cells of onion bulb scales. *Mol. Plant Microbe Interact.* **1999**, *12*, 680–690. [CrossRef]

99. Ueki, S.; Citovsky, V. Spread throughout the plant: Systemic transport of viruses. In *Viral Transport in Plants*, 1st ed.; Waigmann, E., Heinlein, M., Eds.; Springer: Berlin, Germany, 2007; pp. 85–118. ISBN 978-3-540-69967-5.

100. Ding, B. Intercellular protein trafficking through plasmodesmata. In *Plant Molecular Biology: Protein Trafficking in Plant Cells*, 1st ed.; Soll, J., Ed.; Springer: Berlin, Germany, 1998; pp. 279–310. ISBN 978-94-011-5298-3.

101. Oparka, K.J.; Santa Cruz, S. The great escape: Phloem transport and unloading of macromolecules. *Annu. Rev. Plant Physiol. Plant Mol. Biol.* **2000**, *51*, 323–347. [CrossRef] [PubMed]

102. Saito, T.; Yamanaka, K.; Okada, Y. Long-distance movement and viral assembly of tobacco mosaic virus mutants. *Virology* **1990**, *176*, 329–336. [CrossRef]

103. Pallas, V.; Garcia, J.A. How do plant viruses induce disease? Interactions and interference with host components. *J. Gen. Virol.* **2011**, *92*, 2691–2705. [CrossRef] [PubMed]

104. Kao, C.C.; Peng, N.I.; Hema, M.; Huang, X.; Dragnea, B. The coat protein leads the way: An update on basic and applied studies with the brome mosaic virus coat protein. *Mol. Plant Pathol.* **2011**, *12*, 403–412. [CrossRef] [PubMed]

105. Salánki, K.; Kiss, L.; Gellért, A.; Balázs, E. Identification a coat protein region of cucumber mosaic virus (CMV) essential for long-distance movement in cucumber. *Arch. Virol.* **2011**, *156*, 2279–2283. [CrossRef]

106. Requena, A.; Simón-Buela, L.; Salcedo, G.; García-Arenal, F. Potential involvement of a cucumber homolog of phloem protein 1 in the long-distance movement of cucumber mosaic virus particles. *Mol. Plant Microbe Interact.* **2006**, *19*, 734–746. [CrossRef] [PubMed]

107. Bol, J.F. Alfalfa mosaic virus and ilarviruses: Involvement of coat protein in multiple steps of the replication cycle. *J. Gen. Virol.* **1999**, *80*, 1089–1102. [CrossRef] [PubMed]

108. Tenllado, F.; Bol, J.F. Genetic dissection of the multiple functions of alfalfa mosaic virus coat protein in viral RNA replication, encapsidation, and movement. *Virology* **2000**, *268*, 29–40. [CrossRef] [PubMed]
109. Takahashi, H.; Miller, J.; Nozaki, Y.; Sukamuto, J.; Takeda, M.; Shah, J.; Hase, S.; Ikegami, M.; Ehara, Y.; Dinesh-Kumar, S.P. RCY1, an Arabidopsis thaliana RPP8/HRT family resistance gene, conferring resistance to cucumber mosaic virus requires salicylic acid, ethylene and a novel signal transduction mechanism. *Plant J.* **2002**, *32*, 655–667. [CrossRef] [PubMed]
110. Inaba, J.; Kim, B.M.; Shimura, H.; Masuta, C. Virus-induced necrosis is a consequence of direct protein-protein interaction between a viral RNA-silencing suppressor and a host catalase. *Plant Physiol.* **2011**, *156*, 2026–2036. [CrossRef] [PubMed]

International Journal of
*Molecular Sciences*

MDPI

*Article*

# Plant Cell Wall Dynamics in Compatible and Incompatible Potato Response to Infection Caused by *Potato Virus Y* (PVY$^{NTN}$)

Katarzyna Otulak-Kozieł [1],*, Edmund Kozieł [1] and Benham E. L. Lockhart [2]

[1]  Department of Botany, Faculty of Agriculture and Biology, Warsaw University of Life Sciences—SGGW, 159 Nowoursynowska St., 02-776 Warsaw, Poland; edmund_koziel@sggw.pl
[2]  Department of Plant Pathology, University of Minnesota, St. Paul, MN 55108, USA; lockh002@umn.edu
*   Correspondence: katarzyna_otulak@sggw.pl; Tel.: +48-225-932-657

Received: 22 January 2018; Accepted: 13 March 2018; Published: 15 March 2018

**Abstract:** The cell wall provides the structure of the plant, and also acts as a barier against biotic stress. The vein necrosis strain of *Potato virus Y* (PVY$^{NTN}$) induces necrotic disease symptoms that affect both plant growth and yield. Virus infection triggers a number of inducible basal defense responses, including defense proteins, especially those involved in cell wall metabolism. This study investigates the comparison of cell wall host dynamics induced in a compatible (potato cv. Irys) and incompatible (potato cv. Sárpo Mira with hypersensitive reaction gene *Ny-Smira*) PVY$^{NTN}$–host–plant interaction. Ultrastructural analyses revealed numerous cell wall changes induced by virus infection. Furthermore, the localization of essential defensive wall-associated proteins in susceptible and resistant potato host to PVY$^{NTN}$ infection were investigated. The data revealed a higher level of detection of pathogenesis-related protein 2 (PR-2) in a compatible compared to an incompatible (HR) interaction. Immunofluorescence analyses indicated that hydroxyproline-rich glycoproteins (HRGP) (extensin) synthesis was induced, whereas that of cellulose synthase catalytic subunits (CesA4) decreased as a result of PVY$^{NTN}$ infection. The highest level of extensin localization was found in HR potato plants. Proteins involved in cell wall metabolism play a crucial role in the interaction because they affect the spread of the virus. Analysis of CesA4, PR-2 and HRGP deposition within the apoplast and symplast confirmed the active trafficking of these proteins as a step-in potato cell wall remodeling in response to PVY$^{NTN}$ infection. Therefore, cell wall reorganization may be regarded as an element of "signWALLing"—involving apoplast and symplast activation as a specific response to viruses.

**Keywords:** cell wall; cellulose synthase; hypersensitive response; pathogenesis related-protein 2; plant-virus interaction; *Potato virus Y*; ultrastructure

---

## 1. Introduction

The cell wall serves both to define and maintain the structural integrity of the plant as well as to protect from external stress. It is estimated that as much as 15% of the plant genome is associated with cell wall metabolism and biosynthesis. It also functions to adapt the plant to changes during growth and development [1]. This involves modifications in composition and structure of the cell wall through sensing and signaling, intercellular communication as well as exchange interfaces [2,3].

*Potato virus Y* (genus *Potyvirus*) has been listed as the fifth most economically damaging plant virus in the world [4]. The symptoms of PVY infection depend on virus strain and host resistance level. It infects more than 170 species belonging to 34 genera and has a worldwide distribution. The PVY$^{NTN}$, or vein necrosis strain, induces diverse symptoms including necrotic ringspotting which greatly decreases the market quality of harvested tubers [5]. Host response to virus infection is categorized as either compatible (susceptible) or incompatible (resistant) interactions. In the former, the virus spreads systemically throughout the host, whereas in the latter, systemic spread is prevented by localized cell death (hypersensitive response, HR) [6,7].

Viruses, like many other pathogens, trigger a number of inducible defense responses when recognized by plants, inducing the up-regulation of a number of common defensive proteins. Proteins/enzymes involved in cell wall metabolism play a crucial role in this interaction through the influence on cell-to-cell virus spread [3]. Following initial infection, viruses are potentially able to invade adjacent cells, but systemic spread depends on the mechanism involved in cell-to-cell movement [8]. The cell wall may act as a physical barrier to virus infection as well as to restrict intercellular movement [9]. Initial infection of the plant occurs via mechanical wounding or insect feeding that breeches the surface (epidermal) layer [10]. Following replication in initial sites of infection, both intra-cellular and inter-cellular movement of progeny virus depends on the intervention of several host mechanisms, including plasmodesmata modifications [11]. Current information on modifications in composition, structure and function of plant cell walls as related to defense against pathogen invasion has been largely restricted to non-viral pathogens [12–14]. This work has concentrated mainly on pectin, pectic polysaccharide or glycan modifications during plant-microbe interaction, and has focused on cell wall polymers such as callose [15] or lignin [16]. However, knowledge about crucial components in the active cell wall defense strategies relative to responses to other pathogens is lacking. Recent proteomic studies have suggested that certain proteins undergo regulation during plant virus infection. This, as well as other studies on molecular host-virus proteins interaction, suggest that changes in cell wall-associated proteins and cell wall metabolism may play an important function in regulating cell-to-cell trafficking of plant viruses [7,17]. Therefore, this study focuses on ultrastructural cell wall changes and dynamics in potential rearrangements of cell wall proteins in response to PVY$^{NTN}$ infection using differences between susceptible and hypersensitive resistant potato varieties. This study describes for the first time the precise ultrastructural changes occurring in the two types of PVY$^{NTN}$-potato interactions (compatible and incompatible). The findings reveal changes in the regulation of protein production involved in potato cell wall synthesis (e.g., CesA4), and localization of proteins arranged in cell wall remodeling processes (pathogenesis related protein (PR-2) and HRGP (extensin)).

## 2. Results

### 2.1. Ultrastructural Analyses of Adjoining Cell Wall Connections in Compatible and Incompatible PVY$^{NTN}$-Potato Interactions

Changes in cell wall ultrastructure induced by PVY$^{NTN}$ infection were analyzed ten days after mechanical inoculation of potato cvs. Irys (compatible) and Sárpo Mira (hypersensitive). In the susceptible potato-PVY$^{NTN}$ interaction the cell wall folding with loosening structure was mainly observed in mesophyll cells and also in necrotic areas (Figure 1A). In compatible interactions multivesicular bodies connected with plasmodesmata were noticed in associations with PVY$^{NTN}$ cytoplasmic inclusions (Figure 1C). Moreover, modifications of cell wall area were very strongly associated with membrane and vesicular structures and also with presence of PVY particles or inclusions (Figure 1E). Additionally, in mesophyll or parenchyma in the vascular bundle area, complexes of viral cytoplasmic inclusions and plasmodesmata resulted in cell wall protrusions (Figure 1F,G). Phloem tissue typically showed abnormal thickening of cell walls with numerous plasma membrane-derived vesicles (Figure 1B,D).

In incompatible interactions extensive disruptions of epidermal cell wall structure were observed (Figure 2A,B). In sections of mesophyll, the plasma membrane was separated from cell wall (Figure 2C). Irregular thickening and folding of the cell wall was also observed in the hypersensitive response (Figure 2D). Phenolic-like compounds were observed as electron dense areas within cell walls (Figure 2E) or as layers between cell walls and plasmalemma (Figure 2D). Plasmodesmata enlargement in the vicinity of enlarged endoplasmic reticulum membranes were observed primarily in mesophyll tissue, but virus particles were rarely observed (Figure 2E). The deposition of callose-like material was associated with phloem adjacent cells, especially in sieve elements (Figure 2E–H). Multivesicular bodies were detected in phloem tissue and observed changes indicated clearly that cell wall material was being translocated to the vacuoles (Figure 2G). Neither cell wall disruption, nor virus particles and inclusions were observed in comparable healthy tissues (mock-inoculated) (Figure 2I).

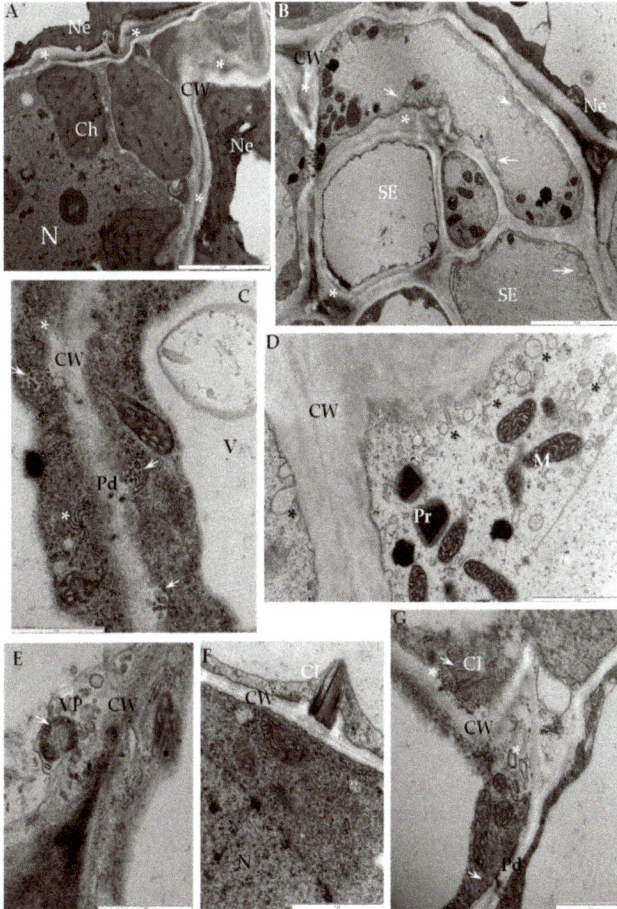

**Figure 1.** Ultrastructure changes in cell wall area from potato Irys–PVY$^{NTN}$ interaction. (**A**) Cell wall loosening and folding (*) between necrotized and non-necrotized mesophyll cells. Bar 5 μm. (**B**) Cell wall thickening (*) with paramural bodies (arrows) in phloem cells. Bar 5 μm. (**C**) Paramural bodies (arrow) connected with plasmodesmata in association with virus cytoplasmic inclusions (*). Bar 1 μm. (**D**) Cell wall thickening with intense vesicles distribution (*). Bar 2 μm. (**E**) PVY particles (arrow) in vesicle derived from cell wall with other paramural bodies in the vicinity of changed cell wall. Bar 1 μm. (**F**) PVY inclusions in associated with highly changed cell wall in mesophyll cell. Bar 1 μm. (**G**) PVY inclusions (arrow) associated with plasma membranes and intense distribution of paramural bodies (asterisk) along the deformed cell wall. Bar 1 μm. Ch—chloroplast, CI—cytoplasmic inclusions, CW—cell wall, ER—endoplasmic reticulum, N—nucleus, Ne—necrosis, Pd—plasmodesmata, SE—sieve element, V—vacuole, VP—virus particles.

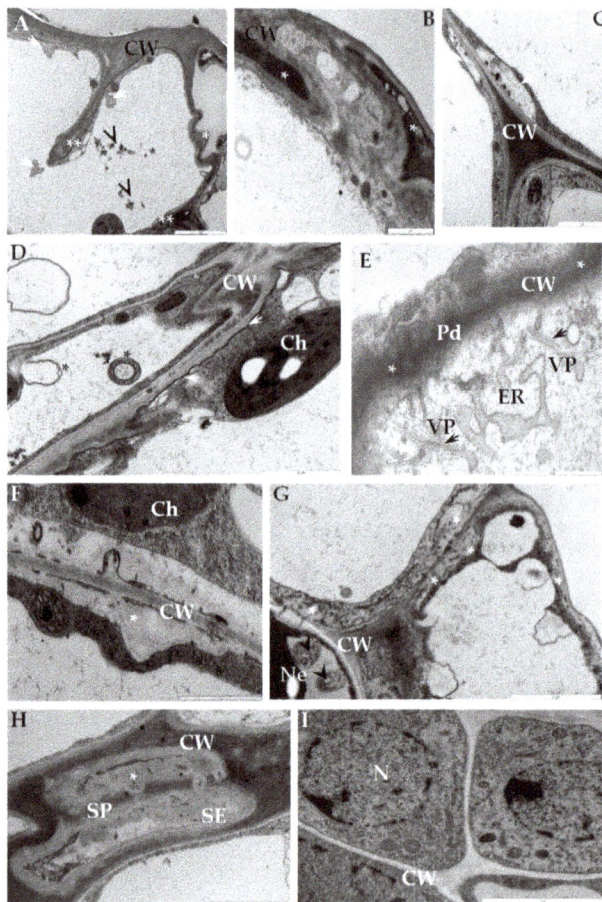

**Figure 2.** Ultrastructure changes in cell wall area from potato Sárpo Mira–PVY$^{NTN}$ interaction.
(**A**) The cell wall folding (*) and invagination (**) in epidermis during HR. Multivesicular bodies
(arrows) and phenolic compounds (arrowhead) presented in vacuole. Bar 2 μm. (**B**) Electron dense
phenolic compounds material in reinforced epidermis cell wall (*). Bar 1 μm. (**C**) Plasma membrane
retrations from cell wall (arrows) between mesophyll cells. Bar 1 μm. (**D**) Reinforced cell wall
between mesophyll cells, phenolic compounds as a layer between plasma membrane and cell wall
(arrows). Multivesicular bodies in vacuole (*). Bar 2 μm. (**E**) Expanded plasmodesmata area, enlarged
endoplasmic reticulum, virus particles associated with membranous structure of ER (arrows). Phenolic
compounds as an electron dense cell wall (asterisk). Bar 2 μm. (**F**) Callose material deposition between
cell wall and plasma membrane in mesophyll cells. Bar 1 μm. (**G**) Callose deposition (asterisk) between
phloem parenchyma cells. Multivesicular bodies in vacuole of necrotized cell (arrowhead). Bar 2 μm.
(**H**) Callose (*) around sieve plate in sieve element. Bar 2 μm. (**I**) Phloem parenchyma cells from
mock-inoculated leaf (control material). Bar 1 μm. Ch—chloroplast, CW—cell wall, ER—endoplasmic
reticulum, Pd—plasmodesmata, SE—sieve element, V—vacuole, VP—virus particles.

### 2.2. Localization of Pathogenesis Related Protein 2 (PR-2, β-1,3 Glucanase, EC 3.2.1.39)

The immunofluorescence localization of PR-2 in susceptible and resistant potato leaves indicated
that PR-2 was induced by PVY$^{NTN}$ infection (Figure 3), but absent in controls (Figure 3A,G).
Green fluorescence due to PR-2 accumulation was noticeably higher in the compatible interaction

(Figure 3B,C), as compared to the hypersensitive response (Figure 3D–F). In cv. Irys leaf tissues green signal developing in mesophyll cells, stomata and vascular bundles, was higher in inoculated as compared to mock-inoculated treatments, especially in xylem (Figure 3B,C). In contrast, PR-2 deposition associated with hypersensitive response occurred mainly in xylem and phloem elements (Figure 3D–F) with scattered symplast deposition in spongy mesophyll cells adjacent to the vasculature (Figure 3E,F). Immunogold labeling revealed that PR-2 occurred in areas of the cell wall (Figure 4B–E,G–I), often adjacent to plasmodesmata (Figure 4B–E,G–H) and also associated with presence of virus inclusions. PR-2 was detected attached to the symplast membranous structure in phloem parenchyma (Figure 4C) and mesophyll cells (Figure 4E,H).

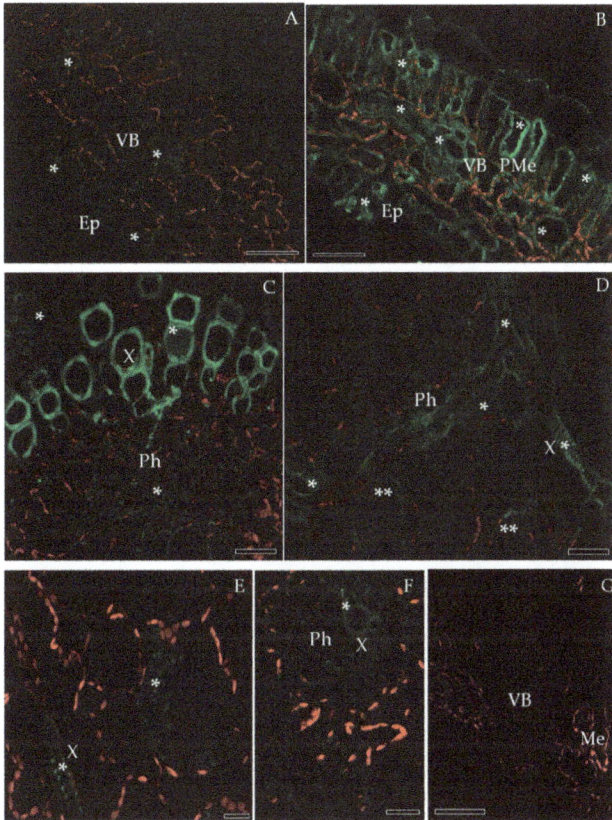

**Figure 3.** Immunofluorescence localization of PR-2 protein in potato-PVY$^{NTN}$ compatible (**B,C**) and incompatible interaction (**D–G**). (**A**) Green fluorescence signal of PR-2 in vascular bundles and epidermis (asterisk) of mock-inoculated potato leaf (control). Bar 200 μm. (**B**) Green fluorescence of PR-2 in palisade mesophyll, vascular bundle and epidermis with stomata (asterisk) of cv. Irys infected with PVY$^{NTN}$. (**C**) Fluorescence PR-2 signal in xylem and phloem elements (asterisk) of potato Irys infected with PVY$^{NTN}$. (**D**) PR-2 signal in phloem & xylem element (*) of Sárpo Mira inoculated with PVY$^{NTN}$, also in the cell wall of spongy parenchyma cells (**). (**E**) PR-2 in the cell wall of xylem elements and in the symplast of spongy mesophyll. (**F**) PR-2 in the cell wall of phloem and xylem elements inside Sárpo Mira leaflets. Bar 200 μm. (**G**) Control—lack of green fluorescence signal in hypersensitive response when primary antibodies were omitted. Bar 200 μm. Ep—epidermis, Me—mesophyll, Ph—phloem, PMe—palisade mesophyll, VB—vascular bundle, X—xylem.

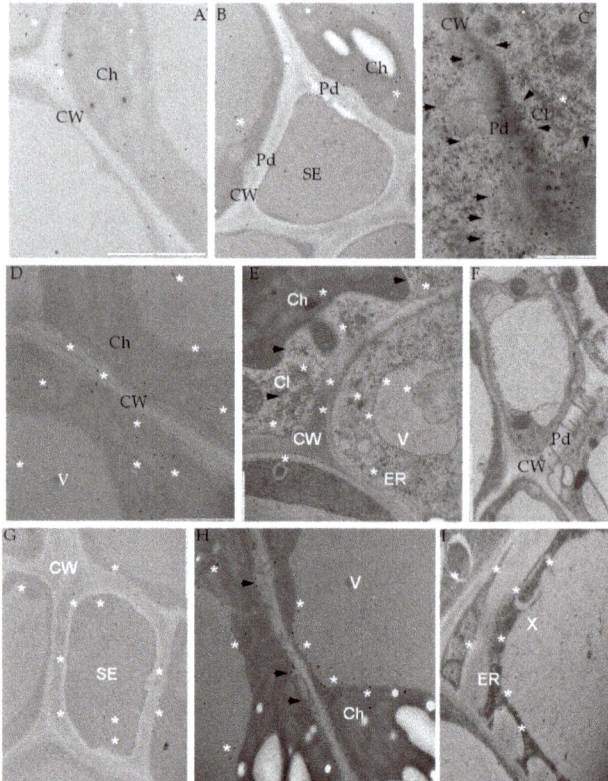

**Figure 4.** Immunogold labeling of PR-2 in potato-PVY$^{NTN}$ compatible (**B–E**) and incompatible interaction (**F–I**). (**A**) Gold deposition in mesophyll of mock-inoculated leaf. Bar 2μm. (**B**) PR-2 deposition in sieve elements in potato Irys inoculated with PVY$^{NTN}$. Bar 1 μm. (**C**) PR-2 deposition in plasmodesmata area and around (\*) membranous structure (arrows) in phloem parenchyma. Virus inclusions associated with plasmodesmata. Bar 1 μm. (**D**) PR-2 localization (\*) in symplast of potato Irys mesophyll cells. Deposition in cell wall in plasmodesmata area, chloroplast and vacuole. Bar 2 μm. (**E**) PR-2 localization along membranous structures (asterisk) [ER, tonoplast and plasmalema]. Virus inclusion associated to membranes (arrows). Bar 1 μm. (**F**) Control—lack of gold deposition in hypersensitive response when primary antibodies were omitted. Bar 2 μm. (**G**) PR-2 location (\*) in sieve elements in HR response. Bar 1 μm. (**H**) PR-2 location along expanded cell wall area in the vicinity of plasmodesmata (arrows). Localization also in vacuole and chloroplasts (asterisk). Bar 1 μm. (**I**) PR-2 deposition inside xylem tracheary elements, along cell wall and along ER or in vacuole. Bar 1 μm. Ch—chloroplast, CI—cytoplasmic inclusions, CW—cell wall, ER—endoplasmic reticulum, Pd—plasmodesmata, SE—sieve element, V—vacuole, X—xylem tracheary element.

Levels of PR-2 antigen detection were slightly lower in mock-inoculated cv. Sárpo Mira than in cv. Irys plants (Figure 5A). Regardless of cultivar, PVY infection caused increase of PR-2 presence in host tissues in comparison to controls (Figures 4A–I and 5A). Increase of PR-2 levels in inoculated relative to mock-inoculated controls was 66.3% for susceptible and 60.3% for resistant cultivars. The highest levels of PR-2 was observed in PVY$^{NTN}$ infected potato plants cv. Irys (mean number of gold particles = 50.5) (Figure 5A). A lower rate of PR-2 localization occurred in PVY$^{NTN}$ resistant cv. Sárpo Mira (mean number of gold particles = 39). The difference between susceptible and resistant potato plants was statistically significant and was approximately 23%. The above results suggest that PR-2 protein is present at higher levels during PVY infection in susceptible than in resistant cultivars exhibiting a hypersensitive response. Quantification of PR-2 antigen by immunogold localization in host cell compartments (Figure 5B) indicated that the localization pattern was in general similar to overall tissue localization of PR-2 (Figure 5A). A lack of localization in trans-Golgi network (TGN) was observed. The highest levels of PR-2 localization were noticed in the susceptible PVY-infected cultivar and was greatest in cell walls and the lowest in cytoplasm (Figure 5B). In contrast, in the resistant potato cultivar the level of detectable PR-2 was highest in the vacuole. Levels of PR-2 in the vacuole in plants undergoing a hypersensitive response exceeded not only that occurring in mock-inoculated plants, but also in cv. Irys plants systemically infected with PVY. Immunogold labeling results showed induction of PR-2 in all cell compartments of PVY-infected potato cultivars cv. Irys and cv. Sárpo Mira compared to mock-inoculated plants. Statistical analyses of PR-2 localization also indicated different pattern of plant reaction to PVY$^{NTN}$ infection depending on the level of resistance. In susceptible cultivars during PVY infection, PR-2 was primarily located in the cell wall. This specific protein localization is directly associated with host cell wall modification induced by the pathogen to increase virus mobility. Similar cell wall modification was also observed by ultrastructural analysis. A different host–plant response was observed in PVY-infected resistant plants, in which PR-2 was strongly sequestered in the vacuole, and in this case storage of PR-2 was seldom observed in the cell wall. During the HR response induced by PVY, PR-2 is likely actively withdrawn or stored in the vacuole.

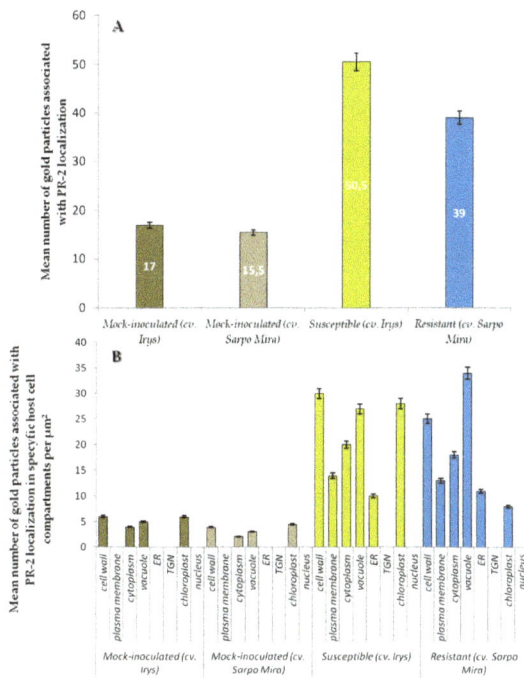

**Figure 5.** Statistical significance assessment of PR-2 epitopes immunogold localization (**A**) in mock-inoculated and PVY inoculated potato tissues of cultivars Irys (susceptible) and Sárpo Mira (resistant) & (**B**) in specific host cell compartments of mock-inoculated and PVY infected potato plants (susceptible and resistant). Figure presents mean numbers of gold particles located in specific compartment per $\mu m^2$. Quantification immunogold localization was prepared using ANOVA method. Mean values of gold particle localization were evaluated at the $p < 0.05$ level of significance using post-hoc Tukey HSD test.

## 2.3. Localization of Cellulose Synthase 4 (CesA4, EC 2.4.1.12)

Immunofluorescence detection of CesA4 antigen gave similar results in both susceptible and resistant potato cultivars (Figure 6B–D) in comparison to controls (Figure 6A,E). CesA4 signal was observed mainly in xylem elements, but also in spongy mesophyll regardless of potato resistance level (Figure 6B–D). In addition, dispersed fluorescence occurred also in epidermis with stomata in the compatible interaction (Figure 6B). The CesA4 signal comes not only from the cell wall, but also from the symplast (Figure 6C).

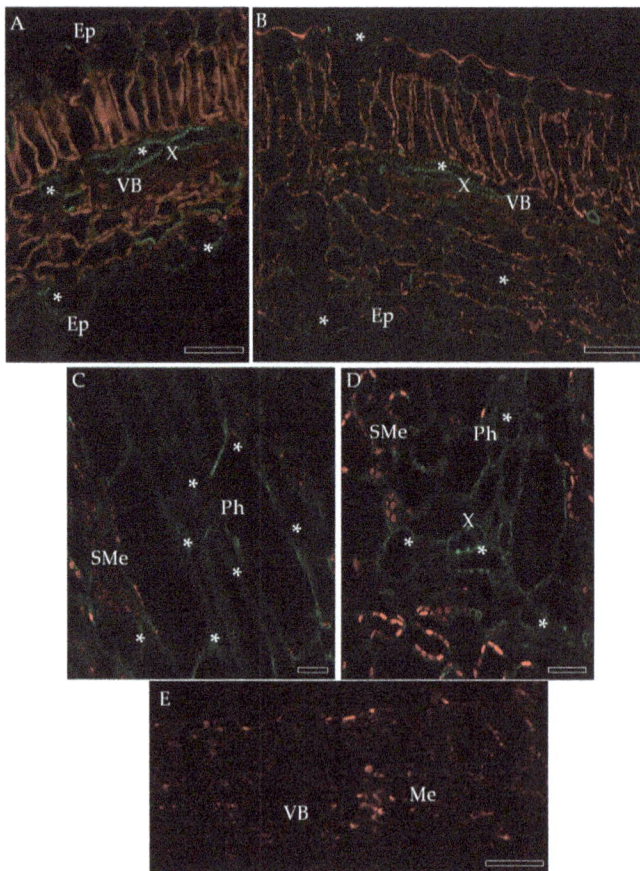

**Figure 6.** Fluorescence detection of CesA4 protein in potato–PVY$^{NTN}$ compatible (**B**) and incompatible interaction (**C–E**). (**A**) CesA4 signal (*) in xylem elements and epidermis of mock-inoculated leaf. (**B**) CesA4 signal (*) in xylem, spongy mesophyll cell wall and epidermis (also in stomata) of Irys leaf inoculated with PVY$^{NTN}$. (**C**) CesA4 green fluorescence in cell wall and symplast of phloem elements in Sárpo Mira leaf (asterisk). Signal also in spongy mesophyll cell wall. (**D**) CesA 4 (*) detection in spongy mesophyll and vasculature of Sárpo Mira. (**E**) Control-lack of green fluorescence in hypersensitive response when primary antibodies were omitted. Bar 200 μm. Ep—epidermis, Ph—phloem, SMe—spongy mesophyll, VB—vascular bundle, X—xylem.

Analysis of CesA4 deposition within the apoplast and symplast (Figure 7) confirmed the active trafficking of this protein as a step-in potato cell wall remodeling in response to PVY$^{NTN}$ infection.

CesA4 localization was observed in all leaf tissues from epidermis to xylem tracheary elements in susceptible potato, particularly in areas where virus cytoplasmic inclusions or particles were present (Figure 7B–E). In the hypersensitive response, significant CesA4 deposition was observed in both vascular tissues (Figure 7G–I).

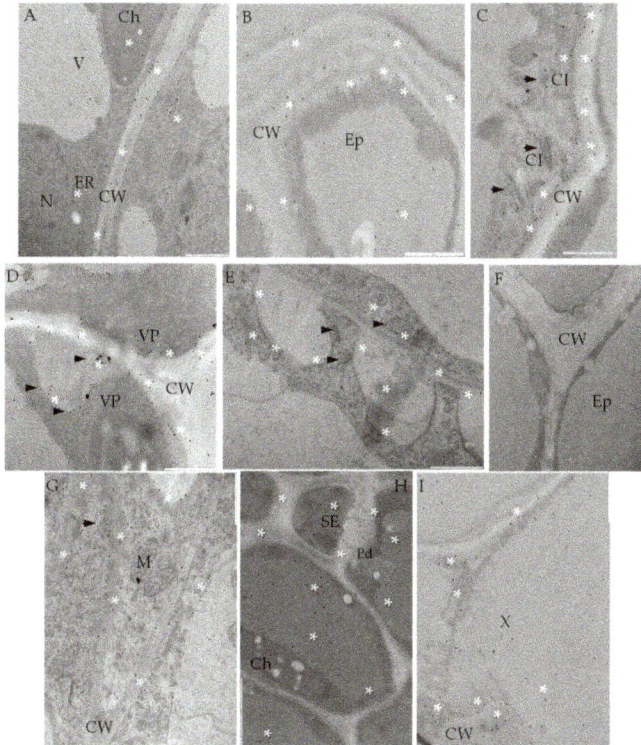

**Figure 7.** Immunogold labeling of CesA4 in potato–PVY$^{NTN}$ compatible (**B–E**) and incompatible interaction (**F–I**). (**A**) Gold deposition (*) of CesA4 along cell wall, plasmolema and ER in mesophyll cells of potato mock-inoculated. Bar 1 μm. (**B**) CesA4 deposition (*) in cell wall and vacuole of epidermis cell. Bar 1 μm. (**C**) CesA4 deposition along cell wall and membranous structures associated with virus cytoplasmic inclusions (arrows). Bar 1 μm. (**D**) CesA4 localization (*) along cell wall in potato Irys mesophyll cells. Virus particles around plasmodesmata, gold deposition in area of protoplast retraction from the cell wall (arrow). Bar 1 μm. (**E**) CesA4 localization (*) in xylem tracheary elements. Deposition along cell wall, membranous structures associated with virus inclusions (arrows) and trans Golgi network. Bar 1 μm. (**F**) Control—lack of gold deposition in hypersensitive response when primary antibodies were omitted. Bar 1 μm. (**G**) CesA4 localization (*) along plasma membrane structures and plasmalema (also in connection with cytoskeleton, arrow). Bar 1 μm. (**H**) Gold deposition (*) in cell wall associated with plasmodesmata in sieve element in hypersensitive reaction. Localization also in symplast, especially in vacuole. Bar 2 μm. (**I**) CesA4 localization inside xylem tracheary elements in hypersensitive reaction. Bar 2 μm. Ch—chloroplast, CI—cytoplasmic inclusions, CW—cell wall, Ep—epidermis, ER—endoplasmic reticulum, M—mitochondria, Pd—plasmodesmata, SE—sieve element, V—vacuole, VP—virus particles, X—xylem tracheary element.

The quantification analyses revealed a statistically significant decrease of CesA4 deposition in both types of PVY infected plants, in comparison to controls (Figures 7A,F and 8A). The estimated reduction of CesA4 based on gold particles localization was aproximately 20%-cv. Irys, and 27.6% in cv. Sárpo Mira. No significant differences in CesA4 localization between mock-inoculated cv. Irys (mean number of gold particles = 90) and cv. Sárpo Mira (mean number of gold particles = 87) were observed (Figure 7A). Decreased deposition of CesA4 in PVY infected plants of both cultivars was statistically valid and was greater in cv. Sárpo Mira (HR) than in cv. Irys (compatible). The CesA4 antigen was detected primarily in membranous structures and organelles, also in the TGN and ER, in both mock-inoculated and inoculated potato plants (Figure 8B). Statistical analyses of CesA4 antigen concentration showed that significant decline in levels of CesA4 accumulation occurred in PVY-infected compared to mock-inoculated potato plants (Figure 8B). High amounts of CesA4 antigen were detected in the plasma membrane and cell wall in both cultivars. ANOVA evaluations of CesA4 antigen levels showed statistically significant differences between localization patterns in relation to PVY resistance level. In resistant plants, similar CesA4 levels founded within both vacuole and plasma membrane and were noticed in lower levels than in mock-inoculated and infected susceptible plants. In susceptible cultivars, levels of CeA4 antigen in plasma membrane, cell wall and ER were higher.

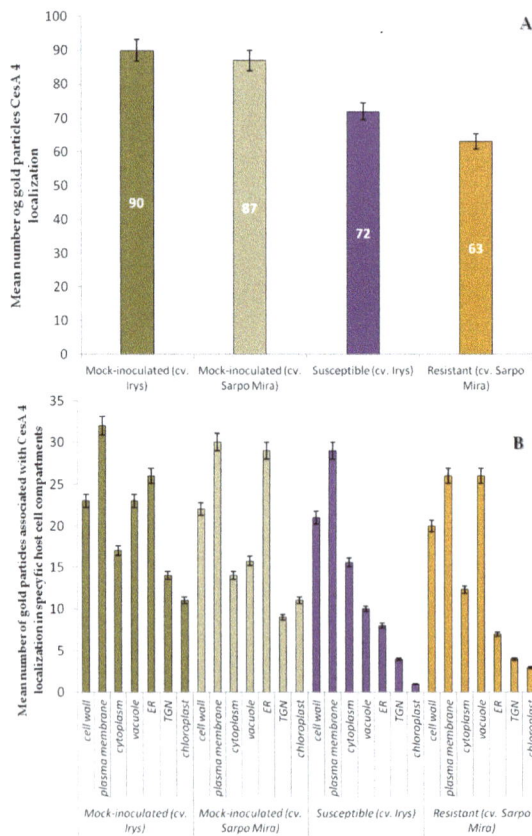

**Figure 8.** Statistical significance assessment of CesA4 epitopes immunogold localization (**A**) in mock-inoculated and PVY inoculated potato tissues of cultivars Irys (susceptible) and Sárpo Mira (resistant) & (**B**) in specific host cell compartments of mock-inoculated and PVY inoculated potato plants (susceptible and resistant). Figure presents mean numbers of gold particles located in specific compartment per $\mu m^2$. Quantification immunogold localization was prepared using ANOVA method. Mean values of gold particle localization were evaluated at the $p < 0.05$ level of significance using post-hoc Tukey honest significant difference (HSD) test.

These data indicate that CesA4 synthesis reduction is greater during viral infection in cv. Sárpo Mira than cv. Irys plants. HR reaction induced by PVY (cv. Sárpo Mira) caused high accumulation of CesA4 inside the vacuole. Such high levels of CesA4 deposition in vacuole were observed neither in mock-inoculated nor infected susceptible plants.

## 2.4. Localization of Extensin (HRGPs) in Susceptible and Resistant Potato-PVY$^{NTN}$ Interactions

Localization of extensin (hydroxyproline-rich glycoproteins, HRGPs), in the PVY$^{NTN}$-infected susceptible (Irys) and resistant (Sárpo Mira) potato cultivars was investigated by immunofluorescence. In compatible interaction, the strong green fluorescence signal of extensin was observed on the surface of epidermial cell walls as well as in both xylem and phloem tissues (Figure 9B,C) in comparison to controls (Figure 9A,F). In the resistant cultivar (HR), HRGPs antigen occurred predominantly in xylem and phloem elements, and was also visible in spongy mesophyll cells (Figure 9D,E). In cv. Irys (compatible interaction) extensin was detected by immunogold labeling in the cell wall area, but primarily in symplast of epidermis (Figure 10B) and mesophyll cells, especially when virus particles or inclusions were deposited in cytoplasm (Figure 10C,D). In the incompatible interaction (HR) the level of HRGPs deposition was visibly greater (Figure 10F–J). The localization occurred in lower and upper epidermis, and also in necrotic areas (Figure 10F,G), mesophyll (Figure 10H,) as well as in xylem tracheary elements (Figure 10J).

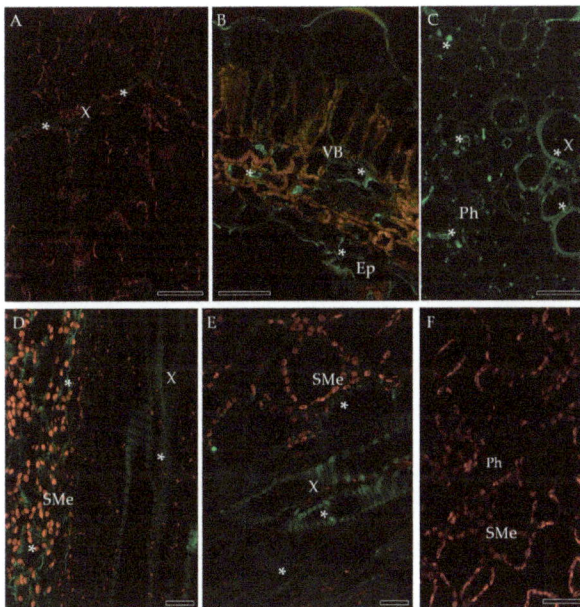

**Figure 9.** Fluorescence detection of extensin (HRGP) in potato–PVY$^{NTN}$ compatible (**B,C**) and incompatible interaction (**D–E**). (**A**) Extensin signal (*) in xylem elements of mock-inoculated leaf. (**B**) Extensin in xylem cell wall, spongy mesophyll symplast as well as in epidermis in Irys leaf cross section inoculated with PVY$^{NTN}$ (asterisk). (**C**) Green fluorescence signal of extension (*) in phloem and xylem elements in Irys leaf blades. (**D**) Fluorescence signal in the cell wall of xylem and spongy mesophyll in HR. (**E**) Fluorescence extension signal in phloem and xylem elements in HR. (**F**) Control—lack of green fluorescence signal in hypersensitive response when primary antibodies were omitted. Bar 200 μm. Ep—epidermis, Ph—phloem, SMe—spongy mesophyll, VB—vascular bundle, X—xylem.

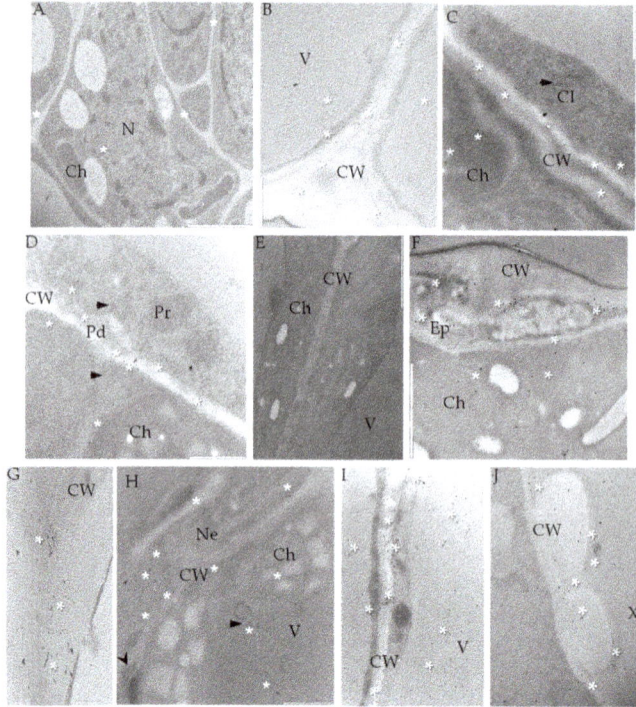

**Figure 10.** Immunogold labeling of extensin (HRGP) in potato–PVY$^{NTN}$ compatible (**B–D**) and incompatible interaction (**E–J**). (**A**) Gold deposition (*) of extensin in phloem parenchyma cells of mock-inoculated potato. Bar 1 μm. (**B**) Extensin localization (*) in epidermis cell wall and vacuole. Bar 1 μm. (**C**) Extensin localization (*) in cell wall and chloroplast in mesophyll cell of potato Irys. Virus cytoplasmic inclusion (arrow) in cytoplasm. Bar 1 μm. (**D**) Extensin deposition (*) around plasmodesmta associated with virus cytoplasmic inclusion (arrow). Bar 1 μm. (**E**) Control—lack of gold deposition in hypersensitive response when primary antibodies were omitted. Bar 2 μm. (**F**) Extensin gold deposition (*) in necrotized epidermis and inside chloroplast in Sárpo Mira mesophyll cell. Bar 3 μm. (**G**) Extensin deposition (*) in epidermis cell wall in Sárpo Mira. Bar 1 μm. (**H**) Extensin localization (*) in necrotized mesophyll area in hypersensitive reaction. Deposition also around multivesicular bodies (arrow) and paramural bodies (arrowhead). Bar 2 μm. (**I**) Extensin deposition (*) in cell wall, cytoplasm and vacuole of mesophyll cell in potato Sárpo Mira. Bar 2 μm. (**J**) Extensin localization (*) in xylem tracheary elements. Bar 1 μm. Ch—chloroplast, CI—cytoplasmic inclusions, CW—cell wall, Ep—epidermis, ER—endoplasmic reticulum, M—mitochondria, N—nucleus, Pd—plasmodesmata, Pr—peroxisome, SE—sieve element, V—vacuole, VP—virus particles, X—xylem tracheary element.

Quantification by immunogold localization of extensin epitopes revealed an increase of HRGPs in potato plants infected by PVY$^{NTN}$. The mean number of gold particles was 61.6 for the compatible interaction and 96.2 for the HR interaction (Figure 11A). The level of detectable extensin in mock-inoculated cv. Sárpo Mira plants was estimated as lower than in cv. Irys. Therefore, the induction of extensin by PVY$^{NTN}$ infection in different cultivars was at various level of intensity. Analyzed data showed that induction of extensin in resistant plants during PVY infection was more rapid and intense. Further analyses of variances in specific cell compartments indicated more interesting results. In all type of controls we neither observed localization of extensin in cell membrane and vacuole, nor in TGN and ER (Figures 10A,E and 11B). Regardless of cultivar resistant level, PVY infection induced presence of extensin in plasma membrane, vacuole or ER in contrast to mock-inoculated plants. The highest level of extensin was detected in resistant cultivars (Figure 11B), with the highest levels in cell wall and cytoplasm, and the lowest in ER. Interestingly, the localization of extensin in vacuoles was significantly greater in infected cv. Irys than in cv. Sárpo Mira. In infected susceptible potato cv. Irys localization pattern was different from observed in cv. Sárpo Mira. Extensin accumulation was largest in the cell wall and vacuole and least in the ER in the compatible interaction. The statistical data indicate that viral infection induced different patterns of extensin localization within potato cell compartments depending on plant-host interaction. It appears possible that in resistant potato cultivars, HRGPs translocation is diverted from vacuoles to cell walls. In contrast, in susceptible cv. Irys plant-PVY$^{NTN}$ interaction extensin accumulation occurred in the vacuole. These data show clearly that location of extensin in the cell wall during HR is crucial for cell wall modification, which is a part of the complex system providing resistance to PVY infection.

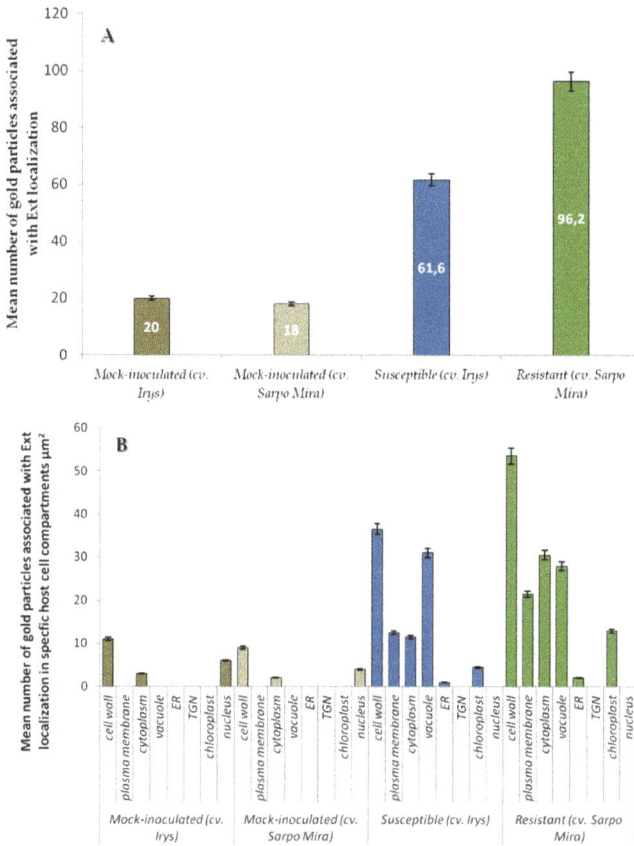

**Figure 11.** Statistical significance assessment of extensin (HRGP) epitopes immunogold localization (**A**) in mock-inoculated and PVY inoculated potato tissues of cultivars Irys (susceptible) and Sárpo Mira (resistant) & (**B**) in specific host cell compartments of mock-inoculated and PVY inoculated potato plants (susceptible and resistant). Figure presents mean numbers of gold particles located in specific compartment per μm². Quantification immunogold localization was prepared using ANOVA method. Mean values of gold particle localization were evaluated at the $p < 0.05$ level of significance using post-hoc Tukey HSD test.

## 3. Discussion

Successful systemic infection of the plant host by plant virus infection depends on its ability to overcome the several barriers to cell-to-cell transport. After cell entry via the host symplast and initiation of replication, systemic infection is successful when progeny virus is able to move freely to adjacent cells. Plasmodesmata act as intercellular connections between plant cells [18] with the help of the host cytoskeleton and endomembranes, which can facilitate virus symplastic movement [19]. The plant cell wall is the first contact area, which plays an important role in initiation as well as regulation processes of defense response. A deeper understanding of the different properties and activities of the plant cell wall is of fundamental importance for understanding the evolution of plant defense strategies. A number of recent publications have proposed models of host–plant cell wall modifications in response to actively penetrating pathogens such as fungi, nematodes and bacteria. Interaction between plant pathogens and the host cell wall depend on the lifestyle of the pathogen. For example, necrotrophs kill cells and macerate dead plant tissues by secreting enzymes, which are able to degrade the cell wall [20]. Bio- and hemi-biotrophic pathogens are dependent on living tissues, and therefore require a different strategy for interaction with the cell wall. Cell wall penetration by fungi or oomycetes utilizing haustoria as feeding structures involves minimum cell wall damage, whereas bacteria potentially delivers effector proteins to the cell wall [21]. Cell wall associated defense mechanisms are designed to retard disease development at an early stage or even to eliminate the pathogen as in the case of a hypersensitive response. In potato two types of resistance genes to PVY have been identified, *Ry* causing symptomless extreme resistance (ER) and *Ny* inducing a hypersensitive response (HR). The potato cv. Sárpo Mira originating in Hungary revealed a very high resistance level to PVY, described as score 9 (in a 1–9 scale accordingly to the European Cultivated Potato Database) [22]. Plants reacted with hypersensitive necrosis after PVY$^{NTN}$ inoculation in detached leaves and the whole plant [23].

Two classes of plant responses to pathogens are currently recognized. The first is the pathogen or microbe associated molecular pattern (PAMP or MAMP) in which plant recognition receptors (PRR) form PAMP- or MAMP-triggered immune responses (described as PTI) [24]. The second class involves intercellular receptors of pathogen virulence called effectors (effector-triggered immunity or ETI) [25]. Viruses do not code for PAMPs or ETIs and antiviral immune responses are triggered by the resistance proteins, which are not yet classified as ETI. At the molecular level many cascades of genetic signaling could be activated in HR to induce expression of defense proteins. Primary signaling cascades are associated with expression or even up-regulation of several defense proteins such as glucanases, chitinases, and other pathogenesis-related proteins [7,26]. Commonly known host responses involved in HR include ion fluxes, oxidative burst-production reactive oxygen species, localized cell death at site of pathogen infection or cross linking of the cell wall proteins [27,28]. Our previous studies determining the hypersensitive response in potato cv. Rywal described ultrastructural alterations with various organelles changes [29,30], but precise ultrastructural analyses of apoplast area rearrangements in PVY infection are still lacking. Our paper concentrated on PVY$^{NTN}$ impact on cell wall ultrastructure modification and localization of proteins associated with structure and/or modification of the cell wall. In a compatible potato-PVY$^{NTN}$ interaction loss of cell wall structure, association of virus cytoplasmic inclusions with plasmodesmata and active formation of paramural bodies (PMB) were observed. It has previously been reported that the viral movement protein (MP) is able to modify the plasmodesmata by changing the size exclusion limit (SEL) [31]. Despite movement proteins in plant-virus interaction, viruses require also some host factors to support movement process. However, the connections between plasmodesmata desmotubule and potyvirus inclusions with associations of vesicular structure and strong deposition of P3N-PIPO protein could be postulated as cell-to-cell translocation complex machinery [32]. Cytoplasmic inclusions (CI) were directed to plasmodesmata through interaction with P3N-PIPO. An important feature of plant response to pathogen infection is the alteration of the host cell walls which leads to the formation of paramural bodies [33] in the form of multivesicular compartments, but usually associated with cell wall and attached to plasma membrane [34]. The role of vesicular structure in virus-associated ultrastructural modifications has been described for *Potato virus M* [35]. These modifications include the formation of vesicles and tubules originating from the border of the cell wall and intruding into the plasmalemma.

In the PVY$^{NTN}$-Sárpo Mira hypersensitive reaction cell wall disruption (invaginations with folding) and thickening were mainly observed. Additionally, typical for potato-PVY$^{NTN}$ HR interaction ultrastructural changes consisted of phenolic compounds in the apoplast and vacuole. Electron dense deposits within and between the wall and plasmolemma were usually found in mesophyll and epidermis. Callose-like deposits were observed around sieve plates and cell walls primarily in phloem cells. According to Chappell et al. [36], the apoplast is the source of, or the route for, signals related to defense response. Induction of phenolic compound accumulation in cell walls may serve as a physical barrier to invading pathogens [37]. In host-plant reaction to infection by powdery mildew, *Blumeria graminis* (Bgh) or *Colleotrichum lindemuthianum*, phenolics were distributed throughout the cells and host-localized material were shown at the side of infected cell [33,34,38]. The partial retraction of the protoplast from the wall, as observed in hypersensitive reaction, especially in mesophyll cells between necrotized and unaffected, was cited, along with the aforementioned phenolic barrier, as an "ultrastructural symptom" of physical separation [6,39]. This phenomenon may possibly limit contact between infected and non-infected cells and may possibly be involved in hypersensitive reaction. Similar to PVM-red kidney bean interaction [35] and PVY$^{NTN}$–potato cv. Rywal interaction [40], the cell wall thickening in hypersensitive reaction PVY$^{NTN}$–potato cv. Sárpo Mira was mainly observed in mesophyll and phloem parenchyma cells. As postulated by O'Brien et al. [41], strengthening of the plant cell wall may play a key role in host-plant defense. Strengthening of the cell wall could be accomplished by deposition of wall material associated with vesicular structure secretion, such as paramural bodies or multivesicular bodies (MVB), as observed in hypersensitive response in Sárpo Mira tissue. These participate either in the degradation process in the vacuole or fuse to the plasma membrane to release their vesicles [42].

Comparative transcriptome profiling and microarray analysis of gene expression in plant virus infection in susceptible as well as resistance hosts indicate that the cell wall could be a kind of target for different plant viruses. It was reported that the cell wall related genes and transcripts are differentially expressed and regulated during the interactions [31,43–45]. Proteins/enzymes involved in plant cell wall organization and metabolism participate in the interaction through which they could influence virus spread [46]. Callose in the apoplast area is directly controlled by the proportion between the actions of two proteins: callose synthase and β-1,3, glucanases that hydrolyze callose [47]. Class II β-1,3 glucanases include pathogenesis-related protein PR-2. It is regarded as cellular factor controlling virus movement [48]. It has also been postulated [49] that TMV infection in tobacco increases the activity of PR-2, which facilitates virus movement. Suppression of β-1,3 glucanase results in increased callose deposition in the cell walls, reducing plasmodesmatal size exclusion limit. This process also restricts short and long-distance transport of PVX (*Potato virus X*), TMV (*Tobacco mosaic virus*) and TNV (*Tobacco necrosis virus*). Alternatively, overexpression not only of β-1,3 glucanase (class II) but also class III glucanase in PVY$^{NTN}$ infection expedited virus spread between cells [46,50]. Our results correspond with analysis of PR-2 in TMV infection. In incompatible interaction we observed callose by ultrastructural analyses. Immunolocalization analyses revealed lower deposition of PR-2 in the hypersensitive than in the compatible interaction. It is generally regarded that glucanases may potentially be localized preferably in vacuole and cell wall [51]. This was confirmed by our results, independently on the type of the interaction, cell wall and vacuole deposition of PR-2 dominated, but also cytoplasm localization was noticed, it indicated probably, that transport of PR-2 in vesicles and membranous structure are possible. Similar phenomena suggested Epel [52]. Furthermore, resistant tobacco plants (VAM) infected with PVY reacted by accumulation of transcript connected with PR proteins [53]. It was summarized that pathogenesis related genes were almost always up-regulated at later stage of infection (late than 2 days after infection). Respectively, in hypersensitive response Sárpo Mira potato plant-PVY$^{NTN}$ interaction, the induction of PR-2 deposition was noticed compared to mock-inoculated control plants. Additionally, also the transcriptome analysis followed by Babler et al. [54] recognized potentiall component of plants responses to virus infection, highlighted that, pathogenesis related PR enzymes involved in cell wall organization were upregulated in potato cv. Rywal hypersensitive to PVY inoculation (starting from 1 day after inoculation). On the other hand, for absolutely different plant virus member of the genus *Phytoreovirus* with quite different range of the host *Rice dwarf virus* (RDV) presented potentially similar reaction [44]. Microarray analyses revealed that about 9 days after inoculation of RDV almost all pathogenesis related protein genes expression was enhanced on rice plants as an example of compatible interaction.

This is the first study on the role cellulose synthase subunit 4 (CesA4) in cell wall structural protein modification in PVY$^{NTN}$-host compatible and incompatible interaction. The β-1,4 glucan (cellulose) is produced at the plasma membrane by large glucan synthase complexes referred to as cellulase synthases CesAs. The proposed model for cellulose synthase complex (CSC) synthesis is that it consists of 36 CESA proteins that produces a cellulose fibril containing 36 glucan chains [55,56]. CESA4 as an element of catalytic subunit of cellulose synthase terminal complex, participated in major mechanisms of the cell wall formation, and is required for β-1,4 glucan microfibril crystallization [57]. It functions in cellulose synthesis during secondary cell wall formation with CESA7 and 8 and is required for xylem cell wall thickening [58,59]. Analyses of localization CesA4 in the HR Sárpo Mir-PVY$^{NTN}$ interaction as well as in the susceptible Irys-PVY$^{NTN}$ interaction revealed a decrease in deposition of CesA4 compared to mock-inoculated controls, based on measurement of relative protein levels detected by immunoassay. These results appear to be compatible with those reported by Humphrey et al. [60] stating that reduction in cellulose content is a feedback response associated with biotic as well as abiotic stress. Hernandez-Blanco et al. [59] postulated that the production of CESA4 mutants 7 or 8 is triggered in response to resistance to two types of pathogenic bacteria and fungi. The above data is supported by research conducted by Burton et al. [61], showing that the putative cellulose synthase gene (*Cesa*) induced gene silencing when inserted into a *Potato virus X* vector in *N. benthamiana*, resulting in a 25% decrease in cellulose content in leaves tissue. From this it was concluded that phenotypic changes, for example changes in cell wall composition, in infected leaves were associated with reduced levels of CesA mRNA [61]. In contrast, transcriptome analyses during rice response to *Rice stripe virus* infection gave ambiguous results. The cellulose synthase genes after virus inoculation were markedly up-regulated in the resistant host but were down-regulated or even unchanged in the susceptible rice cultivar [43]. Interestingly, the microarray analysis conducted starting from 12 days after RSV inoculation revealed cellulose synthase significantly down-regulated. Similarly, in an investigation conducted by Shimizu et al. [44] for *Rice dwarf virus* 9 days after inoculation, there was significant repression of the genes for proteins generally involved in cell wall synthesis. The key enzyme involved in cell wall synthesis and remodeling, cellulose synthase, was suppressed, but this was mainly observed for the compatible interaction. Despite the reduction in immunodetection of CesA4 deposition in resistant and susceptible plants, the localization pattern strongly indicated that plasma membranes and vacuolated membranous structures, rather than cell walls, were the preferable areas for CeaA4 location independent of host resistance. These findings would be compatible with the assumption that cell walls are composed mostly of polysaccharides and are mainly produced in the Golgi, then secreted to the apoplast where they could be modified or join to the growing cell wall [62,63]. It was ascertained that the β-1,4 glucan cellulose is made on the plasma membrane by CesA complexes, which are assembled in the endoplasmic reticulum or Golgi network [15,64]. The process of delivery and endocytosis of the cellulose synthase complex to and from the plasma membrane are essential for regulation of cellulose biosynthesis, also in stress response [56]. This trafficking pathway could be the explanation for increased deposition of CesA4 proteins in the vacuole in HR reaction, and the relatively lower rate of CesA4 protein in the resistant plant. Vacuolar localization may be an effect of sequestration of CesA4 to the vacuole by an active participation of multivesicular bodies or prevacuolar vesicles. These findings suggest that inhibition of CesA4 deposition as a consequence of PVY$^{NTN}$ leads to generate antimicrobial process associated with resistance to the pathogen. This inhibition could be an effect of sequestration of CesA4 to the vacuole thus vesicular structures, in response to HR reaction.

Based on analyses of the localization of extensin (hydroxyproline-rich glycoprotein, HRGP) involved in cell wall structure and modification, it may be concluded that HRGP deposition is induced in both susceptible and resistant cultivars by PVY$^{NTN}$ infection, however, the post-inoculation levels were much higher in the hypersensitive than in the compatible host-pathogen interaction. HRGP epitopes (detected by LM1 antibodies) were located especially in both vascular tissues, moreover, preferably in apoplast region, but in symplast the deposition was also noticed with the special consideration of vacuole, membranous structures and cytoplasm. Our observations and analyses are consistent with data presented by O'Connel et al. [65], who postulated that HRGP protein may be localized not only in the cell wall but is concentrated close to plasma membrane and membranous structures in bean tissues infected with *P. syringae* as well as *Colletotrichum*. Our analyses are consistent with the findings that TMV infection acts as a biotic factor inducing HRGP in tobacco [66]. Similarly, Wycoff et al. [67] examined transgenic tobacco, containing a HRGP-GUS gene and inoculation also with TMV. The authors reported that 27 test plants had an increase of 2.5-fold GUS activity compared to mock-inoculated plants. In our experiment HRGP epitopes were mainly localized in both vascular tissues or even in epidermis as reported by Templeton et al. [68], whose in situ hybridization analyses found accumulation of HRGPs transcripts in the epidermis and in the outer phloem in the vascular area. It was postulated that HRGPs accumulate as a soluble cell wall protein at main stages of development and could be made insoluble by intermolecular cross-linkage formation [69]. According to Raggi [70], the activation by PVY$^N$ infection of HRGPs in the cell wall of tobacco leaves and resultant cross-linking and cell wall lignifications contributes to resistance against *Erysiphe cichoracearum* by reduction in fungal haustoria production. These cross-linking process has been proposed as a mechanism of strengthening the cell wall [71]. Biotic stress induced HRGP (extensin) provides a structural barrier against the pathogen [66,72]. Based on our findings and postulated literature information, the induction of extension (HRGP) in hypersensitive Sárpo Mira clearly indicated that we observed reinforcement of the cell wall and active distribution of HRGP, which are important in preventing the entry-point of PVY in leaf tissues. Moreover, the deep sequencing analysis describing host response to *Rice stripe virus* infection indicated that cell wall strengthening was induced especially in resistance rice cultivars by markedly up-regulation glycine-rich repeat proteins [43].

## 4. Materials and Methods

### 4.1. Plant Material and Virus Inoculation

*Solanum tuberosum* L. cvs. Irys (PVY$^{NTN}$ resistance score 5.5 in a 1–9 scale) [73] and Sárpo Mira (resistance score 9) [22] were obtained from IHAR-PIB, Plant Breeding and Acclimatization Institute, Bonin Research Center. Plants were grown at 20 °C, under 16/8-h light/dark photoperiod and a light intensity of 400 µmol m$^{-2}$ s$^{-1}$, and were inoculated mechanically as previously described [74] at the four leaf stage with the NTN strain of PVY. Potato cv. Sárpo Mira developed a hypersensitive necrotic response visible at 7 days post-inoculation. This reaction is conferred by the *Ny-Smira* gene located on the long arm of the potato IX chromosome [23]. Hypersensitive reaction symptoms on inoculated leaves appeared 7 days post-inoculation. Potato cv. Irys reacted with systemic necrosis visible at 10 days post-inoculation. Leaves from both PVY$^{NTN}$-infected and healthy controls inoculated with phosphate buffer were collected 10 days post-inoculation and tested by DAS-ELISA [75].

### 4.2. Ultrastructural Examinations

Potato (infected and mock-inoculated) leaf samples (2 mm × 2 mm sections) were fixed in 2% (*v*/*v*) paraformaldehyde and 2% (*v*/*v*) glutharaldehyde in 0.1 M cacodylate buffer (pH 7.2) for 2 h and washed 4 times in cacodylate buffer [30]. Samples were post-fixed in 2% (*v*/*v*) OsO$_4$ for 2 h at 4 °C, dehydrated in an ethanol series (10–99%) and propylene oxide, embedded in epoxy resin (EPON,

Fluka, Switzerland) and polymerized at 60 °C for 24 h. Ultra-thin (90 nm) sections were stained with uranyl acetate/lead citrate (Sigma-Aldrich, St. Louis, MO, USA) and examined by transmission electron microscopy [76].

### 4.3. Immunofluorescence Localization

Rat IgG monoclonal LM1 antibodies (HRGP/extensin specific) were purchased from PlantProbes (Leeds, UK). Polyclonal rabbit anti-CesA4 and anti PR-2 antibodies were obtained from Agrisera (Vänäs, Sweden). Potato leaf tissue samples were fixed in 4% ($w/v$) paraformaldehyde in 0.1 M microtubule stabilizing buffer (MSB) pH 6.9 containing 0.1% ($v/v$) Triton X-100 for 2 h at room temperature as described by Gubler [77]. Samples were dehydrated in an ethanol series containing 10 mM dithiothreitol and embedded in butyl-methyl-methacrylate (BMM) resin and polymerized under UV radiation for 20 h at −20 °C. Acetone was used to remove the BMM from 2 μm sections collected on silane slides (Thermo-Fischer Scientific, Warsaw, Poland). Immunofluorescence analysis was carried out after pre-incubation in 3% ($w/v$) bovine serum albumin in PBS for 1 h at room temperature. Sections on slides were incubated for 2 h in a humid chamber with antibodies diluted in PBS (1:50 $v/v$ for polyclonal and 1:10 $v/v$ for monoclonal). Controls consisted of mock-inoculated tissue and pre-immune serum. After rinsing with PBS-Tween20 buffer slides were treated with secondary goat anti-rabbit IgG conjugated with FITC (fluorescein isothiocyanate) goat anti-rat IgG in PBS at RT in the dark for 2 h. An Olympus AX70 Provis (Olympus Poland, Warsaw, Poland) with a UM61002 filter set and equipped with an Olympus SC35 camera was used for fluorescence imaging [30,74].

### 4.4. Immunogold Labeling

Leaf sections (50–70 nm thick) were mounted on Formvar-coated nickel grids and treated with 10% hydrogen peroxide solution for 10 min to remove the resin. The grids were pre-incubated for 1 h in blocking medium containing 2% normal goat serum (Thermo-Fischer Scientific) with 3% BSA in 0.1 M PBS buffer (pH 7.6) as described previously by Otulak and Garbaczewska [76]. Grids were then rinsed three times in buffer containing 0.05% Tween-20 for 10 min and treated for 2 h at room temperature with primary antibodies (as mentioned in above section) in PBS and washed in PBS-Tween 20. The samples were treated for 1 h with gold-conjugated secondary antibody (15 nm, Sigma-Aldrich, Warsaw, Poland) and then rinsed first in PBS and then in distilled water for 5 min. Labeling specificity was checked by incubating grids with material from mock-inoculated plants and by omission of the primary antibody from incubating solution. The grids were then counterstained with 1% uranyl acetate for 5 min and washed 5 × 2 min with distilled water. Immunogold-labeled sections on grids were examined by transmission electron microscope (as described above).

### 4.5. Statistical Quantification of Tissue and Cell Distribution of Immunogold-Labeled Cell Wall-Associated Proteins PR-2, CesA4 and HRGP (Extensin)

Labeling quantification of cell wall associated proteins PR-2, CesA4 and extensin was based on the method proposed by Luschin-Ebengreuth and Zechmann [78], with modifications of the type of statistical method and program used for statistical analyses. Data on gold particle concentration was investigated by analysis of variance (ANOVA) and the post-hoc Tukey HSD test in STATISTICA software (StataSoft and TIBCO Software Inc., Palo Alto, CA, USA, version 13.0). ANOVA was used as an efficient estimator of gold labeling. Each cell wall-associated protein was investigated individually. For statistical estimation of immunogold labeling we compared infected and healthy (mock-inoculated) plants from different cultivars. Gold particles in tissues/cell compartments were counted in forty 10 μm$^2$ fields per image. In each combination (two mock-inoculated, cv. Irys and cv. Sárpo Mira PVY$^{NTN}$ inoculated potato) gold particles from 200 photos were counted for PR-2, CesA4 and HRGP. Data were based on analyses of PR-2 (600 photos), CesA4 (600 photos) and HRGP (600 photos).

*Int. J. Mol. Sci.* **2018**, *19*, 862

## 5. Conclusions

Recently, transcriptome analysis research on responses triggered during plant-pathogen interactions have increased. These investigations delivered new findings relating to up- and/or down-regulation of genes involved in pathogen infection. This study utilizes ultrastructural studies and immunodetection of specific proteins to obtain a better understanding of the involvement of cell wall components in response to biotic stress. The ultrastructural analyses and precise localization and distribution of cell wall proteins/enzymes permitted the collection in situ of new information on structural changes, reorganization and remodeling within the apoplast and symplast associated with plant-host defense response to virus infection. Previous research has concentrated on cell wall changes in different plant-pathogen interaction models of plant pathogenic bacteria or fungal, but not virus-host-plant interactions. In contrast to bacterial and fungal pathogens, viruses are not mechanical destructors of host cell walls. Our studies for the first time have precisely described the ultrastructural changes accompanying the two types of PVY$^{NTN}$-potato interactions, compatible and incompatible. Our findings revealed, that deposition of proteins (e.g., CesA4) involved in potato cell wall synthesis was suppressed 10 days after PVY inoculation in compatible as well as hypersensitive reaction. In contrast, we observed, in inoculated but not in mock-inoculated controls, the induction of proteins (e.g., PR2 and HRGP) involved in cell wall modification processes. Additionally, the deposition and distribution of PR-2 protein was significantly lower in hypersensitive cultivar Sárpo Mira than in the compatible interaction, and as a consequence the deposition of callose-like materials was observed more frequently. In contrast, HRGP (extensin) localization as measured by cell wall thickening was significantly greater in the HR than the compatible reaction. From these observations it may be concluded that cell wall modifications as determined by protein analysis as well as the symplast are involved in apoplast changes in both types of virus-host interaction. The results of the study on cell wall alterations in PVY$^{NTN}$-resistant and susceptible interaction have paved the way for subsequent studies focused on the potential involvement of other proteins/enzymes and their functions in defense response. Further molecular and cellular studies are needed to elucidate the possible role of other cell wall functions (e.g., expansins and xyloglucan metabolism and distribution) that are potentially active in PVY-host-plant interactions.

**Acknowledgments:** The work has been supported by Warsaw University of Life Sciences-SGGW grants number: 505-10-011100-N00208-99 and 505-10-011100-P00122-99. Authors would like to thank A. Przewodowska and Danuta Sekrecka (IHAR-PIB, Bonin, Poland) for kindly provided potato tubers, and inż. E. Znojek for her excellent ultramicrotome work.

**Author Contributions:** Katarzyna Otulak-Kozieł wrote the manuscript, conceived the idea of manuscript, performed experiments presented in the article (microscopy, immunolabeling) and analyzed data. Edmund Kozieł performed quantification of immunogold labeling and analyzed data, Benham E.L. Lockhart helped to write and edit the article, and also offered his experience in plant virology and microscopy in analyzing and making interpretation of data.

**Conflicts of Interest:** The authors declare no conflict of interest.

## References

1. Carpita, N.C. Progress in the biological synthesis of the plant cell wall: New ideas for improving biomass for bioenergy. *Curr. Opin. Biotechnol.* **2012**, *23*, 330–337. [CrossRef] [PubMed]
2. Showalter, A.M. Structure and function of plant cell wall proteins. *Plant Cell* **1993**, *5*, 9–23. [CrossRef] [PubMed]
3. Malinovsky, F.G.; Fangel, J.U.; Willats, W.G.T. The role of the cell wall in plant immunity. *Front. Plant Sci.* **2014**, *5*, 1–12. [CrossRef] [PubMed]
4. Rybicki, E.P. A top ten list for economically important plant viruses. *Arch. Virol.* **2015**, *160*, 17–20. [CrossRef] [PubMed]
5. Chikh Ali, M.; Karasev, A.V.; Furutani, N.; Taniguchi, M.; Kano, Y.; Sato, M.; Natsuaki, T.; Maoka, T. Occurrence of *Potato virus Y* strain PVY NTN in foundation seed potatoes in Japan, and screening for symptoms in Japanese potato cultivars. *Plant Pathol.* **2013**, *62*, 1157–1165. [CrossRef]

6. Dangl, J.L.; Dietrich, R.A.; Richberg, M.H. Death don't have no mercy: cell death programs in plant-microbe interactions. *Plant Cell* **1996**, *8*, 1793–1807. [CrossRef] [PubMed]

7. Alexander, M.M.; Cilia, M. A molecular tug-of-war: Global plant proteome changes during viral infection. *Curr. Plant Biol.* **2016**, *5*, 13–24. [CrossRef]

8. Stavolone, L.; Lionetti, V. Extracellular matrix in plants and animals: Hooks and locks for viruses. *Front. Microbiol.* **2017**, *8*, 1–8. [CrossRef] [PubMed]

9. Lionetti, V.; Cervone, F.; Bellincampi, D. Methyl esterification of pectin plays a role during plant-pathogen interactions and affects plant resistance to diseases. *J. Plant Physiol.* **2012**, *169*, 1623–1630. [CrossRef] [PubMed]

10. Hull, R. *Plant Virology*, 5th ed.; Elsevier Academic Press: London, UK, 2013; pp. 669–754, ISBN 9780123848710.

11. Otulak, K.; Kozieł, E.; Garbaczewska, G. "Seeing is believing". The use of light, fluorescent and transmission electron microscopy in the observation of pathological changes during different plant—Virus interactions. In *Microscopy: Advances in Scientific Research and Education*, 6th ed.; Mendez-Vilas, A., Ed.; Formatex Research Center: Badajoz, Spain, 2014; Volume 1, pp. 367–376, ISBN 978-84-942134-3-4.

12. Nguema-Ona, E.; Vicré-Gibouin, M.; Cannesan, M.A.; Driouich, A. Arabinogalactan proteins in root–microbe interactions. *Trends Plant Sci.* **2013**, *18*, 440–449. [CrossRef] [PubMed]

13. Bethke, G.; Thao, A.; Xiong, G.; Li, B.; Soltis, N.E.; Hatsugai, N.; Hillmer, R.A.; Katagiri, F.; Kliebenstein, D.J.; Pauly, M.; et al. Pectin biosynthesis is critical for cell wall integrity and immunity in Arabidopsis thaliana. *Plant Cell* **2016**, *28*, 537–556. [CrossRef] [PubMed]

14. Zhang, L.; Lilley, C.J.; Imren, M.; Knox, J.P.; Urwin, P.E. The complex cell wall composition of syncytia induced by plant parasitic cyst nematodes reflects both function and host plant. *Front. Plant Sci.* **2017**, *8*, 1–12. [CrossRef] [PubMed]

15. Ellinger, D.; Voigt, C.A. Callose biosynthesis in Arabidopsis with a focus on pathogen response: What we have learned within the last decade. *Ann. Bot.* **2014**, *114*, 1349–1358. [CrossRef] [PubMed]

16. Sattler, S.E.; Funnell-Harris, D.L. Modifying lignin to improve bioenergy feedstocks: Strengthening the barrier against pathogens. *Front. Plant Sci.* **2013**, *4*, 1–8. [CrossRef] [PubMed]

17. Di Carli, M.; Benvenuto, E.; Donini, M. Recent insights into plant–virus interactions through proteomic analysis. *J. Proteome Res.* **2012**, *11*, 4765–4780. [CrossRef] [PubMed]

18. Bruknard, J.O.; Zambryski, P.C. Plasmodesmata enable multicellularity: New insights into their evolution, biogenesis, and functions in development and immunity. *Curr. Opin. Plant Biol.* **2017**, *35*, 76–83. [CrossRef]

19. Harries, P.A.; Schoelz, J.E.; Nelson, R.S. Intracellular transport of viruses and their components: Utilizing the cytoskeleton and membrane highways. *Mol. Plant Microbe Interact.* **2010**, *23*, 1381–1393. [CrossRef] [PubMed]

20. Underwood, W. The plant cell wall: A dynamic barrier against pathogen invasion. *Front. Plant Sci.* **2012**, *3*, 1–6. [CrossRef] [PubMed]

21. Büttner, D.; He, S.Y. Type III protein secretion in plant pathogenic bacteria. *Plant Physiol.* **2009**, *150*, 1656–1664. [CrossRef] [PubMed]

22. The European Cultivated Potato Database. Available online: https://www.europotato.org/quick_search.php (accessed on 24 December 2017).

23. Tomczyńska, I.; Jupe, F.; Hein, I.; Marczewski, W.; Śliwka, J. Hypersensitive response to *Potato virus Y* in potato cultivar Sárpo Mira is conferred by the *Ny-Smira* gene located on the long arm of chromosome IX. *Mol. Breed.* **2014**, *34*, 471–480. [CrossRef] [PubMed]

24. Jones, J.D.; Dangl, J.L. The plant immune system. *Nature* **2006**, *444*, 323–329. [CrossRef] [PubMed]

25. Schwessinger, B.; Ronald, P.C. Plant innate immunity: Perception of conserved microbial signatures. *Annu. Rev. Plant Biol.* **2012**, *63*, 451–482. [CrossRef] [PubMed]

26. Hammond-Kosack, K.E.; Jones, J.D. Resistance gene-dependent plant defense responses. *Plant Cell* **1996**, *8*, 1773–1791. [CrossRef] [PubMed]

27. De León, P. The moss *Physcomitrella patens* as a model system to study interactions between plants and phytopathogenic fungi and Oomycetesinés. *J. Pathog.* **2011**, *2011*, 1–6. [CrossRef] [PubMed]

28. Jayamohan, N.S.; Kumudini, B.S. Host pathogen interaction at the plant cell wall. *Int. Res. J. Pharm. Pharmacol.* **2011**, *1*, 242–249.

29. Otulak, K.; Garbaczewska, G. Cellular localisation of calcium ions during potato hypersensitive response to *Potato virus Y*. *Micron* **2011**, *42*, 381–392. [CrossRef] [PubMed]

30. Otulak, K.; Garbaczewska, G. The participation of plant cell organelles in compatible and incompatible *Potato virus Y*-tobacco and -potato plant interaction. *Acta Physiol. Plant.* **2013**, *36*, 85–99. [CrossRef]

31. Allie, F.; Pierce, E.J.; Okoniewski, M.J.; Rey, C. Transcriptional analysis of South African cassava mosaic virus-infected susceptible and tolerant landraces of cassava highlights differences in resistance, basal defense and cell wall associated genes during infection. *BMC Genom.* **2014**, *14*, 1–30. [CrossRef] [PubMed]

32. Movahed, N.; Patarroyo, C.; Sun, J.; Vali, H.; Laliberté, J.F.; Zheng, H. Cylindrical inclusion protein of Turnip mosaic virus serves as a docking point for the intercellular movement of viral replication vesicles. *Plant Physiol.* **2017**, *175*, 1732–1744. [CrossRef] [PubMed]

33. An, Q.; Ehlers, K.; Kogel, K.H.; van Bel, A.J.E.; Hückelhoven, R. Multivesicular compartments proliferate in susceptible and resistant MLA12-barley leaves in response to infection by the biotrophic powdery mildew fungus. *New Phytol.* **2006**, *172*, 563–576. [CrossRef] [PubMed]

34. An, Q.; Hückelhoven, R.; Kogel, K.H.; van Bel, A.J.E. Multivesicular bodies participate in a cell wall-associated defence response in barley leaves attacked by the pathogenic powdery mildew fungus. *Cell. Microbiol.* **2006**, *8*, 1009–1019. [CrossRef] [PubMed]

35. Tu, J.D.; Hiruki, C. Electron microscopy of cell wall thickening in local lesions of potato virus M-infected red kidney bean. *Phytopathology* **1971**, *6*, 862–868. [CrossRef]

36. Chappell, J.; Levine, A.; Tenhaken, R.; Lusso, M.; Lamb, C. Characterization of a diffusible signal capable of inducing defense gene expression in tobacco. *Plant Physiol.* **1997**, *113*, 621–629. [CrossRef] [PubMed]

37. McLusky, S.R.; Bennett, M.H.; Beale, M.H.; Lewis, M.J.; Gaskin, P.; Mansfield, J.W. Cell wall alterations and localized accumulation of feruloyl-3′-methoxytyramine in onion epidermis at sites of attempted penetration by *Botrytis allii* are associated with actin polarisation, peroxidase activity and suppression of flavonoid biosynthesis. *Plant J.* **1999**, *17*, 523–534. [CrossRef]

38. Bolwell, P.P.; Page, A.; Piślewska, M.; Wojtaszek, P. Pathogenic infection and the oxidative defences in plant apoplast. *Protoplasma* **2001**, *217*, 20–32. [CrossRef] [PubMed]

39. Benhamou, N.; Kloepper, J.W.; Quadt-Hallman, A.; Tuzun, S. Induction of defence-related ultrastructural modifications in pea root tissues inoculated with endophytic bacteria. *Plant Physiol.* **1996**, *112*, 919–929. [CrossRef] [PubMed]

40. Otulak, K.; Garbaczewsk, G. Localisation of hydrogen peroxide accumulation during *Solanum tuberosum* cv. *Rywal* hypersensitive response to Potato virus Y. *Micron* **2010**, *41*, 327–335. [CrossRef] [PubMed]

41. O'Brien, J.A.; Daudi, A.; Butt, V.S.; Bolwell, G.P. Reactive oxygen species and their role in plant defence and cell wall metabolism. *Planta* **2012**, *236*, 765–779. [CrossRef] [PubMed]

42. Tse, Y.C.; Mo, B.; Hillmer, S.; Zhao, M.; Lo, S.W.; Robinson, D.G.; Jiang, L. Identification of multivesicular bodies as prevacuolar compartments in *Nicotiana tabacum* BY-2 cells. *Plant Cell* **2004**, *16*, 672–693. [CrossRef] [PubMed]

43. Zheng, W.; Ma, L.; Zhao, J.; Li, Z.; Sun, F.; Lu, X. Comparative Transcriptome Analysis of Two Rice Varieties in Response to Rice Stripe Virus and Small Brown Planthoppers during Early Interaction. *PLoS ONE* **2013**, *8*, 1–9. [CrossRef] [PubMed]

44. Shimizu, T.; Satoh, K.; Kikuchi, S.; Omura, T. The repression of cell wall and plastid-related genes and the induction of defense-related genes in rice plants infected with Rice dwarf virus. *Mol. Plant-Microbe Interact.* **2007**, *20*, 247–254. [CrossRef] [PubMed]

45. Shimizu, T.; Ogamino, T.; Hiraguri, A.; Nakazono-Nagaoka, E.; Uehara-Ichiki, T.; Nakajima, M.; Akutsu, K.; Omura, T.; Sasaya, T. Strong resistance against *Rice grassy stunt virus* is induced in transgenic rice plants expressing double-stranded RNA of the viral genes for nucleocapsid or movement proteins as targets for RNA interference. *Phytopathology* **2013**, *103*, 513–519. [CrossRef] [PubMed]

46. Bucher, G.L.; Tarina, C.; Heinlein, M.; Di Serio, F.; Meins, F., Jr.; Iglesias, V.A. Local expression of enzymatically active class I beta-1,3-glucanase enhances symptoms of TMV infection in tobacco. *Plant J.* **2001**, *28*, 361–369. [CrossRef] [PubMed]

47. Kauss, H. Callose synthesis. In *Membranes: Specialized Function in Plants*, 2nd ed.; Smallwood, M., Knox, J.P., Bowles, D.J., Eds.; BIOS Scientific: Oxford, UK, 1996; Volume 1, pp. 77–92, ISBN 978-1859962008.

48. Beffa, R.; Meins, F., Jr. Pathogenesis-related functions of plant beta-1,3-glucanases investigated by antisense transformation—A review. *Gene* **1996**, *179*, 97–103. [CrossRef]

49. Iglesias, V.A.; Meins, F., Jr. Movement of plant viruses is delayed in a beta-1,3-glucanase-deficient mutant showing a reduced plasmodesmatal size exclusion limit and enhanced callose deposition. *Plant J.* **2000**, *21*, 157–166. [CrossRef] [PubMed]

50. Dobnik, D.; Baebler, S.; Kogovšek, P.; Pompe-Novak, M.; Štebih, D.; Panter, G.; Janež, N.; Morisset, D.; Žel, J.; Gruden, K. β-1,3-glucanase class III promotes spread of PVY$^{NTN}$ and improves in planta protein production. *Plant Biotechnol. Rep.* **2013**, *7*, 547–555. [CrossRef] [PubMed]

51. Keefe, D.; Hinz, U.; Meins, F., Jr. The effect of ethylene on the cell-type-specific and intracellular localization of β-1,3-glucanase and chitinase in tobacco leaves. *Planta* **1990**, *182*, 43–51. [CrossRef] [PubMed]

52. Epel, B.L. Plant viruses spread by diffusion on ER-associated movement-protein-rafts through plasmodesmata gated by viral induced host beta-1,3-glucanases. *Semin. Cell Dev. Biol.* **2009**, *20*, 1074–1081. [CrossRef] [PubMed]

53. Chen, S.; Li, F.; Liu, D.; Jiang, C.; Cui, L.; Shen, L.; Liu, G.; Yang, A. Dynamic expression analysis of early response genes induced by potato virus Y in PVY-resistant *Nicotiana tabacum*. *Plant Cell Rep.* **2016**, *36*, 297–311. [CrossRef] [PubMed]

54. Baebler, S.; Witek, K.; Petek, M.; Stare, K.; Tušek-Žnidarič, M.; Pompe-Novak, M.; Renaut, J.; Szajko, K.; Strzelczyk-Żyta, D.; Marczewski, W.; et al. Salicylic acid is an indispensable component of the *Ny-1* resistance-gene-mediated response against *Potato virus Y* infection in potato. *J. Exp. Bot.* **2014**, *65*, 1095–1109. [CrossRef] [PubMed]

55. Delmer, D.P. Cellulose biosynthesis: exciting times for a difficult field of study. *Annu. Rev. Plant Physiol. Plant Mol Biol.* **1999**, *50*, 245–276. [CrossRef] [PubMed]

56. Bashline, L.; Li, S.; Gu, Y. The trafficking of the cellulose synthase complex in higher plants. *Ann. Bot.* **2014**, *114*, 1059–1067. [CrossRef] [PubMed]

57. Atanassov, I.I.; Pittman, J.K.; Turner, S.R. Elucidating the mechanisms of assembly and subunit interaction of the cellulose synthase complex of Arabidopsis secondary cell walls. *J. Biol. Chem.* **2009**, *284*, 3833–3841. [CrossRef] [PubMed]

58. Taylor, N.G.; Howells, R.M.; Huttly, A.K.; Vickers, K.; Turner, S.R. Interactions among three distinct CesA proteins essential for cellulose synthesis. *Proc. Natl. Acad. Sci. USA* **2003**, *100*, 1450–1455. [CrossRef] [PubMed]

59. Hernández-Blanco, C.; Feng, D.X.; Hu, J.; Sánchez-Vallet, A.; Deslandes, L.; Llorente, F.; Berrocal-Lobo, M.; Keller, H.; Barlet, X.; Sánchez-Rodríguez, C.; et al. Impairment of cellulose synthases required for Arabidopsis secondary cell wall formation enhances disease resistance. *Plant Cell* **2007**, *19*, 890–903. [CrossRef]

60. Humphrey, T.V.; Bonetta, D.T.; Goring, D.R. Sentinels at the wall: cell wall receptors and sensors. *New Phytol.* **2007**, *176*, 7–21. [CrossRef] [PubMed]

61. Burton, R.A.; Gibeaut, D.M.; Bacic, A.; Findlay, K.; Roberts, K.; Hamilton, A.; Baulcombe, D.C.; Fincher, G.B. Virus-induced silencing of a plant cellulose synthase gene. *Plant Cell* **2000**, *12*, 691–705. [CrossRef] [PubMed]

62. Scheller, H.V.; Ulvskov, P. Hemicelluloses. *Annu. Rev. Plant Biol.* **2010**, *61*, 263–289. [CrossRef] [PubMed]

63. Schneider, R.; Hanak, T.; Persson, S.; Voigt, C.A. Cellulose and callose synthesis and organization in focus, what's new? *Curr. Opin. Plant Biol.* **2016**, *34*, 9–16. [CrossRef] [PubMed]

64. McFarlane, H.E.; Watanabe, Y.; Yang, W.; Huang, Y.; Ohlrogge, J.; Lacey Samuels, A. Golgi- and trans-golgi network-mediated vesicle trafficking is required for wax secretion from epidermal cells. *Plant Physiol.* **2014**, *164*, 1250–1260. [CrossRef] [PubMed]

65. O'Connell, R.J.; Brown, I.R.; Mansfield, J.W.; Bailey, J.A.; Mazau, D.; Rumeau, D.; Esquerré-Tugayé, M.T. Immunocytochemical localization of hydroxyproline-rich glycoproteins accumulating in melon and bean at sites of resistance to bacteria and fungi. *Mol. Plant Microbe Interact.* **1990**, *3*, 33–40. [CrossRef]

66. Benhamou, N.; Mazau, D.; Esquerré-Tugayé, M.T. Immunocytochemical localization of hydroxyproline-rich glycoproteins in tomato root cells infected by *Fusarium oxysporum* f. sp. radicis-lycopersici: Study of a compatible interaction. *Phytopathology* **1990**, *80*, 163–173. [CrossRef]

67. Wycoff, K.L.; Powell, P.A.; Gonzales, R.A.; Corbin, D.R.; Lamb, C.; Dixon, R.A. Stress activation of a bean hydroxyproline-rich glycoprotein promoter is superimposed on a pattern of tissue-specific developmental expression. *Plant Physiol.* **1995**, *109*, 41–52. [CrossRef] [PubMed]

68. Templeton, M.D.; Dixon, R.A.; Lamb, C.J.; Lawton, M.A. Hydroxyproline-rich glycoprotein transcripts exhibit different spatial pat- tems of accumulation in compatible and incompatible interactions between Phaseolus vulgaris and Colletotrichum lindemuthianum. *Plant Physiol.* **1990**, *94*, 1265–1269. [CrossRef] [PubMed]

69. Kieliszewski, M.J.; Lamport, D.T.A. Extensin: Repetitive motifs, functional sites, post-translational codes, and phylogeny. *Plant J.* **1994**, *5*, 157–172. [CrossRef] [PubMed]

70. Raggi, V. Hydroxyproline-rich glycoprotein accumulation in tobacco leaves protected against *Erysiphe cichoracearumby* potato virus Y infection. *Plant Pathol.* **2000**, *49*, 179–186. [CrossRef]

71. Varner, J.E.; Lin, L.S. Plant cell wall architecture. *Cell* **1989**, *56*, 231–239. [CrossRef]

72. Corbin, D.R.; Sauer, N.; Lamb, C. Differential regulation of a hydroxyproline-rich glycoprotein gene family in wounded and infected plants. *Mol. Cell. Biol.* **1987**, *7*, 4337–4344. [CrossRef] [PubMed]

73. Zimmoch-Guzowska, E.; Yin, Z.; Chrzanowska, M.; Flis, B. Sources and effectiveness of potato PVY resistance in IHAR's breeding research. *Am. Potato J.* **2013**, *90*, 21–27. [CrossRef]

74. Otulak, K.; Kozieł, E.; Garbaczewska, G. Ultastructural impact of tobacco rattle virus on tobacco and pepper ovary and anther tissues. *J. Phytopatol.* **2016**, *164*, 226–241. [CrossRef]

75. Chrzanowska, M.; Doroszewska, T. Comparison between PVY isolates obtained from potato and tobacco plants grown in Poland. *Phytopathol. Pol.* **1997**, *13*, 63–71.

76. Otulak, K.; Garbaczewska, G. Ultrastructural events during hypersensitive response of potato cv. Rywal infected with necrotic strains of *Potato virus Y. Acta Physiol. Plant.* **2010**, *32*, 635–644. [CrossRef]

77. Gubler, F. Immunofluorescence localisation of microtubules in plant root tips embedded in butyl-methyl methacrylate. *Cell Biol. Int. Rep.* **1989**, *13*, 137–145. [CrossRef]

78. Luschin-Ebengreuth, N.; Zechmann, B. Compartment-specific investigations of antioxidants and hydrogen peroxide in leaves of Arabidopsis thaliana during dark-induced senescence. *Acta Physiol. Plant.* **2016**, *38*, 1–15. [CrossRef] [PubMed]

International Journal of
*Molecular Sciences*

MDPI

*Article*

# A Comparison of the Effects of *FATTY ACID DESATURASE 7* and *HYDROPEROXIDE LYASE* on Plant–Aphid Interactions

Jiamei Li [1], Carlos A. Avila [2], Denise M. Tieman [3], Harry J. Klee [3] and Fiona L. Goggin [1,*]

[1] Department of Entomology, University of Arkansas, Fayetteville, AR 72701, USA; jxl080@uark.edu
[2] Department of Horticultural Sciences, Texas A&M AgriLife Research, Weslaco, TX 78596, USA; carlos.avila@ag.tamu.edu
[3] Horticultural Sciences Department, University of Florida, Gainesville, FL 32611, USA; dtieman@ufl.edu (D.M.T.); hjklee@ufl.edu (H.J.K.)
* Correspondence: fgoggin@uark.edu; Tel.: +1-479-575-6751

Received: 18 February 2018; Accepted: 1 April 2018; Published: 4 April 2018

**Abstract:** The *spr2* mutation in tomato (*Solanum lycopersicum*), which disrupts function of FATTY ACID DESATURASE 7 (FAD7), confers resistance to the potato aphid (*Macrosiphum euphorbiae*) and modifies the plant's C6 volatile profiles. To investigate whether C6 volatiles play a role in resistance, *HYDROPEROXIDE LYASE* (*HPL*), which encodes a critical enzyme in C6 volatile synthesis, was silenced in wild-type tomato plants and *spr2* mutants. Silencing *HPL* in wild-type tomato increased potato aphid host preference and reproduction on 5-week old plants but had no influence on 3-week old plants. The *spr2* mutation, in contrast, conferred strong aphid resistance at both 3 and 5 weeks, and silencing *HPL* in *spr2* did not compromise this aphid resistance. Moreover, a mutation in the *FAD7* gene in *Arabidopsis thaliana* also conferred resistance to the green peach aphid (*Myzus persicae*) in a genetic background that carries a null mutation in *HPL*. These results indicate that *HPL* contributes to certain forms of aphid resistance in tomato, but that the effects of *FAD7* on aphids in tomato and Arabidopsis are distinct from and independent of *HPL*.

**Keywords:** aphid resistance; *Arabidopsis thaliana*; hydroperoxide lyase; *Macrosiphum euphorbiae*; *Myzus persicae*; *Solanum lycopersicum*; ω-3 fatty acid desaturase

## 1. Introduction

FATTY ACID DESATURASE 7 (FAD7) is an ω-3 fatty acid desaturase (FAD) that is widely similar in sequence throughout the plant kingdom, and that desaturates 16- and 18-carbon fatty acids with two double bonds (C16:2 and C18:2) to generate fatty acids with three double bonds (C16:3 and C18:3) [1,2]. In diverse plant species, expression levels of genes encoding FAD7 and other FADs increase in response to some stresses and decrease in response to others, suggesting that modulation of desaturase activity plays a role in stress adaptation [3–6]. Moreover, artificial manipulation of desaturase activity through silencing or other genetic modifications alters plants' susceptibility to a variety of abiotic and biotic stresses, enhancing resistance to some stresses and compromising resistance to others [7–10]. In short, FADs appear to influence stress resistance.

One form of stress resistance that is negatively correlated with FAD7 activity is resistance to aphids, a group of phloem-feeding insects that include many serious agricultural pests. The *suppressor of prosystemin-mediated responses2* (*spr2*) mutant in tomato, which has a null mutation in *FAD7*, has enhanced resistance to potato aphids (*Macrosiphum euphorbiae*) compared to wild-type plants [10]. Aphid resistance in *spr2* comprises both antixenosis (i.e., decreased host preference) and antibiosis (increased mortality and decreased fecundity). Moreover, population growth of the green peach aphid

(*Myzus persicae*) is significantly lower on *Arabidopsis thaliana* mutants with null mutations in *FAD7* than on wild-type controls [10]. These results indicate that the FAD7 enzyme or its products negatively regulate aphid resistance in more than one plant family.

One way that FAD7 could possibly influence plant defenses against aphids is by affecting the profile of substrates available to the hydrogen peroxide lyase (HPL) pathway. HPL generates six-carbon aldehydes and alcohols (C6 volatiles) from fatty acid hydroperoxides that are produced from C18:2 and C18:3 by 13-lipoxygenase (13-LOX) (Figure 1) [11]. Loss of function of FAD7 results in decreased C18:3 and increased C18:2, and in tomato, this shift in precursors has been shown to result in dramatically altered C6 volatile composition [12,13]. In particular, decreased levels of C18:3 result in significantly lower (Z)-3-hexenal and (Z)-3-hexenol than observed in wild-type plants, and enhanced C18:2 levels result in elevated levels of hexanal and hexanol [12,13]. Conversely, overexpression of the *FAD7* gene in tomato increases the production of C6 volatiles derived from 18:3 and decreases the production of 18:2-derived compounds [14]. The impact of altered FAD7 activity on volatile profiles generated by the HPL pathway could potentially alter aphid host selection and/or survival and fecundity on foliage.

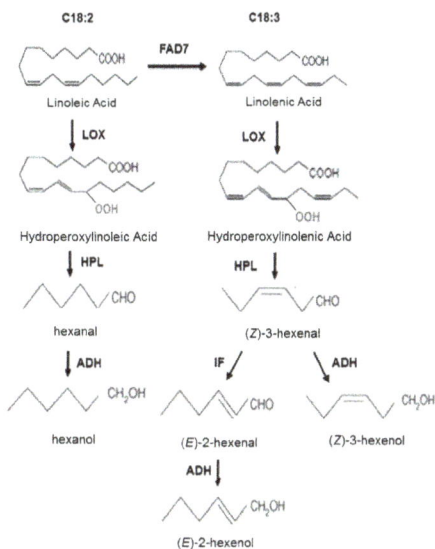

**Figure 1.** Biochemical pathway for synthesis of C6 volatiles in tomato. In tomato, C6 volatiles are synthesized from the polyunsaturated fatty acids linoleic acid (C18:2) and linolenic acid (C18:3) through the successive action of the enzymes lipoxygenase (LOX), hydroperoxide lyase (HPL), alcohol dehydrogenase (ADH), and isomerization factor (IF). FATTY ACID DESATURASE 7 (FAD7) is an omega-3 FAD that desaturates linoleic acid (C18:2) to generate linolenic acid (C18:3).

Several lines of evidence indicate that the HPL pathway can influence direct defenses against insects. In in vitro tests, C6 volatiles including hexanal, (E)-2-hexenal, hexanol, (E)-2-hexenol, and (Z)-3-hexenol have been shown to reduce aphid fecundity [15]. Moreover, artificial manipulation of *HPL* gene expression can influence insect resistance. In potato, antisense suppression of a gene encoding a 13-HPL resulted in increased aphid fecundity [16]. Similarly, a null mutation in an *HPL* homolog rendered rice more susceptible to another piercing-sucking insect, the rice brown plant hopper (*Nilaparvata lugens*) [17]. In Arabidopsis, however, overexpression of *HPL* did not influence the host preference, fecundity, or growth of the green peach aphid, even though it resulted in a fifty-fold increase in C6 volatile production [18]. Furthermore, silencing *HPL* in coyote tobacco (*Nicotiana attenuata*) decreased the feeding behavior and growth of a tobacco hornworm

(*Manduca sexta*) [19], and overexpression of an *HPL* gene from tea (*Camellia sinensis*) in tomato decreased resistance to another chewing insect, *Prodenia litura* [20]. These studies indicate that the HPL pathway influences insect resistance, but that its effects vary in different species combinations, and that further work is needed to understand the role of HPL in specific plant-insect interactions.

The goals of this study were to determine whether the HPL pathway contributes to direct plant defenses against aphids in wild-type tomato, and whether it is required for the enhanced aphid resistance observed in *spr2* plants, which have impaired FAD7 function. We examined the effects of silencing *HPL* in wild-type plants on aphid infestations at two different stages of plant development (3 and 5 weeks after planting), in comparison with the effects of the *spr2* mutation. Silencing *HPL* increased aphid host preference, offspring production, and offspring survival on 5-week old plants, but did not affect aphid infestations on 3-week old plants, and did not influence the survival of adult aphids at either stage of plant development. In contrast, the *spr2* mutation had a strong suppressive effect on adult survival, fecundity, and offspring survival at both 3 and 5 weeks after planting. Thus, the timing and effects of aphid resistance associated with *HPL* differ from those of *spr2*-dependent resistance. We also developed a tomato line (*spr2HPL-RNAi*) that is deficient in both FAD7 function and *HPL* expression in order to determine if loss of function of HPL would compromise aphid resistance associated with *spr2*. The *spr2HPL-RNAi* line showed similar levels of aphid resistance as the *spr2* parent, despite dramatically reduced levels of *HPL* expression. Similarly, bioassays in Arabidopsis indicated that loss of function of FAD7 could confer aphid resistance even in a genotype that carries a null mutation in *HPL* (the *fad7-1* mutant, which we confirmed to be homozygous for the *hpl* mutation). In summary, our results indicate that both *HPL* and *FAD7* influence antixenosis and antibiosis against aphids in tomato, but that the effects of *FAD7* on aphids are distinct from and independent of *HPL*.

## 2. Results

### 2.1. Confirmation of Silencing of HPL in Tomato

A transgenic line in which expression of *HPL* was targeted for silencing by RNA interference (RNAi) [21] was utilized for this study. PCR detection of the kanamycin resistance gene *NPTII* confirmed the presence of the transgene in individuals of the *HPL-RNAi* line, and semi-quantitative PCR confirmed that *HPL* transcript abundance was lower in *HPL-RNAi* plants than in untransformed controls (cv. Flora-Dade) (Figure 2). Analysis of hexanal, (Z)-3-hexenal, (E)-2-hexenal, and (Z)-3-hexen-1-ol levels by gas chromatography (Supplementary Figure S1A–D) also revealed that average C6 volatile production was lower in 5-week-old *HPL-RNAi* plants than in wild-type Flora-Dade.

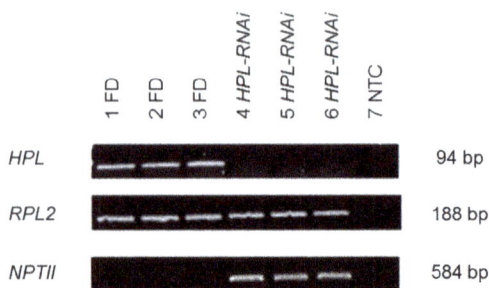

**Figure 2.** Suppression of *HPL* expression by RNAi. Semiquantitative RT-PCR confirmed reduced *HPL* transcript abundance in *HPL-RNAi* plants (lanes 4–6) compared to untransformed wild-type plants (cv. Flora-Dade, FD, lanes 1–3). The housekeeping gene *Rpl2* was used to confirm uniform RNA quantities across samples. The presence of the transgene in the *HPL-RNAi* line was confirmed by PCR detection of the selective marker *NPTII* in genomic DNA samples from the same plants. NTC = no template control.

## 2.2. Influence of HPL and spr2 on Aphid Survival and Fecundity on Tomato

Survival of adult aphids and offspring production were measured in no-choice aphid bioassays to assess the effects of silencing *HPL* on aphid antibiosis in a wild-type tomato cultivar (cv. Flora-Dade). For comparison, antibiotic aphid resistance was also quantified in *spr2* compared to its wild-type control (cv. Castlemart). Assays were performed with more than one age of plant (3- and 5-week old plants) just in case the effects of *HPL* on antibiotic defenses varied with plant age. Based on previous reports, HPL activity and volatile production both can vary over the course of development [22,23]. At both stages of development tested here, adult survival was significantly lower on *spr2* than on the wild-type control Castlemart six days after inoculation (3-week old plants: $p = 0.042$; Figure 3A. 5-week old plants: $p < 0.0001$; Figure 3B). The average number of live offspring per cage, which is a measure of adult fecundity, was also more than 50% lower on *spr2* than on Castlemart at either developmental stage ($p < 0.0001$; Figure 3C,D).

In contrast to *spr2*, which influenced aphid infestations at both 3 and 5 weeks, the effects of silencing *HPL* on aphid populations varied with plant age. The live offspring on *HPL-RNAi* were not significantly different from its control Flora-Dade at 3 weeks ($p = 0.977$; Figure 3C), but were approximately 29% higher than on wild-type at five weeks ($p = 0.005$; Figure 3D). Adult survival did not differ between *HPL-RNAi* and Flora-Dade at either age ($p > 0.05$; Figure 3A,B). These data indicate that *HPL* contributes to antibiotic defenses against juvenile aphids in five-week old plants, whereas the *spr2* mutation promotes antibiosis against adults and juveniles at both 3 and 5 weeks after planting.

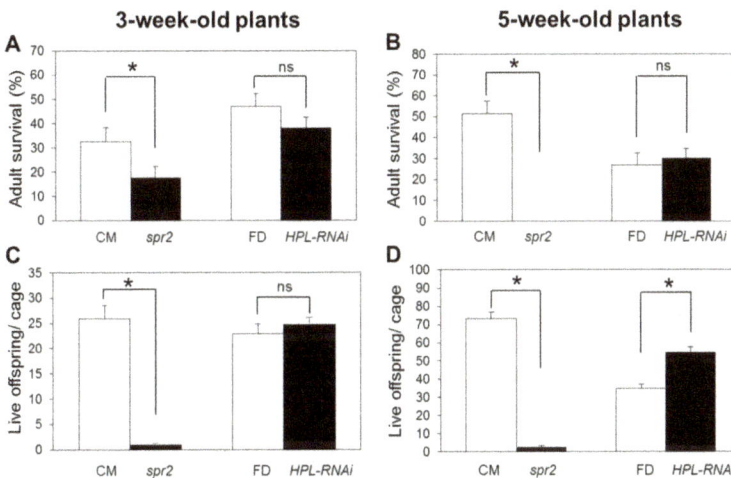

**Figure 3.** Aphid survival and reproduction on tomato. No-choice assays were performed to assess adult survival (**A,B**) and offspring production (**C,D**) of caged potato aphids on 3-week (**A,C**) and 5-week old plants (**B,D**) measured 6 days after inoculation. Asterisks (*) indicate statistically significant differences at $\alpha=0.05$ according to student's *t* test, and error bars represent SEM ($n \geq 10$). Castlemart (CM) and Flora-Dade (FD) are the respective wild-type (WT) controls for *spr2* and *HPL-RNAi*. ns = no significant difference.

## 2.3. Influence of HPL and spr2 on Aphid Host Preference on Tomato

To measure the effects of *HPL* and *spr2* on aphid settling behavior, pair-wise choice tests were performed to compare the *HPL-RNAi* line or *spr2* to their respective wild-type controls. Choice tests were performed with 5-week-old plants, since both *HPL* and *spr2* influence antibiotic defenses at this life stage (Figure 3). In comparisons between *spr2* and wild-type (cv. Castlemart) plants, aphids initially dispersed onto the two genotypes equally, with no significant difference in adult numbers at 1 h after introduction into the choice arena ($p = 0.19$; Figure 4A). Over time, the numbers of adults on *spr2* decreased and the numbers on wild-type controls increased; compared to wild type plants, *spr2* had significantly fewer aphids at 6 h ($p = 0.005$), 24 h ($p < 0.0001$), and 48 h ($p < 0.0001$) after inoculation (Figure 4A). Offspring were first observed at 6 h, and their abundance was significantly lower on *spr2* compared to wild-type control at 24 h ($p < 0.0001$), and 48 h ($p < 0.0001$) (Figure 4B).

When *HPL-RNAi* was compared to its wild-type control (cv. Flora-Dade), the numbers of adult aphids were significantly higher on *HPL-RNAi* at 6 h ($p = 0.038$), 24 h ($p = 0.009$) and 48 h ($p = 0.038$) (Figure 4C); and the numbers of offspring were significantly higher on *HPL-RNAi* at 24 h ($p = 0.044$) and 48 h ($p = 0.030$) (Figure 4D). These results indicate that aphid host preference is enhanced on the *HPL-RNAi* line compared to wild-type plants, whereas the *spr2* mutation decreases aphid host preference. The effects of *spr2* on aphid host preference appeared to be much greater than the effects of *HPL-RNAi*; for example, at 48 h, the number of juveniles on *spr2* was 97% lower than the number on wild-type plants, whereas silencing *HPL* caused only a 37% change in offspring numbers.

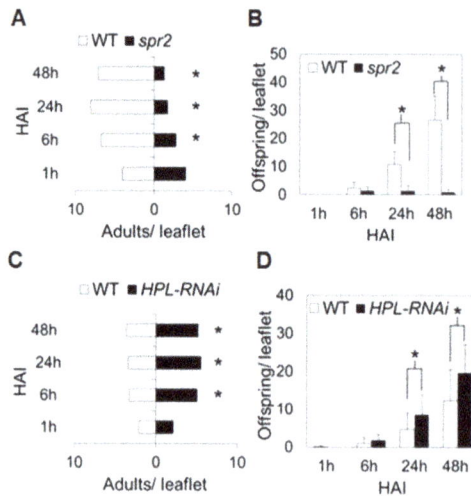

**Figure 4.** Aphid host preference on tomato. Choice assays were performed to compare aphid settling on *spr2* and *HPL-RNAi* with settling behavior on the respective wild-type (WT) controls: Castlemart and Flora-Dade. Adult potato aphids were offered a choice of two plants from different genotypes (14 aphids per pair of plants; 10 pairs of plants for panels (**A,B**), and 15 pairs of plants for panels (**C,D**)). Aphid settling behavior was assessed by recording on which plant the adults were located, and how many offspring they produced at 1 h, 6 h, 24 h and 48 h after inoculation (HAI). Asterisks (*) indicate statistically significant differences at $\alpha = 0.05$ according to Matched pairs one-sided *t*-tests, and error bars represent SEM ($n \geq 10$).

### 2.4. Silencing HPL in the spr2 Mutant

To determine whether the HPL pathway might contribute to enhanced aphid resistance in *spr2*, the *HPL-RNAi* line was crossed with *spr2*, and PCR genotyping was used to identify $F_2$ progeny that were homozygous for the *spr2* mutation and positive for the *HPL-RNAi* construct. RT-qPCR confirmed that *HPL* transcript abundance was significantly reduced in these plants (*spr2HPL-RNAi*), as well as in the *HPL-RNAi* parent line (Figure 5). Compared to the parental lines, the *spr2HPL-RNAi* line had intermediate levels of hexanal, a C6 volatile that accumulates to high levels in *spr2* (Supplementary Figure S1A). No-choice assays were performed to assess aphid performance on *spr2HPL-RNAi* and the parental lines. Bioassays were performed five weeks after planting to focus on a time point when both *spr2* and *HPL-RNAi* impact aphid resistance. As in previous assays, adult survival and live offspring production were significantly lower on *spr2* than on the wild-type control Castlemart (*p* < 0.05; Figure 6A,B), and live offspring production was higher on *HPL-RNAi* than on Flora-Dade (*p* < 0.05; Figure 6B). For both measures of aphid performance, numbers on *spr2HPL-RNAi* were similar to numbers observed on *spr2* (*p* > 0.05, Figure 6A,B), and were significantly lower than numbers observed on the wild-type controls Castlemart and Flora-Dade or on the *HPL-RNAi* parental line (*p* < 0.0001; Figure 6A,B). These data indicate that levels of aphid resistance in *spr2HPL-RNAi* are comparable to levels of resistance in *spr2*, and that silencing *HPL* in an *spr2* background does not compromise aphid resistance mediated by *spr2*.

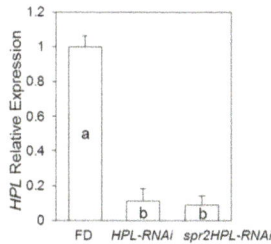

**Figure 5.** *HPL* Expression in *spr2HPL-RNAi*. RT-qPCR was used to compare *HPL* expression in the wild-type cultivar Flora-Dade (FD), a transgenic line in which the *HPL* gene was silenced (*HPL-RNAi*), and progeny of the *spr2* X *HPL-RNAi* cross that were homozygous for the *spr2* mutation in the *FAD7* gene and positive for the *HPL-RNAi* transgene (*spr2HPL-RNAi*). Expression values were normalized using the housekeeping gene *Rpl2* and calculated relative to the wild-type control. Relative expression data were analyzed by one-way ANOVA, and mean separations were performed using Tukey-Kramer HSD. Bars having the same letter are not significantly different at $\alpha = 0.05$, and error bars represent SEM ($n \geq 3$).

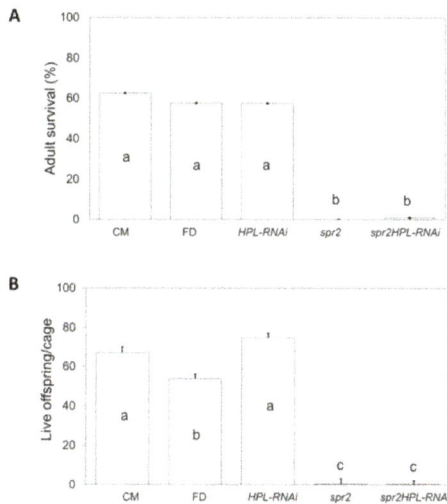

**Figure 6.** The single and combined effects of impairments in *FAD7* and *HPL* on aphid survival and reproduction. A no-choice assay on 5-week-old plants was used to measure aphid performance on $F_2$ progeny of the *spr2* X *HPL-RNAi* cross compared to aphid performance on the parental lines (*spr2* and *HPL-RNAi*) and their respective wild-type controls, Castlemart (CM) and Flora-Dade (FD). All progeny used for this assay were confirmed by PCR to be homozygous for the *spr2* mutation and positive for the *HPL-RNAi* transgene. Data was analyzed by one-way ANOVA and Tukey–Kramer HSD. Bars having the same letter are not significantly different at $\alpha = 0.05$, and error bars represent SEM ($n \geq 10$).

## 2.5. FAD7 and HPL in Arabidopsis

In parallel, we also explored whether the homologous *FAD7* gene in Arabidopsis required a functional copy of the *HPL* gene to influence aphid performance. The Columbia ecotype (Col-0) was previously reported to carry a ten-nucleotide deletion in *HPL* that eliminates the gene's function [24]. PCR genotyping revealed that the *fad7-1* mutant, which originated from Col-0, is also homozygous for the *hpl* mutation (Figure 7A). This *fad7-1 hpl* mutant supported significantly fewer aphids than the wild-type genotypes Columbia and Nossen (*p* < 0.0001; Figure 7B). Thus, aphid resistance conferred by loss of function of the FAD7 protein is independent of HPL in Arabidopsis as well as in tomato.

Unlike Columbia, the Nossen (Nos) ecotype carries a functional allele of the *HPL* gene (Figure 7A) [24]. Aphid population growth did not differ significantly between these two ecotypes (*p* = 0.2293; Figure 7B). These results suggest that *HPL* does not have a major impact on aphid infestations on Arabidopsis, or that its effects are outweighed by other differences between these two genotypes.

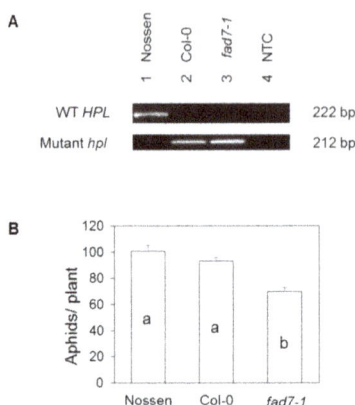

**Figure 7.** Aphid performance on Arabidopsis genotypes with and without mutations in *FAD7* and *HPL*. PCR was used to determine the presence of wild-type and mutant alleles of the *HPL* gene in Columbia (Col-0), Nossen, and *fad7-1* ((**A**), NTC = no template control). A no-choice test was used to assess performance of the green peach aphid on these genotypes ((**B**), *n* = 20). Aphid numbers were analyzed by one-way ANOVA, and mean separations were performed using Tukey's HSD. Bars having the same letter are not significantly different at $\alpha$ = 0.05, and error bars represent SEM.

## 3. Discussion

One objective of this study was to determine whether *HPL* contributes to plant defenses against aphids in wild-type tomato plants. Silencing *HPL* expression in wild-type plants (cv. Flora-Dade) had no effect on aphid infestations three weeks after planting (Figure 3), but resulted in enhanced aphid host preference, reproduction, and survival when infestations occurred about five weeks after planting (Figures 3 and 4). These data suggest that *HPL* contributes to both antixenotic and antibiotic defenses against aphids on tomato, and that these defenses vary between 3- and 5-week old plants. Potentially, the activity of the HPL pathway may vary with plant age; for example, in rice, HPL enzyme activity is low in seedlings and peaks twelve weeks after sowing [22]. Therefore, the fact that silencing *HPL* at three weeks had little effect on aphids could be due to relatively low levels of HPL activity in three-week old plants.

A second objective was to determine if the HPL pathway contributes to aphid resistance in the *spr2* mutant, which is defective in a chloroplast-localized fatty acid desaturase FAD7. Like *HPL*, the *spr2* mutation also impacts C6 volatile synthesis because it alters the relative abundance the fatty acid substrates for volatile synthesis. Because the *spr2* mutation enhances C18:2 accumulation and decreases C18:3 synthesis, it promotes production of C18:2-derived volatiles such as hexanal and inhibits accumulation C18:3 derivatives such as (Z)-3-hexenol [12,13]. However, the results of our aphid bioassays suggest that products of the HPL pathway do not have a causal role in the enhanced levels of aphid resistance observed in *spr2*. Whereas silencing *HPL* did not influence aphid infestations on three-week old plants, the *spr2* mutation has just as strong an impact on aphids in three-week old seedlings as in five-week old plants (Figure 3). Moreover, when *HPL* is silenced in the *spr2* mutant line, it has no effect on aphid performance on this line (Figure 6). These results indicate that *HPL* expression is not essential to aphid resistance in *spr2*.

Consistent with our observations in tomato, the *fad7-1* mutation in Arabidopsis also enhances aphid resistance even in a genetic background that carries the *hpl* mutation (Figure 7), even though this mutation has previously been shown to suppress C6 volatile synthesis [22]. Thus, FAD7 activity in tomato and Arabidopsis modulates direct defenses against aphids independent of HPL activity. Our results also suggest that *HPL* may not have a strong, direct impact on aphid infestations on Arabidopsis, since aphid population growth was similar on ecotypes with (Col-0) and without (Nossen) the *hpl* mutation (Figure 7). This is consistent with a previous report that overexpression of *HPL* had no influence green peach aphid population increase or host preference on Arabidopsis even though overexpression increased C6 volatile levels by over 40-fold [18].

In conclusion, *HPL* contributes to basal aphid resistance in wild-type tomato plants, but enhanced aphid resistance in mutants with impaired FAD7 function is independent of *HPL* gene expression. In Arabidopsis, although *HPL* aids in indirect defenses against aphids by recruiting parasitoids wasps [18], HPL does not appear to contribute significantly to direct defenses against aphids in the *fad7* mutant. These results indicate that fatty acid metabolism in plants can influence plant-aphid interactions through routes independent of C6 volatile synthesis. It is unlikely that loss of function of FAD7 would impact the nutritional quality of plants for aphids, because polyunsaturated fatty acids are naturally rare in the phloem sap on which aphids feed [25]. Our prior work also indicates that aphid resistance in *spr2* is not influenced by jasmonate signaling, although it requires salicylic acid accumulation and *Non-expressor of Pathogenesis Related Proteins* (NPR1) [10]. These findings emphasize the need for further work to understand the mechanisms through which components of primary metabolism including fatty acid desaturation influence plant defense signaling and immunity.

## 4. Materials and Methods

### 4.1. Tomato Culture

Five tomato (*Solanum lycopersicum*) genotypes were used in this study: a mutant line with impaired FAD7 activity called *suppressor of prosystemin-mediated responses2* (*spr2*) [26], a transgenic line silenced for *HPL* (HPLi-1653-3 [21], referred to here as *HPL-RNAi*), a line deficient in both FAD7 function and *HPL* expression (*spr2HPL-RNAi*), and two wild-type cultivars, Castlemart and Flora-Dade. The *spr2* mutant carries a point mutation that results in loss of function of FAD7 [2], and Castlemart is the genetic background that was originally used to develop *spr2*. The creation of the *HPL-RNAi* line in the tomato cultivar Flora-Dade was previously described [21]. In brief, a 330-bp fragment comprising bases 562–881 of the *HPL* open reading frame in the sense orientation and a 595-bp fragment comprising bases 562–1154 of the *HPL* open reading frame in the antisense orientation were expressed in cv. Flora-Dade under the control of the Figwort mosaic virus 35S in order to induce silencing of *HPL*. The authors previously demonstrated that the *HPL RNAi* line used in this study has significantly reduced *HPL* mRNA accumulation and C6 volatile production in the fruits as well as foliage [21]. Since fruits are not typically produced until at least 8 weeks after planting, this data indicated that silencing was persistent in mature plants. The *spr2HPL-RNAi* line was produced by crossing *spr2* and *HPL-RNAi* (described below). Tomato plants (*Solanum lycopersicum*) were grown in LC1 Sunshine potting mix (Sungro Horticulture, Bellevue, WA, USA) with 15-9-12 Osmocote slow-release fertilizer (Scotts-MiracleGro

Company, Marysville, OH, USA) at 23 °C and L16:D8 photoperiod in an environmental growth chamber (Conviron, Winnipeg, MB, Canada), and watered daily with a dilute nutrient solution containing 1000 ppm $CaNO_3$ (Hydro Agri North America, Tampa, FL, USA), 500 ppm $MgSO_4$ (Giles Chemical Corp, Waynesville, NC, USA), and 500 ppm 4-18-38 Gromore fertilizer (Gromore, Gardena, CA, USA).

*4.2. Development of the spr2HPL-RNAi Line*

A tomato line with impairments in both *FAD7* and *HPL* was developed by manually transferring pollen from *HPL-RNAi* to *spr2* and screening the (*spr2* × *HPL-RNAi*) $F_2$ generation for individuals that were positive for the *HPL-RNAi* transgene and homozygous for the *spr2* mutation. Screening for the *spr2* mutation was performed by PCR using two allele-specific primer sets that target a single nucleotide polymorphism as described by Avila et al. [10]. Presence of the *HPL-RNAi* transgene in $F_2$ plants was determined by amplifying the selectable marker *NPTII* (*Neomycin phosphotransferase II*) using forward (5′-GCAATATCACGGGTAGCCAA-3′) and reverse (5′-GCCGTGTTCCGGCTGTCA-3′) primers. *NPTII* PCR was performed at 95 °C for 5 min; 95 °C for 45 s, 50 °C for 45 s, and 72 °C for 45 s (30 cycles); and final extension at 72 °C for 5 min.

*4.3. Arabidopsis Culture and Materials Development*

Arabidopsis plants (*Arabidopsis thaliana*) were grown in a peat, vermiculite, perlite (4:3:2 ratio, Sungro Horticulture, Bellevue, WA, USA) soil mixture supplemented with 15-9-12 Osmocote Plus fertilizer (Scotts-MiracleGro Company, Marysville, OH, USA) at 23 °C and L13:D11 photoperiod in a growth chamber (Conviron, Winnipeg, MB, Canada). The plants were fertilized weekly with Miracle Gro® all-purpose plant food (Scotts-MiracleGro Company, Marysville, OH, USA). Arabidopsis ecotype Columbia (Col-0, CS70000) was obtained from the Arabidopsis Biological Resource Center (Ohio State University, Columbus, OH, USA), and the Nossen ecotype and the *fad7-1gl1* mutant (developed in a Col-0 genetic background) were obtained from Dr. Jyoti Shah (University of North Texas). Because the *fad7-1gl1* mutant carries a mutation (*gl1*) in a gene required for trichome production (*GLABRA1*) in addition to a mutation in *FAD7*, this mutant was crossed with Col-0 to develop another line (*fad7-1*) with impaired FAD7 function but normal trichome development. In the $F_2$ generation, plants with trichomes were screened by PCR with primer sets specific to the mutant and wild-type alleles of *FAD7* to select for plants homozygous for the mutant *fad7-1* allele [27]. Plants that lacked the *gl1* mutant phenotype and that were homozygous for the *fad7-1* mutation were then propagated to generate seeds for subsequent assays. All plants were observed to confirm the presence of trichomes before they were used for experiments.

*4.4. Identification of the HPL Mutation in the Arabidopsis fad7-1 Mutant*

The *fad7-1* mutant in Arabidopsis was screened by PCR for the presence of a 10-bp deletion (from 161 bp to 170 bp) in the *HYDROPEROXIDE LYASE* (*HPL*) gene that occurs naturally in the ecotype Columbia (Col-0), and that results in a non-functional HPL protein [24]. Col-0 was included as a positive control for the mutant allele, and the Nossen ecotype was included as a positive control for the wild-type *HPL* allele. Genomic DNA was extracted using an extraction buffer that was made by diluting Edwards solution (200 mM Tris-HCl (pH 7.5), 250 mM NaCl, 25 mM EDTA, and 0.5% SDS) by 10-fold with TE buffer (10 mM Tris-HCl (pH 8) and 1 mM EDTA) [28,29]. Each sample was used for two separate PCR reactions: one with a primer set that amplifies only the wild-type *HPL* allele (At4g15440.1) (forward 5′-GGACCGTTTAGATTACTTCTGGTT-3′, reverse 5′-CGGAAGTCTCCGATGAGAAC-3′), and another reaction with a primer set that specifically targets the mutant *hpl* allele with the 10 nts deletion (5′-GACCGTTTAGATTCCAAGGAC-3′, reverse 5′-CGGAAGTCTCCGATGAGAAC-3′). PCR amplification was performed at 95 °C for 5 min, followed by 30 cycles of 95 °C for 45 s, 55 °C for 45 s, and 72 °C for 45 s, and a final extension at 72 °C for 5 min. PCR products were separated by electrophoresis on 1% agarose gels.

### 4.5. RNA Isolation and Gene Expression Analysis

For analysis of gene expression, total RNA was extracted from approximately 100 mg of flash-frozen leaf tissue using TRIzol reagent and chloroform (Invitrogen Corp., Carlsbad, CA, USA) using the manufacturer's instructions. cDNA was synthesized from 1 μg of total RNA per sample using Superscript III reverse transcriptase and oligo dT$_{(18)}$ primers in a 20 μL reaction volume (Invitrogen Corp., Carlsbad, CA, USA). Transcript abundance in the cDNA was then quantified by semi-quantitative PCR or real-time PCR. For semi-quantitative PCR, 50 ng of cDNA was used as a template and the final concentration for each primer was 0.4 μM. The PCR program was: 95 °C for 5 min; 95 °C for 30 s, 52 °C for 30 s, and 72 °C for 30 s (22 cycles); and final extension at 72 °C for 5 min. PCR products were detected on a 1% agarose gel. Real-time quantitative PCR was performed on 2 μL of cDNA in a 20 μL reaction volume using the QuantiTect SYBR Green PCR Kit (Qiagen, Inc., Valencia, CA, USA) on a StepOnePlus Real-time PCR system (Applied Biosystems, Foster City, CA, USA). The RT-qPCR program was: 95 °C activation for 10 min, followed by 40 cycles of amplification and quantification (denaturation at 95 °C for 15 s, annealing at 52 °C for 30 s, extension 72 °C for 30 s with a single fluorescence detection). Melting curves were generated at 60–95 °C with a heating rate of 0.3 °C per second. Three biological replicates for each genotype and two technical replicates for each biological replicate were used. Transcript abundance of tomato *HPL* (Solyc07g049690; GenBank accession AF230372.1) was measured using primers previously described by Shen and coworkers [21] (5′-AGCTACGGATTGCCGTTAGT-3′/ 5′-TTTCCATTCTCTTGGTGAAGAA-3′). Data were normalized to the expression levels of the endogenous control *Ribosomal Protein L2 (RPL2)* using primers previously described by Avila and coworkers [10] (5′-GAGGGCGTACTGAGAAACCA-3′/5′-CTTTTGTCCAGGAGGTGCAT-3′). Gene expression was calculated relative to the wild-type control for each genotype comparison using the methodology described by Pfaffl [30].

### 4.6. Aphid Bioassays

#### 4.6.1. Insect Materials

Potato aphids (*Macrosiphum euphorbiae*) were reared on an aphid-susceptible tomato cultivar (cv. BHN876, potato (*Solanum tuberosum* Linnaeus), and Jimson weed (*Datura stramonium* Linnaeus) plants at 20 °C and 16-h light photoperiod. Green peach aphids (*Myzus persicae*) were reared on an aphid-susceptible cabbage cultivar (*Brassica oleracea* var. Joychoi) at ~23 °C and 16-h light photoperiod.

#### 4.6.2. Aphid Survival and Fecundity on Tomato

No-choice assays were performed to evaluate potato aphid survival and fecundity. Wingless adult potato aphids within 24 h of emergence to adulthood were confined to single leaflets of intact plants using clip cages (4 adults per cage, 2 cages per plant, 10–15 replicate plants per genotype), and the numbers of living and dead adults and offspring were recorded at six days after infestation (6-DAI). Both 3- and 5-week-old plants were inoculated to determine if aphid resistance varied with age. The positions of the cages were standardized for all plants in each assay; cages were placed on the terminal leaflet of the 2–3 fully-expanded leaves below the apical meristem. Plants were maintained in a growth chamber at 23 °C and 16L: 8D photoperiod during the bioassay.

#### 4.6.3. Aphid Host Preference on Tomato

Settling behavior of the potato aphid was measured on intact tomato plants by placing adult aphids on choice arenas that allowed them to move back and forth between paired tomato genotypes. Each arena consisted of a Styrofoam platform (15 cm diameter) that was placed underneath two paired leaflets: the terminal leaflet of the third fully expanded leaf below the apical meristem of each of the paired plants. Wingless adult potato aphids within 24 h of emergence to adulthood (14 adults per arena) were placed between the leaflets and confined to the arena using a vented petri dish lid with a soft gasket that prevented damaging the petioles (Supplementary Figure S2). The number of adult aphids on each plant were recorded at 1 h, 6 h, 24 h and 48 h after release. Offspring production was also monitored because it is a well-established marker of host plant acceptance [31]. In each experiment, ten to fifteen replicate pairs per combination of genotypes were tested using five-week-old tomato plants, and each experiment was performed at least twice.

### 4.6.4. Aphid Survival and Fecundity on Arabidopsis

To measure aphid performance on Arabidopsis, plants were inoculated with the green peach aphid (3 wingless newly-emerged adults/plant; 15 plants/genotype) when first flower buds were visible (developmental stage 5.1 according to [32]). After infestation, plants were covered with sleeve cages and maintained for 7 days in a growth chamber (23 °C; 65% relative humidity; L13:D11 photoperiod). The numbers of live and dead adults and offspring aphids on each plant were scored 7 days after infestation (DAI) in this no-choice assay.

### 4.7. Statistical Analysis

All statistical analyses were done with JMP® v.11 (SAS Institute Inc., Cary, NC, USA). Host preference assays were analyzed by matched pairs one sided $t$-tests within each time point, and no-choice assays were analyzed by one-way ANOVA, with $\alpha = 0.05$.

**Supplementary Materials:** The following are available online at http://www.mdpi.com/1422-0067/19/4/1077/s1. Figure S1: Comparison of C6 volatile levels in five-week-old plants with modifications in fatty acid desaturation and/or HYDROPEROXIDE LYASE expression. Figure S2: Design of choice assays to study aphid host preference on tomato.

**Acknowledgments:** We thank Mali C. Sirisena, Carmen S. Padilla, Junhuan Xu, Min Woo Lee, Aravind Galla, Dhaval S. Shah, and Janithri Wickramanayake for assistance in the laboratory. We also thank Gregg Howe at Michigan State University for providing seeds of Castlemart and *spr2*, Jyoti Shah at the University of North Texas for providing Arabidopsis *fad7-1gl1* and Nossen seeds, and Kaleb L. Vaughn for developing the Arabidopsis *fad7-1* line without *gl1*. This research was supported by USDA-NRI (grant number 2015-67013-23412) and the National Science Foundation (grant number IOS-0951287).

**Author Contributions:** Jiamei Li conducted plant hybridization, PCR screening, gene expression analysis, aphid bioassays, and statistical analysis. Carlos A. Avila participated in aphid bioassays. Denise M. Tieman and Harry J. Klee developed and characterized the *HPL-RNAi* transgenic line and conducted volatile analysis. Fiona L. Goggin designed the experiments.

**Conflicts of Interest:** The authors declare no conflict of interest. The funding sponsors had no role in the design of the study; in the collection, analyses, or interpretation of data; and in the decision to publish the results.

### References

1. Browse, J.; McCourt, P.; Somerville, C. A mutant of Arabidopsis deficient in $C_{18:3}$ and $C_{16:3}$ leaf lipids. *Plant Physiol.* **1986**, *81*, 859–864. [CrossRef] [PubMed]

2. Li, C.; Liu, G.; Xu, C.; Lee, G.I.; Bauer, P.; Ling, H.Q.; Ganal, M.W.; Howe, G.A. The tomato *suppressor of prosystemin-mediated responses2* gene encodes a fatty acid desaturase required for the biosynthesis of jasmonic acid and the production of a systemic wound signal for defense gene expression. *Plant Cell* **2003**, *15*, 1646–1661. [CrossRef] [PubMed]

3. Gibson, S.; Arondel, V.; Iba, K.; Somerville, C. Cloning of a temperature-regulated gene encoding a chloroplast omega-3 desaturase from *Arabidopsis thaliana. Plant Physiol.* **1994**, *106*, 1615–1621. [CrossRef] [PubMed]

4. Berberich, T.; Harada, M.; Sugawara, K.; Kodama, H.; Iba, K.; Kusano, T. Two maize genes encoding omega-3 fatty acid desaturase and their differential expression to temperature. *Plant Mol. Biol.* **1998**, *36*, 297–306. [CrossRef] [PubMed]

5. Kodama, H.; Nishiuchi, T.; Seo, S.; Ohashi, Y.; Iba, K. Possible involvement of protein phosphorylation in the wound-responsive expression of *Arabidopsis* plastid $\omega$-3 fatty acid desaturase gene. *Plant Sci.* **2000**, *155*, 153–160. [CrossRef]

6. Dong, C.-J.; Cao, N.; Zhang, Z.-G.; Shang, Q.-M. Characterization of the fatty acid desaturase genes in cucumber: Structure, phylogeny, and expression patterns. *PLoS ONE* **2016**, *11*, e0149917. [CrossRef] [PubMed]

7. Murakami, Y.; Tsuyama, M.; Kobayashi, Y.; Kodama, H.; Iba, K. Trienoic fatty acids and plant tolerance of high temperature. *Science* **2000**, *287*, 476–479. [CrossRef] [PubMed]
8. Im, Y.J.; Kim, M.S.; Yang, K.Y.; Kim, Y.H.; Back, K.; Cho, B.H. Antisense expression of a ω-3 fatty acid desaturase gene in tobacco plants enhances susceptibility against pathogens. *Can. J. Bot.* **2004**, *82*, 297–303. [CrossRef]
9. Yara, A.; Yaeno, T.; Hasegawa, M.; Seto, H.; Montillet, J.-L.; Kusumi, K.; Seo, S.; Iba, K. Disease resistance against *Magnaporthe grisea* is enhanced in transgenic rice with suppression of ω-3 fatty acid desaturases. *Plant Cell Physiol.* **2007**, *48*, 1263–1274. [CrossRef] [PubMed]
10. Avila, C.A.; Arevalo-Soliz, L.M.; Jia, L.; Navarre, D.A.; Chen, Z.; Howe, G.A.; Meng, Q.W.; Smith, J.E.; Goggin, F.L. Loss of function of FATTY ACID DESATURASE 7 in tomato enhances basal aphid resistance in a salicylate-dependent manner. *Plant Physiol.* **2012**, *158*, 2028–2041. [CrossRef] [PubMed]
11. Matsui, K. Green leaf volatiles: Hydroperoxide lyase pathway of oxylipin metabolism. *Curr. Opin. Plant Biol.* **2006**, *9*, 274–280. [CrossRef] [PubMed]
12. Canoles, M.A.; Beaudry, R.M.; Li, C.; Howe, G. Deficiency of linolenic acid in *Lefad7* mutant tomato changes the volatile profile and sensory perception of disrupted leaf and fruit tissue. *J. Am. Soc. Hortic. Sci.* **2006**, *131*, 284–289.
13. Sanchez-Hernandez, C.; Lopez, M.G.; Delano-Frier, J.P. Reduced levels of volatile emissions in jasmonate-deficient *spr2* tomato mutants favour oviposition by insect herbivores. *Plant Cell Environ.* **2006**, *29*, 546–557. [CrossRef] [PubMed]
14. Domínguez, T.; Hernández, M.L.; Pennycooke, J.C.; Jiménez, P.; Martínez-Rivas, J.M.; Sanz, C.; Stockinger, E.J.; Sánchez-Serrano, J.J.; Sanmartín, M. Increasing ω-3 desaturase expression in tomato results in altered aroma profile and enhanced resistance to cold stress. *Plant Physiol.* **2010**, *153*, 655–665. [CrossRef]
15. Hildebrand, D.F.; Brown, G.C.; Jackson, D.M.; Hamilton-Kemp, T.R. Effects of some leaf-emitted volatile compounds on aphid population increase. *J. Chem. Ecol.* **1993**, *19*, 1875–1887. [CrossRef] [PubMed]
16. Vancanneyt, G.; Sanz, C.; Farmaki, T.; Paneque, M.; Ortego, F.; Castanera, P.; Sanchez-Serrano, J.J. Hydroperoxide lyase depletion in transgenic potato plants leads to an increase in aphid performance. *Proc. Natl. Acad. Sci. USA* **2001**, *98*, 8139–8144. [CrossRef] [PubMed]
17. Tong, X.; Qi, J.; Zhu, X.; Mao, B.; Zeng, L.; Wang, B.; Li, Q.; Zhou, G.; Xu, X.; Lou, Y.; et al. The rice hydroperoxide lyase OsHPL3 functions in defense responses by modulating the oxylipin pathway. *Plant J.* **2012**, *71*, 763–775. [CrossRef] [PubMed]
18. Chehab, E.W.; Kaspi, R.; Savchenko, T.; Rowe, H.; Negre-Zakharov, F.; Kliebenstein, D.; Dehesh, K. Distinct roles of jasmonates and aldehydes in plant-defense responses. *PLoS ONE* **2008**, *3*, e1904. [CrossRef] [PubMed]
19. Halitschke, R.; Ziegler, J.; Keinanen, M.; Baldwin, I.T. Silencing of hydroperoxide lyase and allene oxide synthase reveals substrate and defense signaling crosstalk in *Nicotiana attenuata*. *Plant J.* **2004**, *40*, 35–46. [CrossRef] [PubMed]
20. Xin, Z.; Zhang, L.; Zhang, Z.; Chen, Z.; Sun, X. A tea hydroperoxide lyase gene, *CsiHPL1*, regulates tomato defense response against *Prodenia litura* (Fabricius) and *Alternaria alternate* f. sp. *Lycopersici* by modulating green leaf volatiles (GLVs) release and jasmonic acid (JA) gene expression. *Plant Mol. Biol. Report.* **2014**, *32*, 62–69. [CrossRef]
21. Shen, J.; Tieman, D.; Jones, B.; Taylor, G.; Schmelz, E.; Huffaker, A.; Bies, D.; Chen, K.; Klee, H.J. A 13-lipoxygenase, TomloxC, is essential for synthesis of C5 flavour volatiles in tomato. *J. Exp. Bot.* **2014**, *65*, 419–428. [CrossRef] [PubMed]
22. Liu, X.; Li, F.; Tang, J.; Wang, W.; Zhang, F.; Wang, G.; Chu, J.; Yan, C.; Wang, T.; Chu, C.; et al. Activation of the jasmonic acid pathway by depletion of the hydroperoxide lyase OsHPL3 reveals crosstalk between the HPL and AOS branches of the oxylipin pathway in rice. *PLoS ONE* **2012**. [CrossRef] [PubMed]
23. Hanley, M.E.; Girling, R.D.; Felix, A.-E.; Olliff, E.D.; Newland, P.L.; Poppy, G.M. Olfactory selection of *Plantago lanceolata* declines with seedling age. *Ann. Bot.* **2013**, *112*, 671–676. [CrossRef] [PubMed]
24. Duan, H.; Huang, M.-Y.; Palacio, K.; Schuler, M.A. Variations in CYP74B2 (hyperoxide lyase) gene expression differentially affect hexenal signaling in the Columbia and Landsberg erecta ecotypes of *Arabidopsis*. *Plant Physiol.* **2005**, *139*, 1529–1544. [CrossRef] [PubMed]
25. Madey, E.; Nowack, L.; Thompson, J. Isolation and characterization of lipid in phloem sap of canola. *Planta* **2002**, *214*, 625–634. [CrossRef] [PubMed]

26. Howe, G.A.; Ryan, C.A. Suppressors of systemin signaling identify genes in the tomato wound response pathway. *Genetics* **1999**, *153*, 1411–1421. [PubMed]

27. Vaughn, K.L.; Avila, C.A.; Padilla-Marcia, C.S.; Goggin, F.L. Development of *fad7-1* single mutant *Arabidopsis thaliana* plants that are resistant to aphids. *Discovery* **2014**, *15*, 94–99.

28. Edwards, K.; Johnstone, C.; Thompson, C. A simple and rapid method for the preparation of plant genomic DNA for PCR analysis. *Nucleic Acids Res.* **1991**, *19*, 1349. [CrossRef] [PubMed]

29. Kasajima, I.; Ide, Y.; Ohkama-Ohtsu, N.; Hayashi, H.; Yoneyama, T.; Fujiwara, T. A protocol for rapid DNA extraction from *Arabidopsis thaliana* for PCR analysis. *Plant Mol. Biol. Report.* **2004**, *22*, 49–52. [CrossRef]

30. Pfaffl, M.W. A new mathematical model for relative quantification in real-time RT-PCR. *Nucleic Acids Res.* **2001**, *29*, e45. [CrossRef] [PubMed]

31. Powell, G.; Tosh, C.R.; Hardie, J. Host plant selection by aphids: Behavioral, evolutionary, and applied perspectives. *Annu. Rev. Entomol.* **2006**, *51*, 309–330. [CrossRef] [PubMed]

32. Boyes, D.C.; Zayed, A.M.; Ascenzi, R.; McCaskill, A.J.; Hoffman, N.E.; Davis, K.R.; Görlach, J. Growth stage-based phenotypic analysis of Arabidopsis: A model for high throughput functional genomics in plants. *Plant Cell* **2001**, *13*, 1499–1510. [CrossRef] [PubMed]

International Journal of
*Molecular Sciences*

MDPI

*Article*

# Expressing *OsMPK4* Impairs Plant Growth but Enhances the Resistance of Rice to the Striped Stem Borer *Chilo suppressalis*

Xiaoli Liu, Jiancai Li, Liping Xu, Qi Wang and Yonggen Lou *

State Key Laboratory of Rice Biology & Ministry of Agriculture Key Lab of Agricultural Entomology, Institute of Insect Sciences, Zhejiang University, Hangzhou 310058, China; 21216107@zju.edu.cn (X.L.); jiancai.2007@163.com (J.L.); 21616161@zju.edu.cn (L.X.); q.wang302@gmail.com (Q.W.)
* Correspondence: yglou@zju.edu.cn; Tel.: +86-571-8898-2622

Received: 3 March 2018; Accepted: 10 April 2018; Published: 13 April 2018

**Abstract:** Mitogen-activated protein kinases (MPKs) play a central role not only in plant growth and development, but also in plant responses to abiotic and biotic stresses, including pathogens. Yet, their role in herbivore-induced plant defenses and their underlying mechanisms remain largely unknown. Here, we cloned a rice MPK gene, *OsMPK4*, whose expression was induced by mechanical wounding, infestation of the striped stem borer (SSB) *Chilo suppressalis*, and treatment with jasmonic acid (JA), but not by treatment with salicylic acid (SA). The overexpression of *OsMPK4* (oe-MPK4) enhanced constitutive and/or SSB-induced levels of JA, jasmonoyl-L-isoleucine (JA-Ile), ethylene (ET), and SA, as well as the activity of elicited trypsin proteinase inhibitors (TrypPIs), and reduced SSB performance. On the other hand, compared to wild-type plants, oe-MPK4 lines in the greenhouse showed growth retardation. These findings suggest that *OsMPK4*, by regulating JA-, ET-, and SA-mediated signaling pathways, functions as a positive regulator of rice resistance to the SSB and a negative regulator of rice growth.

**Keywords:** rice; *Chilo suppressalis*; mitogen-activated protein kinase 4; jasmonic acid; salicylic acid; ethylene; herbivore-induced defense response

## 1. Introduction

In their natural habitats, plants often face infestations by various herbivores. To protect themselves, plants have developed constitutive and induced defensive mechanisms [1–3]. Induced defenses are made when a plant recognizes damage-related signals from herbivores. In response to the signals, plants activate a defense-related signaling network consisting mainly of jasmonic acid (JA)-, salicylic acid (SA)-, and ethylene (ET)-mediated pathways, and induce the production of defensive compounds; as a result, herbivore resistance in plants increases [4–7]. During this process, mitogen-activated protein kinase (MPK) cascades play vital roles in the signaling network, functioning upstream and downstream of defense-related signaling pathways [8–11].

A typical MPK cascade comprises three modules: MPKs, MPK kinases (MPKKs or MEKs), and MPKK kinases (MPKKKs or MEKKs), all of which are evolutionarily conserved in all eukaryotes [12]. These modules are sequentially activated by the dual phosphorylation of threonine and tyrosine residues in a TxY (x = D or E) motif located in their kinase catalytic activation loop. By activating the module MPKs, which subsequently phosphorylate some transcription factors and enzymes, and thereby trigger downstream signaling components or pathways, MPK cascades have been reported to play central roles in plant growth, development, and responses to abiotic and biotic stresses, including defense responses to pathogens and herbivores [8,13–17]. In *Arabidopsis thaliana*, for instance, MPK3 and MPK6 directly enhance the activity of 1-amino-cyclopropane-1-carboxylic acid synthase 6 (ACS6) and

ACS2 at both transcriptional and protein levels, and thereby lead to the increase in the production of ET [18,19]. In *Nicotiana attenuata*, the wound-induced protein kinase (WIPK) and salicylic acid-induced protein kinase (SIPK) (orthologs of *Arabidopsis thaliana* MPK3 (AtMPK3) and AtMPK6, respectively) regulate JA- and SA-mediated signaling pathways, as well as herbivore-induced defense responses [9]. In rice, silencing *OsMPK3* (an ortholog of *AtMPK3*, also known as *OsBIMK1*) reduces the level of herbivore-elicited JA and the activity of trypsin protease inhibitors (TrypPIs), which enhances its resistance to the striped stem borer (SSB) *Chilo suppressalis* [10]. Moreover, OsMPK3 has also been found to negatively regulate the resistance of rice to *Magnaporthe oryzae* and positively regulate the plant's tolerance to both drought and submergence [20,21].

MPK4, which acts downstream of the MEKK1–MEK1/MEK2 cascade, is one of the best-characterized MPKs. In Arabidopsis, MPK4 exists in the nucleus in a ternary complex with MAP KINASE 4 SUBSTRATE1 (MKS1) and the WRKY transcription factor WRKY33; these inhibit the extent to which WRKY33 can act as a transcription factor. Upon stimulation, MPK4 is activated and MKS1 is phosphorylated, and, subsequently, both MKS1 and WRKY33 are released from MPK4; the released WRKY33 then regulates the expression of target genes [22,23]. MPK4 is believed to act as a negative regulator of plant immunity to pathogens and a positive regulator of plant growth [24–27]. In Arabidopsis, for example, the *mpk4* mutant is severely dwarfed, and exhibits elevated SA levels and enhanced resistance to biotrophic pathogens [28–30]. Similarly, in soybean (*Glycine max*), silencing *GmMPK4* enhances SA and $H_2O_2$ accumulation, and plant resistance to downy mildew and to the soybean mosaic virus, but reduces plant growth [26]. By contrast, in rice, OsMPK4 has been reported to act both as an activator and a repressor of plant resistance to *Xanthomonas oryzae* pv *oryzae* (*Xoo*) [31]. Recently, a few studies have revealed that MPK4 also plays an important role in herbivore-induced defense responses. In *N. attenuata*, silencing *MPK4* elevates elicited levels of JA and jasmonoyl-L-isoleucine (JA-Ile); silencing *MPK4* also activates a JA-independent defense pathway, which in turn increases the resistance of plants to *Manduca sexta* [32]. In summary, these new findings shed light on the role of MPK4 in herbivore-induced plant defenses, a role which has until recently remained largely unexplored.

Rice, a staple food worldwide, is attacked by many insect pests [33], among which SSB is one of the most serious. Previous studies in rice have revealed that SSB infestation induces the biosynthesis of JA, JA-Ile, SA, and ET; these compounds subsequently modulate defense responses, including the production of herbivore-induced volatiles and an increase in the activity of TrypPIs [34–38]. Given the key role of MPK4 in plant defenses, we isolated the rice MPK4 gene, *OsMPK4* (TIGR ID: *Os10g38950*, the homologue of Arabidopsis *MPK4* and tobacco *MPK4*; also known as *OsMPK6*), which belongs to subgroup B of the MPK family and harbors a well-conserved TEY motif, as well as the evolutionarily conserved C-terminal common docking domain [39], and we characterized its role in herbivore-induced defense responses. By combining molecular biology, inverse genetics, chemical analysis, and bioassays, we show that *OsMPK4* is induced by mechanical wounding and herbivore attack. The overexpression of *OsMPK4* increases basal and/or SSB-induced levels of JA, JA-Ile, ET, and SA, as well as the activity of TrypPIs; in response, the performance of SSB larvae is reduced. Moreover, lines overexpressing *OsMPK4* exhibit reduced size. All these findings suggest that OsMPK4 acts as a positive modulator of herbivore-elicited defense responses and a negative mediator of plant growth in rice.

## 2. Results

### 2.1. Mechanical Wounding, Striped Stem Borer (SSB) Infestation, and Jasmonic Acid (JA) Treatment Induce Expression of OsMPK4

We screened rice plants for herbivore-induced transcripts using rice microarrays and found that one MPK gene, *OsMPK4*, was upregulated after SSB infestation [40]. The full-length cDNA of the *OsMPK4*, including an open reading frame (ORF) of 1131 bp, was obtained by reverse transcription polymerase chain reaction (PCR) (Figure S1). Phylogenetic analysis of some MPKs in subgroup B from different species revealed that OsMPK4 is homologous to BdMPK6 in *Brachypodium distachyon*, to

SiMPK6 in *Setaria italica* and to SbMPK6 in *Sorghum bicolor*; all of these proteins share more than 94% amino acid sequence identity with OsMPK4. OsMPK4 also shows homologous to well characterized MPK4 in *Arabidopsis thaliana* (AtMPK4) [22], *Brassica napus* (BnMPK4) [41], and *Nicotiana tabacum* (NtMPK4) [42] (Figure S2), whose amino acid sequence similarities to OsMPK4 were 81.12%, 82.18%, and 81.91%, respectively.

Quantitative real-time PCR (qRT-PCR) analysis revealed low constitutive levels of *OsMPK4* transcripts. Mechanical wounding, SSB infestation, and JA treatment markedly enhanced transcript levels of *OsMPK4*, whereas SA treatment did not (Figure 1). Moreover, unlike mechanical wounding, which quickly induced the expression of *OsMPK4*, SSB infestation and, especially, JA treatment increased the transcript level of *OsMPK4* only at later treatment stages. These data indicate that OsMPK4 might be involved in defense responses of rice to SSB.

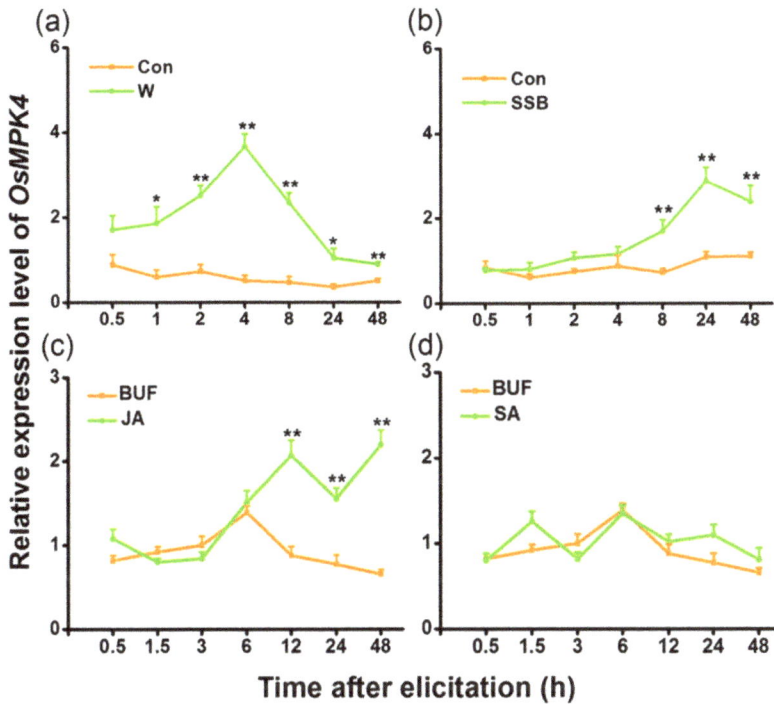

**Figure 1.** Relative transcript levels of *OsMPK4* in rice after different treatments. Mean transcript levels (+standard error (SE), $n = 5$) of *OsMPK4* in rice stems that were mechanically wounded (W) (**a**), infested by striped stem borer (SSB) (**b**), or treated with jasmonic acid (JA) (**c**) or salicylic acid (SA) (**d**). Transcript levels were analyzed by quantitative real-time PCR (qRT-PCR). Con, non-manipulated plant, BUF, buffer. Asterisks indicate significant differences in transcript levels between treatments and controls (*, $p < 0.05$; and **, $p < 0.01$, Student's *t*-tests).

## 2.2. Overexpression of OsMPK4

To explore the function of OsMPK4 in herbivore-induced defenses, we obtained two homozygous single-insertion lines with over-expressed *OsMPK4* (oe-MPK4-43 and oe-MPK4-59) (Figure 2a). Transcript analysis found that constitutive transcript levels of *OsMPK4* in two oe-MPK4 lines, oe-MPK4-43 and oe-MPK4-59, were 20.4- and 25.0-fold higher, respectively, than levels of *OsMPK4* in wild type (WT) plants (Figure 2b); after mechanical wounding, transcript levels of *OsMPK4* in the two oe-MPK4 lines were still significantly higher than levels in WT plants: 4 h after mechanical wounding, for example, transcript levels of *OsMPK4* in the two oe-MPK4 lines were 5.2- and 6.0-fold higher than those in WT plants (Figure 2b). Lines with overexpressed *OsMPK4* showed growth retardation, especially at later growth stages (Figure 3a–d,f). At 55 days, for instance, the root length of the two oe-MPK4 lines, oe-MPK4-43 and oe-MPK4-59, decreased by approximately 31.85% and 25.35%, respectively, compared to that of WT plants (Figure 3b), and the total root mass of the two oe-MPK4 lines decreased by approximately 35.56% and 24.47% (Figure 3d); moreover, the mass of the aboveground part of oe-MPK4 plants decreased by approximately 18.96% and 12.46% (Figure 3c). Unlike the growth phenotype, the chlorophyll content in oe-MPK4 lines was a bit higher (about 4%) than in WT plants (Figure 3e). These results suggest that OsMPK4 is involved in rice growth and development.

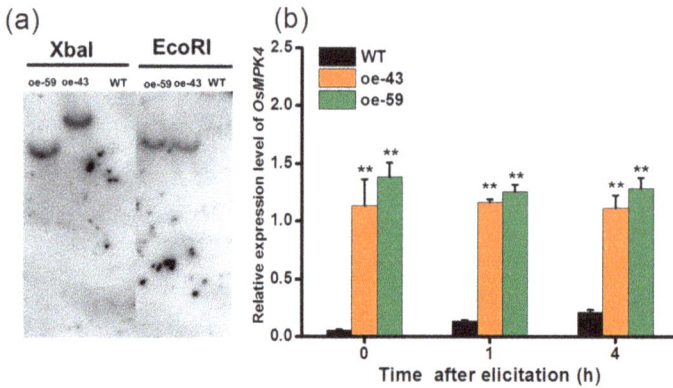

**Figure 2.** DNA gel-blot analysis of oe-MPK4 and expression levels of *OsMPK4* in transgenic lines and wild-type (WT) plants. (**a**) Genomic DNA was digested with *EcoR* I or *Xba* I. The blot was hybridized with a probe specific for the reporter gene encoding β-glucuronidase. Two transgenic lines (oe-43 and oe-59) have a single insertion of the transgene. The magnification is 1; (**b**) Mean transcript levels (+SE, $n = 5$) of *OsMPK4* in oe-MPK4 lines and WT plants that were individually wounded for 1 h and 4 h. Transcript levels were analyzed by qRT-PCR. Asterisks indicate significant differences in oe-MPK4 lines compared to WT plants (**, $p < 0.01$, Student's *t*-tests).

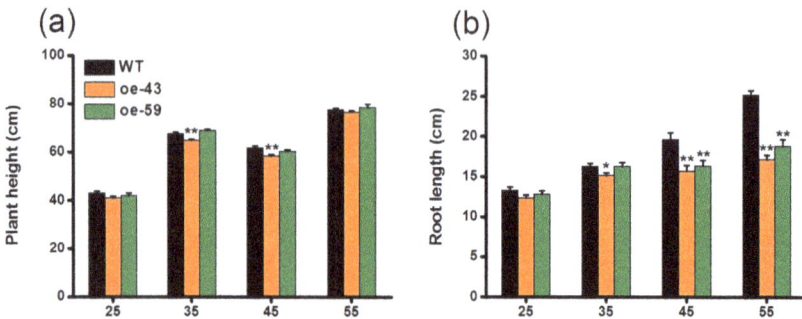

**Figure 3.** *Cont.*

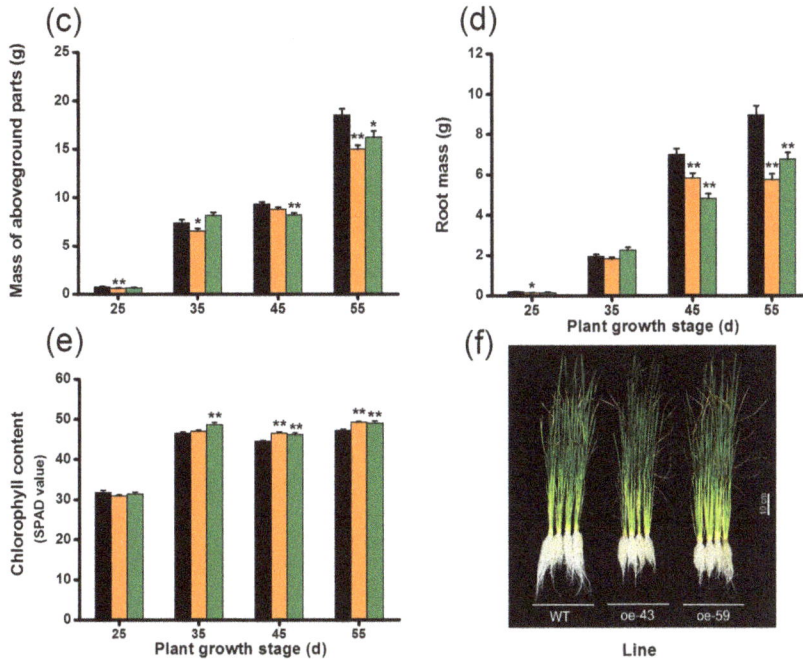

**Figure 3.** Growth phenotypes of oe-MPK4 lines and wild-type (WT) plants in the greenhouse. (**a–d**) Mean plant height (**a**) and root length (**b**) (+SE, $n = 10$) as well as mean mass (+SE, $n = 10$) of the aboveground part of plants (**c**) and roots (**d**) of oe-MPK4 lines and WT plants at 25, 35, 45, and 55 days; (**e**) Mean chlorophyll content (+SE, $n = 30$), measured by soil-plant analysis development (SPAD) meter, of oe-MPK4 lines and WT plants at 25, 35, 45, and 55 days; (**f**) Growth phenotype of 55 day old seedlings of oe-MPK4 lines and WT plants in the greenhouse. Asterisks indicate significant differences in oe-MPK4 lines compared to WT plants (*, $p < 0.05$; **, $p < 0.01$, Student's *t*-tests).

*2.3. Overexpressing OsMPK4 Increases Basal and/or Elicited Levels of JA, JA-Ile, Ethylene (ET), and Salicylic Acid (SA)*

Signaling pathways mediated by plant hormones, such as JA, ET, and SA, play key roles in herbivore-elicited defense responses in various plant species, including rice [6,34–38,43–45]. To evaluate whether *OsMPK4* affects the production of these plant hormones, and thus, regulates herbivore-induced defenses in rice, levels of these phytohormones were quantified in oe-MPK4 lines and WT plants. Basal JA and JA-Ile levels were similar between oe-MPK4 lines and WT plants, whereas levels of these two signals in oe-MPK4 lines were significantly increased compared to levels in WT plants after infestation by SSB: levels of JA and JA-Ile in the two oe-MPK4 lines were 2.48- and 2.32-fold, as well as 5.83- and 4.82-fold, respectively, higher than levels in WT plants at 1.5 h after SSB infestation (Figure 4a,b). Similarly, elicited levels of ET in oe-MPK4 lines were also higher than those in WT plants: 24 h after SSB infestation, ET accumulation in oe-MPK4 lines was increased by 25–28% compared to that in WT plants (Figure 4c). The overexpression of *OsMPK4* significantly enhanced constitutive levels of SA: levels in the two oe-MPK4 lines were 3.84- and 2.82-fold higher than levels in WT plants; after SSB infestation, the difference in the SA levels between oe-MPK4 lines and WT plants diminished, and only one oe-MPK4 line (oe-43) had higher elicited SA levels than did WT plants (Figure 4d). These results suggest that OsMPK4 modulates the biosynthesis of basal or SSB-elicited JA, JA-Ile, ET, and SA.

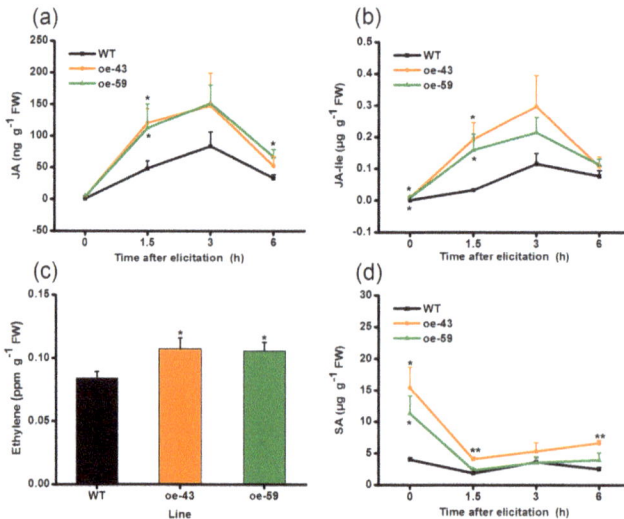

**Figure 4.** *OsMPK4* mediates SSB-induced JA, JA-Ile, ethylene, and SA accumulation. Mean levels (+SE, *n* = 5) of JA (**a**), JA-Ile (**b**), ethylene (**c**), and SA (**d**) in oe-MPK4 lines and wild-type (WT) plants that were individually infested by a third-instar SSB larva. Asterisks indicate significant differences in oe-MPK4 lines compared to WT plants (*, $p < 0.05$; **, $p < 0.01$, Student's *t*-tests). FW, fresh weight.

*2.4. Overexpression of OsMPK4 Enhances SSB-Induced Levels of Trypsin Protease Inhibitors (TrypPIs) and Resistance to SSB*

TrypPIs, important direct defense compounds, help rice resist herbivores, especially chewing herbivores, and their production is positively regulated by both JA and ET signaling pathways [10,37,38]. Thus, we investigated if OsMPK4 modulates the production of TrypPIs and, eventually, the resistance of rice to the SSB. As expected, the activity of SSB-induced TrypPIs was higher in the two oe-MPK4 lines than that in WT plants (Figure 5a). Consistent with this, SSB caterpillars fed on oe-MPK4 lines gained significantly less mass compared with those fed on WT plants: the mass of larvae fed on the oe-MPK4 lines was only 52–59% of the mass of larvae fed on WT plants (Figure 5b). These data demonstrate that *OsMPK4* positively regulates SSB-induced TrypPIs and, thus, rice resistance to SSB.

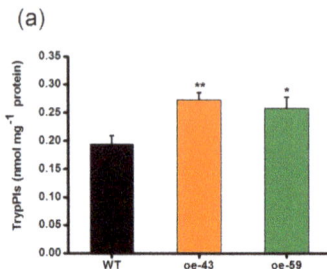

**Figure 5.** *Cont.*

(b)

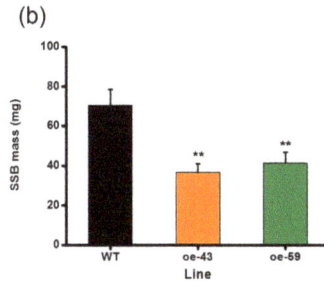

**Figure 5.** *OsMPK4* positively regulates SSB-induced trypsin protease inhibitor (TrypPI) accumulation and the resistance of rice to SSB. (**a**) Mean TrypPI activity (+SE, $n = 5$) in oe-MPK4 lines and wild-type (WT) plants that were individually infested by a third-instar SSB larva for 3 days; (**b**) Mean larval mass (+SE, $n = 60$) of SSB fed on oe-MPK4 lines and WT plants for 12 days. Asterisks indicate significant differences in oe-MPK4 lines compared to WT plants (*, $p < 0.05$; **, $p < 0.01$, Student's *t*-tests).

## 3. Discussion

In the study, we found that *OsMPK4* could be induced by mechanical wounding, SSB infestation and JA treatment, but not by SA treatment (Figure 1). The overexpression of *OsMPK4* enhanced SSB-elicited and/or basal levels of JA, JA-Ile, ET, and SA in rice (Figure 4), which in turn, increased the activity of TrypPIs and the resistance of rice to the SSB (Figure 5a,b). In addition, the overexpression of *OsMPK4* impaired plant growth, shortening root lengths, and reducing stem and root biomasses (Figure 3b–d). The data demonstrate that *OsMPK4* functions as a negative regulator of plant growth but a positive regulator of SSB resistance in rice.

MPK4 has been reported to play an important but variable role in regulating defense-related signaling pathways in plants. In Arabidopsis, for instance, AtMPK4 negatively regulates SA levels, but positively modulates JA- and ET-mediated signaling pathways [28,29]. Similarly, NtMPK4 in *N. tabacum* inhibits the production of ozone-induced SA, but is required for the activation of the JA pathway [42]. By contrast, silencing *MPK4* in soybean enhances levels of pathogen-induced SA and simultaneously activates JA signaling [26]; in *N. attenuata*, NaMPK4 does not influence the biosynthesis of constitutive and herbivore-elicited SA, but does positively regulate the production of herbivore-induced JA [32]. Here, we found that SSB infestation induced the expression of *OsMPK4* (Figure 1b), and the overexpression of *OsMPK4* increased basal and/or herbivore-induced levels of SA, JA, JA-Ile, and ET in rice (Figure 4). This demonstrates that the function of OsMPK4 in rice defenses is different from its functions in other plant species. *OsMPK4* expression has previously been reported to be induced by *Xoo* infection and the overexpression of *OsMPK4* enhances constitutive and pathogen-elicited SA levels and JA levels in rice [31]. Moreover, JA treatment enhanced *OsMPK4* transcript levels only at late treatment stages, and SA treatment did not induce *OsMPK4* expression (Figure 1c,d). These findings indicate that OsMPK4 plays an important role in plant defenses against both pathogens and herbivores by functioning upstream of defense-related signaling pathways in a similar manner. That JA induces *OsMPK4* suggests that JA may also regulates the activity of this enzyme via positive feedback loops.

As stated above, many studies have demonstrated that MPKs can regulate levels of JA, SA, and ET in plants by influencing the activity of the enzymes related to signal biosynthesis [9,10,18,19]. Moreover, in Arabidopsis, AtMPK4 modulates SA-, JA-, and ET-dependent responses by suppressing the activity of enhanced disease susceptibility 1 (EDS1) and phytoalexin deficient 4 (PAD4), both of which act as activators of the SA pathway and repressors of the ET/JA pathway [29]. Thus, in rice, the regulation of OsMPK4 on these defense-related signaling pathways may occur via similar mechanisms. Interestingly, OsMPK3 has also been found to mediate the biosynthesis of herbivore-induced JA and the resistance of rice to herbivores [10]. This indicates that there is an overlap in the function of the two MPKs in the herbivore-induced defense responses in rice. Future research should elucidate how the two MPKs regulate rice defenses, and what the similarities and the differences are between them.

Accumulated evidence demonstrates that there is antagonism between the JA and SA pathways in many plant species, including rice [38,46,47]. However, we found that the overexpression of *OsMPK4* enhanced basal and/or herbivore-elicited levels of both SA and JA (Figure 4); this finding was consistent with the previously reported result that rice lines that overexpressed *OsMPK4* had higher basal and/or pathogen-induced levels of SA and JA than did WT plants [31]. A similar result was observed in soybean [26]. Complicating the results, knocking out or silencing *OsMPK4* enhances constitutive and pathogen-elicited SA levels, but slightly decreases pathogen-elicited JA levels, suggesting there is antagonism between the SA and JA pathways [31,48]. The complexity of the findings indicates that, in addition to acting as a regulator of SA–JA antagonism, MPK4 may also function as a regulator of SA and JA synergism in some plant species or under some conditions. Further research should elucidate these mechanisms.

JA and ET signaling pathways have been reported to positively mediate the production of TrypPIs, a kind of direct defense compound that protects rice plants against chewing herbivores, including the SSB [37,38,49]. Therefore, enhanced resistance of oe-MPK4 lines to SSB is probably because of higher levels of JA, JA-Ile, and ET in these lines compared to WT plants; these signals led to increased activity among defensive compounds, including TrypPIs in oe-MPK4 lines.

In addition to its role in plant defense responses, OsMPK4 plays a role in plant growth. In Arabidopsis and soybean (*G. max*), for instance, the *mpk4* mutant is severely dwarfed [26,27]. In the wild tobacco species *N. attenuata*, silencing *NaMPK4* causes plants to be somewhat small in stature [50]. Consistent with these data, the overexpression of *BnMPK4* in *B. napus* increases plant size [41]. Research into the mechanisms underlying the growth phenotype of *mpk4* mutants revealed that elevated constitutive defense responses [28], ROS homeostasis deregulation, and photosynthesis damage [27], auxin repression [26], or rapid water loss [50] may explain why these mutants are so small. Unlike the positive effect of MPK4 on plant growth found in other species, we found that the overexpression of *OsMPK4* retarded rice growth moderately (Figure 3). This result may be related to the function of OsMPK4 in rice, which is different from that of MPK4 in other plant species, as stated above: the overexpression of *OsMPK4* enhanced constitutive SA levels (Figure 4d). Given that the growth retardation of the *mpk4* mutant in Arabidopsis is partly related to the accumulation of SA [28], the possibility could not be ruled out that the reduced size of mutants might be, at least in part, due to their higher SA levels than WT plants.

In summary, our findings demonstrate that OsMPK4 plays an important role in plant growth and herbivore-induced defense responses of rice by regulating SA-, JA-, and ET-mediated signaling pathways, and that the function of OsMPK4 in rice, a positive regulator of plant herbivore resistance and a negative regulator of plant growth, is different from the function of MPK4 reported in other plant species, such as Arabidopsis and soybean. Moreover, OsMPK4 may also mediate SA–JA/ET interactions in rice, depending on the stimulus: a stimulus that suppresses the activity of OsMPK4 will cause SA–JA/ET antagonism, whereas a stimulus that induces the activity of OsMPK4 will result in SA–JA/ET synergism; in other words, OsMPK4 may also play an important role in modulating appropriate and specific defense responses in plants to different stresses.

## 4. Materials and Methods

### 4.1. Plant Growth

The rice (*Oryza sativa*) genotypes used in this study were cv Xiushui 11 wild-type (WT) and two oe-MPK4 transgenic lines (see below). Pre-germinated seeds of WT plants and transgenic lines were cultured in plastic cups (diameter, 8 cm; height, 10 cm) in a climate incubator at $28 \pm 2$ °C, with a 14 h light phase. Twelve day old seedlings were transferred to 25 L hydroponic boxes with a rice nutrient solution [51]. After 30–35 days, plants were transplanted in individual 500 mL hydroponic plastic pots. Plants were used for experiments 4–5 days after transplanting.

### 4.2. Insects

An SSB colony was originally obtained from a rice field in Hangzhou, China, and maintained on TN1 (a rice variety that is susceptible to SSB) rice seedlings in a controlled climate chamber at $26 \pm 2\,°C$, with a 14 h light phase and 65% relative humidity.

### 4.3. Isolation and Characterization of OsMPK4 cDNA

The full-length cDNA of *OsMPK4* was PCR-amplified. The primers MPK4-F (5′-ATAGTCG ACTCCACCTCGTCCT-3′) and MPK4-R (5′-ATATCTAGAAGGGGATTTGGCTTT-3′) were designed based on the sequence of *OsMPK4*. The PCR products were cloned into the pMD19-T vector (TaKaRa, Tokyo, Japan) and sequenced.

### 4.4. Phylogenetic Analysis

For the phylogenetic analysis, the program MEGA 6.0 [52] was used. Protein sequences were downloaded from GenBank (https://www.ncbi.nlm.nih.gov/genbank/) and were aligned using the ClustalW algorithm in MEGA 6.0 (Tempe, AZ, USA). This alignment was then used to generate an unrooted tree using the maximum-likelihood approach [53] (1000 replications).

### 4.5. Generation and Characterization of Transgenic Plants

The full-length cDNA sequence of *OsMPK4* was cloned into the pCAMBIA1301 transformation vector to yield an overexpressed construct containing the hygromycin resistance gene *hph* and the reporter gene encoding β-glucuronidase (GUS) (GenBank: AF234297) as selectable markers (Figure S3). The vector was used for transforming the rice variety Xiushui 11 by *Agrobacterium*-mediated transformation system. Homozygous T2 transgenic plants were selected using GUS staining or hygromycin resistance screening [38]. Two T2 homozygous lines (oe-43 and oe-59), each harboring a single insertion (Figure 2a), were used in subsequent experiments.

### 4.6. Plant Treatments

For mechanical wounding, plants were individually pierced with a needle on the lower part of leaf sheaths (about 2 cm long), each 200 times (W). Control plants were not manipulated (C). For SSB treatment, plants were individually infested with a third-instar SSB larva that had been starved for 2 h. Non-infested plants were used as controls (C). For JA and SA treatment, each plant was sprayed with 2 mL of JA (100 µg·mL$^{-1}$) or SA (70 µg·mL$^{-1}$) in 50 mM sodium phosphate buffer (pH 8, with 0.01% Tween). Plants sprayed with 2 mL of the buffer (BUF) were used as controls. Each treatment at each time point was replicated five times.

### 4.7. Measurement of Plant Growth Parameters

Plant growth parameters, including plant height, root length, mass of the above- and belowground part of a plant, and chlorophyll content, of oe-WPK4 lines and WT plants, were measured at 25, 35, 45, and 55 days after plant growth. Plant height and root length were defined as the part of a plant from the stem base to the longest leaf apex and that from the stem base to the longest root tip, respectively. Plants were cut off from the stem base and then the mass of aboveground and belowground (roots) part of plants was measured. For chlorophyll content determination, three leaves with identical leaf positions from each plant and three identical locations from each leaf were measured by a soil-plant analysis development (SPAD) meter. Every experiment at each stage was replicated ten times.

### 4.8. JA, JA-Ile, and SA Analysis

For JA, JA-Ile, and SA analysis, plants of the different genotypes were randomly assigned to SSB and control treatments. Plant stems were harvested at different time points after the start of treatment (see details in Figure 4). Samples (each with about 150 mg) were ground in liquid nitrogen, and JA, JA-Ile, and SA were extracted with ethyl acetate spiked with labeled internal standards (D6-JA, D6-JA-Ile, D4-SA) and analyzed with an HPLC/mass spectrometry/mass spectrometry system following the method described in Lu et al. [54]. Each treatment at each time point was replicated five times.

### 4.9. Ethylene Analysis

Plants of WT and transgenic lines were randomly assigned to SSB and control treatments, and were individually confined with sealed glass cylinders (diameter 4 cm, height 50 cm). The production of ethylene was determined following the same method as described in Lu et al. [55]. Each treatment at each time point was replicated five times.

### 4.10. TrypPI Activity Analysis

Plants of different genotypes were randomly assigned to SSB and control treatments. Stems of oe-MPK4 lines and WT plants (about 0.3 g per sample) were individually harvested 3 days after the start of the treatment. TrypPI concentrations were measured using a radial diffusion assay as described by van Dam et al. [56]. Each treatment was replicated five times.

### 4.11. SSB Performance Measurement

The performance of SSB larvae on oe-MPK4 lines and WT plants was determined by releasing one freshly hatched larva onto each plant. Larval mass (to an accuracy of 0.1 mg) was measured and recorded 12 days after the release of the herbivore. Each treatment was replicated thirty times.

### 4.12. Data Analysis

Differences in data in different lines or treatments were determined by analyzing variance followed by Duncan's multiple range test (or Student's *t*-test for comparing two treatments). When necessary, data were log-transformed or arcsine-transformed to meet requirements for the homogeneity of variance. All tests were carried out with SPSS software version 20.

**Supplementary Materials:** Supplementary materials can be found at http://www.mdpi.com/1422-0067/19/4/1182/s1..

**Acknowledgments:** We thank Emily Wheeler for editorial assistance. The study was jointly sponsored by the Fund for Agro-scientific Research in the Public Interest (201403030), the National Natural Science Foundation of China (31330065, 31272052), the National Program of Transgenic Variety Development of China (2016ZX08001-001) and the earmarked fund for China Agriculture Research System (CARS-01-40).

**Author Contributions:** Yonggen Lou and Xiaoli Liu conceived and designed the experiments; Xiaoli Liu, Jiancai Li and Liping Xu performed the experiments; Xiaoli Liu and Jiancai Li analyzed the data; Qi Wang contributed the transgenic lines; Yonggen Lou and Xiaoli Liu wrote the paper.

**Conflicts of Interest:** The authors declare no conflict of interest.

### Abbreviations

| | |
|---|---|
| MPKs | mitogen-activated protein kinases |
| JA | jasmonic acid |
| JA-Ile | jasmonoyl-L-isoleucine |
| SA | salicylic acid |
| ET | ethylene |
| TrypPIs | trypsin proteinase inhibitors |
| SSB | striped stem borer |

## References

1. Kempel, A.; Schadler, M.; Chrobock, T.; Fischer, M.; van Kleunen, M. Tradeoffs associated with constitutive and induced plant resistance against herbivory. *Proc. Natl. Acad. Sci. USA* **2011**, *108*, 5685–5689. [CrossRef] [PubMed]

2. Mithofer, A.; Boland, W. Plant defense against herbivores: Chemical aspects. *Annu. Rev. Plant Biol.* **2012**, *63*, 431–450. [CrossRef] [PubMed]

3. Bosch, M.; Wright, L.P.; Gershenzon, J.; Wasternack, C.; Hause, B.; Schaller, A.; Stintzi, A. Jasmonic acid and its precursor 12-oxophytodienoic acid control different aspects of constitutive and induced herbivore defenses in tomato. *Plant Physiol.* **2014**, *166*, 396–410. [CrossRef] [PubMed]

4. Zipfel, C. Plant pattern-recognition receptors. *Trends Immunol.* **2014**, *35*, 345–351. [CrossRef] [PubMed]

5. Hilker, M.; Fatouros, N.E. Resisting the onset of herbivore attack: Plants perceive and respond to insect eggs. *Curr. Opin. Plant Biol.* **2016**, *32*, 9–16. [CrossRef] [PubMed]

6. Erb, M.; Meldau, S.; Howe, G.A. Role of phytohormones in insect-specific plant reactions. *Trends Plant Sci.* **2012**, *17*, 250–259. [CrossRef] [PubMed]

7. Venkatesan, R. Biosynthesis and regulation of herbivore-induced plant volatile emission. *J. Indian Inst. Sci.* **2015**, *95*, 25–34.

8. Hettenhausen, C.; Schuman, M.C.; Wu, J. MAPK signaling: A key element in plant defense response to insects. *Insect Sci.* **2015**, *22*, 157–164. [CrossRef] [PubMed]

9. Wu, J.; Hettenhausen, C.; Meldau, S.; Baldwin, I.T. Herbivory rapidly activates MAPK signaling in attacked and unattacked leaf regions but not between leaves of *Nicotiana attenuata*. *Plant Cell* **2007**, *19*, 1096–1122. [CrossRef] [PubMed]

10. Wang, Q.; Li, J.; Hu, L.; Zhang, T.; Zhang, G.; Lou, Y. *OsMPK3* positively regulates the JA signaling pathway and plant resistance to a chewing herbivore in rice. *Plant Cell Rep.* **2013**, *32*, 1075–1084. [CrossRef] [PubMed]

11. Guo, H.; Peng, X.; Gu, L.; Wu, J.; Ge, F.; Sun, Y. Up-regulation of MPK4 increases the feeding efficiency of the green peach aphid under elevated $CO_2$ in *Nicotiana attenuata*. *J. Exp. Bot.* **2017**, *68*, 5923–5935. [CrossRef] [PubMed]

12. Vidhyasekaran, P. Mitogen-activated protein kinase cascades in plant innate immunity. In *PAMP Signals in Plant Innate Immunity: Signal Perception and Transduction*, 1st ed.; Springer: Dordrecht, The Netherlands, 2014; Volume 21, pp. 331–374.

13. Xu, J.; Zhang, S. Mitogen-activated protein kinase cascades in signaling plant growth and development. *Trends Plant Sci.* **2015**, *20*, 56–64. [CrossRef] [PubMed]

14. Taj, G.; Agarwal, P.; Grant, M.; Kumar, A. MAPK machinery in plants: Recognition and response to different stresses through multiple signal transduction pathways. *Plant Signal. Behav.* **2010**, *5*, 1370–1378. [CrossRef] [PubMed]

15. Cheong, Y.-H.; Kim, M.-C. Functions of MAPK cascade pathways in plant defense signaling. *Plant Pathol. J.* **2010**, *26*, 101–109. [CrossRef]

16. Rodriguez, M.C.; Petersen, M.; Mundy, J. Mitogen-activated protein kinase signaling in plants. *Annu. Rev. Plant Biol.* **2010**, *61*, 621–649. [PubMed]

17. Schwessinger, B.; Ronald, P.C. Plant innate immunity: Perception of conserved microbial signatures. *Annu. Rev. Plant Biol.* **2012**, *63*, 451–482. [CrossRef] [PubMed]

18. Han, L.; Li, G.J.; Yang, K.Y.; Mao, G.; Wang, R.; Liu, Y.; Zhang, S. Mitogen-activated protein kinase 3 and 6 regulate *Botrytis cinerea*-induced ethylene production in Arabidopsis. *Plant J.* **2010**, *64*, 114–127. [CrossRef] [PubMed]

19. Li, G.; Meng, X.; Wang, R.; Mao, G.; Han, L.; Liu, Y.; Zhang, S. Dual-level regulation of ACC synthase activity by MPK3/MPK6 cascade and its downstream WRKY transcription factor during ethylene induction in Arabidopsis. *PLoS Genet.* **2012**, *8*, e1002767. [CrossRef] [PubMed]

20. Song, F.M.; Goodman, R.M. *OsBIMK1*, a rice MAP kinase gene involved in disease resistance responses. *Planta* **2002**, *215*, 997–1005. [CrossRef] [PubMed]

21. Singh, P.; Sinha, A.K. A positive feedback loop governed by SUB1A1 interaction with mitogen-activated protein kinase3 imparts submergence tolerance in rice. *Plant Cell* **2016**, *28*, 1127–1143. [CrossRef] [PubMed]

22. Qiu, J.L.; Fiil, B.K.; Petersen, K.; Nielsen, H.B.; Botanga, C.J.; Thorgrimsen, S.; Palma, K.; Suarez-Rodriguez, M.C.; Sandbech-Clausen, S.; Lichota, J.; et al. Arabidopsis MAP kinase 4 regulates gene expression through transcription factor release in the nucleus. *EMBO J.* **2008**, *27*, 2214–2221. [CrossRef] [PubMed]

23. Qiu, J.L.; Zhou, L.; Yun, B.W.; Nielsen, H.B.; Fiil, B.K.; Petersen, K.; Mackinlay, J.; Loake, G.J.; Mundy, J.; Morris, P.C. Arabidopsis mitogen-activated protein kinase kinases MKK1 and MKK2 have overlapping functions in defense signaling mediated by MEKK1, MPK4, and MKS1. *Plant Physiol.* **2008**, *148*, 212–222. [CrossRef] [PubMed]

24. Berriri, S.; Garcia, A.V.; Frei dit Frey, N.; Rozhon, W.; Pateyron, S.; Leonhardt, N.; Montillet, J.L.; Leung, J.; Hirt, H.; Colcombet, J. Constitutively active mitogen-activated protein kinase versions reveal functions of *Arabidopsis* MPK4 in pathogen defense signaling. *Plant Cell* **2012**, *24*, 4281–4293. [CrossRef] [PubMed]

25. Li, B.; Jiang, S.; Yu, X.; Cheng, C.; Chen, S.; Cheng, Y.; Yuan, J.S.; Jiang, D.; He, P.; Shan, L. Phosphorylation of trihelix transcriptional repressor ASR3 by MAP KINASE4 negatively regulates Arabidopsis immunity. *Plant Cell* **2015**, *27*, 839–856. [CrossRef] [PubMed]

26. Liu, J.Z.; Horstman, H.D.; Braun, E.; Graham, M.A.; Zhang, C.; Navarre, D.; Qiu, W.L.; Lee, Y.; Nettleton, D.; Hill, J.H.; et al. Soybean homologs of MPK4 negatively regulate defense responses and positively regulate growth and development. *Plant Physiol.* **2011**, *157*, 1363–1378. [CrossRef] [PubMed]

27. Gawroński, P.; Witoń, D.; Vashutina, K.; Bederska, M.; Betliński, B.; Rusaczonek, A.; Karpiński, S. Mitogen-activated protein kinase 4 is a salicylic acid-independent regulator of growth but not of photosynthesis in *Arabidopsis*. *Mol. Plant* **2014**, *7*, 1151–1166. [CrossRef] [PubMed]

28. Petersen, M.; Brodersen, P.; Naested, H.; Andreasson, E.; Lindhart, U.; Johansen, B.; Nielsen, H.B.; Lacy, M.; Austin, M.J.; Parker, J.E.; et al. Arabidopsis MAP kinase 4 negatively regulates systemic acquired resistance. *Cell* **2000**, *103*, 1111–1120. [CrossRef]

29. Brodersen, P.; Petersen, M.; Bjorn Nielsen, H.; Zhu, S.; Newman, M.A.; Shokat, K.M.; Rietz, S.; Parker, J.; Mundy, J. Arabidopsis MAP kinase 4 regulates salicylic acid- and jasmonic acid/ethylene-dependent responses via EDS1 and PAD4. *Plant J.* **2006**, *47*, 532–546. [CrossRef] [PubMed]

30. Kong, Q.; Qu, N.; Gao, M.; Zhang, Z.; Ding, X.; Yang, F.; Li, Y.; Dong, O.X.; Chen, S.; Li, X.; et al. The MEKK1-MKK1/MKK2-MPK4 kinase cascade negatively regulates immunity mediated by a mitogen-activated protein kinase kinase kinase in Arabidopsis. *Plant Cell* **2012**, *24*, 2225–2236. [CrossRef] [PubMed]

31. Shen, X.; Yuan, B.; Liu, H.; Li, X.; Xu, C.; Wang, S. Opposite functions of a rice mitogen-activated protein kinase during the process of resistance against *Xanthomonas oryzae*. *Plant J.* **2010**, *64*, 86–99. [CrossRef] [PubMed]

32. Hettenhausen, C.; Baldwin, I.T.; Wu, J. *Nicotiana attenuata* MPK4 suppresses a novel jasmonic acid (JA) signaling-independent defense pathway against the specialist insect *Manduca sexta*, but is not required for the resistance to the generalist *Spodoptera littoralis*. *New Phytol.* **2013**, *199*, 787–799. [CrossRef] [PubMed]

33. Lou, Y.G.; Zhang, G.R.; Zhang, W.Q.; Hu, Y.; Zhang, J. Biological control of rice insect pests in China. *Biol. Control* **2013**, *67*, 8–20. [CrossRef]

34. Li, R.; Zhang, J.; Li, J.; Zhou, G.; Wang, Q.; Bian, W.; Erb, M.; Lou, Y. Prioritizing plant defence over growth through WRKY regulation facilitates infestation by non-target herbivores. *eLife* **2015**, *4*, e04805. [CrossRef] [PubMed]

35. Hu, L.; Ye, M.; Li, R.; Zhang, T.; Zhou, G.; Wang, Q.; Lu, J.; Lou, Y. The rice transcription factor WRKY53 suppresses herbivore-induced defenses by acting as a negative feedback modulator of mitogen-activated protein kinase activity. *Plant Physiol.* **2015**, *169*, 2907–2921. [PubMed]

36. Lu, J.; Li, J.; Ju, H.; Liu, X.; Erb, M.; Wang, X.; Lou, Y. Contrasting effects of ethylene biosynthesis on induced plant resistance against a chewing and a piercing-sucking herbivore in rice. *Mol. Plant* **2014**, *7*, 1670–1682. [CrossRef] [PubMed]

37. Lu, J.; Ju, H.; Zhou, G.; Zhu, C.; Erb, M.; Wang, X.; Wang, P.; Lou, Y. An EAR-motif-containing ERF transcription factor affects herbivore-induced signaling, defense and resistance in rice. *Plant J.* **2011**, *68*, 583–596. [CrossRef] [PubMed]

38. Zhou, G.; Qi, J.; Ren, N.; Cheng, J.; Erb, M.; Mao, B.; Lou, Y. Silencing *OsHI-LOX* makes rice more susceptible to chewing herbivores, but enhances resistance to a phloem feeder. *Plant J.* **2009**, *60*, 638–648. [CrossRef] [PubMed]

39. Reyna, N.S.; Yang, Y. Molecular analysis of the rice map kinase gene family in relation to *Magnaporthe grisea* infection. *Mol. Plant-Microbe Interact.* **2006**, *19*, 530–540. [CrossRef] [PubMed]

40. Zhou, G.; Wang, X.; Yan, F.; Wang, X.; Li, R.; Cheng, J.; Lou, Y. Genome-wide transcriptional changes and defence-related chemical profiling of rice in response to infestation by the rice striped stem borer *Chilo suppressalis*. *Physiol. Plant.* **2011**, *143*, 21–40. [CrossRef] [PubMed]

41. Wang, Z.; Mao, H.; Dong, C.; Ji, R.; Cai, L.; Fu, H.; Liu, S. Overexpression of *Brassica napus* MPK4 enhances resistance to *Sclerotinia sclerotiorum* in oilseed rape. *Mol. Plant-Microbe Interact.* **2009**, *22*, 235–244. [CrossRef] [PubMed]

42. Gomi, K.; Ogawa, D.; Katou, S.; Kamada, H.; Nakajima, N.; Saji, H.; Soyano, T.; Sasabe, M.; Machida, Y.; Mitsuhara, I.; et al. A mitogen-activated protein kinase NtMPK4 activated by SIPKK is required for jasmonic acid signaling and involved in ozone tolerance via stomatal movement in tobacco. *Plant Cell Physiol.* **2005**, *46*, 1902–1914. [CrossRef] [PubMed]

43. Tian, D.; Peiffer, M.; de Moraes, C.M.; Felton, G.W. Roles of ethylene and jasmonic acid in systemic induced defense in tomato (*Solanum lycopersicum*) against *Helicoverpa zea*. *Planta* **2014**, *239*, 577–589. [CrossRef] [PubMed]

44. Fragoso, V.; Rothe, E.; Baldwin, I.T.; Kim, S.G. Root jasmonic acid synthesis and perception regulate folivore-induced shoot metabolites and increase *Nicotiana attenuata* resistance. *New Phytol.* **2014**, *202*, 1335–1345. [CrossRef] [PubMed]

45. Li, R.; Afsheen, S.; Xin, Z.; Han, X.; Lou, Y. *OsNPR1* negatively regulates herbivore-induced JA and ethylene signaling and plant resistance to a chewing herbivore in rice. *Physiol. Plant.* **2013**, *147*, 340–351. [CrossRef] [PubMed]

46. Laurie-Berry, N.; Joardar, V.; Street, I.H.; Kunkel, B.N. The *Arabidopsis thaliana JASMONATE INSENSITIVE* 1 gene is required for suppression of salicylic acid-dependent defenses during infection by *Pseudomonas syringae*. *Mol. Plant-Microbe Interact.* **2006**, *19*, 789–800. [CrossRef] [PubMed]

47. Bruessow, F.; Gouhier-Darimont, C.; Buchala, A.; Metraux, J.P.; Reymond, P. Insect eggs suppress plant defence against chewing herbivores. *Plant J.* **2010**, *62*, 876–885. [CrossRef] [PubMed]

48. Yuan, B.; Shen, X.; Li, X.; Xu, C.; Wang, S. Mitogen-activated protein kinase OsMPK6 negatively regulates rice disease resistance to bacterial pathogens. *Planta* **2007**, *226*, 953–960. [CrossRef] [PubMed]

49. Qi, J.; Zhou, G.; Yang, L.; Erb, M.; Lu, Y.; Sun, X.; Cheng, J.; Lou, Y. The chloroplast-localized phospholipases D $\alpha$4 and $\alpha$5 regulate herbivore-induced direct and indirect defenses in rice. *Plant Physiol.* **2011**, *157*, 1987–1999. [CrossRef] [PubMed]

50. Hettenhausen, C.; Baldwin, I.T.; Wu, J. Silencing *MPK4* in *Nicotiana attenuata* enhances photosynthesis and seed production but compromises abscisic acid-induced stomatal closure and guard cell-mediated resistance to *Pseudomonas syringae* pv *tomato* DC3000. *Plant Physiol.* **2012**, *158*, 759–776. [CrossRef] [PubMed]

51. Yoshida, S.; Forno, D.A.; Cock, J.H. *Laboratory Manual for Physiological Studies of Rice*, 3rd ed.; International Rice Research Institute: Manila, Philippines, 1976; pp. 1–83.

52. Tamura, K.; Stecher, G.; Peterson, D.; Filipski, A.; Kumar, S. MEGA6: Molecular evolutionary genetics analysis version 6.0. *Mol. Biol. Evol.* **2013**, *30*, 2725–2729. [CrossRef] [PubMed]

53. Whelan, S.; Goldman, N. A general empirical model of protein evolution derived from multiple protein families using a maximum-likelihood approach. *Mol. Biol. Evol.* **2001**, *18*, 691–699. [CrossRef] [PubMed]

54. Lu, J.; Robert, C.A.; Riemann, M.; Cosme, M.; Mene-Saffrane, L.; Massana, J.; Stout, M.J.; Lou, Y.; Gershenzon, J.; Erb, M. Induced jasmonate signaling leads to contrasting effects on root damage and herbivore performance. *Plant Physiol.* **2015**, *167*, 1100–1116. [CrossRef] [PubMed]

55. Lu, Y.; Wang, X.; Lou, Y.; Cheng, J. Role of ethylene signaling in the production of rice volatiles induced by the rice brown planthopper *Nilaparvata lugens*. *Chin. Sci. Bull.* **2006**, *51*, 2457–2465. [CrossRef]

56. Van Dam, N.M.; Horn, M.; Mares, M.; Baldwin, I.T. Ontogeny constrains systemic protease inhibitor response in *Nicotiana attenuata*. *J. Chem. Ecol.* **2001**, *27*, 547–568. [CrossRef] [PubMed]

International Journal of
*Molecular Sciences*

MDPI

*Article*

# Two New Polyphenol Oxidase Genes of Tea Plant (*Camellia sinensis*) Respond Differentially to the Regurgitant of Tea Geometrid, *Ectropis obliqua*

Chen Huang [1,2,3], Jin Zhang [1,3], Xin Zhang [1,3], Yongchen Yu [1], Wenbo Bian [1,3], Zhongping Zeng [2], Xiaoling Sun [1,3,*] and Xinghui Li [2,*]

[1]  Tea Research Institute, Chinese Academy of Agricultural Sciences, Hangzhou 310008, China; yellowchen92@163.com (C.H.); zhangjin7981@163.com (J.Z.); xinzhang@tricaas.com (X.Z.); w935369897@163.com (Y.Y.); bwbkg21@163.com (W.B.)
[2]  Tea Research Institute, College of Horticulture, Nanjing Agricultural University, Nanjing 210095, China; zpzeng@njau.edu.cn
[3]  Key Laboratory of Tea Biology and Resources Utilization, Ministry of Agriculture, Hangzhou 310008, China
*  Correspondence: xlsun@tricaas.com (X.S.); lxh@njau.edu.cn (X.L.); Tel.: +86-571-86650350 (X.S.); +86-025-84396651 (X.L.)

Received: 15 July 2018; Accepted: 13 August 2018; Published: 16 August 2018

**Abstract:** Polyphenol oxidases (PPOs) have been reported to play an important role in protecting plants from attacks by herbivores. Though PPO genes in other plants have been extensively studied, research on PPO genes in the tea plant (*Camellia sinensis*) is lacking. In particular, which members of the PPO gene family elicit the defense response of the tea plant are as yet unknown. Here, two new PPO genes, *CsPPO1* and *CsPPO2*, both of which had high identity with PPOs from other plants, were obtained from tea leaves. The full length of *CsPPO1* contained an open reading frame (ORF) of 1740 bp that encoded a protein of 579 amino acids, while *CsPPO2* contained an ORF of 1788 bp that encoded a protein of 595 amino acids. The deduced CsPPO1 and CsPPO2 proteins had calculated molecular masses of 64.6 and 65.9 kDa; the isoelectric points were 6.94 and 6.48, respectively. The expression products of recombinant CsPPO1 and CsPPO2 in *Escherichia coli* were about 91 and 92 kDa, respectively, but the recombinant proteins existed in the form of an inclusion body. Whereas *CsPPO1* is highly expressed in stems, *CsPPO2* is highly expressed in roots. Further results showed that the expression of *CsPPO1* and *CsPPO2* was wound- and *Ectropis obliqua*-induced, and that regurgitant, unlike treatment with wounding plus deionized water, significantly upregulated the transcriptional expression of *CsPPO2* but not of *CsPPO1*. The difference between regurgitant and wounding indicates that *CsPPO2* may play a more meaningful defensive role against *E. obliqua* than *CsPPO1*. Meanwhile, we found the active component(s) of the regurgitant elicited the expression of *CsPPO* may contain small molecules (under 3-kDa molecular weight). These conclusions advance the understanding of the biological function of two new PPO genes and show that one of these, *CsPPO2*, may be a promising gene for engineering tea plants that are resistant to *E. obliqua*.

**Keywords:** polyphenol oxidase; *Camellia sinensis*; *Ectropis obliqua*; wounding; regurgitant

## 1. Introduction

Plant polyphenol oxidases (PPOs), which are ubiquitous, dinuclear, copper-containing metalloproteins, contribute to the lignification, pigmentation, and, in higher plant species, defense against pathogens or herbivores [1–5]. PPOs utilize molecular oxygen to oxidize various phenolic precursors to their corresponding quinines [6], and these quinones are responsible for the enzymatic browning of many fruits, vegetables, and grains. Such browning often accompanies senescence, mechanical damage, and attack by pathogens or herbivores [7–9]. The negative effect of PPOs on the appearance and nutritional quality of products has prompted numerous ecological and molecular studies, as has the role of PPOs in plant defense against herbivores and pathogens [10–15]. For instance, an inverse correlation has been found between the performance of cotton bollworm (*Helicoverpa armigera* (Hübner)), beet armyworm (*Spodoptera exigua* (Hübner)), and PPO levels [12,16], and the antisense suppression of potato (*Solanum tuberosum* L.) *StPPO* has been shown to increase susceptibility, and the overexpression of PPO cDNAs has been demonstrated in tomato (*Lycopersicon esculentum* L.) to increase resistance to *Pseudomonas syringae* pv. *tomato* and to *Spodoptera litura* (Fabricius) [11,17,18]. In addition, *PPO* genes are frequently found to be differentially induced in response to injuries inflicted by wounding, pathogens, or herbivores from various plant species, and also to signaling molecules (jasmonic acid (JA), methyl jasmonate (MeJA), salicylic acid (SA), ethylene (ET)), suggesting that these genes have a defensive role [19–23].

To our knowledge, *PPO* gene families have been described in more than 26 plant species [22,24–26]: the *PPO* gene family of the tomato consists of seven members [27] and the family of *Salvia miltiorrhiza* comprises 19 *PPO* genes [22]. Until now, according to reports, only one full-length genomic DNA sequence of *PPO* has been cloned from *Camellia sinensis* cv. *Longjing 43* (EF635860.1), although the *PPO* gene family of *C. sinensis* (L.) O. Kuntzeis is thought to have from five to six members [28].

The tea plant, *C. sinensis*, is not only one of the world's most important woody-plantation crops but is also valued as a source of secondary metabolic products, including phyto-oxylipins [29]. Tender tea buds and leaves are the raw material for commercial tea, one of the most popular nonalcoholic drinks worldwide [30]. Developing tea shoots and leaves may be damaged by numerous pests, such as the tea geometrid *Ectropis obliqua* (Prout), whose larvae seriously affect the yield and quality of tea [31], and the tea green leafhopper, *Empoasca onukii* (Matsuda). Herbivore-induced plant defenses are induced both by wounding, which is caused by the herbivore mouthparts involved in chewing/piercing, and by the elicitors/effectors that come from the insect's oral or oviduct secretions [32,33]. Oral secretions (OS) are the key factors according to which the plant distinguishes between mechanical damage and herbivore feeding, and then responds, as different responses are elicited by different herbivore species [32–37]. Previously, we found that PPOs were an important antiherbivore factor in tea plants, defending them directly against *E. obliqua* larvae [4]. Though mechanical damage and JA treatment can upregulate PPO activity in tea leaves, both the infestation of the tea geometrids and wounding plus the regurgitant significantly suppressed wound-induced PPO activity, from which we inferred that *E. obliqua* larvae have evolved to be able to elude the tea plant's defenses by inhibiting the production of PPOs [4]. Unfortunately, which genes are responsible for PPO activity induced or inhibited by the exogenous application of JA or the infestation of *E. obliqua* remains unknown.

To elucidate the *CsPPOs* responsible for the defense/coevolutionary response of the tea plant, we first isolated and characterized two new full-length cDNA sequences of *CsPPO* genes from *C. sinensis* cv. *Longjing 43*. Second, the phylogenetic relationship was analyzed by DNAMAN software. Third, we analyzed the transcriptional expression characteristics of these two genes in different tissues and in response to mechanical damage, *E. obliqua* infestation, and treatment involving mechanical damage plus regurgitant or exogenous application of JA. Finally, *CsPPO1* and *CsPPO2* were used as target genes to screen the active components of the regurgitant by detecting transcriptional expression levels of leaves that were treated with three separate compounds of the regurgitant. Our results will help clarify the interaction between tea plants and tea geometrids, and provide candidate defensive gene resources for breeding molecular resistance to insects in tea.

## 2. Results

### 2.1. cDNA Cloning and Sequence Analysis

The full length of the *CsPPO1* contained an ORF of 1740 bp that encoded a 579-amino acid residue, while the *CsPPO2* contained an ORF of 1788 bp that encoded a 595-amino acid residue. The deduced

CsPPO1 and CsPPO2 proteins had calculated the molecular weight (Mw) of 64.6 kDa and 65.9 kDa, and the isoelectric points (pI) were 6.94 and 6.48, respectively. Compared to the published *CsPPO* (EF635860.1), the putative conserved domains of CsPPOs were predicted on a protein-blast website, and the results indicated that the *CsPPOs* we isolated are new genes encoding two new putative PPOs of the tea plant. The three cDNA sequences share 71–76% and 68–72% pairwise identity at the nucleic-acid and amino-acid levels, respectively (Table 1). The 80–90 residues of the derived amino-acid sequences of CsPPOs in the N-terminal region (Figure 1) show many typical features of a chloroplast transit peptide. The proteins had a conserved tyrosinase superfamily motif and two copper ion-binding sites. The PPO-DWL supermotif was the conserved domain of PPO that contained approximately 50 amino acids. The PPO-KFDV superfamily, whose function has not yet been studied, was the C-terminal domain of these oxidases (Figure 1). Interestingly, sequence analysis revealed only one copper-binding site in CsPPO1, TpPPO, and AmAS1, and two copper-binding sites in CsPPO2, CsPPO, GhPPO, NtaPPO, SlyPPOA, SmePPO1, PtdPPO1, and VvPPO (Figure 1).

**Table 1.** Percentage of similarity among three tea-plant PPO sequences, calculated for both nucleotide and amino-acid sequences using MEGALIGN (DNAStar). Protein identities are in bold.

| Name | CsPPO1 | CsPPO2 | CsPPO |
|---|---|---|---|
| CsPPO1 | 100.0 | **68.45** | **69.83** |
| CsPPO2 | 71.68 | 100.0 | **71.14** |
| CsPPO | 73.27 | 75.27 | 100.0 |

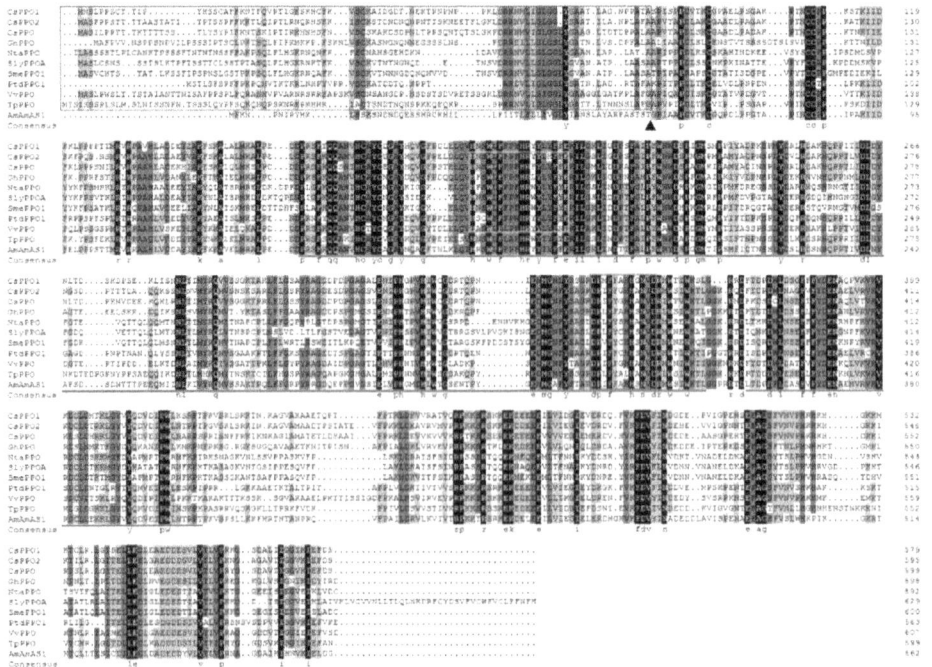

**Figure 1.** Amino acid sequence alignment of three polyphenol oxidases (PPOs) in *C. sinensis* and other plant PPOs. The region corresponding to the chloroplast transit peptide is boxed, and the thylakoid peptidase-processing site is indicated by ▲. The three domains: red underline, tyrosinase domain; blue underline, PPO1_DWL domain; yellow underline, PPO1_KFDV domain. The green box refers to the CuA binding site; the purple box to the CuB binding site.

## 2.2. The Phylogenetic Analysis

On the basis of the alignment of the amino-acid sequences of several plant PPOs, a phylogenetic tree was generated using MEGA 6. The phylogenetic tree (Figure 2) showed that CsPPO1 and CsPPO2 were clustered on the same branch, suggesting that CsPPO1 and CsPPO2 have higher similarity to each other than to CsPPO. The CsPPOs are much closer to PPOs from red clover (*Trifolium pretense*), lotus flower (*Nelumbo nucifera*), and pokeweed (*Phytolacca americana*) than to PPOs from eggplant (*Solanum melongena*), tomato, and tobacco (*Nicotiana tabacum*). The PPO members of the nightshade family are clustered on the same branch, close to the NtasPPO. The rest of the PPO members from the same plant in the PPO family are relatively closer than others. Most of the PPO members from different plants are not inducible, and the inducible PPO members from the same plant are closer to each other than to the others.

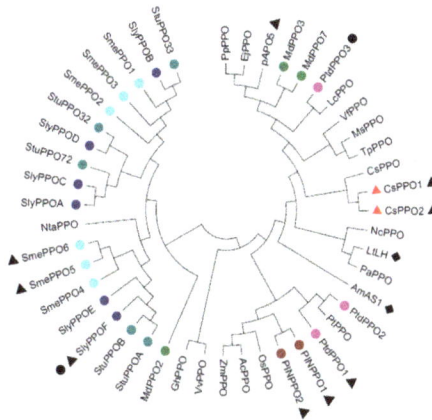

**Figure 2.** Phylogenetic analysis of PPO protein squences. The numbers on the tree branches represent bootstrap confidence values, as "Bootstrap" is 1000. The black triangles indicate wound-inducible; black dots indicate pathogen-inducible; black rhombuses indicate PPOs with biosynthetic functions. All sequence data may be found in GenBank, and differently colored bullet points represent different *PPO* gene families. *Malus × domestica* (*pAPO5*, P43309; *MdPPO2*, AAK56323; *MdPPO3*, BAA21676; and *MdPPO7*, BAA21677); *Pyrus pyrifolia* (*PpPPO*, AB056680); *Eriobotrya japonica* (*EjPPO*, AFO55217); *Vicia faba* (*VfPPO*, CAA77764.1); *Medicago sativa* (*MsPPO*, AAP33165.1); *Trifolium pretense* (*TpPPO*, AAK13244.1); *Populus trichocarpa × P. deltoids* (*PtdPPO1*, AAG21983.1; *PtdPPO2*, AAU12256.1; and *PtdPPO3*, AAU12257.1); *Litchi chinensis* (*LcPPO*, AEQ30073); *Phytolacca American* (*PaPPO*, D45386.1); *Larrea tridentata* (*LtLH*, AAQ67412); *Nelumbo nucifera* (*NcPPO*, ADP89908.1); *Vitis vinifera* (*VvPPO*, AB871370); *Camellia sinensis* (*CsPPO*, EF635860.1); *Populus trichocarpa* (*PtPPO*, AEH41424.1); *Ananas comosus* (*PINPPO1*, AAO16863.1; and *PINPPO2*, AAO16865.1); *Annona cherimola* (*AcPPO*, ABJ90144.1); *Oryza sativa* (*OsPPO*, DQ532396); *Zea mays* (*ZmPPO*, ACG28948.1); *Antirrhinum majus* (*AmAS1*, BAB20048.1); *Gossypium hirsutum* (*GhPPO*, AFC36521.1); *Solanum tuberosum* (*StuPPO32*, AAA85121.1; *StuPPO33*, AAA85122.1; *StuPPO72*, AAA85123.1; *StuPPOA*, AAA02877.1; and *StuPPOB*, AAA02879.1); *Solanum lycopersicum* (*SlyPPOA*, Q08303; *SlyPPOB*, Q08304; *SlyPPOC*, Q08305; *SlyPPOD*, Q08306; *SlyPPOE*, Q08307; and *SlyPPOF*, Q08296); *Solanum melongena* (*SmePPO*, ACR61399.1) and *Nicotiana tabacum* (*NtaPPO*, CAA73103.1).

## 2.3. Expression of the Recombinant Protein in E. coli

The expression vectors pGEX-4T-2/CsPPO1 and pGEX-4T-2/CsPPO2 were constructed and transferred into *E. coli* BL21 (DE3). *CsPPO1* and *CsPPO2* genes were highly expressed in *E. coli* cells. Results from a solubility analysis indicate that the recombinant proteins of pGEX-4T-2/CsPPO1 and pGEX-4T-2/CsPPO2 exist in the form of an inclusion body (Figure 3). The expression products of recombinant CsPPO1 and CsPPO2 in *E. coli* are about 91 and 92 kDa, respectively.

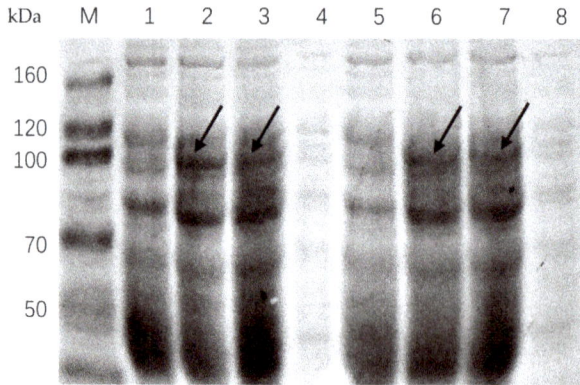

**Figure 3.** SDS-PAGE analysis of recombinant CsPPO1 and CsPPO2 proteins expressed in *E. coli*. The lanes M and 1–8 in the graph are as follows: M, protein molecular mass marker. 1, pGEX-4T-2/CsPPO1 without isopropyl-β-D-thiogalactopyranoside (IPTG) induction. 2, pGEX-4T-2/CsPPO1 with IPTG induction. 3, precipitation of pGEX-4T-2/CsPPO1 with IPTG induction after sonication. 4, supernatant of pGEX-4T-2/CsPPO1 with IPTG induction after sonication. 5, pGEX-4T-2/CsPPO2 without IPTG induction. 6, pGEX-4T-2/CsPPO2 with IPTG induction. 7, precipitation of pGEX-4T-2/CsPPO2 with IPTG induction after sonication. 8, supernatant of pGEX-4T-2/CsPPO2 with IPTG induction after sonication. The bands of pGEX-4T-2/CsPPO1 and pGEX-4T-2/CsPPO2 were pointed out by arrows in lanes 2, 3, 6, 7.

## 2.4. Expression of CsPPO1 and CsPPO2 in Different Tissues

The level of transcriptional expression of *CsPPO1* in the stem is significantly higher than that in the roots, leaves, and flowers (Figure 4A). The expression level of *CsPPO2* in the roots is significantly higher than that in the leaves, flowers, and stems, while the expression of *CsPPO2* in the stems is significantly lower than that in the leaves and flowers (Figure 4B).

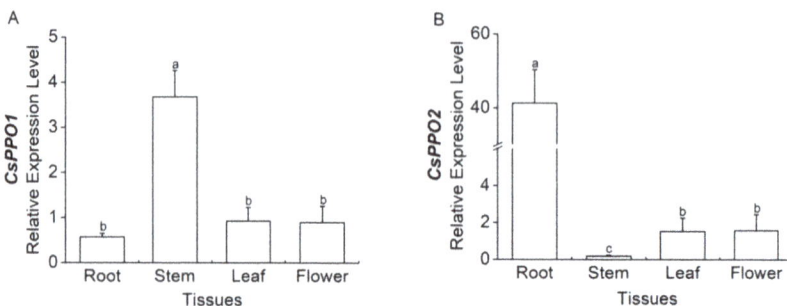

**Figure 4.** Mean levels of transcriptional expression (±SE) of (**A**) *CsPPO1* and (**B**) *CsPPO2* in tissues from the roots, stems, leaves, and flowers of *Camellia sinensis*. *CsGAPDH* was used as a reference gene. For each column, different letters indicate significant differences among tissues ($p < 0.05$, Duncan's multiple range test, $n = 5$).

## 2.5. Jasmonic Acid Elicits the Expression of CsPPO1 and CsPPO2 Differentially

Exogenous application of JA significantly elicited the expression of *CsPPO2* at 6 and 48 h after the start of treatment, but not of *CsPPO1* (Figure 5).

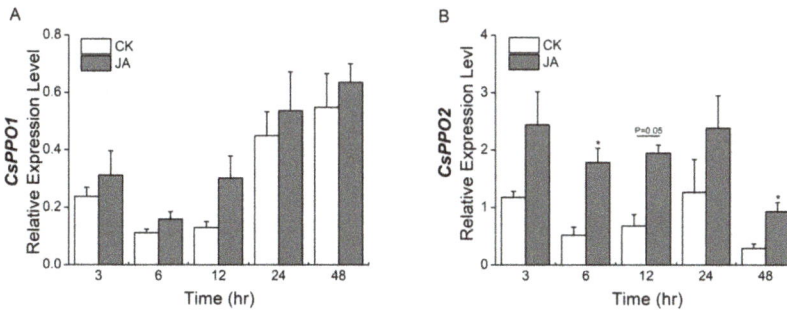

**Figure 5.** Mean levels of transcriptional expression (±SE) of (**A**) *CsPPO1* and (**B**) *CsPPO2* in tea leaves of jasmonic acid (JA)-treated plants and control plants. *CsGAPDH* was used as a reference gene. The asterisks indicate significant differences between treatments and controls (* $p < 0.05$, Student's *t*-test, $n = 5$).

## 2.6. Infestation of Caterpillars Elicit the Expression of CsPPO1 and CsPPO2

The infestation of *E. obliqua* larvae significantly elicited the expression of *CsPPO1* and *CsPPO2* (Figure 6). Levels of *CsPPO1* and *CsPPO2* in caterpillar-infested tea plants at 24 and 48 h after the start of treatment were significantly higher than those in control plants: 7.8- and 11.6-fold higher, and 5.8- and 3.7-fold higher, respectively.

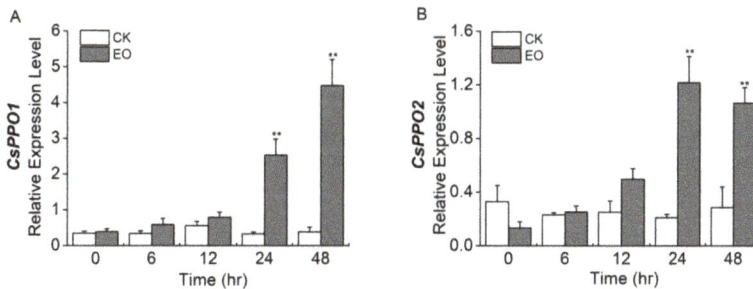

**Figure 6.** Mean levels of transcriptional expression (±SE) of (**A**) *CsPPO1* and (**B**) *CsPPO2* elicited in leaves of *Camellia sinensis* by the infestation of *Ectropis obliqua*. *CsGAPDH* was used as a reference gene. The asterisks indicate significant differences between treatments and controls (* $p < 0.05$, ** $p < 0.01$, Student's *t*-test, $n = 5$).

## 2.7. Regurgitant Elicits the Expression of CsPPO1 and CsPPO2 Differentially

Wounding plus regurgitant and wounding plus deionized water significantly elicited the expression of *CsPPO1* and *CsPPO2* (Figure 7). The expression of *CsPPO1* was significantly induced in plants by treatment involving wounding plus deionized water or wounding plus regurgitant at 12 h after the start of treatment, compared to the expression in intact plants, whereas the expression of *CsPPO2* was only significantly induced by wounding plus regurgitant at 12 h. Moreover, plants treated with wounding plus regurgitant significantly upregulated the expression of *CsPPO2* compared to plants treated with wounding plus deionized water at 12 and 24 h after the start of treatment

(Figure 7B), approximately 2.22 and 3.73 times, respectively. But, the expression of *CsPPO1* did not differ significantly regardless of whether or not plants had been treated by wounding plus deionized water or wounding plus regurgitant at 12 and 24 h after the start of treatment (Figure 7A).

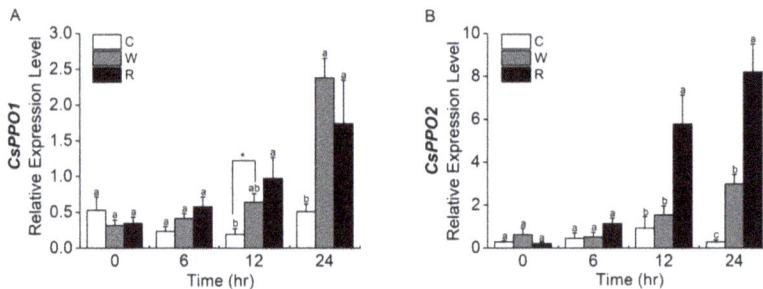

**Figure 7.** Mean levels of transcriptional expression (±SE) of (**A**) *CsPPO1* and (**B**) *CsPPO2* in treated tea leaves elicited by mechanical wounding supplemented with distilled water or regurgitant and intact control plants (C). *CsGAPDH* was used as a reference gene. For each time point, different letters indicate significant differences among treatments ($p < 0.05$, Duncan's multiple range test, $n = 5$). The asterisks indicate significant differences between treatments and controls (* $p < 0.05$, Student's *t*-test, $n = 5$).

### 2.8. Separated Compounds of Regurgitant Elicit the Expression of CsPPO1 and CsPPO2 Differentially

The expression of *CsPPO1* was only significantly induced by the wounding plus diluted regurgitant treatment at 24 h after the start of treatment, compared to the expression in intact plants, whereas the expression of *CsPPO2* was significantly induced by all treatments at 24 h; in addition, the plants treated with wounding plus diluted regurgitant and sterile extract significantly upregulated the expression of *CsPPO2* compared to plants treated with wounding plus deionized water (Figure 8). Under treatment of the three separate compounds of regurgitant, the expressions of *CsPPO1* between wounding plus diluted regurgitant treatment and wounding plus sterile extract removed 3-kDa molecular weight off differ (Figure 8), whereas the expression levels of *CsPPO2* between wounding plus diluted regurgitant treatment and wounding plus sterile extract treatment are the same; but, the expression of both had significant differences with the treatment under wounding plus sterile extract without 3-kDa molecules (Figure 8).

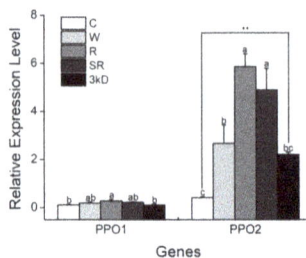

**Figure 8.** Mean levels of transcriptional expression (±SE) of *CsPPO1* and *CsPPO2* in treated tea leaves elicited by mechanical wounding supplemented with distilled water (W), diluted regurgitant (R), sterile extract (SR), sterile extract without 3-kDa molecules(3kD) and intact control plants (C). *CsGAPDH* was used as a reference gene. For each time point, different letters indicate significant differences among treatments ($p < 0.05$, Duncan's multiple range test, $n = 5$). The asterisks indicate significant differences between treatments and controls, ** $p < 0.01$, Student's *t*-test, $n = 5$).

## 3. Discussion

PPOs play multiple roles in *C. sinensis*, such as the elicitation of defenses against *E. obliqua* and the oxidation of flavanols to theaflavins and thearubigins during tea processing [4,38,39]. However, which members of the *CsPPO* gene family are involved in the defense response of tea plants have not yet been identified. In the present study, two new *PPO* genes, *CsPPO1* and *CsPPO2*, both of which had high similarity with *PPOs* from other plants, were obtained by rapid amplification of cDNA ends PCR (RACE-PCR) from leaves of *C. sinensis*. The transit peptides were predicted in the N-terminal region of CsPPO1 and CsPPO2, indicating that the tea plant's PPO is a chloroplastic enzyme, which is consistent with the report of Halder et al. [40] and similar to other plant PPO proteins [41–43]. Moreover, the deduced amino-acid sequences of CsPPO1 and CsPPO2 were found to contain three conserved domains (Tyrosinase, PPO1-DWL, and PPO-KFDV), which are considered to be expression sequence tags of PPOs [44]. That CsPPO1 has only one copper-binding site and CsPPO2 contains two (Figure 1) suggests that they may have different functions, similar to *PPO* genes of strawberry (*Fragaria vesca*) [45]. Furthermore, results from 25 different plants involving 46 *PPO* genes in a phylogenetic analysis suggest that *CsPPOs* have a common ancestor with the red clover, lotus flower, and pokeweed, as these species were clustered on the adjoining branch (Figure 2). Results of phylogenetic and gene-structure analysis indicate that *PPO* genes are relatively conserved across different species. *SmePPO5*, *SmePPO6*, *SlyPPOF*, *PINPPO1*, *PINPPO2*, *PtdPPO1*, *PtdPPO3*, *pAPO5*, *CsPPO1*, and *CsPPO2* were wound-induced, while *SlyPPOF* and *PtdPPO3* were induced by pathogens. *AmAS1* has been unambiguously demonstrated to play a role in the biosynthesis of chalcone-derived yellow-colored aurone pigments in *Antirrhinum majus* (snapdragon) [46], and *LtLH* is an enantiospecific PPO involved in the biosynthesis of linked lignins that have been isolated and characterized in *Larrea tridentata* (creosote bush) [47–49]. Genes with functions similar to those of PPOs from different species do not cluster together in the phylogenetic analysis (Figure 2), suggesting that the adaptation of *PPO* genes for defense evolved independently in different plants. This result might explain the differences, which are consistent with those reported by Schmidt et al. [50]. Wu et al. expressed a reported *CsPPO* in *E. coli* and found that recombinant CsPPO appears as an inclusion body upon expression in *E. coli* despite the removal of chloroplast-targeting transit peptide; subsequently, they attempted to solubilize inclusion bodies in a suitable buffer; the specific activity of PPO was only 19.01 U/mg protein [51]. Our results show that the *CsPPO1* and *CsPPO2* genes highly express in *E. coli* cells, but the recombinant proteins exist in the form of an inclusion body (Figure 3), which are similar to the results of Liu et al. [29]. In a follow-up study, we need to solubilize inclusion bodies in a suitable buffer to measure the specific enzymatic activity of CsPPO1 and CsPPO2.

Previous studies have reported that different *PPO* gene members have distinct expression levels in various tissues [22,45,47,52]. For instance, although three *PPO* genes have been found in a poplar hybrid (*Populus trichocarpa* × *P. deltoides*), *PtdPPO1* is exclusively expressed in damaged leaves, whereas *PtdPPO2* and *PtdPPO3* are predominantly expressed in stems, petioles, or roots [52]. Here, we found that *CsPPO1* was highly expressed in stems, and *CsPPO2* was highly expressed in roots. In other plants' *PPO* gene families, the expression patterns of some members are similarly organ-specific with *CsPPO1* and *CsPPO2*, such as *SmPPO1*, *SmPPO16*, *FaPPO4*, *PtrPPO2*, and *PtrPPO3*, which are highly expressed in roots, and *SmPPO11*, which is highly expressed in stems [22,45,52]. The diverse expression profiles identified for *PPO* genes in different tissues suggest that they may have diverse functional roles. For example, *AmAS1* from snapdragon is localized in vacuoles and specifically catalyzes the formation of aurones from chalcones [46]; Li et al. [22] reported that eight *SmPPOs* that were expressed in *S. miltiorrhiza* roots have potential in lithospermic acid B biosynthesis and metabolism. The roots of tea plants are a component of traditional Chinese medicine, and have been reported to have pharmacological effects [53]; *CsPPO2* was highly expressed in the roots, suggesting that it may be involved in the biosynthesis and metabolism of phenolic acid in roots—a hypothesis that needs to be verified. However, our study mainly focused on *CsPPOs* with defensive roles against leaf-feeding pests, so we paid attention to the expression of *CsPPOs* in leaves but not to that in roots.

What the biological meaning of *CsPPO2* is, which was highly expressed in the roots but plays important role in defending against *E. obliqua*, needs to be investigated further.

Strong evidence has shown that PPOs play defensive roles in tomato, poplar (*Populus trichocarpa*), strawberry, and some other plants [12,13,18,45,52,54–56]. Among the 15 members of the poplar *PPO* gene family, levels of transcriptional expression of *PtrPPO1*, *PtrPPO3*, and *PtrPPO11* were significantly induced by mechanical damage, the exogenous application of MeJA, or the infection of *Melampsora laricipopulina*, while others are developmentally regulated [48,57]. Similar results have also been reported in tomato, apple (*Malus domestica*), and pineapple (*Ananas comosus*) [58–60]. Our current results showed that the expression of *CsPPO1* and *CsPPO2* was wound- and *E. obliqua*-induced (Figure 6A,B), indicating that both were probably promoted to play defensive roles in *C. sinensis*. In contrast, JA treatment induced only the expression of *CsPPO2*, not of *CsPPO1* (Figure 5). The situation is similar in poplar [48], apple [58], and pineapple [59], where *PtrPPO5*, *pAPO5*, *PINPPO1*, and *PINPPO2* were found to be the only wound-induced *PPO*s. In other plants, most *PPO* genes are not wound- or herbivore-induced [48]. Nevertheless, the infestation of herbivores not only mechanically damaged plants but also introduced OS into the wounding sites [32,36,61]. Musser et al. [36] found that the caterpillar labial saliva of cotton bollworm (*Helicoverpa zea*) alters gene expression in the tomato plant. Our results showed that regurgitant could upregulate the expression level of *CsPPO2* but not of *CsPPO1* more significantly than treating the plant with wounding plus deionized water at 12 and 24 h after the start of the experiment (Figure 7). This result indicates that the accumulation of *CsPPO2* can be upregulated by regurgitant and may constitute a more meaningful defense against *E. obliqua* than the accumulation of *CsPPO1*. The disparity between the transcripts of *CsPPO1* and of *CsPPO2* may have resulted from their different structural attributes, as stated above. Similar disparities were previously found in other plants as well [22,57,59].

Several examples have shown that the components of OS can interfere with, or even suppress, the activation of defensive responses in plants [62]. Among the known herbivore-associated molecular patterns (HAMPs) [63], fatty acid–amino acid conjugates are widely distributed in the OS of lepidopteran insects and elicit specific responses in various plants; furthermore, $H_2O_2$, which is produced by glucoseoxidase in OS, is also believed to take part in activating insect feeding-induced defensive reactions [64]. In the present study, we found that *CsPPO2* was significantly amplified by diluted regurgitant, but its expression level did not differ significantly between wounding plus sterile extract and wounding plus diluted regurgitant, suggesting the active component(s) was not microbial (Figure 8). Chung et al. [32] found the Colorado potato beetles (CPB, *Leptinotarsa decemlineata*) can exploit orally secreted bacteria to suppress plant defenses, and Wang et al. (2016) found different microbes in insects can have species-specific effects on different host plants. The microbes in the regurgitant of *E. obliqua* may have effects on other defense genes that need to be verified. Treatments of wounding plus diluted regurgitant and sterile extract significantly elicited the expression of *CsPPO2* compared with its expression in wounding plus deionized water, and wounding plus sterile extract removed small molecules off and intact plants (Figure 8), which suggests the active component(s) of the regurgitant that elicited the expression of *CsPPO2* may contain in small molecules (within 3-kDa molecular weight). Similarly, the difference in responses between the expression of *CsPPO1* under wounding plus diluted regurgitant and wounding plus sterile extract without small molecules supported this result (Figure 8).

Previously, our group noted the infestation of *E. obliqua* or wounding plus the regurgitant significantly suppressed wound-induced PPO activities [4]. Given our earlier results and our results here, we hypothesize that there are other *PPO* gene members that are particularly responsive to the infestation of *E. obliqua* in the tea plant; the assumption needs to be further investigated. Meanwhile, microRNAs (miRNAs) represent key post-transcriptional regulators of eukaryotic gene expression and play important roles in stress responses. The prediction of miRNAs in *CsPPO1* and *CsPPO2* indicates that miRNAs are involved the post-transcriptional regulation of *PPO*s, which may lead to the difference between our earlier results and results here.

The regurgitant from *E. obliqua* was separated into its components for the first time in this study, and we found the active component(s) may elicit the expression of *CsPPO2* contained in small molecules (within 3-kD molecular weight). Future studies are planned identifying the active components in regurgitant that launch the induced defense response of the tea plant against this herbivore.

## 4. Materials and Methods

### 4.1. Insects

*E. obliqua* eggs were originally obtained from Plantation Centre of Tea Research Institute of Chinese Academy of Agricultural Sciences (CAAS), Hangzhou, Zhejiang, and maintained in an insectary. The newly hatched larvae were raised in net cages (75 × 75 × 75 cm) with potted fresh tea shoots and kept in a controlled climate room programmed at 26 ± 2 °C, 70 ± 5% RH (relative humidity), and 12-h photophase. After one generation, *E. obliqua* caterpillars were used for the experiments.

### 4.2. Regurgitant Collection and Separation

Regurgitant was collected from the oral cavity of 4th-instar *E. obliqua* with a P200 Pipetteman (Gilson, Middleton, WI, USA). The collected regurgitant was centrifuged for 5 min at 10,000× $g$, after which the supernatant was collected and stored at −80 °C for leaf treatment. Preliminary separation of the regurgitant was carried out according to the method described in Ray et al. [65] (2015) with minor revision. The collected regurgitant was diluted 1:4 $v/v$ with water and filtered through Miracloth (EMD Millipore, Billerica, MA, USA) to remove debris. Then, the diluted regurgitant was sterilized in 0.2-μm filters (EMD Millipore), and the sterile regurgitant was concentrated using centrifugal columns with a 3-kDa molecular weight cut-off (Pall Life Sciences, Louisville, KY, USA) to remove small molecules present in the extract.

### 4.3. Tea Plants and Treatments

One-year-old Longjing 43 tea plants were planted individually in plastic pots (14 cm diameter × 15 cm high) and grown in a greenhouse (26 ± 2 °C, relative humidity of 70–80%, 12 h photophase), irrigated once every other day and fertilized with rapeseed cake once a month. The light intensity for plants was about 450 μmol·m$^{-2}$·s$^{-1}$ during the photophase. Three-year-old plants were used for experiments. Potted plants were washed under running water and then air-dried. Samples and plant treatments were prepared as follows:

#### 4.3.1. Different Tissues

Different tissues, including roots (Figure 9A), tender internodes between the fourth to seventh (Figure 9B), leaves (the second to third intact leaves away from the terminal growing point, Figure 9C) and half-open flowers (Figure 9D) were harvested from the same plant, immediately frozen in liquid nitrogen and then stored at −80 °C. Five replications were carried out.

**Figure 9.** Characteristic photographs of treatments. (**A**): Root. (**B**): Stem. (**C**): Leaf. (**D**): Flower. (**E**): Caterpillar infestation. (**F**): Mechanical damage with deionized water. (**G**): Mechanical damage with regurgitant.

### 4.3.2. JA Treatment

JA (Sigma Chemical Co., St. Louis, MO, USA) was dissolved in a small amount of ethanol and made up to a concentration of 150 µg/mL in 50 mM sodium-phosphate buffer (titrated with 1 M citric acid until pH 8). Plants were individually sprayed with 8 mL of JA solution. Control plants were individually sprayed with 8 mL of the buffer. Plants were treated at 9 am and then placed in a controlled-climate room that was maintained at $26 \pm 2 \, ^\circ C$, 12-h photophase, and 80% RH. The second leaves were harvested at 0, 3, 6, 12, 24, and 48 h after the start of treatment. Five replications were carried out.

### 4.3.3. Caterpillar Infestation

The second leaf of each plant was covered with a fine-mesh sleeve into which 2 3rd-instar caterpillars that had been starved for 10 h were introduced. Plants with only their second leaves covered with fine-mesh sleeves were used as controls. The second leaves were harvested at 0, 6, 12, 24, and 48 h after the start of treatment (Figure 9E). Five replications were carried out.

### 4.3.4. Regurgitant Induction

The mechanical damage was made by a fabric pattern wheel following the method described in Lou and Baldwin [66]. Each leaf was rolled 6 times, and 15 µL whole regurgitant from 4th-instar larvae was added to the puncture wounds on each leaf (Figure 9G). 15 µL of deionized water was added to the damaged leaves for wounding treatment (Figure 9F). Intact second leaves were used as controls. All the second leaves were harvested at 0, 6, 12, and 24 h after the start of treatment. Five replications were carried out.

### 4.3.5. Separated Regurgitant Induction

15 µL of deionized water, diluted regurgitant, sterile extract, and sterile extract without 3 kDa molecular weight were added to damaged leaves. Intact second leaves were used as controls. Second leaves were harvested at 24 h after the start of treatment. Five replications were carried out.

### 4.4. RNA Extraction and cDNA Synthesis

Total RNA was isolated with TRIzol™ kit, according to the manufacturer's instructions (TIANGEN, Beijing, China). Quality and concentration were checked by agarose-gel electrophoresis and spectrophotometer analysis, and the RNA was stored at −80 °C until use. First-strand cDNA was synthesized from total RNA using a PrimerScript® RT Reagent Kit (Takara, Dalian, China) according to the manufacturer's instructions. After reverse transcription, the synthesized cDNA was stored at −20 °C for future use.

### 4.5. Cloning Full-Length CsPPO Gene Sequences

The cDNA fragments were obtained from the previous transcriptome of Longjing 43. The 5′ and 3′ sequences of *CsPPO1* and *CsPPO2* were acquired by rapid amplification of RACE-PCR using the manufacturer's protocol (SMARTer® RACE 5′/3′Kit, Clontech Lab, Inc., Mountain View, CA, USA). The PCR products were ligated into a pMD-19T easy vector (Takara, Dalian, China) after purification. After the 3′ and 5′ sequences were obtained, the full-length cDNA sequences of *CsPPO1* and *CsPPO2* were cloned by RT-PCR reaction using primers PPO1-full1, PPO1-full2, PPO2-full1, and PPO2-full2. All the primers used in RACE are listed in Table 2.

**Table 2.** Primers used for cloning and analysis of *CsPP1* and *CsPPO2*.

| Primers | Purpose | Primer Sequence (5′-3′) |
|---------|---------|-------------------------|
| GSP1-3 | 3′-RACE | AATGTGGATCGGATGTGG |
| NGSP1-3 | 3′-RACE | CAGAGACCGAAGAAATCAAG |
| GSP2-3 | 3′-RACE | GGTTTGTGTTCTATGATGAG |
| NGSP2-3 | 3′-RACE | AGAAGGATGATGAAGAGGAG |
| GSP1-5 | 5′-RACE | GAGCCATGAGTTGTGGACTTGAAG |
| NGSP1-5 | 5′-RACE | TGGTGATAAGCTCCGTCGCAATAA |
| GSP2-5 | 5′-RACE | ATCCGACCCGTTGTAATCCA |
| NGSP2-5 | 5′-RACE | CGAAGAAGTGGAGGTAGTAT |
| UPM | RACE | Long primer: CTAATACGACTCACTATAGGGCAAGCAGTGGTATCAACGCAGAGT<br>Short primer: CTAATACGACTCACTATAGGGC |
| PPO1-full1 | Clone | ATGAATTCTCTTCCACCATCA |
| PPO1-full2 | Clone | TTAGGAATCAAACTCAATCTTG |
| PPO2-full1 | Clone | ATGGCTTCTTTTCCACCTTC |
| PPO2-full2 | Clone | TCAAGAATCAAACTCTATCTTGA |
| PPO1-PF | protein | CTGGTTCCGCGTGGATCCATGAATTCTCTTCCACCATCATGCA |
| PPO1-PR | protein | CGCTCGAGTCGACCCGGGTTAGGAATCAAACTCAATCTTG |
| PPO2-PF | protein | CTGGTTCCGCGTGGATCCATGGCTTCTTTTCCACCTTC |
| PPO2-PR | protein | CGCTCGAGTCGACCCGGGTCAAGAATCAAACTCTATCTTGA |
| GAPDH-F | QPCR | GACTGGAGAGGTGGAAGAGC |
| GAPDH-R | QPCR | AGCCATTCCAGTCAATTTCC |
| PPO1-RTF | QPCR | CCATCTGGAAGAGTTTGGGT |
| PPO1-RTR | QPCR | CCTTCACTTTGACAGGCTGA |
| PPO2-RTF | QPCR | CGGAATGCCAATGCCTGCAA |
| PPO2-RTR | QPCR | AGTTGGATCCGACCCGTTGT |

### 4.6. Bioinformatic and Phylogenetic Analysis

The BLAST program at the NCBI Web site (https://blast.ncbi.nlm.nih.gov/Blast.cgi) was used to compare the amino-acid sequences of CsPPO1 and CsPPO2. The ORF was analyzed using the ORF finder (https://www.ncbi.nlm.nih.gov/orffinder/). The molecular weight (Mw) and theoretical isoelectric (pI) points of *CsPPO1/CsPPO2* were calculated using the online computer pI/Mw tool (http://cn.expasy.org/tools). The putative domains were identified using the InterPro database (http://www.ebi.ac.uk/interpro/), and the Cu-binding sites were analyzed with the website (http://www.expasy.ch/prosite). ClustalW2 software (Lynnon Biosoft, Los Angeles, CA, USA) was used to align the CsPPO1 and CsPPO2 sequences with other PPO sequences from related species. The phylogenetic tree was constructed by MEGA6 (http://www.megasoftware.net/) using the neighbor-joining method with the bootstrap values calculated from 1000 replicates.

### 4.7. Recombinant Protein Expression in Escherichia coli

Plasmid vector pGEX-4T-2 was used to produce recombinant protein. The plasmids pTA-CsPPO1 and pTA-CsPPO2 were cloned by the forward primers with *BamH I* site and the reverse primers with *Sma I* site (Table 2). The pGEX-4T-2 vector, plasmid pTA-CsPPO1, and pTA-CsPPO2 were digested with the restriction enzymes *BamH I* and *Sma I* and ligated to obtain the expression plasmid pGEX-4T-2/CsPPO1 and pGEX-4T-2/CsPPO2. After being sequenced to confirm the cloned fragments, the constructs were transformed into the *E. coli* BL21 (DE3). Cells were grown at 37 °C overnight in LB media containing ampicillin (50 $\mu$g·mL$^{-1}$). Following centrifugation, *E. coli* cells were adjusted to OD$_{600}$ = 0.5–0.8, and the production of recombinant protein was induced by adding 1 mM isopropyl-$\beta$-D-thiogalactopyranoside (IPTG) to a final concentration of 0.2 mM. No IPTG induction samples were treated as controls. Incubation was continued for 6 h at 18 °C and finally analyzed by sodium dodecylsulfate-polyacrylamide gel electrophoresis (SDS-PAGE).

### 4.8. Real-Time PCR (qPCR) Analysis

RT–qPCR was carried out to investigate the expression profiles of *CsPPO1* and *CsPPO2* in different tissues and under different treatments. Five independent biological samples were used, and the cDNA from different tissues or different treatments was used as a template. The *GAPDH* (Accession No. GE651107, Deng et al., 2016) was used as the internal standard. The RT-qPCR primers were designed by primer premier 5.0 software (PREMIER Biosoft, Palo Alto, CA, USA), and all the primers are listed in Table 1. The RT-qPCR was performed on a LightCycler 480 system (Roche Diagnostics, Mannheim, Germany) using a Premix Ex Taq kit (TaKaRa, Dalian, China) with a 20 $\mu$L reaction mixture containing 1 $\mu$L cDNA, 10 $\mu$L qPCR Master Mix (Mountain View, CA, USA), 1 $\mu$L of each specific primer (0.2 mM), and 7 $\mu$L nuclease-free water. The qPCR program included a preliminary step at 95 °C for 30 s, 40 cycles of a denaturation at 95 °C for 10 s, and an annealing and extension step at 58 °C for 1 min. The relative expressions were calculated by $2^{-\Delta\Delta Ct}$ method.

### 4.9. Statistical Analysis

All statistical analyses were performed by using the Statistica (Statistica, SAS, Institute Inc., Cary, NC, USA). The differences among the transcriptional levels of *CsPPO1* and *CsPPO2* expressed in different tissues, and among tissue from intact plants, and from plants subjected to wounding and regurgitant induction were analyzed by using one-way ANOVAs with $p < 0.05$ indicating statistical significance. Differences in the levels of gene expression between *E. obliqua*-infested and intact tea plants, and JA-treated and intact tea plants were determined by Student's *t*-test.

**Author Contributions:** C.H., X.S., and X.L. conceived and designed the experiments; X.S. and X.L. supported the experimental materials; C.H., J.Z. and X.Z. performed the experiments; C.H., Y.Y., and W.B. analyzed the data; C.H. wrote the paper; and the manuscript was modified by X.Z., Z.Z., and X.S.

**Funding:** The study was sponsored by the Special Fund for Agro-scientific Research in the Public Interest (201403030), the National Natural Science Foundation of China (31471784; 31272053), the China Earmarked Fund for Modern Agro-industry Technology Research System (CARS-19), and the Key Research and Development Program in Guangxi (Demonstration and Application of Standardized Production Technology in Organic Liu Bao tea Garden).

**Acknowledgments:** We gratefully acknowledge Emily Wheeler, Boston, for editorial assistance.

**Conflicts of Interest:** The authors declare no conflict of interest.

## Abbreviations

| | |
|---|---|
| PPOs | Polyphenol oxidases |
| ORF | Open reading frame |
| JA | Jasmonic acid |
| MeJA | Methyl jasmonate |
| SA | Salicylic acid |
| ET | Ethylene |
| OS | Oral secretions |
| HAMPs | Herbivore-associated molecular patterns |
| CPB | Colorado potato beetles |
| CAAS | Chinese Academy of Agricultural Sciences |
| Mw | Molecular weight |
| pI | Isoelectric point |
| IPTG | Isopropyl-β-D-thiogalactopyranoside |
| SDS–PAGE | Sodium dodecylsulfate-polyacrylamide gel electrophoresis |

## References

1. Joy, R.W., IV; Sugiyama, M.; Fukuda, H.; Komamine, A. Cloning and characterization of polyphenol oxidase cDNAs of *Phytolacca americana*. *Plant Physiol.* **1995**, *107*, 1083–1089. [CrossRef] [PubMed]
2. Nakayama, T.; Yonekura-Sakakibara, K.; Sato, T.; Kikuchi, S.; Fukui, Y.; Fukuchi-Mizutani, M.; Ueda, T.; Nakao, M.; Tanaka, Y.; Kusumi, T. Aureusidin synthase: A polyphenol oxidase homolog responsible for flower coloration. *Science* **2000**, *290*, 1163–1166. [CrossRef] [PubMed]
3. Ralph, J.; Brunow, G.; Harris, P.J.; Dixon, R.A.; Schatz, P.F.; Boerjan, W. Lignification: Are lignins biosynthesized via simple combinatorial chemistry or via proteinaceous control and template replication. *Recent Adv. Polyphen. Res.* **2008**, *1*, 36–66.
4. Yang, Z.-W.; Duan, X.-N.; Jin, S.; Li, X.-W.; Chen, Z.-M.; Ren, B.-Z.; Sun, X.-L. Regurgitant derived from the tea geometrid *Ectropis obliqua* suppresses wound-induced polyphenol oxidases activity in tea plants. *J. Chem. Ecol.* **2013**, *39*, 744–751. [CrossRef] [PubMed]
5. Du, M.-H.; Yan, Z.-W.; Hao, Y.-J.; Yan, Z.-T.; Si, F.-L.; Chen, B.; Qiao, L. Suppression of *Laccase 2* severely impairs cuticle tanning and pathogen resistance during the pupal metamorphosis of *Anopheles sinensis* (Diptera: Culicidae). *Parasites Vectors* **2017**, *10*, 171. [CrossRef] [PubMed]
6. Mayer, A.; Harel, E. Phenoloxidases and their significance in fruit and vegetables. *Food Enzym.* **1991**, *1*, 373–398.
7. Si, H.; Zhou, Z.; Wang, X.; Ma, C. A novel molecular marker for the polyphenol oxidase gene located on chromosome 2B in common wheat. *Mol. Breed.* **2012**, *30*, 1371–1378. [CrossRef]
8. Jukanti, A.K.; Bhatt, R. Eggplant (*Solanum melongena* L.) polyphenol oxidase multi-gene family: A phylogenetic evaluation. *Biotech* **2015**, *5*, 93–99. [CrossRef] [PubMed]
9. Wang, J.; Liu, B.; Xiao, Q.; Li, H.; Sun, J. Cloning and expression analysis of litchi (*Litchi chinensis* sonn.) polyphenol oxidase gene and relationship with postharvest pericarp browning. *PLoS ONE* **2014**, *9*, e93982. [CrossRef] [PubMed]
10. Felton, G.; Donato, K.; Broadway, R.; Duffey, S. Impact of oxidized plant phenolics on the nutritional quality of dietar protein to a noctuid herbivore, *Spodoptera exigua*. *J. Insect Physiol.* **1992**, *38*, 277–285. [CrossRef]
11. Thipyapong, P.; Hunt, M.D.; Steffens, J.C. Antisense downregulation of polyphenol oxidase results in enhanced disease susceptibility. *Planta* **2004**, *220*, 105–117. [CrossRef] [PubMed]
12. Bhonwong, A.; Stout, M.J.; Attajarusit, J.; Tantasawat, P. Defensive role of tomato polyphenol oxidases against cotton bollworm (*Helicoverpa armigera*) and beet armyworm (*Spodoptera exigua*). *J. Chem. Ecol.* **2009**, *35*, 28–38. [CrossRef] [PubMed]
13. Bosch, M.; Berger, S.; Schaller, A.; Stintzi, A. Jasmonate-dependent induction of polyphenol oxidase activity in tomato foliage is important for defense against *Spodoptera exigua* but not against *Manduca Sexta*. *BMC Plant Biol.* **2014**, *14*, 257. [CrossRef] [PubMed]
14. Redondo, D.; Venturini, M.E.; Oria, R.; Arias, E. Inhibitory effect of microwaved thinned nectarine extracts on polyphenol oxidase activity. *Food Chem.* **2016**, *197*, 603–610. [CrossRef] [PubMed]

15. Zhou, L.; Liu, W.; Xiong, Z.; Zou, L.; Chen, J.; Liu, J.; Zhong, J. Different modes of inhibition for organic acids on polyphenol oxidase. *Food Chem.* **2016**, *199*, 439–446. [CrossRef] [PubMed]

16. Felton, G.; Donato, K.; Del Vecchio, R.; Duffey, S. Activation of plant foliar oxidases by insect feeding reduces nutritive quality of foliage for noctuid herbivores. *J. Chem. Ecol.* **1989**, *15*, 2667–2694. [CrossRef] [PubMed]

17. Li, L.; Steffens, J.C. Overexpression of polyphenol oxidase in transgenic tomato plants results in enhanced bacterial disease resistance. *Planta* **2002**, *215*, 239–247. [CrossRef] [PubMed]

18. Mahanil, S.; Attajarusit, J.; Stout, M.J.; Thipyapong, P. Overexpression of tomato polyphenol oxidase increases resistance to common cutworm. *Plant Sci.* **2008**, *174*, 456–466. [CrossRef]

19. Haruta, M.; Pedersen, J.A.; Constabel, C.P. Polyphenol oxidase and herbivore defense in trembling aspen (*Populus tremuloides*): cDNA cloning, expression, and potential substrates. *Physiol. Plant.* **2001**, *112*, 552–558. [CrossRef] [PubMed]

20. Webb, K.J.; Cookson, A.; Allison, G.; Sullivan, M.L.; Winters, A.L. Gene expression patterns, localization, and substrates of polyphenol oxidase in red clover (*Trifolium pratense* L.). *J. Agric. Food Chem.* **2013**, *61*, 7421–7430. [CrossRef] [PubMed]

21. Zhu, X.Z.; Qiao-Ying, M.A.; Zhang, S.; Lv, L.M.; Luo, J.Y.; Wang, C.Y.; Cui, J.J. Cloning of a polyphenol oxidase gene (*GhPPO1*) of *Gossypium hirsutum* and its role in cotton after *Helicoverpa armigera* feeding. *Sci. Agric. Sin.* **2014**, *47*, 3174–3183.

22. Li, C.; Li, D.; Li, J.; Shao, F.; Lu, S. Characterization of the polyphenol oxidase gene family reveals a novel microRNA involved in posttranscriptional regulation of PPOs in *Salvia miltiorrhiza*. *Sci. Rep.* **2017**, *7*, 44622. [CrossRef] [PubMed]

23. Liang, X.; Chen, Q.; Lu, H.; Wu, C.; Lu, F.; Tang, J. Increased activities of peroxidase and polyphenol oxidase enhance cassava resistance to *Tetranychus urticae*. *Exp. Appl. Acarol.* **2017**, *71*, 195–209. [CrossRef] [PubMed]

24. Thipyapong, P.; Stout, M.J.; Attajarusit, J. Functional analysis of polyphenol oxidases by antisense/sense technology. *Molecules* **2007**, *12*, 1569–1595. [CrossRef] [PubMed]

25. Winters, A.; Heywood, S.; Farrar, K.; Donnison, I.; Thomas, A.; Webb, K.J. Identification of an extensive gene cluster among a family of PPOs in *Trifolium pratense* L. (red clover) using a large insert BAC library. *BMC Plant Biol.* **2009**, *9*, 94. [CrossRef] [PubMed]

26. Tran, L.T.; Taylor, J.S.; Constabel, C.P. The polyphenol oxidase gene family in land plants: Lineage-specific duplication and expansion. *BMC Genom.* **2012**, *13*, 395. [CrossRef] [PubMed]

27. Newman, S.M.; Eannetta, N.T.; Yu, H.; Prince, J.P.; de Vicente, M.C.; Tanksley, S.D.; Steffens, J.C. Organisation of the tomato polyphenol oxidase gene family. *Plant Mol. Biol.* **1993**, *21*, 1035–1051. [CrossRef] [PubMed]

28. Zhao, D.; Liu, Z.S.; Biao, X.I. Cloning and alignment of polyphenol oxidase cDNA of tea plant. *J. Tea Sci.* **2001**, *2*, 94–98.

29. Liu, S.; Han, B. Differential expression pattern of an acidic 9/13-lipoxygenase in flower opening and senescence and in leaf response to phloem feeders in the tea plant. *BMC Plant Biol.* **2010**, *10*, 228. [CrossRef] [PubMed]

30. Mu, D.; Cui, L.; Ge, J.; Wang, M.X.; Liu, L.F.; Yu, X.P.; Zhang, Q.H.; Han, B.Y. Behavioral responses for evaluating the attractiveness of specific tea shoot volatiles to the tea green leafhopper, *Empoaca vitis*. *Insect Sci.* **2012**, *19*, 229–238. [CrossRef]

31. Chen, Y. Biological control progress of *Ectropis obliqua* prout. *Nat. Enemies Insects* **2001**, *23*, 181–184.

32. Chung, S.H.; Rosa, C.; Scully, E.D.; Peiffer, M.; Tooker, J.F.; Hoover, K.; Luthe, D.S.; Felton, G.W. Herbivore exploits orally secreted bacteria to suppress plant defenses. *Proc. Natl. Acad. Sci. USA* **2013**, *110*, 15728–15733. [CrossRef] [PubMed]

33. Wang, J.; Chung, S.H.; Peiffer, M.; Rosa, C.; Hoover, K.; Zeng, R.; Felton, G.W. Herbivore oral secreted bacteria trigger distinct defense responses in preferred and non-preferred host plants. *J. Chem. Ecol.* **2016**, *42*, 463–474. [CrossRef] [PubMed]

34. Baldwin, I.T.; Halitschke, R.; Kessler, A.; Schittko, U. Merging molecular and ecological approaches in plant-insect interactions. *Curr. Opin. Plant Biol.* **2001**, *4*, 351–358. [CrossRef]

35. Musser, R.O.; Farmer, E.; Peiffer, M.; Williams, S.A.; Felton, G.W. Ablation of caterpillar labial salivary glands: Technique for determining the role of saliva in insect–plant interactions. *J. Chem. Ecol.* **2006**, *32*, 981–992. [CrossRef] [PubMed]

36. Musser, R.O.; Hum-Musser, S.M.; Lee, H.K.; DesRochers, B.L.; Williams, S.A.; Vogel, H. Caterpillar labial saliva alters tomato plant gene expression. *J. Chem. Ecol.* **2012**, *38*, 1387–1401. [CrossRef] [PubMed]

37. Suzuki, H.; Dowd, P.F.; Johnson, E.T.; Hum-Musser, S.M.; Musser, R.O. Effects of elevated peroxidase levels and corn earworm feeding on gene expression in tomato. *J. Chem. Ecol.* **2012**, *38*, 1247–1263. [CrossRef] [PubMed]

38. Tram, N.N. Change of polyphenol oxidase activity during oolong tea process. *J. Food Nutr. Sci.* **2015**, *3*, 88–93. [CrossRef]

39. Teng, J.; Gong, Z.; Deng, Y.; Chen, L.; Li, Q.; Shao, Y.; Lin, L.; Xiao, W. Purification, characterization and enzymatic synthesis of theaflavins of polyphenol oxidase isozymes from tea leaf (*Camellia sinensis*). *LWT Food Sci. Technol.* **2017**, *84*, 263–270. [CrossRef]

40. Halder, J.; Tamuli, P.; Bhaduri, A.N. Isolation and characterization of polyphenol oxidase from Indian tea leaf (*Camellia sinensis*). *J. Nutr. Biochem.* **1998**, *9*, 75–80. [CrossRef]

41. Wahler, D.; Gronover, C.S.; Richter, C.; Foucu, F.; Twyman, R.M.; Moerschbacher, B.M.; Fischer, R.; Muth, J.; Prüfer, D. Polyphenol oxidase silencing affects latex coagulation in taraxacum species. *Plant Physiol.* **2009**, *151*, 334–346. [CrossRef] [PubMed]

42. Quarta, A.; Mita, G.; Durante, M.; Arlorio, M.; De Paolis, A. Isolation of a polyphenol oxidase (*PPO*) cDNA from artichoke and expression analysis in wounded artichoke heads. *Plant Physiol. Biochem.* **2013**, *68*, 52–60. [CrossRef] [PubMed]

43. Li, D.; Deng, Z.; Liu, C.; Zhao, M.; Guo, H.; Xia, Z.; Liu, H. Molecular cloning, expression profiles, and characterization of a novel polyphenol oxidase (*PPO*) gene in *Hevea brasiliensis*. *Biosci. Biotechnol. Biochem.* **2014**, *78*, 1648–1655. [CrossRef] [PubMed]

44. Taranto, F.; Pasqualone, A.; Mangini, G.; Tripodi, P.; Miazzi, M.M.; Pavan, S.; Montemurro, C. Polyphenol oxidases in crops: Biochemical, physiological and genetic aspects. *Int. J. Mol. Sci.* **2017**, *18*, 377. [CrossRef] [PubMed]

45. Jia, H.; Zhao, P.; Wang, B.; Tariq, P.; Zhao, F.; Zhao, M.; Wang, Q.; Yang, T.; Fang, J. Overexpression of polyphenol oxidase gene in strawberry fruit delays the fungus infection process. *Plant Mol. Boil. Report.* **2016**, *34*, 592–606. [CrossRef]

46. Nakayama, T.; Sato, T.; Fukui, Y.; Yonekura-Sakakibara, K.; Hayashi, H.; Tanaka, Y.; Kusumi, T.; Nishino, T. Specificity analysis and mechanism of aurone synthesis catalyzed by aureusidin synthase, a polyphenol oxidase homolog responsible for flower coloration. *FEBS Lett.* **2001**, *499*, 107–111. [CrossRef]

47. Shetty, S.M.; Chandrashekar, A.; Venkatesh, Y.P. Eggplant polyphenol oxidase multigene family: Cloning, phylogeny, expression analyses and immunolocalization in response to wounding. *Phytochemistry* **2011**, *72*, 2275–2287. [CrossRef] [PubMed]

48. Tran, L.T.; Constabel, C.P. The polyphenol oxidase gene family in poplar: Phylogeny, differential expression and identification of a novel, vacuolar isoform. *Planta* **2011**, *234*, 799. [CrossRef] [PubMed]

49. Cho, M.H.; Moinuddin, S.G.; Helms, G.L.; Hishiyama, S.; Eichinger, D.; Davin, L.B.; Lewis, N.G. (+)-Larreatricin hydroxylase, an enantio-specific polyphenol oxidase from the creosote bush (*Larrea tridentata*). *Proc. Natl. Acad. Sci. USA* **2003**, *100*, 10641–10646. [CrossRef] [PubMed]

50. Schmidt, D.D.; Voelckel, C.; Hartl, M.; Schmidt, S.; Baldwin, I.T. Specificity in ecological interactions. Attack from the same lepidopteran herbivore results in species-specific transcriptional responses in two solanaceous host plants. *Plant Physiol.* **2005**, *138*, 1763–1773. [CrossRef] [PubMed]

51. Wu, Y.L.; Pan, L.P.; Yu, S.L.; Li, H.H. Cloning, microbial expression and structure-activity relationship of polyphenol oxidases from *Camellia sinensis*. *J. Biotechnol.* **2010**, *145*, 66–72. [CrossRef] [PubMed]

52. Wang, J.; Constabel, C.P. Three polyphenol oxidases from hybrid poplar are differentially expressed during development and after wounding and elicitor treatment. *Physiol. Plant.* **2004**, *122*, 344–353. [CrossRef]

53. Yu, X.; Zhou, J.; Chen, M.L.; Ou Yang, Y.; Luo, X.Q. Research on the anti-inflammatory and analgesic actions of tea plant root. *Res. Pract. Chin. Med.* **2012**, *4*, 39–41.

54. Castañera, P.; Steffens, J.; Tingey, W. Biological performance of colorado potato beetle larvae on potato genotypes with differing levels of polyphenol oxidase. *J. Chem. Ecol.* **1996**, *22*, 91–101. [CrossRef] [PubMed]

55. Constabel, C.P.; Bergey, D.R.; Ryan, C.A. Polyphenol oxidase as a component of the inducible defense response in tomato against herbivores. *Recent Adv. Phytochem.* **1996**, *30*, 231–252.

56. He, J.; Chen, F.; Chen, S.; Lv, G.; Deng, Y.; Fang, W.; Liu, Z.; Guan, Z.; He, C. Chrysanthemum leaf epidermal surface morphology and antioxidant and defense enzyme activity in response to aphid infestation. *J. Plant Physiol.* **2011**, *168*, 687–693. [CrossRef] [PubMed]

57. Miranda, M.; Ralph, S.G.; Mellway, R.; White, R.; Heath, M.C.; Bohlmann, J.; Constabel, C.P. The transcriptional response of hybrid poplar (*Populus trichocarpa* × *P. deltoids* to infection by melampsora medusae leaf rust involves induction of flavonoid pathway genes leading to the accumulation of proanthocyanidins. *Mol. Plant Microbe Interact.* **2007**, *20*, 816–831.

58. Boss, P.K.; Gardner, R.C.; Janssen, B.-J.; Ross, G.S. An apple polyphenol oxidase cdna is up-regulated in wounded tissues. *Plant Mol. Biol.* **1995**, *27*, 429–433. [CrossRef] [PubMed]

59. Stewart, R.J.; Sawyer, B.J.; Bucheli, C.S.; Robinson, S.P. Polyphenol oxidase is induced by chilling and wounding in pineapple. *Funct. Plant Biol.* **2001**, *28*, 181–191. [CrossRef]

60. Thipyapong, P.; Steffens, J.C. Differential expression and turnover of the tomato polyphenol oxidase gene family during vegetative and reproductive development. *Plant Physiol.* **1997**, *113*, 707–718. [CrossRef] [PubMed]

61. Wang, Y.N.; Tang, L.; Hou, Y.; Wang, P.; Yang, H.; Wei, C.L. Differential transcriptome analysis of leaves of tea plant (*Camellia sinensis*) provides comprehensive insights into the defense responses to *Ectropis obliqua* attack using RNA-seq. *Funct. Integr. Genom.* **2016**, *16*, 383–393. [CrossRef] [PubMed]

62. Musser, R.O.; Hummusser, S.M.; Eichenseer, H.; Peiffer, M.; Ervin, G.; Murphy, J.B.; Felton, G.W. Herbivory: Caterpillar saliva beats plant defences. *Nature* **2002**, *416*, 599–600. [CrossRef] [PubMed]

63. Erb, M.; Glauser, G.; Robert, C.A. Induced immunity against belowground insect herbivores-activation of defenses in the absence of a jasmonate burst. *J. Chem. Ecol.* **2012**, *38*, 629–640. [CrossRef] [PubMed]

64. Wu, J.; Baldwin, I.T. New insights into plant responses to the attack from insect herbivores. *Annu. Rev. Genet.* **2010**, *44*, 1–24. [CrossRef] [PubMed]

65. Ray, S.; Gaffor, I.; Acevedo, F.E.; Helms, A.; Chuang, W.P.; Tooker, J.; Felton, G.W.; Luthe, D.S. Maize plants recognize herbivore-associated cues from caterpillar frass. *J. Chem. Ecol.* **2015**, *41*, 781–792. [CrossRef] [PubMed]

66. Lou, Y.; Baldwin, I.T. Nitrogen supply influences herbivore-induced direct and indirect defenses and transcriptional responses in *Nicotiana attenuata*. *Plant Physiol.* **2004**, *135*, 496–506. [CrossRef] [PubMed]

International Journal of
*Molecular Sciences*

MDPI

*Article*

# Overexpression of *OsGID1* Enhances the Resistance of Rice to the Brown Planthopper *Nilaparvata lugens*

Lin Chen [†], Tiantian Cao [†], Jin Zhang and Yonggen Lou *

State Key Laboratory of Rice Biology & Ministry of Agriculture Key Lab of Agricultural Entomology,
Institute of Insect Sciences, Zhejiang University, Hangzhou 310058, China; chenlin88@126.com (L.C.);
caotiantiande@sina.cn (T.C.); zhangjin1369@tricaas.com (J.Z.)
* Correspondence: yglou@zju.edu.cn; Tel.: +86-0571-8898-2622
† These authors contributed equally to this work.

Received: 18 July 2018; Accepted: 6 September 2018; Published: 13 September 2018

**Abstract:** Gibberellins (GAs) play pivotal roles in plant growth and development, and in defenses against pathogens. Thus far, how the GA-mediated signaling pathway regulates plant defenses against herbivores remains largely unknown. In this study, we cloned the rice GA receptor gene *OsGID1*, whose expression was induced by damage from the brown planthopper (BPH) *Niaparvata lugens*, mechanical wounding, and treatment with salicylic acid (SA), but not jasmonic acid. The overexpression of *OsGID1* (oe-GID1) decreased BPH-induced levels of SA, $H_2O_2$, and three SA-pathway-related WRKY transcripts, but enhanced BPH-induced levels of ethylene. Bioassays in the laboratory revealed that gravid BPH females preferred to feed and lay eggs on wild type (WT) plants than on oe-GID1 plants. Moreover, the hatching rate of BPH eggs on oe-GID1 plants was significantly lower than that on WT plants. In the field, population densities of BPH adults and nymphs were consistently and significantly lower on oe-*OsGID1* plants than on WT plants. The increased resistance in oe-GID1 plants was probably due to the increased lignin level mediated by the GA pathway, and to the decrease in the expression of the three WRKY genes. Our findings illustrated that the *OsGID1*-mediated GA pathway plays a positive role in mediating the resistance of rice to BPH.

**Keywords:** rice; *OsGID1*; gibberellin; herbivore-induced plant defenses; *Nilaparvata lugens*

---

## 1. Introduction

In natural ecosystems, the evolutionary arms race between plants and insects has been ongoing for more than 400 million years, and along the way plants have evolved diverse strategies to protect themselves [1–3]. When facing herbivore attack, plants identify the type of damage caused by herbivores and then trigger a signaling network comprising calcium signaling, mitogen-activated protein kinase (MAPK) cascades, and signaling pathways mediated by phytohormones, such as jasmonic acid (JA), salicylic acid (SA), and ethylene (ET) [4–6]. Once activated, this signaling network elicits plants' production of defense compounds and enhances their resistance to herbivores. Recent research advances have highlighted how some growth-related phytohormones, such as auxins, brassinosteroids, cytokinins, and gibberellins (GAs), also regulate plant defenses [7].

GAs are a series of tetracyclic diterpenoid phytohormones, which play crucial roles in numerous processes underlying plant growth and development, including cell expansion and division, seed germination, and floral development [8]. By combining genetics, molecular biology, and plant physiology, key components acting in the regulation of GA biosynthesis, signal perception, and transduction have been identified [9], such as the GA receptor GID1 (GA-INSENSITIVE DWARF1) [10,11], DELLA proteins [12], and the F-box protein GID2 (GA-INSENSITIVE DWARF2) in rice [13]. Moreover, the GA–GID1–DELLA regulatory module, which controls GA signaling is well established [14]. In this module, GA signaling starts by binding bioactive GAs to the GA receptor GID1 to form a GA-GID1 complex [10,15]. Together with E3 ubiquitin-ligase $SCF^{SLY1/GID2}$, the complex then stimulates the rapid degradation of DELLA proteins, the master repressors of the GA-signaling pathway, and activate the GA pathway [14,16]. As a key component of the gibberellin-signaling pathway, GA receptors have been cloned in many higher plant species and characterized using genetic approaches [17]. In rice, there is only one GA receptor gene, *OsGID1*, and it encodes a soluble hormone-sensitive lipase-like protein, whereas in *Arabidopsis*, there are three GA receptor genes, *AtGID1a*, *AtGID1b*, and *AtGID1c*, and these genes function redundantly [15,18].

In addition to its central role in regulating plant growth and development, GA-mediated signaling has also been reported to play a pivotal role in plant defenses against pathogens by directly or indirectly interacting with defense-related signaling pathways. In *Arabidopsis*, for instance, DELLAs were found to negatively mediate SA biosynthesis and signaling; such regulation enhances the plant's resistance to necrotrophs and susceptibility to biotrophs [19]. In rice, the DELLA protein Slender Rice1 (OsSLR1) promotes defenses mediated by both SA and JA, and these enhance rice resistance to hemibiotrophic pathogens [20]. In addition, Tanaka et al. [21] reported that *gid1*, a gibberellin-insensitive dwarf mutant, showed increased resistance to blast fungus due to high levels of a probenazole-inducible protein (PBZ1). In contrast, less attention has been paid to the role of the GA-mediated signaling pathway in herbivore-induced plant defenses. Lan et al. [22] found that in *Arabidopsis*, DELLA proteins optimize and fine-tune plant defense and growth by suppressing plant defense hormones in the labial salivary secretions of caterpillars. In rice, the GA-mediated pathway has been found to positively regulate resistance to the brown planthopper (BPH) *Nilaparvata lugens* (Stål) [23]. Moreover, Zhang et al. [24] reported that silencing the DELLA gene *OsSLR1* enhances the resistance of rice to BPH in the laboratory and field by impairing JA and ethylene pathways, and by enhancing the levels of lignin. These new findings shed light on the role of GA-mediated signaling in plant defenses against herbivores. However, how the GA-mediated pathway regulates herbivore-induced defenses remains largely unknown.

Rice, one of the most important food crops in the world and eaten by more than half of the world's population, suffers severely from many insect pests, including the phloem-sucking insect, brown planthopper (BPH) *Nilaparvata lugens* [25]. Previous studies of rice have revealed that the infestation of BPH gravid females changes the biosynthesis of many defense-related signals, such as JA, JA-Ile, SA, ET, and $H_2O_2$; in turn, these compounds modulate defense responses, for example, by influencing the production of herbivore-induced volatiles and the activity of TrypPIs [26–28]. Given the role of GA-mediated signaling in plant defenses, we cloned the rice GA receptor gene *OsGID1* (TIGR ID Os05g33730) and characterized its role in herbivore-induced defenses in rice. By combining molecular biology, reverse genetics, chemical analysis, and bioassays, we found that the overexpression of *OsGID1* decreases the BPH-induced accumulation of three WRKY transcription factor (TF) mRNAs, and of SA and $H_2O_2$, but increases ethylene levels. Moreover, the overexpression of *OsGID1* enhances the resistance of rice to BPH, suggesting the *OsGID1*-mediated GA signaling pathway also plays a role in resistance.

## 2. Results

### 2.1. Isolation and Expression Patterns of OsGID1

Based on the cDNA sequence of *OsGID1* in the MSU Rice Genome Annotation Project database, we cloned the *OsGID1* gene from a cDNA library of rice variety XS110 using PCR. The *OsGID1* gene contains an open reading frame of 1065 nucleotides, of which it consists of two exons and one intron (Supplemental Figure S1) and encodes a soluble GA receptor of 354 amino acids, which is mainly localized in the nucleus [10]. There are four single nucleotide polymorphisms (SNPs) in the coding sequence (CDS) of *OsGID1*, two of which are missense substitutions and the others are synonymous nucleotide substitutions (Supplemental Figure S1). To analyze the evolutionary history of OsGID1 and its orthologs from higher plant species, a phylogenetic tree was constructed using neighbor-joining analysis. According to the phylogenetic tree, OsGID1 shows the highest homology with GID1s from four monocotyledonous species: ZmGID1 in *Zea mays*, SiGID1 in *Setaria italic*, SbGID1 in *Sorghum bicolor*, and SoGID1 in *Saccharum officinarum*, sharing 80.79%, 84.46%, 82.91%, and 81.97% amino acid sequence identity with OsGID1 (Supplemental Figure S2).

Relative expression levels of *OsGID1* in response to different treatments were analyzed by quantitative real-time PCR (qRT-PCR). The expression of *OsGID1* was induced by gravid BPH females, at 24 and 48 h after infestation (Figure 1a); after mechanical wounding, expression levels increased rapidly and peaked at 4 h (Figure 1b). JA treatment had almost no influence on levels of *OsGID1* transcripts, whereas at some time points SA treatment significantly enhanced the expression of *OsGID1* (Figure 1c,d). These data indicate that *OsGID1* might mediate BPH-induced defenses in rice.

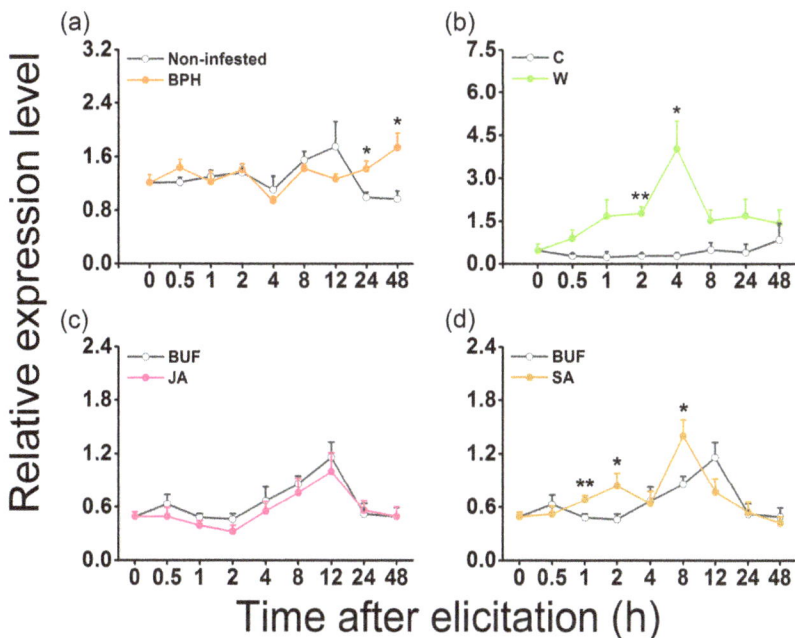

**Figure 1.** Quantitative real-time PCR (qRT-PCR) analysis of *OsGID1* transcript levels in rice after different treatments. Relative levels (means + SE, $n = 5$) of *OsGID1* transcripts in rice plants that were infested by gravid brown planthopper (BPH) females (**a**) or treated with mechanically wounding (**b**), jasmonic acid (JA) (**c**) or salicylic acid (SA) (**d**). Asterisks indicate significant differences between treatments and controls (* $p < 0.05$, ** $p < 0.01$, Student's *t*-test).

### 2.2. Overexpression of OsGID1

To further investigate the function of *OsGID1* in herbivore-induced plant defenses, we generated rice plants with overexpressed *OsGID1* (oe-GID1), under the control of the cauliflower mosaic virus 35S promoter. Two T2 homozygous lines with a single insertion (Supplemental Figure S3), G1 and G11, were selected for further investigation. qRT-PCR analysis showed that levels of *OsGID1* transcripts in plants of the two oe-GID1 lines, G1 and G11, were 81.3–89.7 fold and 34.9–23.7 fold higher than those levels in wild type (WT) plants 0 and 4 h after mechanical wounding (Figure 2a). Similar to the results reported by Ueguchi-Tanaka et al. [10], transgenic plants from the lines expressing *OsGID1* exhibited a GA overdose phenotype: oe-GID1 plants were taller and had longer leaves, compared to WT plants; moreover, in both 10- and 38-day-old seedlings of oe-GID1 plants, the second leaf sheath was significantly longer than the same sheath in WT plants (Figure 2b–d). At 38 days, the mass of the above-ground part of oe-GID1 lines, G1 and G11, increased by approximately 15% and 24%, respectively, compared to that of WT plants (Supplemental Figure S4).

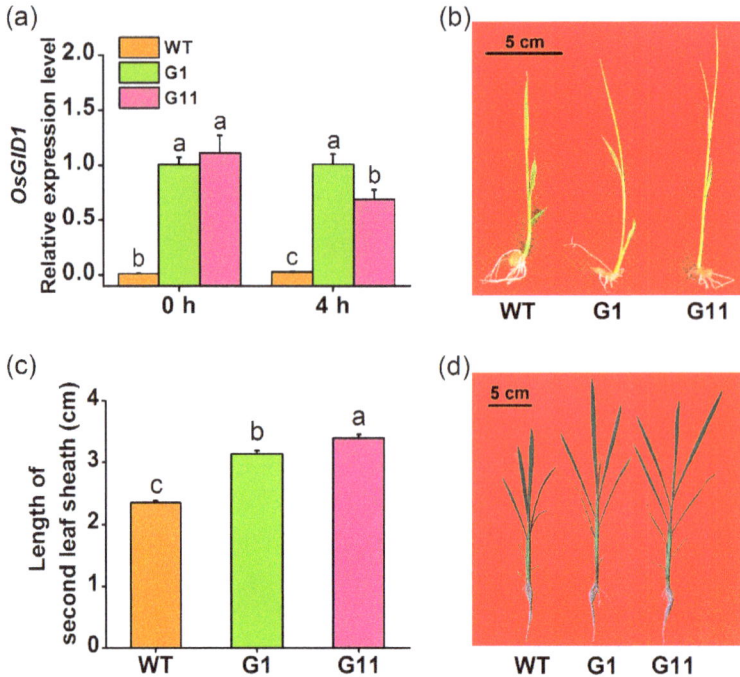

**Figure 2.** Relative expression levels of *OsGID1* in wild type (WT) plants, and in plants from lines expressing *OsGID1* and their growth phenotypes. (**a**) Relative levels (means + SE, $n = 5$) of *OsGID1* transcripts in oe-GID1 and WT plants 0 and 4 h after they were wounded (these two experiments (two time points) were performed separately); (**b,d**) Growth phenotypes of oe-GID1 and WT plants at day 10 and day 38 in the greenhouse, bar = 5 cm; (**c**) Mean length (+ SE, $n = 30$) of the second leaf sheath of 10-day-old oe-GID1 and WT plants. Different letters indicate significant differences between oe-GID1 lines and WT plants (Duncan's multiple range test, $p < 0.05$).

## 2.3. Overexpressing OsGID1 Enhances the Resistance of Rice to BPH

Previous studies have shown that GA signaling is involved in plant defense [7,24]; therefore, we examined whether the overexpression of *OsGID1* affects the resistance of rice to BPH. In a two-choice experiment, gravid BPH females preferred to feed and lay eggs on WT plants than on oe-GID1 plants (Figure 3a,b). Moreover, the hatching rate of BPH eggs on WT plants was dramatically higher than on oe-GID1 plants (Figure 3c). Finally, the developmental duration of BPH eggs on oe-GID1 plants showed longer than that on WT plants, although the difference between WT plants and G1 plants was not significant (Figure 3d). These results indicated that the overexpression of *OsGID1* enhanced the resistance of rice to BPH.

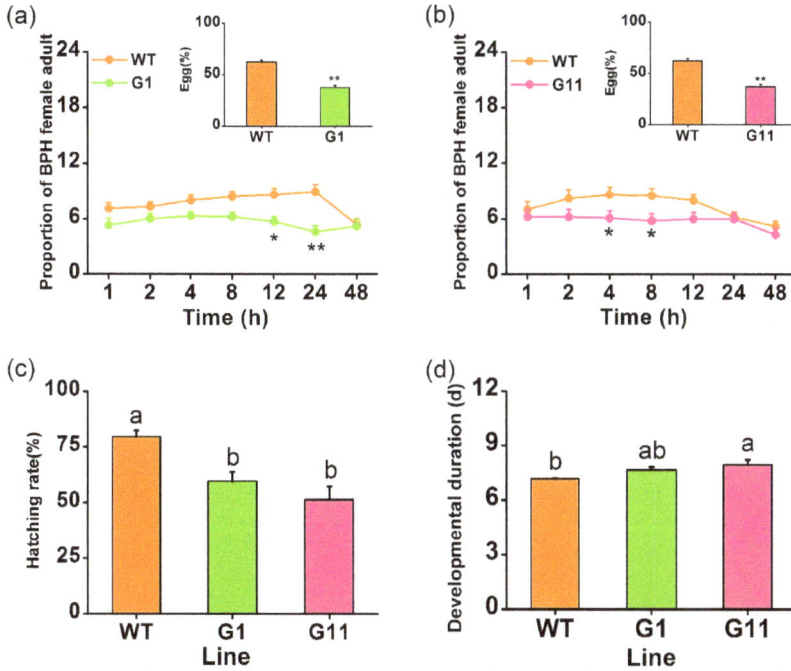

**Figure 3.** *OsGID1* positively mediates rice resistance to BPH. (**a,b**) Mean number of gravid BPH females per plant (+SE, $n = 10$) on pairs of plants (WT versus G1 or G11) in two-choice assays, 1–48 h after pairs were exposed to the herbivore. Inserts: Mean percentage of BPH eggs per plant (+SE, $n = 10$) on pairs of plants as stated above, 48 h after the release of BPH. (**c**) Mean hatching rate (+SE, $n = 10$) of BPH eggs on WT and oe-GID1 plants; (**d**) Mean developmental duration (+SE, $n = 10$) of BPH eggs on WT and oe-GID1 plants. Asterisks indicate significant differences in oe-GID1 plants, compared with WT plants (* $p < 0.05$, ** $p < 0.01$, chi-square test); different letters indicate significant differences between oe-GID1 lines and WT plants (Duncan's multiple range test, * $p < 0.05$).

### 2.4. Overexpression of OsGID1 Suppressed the Expression of Three WRKY Transcription Factors

Three WRKY TFs—*OsWRKY13*, *OsWRKY30*, *OsWRKY33*—have been reported to be involved in the SA-mediated signaling pathway, and thus regulate rice defense [29–31]. Hence, we monitored mRNA levels of these TFs in plants of oe-GID1 lines, and in WT plants after infestation with gravid BPH females. The transcript levels of *OsWRKY13* did not differ between WT and oe-GID1 plants at 0 h and 0.5 h post BPH infestation. However, from 1 h to 12 h post BPH infestation, the transcript level of *OsWRKY13* was significantly lower in oe-GID1 plants than in WT plants (Figure 4a). Similarly, mRNA levels of *OsWRKY30* (Figure 4b) and *OsWRKY33* (Figure 4c) in oe-GID1 plants decreased from 0.5 to 12 h and from 1 to 6 h, respectively, post BPH infestation. These data suggest that *OsGID1* negatively regulated the transcript levels of the three WRKY genes.

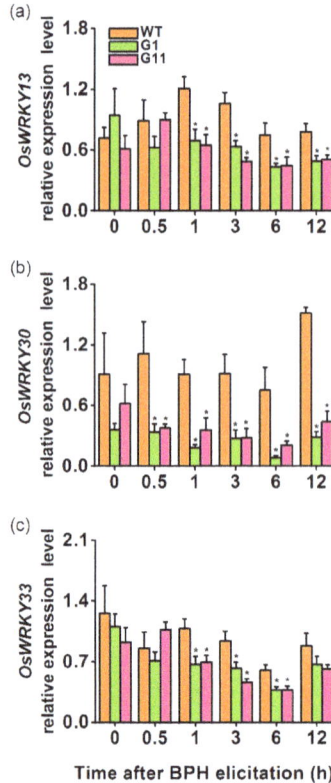

**Figure 4.** Overexpression of *OsGID1* decreases transcript levels of three WRKY TF genes. Relative levels (means + SE, *n* = 4–5) of *OsWRKY13* (**a**), *OsWRKY30* (**b**), and *OsWRKY33* (**c**) transcripts in oe-GID1 and WT plants after infestation by gravid BPH females. Asterisks indicate significant differences in oe-GID1 plants, compared with WT plants (Duncan's multiple range test; * *p* < 0.05).

## 2.5. Overexpression of OsGID1 Alters Accumulation of BPH-induced $H_2O_2$, SA and Ethylene

JA, JA-Ile, SA, $H_2O_2$, and ethylene have been reported to play important roles in plant defenses against herbivores [32–34]. Moreover, as reported previously, these signals have been implicated in defense responses of rice to BPH [35–37]. To test if JA, JA-Ile, SA, $H_2O_2$, or ethylene were responsible for *OsGID1*-mediated rice resistance to BPH, we investigated their levels in oe-GID1 and in WT plants, after infestation by gravid BPH females. Although JA levels accumulated over time following infestation, differences between oe-GID1 and WT plants were slight: Only at 0 h post BPH infestation, JA levels were higher in G11 plants than in WT plants (Figure 5a). Similarly, small differences were observed in JA-Ile: JA-Ile levels in oe-GID1 lines at 24 h after BPH infestation were significantly lower than those in WT plants, whereas, levels of JA-Ile in G1 plants at 0 h and 3 h post BPH infestation were higher than those in WT plants (Figure 5b). The level of $H_2O_2$ in oe-GID1 plants was generally lower than that in WT plants at different time points after infestation (Figure 5c). At 3 h post BPH infestation, SA levels in oe-GID1 plants started to decrease, and this decline continued for 48 h (Figure 5d). Transcript levels of the gene encoding isochorismate synthase (OsICS1), a key enzyme in SA biosynthesis in rice, were consistently lower in plants from G1 and G11 lines, compared to those in WT plants (Figure 5e). In contrast, elicited and constitutive levels of ethylene were higher in plants from G1 and G11 lines than in WT plants (Figure 5f). As oe-GID1 plants were bigger than WT plants (Supplemental Figure S4), levels of ethylene emitted from per g biomass (above-ground parts) of oe-GID1 and WT plants, showed no significant difference (Figure 5f, insert). Consistent with this, relative expression levels of 1-aminocyclopropane-1-carboxylic acid synthase 2 (*OsACS2*), an ethylene biosynthesis-related gene, did not differ over time between oe-GID1 and WT plants, following BPH infestation (Figure 5g).

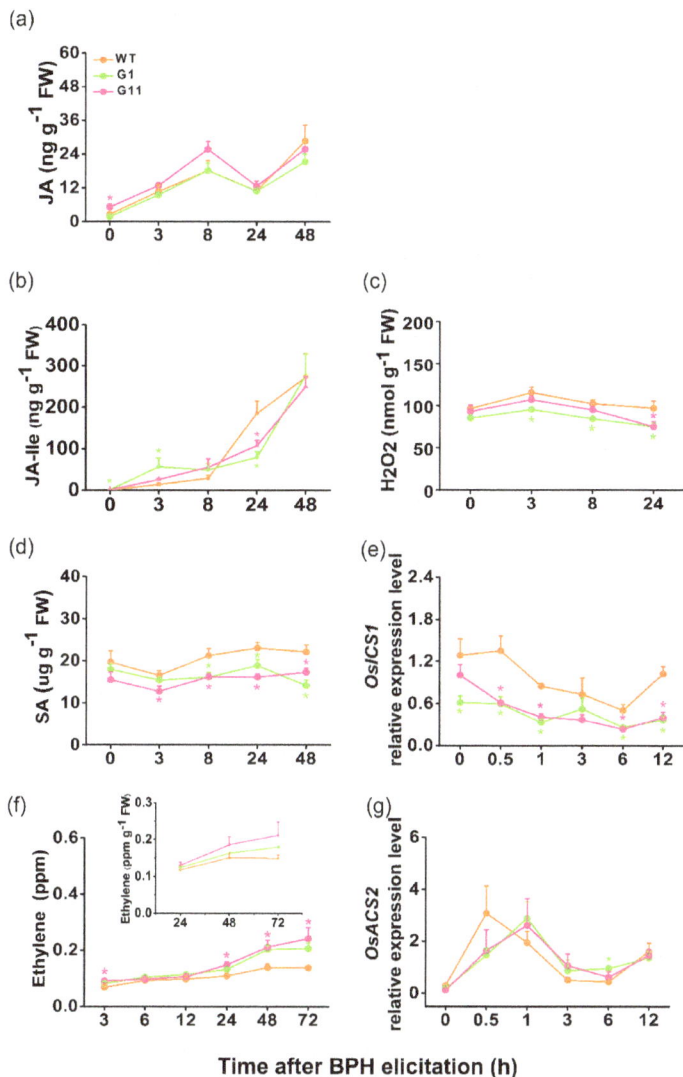

**Figure 5.** *OsGID1* regulates the accumulation of BPH-induced JA, JA-Ile, $H_2O_2$, SA, and ethylene. (a–d,f) Levels (means + SE, $n$ = 4–8) of JA (a), JA-Ile (b), $H_2O_2$ (c), SA (d), and ethylene (f, insert: ethylene levels per g biomass) in/from oe-GID1 and WT plants at different times after they were infested by 15 gravid BPH females; (e,g) Relative levels (means + SE, $n$ = 5) of *OsICS1* (e) and *OsACS2* (g) in oe-GID1 and WT plants at different times after they were infested by 15 gravid BPH females. Asterisks indicate significant differences in oe-GID1 plants, compared with WT plants (Duncan's multiple range test; * $p < 0.05$; ** $p < 0.01$).

### 2.6. Overexpression of OsGID1 Decreases the Population Density of Rice Planthoppers and Spiders in the Field

Since plants with overexpression of *OsGID1* showed enhanced rice resistance to BPH in the laboratory, we questioned whether the overexpression of *OsGID1* also affected the population density of rice planthoppers, and their natural enemies in the field. To answer this question, a field survey of the population dynamics of BPH, white-backed planthopper (WBPH) *Sogatella furcifera*, and predatory spiders was conducted. The population density of BPH adults and nymphs was significantly lower on oe-GID1 plants, compared to those on WT plants (Figure 6a,b). The population density of WBPH adults and nymphs was also lower on oe-GID1 plants than on WT plants; especially on 16 August, the population density of WBPH adults and nymphs on oe-GID1 plants was significantly lower than the population density of WBPH adults and nymphs on WT plants (Figure 6c,d). In the field, the dominant predatory spider species were *Pirata subpiraticus*, *Misumenops tricuspidatus*, *Pardosa pseudoannulata*, and *Tetragnatha maxillosa*. There were fewer spiders on oe-GID1 plants than on WT plants, especially on 23 August, 6 September, and 3 October (Figure 6e).

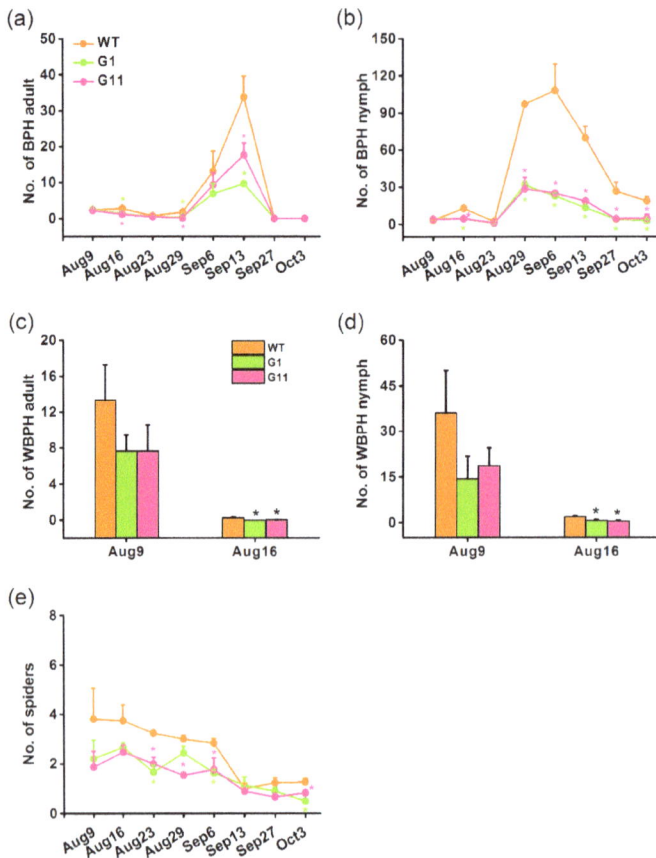

**Figure 6.** Population dynamics of brown planthopper (BPH), white-backed planthopper (WBPH), and spiders on oe-GID and WT plants in the field. Mean number (+SE, $n = 3$) of BPH adults (**a**), BPH nymphs (**b**), WBPH adults (**c**), WBPH nymphs (**d**), and spiders (**e**) per hill on G1, G11, and WT plants. Asterisks indicate significant differences in oe-GID1 plants, compared with WT plants (Duncan's multiple range test; * $p < 0.05$).

## 2.7. Overexpression of OsGID1 Reduces Rice Yield in the Field

We also investigated the effect of overexpressing OsGID1 on rice yield in the field. Consistent with the growth phenotype reported in Ueguchi-Tanaka et al. [10], oe-GID1 plants showed taller and longer leaves, and fewer tillers in the field, compared with WT plants (Figure 7a). Plants with overexpression of *OsGID1* had far fewer panicles per hill than WT plants (Figure 7b). Although the seed setting rate was similar between WT plants and G11 plants, statistically, it was significantly lower in G1 plants (Figure 7c). The mean weight of mature seeds per WT plant and the 1000-seed weight from WT plants, were clearly heavier than those from oe-GID1 plants (Figure 7d,e). Taken together, these results suggest that the overexpression of *OsGID1* reduces rice yield.

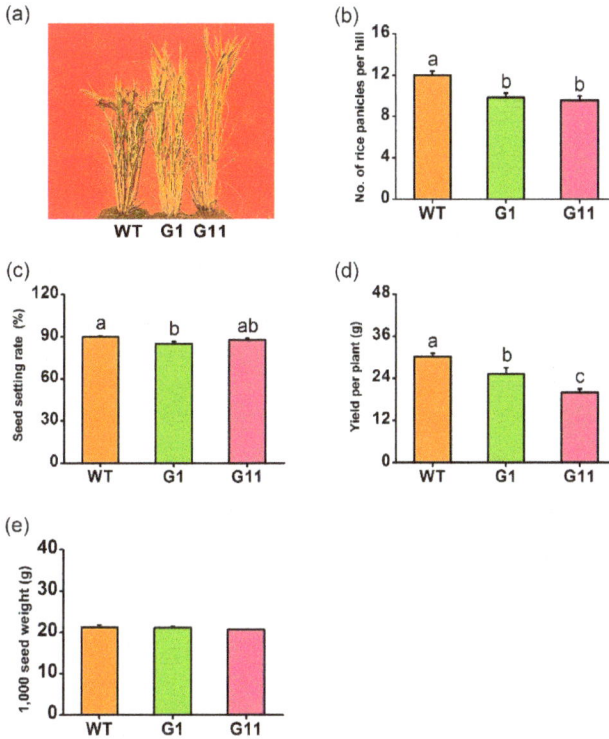

**Figure 7.** Overexpression of *OsGID1* reduces rice yield in the field. (**a**) Growth genotype of oe-GID1 and WT plants in the field; (**b**) Mean number (+SE, $n = 30$) of panicles per plant of oe-GID1 and WT plants; (**c**) Mean seed setting rate (+SE, $n = 30$) of oe-GID1 and WT plants; (**d**) Mean yield of mature seeds per plant (+SE, $n = 30$) of oe-GID1 and WT plants; (**e**) Mean weight (+SE, $n = 30$) of 1000 seeds of oe-GID1 and WT plants. Different letters indicate significant differences between oe-GID1 lines and WT plants (Duncan's multiple range test, * $p < 0.05$).

## 3. Discussion

We found that *OsGID1*, in addition to its pivotal role in plant growth and development, also plays an important role in rice defenses. Several lines of evidence support this statement. First, transcript levels of *OsGID1* were induced by BPH infestation, wounding, and SA treatment but not by JA treatment (Figure 1). Second, *OsGID1*-overexpression repressed BPH-elicited levels of three WRKY gene transcripts, SA, and $H_2O_2$, but had a small or no effect on constitutive and elicited levels of ethylene, JA, and JA-Ile (Figures 4 and 5). Since defenses in rice can be induced only by BPH oviposition and not by BPH feeding [38], with no difference found in the number of eggs laid by gravid BPH females on oe-GID1 and WT plants for 12 h (Supplemental Figure S5), the difference in induced defenses stated above between oe-GID1 lines and WT plants should have resulted from the modulation of OsGID1, not from the difference in damage levels. Third, oe-GID1 lines exhibited a GA overdose phenotype and had higher constitutive levels of lignin, a GA responsive marker [24], than WT plants (SupplementalFigure S6), suggesting that the overexpression of *OsGID1* activates the GA-mediated pathway. Fourth, the overexpression of *OsGID1* enhances the resistance of rice to BPH in the laboratory and field (Figures 3 and 6). These findings suggest that *OsGID1*-mediated GA signaling, as we previously reported in References [23,24], positively regulates rice resistance to BPH.

The single GA receptor in rice [10] shows high homology with GID1s from other Poaceae species (Supplemental Figure S2). It has been reported that drought, submergence, and NaCl treatment can rapidly and strongly induce the expression of *OsGID1* [39]. In this study, we observed that infestation by gravid BPH females and wounding, could also induce the expression of *OsGID1*. These data demonstrate that *OsGID1*-mediated GA signaling is influenced by both abiotic and biotic stresses. The trade-off between GA-mediated plant growth and JA- or SA-mediated plant defense has been well documented [7,40,41]. We found that the exogenous application of JA to rice plants had almost no effect on the transcript level of *OsGID1*, and SA treatment had only a weak effect (Figure 1). These findings indicate that the inhibition of plant growth by JA- or SA-mediated plant defenses, may not occur by suppressing transcript levels of the GA receptor gene *OsGID1*.

The GA-signaling pathway is known to influence early events in plant defenses [7,34,42]. For instance, DELLAs can interact directly with the JA ZIM-domain 1 (JAZ1) protein, a key repressor of JA signaling, and EIN3, and thus influence defense responses mediated by JA and ethylene [41,43]. DELLAs can also suppress the accumulation of reactive oxygen species (ROS) and the action of SA, but rather to strengthen JA action in *Arabidopsis* [19,44]. Recently, the DELLA protein in rice OsSLR1 has been observed to enhance transcriptional levels of two herbivore-induced MAPK genes, *OsMPK3* and *OsMPK20-5*, and four WRKY genes, *OsWRKY53*, *OsWRKY70*, *OsWRKY45*, and *OsWRKY24*, as well as the biosynthesis or action of elicited JA, SA, ethylene, and $H_2O_2$ [24,42]. These OsSLR1 actions were thought to occur both directly, by interacting with components of these pathways, and indirectly, through cross-talk among pathways [24]. We did not investigate the effect of the overexpression of *OsGID1* on transcript levels of the MAPK and WRKY genes described above, but we did find that the overexpression of *OsGID1* decreased BPH-induced mRNA levels of three WRKY genes, *OsWRKY13*, *OsWRKY30*, and *OsWRKY33*, as well as induced levels of SA and $H_2O_2$, but increased levels of BPH-elicited ethylene (Figures 4 and 5). Given the close connection between OsGID1, a GA receptor, and OsSLR1, a GA pathway repressor, it is reasonable to think that the same mechanisms may be used to regulate the GA-signaling pathways in plants of both OsGID1 and OsSLR1. Further research should elucidate whether OsGID1 also regulates defense-related MAPK cascades and other WRKYs.

In rice, SA- and $H_2O_2$-mediated signaling pathways positively regulate the resistance of rice to BPH [26,28]; whereas, the ethylene- and JA-mediated signaling pathway negatively regulates rice resistance [27]. However, in this study, we found that the overexpression of *OsGID1* reduced the level of BPH-induced SA and $H_2O_2$ but had a small or no effect on elicited levels of ethylene, JA, and JA-Ile. Thus, the enhanced resistance in oe-GID1 plants to BPH is probably due to factors other than the SA-, $H_2O_2$-, JA- and ethylene-mediated pathways. One of these factors might be lignin, whose levels in plants are positively regulated by the GA pathway [24,45]. Lignin not only strengthens the mechanical hardness of plant tissues and thereby impairs the feeding of herbivores, but it also decreases the digestibility of the food herbivores consume [46]. In rice, it has been reported that silencing *OsSLR1* enhances lignin levels in plants, which then increases the resistance of rice to BPH [24]. Here, we also found that the constitutive level of lignin was higher in oe-GID1 lines than in WT plant (Supplemental Figure S6). The enhanced resistance in plants to BPH may also be due to WRKY TFs, which are modulated by OsGID1, as these have been reported to both positively and negatively regulate plant defenses against herbivores [23,35,47,48]. For example, OsWRKY70 was found to positively modulate the resistance of rice to the striped stem borer (SSB) *Chilo suppressalis* [23], whilst OsWRKY53 negatively modulated rice resistance to SSB [47]. Moreover, silencing *OsWRKY45*, a SA-responsive transcription factor, increased the resistance of rice to BPH, including the decrease in the hatching rate of BPH eggs, suggesting the SA-signaling pathway may negatively affect rice BPH resistance [35]. We observed that BPH-induced mRNA levels of three WRKY genes, *OsWRKY13*, *OsWRKY30*, and *OsWRKY33*, were significantly lower in oe-GID1 plants than in WT plants (Figure 3). These WRKYs are all involved in the activation of the SA-signaling pathway, and in increased plant resistance to pathogens in rice [29–31]. *OsWRKY13*, for example, functions as a positive regulator of the SA pathway and a negative regulator of the JA pathway by directly or indirectly modulating the transcription of a subset of genes, which act both upstream and downstream of SA and JA [31]. OsWRKY30, which positively regulates disease resistance in rice, is an SA-inducible gene and can up-regulate *OsWRKY45* [29]. Thus, the decrease in transcript levels of these three WRKY genes may also contribute to the increase in the resistance of oe-GID1 plants to BPH. Further research should investigate the roles of lignin and these WRKYs in the resistance of rice to BPH.

In the laboratory and in the field, oe-GID1 plants showed taller, longer leaves, fewer tillers, and poor yield, compared with WT plants (Figure 7). These observations are consistent with the growth phenotype observed in plants with high levels of GA [10], confirming that the GA-signaling pathway plays a pivotal role in plant growth, development, and reproduction. Interestingly, in the field, we also observed that the population density of spiders on oe-GID1 plants was lower than on WT plants (Figure 7). This difference is probably because most BPH were found on WT plants, and thus a high prey population density will support a high predator population density.

In summary, our data show that the *OsGID*-mediated GA signaling pathway is an important positive regulator of rice resistance to BPH. Our previous results found that infestation by gravid BPH females suppresses the expression of *OsSLR1*, and thus activates the GA-mediated signaling pathway, which in turn induces the production of lignin, and enhances the resistance of rice to BPH [24]. We observed that infestation also activated the GA-signaling pathway by inducing the expression of the GA receptor gene *OsGID1* (Figure 1). Hence, by recognizing signals associated with gravid BPH females, rice can activate the GA pathway in multiple ways, thereby increasing its resistance. Interestingly, compared to mechanical wounding, gravid BPH female infestation induced the expression of OsGID1, weakly and slowly (Figure 1a). This suggested that BPH may suppress the resistance of rice by inhibiting the activation of GA pathway, an interesting question worth studying in the future. Given that the GA-signaling pathway has a negative effect on rice yield, how to balance GA levels so that rice has both a high yield and sufficient resistance to herbivores is also an important topic for future study.

## 4. Materials and Methods

### 4.1. Plant and Insects

The rice genotypes used in this study were Xiushui 110 (XS110; wild type (WT)) (*Oryza sativa* L.) and plants overexpressing *OsGID1*, G1 and G11 (see below). The seedlings of plants from transgenic lines and WT plants were transplanted 10 days after pre-germination to 30 L hydroponic boxes containing a rice nutrient solution [49]. After 4 weeks, seedlings were transferred to 500 mL hydroponic plastic pots, each of which contained one or two plants. Rice plants were used for the experiments 4–5 days after transplanting. The brown planthopper (BPH) *Nilaparvata lugens* colony was originally collected from a rice field in Hangzhou, China, and maintained in growth chambers as previously described in Reference [23].

### 4.2. Structure of OsGID1 and Phylogenetic Analysis

The structure of *OsGID1* was analyzed using a Gene Structure Display Server (GSDS2.0) (http://gsds.cbi.pku.edu.cn). The SNPs in the CDS of *OsGID1* were analyzed after submitting its accession number to the National Center of Biotechnology Information (NCBI, https://www.ncbi.nlm.nih.gov/SNP/). The protein sequences of OsGID1, and its homologs from different plant species were downloaded from the NCBI (http://www.ncbi.nlm.nih.gov). Sequences were then aligned by ClustalW method, using the default parameters in MEGA X, as described in Reference [50]. The neighbor-joining method was used to construct the phylogenetic tree using bootstrap analysis with 1000 replicates through MEGA X [51].

### 4.3. Generating Transgenic Plants

The full-length coding sequence of *OsGID1* (TIGR ID Os05g33730) was PCR-amplified using a pair of primers, GID1-F (5′-GTCGACAATCTCCTCCCTTCTCGA-3′) and GID1-R (5′-ACTAGTCTAGTAGTAGAGGTTA-3′), from a cDNA library of rice variety XS110. The sequence-confirmed products were cloned into the binary vector pCAMBIA1301, yielding an overexpression transformation vector pCAMBIA1301-*OsGID1*. The expressing vector was introduced into rice variety XS110 via *Agrobacterium tumefaciens*-mediated transformation. Screening of homozygous T2 plants and Southern blotting were performed as described in Reference [28]. Two homozygous *GID1*-overexpressing lines of the T2 generation with a single insertion (G1 and G11) were used for further experiments.

### 4.4. RNA Extraction and Quantitative Real-time PCR

For gene expression analysis, five independent biological samples were used. Total RNA was extracted from 100 mg of rice stem tissues using SV Total RNA Isolation System (Promega, Madison, WI, USA), and RNA was reverse-transcribed using a PrimeScript™ RT reagent Kit (Perfect Real Time) (TaKaRa, Dalian, China), following the manufacturer's protocols. Quantitative real-time PCR (qRT-PCR) was carried out using a CFX96TM Real-Time System (Bio-Rad, Hercules, CA, USA) with Premix Ex Taq™ Kit (TaKaRa, Dalian, China). A linear standard curve, threshold cycle number versus log (designated transcript level), was built using a series concentrations of a specific cDNA standard. Relative levels of the transcript of the target gene in tested samples were calculated according to the standard curve. A rice actin gene *OsACT* (TIGR ID: Os03g50885) was used as an internal standard to normalize cDNA concentrations. Specific primers and probe sequences used for qRT-PCR are provided in Table S1.

### 4.5. Plant Treatments

For mechanical wounding treatments, the lower sections of rice stems were pierced 200 times with a fine needle. Control plants were left untreated. For BPH treatment, individual plants were infested with 15 gravid BPH females, enclosed in a glass cylinder (4 × 8 cm, with 48 small holes, 0.8 mm in diameter). An identical glass cylinder without BPH was enclosed for control plants (non-infested). For JA and SA treatments, rice plants were sprayed individually with 2 mL of JA (100 µg mL$^{-1}$) or SA (70 µg mL$^{-1}$) in 50 mM sodium phosphate buffer. Rice plants sprayed with 2 mL of the buffer were used as controls. Five replications for each treatment, at each time point, were carried out.

### 4.6. JA, JA-Ile, SA and Ethylene Measurements

Under JA, JA-Ile, and SA analysis, oe-GID1 and WT plants were randomly assigned to BPH and control (non-infestation) treatments. The leaf sheaths of each plant were harvested at 0 h, 3 h, 8 h, 24 h, and 48 h, post BPH infestation. Samples were mixed with ethyl acetate containing labelled internal standards, and then analyzed using a HPLC/mass spectrometry/mass spectrometry system to quantify JA, JA-Ile, and SA levels in each plant as previously described in Reference [24]. For ethylene analysis, plants from each line were individually covered with a sealed glass cylinder (4 cm in diameter and 50 cm in height), into which 15 gravid BPH females were released. Ethylene production was measured at 3 h, 6 h, 12 h, 24 h, 48 h, and 72 h post BPH infestation, following the same method as described in Lu et al. [36]. Experiments for each treatment, at each time point, were replicated five times.

### 4.7. Hydrogen Peroxide and Lignin Content Analysis

WT and oe-GID1 plants were randomly assigned to treatment with BPH infestation or non-infestation. The leaf sheaths of each plant were harvested at 0 h, 3 h, 8 h, and 24 h, post BPH infestation. Samples were stored at $-80$ °C until use. Hydrogen peroxide concentrations were measured using a Amplex® Red Hydrogen Peroxide/Peroxidase Assay Kit (Invitrogen, Eugene, OR, USA), following the manufacturer's protocol. The leaf sheaths of 4-week-old rice plants from oe-GID1 and WT plants were harvested, and lignin contents were determined, as described in Xu et al. [51]. Experiments for each treatment, at each time interval, were replicated five times.

### 4.8. BPH Performance Assay

To investigate the effect of oe-GID1 plants on the feeding and oviposition preference of gravid BPH females, stems of two plants, either a G1 or G11 plant and a WT plant, were covered with a glass cylinder (4 × 8 cm, with 48 small holes, 0.8 mm in diameter), into which 15 gravid BPH female adults were introduced. The number of BPH adults on each plant was counted at 1, 2, 4, 8, 24, and 48 h, after the insects had been released. Adults were removed 48 h after infestation, and the eggs laid on each plant were examined with a microscope. The experiment was replicated ten times.

To determine the influence of oe-GID1 plants on the hatching rate of BPH eggs, 15 gravid BPH females were allowed to oviposit on individual plants of oe-GID1 and WT lines for 12 h. Every day newly hatched BPH nymphs on each plant were recorded, until no new nymphs were observed for three consecutive days. Unhatched eggs were examined with a microscope. Based on these data, the developmental duration and hatching rate of BPH eggs were calculated. Each line was replicated ten times.

### 4.9. Field Experiment

To assess the resistance of *OsGID1*-overexpressing plants and their corresponding WT plants to insects in a natural environment, a field experiment was performed in Changxing, Zhejiang, China. The field plot consisted of 9 blocks (6.5 m in length and 3.5 m in width), each of which was surrounded by a 0.5 m buffer zone. On the plot, plants from three lines were randomly assigned to 9 blocks; each line with three independent replicate blocks. The number of adults and nymphs of BPH and WBPH, as well as the number of spiders, the main predators of herbivores, were recorded once a week from June to October in 2014. To collect these data, we randomly sampled 10 hills of plants, in each plot, at each time interval.

To evaluate the effect of overexpressing *OsGID1* on rice yield in the field, 10 hills of mature plants per block were randomly harvested. After being sun dried, the productivity-related parameters—the number of rice panicles per hill, seed setting rate, yield per plant, and 1000-seed weight—were recorded in the laboratory. The seed setting rate was assessed as the ratio of mature seeds to the total number of grains (including abortive grains); yield per plant equaled to the total weight of mature seeds per plant; 1000-seed weight was calculated as follows: 1000-seed weight = total weight of mature seeds/total number of mature seeds × 1000.

### 4.10. Statistical Analysis

Differences in the relative expression levels of *OsGID1* in response to different treatments were determined using the Student's *t*-test. Differences in the feeding and oviposition preference of gravid BPH females were determined by the chi-square test. Differences in other data were analyzed by one-way ANOVA; if the ANOVA analysis was significant ($p < 0.05$), Duncan's multiple range test was used to detect significant differences among groups. When necessary, data were log-transformed or arcsine-transformed to meet requirements for the homogeneity of variance. All tests were carried out with Statistica (SAS Institute, Inc., Cary, NC, USA).

**Supplementary Materials:** Supplementary materials can be found at http://www.mdpi.com/1422-0067/19/9/2744/s1. Supplemental Figure S1: The structure of *OsGID1* and single nucleotide polymorphisms in its coding sequence; Supplemental Figure S2: Phylogenetic analysis of GID1 proteins from different plant species; Supplemental Figure S3: DNA gel-blot analysis of oe-GID1 lines and WT plants; Supplemental Figure S4: Mean mass (+SE, *n* = 20) of the above-ground part of 38-day-old plants of oe-GID1 and WT lines; Supplemental Figure S5: Mean number (+SE, *n* = 5) of eggs laid by 15 gravid BPH females on oe-GID1 and WT plants for 12 h. Supplemental Figure S6: Constitutive levels of Lignin in WT and oe-GID1 plants; Table S1: Primers and probes used for qRT-PCR of target genes.

**Author Contributions:** Y.L., T.C. and L.C. designed the experiments; T.C., L.C. and J.Z. performed the experiments; Y.L., L.C. and T.C. analyzed the data; Y.L. and L.C. prepared and wrote the paper.

**Funding:** This work was jointly supported by the Special Fund for Agro-scientific Research in the Public Interest (201403030), the National Natural Science Foundation of China (31330065), the National Program of Transgenic Variety Development of China (2016ZX08001-001), the earmarked fund for China Agriculture Research System (CARS-01-40).

**Acknowledgments:** We gratefully thank Guilan Dong and Lianggen Xie for their assistance with plant growth and insect rearing. We thank Emily Wheeler for editorial assistance.

**Conflicts of Interest:** The authors declare no conflict of interest.

## References

1. Furstenberg-Hagg, J.; Zagrobelny, M.; Bak, S. Plant defense against insect herbivores. *Int. J. Mol. Sci.* **2013**, *14*, 10242–10297. [CrossRef] [PubMed]
2. Howe, G.A.; Jander, G. Plant immunity to insect herbivores. *Annu. Rev. Plant Biol.* **2008**, *59*, 41–66. [CrossRef] [PubMed]
3. Wu, J.; Baldwin, I.T. New insights into plant responses to the attack from insect herbivores. *Annu. Rev. Genet.* **2010**, *44*, 1–24. [CrossRef] [PubMed]
4. Erb, M.; Meldau, S.; Howe, G.A. Role of phytohormones in insect-specific plant reactions. *Trends Plant Sci.* **2012**, *17*, 250–259. [CrossRef] [PubMed]
5. Pieterse, C.M.; Van der Does, D.; Zamioudis, C.; Leon-Reyes, A.; Van Wees, S.C. Hormonal modulation of plant immunity. *Annu. Rev. Cell Dev. Biol.* **2012**, *28*, 489–521. [CrossRef]
6. Schuman, M.C.; Baldwin, I.T. The layers of plant responses to insect herbivores. *Annu. Rev. Entomol.* **2016**, *61*, 373–394. [CrossRef]
7. De Bruyne, L.; Hofte, M.; De Vleesschauwer, D. Connecting growth and defense: the emerging roles of brassinosteroids and gibberellins in plant innate immunity. *Mol. Plant* **2014**, *7*, 943–959. [CrossRef]
8. Yamaguchi, S. Gibberellin metabolism and its regulation. *Annu. Rev. Plant Biol.* **2008**, *59*, 225–251. [CrossRef] [PubMed]
9. Daviere, J.M.; Achard, P. Gibberellin signaling in plants. *Development* **2013**, *140*, 1147–1151. [CrossRef] [PubMed]
10. Ueguchi-Tanaka, M.; Ashikari, M.; Nakajima, M.; Itoh, H.; Katoh, E.; Kobayashi, M.; Chow, T.Y.; Hsing, Y.I.; Kitano, H.; Yamaguchi, I.; et al. GIBBERELLIN INSENSITIVE DWARF1 encodes a soluble receptor for gibberellin. *Nature* **2005**, *437*, 693–698. [CrossRef] [PubMed]
11. Shimada, A.; Ueguchi-Tanaka, M.; Nakatsu, T.; Nakajima, M.; Naoe, Y.; Ohmiya, H.; Kato, H.; Matsuoka, M. Structural basis for gibberellin recognition by its receptor GID1. *Nature* **2008**, *456*, 520–523. [CrossRef] [PubMed]
12. Silverstone, A.L.; Ciampaglio, C.N.; Sun, T.P. The *Arabidopsis* RGA gene encodes a transcriptional regulator repressing the gibberellin signal transduction pathway. *Plant Cell* **1998**, *10*, 155–169. [CrossRef] [PubMed]
13. Sasaki, A.; Itoh, H.; Gomi, K.; Ueguchi-Tanaka, M.; Ishiyama, K.; Kobayashi, M.; Jeong, D.H.; An, G.; Kitano, H.; Ashikari, M.; et al. Accumulation of phosphorylated repressor for gibberellin signaling in an F-box mutant. *Science* **2003**, *299*, 1896–1898. [CrossRef] [PubMed]
14. Sun, T.P. The molecular mechanism and evolution of the GA-GID1-DELLA signaling module in plants. *Curr. Biol.* **2011**, *21*, R338–R345. [CrossRef] [PubMed]

15. Griffiths, J.; Murase, K.; Rieu, I.; Zentella, R.; Zhang, Z.L.; Powers, S.J.; Gong, F.; Phillips, A.L.; Hedden, P.; Sun, T.P.; et al. Genetic characterization and functional analysis of the GID1 gibberellin receptors in *Arabidopsis*. *Plant Cell* **2006**, *18*, 3399–3414. [CrossRef] [PubMed]

16. Hirano, K.; Ueguchi-Tanaka, M.; Matsuoka, M. GID1-mediated gibberellin signaling in plants. *Trends Plant Sci.* **2008**, *13*, 192–199. [CrossRef] [PubMed]

17. Ueguchi-Tanaka, M.; Matsuoka, M. The perception of gibberellins: clues from receptor structure. *Curr. Opin. Plant Biol.* **2010**, *13*, 503–508. [CrossRef] [PubMed]

18. Nakajima, M.; Shimada, A.; Takashi, Y.; Kim, Y.C.; Park, S.H.; Ueguchi-Tanaka, M.; Suzuki, H.; Katoh, E.; Iuchi, S.; Kobayashi, M.; et al. Identification and characterization of arabidopsis gibberellin receptors. *Plant J.* **2006**, *46*, 880–889. [CrossRef] [PubMed]

19. Navarro, L.; Bari, R.; Achard, P.; Lison, P.; Nemri, A.; Harberd, N.P.; Jones, J.D. DELLAs control plant immune responses by modulating the balance of jasmonic acid and salicylic acid signaling. *Curr. Biol.* **2008**, *18*, 650–655. [CrossRef] [PubMed]

20. De Vleesschauwer, D.; Seifi, H.S.; Filipe, O.; Haeck, A.; Huu, S.N.; Demeestere, K.; Hofte, M. The della protein slr1 integrates and amplifies salicylic acid—And jasmonic acid—Dependent innate immunity in rice. *Plant Physiol.* **2016**, *170*, 1831–1847. [CrossRef] [PubMed]

21. Tanaka, N.; Matsuoka, M.; Kitano, H.; Asano, T.; Kaku, H.; Komatsu, S. *gid1*, a gibberellin-insensitive dwarf mutant, shows altered regulation of probenazole-inducible protein (PBZ1) in response to cold stress and pathogen attack. *Plant Cell Environ.* **2006**, *29*, 619–631. [CrossRef] [PubMed]

22. Lan, Z.; Krosse, S.; Achard, P.; van Dam, N.M.; Bede, J.C. DELLA proteins modulate *Arabidopsis* defences induced in response to caterpillar herbivory. *J. Exp. Bot.* **2014**, *65*, 571–583. [CrossRef] [PubMed]

23. Li, R.; Zhang, J.; Li, J.; Zhou, G.; Wang, Q.; Bian, W.; Erb, M.; Lou, Y. Prioritizing plant defence over growth through WRKY regulation facilitates infestation by non-target herbivores. *eLife* **2015**, *4*, e04805. [CrossRef] [PubMed]

24. Zhang, J.; Luo, T.; Wang, W.; Cao, T.; Li, R.; Lou, Y. Silencing *OsSLR1* enhances the resistance of rice to the brown planthopper *Nilaparvata lugens*. *Plant Cell Environ.* **2017**, *40*, 2147–2159. [CrossRef] [PubMed]

25. Lou, Y.-G.; Zhang, G.-R.; Zhang, W.-Q.; Hu, Y.; Zhang, J. Biological control of rice insect pests in China. *Biol. Control* **2013**, *67*, 8–20. [CrossRef]

26. Lu, J.; Ju, H.; Zhou, G.; Zhu, C.; Erb, M.; Wang, X.; Wang, P.; Lou, Y. An EAR-motif-containing ERF transcription factor affects herbivore-induced signaling, defense and resistance in rice. *Plant J.* **2011**, *68*, 583–596. [CrossRef] [PubMed]

27. Lu, J.; Li, J.C.; Ju, H.P.; Liu, X.L.; Erb, M.; Wang, X.; Lou, Y.G. Contrasting effects of ethylene biosynthesis on induced plant resistance against a chewing and a piercing-sucking herbivore in rice. *Mol. Plant* **2014**, *7*, 1670–1682. [CrossRef] [PubMed]

28. Zhou, G.; Qi, J.; Ren, N.; Cheng, J.; Erb, M.; Mao, B.; Lou, Y. Silencing OsHI-LOX makes rice more susceptible to chewing herbivores, but enhances resistance to a phloem feeder. *Plant J.* **2009**, *60*, 638–648. [CrossRef] [PubMed]

29. Han, M.; Ryu, H.-S.; Kim, C.-Y.; Park, D.-S.; Ahn, Y.-K.; Jeon, J.-S. *OsWRKY30* is a transcription activator that enhances rice resistance to the *Xanthomonas oryzae* pathovar *oryzae*. *J. Plant Biol.* **2013**, *56*, 258–265. [CrossRef]

30. Koo, S.C.; Moon, B.C.; Kim, J.K.; Kim, C.Y.; Sung, S.J.; Kim, M.C.; Cho, M.J.; Cheong, Y.H. OsBWMK1 mediates SA-dependent defense responses by activating the transcription factor OsWRKY33. *Biochem. Biophys. Res. Commun.* **2009**, *387*, 365–370. [CrossRef] [PubMed]

31. Qiu, D.Y.; Xiao, J.; Ding, X.H.; Xiong, M.; Cai, M.; Cao, C.L.; Li, X.H.; Xu, C.G.; Wang, S.P. OsWRKY13 mediates rice disease resistance by regulating defense-related genes in salicylate- and jasmonate-dependent signaling. *Mol. Plant Microbe Interact.* **2007**, *20*, 492–499. [CrossRef] [PubMed]

32. Arimura, G.I.; Ozawa, R.; Maffei, M.E. Recent advances in plant early signaling in response to herbivory. *Int. J. Mol. Sci.* **2011**, *12*, 3723–3739. [CrossRef] [PubMed]

33. Lou, Y.G.; Baldwin, I.T. Silencing of a germin-like gene in *Nicotiana attenuata* improves performance of native herbivores. *Plant Physiol.* **2006**, *140*, 1126–1136. [CrossRef] [PubMed]

34. Yang, D.L.; Yang, Y.N.; He, Z.H. Roles of plant hormones and their interplay in rice immunity. *Mol. Plant* **2013**, *6*, 675–685. [CrossRef] [PubMed]

35. Huangfu, J.; Li, J.; Li, R.; Ye, M.; Kuai, P.; Zhang, T.; Lou, Y. The transcription factor OsWRKY45 negatively modulates the resistance of rice to the brown planthopper *Nilaparvata lugens*. *Int. J. Mol. Sci.* **2016**, *17*, 697. [CrossRef] [PubMed]

36. Lu, Y.; Wang, X.; Lou, Y.; Cheng, J. Role of ethylene signaling in the production of rice volatiles induced by the rice brown planthopper *Nilaparvata lugens*. *Chin. Sci. Bull.* **2006**, *51*, 2457–2465. [CrossRef]

37. Wang, X.; Zhou, G.; Xiang, C.; Du, M.; Cheng, J.; Liu, S.; Lou, Y. β-glucosidase treatment and infestation by the rice brown planthopper *Nilaparvata lugens* elicit similar signaling pathways in rice plants. *Chin. Sci. Bull.* **2008**, *53*, 53–57. [CrossRef]

38. Li, J.C. Comparison of Defense responses in Rice Induced by Feeding and Oviposition of the Brown Planthopper *Nilaparvata lugens* and Their Underlying Mechanisms. Ph.D. Thesis, Zhejiang University, Hangzhou, China, 2015.

39. Du, H.; Chang, Y.; Huang, F.; Xiong, L. GID1 modulates stomatal response and submergence tolerance involving abscisic acid and gibberellic acid signaling in rice. *J. Integr. Plant Biol.* **2015**, *57*, 954–968. [CrossRef] [PubMed]

40. Huot, B.; Yao, J.; Montgomery, B.L.; He, S.Y. Growth-defense tradeoffs in plants: A balancing act to optimize fitness. *Mol. Plant* **2014**, *7*, 1267–1287. [CrossRef] [PubMed]

41. Yang, D.L.; Yao, J.; Mei, C.S.; Tong, X.H.; Zeng, L.J.; Li, Q.; Xiao, L.T.; Sun, T.P.; Li, J.; Deng, X.W.; et al. Plant hormone jasmonate prioritizes defense over growth by interfering with gibberellin signaling cascade. *Proc. Natl. Acad. Sci. USA* **2012**, *109*, E1192–E1200. [CrossRef] [PubMed]

42. Daviere, J.M.; Achard, P. A pivotal role of DELLAs in regulating multiple hormone signals. *Mol. Plant* **2016**, *9*, 10–20. [CrossRef] [PubMed]

43. Hou, X.L.; Lee, L.Y.C.; Xia, K.F.; Yen, Y.Y.; Yu, H. DELLAs modulate jasmonate signaling via competitive binding to JAZs. *Dev. Cell* **2010**, *19*, 884–894. [CrossRef] [PubMed]

44. Achard, P.; Renou, J.P.; Berthome, R.; Harberd, N.P.; Genschik, P. Plant DELLAs restrain growth and promote survival of adversity by reducing the levels of reactive oxygen species. *Curr. Biol.* **2008**, *18*, 656–660. [CrossRef] [PubMed]

45. Okuno, A.; Hirano, K.; Asano, K.; Takase, W.; Masuda, R.; Morinaka, Y.; Ueguchi-Tanaka, M.; Kitano, H.; Matsuoka, M. New approach to increasing rice lodging resistance and biomass yield through the use of high gibberellin producing varieties. *PLoS ONE* **2014**, *9*, e86870. [CrossRef] [PubMed]

46. Santiago, R.; Barros-Rios, J.; Malvar, R.A. Impact of cell wall composition on maize resistance to pests and diseases. *Int. J. Mol. Sci.* **2013**, *14*, 6960–6980. [CrossRef] [PubMed]

47. Hu, L.F.; Ye, M.; Li, R.; Zhang, T.F.; Zhou, G.X.; Wang, Q.; Lu, J.; Lou, Y.G. The rice transcription factor WRKY53 suppresses herbivore-induced defenses by acting as a negative feedback modulator of mitogen-activated protein kinase activity. *Plant Physiol.* **2015**, *169*, 2907–2921. [PubMed]

48. Li, G.J.; Meng, X.Z.; Wang, R.G.; Mao, G.H.; Han, L.; Liu, Y.D.; Zhang, S.Q. Dual-level regulation of ACC synthase activity by MPK3/MPK6 cascade and its downstream WRKY transcription factor during ethylene induction in *Arabidopsis*. *PLoS Genet.* **2012**, *8*, e1002767. [CrossRef] [PubMed]

49. Yoshida, S.; Forno, D.A.; Cock, J.H. *Laboratory Manual for Physiological Studies of Rice*, 3rd ed.; International Rice Research Institute: Los Baños, Philippines, 1976.

50. Kumar, S.; Stecher, G.; Li, M.; Knyaz, C.; Tamura, K. MEGA X: Molecular evolutionary genetics analysis across computing platforms. *Mol. Biol. Evol.* **2018**, *35*, 1547–1549. [CrossRef] [PubMed]

51. Xu, Q.; Yin, X.R.; Zeng, J.K.; Ge, H.; Song, M.; Xu, C.J.; Li, X.; Ferguson, B.I.; Chen, K.S. Activator- and repressor-type MYB transcription factors are involved in chilling injury induced flesh lignification in loquat via their interactionswith the phenylpropanoid pathway. *J. Exp. Bot.* **2014**, *65*, 4349–4359. [CrossRef] [PubMed]

International Journal of
*Molecular Sciences*

MDPI

*Article*

# The Commonly Used Bactericide Bismerthiazol Promotes Rice Defenses against Herbivores

Pengyong Zhou [†], Xiaochang Mo [†], Wanwan Wang, Xia Chen and Yonggen Lou *

State Key Laboratory of Rice Biology & Ministry of Agriculture Key Lab of Molecular Biology of Crop Pathogens and Insects, Institute of Insect Sciences, Zhejiang University, Hangzhou 310058, China; zhoupy1991@163.com (P.Z.); camelotmo@zju.edu.cn (X.M.); wan456wan@126.com (W.W.); 21317040@zju.edu.cn (X.C.)
* Correspondence: yglou@zju.edu.cn; Tel.: +86-571-8898-2622
† These authors contributed equally to this work.

Received: 10 April 2018; Accepted: 20 April 2018; Published: 24 April 2018

**Abstract:** Chemical elicitors that enhance plant resistance to pathogens have been extensively studied, however, chemical elicitors that induce plant defenses against insect pests have received little attention. Here, we found that the exogenous application of a commonly used bactericide, bismerthiazol, on rice induced the biosynthesis of constitutive and/or elicited jasmonic acid (JA), jasmonoyl-isoleucine conjugate (JA-Ile), ethylene and $H_2O_2$ but not salicylic acid. These activated signaling pathways altered the volatile profile of rice plants. White-backed planthopper (WBPH, *Sogatella furcifera*) nymphs and gravid females showed a preference for feeding and/or oviposition on control plants: survival rates were better and more eggs were laid than on bismerthiazol-treated plants. Moreover, bismerthiazol treatment also increased both the parasitism rate of WBPH eggs laid on plants in the field by *Anagrus nilaparvatae*, and also the resistance of rice to the brown planthopper (BPH) *Nilaparvata lugens* and the striped stem borer (SSB) *Chilo suppressalis*. These findings suggest that the bactericide bismerthiazol can induce the direct and/or indirect resistance of rice to multiple insect pests, and so can be used as a broad-spectrum chemical elicitor.

**Keywords:** bismerthiazol; rice; induced defense responses; chemical elicitors; *Sogatella furcifera*; defense-related signaling pathways

## 1. Introduction

To protect themselves from damage by herbivores, plants have constitutive and induced defense systems. Constitutive defenses are defense traits that exist in plants whether or not herbivore infestation occurs. Induced defenses are defense traits that appear only when plants are exposed to herbivory [1,2]. Induced defenses in plants are activated by herbivore-associated signals and regulated by a complex signaling network, which mainly includes mitogen-activated protein kinase cascades and pathways mediated by jasmonic acid (JA), jasmonoyl-isoleucine (JA-Ile), salicylic acid (SA), ethylene (ET) and reactive oxygen species [3–6]. Regulation in plants leads to large changes in transcriptomes, proteomes and metabolomes; these changes may reduce the palatability of herbivores and increase their attractiveness to the natural enemies of herbivores, thereby increasing the direct and/or indirect resistance of plants to herbivores [3].

Induced plant defenses can also be triggered by applying chemicals that elicit plant resistance to herbivores or pathogens [7,8]. These chemicals are called chemical elicitors. Thus far, many potential chemical elicitors that induce the resistance of plants to pathogens have been reported, and several such elicitors—for example, benzo (1,2,3) thiadiazole-7-carbothioic acid *S*-methyl ester (BTH)—have been used commercially [9,10]. However, chemical elicitors that induce plant defenses against insect pests have received little attention. So far, JA and methyl ester of JA (MeJA) have been the most studied and are considered the most active chemical elicitors. They can activate diverse defensive compounds in plants, including polyphenol oxidase, proteinase inhibitors and volatiles [11–14], thereby leading to the direct and indirect resistance of plants to herbivores [1,15–17]. Recently, several JA analogues, such as coronalon, 6-ethyl indanoyl isoleucine conjugate, *cis*-jasmone JA-Ile and JA–amino acid conjugates, have been synthesized and found to induce herbivore resistance in plants [18–21]. Xin et al. [8] reported that the exogenous application of an herbicide, 2,4-dichlorophenoxyacetic acid (2,4-D), can induce the production of trypsin proteinase inhibitors (TrypPIs) and volatiles in rice, which in turn enhance the resistance of rice to the striped stem borer (SSB) *Chilosuppressalis* and the attractiveness of rice to brown planthopper (BPH) *Nilaparvata lugens* (Stål) and its egg parasitoid *Anagrus nilaparvatae*. In addition, several chemical elicitors, such as BTH, laminarin, tiadinil and *cis*-jasmone, have been reported to elicit indirect defenses in plants [22–25].

Bismerthiazol is a commonly used bactericide for the control of rice diseases caused by *Xanthomonas oryzae pv. Oryzae* (*Xoo*), such as bacterial rice leaf blight and bacterial leaf streak, and citrus canker caused by *X. campestris pv. Citri* [26–28]. In addition to having direct anti-microbial properties, bismerthiazol can also enhance $H_2O_2$ production, cellular defense responses and defense-related gene expression in *Xoo*-inoculated rice leaves [29].

Rice, one of the most important crops in the world, is infested by many insect pests, such as the white-backed planthopper (WBPH) *Sogatella furcifera*, BPH and SSB [30]. Previous studies with rice have shown that herbivore attack induces the biosynthesis of a variety of defense-related signals including JA, JA-Ile, SA, $H_2O_2$ and ET; these, in turn, regulate defense responses, such as the release of herbivore-induced volatiles, the accumulation of TrypPIs and peroxidase [4,8,31–36]. In this study, we found that bismerthiazol was not directly toxic to herbivores but enhanced levels of constitutive and herbivore-induced JA, JA-Ile, ET and $H_2O_2$, and these subsequently induced the resistance of rice to WBPH, BPH and SSB. Moreover, the exogenous application of bismerthiazol altered the volatile profile in rice and increased the attractiveness of plants to *A. nilarpavatae*, an egg parasitoid of rice planthoppers. The results demonstrate that bismerthiazol can act as a chemical elicitor to increase the resistance of rice to herbivores.

## 2. Results

### 2.1. Bismerthiazol Treatment Enhances the Direct Resistance of Rice to WBPH and Slightly Impairs Plant Growth

The survival rates of WBPH nymphs fed on plants that were grown in a nutrient solution with a concentration of 10, 20 or 50 mg $L^{-1}$ bismerthiazol decreased by 26.9%, 28.1% and 94%, respectively, relative to untreated plants (Figure 1a). Moreover, lower hatching rates of WBPH eggs and smaller numbers of WBPH eggs laid by gravid females over 12 h on bismerthiazol-treated plants compared with rates and numbers on control plants were observed (Figure 1b,c). In a choice experiment, both gravid females and nymphs preferred to feed on control plants rather than on bismerthiazol-treated plants (Figure 1d,e); and gravid WBPH females laid more eggs on control plants than on bismerthiazol-treated plants (Figure 1d). Like plants whose roots had been treated with bismerthiazol, plants whose above-ground parts had been sprayed with the compound also had enhanced resistance to WBPH: the survival rate of WBPH nymphs fed on rice plants that had been individually sprayed with 4 mL of 50 or 100 mg $L^{-1}$ bismerthiazol decreased by 12.94% and 49.41%, compared to the survival rate of nymphs fed on control plants (Figure 2). We found that the presence of bismerthiazol did not cause contact or stomach-poisoning toxicity in WBPH nymphs (Figure 3a,b). The data demonstrate that bismerthiazol probably affects the performance of WBPH via its induction of defense responses in rice.

We also examined the effect of bismerthiazol treatment on rice growth. Results showed that bismerthiazol has only a slight effect on rice growth: compared to plants growing in the nutrient solution without bismerthiazol, only 30-day-old plants growing in nutrient solution with a concentration of 50 mg $L^{-1}$ bismerthiazol for 10 days exhibited slightly shorter roots (Figure 3c,d).

No difference was observed in plant height, root mass and above-ground part mass between bismerthiazol-treated plants and control plants (Figure S1).

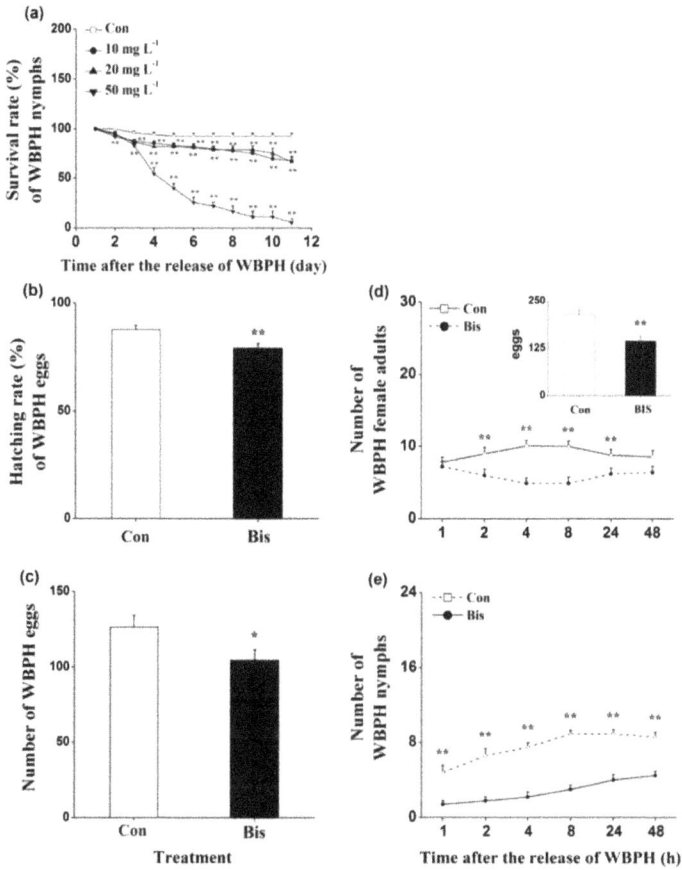

**Figure 1.** Bismerthiazol induces the direct resistance of rice to WBPH. (**a**) Mean survival rates (+ SE, $n = 6$) of 15 newly hatched WBPH nymphs fed on plants that had been grown in nutrient solution with either different concentrations of bismerthiazol (Bis) or without bismerthiazol (Con) for 24 h, 1–11 days after exposure; (**b**) Mean hatching rates (+SE, $n = 10$) of eggs laid over 12 h by 15 gravid WBPH females on plants that had been grown in nutrient solution with 50 mg $L^{-1}$ bismerthiazol (Bis) or without bismerthiazol (Con) for 24 h; (**c**) Mean numbers (+SE, $n = 10$) of eggs laid over 12 h by 15 female WBPH adults fed on plants that had been grown in nutrient solution with 50 mg $L^{-1}$ bismerthiazol (Bis) or without bismerthiazol (Con) for 24 h; (**d**) Mean number of gravid WBPH females (+SE, $n = 10$) fed on plants that had been grown in nutrient solution with 50 mg $L^{-1}$ bismerthiazol (Bis) or without bismerthiazol (Con) for 24 h, 1–48 h after exposure. Inset shows the mean percentage (+SE, $n = 10$) of WBPH eggs per plant on pairs of plants as stated above, 48 h after the release of WBPH; (**e**) Mean number of newly hatched WBPH nymphs (+SE, $n = 10$) on plants that had been grown in nutrient solution with 50 mg $L^{-1}$ bismerthiazol (Bis) or without bismerthiazol (Con) for 24 h, 1–48 h after exposure. For survival rate data, asterisks indicate significant differences in bismerthiazol-treated plants compared with non-treated plants (** $p < 0.01$; Duncan's multiple-range test); for other data, asterisks indicate significant differences between members of a pair (* $p < 0.05$; ** $p < 0.01$; Student's *t*-test).

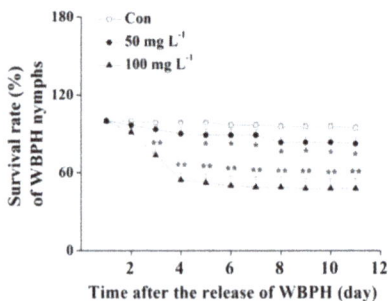

**Figure 2.** Spraying with bismerthiazol induces resistance of rice to WBPH. Mean survival rates (+SE, $n = 6$) of 15 newly hatched WBPH nymphs fed on plants that had been individually sprayed with 4 mL of 50 or 100 mg $L^{-1}$ bismerthiazol (Bis) or with buffer only (Con) for 24 h, 1–11 days after exposure. Asterisks indicate significant differences in bismerthiazol-treated plants compared with non-treated plants (* $p < 0.05$; ** $p < 0.01$; Duncan's multiple-range test).

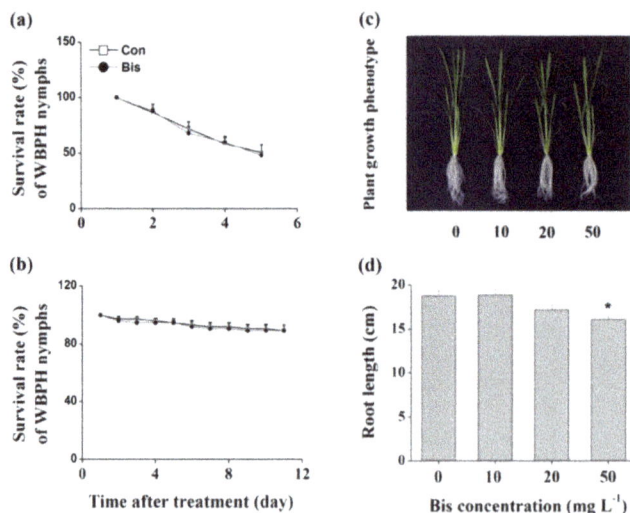

**Figure 3.** Bismerthiazol has no direct toxicity to WBPH but does slightly impair plant growth. (**a**) Mean survival rates (+SE, $n = 6$) of 15 newly hatched WBPH nymphs fed on artificial diets with 50 mg $L^{-1}$ of bismerthiazol (Bis) or without bismerthiazol (Con); (**b**) Mean survival rates (+SE, $n = 6$) of 15 newly hatched WBPH nymphs that were exposed to 50 mg $L^{-1}$ of bismerthiazol (Bis) or to buffer only (Con); (**c**) Growth phenotypes of plants grown in nutrient solutions with different concentrations of bismerthiazol for 10 days; (**d**) Mean root lengths (+ SE, $n = 6$) of plants grown in nutrient solutions with different concentrations of bismerthiazol for 10 days. Asterisks indicate significant differences between treatments and controls (* $p < 0.05$; Student's *t*-test).

*2.2. Bismerthiazol Induces the Biosynthesis of Constitutive and/or Elicited JA, JA-Ile, ET and $H_2O_2$, but Not SA*

JA, JA-Ile, SA, ET and $H_2O_2$ are known to play a pivotal role in mediating plant defense responses [5,37,38]. Therefore, we examined whether bismerthiazol treatment induced the biosynthesis of these signaling molecules in plants. Since feeding on plants whose roots were treated with 50 mg $L^{-1}$ bismerthiazol had the most lethal effect on WBPH nymphs, we used this concentration of bismerthiazol and root treatment to carry out experiments. Compared to non-infested plants, plants infested with gravid WBPH females displayed higher levels of JA, JA-Ile, ET and $H_2O_2$ (Figure 4), suggesting that exposure to infestation by gravid WBPH females could induce the production of all four of these signaling molecules. Bismerthiazol treatment alone significantly enhanced JA levels at 8 and 24 h after treatment, and the levels of JA in bismerthiazol-treated plants were 2.13-fold higher than those in control plants at 8 h (Figure 4a). Additionally, bismerthiazol treatment also enhanced JA levels in plants at 8 h after WBPH infestation (Figure 4a). Bismerthiazol treatment alone did not alter levels of JA-Ile; however, only when bismerthiazol-treated plants were exposed to infestation by gravid WBPH females did they display higher JA-Ile levels than those in control plants: JA-Ile levels in bismerthiazol-treated plants were approximately 2.63- and 2.25-fold higher than those in plants without bismerthiazol treatment at 3 and 8 h, respectively, following exposure to WBPH infestation (Figure 4b). Like the JA levels, the transcript levels of *OsAOS1* and *OsAOS2*, genes related to JA biosynthesis [39,40], increased at 8 h after treatment with bismerthiazol (Figure 4c,d). Bismerthiazol treatment only slightly enhanced SA levels in rice at 3 h after treatment and did not influence the production of SA in plants that had also been exposed to WBPH infestation plants (Figure S2).

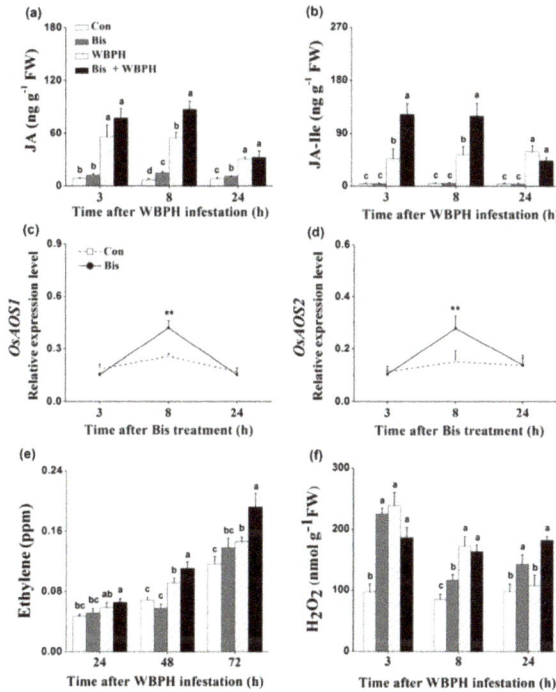

**Figure 4.** Bismerthiazol elicits JA, JA-Ile, ET and $H_2O_2$ biosynthesis. (**a,b**) Mean levels (+SE, $n = 5$) of JA (**a**) and JA-Ile (**b**) in leaf sheaths of rice plants that had been differently treated for 24 h, at 3, 8 and 24 h after being exposed to WBPH infestation. Con, control plants; Bis, bismerthiazol-treated plants; WBPH, WBPH-infested plants; Bis + WBPH, bismerthiazol + WBPH-treated plants. These treatment methods are described in Materials and Methods; (**c,d**) Mean relative expression levels (+SE, $n = 5$) of *OsAOS1* (**c**) and *OsAOS2* (**d**) in leaf sheaths of plants 3–24 h after plants were grown in nutrient solution with 50 mg $L^{-1}$ bismerthiazol (Bis) or not (Con); (**e**) Mean levels (+SE, $n = 10$) of ethylene released from plants that had received different treatments for 24 h, at 24, 48 and 72 h after exposure to WBPH infestation. Treatments are as stated above; (**f**) Mean levels (+SE, $n = 5$) of $H_2O_2$ in leaf sheaths of plants that had received different treatments for 24 h, at 3, 8 and 24 h after exposure to WBPH infestation. Treatments are as stated above. Letters indicate significant differences between different treatments ($p < 0.05$, Duncan's multiple-range test). Asterisks indicate significant differences between treatments and controls at each time point (** $p < 0.01$; Student's $t$-test).

As was observed in JA-Ile accumulation, bismerthiazol treatment alone did not induce the ET production in plants. Only when treated plants were exposed to WBPH infestation were higher ET levels in bismerthiazol-treated plants observed: levels from bismerthiazol-treated plants 48 and 72 h after exposure to infestation by WBPH were 121% and 132% of those in untreated plants that had been exposed to the infestation (Figure 4e). Bismerthiazol treatment alone enhanced $H_2O_2$ levels in plants: levels in plants treated with bismerthiazol at 3, 8 and 24 h, respectively, increased by 2.31-, 1.37- and 1.46-fold compared to those in control plants (Figure 4f). Twenty-four h after exposure to WBPH infestation, the $H_2O_2$ level in bismerthiazol-treated plants was 1.69-fold higher than the level in exposed plants without bismerthiazol treatment (Figure 4f). These findings suggest that bismerthiazol treatment was able to induce the production of constitutive and/or WBPH-elicited JA, JA-Ile, ET and $H_2O_2$ but not of SA.

### 2.3. Bismerthiazol Treatment Alters the Volatile Chemical Profile of Rice

As reported in previous results [8], non-manipulated rice plants released only a few volatile chemicals (Table 1). When treated with bismerthiazol, rice plants released more volatiles. The total amount of volatiles from bismerthiazol-treated plants was about 178% of the amount from control plants. Moreover, bismerthiazol treatment induced the production of 6 individual compounds—α-thujene, unknown 1, sesquithujene, (E)-α-bergamotene, α-curcumene and β-sesquiphellandrene—and increased levels of two chemicals, myrcene and methyl salicylate (MeSA) (Table 1, Figure S3). Exposure to infestation by gravid WBPH females drastically induced the production of volatiles in rice: the level of the total volatiles released from WBPH-infested plants was 14.6-fold higher than the level from non-infested plants. In addition, 11 chemicals that were not detected in non-infested plants—n-thujene, unknown 1, unknown 2, α-copaene, sesquithujene, (E)-α-bergamotene, sesquisabinene A, (E)-β-farnesene, α-curcumene, β-sesquiphellandrene and (E)-γ-bisabolene—were induced; and levels of 9 chemicals—2-heptanone, 2-heptanol, (+)-limonene, (E)-linalool oxide, linalool, MeSA, (E)-β-caryophyllene, zingiberene and β-bisabolene—were increased by exposure to WBPH infestation (Table 1). Interestingly, bismerthiazol treatment decreased the total amount of volatiles emitted from WBPH-infested plants, which was only 47.79% of the total amount of volatiles from plants exposed to WBPH infestation but not treatment. Furthermore, levels of 11 chemicals—2-heptanone, 2-heptanol, (E)-linalool oxide, linalool, unknown 1, sesquithujene, sesquisabinene A, zingiberene, β-bisabolene and β-sesquiphellandrene—in WBPH-infested plants were significantly decreased by bismerthiazol treatment (Table 1, Figure S3).

**Table 1.** Comparison of volatile compounds (mean ± SE, $n = 6$) emitted from control plants, bismerthiazol-treated plants, WBPH-infested plants and bismerthiazol + WBPH plants. These treatment methods were described in the section of Materials and Methods.

| No | Compound | Con | Bis | WBPH | Bis + WBPH |
|----|----------|-----|-----|------|------------|
| 1 | 2-Heptanone | 4.14 ± 1.59 [b] | 5.70 ± 1.69 [b] | 15.77 ± 1.23 [a] | 6.64 ± 0.85 [b] |
| 2 | 2-Heptanol | 0.49 ± 0.22 [b] | 0.50 ± 0.15 [b] | 4.13 ± 0.44 [a] | 0.59 ± 0.16 [b] |
| 3 | α-Thujene | - | 0.29 ± 0.17 [a] | 0.68 ± 0.17 [a] | 0.44 ± 0.04 [a] |
| 4 | α-Pinene | 0.53 ± 0.32 [a] | 0.94 ± 0.43 [a] | 0.81 ± 0.23 [a] | 0.71 ± 0.17 [a] |
| 5 | Myrcene | 1.20 ± 0.26 [b] | 2.12 ± 0.26 [a] | 1.72 ± 0.33 [ab] | 1.19 ± 0.26 [b] |
| 6 | (+)-Limonene | 0.86 ± 0.19 [b] | 1.69 ± 0.51 [b] | 11.95 ± 3.02 [a] | 10.70 ± 0.89 [a] |
| 7 | (E)-Linalool oxide | 0.14 ± 0.03 [c] | 0.32 ± 0.09 [c] | 2.98 ± 0.62 [a] | 0.93 ± 0.23 [b] |
| 8 | Linalool | 0.56 ± 0.07 [c] | 0.57 ± 0.23 [c] | 63.74 ± 15.76 [a] | 26.91 ± 4.48 [b] |
| 9 | Methyl salicylate | 0.15 ± 0.02 [d] | 0.41 ± 0.16 [c] | 6.05 ± 1.04 [a] | 2.06 ± 0.45 [b] |
| 10 | Unknown 1 | - | 1.02 ± 0.28 [a] | 1.34 ± 0.26 [a] | 0.54 ± 0.10 [b] |
| 11 | Unknown 2 | - | - | 1.28 ± 0.41 [a] | 0.33 ± 0.16 [a] |
| 12 | α-Copaene | - | - | 1.15 ± 0.37 [a] | 1.24 ± 0.04 [a] |

Table 1. *Cont.*

| No | Compound | Con | Bis | WBPH | Bis + WBPH |
|----|----------|-----|-----|------|------------|
| 13 | Sesquithujene | - | 0.37 ± 0.10 [b] | 1.05 ± 0.21 [a] | 0.50 ± 0.02 [b] |
| 14 | α-Cedrene | 0.32 ± 0.04 [a] | 0.59 ± 0.14 [a] | 0.75 ± 0.16 [a] | 0.55 ± 0.06 [a] |
| 15 | (E)-β-Caryophyllene | 0.65 ± 0.07 [b] | 1.11 ± 0.39 [b] | 2.41 ± 0.46 [a] | 1.67 ± 0.14 [ab] |
| 16 | (E)-α-Bergamotene | - | 0.20 ± 0.08 [b] | 3.35 ± 0.92 [a] | 1.67 ± 0.10 [a] |
| 17 | Sesquisabinene A | - | - | 3.30 ± 0.87 [a] | 1.71 ± 0.12 [b] |
| 18 | (E)-β-Farnesene | - | - | 2.91 ± 0.83 [a] | 1.30 ± 0.10 [a] |
| 19 | α-Curcumene | - | 0.94 ± 0.34 [a] | 2.79 ± 0.81 [a] | 1.33 ± 0.16 [a] |
| 20 | Zingiberene | 1.13 ± 0.27 [c] | 1.41 ± 0.27 [c] | 10.33 ± 2.74 [a] | 5.08 ± 0.48 [b] |
| 21 | β-Bisabolene | 0.40 ± 0.11 [c] | - | 4.51 ± 1.27 [a] | 1.95 ± 0.20 [b] |
| 22 | β-Sesquiphellandrene | - | 0.48 ± 0.07 [c] | 7.70 ± 2.19 [a] | 3.72 ± 0.34 [b] |
| 23 | (E)-γ-Bisabolene | - | - | 3.78 ± 1.09 [a] | 2.09 ± 0.25 [a] |
| Total | | 10.57 ± 3.17 [c] | 18.80 ± 5.36 [c] | 154.48 ± 35.44 [b] | 73.83 ± 9.78 [b] |

Letters in the same row indicate significant differences between treatments ($p < 0.05$, Duncan's multiple-range test).

## 2.4. Bismerthiazol Treatment Enhances the Indirect Resistance of Rice to WBPH

Because bismerthiazol treatment changed the volatile profile of rice (Table 1), and plant volatiles play an important role in the host/prey-searching behavior of natural enemies of herbivores [41], we wanted to explore the effects of the bactericide on egg parasitism in the field. We found that bismerthiazol treatment affected the parasitism rate of WBPH eggs: eggs laid on bismerthiazol-treated plants by *A. nilaparvatae* were parasitized at rates 2.3-fold higher than eggs laid on plants without bismerthiazol treatment (Figure 5). This result indicates that bismerthiazol treatment was also able to increase the indirect resistance of rice to WBPH.

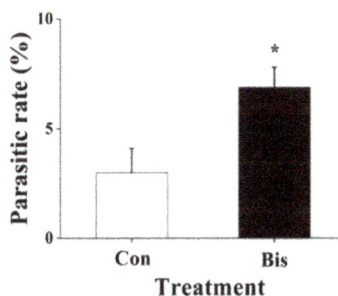

**Figure 5.** Mean parasitism rates (+ SE, $n = 15$) of WBPH eggs by *A. nilaparvatae* on the plants that had been grown for 1 day in nutrient solution with 50 mg $L^{-1}$ bismerthiazol (Bis) or without bismerthiazol, 2 days after plants were placed into the field. Asterisks indicate significant differences between members of a pair (* $p < 0.05$; Student's *t*-test).

## 2.5. Bismerthiazol Treatment Also Enhances the Resistance of Rice to BPH and SSB

To determine whether bismerthiazol induces the broad-spectrum resistance of rice to herbivores, we also measured the influence of bismerthiazol treatment on the performance of another piercing and sucking herbivore, BPH, and a chewing herbivore, SSB. The survival rate of BPH nymphs fed on plants that were grown in the nutrient solution with a concentration of 10, 20 or 50 mg $L^{-1}$ bismerthiazol decreased by 66.3%, 79.1% and 93.0%, respectively, compared to the survival rate of BPH nymphs fed on plants that were grown in the control solution (Figure 6a). Similarly, the larval mass of SSBs fed on bismerthiazol-treated plants for 12 days decreased by 35.45% relative to the larval mass of those fed on untreated plants (Figure 6b).

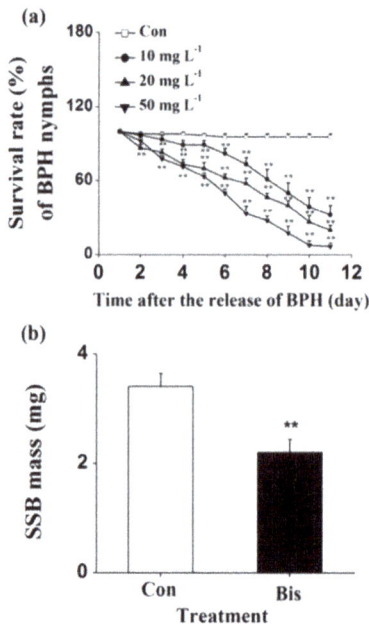

**Figure 6.** Bismerthiazol induces the resistance of rice to BPH and SSB. (**a**) Mean survival rates (+SE, $n = 6$) of 15 newly hatched BPH nymphs fed on plants that had been grown in nutrient solution with bismerthiazol (Bis) or without bismerthiazol (Con) for 24 h, 1–11 days after exposure; (**b**) Mean larval mass (+SE, $n = 30$) of SSB fed for 15 days on plants that had been grown in nutrient solution with 50 mg $L^{-1}$ bismerthiazol (Bis) or without bismerthiazol (Con) for 24 h. For BPH data, asterisks indicate significant differences in bismerthiazol-treated plants compared with non-treated plants (** $p < 0.01$; Duncan's multiple-range test); for SSB data, asterisks indicate significant differences between members of a pair (** $p < 0.01$; Student's *t*-test).

## 3. Discussion

In the study, we discovered that bismerthiazol treatment both induced the biosynthesis of constitutive and/or WBPH-elicited JA, JA-Ile, ET and $H_2O_2$, and altered the volatile profile of rice (Figure 4 and Figure S3, Table 1). These changes enhanced the direct resistance of rice to WBPH, BPH and SSB, as well as the indirect resistance of rice to planthoppers, by attracting *A. nilaparvatae*, the egg parasitoid of rice planthoppers (Figures 1, 2, 5 and 6). These findings demonstrate that bismerthiazol can act as a chemical elicitor that increases the resistance of rice to herbivores.

Signaling pathways mediated by JA, JA-Ile, ET, SA and $H_2O_2$ play a central role in regulating plant defense responses to herbivores [3–6]. Bismerthiazol has been reported to induce defense response in citrus by activating the SA signaling pathway but not the JA signaling pathway [42]. In rice, the exogenous application of bismerthiazol enhances $H_2O_2$ levels, up-regulates expression of defense-related genes, and induces both callose deposition and the hypersensitive response-like cell death in leaves infected with *Xoo* strain ZJ173 but not in non-infected leaves [29]. Here, we found that the exogenous application of bismerthiazol on rice induced the production of constitutive and/or elicited JA, JA-Ile, ET and $H_2O_2$ (Figure 4); however, the application had little or no effect on the production of SA (Figure S2). These discrepancies might be related to the different plant species and genotypes used. Further research should elucidate how biomerthiazol elicits these signaling pathways and how a plant's genetic background influences induction.

In rice, both JA- and ET-mediated pathways positively regulate the plant's resistance to chewing herbivores, such as SSB and leaf folder *Cnaphalocrocis medinalis*, by eliciting the accumulation of corresponding defensive compounds, such as TrypPIs [33,43,44]. In contrast, the JA-mediated pathway (probably by inhibiting the $H_2O_2$ pathway) and the ET-mediated pathway negatively modulate resistance of rice to piercing and sucking herbivores, such as BPH [8,33,34,45,46]; moreover, the $H_2O_2$-mediated pathway positively regulates the plant's resistance to BPH [4,33,46–48]. Some researchers have shown that the activation of the SA pathway increases the resistance of rice to BPH [4,33,49]; however, it has been reported recently that the OsWRKY45-dependent SA pathway negatively modulates the resistance of rice to BPH [50]. Thus, higher elicited levels of $H_2O_2$ and of JA, JA-Ile and ET in bismerthiazol-treated plants compared to in untreated plants contributed at least partially to plant's enhanced resistance to WBPH and BPH, as well as to SSB.

Volatile analysis found that the exogenous application of bismerthiazol on rice plants could induce the production of six volatile chemicals and enhance levels of two chemicals, although the amount of total volatiles was not much increased. In contrast, the exogenous application of bismerthiazol decreased production in plants exposed to WBPH infestation: levels of 11 chemicals were significantly decreased, and the amount of the total volatiles was only about half the amount in untreated plants that have exposed to WBPH infestation (Table 1). In rice, the exogenous application of JA induces the release of rice volatiles [51]; however, surprisingly, the ethylene-mediated pathway positively regulates the production of SSB-induced volatiles but negatively regulates the production of BPH-induced volatiles [44]. We found that bismerthiazol treatment induced the accumulation of JA, JA-Ile and ET in rice plants exposed to infestation by gravid WBPH females, whereas it induced only the biosynthesis of JA in non-infested plants (Figure 4). Thus, the way that the induction of biomerthiazol affects the production of volatiles in non-infested plants is probably related to how the induction regulates JA biosynthesis. The decrease in volatiles in bismerthiazol-treated plants that had been exposed to infestation compared to in exposed but untreated plants suggests that the ratio of JA to ethylene (or even the ratio of JA to other signals) is crucial for the production of rice volatiles. Interestingly, we found that bismerthiazol treatment decreased the level of MeSA in WBPH-infested plants but increased it in non-infested plants, compared to their corresponding control plants, although bismerthiazol treatment had little or no effect on SA levels in these plants (Figure S1). MeSA is produced via methylation of SA catalyzed by salicylic acid carboxyl methyltransferases (SAMTs) [52]. Therefore, these findings suggest that the effect of bismerthiazol treatment on the production of MeSA in rice plants might mainly be via its influence on the activity of SAMTs, while not on the biosynthesis of SA. Moreover, the data demonstrate that the activity of SAMTs is also affected by WBPH infestation and the interaction between bismerthiazol treatment and WBPH infestation. Further research should elucidate these issues.

Plant volatiles affect not only the host-searching behavior of insects but also their performance [32,53,54]. In addition to negatively affecting the performance of WBPH, bismerthiazol treatment also decreased the feeding and/or oviposition preference of gravid WBPH female adults and nymphs on plants (Figure 1d,e). Compared to plants without bismerthiazol treatment, treated plants released more volatiles, including 11 chemicals that were not detected in untreated plants and 2 chemicals whose levels were increased (Table 1). Moreover, one of these chemicals, methyl salicylate, was found to repel BPH [55]. Thus, apart from the other defenses mediated by enhanced signaling pathways, the change of volatiles in bismerthiazol-treated plants also partially contributed to the enhanced direct resistance to herbivores, reducing the performance and altering the preference of herbivores.

Interestingly, unlike the result that WBPH nymphs and gravid females did not prefer to feed and/or lay eggs on bismerthiazol-treated plants, the parasitoid *A. nilaparvatae* did prefer to parasitize WBPH eggs laid on bismerthiazol-treated plants over eggs laid on untreated plants: in the field, the parasitism of WBPH eggs by *A. nilaparvatae* on treated plants was 2.3-fold higher than the parasitism on untreated plants (Figure 5). Plant growth phenotype can also influence the host-searching behavior of parasitoids [56]. However, we could exclude the effect of plant growth phenotype on the parasitism rate of WBPH eggs as there was only a small difference in the growth phenotypes of bismerthiazol-treated and non-treated plants even 10 days after treatment (Figure 3c,d), and in this experiment, we treated plants only for 1 day. Thus, the observed difference in the amount of parasitism of WBPH eggs between bismerthiazol-treated and untreated plants is probably due mainly to the difference in volatile amounts between the two groups of plants. Surprisingly, bismerthiazol treatment decreased the production of all detected volatiles from WBPH-infested plants, including the three chemicals attractive to the parasitoid: linalool, MeSA and (E)-β-caryophyllene [41,55,57] (Table 1). Similarly, in maize, treatment with BTH or laminarin is reported to reduce herbivore-induced volatile emissions but increase the plant's attractiveness to parasitoids [24]. These findings demonstrate that the ratio of volatile chemicals or as-yet undetectable and unidentified chemicals may be of great importance for attracting *A. nilaparvatae*. Another possibility is that the volatiles that negatively affect *A. nilaparvatae* may also be the ones that are reduced. The exact mechanisms underlying this biological phenomenon should be clarified in future research.

In summary, our results demonstrate that bismerthiazol can enhance constitutive, especially herbivore-elicited, levels of defense-related signals, such as JA, JA-Ile, ET and $H_2O_2$. These activated signaling pathways alter the chemical profile of plants, and the changes, in turn, increase both direct and indirect resistance of rice to herbivores. As a commonly used bactericide, bismerthiazol is generally used to control bacterial diseases in rice and its amount applied in the field is about 375 g active ingredient per hectare [58]. The number of rice plants is about 2.5–3.0 million per hectare. Thus, the amount of bismerthiazol received by one plant (about 0.25–0.30 mL of 500 mg $L^{-1}$ bismerthiazol per plant) in the field is comparable to the amount received by one plant that was sprayed with 4 mL of 100 mg $L^{-1}$ bismerthiazol in this study (we suppose that about half of the amount of bismerthiazol used reached the plant). This indicates that bismerthiazol has a potential to be used in the field as a chemical elicitor for insect pest management. Further researches should investigate the control effect of bismerthiazol on rice insect pests in the field. Moreover, given that bismerthiazol is an efficient bactericide and that bismerthiazol can induce plant defenses against pathogens [29,59], these findings also open a pathway for controlling diseases and insect pests by designing and exploiting chemical elicitors that have such characteristic structures.

## 4. Materials and Methods

### 4.1. Plant Material and Growth Conditions

The rice (*Oryza sativa*) genotypes used in this study was Xiushui 110. Seeds were germinated in water for 1 day and cultured in plastic bottles (diameter 8 cm, height 10 cm) in a greenhouse (28 ± 2 °C, 14 h light, 10 h dark). Ten-day-old seedlings were transferred to 20 L communal hydroponic boxes with rice nutrient solution [60]. After 30–35 days, seedlings were transferred to individual 320 mL hydroponic plastic pots. Rice plants were used for experiments 3 days after transplanting.

### 4.2. Compounds

Bismerthiazol (purity, 98%) was provided by Longwan Agrichemical Co. Ltd., Wenzhou, China. *N,N*-Dimethylformamide and Tween-60 were purchased from Sinopharm Chemical Reagent Co. Ltd., Shanghai, China.

### 4.3. Insects

Colonies of WBPH, BPH and SSB were originally obtained from rice fields in Hangzhou, China, and maintained on TN 1 (a variety susceptible to the three herbivore species) rice seedlings in a controlled climate room (26 ± 2 °C, 14 h light, 10 h dark, 80% relative humidity).

## 4.4. Plant Treatments

For bimerthiazol treatment, two methods were used: root treatment and spray treatment. For root treatment, plants were grown in nutrient solution, and bimerthiazol (first dissolved in *N,N*-dimethylformamide, 1 mg per mL of *N,N*-dimethylformamide)—at final concentrations ranging from 10 to 50 mg L$^{-1}$—was added (with 0.01% Tween-60). Control plants (Con) were grown in nutrient solution without bismerthiazol but with an equal volume of *N,N*-dimethylformamide and Tween-60. For spray treatment, bismerthiazol was first dissolved in *N,N*-dimethylformamide (1 mg per mL of *N,N*-dimethylformamide) and then diluted in distilled water (with 0.01% Tween-60) at concentrations ranging from 50 to 100 mg L$^{-1}$. The above-ground parts of each plant were sprayed using an atomizer with 2 mL of bismerthiazol solution twice at interval of 4 h. Control plants were sprayed with distilled water containing the same volume of *N,N*-dimethylformamide and Tween-60. For WBPH treatment, individual plants were infested with 15 WBPH nymphs or gravid females that were confined within a glass cylinder (diameter 4 cm, height 8 cm, with 48 small holes, diameter 0.8 mm), and the top of the cylinder was covered with a piece of sponge. Empty cylinders were attached to control plants (non-infested plants).

## 4.5. Measurement of Plant Growth Parameters

Plant growth parameters, including plant height, root length, and mass of above- and below-ground parts of control plants and bismerthiazol-treated plants (grown in nutrient solution with different concentrations of bismerthiazol) were measured at 10 days after bismerthiazol treatment. Plant height and root length were defined as the part of a plant from the stem base to the longest leaf apex and that from the stem base to the longest root tip, respectively. Plants were cut off from the stem base and then the mass of above-ground and below-ground (roots) part of plants was measured. The experiment was replicated 6 times.

## 4.6. qRT-PCR

For qRT-PCR analysis, five independent biological samples were used. Total RNA was isolated using the SV Total RNA Isolation System (Promega, Madison, WI, USA) following the manufacturer's instructions. One microgram of each total RNA sample was reverse-transcribed using the PrimeScript RT-PCR Kit (TaKaRa, Shiga, Japan). The QRT-PCR assay was performed on a CFX96TM Real-Time system (Bio-Rad, Hercules, CA, USA) using a Premix Ex TaqTM Kit (TaKaRa, Shiga, Japan). A linear standard curve, threshold cycle number versus log (designated transcript level), was built using a series concentrations of a specific cDNA standard. Relative levels of the transcript of the target gene in tested samples were calculated according to the standard curve. A rice actin gene *OsACT* (TIGR ID: Os03g50885) was used as an internal standard to normalize cDNA concentrations. The primers and probes used for qRT-PCR for all tested genes are listed in Table S1.

## 4.7. WBPH and BPH Performance Measurement

The survival rates of WBPH and BPH nymphs on bismerthiazol-treated plants and control plants were investigated. For WBPH bioassays, two bismerthiazol treatments, root and spray, were used. Twenty-four hours after plants had been treated with bismerthiazol, the basal stem of each plant was covered in a glass cylinder into which 15 newly hatched WBPH or BPH nymphs were introduced. Every day, the number of surviving WBPH or BPH nymphs and emerging adults on each plant was recorded until all the nymphs had become adults. The experiment was repeated 6–10 times.

The hatching rates of WBPH eggs on bismerthiazol-treated plants (plants that had been grown in nutrient solution containing bismerthiazol for 24 h) and control plants were determined. Fifteen gravid WBPH females were exposed to individual plants for 12 h, and then all the insects were removed. The number of freshly hatched WBPH nymphs on plants was recorded every day until no new nymphs occurred for three consecutive days. Unhatched eggs and total eggs were counted to determine the hatching rate on each plant. The experiment was repeated 6 times.

To determine the effect of bismerthiazol treatment on the colonization and/or oviposition preferences of gravid WBPH females or nymphs, two plants (one control plant and one bismerthiazol-treated plant, as stated above) were confined within a glass cylinder into which 15 gravid females or nymphs were introduced. The number of gravid WBPH females or nymphs on each plant was counted 1, 2, 3, 8, 24 and 48 h after the release of WBPH; after 48 h, WBPH were removed and the eggs on each plant (for experiments with gravid females) were counted. The experiment was repeated 10 times.

To detect the direct effect of bismerthiazol on the survival rate of newly hatched WBPH nymphs, the contact toxicity and the stomach-poisoning toxicity of bismerthiazol were measured. For the contact toxicity measurement, 15 newly hatched WBPH nymphs were placed into a Petri dish (diameter 9 cm, height 2 cm) lined with filter paper wetted with 0.2 mL of 50 mg $L^{-1}$ bismerthiazol. Controls were lined with filter paper wetted with 0.2 mL of distilled water. After 2 h, WBPH nymphs were transferred onto the basal stem of each plant and covered in a glass cylinder. The experiment was repeated 6 times. For stomach-poisoning toxicity measurements, 15 newly hatched WBPH nymphs were fed on artificial diet with a concentration of 50 mg $L^{-1}$ bismerthiazol in a 30 mL glass cylinder with two ends open (diameter 9 cm, height 2 cm) as described in [61]. Controls were fed on artificial diet without bismerthiazol. Each treatment was replicated 5 times. Every day, the number of surviving WBPH nymphs and emerging adults on each plant in both experiments was recorded until all the nymphs had become adults.

*4.8. SSB Performance Measurement*

To measure the effect of bismerthiazol treatment on the performance of SSB, individual bismerthiazol-treated plants (as stated in Section 4.6) or control plants were infested with a newly hatched SSB larva. Larval mass was measured 15 days after the release of the herbivore. The experiment was repeated 30 times.

*4.9. JA, JA-Ile, SA, ET and $H_2O_2$ Analysis*

For JA, JA-Ile, and SA analysis, both bismerthiazol-treated plants (as stated in Section 4.6) and control plants were individually infested by 15 gravid WBPH females. Plant leaf sheaths were harvested at 3, 8 and 24 h after the start of WBPH infestation. Samples were ground in liquid nitrogen, and JA and JA-Ile were extracted with ethyl acetate spiked with labeled internal standards ($^2$D2-JA, $^2$D6-JA-Ile and $^2$D4-SA) and analyzed with HPLC/mass spectrometry/mass spectrometry following the method described [36]. Each treatment at each time interval was replicated five times.

For ET analysis, bismerthiazol-treated plants (as stated in Section 4.6) and control plants were individually infested by 15 gravid WBPH females. Plants were then individually covered with sealed glass cylinders (diameter 4 cm, height 50 cm). ET production was determined at 24, 48 and 72 h after the start of WBPH infestation by using the method described previously [62]. Each treatment at each time interval was replicated 10 times.

For $H_2O_2$ analysis, both bismerthiazol-treated plants (as stated in Section 4.6) and control plants were individually infested by 15 gravid WBPH females. Plant leaf sheaths were harvested at 3, 8 and 24 h after WBPH infestation. The $H_2O_2$ concentrations were determined by using the Amplex® Red Hydrogen Peroxide/Peroxidase Assay Kit (Invitrogen, available online: http://www.invitrogen.com/) as described previously [4]. Each treatment at each time interval was replicated five times.

*4.10. Collection, Isolation and Identification of Volatile Compounds*

The collection, isolation and identification of rice volatiles were carried out using the same method as our previously described [32,63]. Volatiles emitted from individual plants with different treatments—bismerthiazol-treated plants, WBPH-infested plants, bismerthiazol + WBPH-treated

*Int. J. Mol. Sci.* **2018**, *19*, 1271

plants and control plants—were collected and lasted for 8 h. For control plants, plants were grown in the control solution for 1 day, followed by non-infestation for 12 h. For the WBPH treatment, plants were grown in the control solution for 1 day, followed by infestation by 15 gravid WBPH females for 12 h. For bismerthiazol-treated plants, plants were grown in nutrient solution with 50 mg $L^{-1}$ bismerthiazol for 1 day, followed by non-infestation for 12 h. For bismerthiazol + WBPH-treated plants, plants were grown in nutrient solution with 50 mg $L^{-1}$ bismerthiazol for 1 day, followed by infestation by 15 gravid WBPH females for 12 h. Each treatment was replicated five times. The detected compounds were expressed as the percentage of peak areas relative to the internal standard (IS, diethyl sebacate) per 8 h of trapping for one plant.

*4.11. Field Experiment*

To evaluate the effect of bismerthiazol treatment on the host-searching behavior of the parasitoid *A. nilaparvatae* in the field, experiments were carried out in November 2016 in a rice field in Lin'an, Hangzhou, China. Gravid WBPH females laid eggs on both bismerthiazol-treated plants (as stated in Section 4.6) and control plants for 24 h, and plants were then transferred to fields. Pairs of potted plants (one bismerthiazol-treated plant and one control plant), 20 cm apart each other, were placed at 15 locations (3 m apart) parallel to the ridge (1 m from the ridge) in a rice field. Plants were collected 2 days after plants were introduced into the field. The plants were transferred to the controlled climate room (28 ± 2 °C, 14 h light, 10 h dark, 80% relative humidity), and each pot was confined in a plastic cage (diameter 11 cm, height 10 cm) (herbivores, predators, and parasitoids on plants were all removed). Five days later, the plants were cut off at the soil level and dissected under a microscope to record the total number of WBPH eggs and the number of parasitized eggs.

**Supplementary Materials:** Supplementary materials can be found at http://www.mdpi.com/1422-0067/19/5/1271/s1. Table S1: Primers and probes used for QRT-PCR of target genes; Figure S1: Growth phenotypes of bismerthiazol-treated and control rice plants; Figure S2: Mean levels (+ SE, *n* = 5) of SA in leaf sheaths of rice plants with different treatments; Figure S3: Typical chromatograms obtained by head space collections from plants with different treatments.

**Author Contributions:** Yonggen Lou and Xiaochang Mo conceived and designed the experiments; Xiaochang Mo, Pengyong Zhou, Wanwan Wang and Xia Cheng performed the experiments; Xiaochang Mo, Pengyong Zhou analyzed the data; Yonggen Lou and Pengyong Zhou wrote the paper.

**Funding:** The study was jointly sponsored by the Fund for Agro-scientific Research in the Public Interest (201403030) and the earmarked fund for China Agriculture Research System (CARS-01-40).

**Acknowledgments:** We thank Emily Wheeler, Boston, for editorial assistance.

**Conflicts of Interest:** The authors declare no conflict of interest.

## Abbreviations

| | |
|---|---|
| JA | Jasmonic acid |
| JA-Ile | Jasmonoyl-isoleucine conjugate |
| SA | Salicylic acid |
| ET | Ethylene |
| BTH | Benzo (1,2,3) thiadiazole-7-carbothioic acid *S*-methyl ester |
| MeJA | Methyl ester of JA |
| 2,4-D | 2,4-dichlorophenoxyacetic acid |
| TrypPIs | Trypsin proteinase inhibitors |
| SSB | Striped stem borer |
| BPH | Brown planthopper |
| *Xoo* | *Xanthomonas oryzae pv. Oryzae* |
| WBPH | white-backed planthopper |
| MeSA | Methyl salicylate |
| SAMT | Salicylic acid carboxyl methyltransferase |

## References

1. Howe, G.A.; Jander, G. Plant immunity to insect herbivores. *Annu. Rev. Plant Biol.* **2008**, *59*, 41–66. [CrossRef] [PubMed]
2. Hu, X.; Makita, S.; Schelbert, S.; Sano, S.; Ochiai, M.; Tsuchiya, T.; Hasegawa, S.F.; Hortensteiner, S.; Tanaka, A.; Tanaka, R. Reexamination of chlorophyllase function implies its involvement in defense against chewing herbivores. *Plant Physiol.* **2015**, *167*, 660–670. [CrossRef] [PubMed]
3. Wu, J.; Baldwin, I.T. New insights into plant responses to the attack from insect herbivores. *Annu. Rev. Genet.* **2010**, *44*, 1–24. [CrossRef] [PubMed]
4. Lu, J.; Ju, H.; Zhou, G.; Zhu, C.; Erb, M.; Wang, X.; Wang, P.; Lou, Y. An EAR-motif-containing ERF transcription factor affects herbivore-induced signaling, defense and resistance in rice. *Plant J.* **2011**, *68*, 583–596. [CrossRef] [PubMed]
5. Erb, M.; Meldau, S.; Howe, G.A. Role of phytohormones in insect-specific plant reactions. *Trends Plant Sci.* **2012**, *17*, 250–259. [CrossRef] [PubMed]
6. Lou, Y.; Lu, J.; Li, J. Herbivore-Induced Defenses in Rice and Their Potential Application in Rice Planthopper Management. In *Rice Planthoppers*; Heong, K.L., Cheng, J., Escalada, M.M., Eds.; Zhejiang University Press: Hangzhou, China, 2015; pp. 91–116.
7. Bektas, Y.; Eulgem, T. Synthetic plant defense elicitors. *Front. Plant Sci.* **2014**, *5*, 804. [CrossRef] [PubMed]
8. Xin, Z.; Yu, Z.; Erb, M.; Turlings, T.C.; Wang, B.; Qi, J.; Liu, S.; Lou, Y. The broad-leaf herbicide 2,4-dichlorophenoxyacetic acid turns rice into a living trap for a major insect pest and a parasitic wasp. *New Phytol.* **2012**, *194*, 498–510. [CrossRef] [PubMed]
9. Beckers, G.J.; Conrath, U. Priming for stress resistance: From the lab to the field. *Curr. Opin. Plant Biol.* **2007**, *10*, 425–431. [CrossRef] [PubMed]
10. LaMondia, J.A. Actigard Increases Fungicide Efficacy against Tobacco Blue Mold. *Plant Dis.* **2008**, *92*, 1463–1467. [CrossRef]
11. Thaler, J.S.; Stout, M.J.; karban, R.; Duffey, S.S. Exogenous jasmonates simulate insect wounding in tomato plants (*Lycopersicon esculentum*) in the laboratory and field. *J. Chem. Ecol.* **1996**, *22*, 1767–1781. [CrossRef] [PubMed]
12. Cipollini, D.F.; Redman, A.M. Age-dependent effects of jasmonic acid treatment and wind exposure on foliar oxidase activity and insect resistance in tomato. *J. Chem. Ecol.* **1999**, *5*, 271–281. [CrossRef]
13. Thaler, J.S. Jasmonate-inducible plant defences cause increased parasitism of herbivores. *Nature* **1999**, *399*, 686–688. [CrossRef]
14. Pauwels, L.; Morreel, K.; Witte, E.D.; Lammertyn, F.; Montagu, M.V.; Boerjan, W.; Inzé, D.; Goossens, A. Mapping methyl jasmonate-mediated transcriptional reprogramming of metabolism and cell cycle progression in cultured *Arabidopsis* cells. *Proc. Natl. Acad. Sci. USA* **2007**, *105*, 1380–1385. [CrossRef] [PubMed]
15. Felton, G.W.; Duffey, S.S. Ascorbate Oxidation Reduction in *Helicoverpa zea* as a Scavenging System Against Dietary Oxidants. *Arch. Insect Biochem. Physiol.* **1992**, *19*, 29–37. [CrossRef]
16. Stout, M.J.; Duffey, S.S. Characterization of induced resistance in tomato plants. *Entomol. Exp. Appl.* **1996**, *79*, 273–283. [CrossRef]
17. Li, R.; Schuman, M.C.; Wang, Y.; Llorca, L.C.; Bing, J.; Bennion, A.; Halitschke, R.; Baldwin, I.T. Jasmonate signaling makes flowers attractive to pollinators and repellant to florivores in nature. *J. Integr. Plant Biol.* **2018**, *60*, 190–194. [CrossRef] [PubMed]
18. Kramell, R.; Atzorn, R.; Schneider, G.; Miersch, O.; Bruckner, C.; Schmidt, J.; Sembdner, G.; Parthier, B. Occurrence and Identification of Jasmonic Acid and Its Amino Acid Conjugates Induced by Osmotic Stress in Barley Leaf Tissue. *J. Plant Growth Regul.* **1995**, *14*, 29–36. [CrossRef]
19. Wasternack, C. Jasmonates: An update on biosynthesis, signal transduction and action in plant stress response, growth and development. *Ann. Bot.* **2007**, *100*, 681–697. [CrossRef] [PubMed]
20. Fonseca, S.; Chini, A.; Hamberg, M.; Adie, B.; Porzel, A.; Kramell, R.; Miersch, O.; Wasternack, C.; Solano, R. (+)-7-*iso*-Jasmonoyl-L-isoleucine is the endogenous bioactive jasmonate. *Nat. Chem. Biol.* **2009**, *5*, 344–350. [CrossRef] [PubMed]
21. Jimenez-Aleman, G.H.; Machado, R.A.R.; Baldwin, I.T.; Boland, W. JA-Ile-macrolactones uncouple growth and defense in wild tobacco. *Org. Biomol. Chem.* **2017**, *15*, 3391–3395. [CrossRef] [PubMed]

22. Birkett, M.A.; Campbell, C.A.M.; Chamberlain, K.; Guerrieri, E.; Hick, A.J.; Martin, J.L.; Matthes, M.; Napier, J.A.; Pettersson, J.; Pickett, J.A.; et al. New roles for *cis*-jasmone as an insect semiochemical and in plant defense. *Proc. Natl. Acad. Sci. USA* **2000**, *97*, 9329–9334. [CrossRef] [PubMed]

23. Maeda, T.; Ishiwari, H. Tiadinil, a plant activator of systemic acquired resistance, boosts the production of herbivore-induced plant volatiles that attract the predatory mite *Neoseiulus womersleyi* in the tea plant *Camellia sinensis*. *Exp. Appl. Acarol.* **2012**, *58*, 247–258. [CrossRef] [PubMed]

24. Sobhy, I.S.; Erb, M.; Sarhan, A.A.; El-Husseini, M.M.; Mandour, N.S.; Turlings, T.C. Less is more: Treatment with BTH and laminarin reduces herbivore-induced volatile emissions in maize but increases parasitoid attraction. *J. Chem. Ecol.* **2012**, *38*, 348–360. [CrossRef] [PubMed]

25. Sobhy, I.S.; Erb, M.; Turlings, T.C. Plant strengtheners enhance parasitoid attraction to herbivore-damaged cotton via qualitative and quantitative changes in induced volatiles. *Pest Manag. Sci.* **2015**, *71*, 686–693. [CrossRef] [PubMed]

26. Ma, Z.; Zhou, M.; Ye, Z. The action mode of saikuzuo against *xanthomon as oryzae pv. oryzae*. *Acta Phytopathol. Sin.* **1997**, *27*, 237–241.

27. Shen, G.; Zhou, M. Resistance monitoring of *Xanthomonas oryzae pv. axyzae* to saikuzuo. *Plant Prot.* **2002**, *28*, 9–11.

28. Huang, Q.; Zhou, M.; Ye, Z. Physiological characters of the resistant mutant of *Xanthomonas citri* to amicarthiazol. *Acta Phytopathol. Sin.* **2003**, *33*, 63–66.

29. Liang, X.; Yu, X.; Dong, W.; Guo, S.; Xu, S.; Wang, J.; Zhou, M. Two thiadiazole compounds promote rice defence against *Xanthomonas oryzae pv. oryzae* by suppressing the bacterium's production of extracellular polysaccharides. *Mol. Plant Pathol.* **2015**, *16*, 882–892. [CrossRef] [PubMed]

30. Chen, H.; Stout, M.J.; Qian, Q.; Chen, F. Genetic, Molecular and Genomic Basis of Rice Defense against Insects. *Crit. Rev. Plant Sci.* **2012**, *31*, 74–91. [CrossRef]

31. Lou, Y.G.; Ma, B.; Cheng, J.A. Attraction of the parasitoid *Anagrus nilaparvatae* to rice volatiles induced by the rice brown planthopper *Nilaparvata lugens*. *J. Chem. Ecol.* **2005**, *31*, 2357–2372. [CrossRef] [PubMed]

32. Lou, Y.G.; Du, M.H.; Turlings, T.C.J.; Cheng, J.A.; Shan, W.F. Exogenous application of jasmonic acid induces volatile emissions in rice and enhances parasitism of *Nilaparvata lugens* eggs by the parasitoid *Anagrus nilaparvatae*. *J. Chem. Ecol.* **2005**, *31*, 1985–2002. [CrossRef] [PubMed]

33. Zhou, G.; Qi, J.; Ren, N.; Cheng, J.; Erb, M.; Mao, B.; Lou, Y. Silencing OsHI-LOX makes rice more susceptible to chewing herbivores, but enhances resistance to a phloem feeder. *Plant J.* **2009**, *60*, 638–648. [CrossRef] [PubMed]

34. Wang, Q.; Li, J.; Hu, L.; Zhang, T.; Zhang, G.; Lou, Y. OsMPK3 positively regulates the JA signaling pathway and plant resistance to a chewing herbivore in rice. *Plant Cell Rep.* **2013**, *32*, 1075–1084. [CrossRef] [PubMed]

35. Hu, L.; Ye, M.; Li, R.; Zhang, T.; Zhou, G.; Wang, Q.; Lu, J.; Lou, Y. The Rice Transcription Factor WRKY53 Suppresses Herbivore-Induced Defenses by Acting as a Negative Feedback Modulator of Mitogen-Activated Protein Kinase Activity. *Plant Physiol.* **2015**, *169*, 2907–2921. [CrossRef] [PubMed]

36. Lu, J.; Robert, C.A.; Riemann, M.; Cosme, M.; Mene-Saffrane, L.; Massana, J.; Stout, M.J.; Lou, Y.; Gershenzon, J.; Erb, M. Induced jasmonate signaling leads to contrasting effects on root damage and herbivore performance. *Plant Physiol.* **2015**, *167*, 1100–1116. [CrossRef] [PubMed]

37. Mewis, I.; Appel, H.M.; Hom, A.; Raina, R.; Schultz, J.C. Major signaling pathways modulate Arabidopsis glucosinolate accumulation and response to both phloem-feeding and chewing insects. *Plant Physiol.* **2005**, *138*, 1149–1162. [CrossRef] [PubMed]

38. Ahuja, I.; Kissen, R.; Bones, A.M. Phytoalexins in defense against pathogens. *Trends Plant Sci.* **2012**, *17*, 73–90. [CrossRef] [PubMed]

39. Delker, C.; Stenzel, I.; Hause, B.; Miersch, O.; Feussner, I.; Wasternack, C. Jasmonate biosynthesis in *Arabidopsis thaliana*—Enzymes, products, regulation. *Plant Biol.* **2006**, *8*, 297–306. [CrossRef] [PubMed]

40. Riemann, M.; Riemann, M.; Takano, M. Rice *JASMONATE RESISTANT 1* is involved in phytochrome and jasmonate signalling. *Plant Cell Environ.* **2008**, *31*, 783–792. [CrossRef] [PubMed]

41. Xiao, Y.; Wang, Q.; Erb, M.; Turlings, T.C.; Ge, L.; Hu, L.; Li, J.; Han, X.; Zhang, T.; Lu, J.; et al. Specific herbivore-induced volatiles defend plants and determine insect community composition in the field. *Ecol. Lett.* **2012**, *15*, 1130–1139. [CrossRef] [PubMed]

42. Yu, X.; Armstrong, C.M.; Zhou, M.; Duan, Y. Bismerthiazol Inhibits *Xanthomonas citri* subsp. *citri* Growth and Induces Differential Expression of Citrus Defense-Related Genes. *Phytopathology* **2016**, *106*, 693–701. [CrossRef] [PubMed]

43. Tong, X.; Qi, J.; Zhu, X.; Mao, B.; Zeng, L.; Wang, B.; Li, Q.; Zhou, G.; Xu, X.; Lou, Y.; et al. The rice hydroperoxide lyase OsHPL3 functions in defense responses by modulating the oxylipin pathway. *Plant J.* **2012**, *71*, 763–775. [CrossRef] [PubMed]

44. Lu, J.; Li, J.; Ju, H.; Liu, X.; Erb, M.; Wang, X.; Lou, Y. Contrasting effects of ethylene biosynthesis on induced plant resistance against a chewing and a piercing-sucking herbivore in rice. *Mol. Plant* **2014**, *7*, 1670–1682. [CrossRef] [PubMed]

45. Zhou, G.; Wang, X.; Yan, F.; Wang, X.; Li, R.; Cheng, J.; Lou, Y. Genome-wide transcriptional changes and defence-related chemical profiling of rice in response to infestation by the rice striped stem borer *Chilo suppressalis*. *Physiol. Plant* **2011**, *143*, 21–40. [CrossRef] [PubMed]

46. Zhang, J.; Luo, T.; Wang, W.; Cao, T.; Li, R.; Lou, Y. Silencing *OsSLR1* enhances the resistance of rice to the brown planthopper *Nilaparvata lugens*. *Plant Cell Environ.* **2017**, *40*, 2147–2159. [CrossRef] [PubMed]

47. Qi, J.; Li, J.; Han, X.; Li, R.; Wu, J.; Yu, H.; Hu, L.; Xiao, Y.; Lu, J.; Lou, Y. Jasmonic acid carboxyl methyltransferase regulates development and herbivory-induced defense response in rice. *J. Integr. Plant Biol.* **2016**, *58*, 564–576. [CrossRef] [PubMed]

48. Hu, L.; Ye, M.; Li, R.; Lou, Y. OsWRKY53, a versatile switch in regulating herbivore-induced defense responses in rice. *Plant Signal. Behav.* **2016**, *11*, e1169357. [CrossRef] [PubMed]

49. Li, J. Comparison of Defense Responses in Rice Induced by Feeding and Oviposition of the Brown Planthopper *Nilaparvata lugens* and Their Underlying Mechanisms. Ph.D. Thesis, Zhejiang University, Hangzhou, China, 2015.

50. Huangfu, J.; Li, J.; Li, R.; Ye, M.; Kuai, P.; Zhang, T.; Lou, Y. The Transcription Factor OsWRKY45 Negatively Modulates the Resistance of Rice to the Brown Planthopper *Nilaparvata lugens*. *Int. J. Mol. Sci.* **2016**, *17*, 697. [CrossRef] [PubMed]

51. Lou, Y.; Baldwin, I.T. Silencing of a germin-like gene in *Nicotiana attenuata* improves performance of native herbivores. *Plant Physiol.* **2006**, *140*, 1126–1136. [CrossRef] [PubMed]

52. Deng, W.W.; Wang, R.; Yang, T.; Jiang, L.; Zhang, Z.Z. Functional Characterization of Salicylic Acid Carboxyl Methyltransferase from *Camellia sinensis*, Providing the Aroma Compound of Methyl Salicylate during the Withering Process of White Tea. *J. Agric. Food Chem.* **2017**, *65*, 11036–11045. [CrossRef] [PubMed]

53. Yuan, J.S.; Kollner, T.G.; Wiggins, G.; Grant, J.; Degenhardt, J.; Chen, F. Molecular and genomic basis of volatile-mediated indirect defense against insects in rice. *Plant J.* **2008**, *55*, 491–503. [CrossRef] [PubMed]

54. Christensen, S.A.; Nemchenko, A.; Borrego, E.; Murray, I.; Sobhy, I.S.; Bosak, L.; DeBlasio, S.; Erb, M.; Robert, C.A.; Vaughn, K.A.; et al. The maize lipoxygenase, *ZmLOX10*, mediates green leaf volatile, jasmonate and herbivore-induced plant volatile production for defense against insect attack. *Plant J.* **2013**, *74*, 59–73. [CrossRef] [PubMed]

55. Wang, P. Screening and Primary Field Evaluation of Infochemicals Manipulating Behavioral Responses of Rice Brown Planthopper *Nilaparvata lugens* (Stål) and Its Egg Parasitoid *Anagrus nilaparvatae*. Ph.D. Thesis, Zhejiang University, Hangzhou, China, 2011.

56. Ma, B.; Lou, Y.; Chen, J. Effects of some biotic factors on activities of the volatiles emitted from rice plants infested by the rice brown planthopper, *Nilaparvata lugens* (Stål). *J. Zhejiang Univ.* **2004**, *30*, 589–595.

57. Wang, Q.; Xin, Z.; Li, J.; Hu, L.; Lou, Y.; Lu, J. (*E*)-β-caryophyllene functions as a host location signal for the rice white-backed planthopper *Sogatella furcifera*. *Physiol. Mol. Plant Pathol.* **2015**, *91*, 106–112. [CrossRef]

58. Koo, Y.J.; Kim, M.A.; Kim, E.H.; Song, J.T.; Jung, C.; Moon, J.K.; Kim, J.H.; Seo, H.S.; Song, S.I.; Kim, J.K.; et al. Overexpression of salicylic acid carboxyl methyltransferase reduces salicylic acid-mediated pathogen resistance in *Arabidopsis thaliana*. *Plant Mol. Biol.* **2007**, *64*, 1–15. [CrossRef] [PubMed]

59. Liang, X.; Yu, X.; Pan, X.; Wu, J.; Duan, Y.; Wang, J.; Zhou, M. A thiadiazole reduces the virulence of *Xanthomonas oryzae* pv. *oryzae* by inhibiting the histidine utilization pathway and quorum sensing. *Mol. Plant Pathol.* **2016**, *19*, 116–128. [CrossRef] [PubMed]

60. Yoshida, S.; Forno, D.A.; Cock, J.H.; Gomez, K.A. *Laboratory Manual for Physiological Studies of Rice*, 3rd ed.; International Rice Research Institute: Los Baños, Philippines, 1976.

*Int. J. Mol. Sci.* **2018**, *19*, 1271

61. Fu, Q.; Zhang, Z.; Hu, C.; Lai, F.; Sun, Z. A chemically defined diet enables continuous rearing of the brown planthopper, *Nilaparvata lugens* (Stål) (Homoptera: Delphacidae). *Appl. Entomol. Zool* **2001**, *36*, 111–116. [CrossRef]

62. Lu, Y.; Wang, X.; Lou, Y.; Cheng, J. Role of ethylene signaling in the production of rice volatiles induced by the rice brown planthopper *Nilaparvata lugens*. *Chin. Sci. Bull.* **2006**, *51*, 2457–2465. [CrossRef]

63. Qi, J.; Zhou, G.; Yang, L.; Erb, M.; Lu, Y.; Sun, X.; Cheng, J.; Lou, Y. The chloroplast-localized phospholipases D α4 and α5 regulate herbivore-induced direct and indirect defenses in rice. *Plant Physiol.* **2011**, *157*, 1987–1999. [CrossRef] [PubMed]

International Journal of
*Molecular Sciences*

MDPI

*Article*

# A Biocontrol Strain of *Bacillus subtilis* WXCDD105 Used to Control Tomato *Botrytis cinerea* and *Cladosporium fulvum* Cooke and Promote the Growth of Seedlings

Hui Wang [1], Yuying Shi [2,3], Doudou Wang [1], Zhongtong Yao [4], Yimei Wang [1], Jiayin Liu [5], Shumei Zhang [6] and Aoxue Wang [1,2,3,*]

[1]    College of Life Sciences, Northeast Agricultural University, 150030 Harbin, China;
       wh571080005@163.com (H.W.); wangdou123dou@163.com (D.W.); Jessica0504wym@163.com (Y.W.)
[2]    College of Horticulture and Landscape Architecture, Northeast Agricultural University,
       150030 Harbin, China; 1144869388@126.com
[3]    Key Laboratory of Biology and Genetic Improvement of Horticultural Crops (Northeast Region),
       Ministry of Agriculture, Northeast Agricultural University, 150030 Harbin, China
[4]    College of Agronomy, Northeast Agricultural University, 150030 Harbin, China; yaozhong123tong@163.com
[5]    College of Science, Northeast Agricultural University, 150030 Harbin, China; 13040216@163.com
[6]    Institute of Microbiology Heilongjiang Academy of Sciences, 150001 Harbin, China;
       shumeizhang@yahoo.com
*    Correspondence: axwang@neau.edu.cn; Tel.: +86-451-551-91671

Received: 5 April 2018; Accepted: 30 April 2018; Published: 4 May 2018

**Abstract:** In this study, a strain named WXCDD105, which has strong antagonistic effects on *Botrytis cinerea* and *Cladosporium fulvum* Cooke, was screened out from the rhizosphere of healthy tomato plants. The tomato plants had inhibition diameter zones of 5.00 mm during the dual culture for four days. Based on the morphological and physiological characteristics, the 16S rDNA sequence, and the *gyrB* gene sequence analysis, the strain WXCDD105 was identified as *Bacillus subtilis* suBap. *subtilis*. The results of the mycelial growth test showed that the sterile filtrate of the strain WXCDD105 could significantly inhibit mycelial growth of *Botrytis cinerea* and *Cladosporium fulvum* Cooke. The inhibition rates were 95.28 and 94.44%, respectively. The potting experiment showed that the strain WXCDD105 made effective the control of tomato gray mold and tomato leaf mold. The control efficiencies were 74.70 and 72.07%. The antagonistic test results showed that the strain WXCDD105 had different degrees of inhibition on 10 kinds of plant pathogenic fungi and the average inhibition rates were more than 80%. We also found that the strain WXCDD105 stimulated both the seed germination and seedling growth of tomatoes. Using the fermentation liquid of WXCDD105 ($10^8$ cfu·mL$^{-1}$) to treat the seeds, the germination rate and radicle length were increased. Under the treatment of the fermentation liquid of the strain WXCDD105 ($10^6$ cfu·mL$^{-1}$), nearly all physiological indexes of tomato seedlings were significantly higher than that of the control groups. This could not only keep the nutritional quality of tomato fruits but also prevent them from rotting. This study provided us with an excellent strain for biological control of tomato gray mold, tomato leaf mold, and tomato growth promotion. This also laid the technical foundation for its application.

**Keywords:** tomato gray mold; tomato leaf mold; *Bacillus subtilis*; biological control

## 1. Introduction

The tomato is native to Peru and Ecuador. It is a solanaceae herbaceous plant that belongs to the solanaceae family. It can be grown for many years under suitable conditions and is one of the most widely grown and highly consumed vegetable crops in the world [1]. Tomato gray mold and leaf mold are the main diseases during tomato cultivation, which is caused by *Botrytis cinerea* and *Cladosporium fulvum* Cooke [2,3]. Gray mold harms tomato leaves, flowers, and fruits on the ground. The flower period and fruit enlargement period suffer the most [4]. The leaf mold disease mainly harms the leaves, but can also infect the stem and fruits. This usually happens in the middle and late stages of production. Because of the lack of resistant germplasm resources and the loss of variety in resistance, the plant has not been screened for ideal tomato gray mold and leaf mold disease resistance resource materials worldwide. The breeding progress of disease resistance is slow. At present, chemical pesticides are still mainly used for preventing and treating tomato gray mold and leaf mold diseases [5,6]. Chemical pesticides have played a huge role in the control of plant diseases, but the disadvantages are also becoming more prominent. Using chemical pesticides for a long time can increase pathogens' resistance to chemicals. In addition, while it kills germs, it can also threaten beneficial microorganisms, destroy the agricultural ecosystem, and endanger human health. In recent years, biological control has attracted more attention [7]. Biological control can control the disease to the lowest level. Its advantages include low toxicity, low or no residual levels, low pollution, safety, and efficiency. In the background of the sustainable development of agriculture, food safety and environmental protection, the need to research and develop new biological pesticides that will gradually replace chemical pesticides has become an important target for agricultural production. There are many successful cases of the application of antagonistic microorganisms to control tomato gray mold and leaf mold diseases [8,9]. The prophylaxis and treatment of gray mold disease is more than that of leaf mold [10]. According to related reports, the main antagonistic microorganisms of tomato gray mold and leaf mold diseases include bacterial, fungal, yeasts and actinomycetes [11,12]. *Bacillus subtilis* is widely used in biological control as a variety of antibiotics [13]. However, most studies are only screening the microorganism for one of the diseases and the two diseases often occur simultaneously to the actual form of production. The anti-biological inoculant and antimicrobial spectrum of the microorganisms were narrow, which severely restricted its development and application.

Physiological and biochemical characteristics of *Bacillus subtilis*, a kind of gram positive bacterium, have been studied for decades [14,15]. Hundreds of *Bacillus subtilis* have been purified. There are also many kinds of *Bacillus subtilis* that can produce antibiotics. *Bacillus subtilis* can also affect other microbial communities to a certain extent [16]. Therefore it has been increasingly used in the research on biological control.

This study mainly focused on screening new active bacteria and their biocontrol effects. The *Bacillus subtilis* strain WXCDD105 had strong antagonism to the tomato gray mold disease and the leaf mold disease, which were isolated from healthy tomato plants' rhizosphere soil and identified as *Bacillus subtilis* suBap. *subtilis*. The effects of biological control on other related diseases and its effect on growth was studied. We provided and developed a potential bacterial strain that can prevent tomato gray mold and tomato leaf mold diseases. This also provides a satisfactory technical basis for developing one kind of new biocontrol agent.

## 2. Results

### 2.1. The Separation, Purification, and Screening of Antagonist Strains

A total of 126 strains of bacteria were isolated and purified from rhizosphere soils of different plants. Five antagonistic strains that have different degrees of inhibition effect on *Botrytis cinerea* WD1 and *Myceliophthora* sp. WD2 were obtained by dual-culture screening and were numbered 105, 51, 98, 15, and 109, respectively. The test results of the dual culture are shown in Table 1 below. It can be seen from Table 1 that the strain WXCDD105 isolated from tomato rhizosphere soil has the strongest bacteriostatic activity and the bacteriostatic effects on the strains *Botrytis cinerea* WD1 and *Cladosporium fulvum* Cooke WD2 are, respectively, 71.57 and 71.64%. The antibacterial circle method was used for re-screening and the results are shown in Figure 1. It can be seen from Figure 1 that the strain WXCDD105 has a more stable antibacterial effect on *Botrytis cinerea* WD1 and on *Mycoplasma* WD2. The strain WXCDD105 was, therefore, selected for a follow-up test and it was named WXCDD105.

**Table 1.** Screening of biocontrol bacteria against tomato gray mold disease and tomato leaf mold disease.

| Strain Number | *Botrytis cinereal* WD1 | | *Cladosporium fulvum* Cooke WD2 | |
|---|---|---|---|---|
| | Inhibition Zone/mm | Inhibition Rate/% | Inhibition Zone/mm | Inhibition Rate/% |
| 105 | $4.99 \pm 0.56$ a | 71.57 a | $5.01 \pm 0.75$ a | 71.64 a |
| 51 | $4.00 \pm 0.63$ a | 69.13 a | $4.00 \pm 0.51$ a | 69.13 a |
| 98 | $0.93 \pm 0.12$ c | 61.56 c | $0.92 \pm 0.14$ c | 61.53 c |
| 15 | $1.61 \pm 0.41$ b | 63.23 b | $1.62 \pm 0.32$ b | 63.26 b |
| 109 | $1.00 \pm 0.11$ a | 61.74 a | $1.00 \pm 0.11$ a | 61.74 a |

Notes: the different normal letters in the same column indicate significant difference among treatments at 0.05 level ($n = 3$).

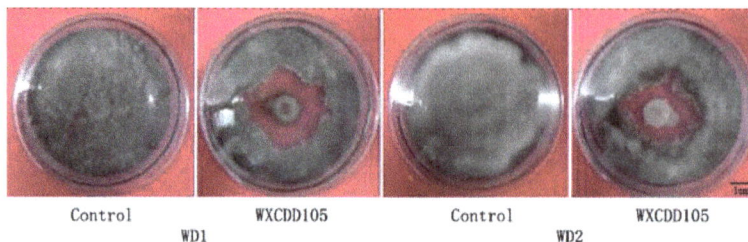

Control   WXCDD105   Control   WXCDD105

WD1         WD2

**Figure 1.** Antagonistic effects of strain WXCDD105 against *Botrytis cinerea* WD1 and *Cladosporium fulvum* Cooke WD2.

## 2.2. Identification of Strain WXCDD105

### 2.2.1. Strain WXCDD105 Morphology Detection

The colony of strain WXCDD105 was pale yellow, suborbicular, and had a dry surface. It also had a flat, irregular edge, which is shown in Figure 2A after being cultured for 18 h on Luria-Bertani (LB) media at 37 °C. Under optical microscopes, which is shown in Figure 2B, the shape of the bacteria was rod-like and the size of the bacteria was 0.7–0.8 μm × 1.8–3.3 μm. The bacteria were arranged individually or in pairs and appeared as columnar, intermediate, or near-terminal spores whose cyst enlargement was not obvious. They were identified as gram positive. Under a transmission electron microscope, the the bacteria was rod-shaped, round, and had a peri-flagellum, which is shown in Figure 2C.

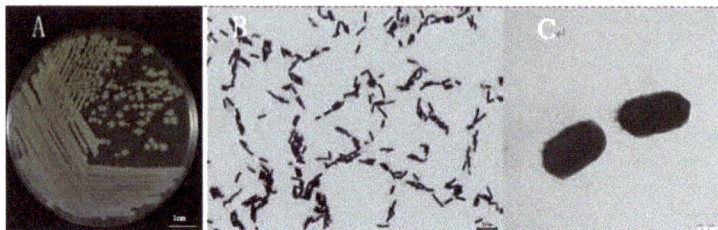

**Figure 2.** Morphological characteristics of strain WXCDD105 on LB medium plate culture (**A**), and observed under an optical microscopes (**B**) and transmission electron microscopy (**C**).

### 2.2.2. Molecular Identification of Bacterial Strain WXCDD105

The strain WXCDD105 consists of aerobic bacteria and ferments to produce acid. The nitrate reduction reaction, starch hydrolysis, contact enzyme reaction, gelatin liquefaction, and the citrate utilization test were positive. The compact automatic bacterial identification system of the WXCDD105 strain is shown in Table 2. According to the detection results of Bergey's Manual of Determinative Bacteriology, the bacterial strain WXCDD105 was identified as *Bacillus subtilis* suBap. *Subtilis*.

**Table 2.** The results of the bacterial automatic identification system of the strain WXCDD105.

| The Bacterial Strain WXCDD105 | | | | | | | |
|---|---|---|---|---|---|---|---|
| Identification Index | Result | Identification Index | Result | Identification Index | Result | Identification Index | Result |
| BXYL | + | LysA | − | AspA | − | LeuA | + |
| PheA | − | ProA | − | BGAL | + | PyrA | + |
| AGAL | + | AlaA | − | TyrA | − | BNAG | − |
| APPA | + | CDEX | − | dGAL | − | GLYG | − |
| INO | + | MdG | + | ELLM | − | MdX | − |
| AMAN | − | MTE | + | GlyA | + | dMAN | + |
| dMNE | + | dMLZ | − | NAG | − | PLE | + |
| IRHA | − | BGLU | + | BMAN | − | PHC | − |
| PVATE | + | AGLU | − | dTAG | − | dTRE | + |
| INU | + | dGLU | + | dRIB | + | PSCNa | − |
| NaCl 6.5% | + | KAN | − | OLD | − | ESC | + |
| TTZ | + | POLYB_R | − | | | | |

Notes: "+" means masculine, "−" means feminine.

According to 16S rRNA sequencing results, the most homologous strains of WXCDD105 were *Bacillus* sp., which had the highest similarity with *Bacillus subtilis* suBap. *inaquosorum* (AMXN01000021) and *Bacllus tequilensis* (AYTO01000043). All of these strains were 99.93% similar. The similarity of the *Bacillus amyloedfaciens* suBap. *amylol* (FN597644) in the same branch was 99.34%. In order to improve the species precision, the strain WXCDD105 was analyzed for the *gyrB* gene sequence, the strain WXCDD105 was analyzed for the *gyrB* gene sequence, which is shown in Figure 3.

The *Bacillus subtilis* suBap. *Subtilis* BCRC 10255T (DQ309293) was in the same branch as the strain WXCDD105, which had the highest similarity in 99.1%. The similarity with other model strains was less than 95%. After morphological observation as well as the physiological and biochemical characterization of the 16S rDNA sequence and *gyrB* gene sequence analysis, the strain WXCDD105 was identified as *Bacillus subtilis* suBap. *subtilis*.

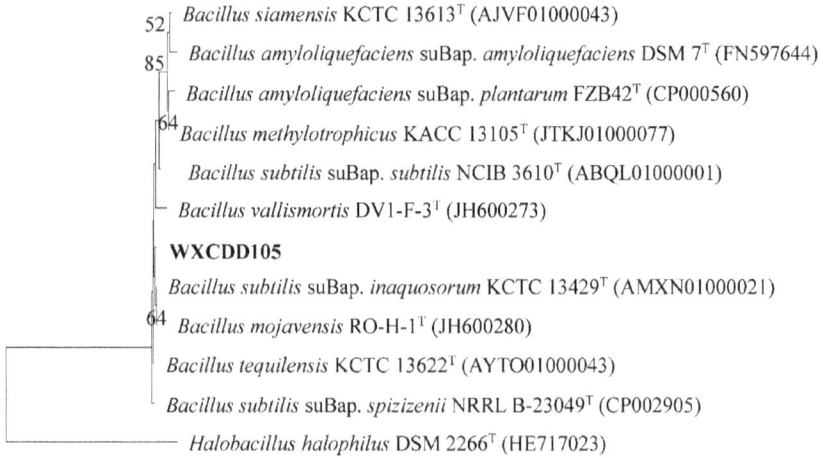

0.01                  (A)

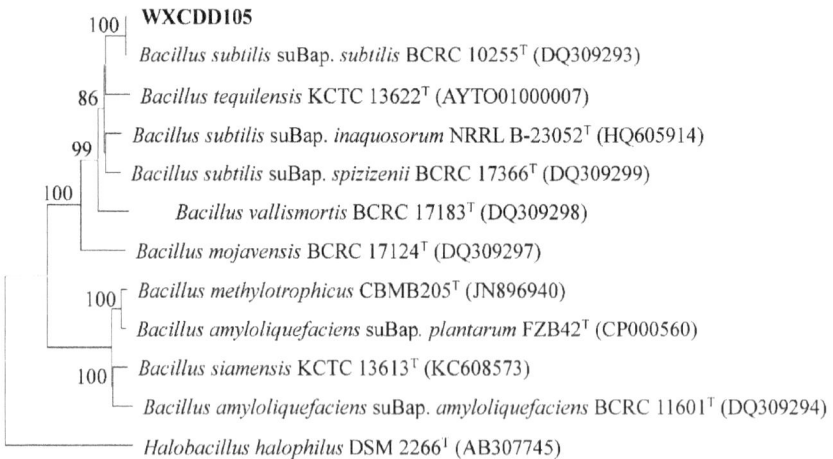

0.05                  (B)

**Figure 3.** Phylogenetic tree based on 16S rDNA (**A**) and *gyrB* (**B**) sequences of strain WXCDD105. The nodes of the phylogenetic tree where the value of bootstrap is greater than 50%, which will be noted in the graph, and the superscript "T" indicates the model strain.

### 2.3. Effect of Strain WXCDD105 Aseptic Filtrate on the Growth of Mycelia of Botrytis cinerea and Fuliva fulva

The bacterial strain WXCDD105 aseptic filtrate showed a significant inhibitory effect on the growth of the pathogens of tomato gray mold disease and tomato leaf mold disease. The hypha of *Botrytis cinerea* WD1 and the *Cladosporium fulvum* Cooke WD2 grow rarely in the strain WXCDD105 solution in contrast to the ddH₂O solution as a control, which is shown in Figure 4. The inhibitory rate of strain WXCDD105 on the growth of *Botrytis cinerea* WD1 mycelia and the growth of *Cladosporium fulvum*

Cooke WD2 mycelia was 95.28 and 94.44%, respectively. Additionally, there was no significant difference between them, which is shown in Table 3 below. The results indicated that the strain WXCDD105 can produce certain bacteriostatic substances and inhibit the growth of the tomato gray mold and leaf mold pathogens.

**Table 3.** Inhibitory effects of sterile filtrate of strain WXCDD105 on mycelial growth of pathogenic fungi.

| Treatment | *Botrytis cinerea* WD1 | | *Cladosporium fulvum* Cooke WD2 | |
|---|---|---|---|---|
| | Mycelium Dry Weight/g | Inhibition Rate/% | Mycelium Dry Weight/g | Inhibition Rate/% |
| WXCDD105 | $0.024 \pm 0.009$ | 95.28 | $0.029 \pm 0.006$ | 94.44 |
| Control | $0.508 \pm 0.014$ | - | $0.522 \pm 0.025$ | - |

| Control | WXCDD105 | Control | WXCDD105 |
| (a) WD1 | | (b) WD2 | |

**Figure 4.** Inhibitory effects of sterile filtrate of strain WXCDD105 on mycelial growth of *Botrytis cinerea* WD1 (**a**) and *Cladosporium fulvum* Cooke WD2 (**b**).

### 2.4. The Biocontrol Effect of Strain WXCDD105 on Tomato Gray Mold and Tomato Leaf Mold

A healthy tomato seedling grown to five real leaves was tested for efficacy. It was set for three repetitions with 10 seedlings repeating in temperature 25 °C and humidity 90% condition culture. After seven days of inoculation, the disease was observed day by day and the disease index and control effect were calculated. After seven days of inoculation of mildew and leaf mold, the incidence of individual leaves was increased and the number of leaves was gradually increased after 10 days. The symptoms on day 15 were obvious. As shown in Figure 5A,B. At this time, the disease index of tomato gray mold and leaf mold diseases were 74.70 and 72.07%, respectively. In addition, compared to the control of ddH$_2$O, the disease index by chemical pesticides and the strain WXCDD105 treatment was significantly lower. The protective effect of 40% pyrimethanil (chemical) on the tomato gray mold disease was 67.77% and the protective effect of 10% polyoxin (chemical) on the tomato leaf mold disease was 70.09%. WXCDD105 has a good biocontrol effect on *Botrytis cinerea* and *Cladosporium fulvum* Cooke simultaneously, and the control effects were respectively 74.70% and 72.07%, which are better than chemical pesticides of 40% ethyl. Pyrimethanil (preventing gray mold) and 10% polyoxin (preventing leaf mold), as shown in Table 4.

(A)                                                                    (B)

**Figure 5.** The biocontrol effects of tomato gray mold disease (**A**) and tomato leaf mold disease (**B**) infected tomato leaves.

**Table 4.** The control effects of strain WXCDD105 against tomato gray mold disease and tomato leaf mold disease on seedlings in the pot experiment.

| Treatment | Tomato Gray Mold Disease | | Tomato Leaf Mold Disease | |
|---|---|---|---|---|
| | Disease Index | Control Efficiency/% | Disease Index | Control Efficiency/% |
| WXCDD105 | 10.73 ± 2. 41 b | 74.70 | 12.09 ± 2.89 b | 72.07 |
| 40% Pyrimethanil | 13.67 ± 2.63 b | 67.77 | - | - |
| 10% Polyoxin | - | - | 12.95 ± 2.74 b | 70.09 |
| Control | 42.41 ± 1.58 a | - | 43.29 ± 1.53 a | - |

Notes: the different normal letters in the same column indicate significant difference among treatments at 0.05 level (*n* = 3).

### 2.5. Determination of Bacteriostasis Spectra of Strain WXCDD105

The results of the antibacterial spectrum determination are shown in Table 5. It can be observed from this table that the bacteriostasis spectrum of strain WXCDD105 is wide. The strain WXCDD105 has an inhibitory effect on 10 plant pathogenic fungi, which cause root and leaf diseases of crops. Among them, the inhibition rate of *Botrytis cinerea* was the highest at 97.37%. The inhibition rate of the *Fusarium oxysporum* f. sp. *niveum* disease was the lowest at 77.45%. The inhibition rate of other diseases was between 77 and 95%.

**Table 5.** Inhibitory effect of strain WXCDD105 on plant pathogenic fungi.

| Pathogenic Fungi | Colony Diameter (mm) | | Inhibition Rate/% |
|---|---|---|---|
| | Control | Treatment | |
| *Corynespora cassiicola* | 88.13 ± 0.78 | 11.49 ± 0.25 | 94.47 ab |
| *Botrytis cinerea* | 88.56 ± 0.99 | 9.13 ± 0.42 | 97.37 a |
| *Sclerotiniasclerotiorum* | 88.51 ± 0.55 | 16.79 ± 0.25 | 87.99 c |
| *Setosphaeria turcica* | 88.36 ± 2.70 | 13.24 ± 1.00 | 92.33 bc |
| *Fusarium oxysporum* f. sp. *cucumerinum* | 88.83 ± 0.38 | 16.17 ± 1.72 | 88.79 c |
| *Fusarium oxysporum* f. sp. *niveum* | 88.33 ± 3.85 | 25.34 ± 5.68 | 77.45 d |
| *Septoria lycopersici Speg.* | 88.15 ± 0.45 | 14.86 ± 0.49 | 90.31 bc |
| *Fusarium oxysporium Schelcht* | 88.11 ± 1.57 | 23.98 ± 3.18 | 79.07 d |
| *Fusarium oxysporum* f. sp. *lycopersici* | 88.89 ± 2.15 | 16.04 ± 0.02 | 88.96 c |
| *Fusarium oxysporum* f. sp. *melonis* | 88.50 ± 1.57 | 22.41 ± 0.83 | 81.08 d |

Notes: the different normal letters in the same column indicate significant difference among treatments at 0.05 level (*n* = 3).

## 2.6. Growth-Promoting Effects of Strain WXCDD105

Tomato seeds were soaked in the strain WXCDD105 with a concentration of $10^8$ cfu·mL$^{-1}$ for 3 h. The germination rate and the root growth of the radicle were measured and the result is shown in Figure 6. The strain WXCDD105 can promote the growth of tomato seedlings to a certain extent. The germination rate of tomato seedlings which were treated with the strain WXCDD105 was 79.00% at the fourth day after treatment, while the germination rates of tomato seedlings treated with the *Bacillus subtilis* WZ-1 and clear water were 72.00 and 71.00%, respectively, during the fourth day after treatment. The strain WZ-1 was isolated in our lab. The effects of three treatments of the radicle length of the sprouting tomato had no significant difference in the first three days. But during the fourth day, the radicle lengths of the tomato seed soaked with strain WXCDD10, WZ-1, and water were 14.52, 19.34, and 13.26 mm, respectively. During the fifth day, the radicle lengths of the tomato seed soaked with the strain WXCDD10, WZ-1, and water were 42.78, 43.94, and 36.08 mm, respectively. The radicle length of tomato seeds soaked with the strain WXCDD10 increased 18.57% over the control group soaked with water and was not significantly different to the control group soaked with strain WZ-1.

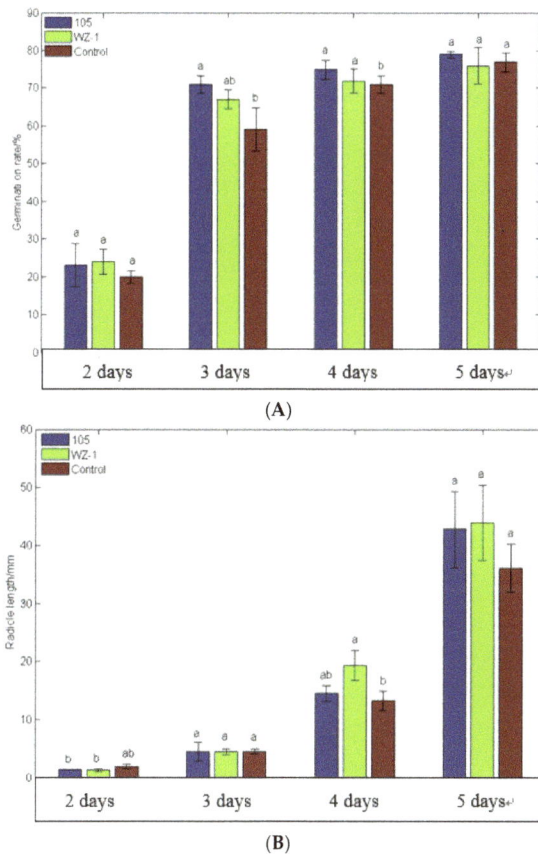

**Figure 6.** Effects of the strain WXCDD105, the strain WZ-1 and the water control on seed germination (**A**) and radicle growth (**B**) of tomato. The different normal letters in the same point indicate significant difference among treatments at 0.05 level (*n* = 3).

Table 6 below shows the effect of the strain WXCDD105 on the growth promotion of tomato seedlings. It can be seen from the table that different concentrations of strain WXCDD105 have certain effects on the growth of tomato seedlings at the 4–5 leaf stage. In addition, the bacterial liquid with a concentration of $10^6$ cfu·mL$^{-1}$ had the best promoting effect on tomato seedlings. The plant height, stem diameter, main root length, and whole plant fresh weight of tomato seedlings treated with bacteria liquid with a concentration of $10^7$ cfu·mL$^{-1}$, $10^6$ cfu·mL$^{-1}$ were significantly higher than the control. Compared with the water control group, the plant height, stem diameter, main root length, and whole plant fresh weight of tomato seedlings treated by bacterial liquid with a concentration of $10^6$ cfu·mL$^{-1}$ increased by 66.34, 41.82, 112.33, and 145.22%, respectively. Additionally, there were differences between different concentrations of bacterial liquid.

**Table 6.** The growth-promoting effect of single strain WXCDD105 on tomato seedlings.

| Processing Method | Plant Height/cm | Stem Diameter/mm | Main Root Length/cm | Whole Plant Fresh Weight/g |
|---|---|---|---|---|
| 10 times dilution | 15.30 ± 0.72 b | 3.21 ± 0.04 b | 13.13 ± 1.78 b | 3.52 ± 0.91 b |
| 100 times dilution | 17.30 ± 1.31 a | 3.90 ± 0.14 a | 17.90 ± 1.84 a | 5.64 ± 0.14 a |
| Control | 10.40 ± 0.85 c | 2.75 ± 0.09 c | 8.43 ± 1.81 c | 2.30 ± 0.84 b |

Notes: the different normal letters in the same column indicate significant difference among treatments at 0.05 level (*n* = 3).

### 2.7. The Effect of Strain WXCDD105 on the Physiological Quality of Tomato Fruit

With the extension of storage time, there was not much difference between the changes in storage from three to nine days. The mass loss rate increased sharply and the control increased significantly. However, the quality loss of fruit treated by strain WXCDD105 was lower than the control (Figure 7). After 15 days of storage, the rotten rate of processing the tomato fruit treated by strain WXCDD105 bacterial suspension was significantly lower than the control group. The strain WXCDD105-treated decay rate was 29.45% and the biocontrol bacteria strain WZ-1 decay rate was 38.14%, while the control group had a decay rate of 57.07% (Table 7). The strain WXCDD105 has a superior antiseptic effect and strong effects on storage of the biocontrol bacterium WZ-1. After 15 days of storage, the control fruit became soft compared with the biocontrol bacteria strain WXCDD105, and WZ-1-treatment fruit firmness was 0.68, 0.70, and 0.56 kg·cm$^{-2}$, respectively. The biocontrol agent strain WXCDD105 can delay fruit softening, but has no significant difference with the biocontrol agent strain WZ-1. At the end of storage for different groups of soluble solids, the titratable acid content was determined. The results showed that there was no significant difference in the nutrient content between the treatments (see Table 7).

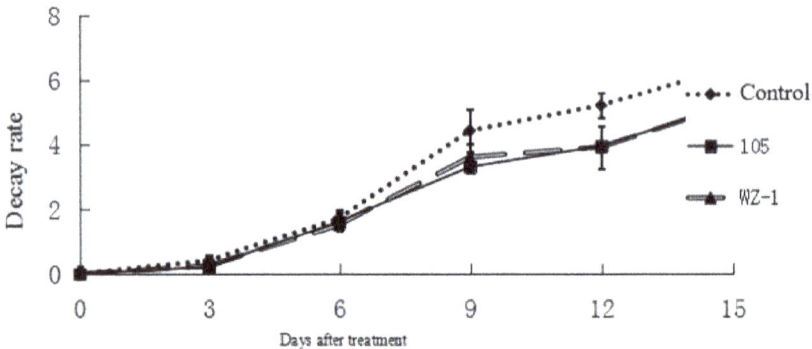

**Figure 7.** Effects of different treatments on weight-loss rate of tomato.

**Table 7.** Effects of different treatments on fruit quality of tomato.

| Treatment | Rotting Rate/% | Firmness/kg·cm$^{-2}$ | Soluble Solids Content/% | Titratable Acid Content/% |
|---|---|---|---|---|
| Control | 56.67 ± 1.21 a | 0.57 ± 0.06 b | 6.24 ± 0.21 a | 0.32 ± 0.13 a |
| WZ-1 | 38.34 ± 1.13 b | 0.71 ± 0.02 a | 6.25 ± 0.13 a | 0.33 ± 0.04 a |
| WXCDD105 | 28.15 ± 1.02 c | 0.71 ± 0.03 a | 6.27 ± 0.22 a | 0.35 ± 0.06 a |

Notes: the different normal letters in the same column indicate significant difference among treatments at 0.05 level (*n* = 3).

## 3. Discussion

In recent years, the use of *Bacillus subtilis* and biological control of the plant have been carried out extensively [17]. In these studies, some strains can produce Idole acetic acid (IAA) and some strains can produce iron cells [18,19]. The *Bacillus* strains can significantly inhibit the growth of pathogenic bacteria and pathogenic bacteria biofilm synthesis of the cell wall that will cause severe damage to plants and lead to cell lysis of pathogens. Additionally, the interaction of rhizosphere bacteria and plants had the effect of promoting plant growth [20]. Previous research showed that *Bacillus subtilis* during the process of phosphorylation has an ability to dissolve inorganic phosphate and can increase the utilization rate of phosphorus in the soil. *Bacillus* resistance can lead to the secretion of a spectrum of the compounds or the induced resistance of plants to increase system capacity [21]. *Bacillus* produces siderophores at the same time in order to adhere to the surface of non-biological biofilm formation, which makes the *Bacillus* secrete broad-spectrum resistance compounds to a certain extent and increase the spectrum resistance of plants [22]. The tomato plant has attracted increasing attention because of its special flavor, rich vitamins, and its excellent antioxidation effects. Tomato plants are planted in many areas in China but are vulnerable to many diseases such as fungi, bacteria, and nematodes both in the greenhouse and field [23,24]. Tomato gray mold and tomato leaf mold diseases are the main causes of large area reductions in tomato cultivation. These two diseases have mainly occurred in the aboveground parts of plants, which seriously threatens plant growth and development [25]. Boukaewet reported that *Streptomyces* is a genus known for its ability to protect plants against many pathogens and various strains of this bacteria have been used as biological control agents [26]. At present, the prevention and treatment of these two diseases are mainly based on chemical control, but the long-term use of chemical fungicides will lead to food safety, environmental pollution, and pathogenic bacterial resistance. A series of problems such as a single control effect has serious harm. A growing number of researchers are paying attention to these problems. More people began to use biological control methods in the control of disease and made great progress. Biocontrol agents mainly *Bacillus*, *Trichoderma*, *Pseudomonas*, nonpathogenic *Fusarium*, and *Penicillium* strains were evaluated to control *Fusarium* wilt, but still this lethal disease could not be controlled completely [23]. Biological or biological metabolites provide effective protection for animal and plant control. The biggest feature of this form of protection is not the fact that it will not destroy the natural environment and will not hurt people, animals, and plants, but also that it will not pollute or destroy the normal natural environment and has the advantages of controllability and stability. In recent years, the application of *Bacillus* spore to control plant diseases has been reported worldwide. The application includes the effect of *Bacillus subtilis* on the growth immune function and intestinal morphology in weaned piglets. The present study compared the effects of soybean meal fermented by three different probiotics organisms with a non-fermented soybean meal on the growth performance, serum parameters, immune chemistry, and intestinal morphology of weaned piglets [23]. The *Bacillus subtilis* strain WXCDD105 obtained in this study can prevent *Corynesporacassiicola*, *Botrytiscinerea*, *Sclerotiniasclerotiorum*, *Setosphaeria turcica*, among others. The control effect is more than 80%. In particular, the effect of the prevention of gray mold was 10.26% higher than that of 40% Pyrimethanil. Additionally, it was higher than for the control of tomato leaf mold by 10% Polyoxin. In this study, 126 strains of bacteria were isolated from soil. Five of them were 105, 51, 98, 15, and 109, which had some control effect of *Botrytis cinerea* and *Cladosporium fulvum* Cooke, while 105 had the best control effect against both diseases. The mechanism of *Bacillus subtilis* disease prevention has been studied. Asaka reported that *Bacillus subtilis* secreted antibiotics iturin A and surfactin for controlling plant diseases [27]. *Bacillus subtilis* enhanced the expression of the chitinase promoter [28]. Such bacteria indicated expression of both chiS and chiL chitinases under the control of a strong promoter. The activity of this promoter was evaluated in *B. subtilis* as a host organism. Our study found that 105 of the aseptic filtrate had a significant control effect on tomato gray mold and tomato leaf mold mycelia growth and all of them reached over 90%. This also indicates that these 105 can produce some antibacterial substances during the growth process, which stops plant diseases. Al-Masri studied this in only 73% of cases [29].

On the general antibacterial range, especially in terms of the *Botrytis cinerea* and tomato *Cladosporium fulvum* Cooke, the significant control effect not only showed that strain WXCDD105 has a certain promoting effect on tomato seeds and seedlings, but also the characteristics necessary to maintain the quality of the tomato fruit. The firmness compared to the control group increased 21.42% and greatly reduced the decay rate of the tomato. However, tomato firmness, soluble solids content, and titratable acid content had no significant difference when compared to those with water in the control group. Fresh weight may be the cause of the seedling test, where there is no big difference; but as the plant grows, there may be large differences. We also found that the biocontrol strain WZ-1-treated tomato fruit reached the approximate treatment effect. Whether the biocontrol affected the tomato fruit and produced a similar biocontrol substance, and whether this is the source of plant-induced disease resistance, this is how the role of the antimicrobial substance in the tomato induced the overexpression of the tomato gene and its mechanism of action. This mechanism needs to be studied further.

## 4. Materials and Methods

### 4.1. Materials

Pathogens *Botrytis cinerea* WD1, *Cladosporium fulvum* Cooke WD2, *Corynespora cassiicola* HH, *Botrytis cinerea* WS1, *Sclerotinia sclerotiorum* WS2, *Setosphaeria turcica* YD, *Fusarium oxysporum* f. sp. *cucumerinum* WS3, *Fusarium oxysporum* f. sp. *niveum* XK, *Septoria lycopersici* Speg. FB, *Fusarium oxysporium* Schelcht SL, *Fusarium oxysporum* f. sp. *lycopersici* FK, and *Fusarium oxysporum* f. sp. *melonis* TK were kept and propagated at 25 °C on potato dextrose agar (PDA) plates in our lab.

Additionally, 126 strains of bacteria were isolated from the rhizosphere soil of a carrot, *Portulaca*, *Solanum nigrum*, a radish, wheat, soybean, Chinese cabbage, and a tomato. The biocontrol bacteria strain WXCDD105 was screened for 126 strains and originated from tomato rhizosphere soil. *Bacillus* sp. X was maintained in our lab. Previous studies have shown that it provided effective control of tomato gray mold, promoted tomato plant growth, and had a preservation effect. All bacteria were cultured at 37 °C on LB medium.

The tomato cv. Dongnong 713 was provided by the Tomato Research Institute of Northeast Agricultural University in China. It has no resistance to *Botrytis cinerea* and *Cladosporium fulvum* Cooke. The plants were grown from seeds in growth chambers at 25/22 (day/night), 75% relative humidity, and a 16/8 h (day/night) photoperiod. Healthy tomato fruit and leaves with similar sizes and colors were used in this study. The fruit and leaves were free from visible wounds and diseases. Before each treatment, the fruit was washed with water and then in 75% ethanol for 30 s, rinsed in sterile water, and air dried.

### 4.2. Isolation and Screening of Antagonistic Bacteria

Soil samples were collected from different fields. The sterile water was mixed with one gram of soil samples with 9 mL of mixing and dilution ($10^{-1}$–$10^{-6}$). 100 µL of sample from $10^{-6}$ dilution were plated and spread separately onto LB medium and incubated at 37 °C for 24 h. The same morphology, color, and size of colonies was picked up from the LB medium.

The pathogenic fungi (*Botrytis cinerea* WD1 and *Cladosporium fulvum* Cooke WD2) were inoculated on potato dextrose agar (PDA) medium for 1 week. A mycelia disk (7 mm diameter) of each pathogen was transferred to the center of the freshly prepared PDA medium on different plates and the bacteria were inoculated onto the areas 20 mm from the central pathogen colony, which was followed by incubation at 28 °C for 7 days. Each experiment was replicated three times and the average zone of inhibition was calculated.

Antagonistic strains were screened repeatedly by using the inhibition zone method. 50 μL of the spore suspension of *Botrytis cinerea* WD1 and *Cladosporium fulvum* Cooke WD2 ($10^5$ cfu·mL$^{-1}$) were plated and spread separately onto PDA medium and incubated at 25 °C for 14 days. The Oxford cup was transferred to the center and the antagonistic bacteria fermentation broth was added to the Oxford cup. Each experiment was replicated three times and the average zone of effects was measured.

### 4.3. Morphological and Biochemical Characteristics of Strain WXCDD105

According to Bergey's Manual of Determinative Bacteriology [12], the morphological characteristics and physiological and biochemical characteristics of the strain WXCDD105 were identified by conventional methods.

Genomic DNA extraction of strain WXCDD105 was performed as described by the CTAB method and PCR amplification of the 16S rRNA gene was carried out with universal primers (8F: 5′-AGAGTTTGATCMTGGCTCAG-3′, 1492R: 5′-ACGGATACCTTGTTACGACTT-3′). The conditions used for the thermal cycler (TaKaRa, Dalian, China) were measured according to the following amplification profile: 94 °C for 3 min followed by 30 cycles consisting of denaturation at 94 °C for 30 s, primer annealing at 55 °C for 30 s and primer extension at 72 °C for 1 min. At the end of the cycles, the reaction mixture was kept at 72 °C for 10 min and then cooled to 4 °C. PCR amplification of the *gayB* gene was carried out with universal primers (UP-1: 5′-GAAGTCATCATGACCGTTCTGCAYGCNGGNGGNAARTTYGA-3′, UP-2r: 5′-AGCAGGGTACGGATGTGCGAGCCRTCNACRTCNGCRTCNGCRTCNGTCAT). The conditions used for the thermal cycler (TaKaRa) were based on the following amplification profile: 96 °C for 5 min followed by 30 cycles consisting of denaturation at 94 °C for 30 s, primer annealing at 63 °C for 90 s, and primer extension at 73 °C for 90 s. At the end of the cycles, the reaction mixture was kept at 72 °C for 10 min and then cooled to 4 °C. The extracted product was recovered and purified by a gel recovery kit and was sequenced by the Beijing Genomics Institute (BGI) (Beijing, China). The 16S rRNA and *gayB* gene sequences determined were compared with the GenBank database by using the BLAST search program. The phylogenetic and molecular evolutionary analyses were conducted using the software included in the MEGA 6.06 package. The 16S rRNA and *gyrB* gene sequences of the strain WXCDD105 and 11 types of model strains were aligned using the ClustalX 1.83 program [30]. The evolutionary tree was inferred by using the neighbor-joining method [31] from the evolutionary distance data corrected by Kimura's two-parameter model [32], and the topology of the phylogenetic tree was evaluated by the bootstrap resampling method of 1000 replicates [33].

### 4.4. The Inhibiting Effect of the Biocontrol Bacterium WXCDD105 Fermentation Filtrate on the Growth of Tomato-Grown Mold and Leaf Mold

We transferred the bacterial strain WXCDD105 to LB broth, 30 °C, 200 rpm shake culture for 72 h, the 10,000 rpm centrifuge for 10 min, and took on a clear liquid with 0.22 μm aseptic filtrate microporous membrane filter. The strain WXCDD105 aseptic filtrate was added to the same amount of aseptic 2-fold concentrated PDA medium to produce the medium. Afterward, 50 mL containing medium was added to 250 mL triangular bottles and inoculated with two strains of *Botrytis cinerea* WD1 and *Cladosporium fulvum* Cooke WD2 (7 mm in diameter). This was maintained in sterile water for contrast, trained for 5 days in the condition of 28 °C, 120 rpm, filtered by gauze, and then the hyphae were collected, weighed, and the inhibition rate calculated. Each process was repeated three times.

*4.5. Antifungal Activity Assay Using Strain WXCDD105 in the Greenhouse*

Tomato seeds were surface-sterilized for 15 min in 2% sodium hypochlorite followed by a brief rinse with 75% ethanol. Seeds were sown on a seedling tray filled with commercially available horticulture soil. The soil used in the experiment was steam-sterilized at 121 °C for 30 min three times. The seedlings were grown in the greenhouse. After four weeks, individual seedlings were transplanted into plastic pots (10 cm in diameter) and filled with the same sterile soil. Plants were watered twice a day. Plants were maintained at 20–30 °C in the greenhouse for two months. *Botrytis cinerea* WD1, *Cladosporium fulvum* Cooke WD2 were cultured on a PDA plate at 20 °C for two weeks. A conidial suspension was diluted in potato dextrose broth after incubation to reach the final concentration of $10^5$ cfu·mL$^{-1}$, which was determined with a hemocytometer. Tomato seedlings were treated with the fermentation liquid of strain WXCDD105 ($10^8$·cfu·mL$^{-1}$) for three consecutive weeks (once a week) by spray inoculation. Then *Botrytis cinerea* WD1 and *Cladosporium fulvum* Cooke WD2 were inoculated. Pyrimethanil (40% wettable powder) and polyoxin (10% wettable powder) were used as positive controls for biocontrol and fungicide effects, respectively, and water was used as the negative control and applied at the same time as the antagonists. The inoculated plants were kept in a moist chamber and maintained at approximately 90% humidity at 25 °C. Disease severity was evaluated at 14 days after inoculation on a scale of 0 to 4. Level 0 stands for no disease, level 1 represents morbidity $\leq 25\%$, level 2 represents morbidity 25.1–50%, level 3 represents morbidity 50.1–75%, level 4 represents morbidity $\geq 75.1\%$. There were three replications for each treatment. All 30 plants per treatment were used for disease symptom investigation. The plant disease index (DI), which would represent both disease incidence and symptom severity, can be calculated as: DI = ($\Sigma$Di × Dd)/(Mi × Md) × 100 where i means a 0–4 disease level and Mi means plant number of reaction i, Dd means total number of leaves investigated.

*4.6. Determination of Biocontrol Bacterial Antibacterial Spectrum*

The antimicrobial activity of the bacterial strain WXCDD105 on 10 plant pathogenic fungi was determined by using the mycelial growth rate method. We took 100 μm bioantibacterial fluid ($10^8$ cfu·mL$^{-1}$) on the PDA plate and inoculated the pathogenic bacteria cake (7 mm in diameter) at the center of the tablet with the sterile LB medium as the control at 28 °C. When the pathogens in the control group were covered the whole petri dish, we measured the colony diameter and calculated the inhibition rate. Each experiment was repeated three times. Bacteriostatic rate (%) = [(control group pathogenic bacteria diameter − experimental group pathogenic bacteria diameter)/(control group pathogenic bacteria diameter − 7)] × 100.

*4.7. Effects of Biocontrol Bacteria on the Growth of Tomato Seeds and Seedlings*

Same-size grain tomato seeds were selected and added to 2% sodium hypochlorite ($v/v$) for 15 min, 75% alcohol, 30 s treatment, then washed in a strain WXCDD105 bacterial liquid of 10 mL after washing 3 times in sterile water (concentration of $10^8$ cfu·mL$^{-1}$). After soaking for 3 h, they were dried and placed in a sterile Petri dish. The filter paper held 50 grains per dish and included 5 dishes. The cultivation of the moisturizing box was maintained at a temperature of 25 °C and water treatment, and the strain WZ-1 was used as the control. Records of the tomato seed germination rate and radicle length were reported at 2–5 days. Each treatment was repeated 3 times.

Tomato seeds were seeded in plantlets containing sterilized soil and managed in a greenhouse after disinfection. Tomato seedlings with the same growth were selected after 10 days of seeding and were treated with strain WXCDD105 liquid, respectively. $10^6$ cfu·mL$^{-1}$ mixed soil was treated with clear water as the control. The optimum concentration was screened and seed soaking was used. After soaking tomato seeds with the best dilution ratio of bacteria for 3 h, the seeds were seeded on a seedling tray until the seedlings grew to 2 leaves and 1 heart. This was transplanted into a nutritious bowl. Soaking seeds in sterile water was used as a control. Seeds were disinfected and seeded in a seeding tray. Uniformly grown seedlings were positioned in the pot with the best dilution broth with root treatment. Each bowl held 20 mL and sterile water irrigation treatment was used as a control. Seed disinfection after sowing to seedling trays was used to grow to 2 leaves and 1 heart. Uniform grown seedlings were transplanted into pots with each bowl containing an equal amount of nutrient soil mixed with 20 mL optimum dilution of bacteria liquid with sterile water and soil treatment used as a control. After the completion of each treatment in the greenhouse under routine management, when the plants reached the 4–5 leaf stage they underwent a determination of the plant height, stem diameter and root length.

### 4.8. The Effects of Bacteria Strain WXCDD105 on the Physiological Quality of Tomato Fruits

The fruits of the tomato that had no damage and disease with the same size maturity were graded, weighed and washed with sterile water. They were soaked for 4 min in strain WXCDD105 bacterial suspension liquid ($10^7$ cfu·mL$^{-1}$). In addition, they were dredged to dry and store in a fresh bag at room temperature. The condition of being putrefied was observed and weighed every 3 days and the weight loss rate was calculated. After the storage period, some indexes were tested including firmness, soluble solid, and titratable acid content. The test was treated and contrasted with aseptic water and the antibacterial X was treated as a positive control. In total, 30 fruits were processed in each group and each treatment was repeated 3 times.

Fruit rot rate: fruit rot rate of tomato (%) = rotten number of fruits/total number of fruits × 100, Weight loss rate: fruit weightlessness (%) = (initial fruit weight − survey fruit weight)/initial fruit weight × 100. Determination of the firmness of the fruit: using the GY-4 fruit firmness tester to record readings (Zhejiang Top Instrument Co., LTD., Hangzhou, China). Determination of soluble solids: using a hand-held refractometer to record readings. Titration acid was determined by acid base titration.

Excel 2013 (Microsoft) software was used to carry out the statistical analysis, DPS (Data processing system) V15.10 (Hangzhou Ruifeng Information Technology Co. LTD., Hangzhou, China) to carry out data analysis, and Duncan's test was performed.

### 5. Conclusions

We isolated a bacterium that had a good control effect on *Botrytis cinerea* and *Cladosporium fulvum* Cooke from the rhizosphere soil of a healthy plant and identified it as *Bacillus subtilis*. The strain not only showed significant potential for biological control but also promoted the growth of tomato seedlings.

**Author Contributions:** A.W. conceived and designed the experiments; Y.S., D.W., H.W. and Y.W. performed the experiments; H.W., J.L. and S.Z. analyzed the data; Z.Y. contributed reagents/materials/analysis tools; A.W. and H.W. and Y.S. wrote the paper; H.W., Y.S. contributed equally to this study and share first authorship.

**Acknowledgments:** This work was financially supported by research fund of Harbin Municipal Science and Techonology Bureau for leaders in specific discipines (2016RAXXJ051) and Heilongjiang Provincial Science Fund for Distinguished Young Scholars (JC2015004) the China Postdoctoral Science Foundation (No. 2017M621235) to A.W., the Postdoctoral Science Foundation of Heilongjiang Province (No. LBH-Z17034), the "Young Talents" Project of North-East Agricultural University (No. 17Q26) to J.L.

**Conflicts of Interest:** The authors declare no conflict of interest.

### Abbreviations

| | |
|---|---|
| WD1 | *Botrytis cinerea* |
| WD2 | *Cladosporium fulvum* Cooke |
| LB | Luria–Bertani |
| PDA | Potato dextrose agar |
| PDB | Potato dextrose broth |
| DI | Disease index |

| | |
|---|---|
| BXYL | Beta-XYLOSIDASE |
| LysA | L-Lysine-ARYLAMIDASE |
| AspA | Aspartic-ARYLAMIDASE |
| LeuA | Leucine-ARYLAMIDASE |
| PheA | Phenylalanine ARYLAMIDASE |
| ProA | Proline ARYLAMIDASE |
| BGAL | BETA-GALACTOSIDASE |
| PyrA | Pyrrolydonyl ARYLAMIDASE |
| AGAL | ALPHA-GALACTOSIDASE |
| AlaA | Alanine ARYLAMIDASE |
| TyrA | Tyrosine ARYLAMIDASE |
| BNAG | BETA-N-ACETYL-GLUCOSAMINIDASE |
| APPA | Ala-Phe-Pro-ARYLAMIDASE |
| CDEX | Alanine-phenylalanine-proline ARYLAMIDASE |
| dGAL | D-GALACTOSE |
| GLYG | Glycogen |
| INO | Inositol |
| MdG | Methyl glucoside acidification |
| ELLM | ELLMAN |
| MdX | Methyl-d-xyloside |
| AMAN | Alpha mannosidase |
| MTE | Germination three sugar |
| GlyA | Glycine ARYLAMIDASE |
| dMAN | D-MANNITOL |
| dMNE | D-MANNOSE |
| dMLZ | D-MELEZITOSE |
| NAG | N-ACETYL-GLUCOSAMINE |
| PLE | PALATINOSE |
| IRHA | L-rhamnose monohydrate |
| BGLU | BETA-GLUCOSIDASE |
| INU | Inulin |
| dGLU | D-GLUCOSE |
| dRIB | Ribose |
| PSCNa | Decomposing ammonia assimilation |
| OLD | Old sugar |
| ESC | ESCULIN hydrolyse |
| TTZ | Polymyxin B resistance |
| POLYB_R | Polymyxin B tolerance |

## References

1. Al-Saleh, M.A. Pathogenic variability among five bacterial isolates of *Xanthomonas campestris* pv. vesicatoria, causing spot disease on tomato and their response to salicylic acid. *J. Saudi Soc. Agric. Sci.* **2011**, *10*, 47–51. [CrossRef]
2. Babadoost, M. Leaf mold (*Fulvia fulva*), a serious threat to high tunnel tomato production in illinois. *Acta Hortic.* **2011**, *914*, 93–96. [CrossRef]
3. Williamson, B.; Tudzynski, B.P.; Kan, J.A.L.V. *Botrytis cinerea*: The cause of grey mould disease. *Mol. Plant Pathol.* **2007**, *8*, 561–580. [CrossRef] [PubMed]
4. Hong, S.J. National Academy of Agricultural Science, RDA, Suwon, Republic of Korea, Study on the Control of Leaf Mold, Powdery Mildew and Gray Mold for Organic Tomato Cultivation. *Korean J. Organ. Agric.* **2012**, *20*, 655–668. [CrossRef]
5. Wahab, S. Biotechnological approaches in the management of plant pests, diseases and weeds for sustainable agriculture. *J. Biopestic.* **2009**, *2*, 115–134.

6. Correa, O.S.; Soria, M.A. Potential of Bacilli for Biocontrol and Its Exploitation in Sustainable Agriculture. *Microbiol. Monogr.* **2010**, *18*, 197–209.

7. Medeiros, F.H.V.D.; Martins, S.J.; Zucchi, T.D.; Melo, I.S.D.; Batista, L.R.; Machado, J.D.C. Controle biológico de fungos de armazenamento produtores de micotoxinas. *Ciênc Agrotec* **2012**, *36*, 483–497. [CrossRef]

8. Smith, S.M. Biological control with Trichogramma: Advances, successes, and potential of their use. *Annu. Rev. Entomol.* **1996**, *41*, 375–406. [CrossRef] [PubMed]

9. Yuan, Y.; Feng, H.; Wang, L.; Li, Z.; Shi, Y.; Zhao, L.; Feng, Z.; Zhu, H. Potential of Endophytic Fungi Isolated from Cotton Roots for Biological Control against Verticillium Wilt Disease. *PLoS ONE* **2017**, *12*, e0170557. [CrossRef] [PubMed]

10. Heimpel, G.E.; Ragsdale, D.W.; Venette, R.; Hopper, K.R.; O'Neil, R.J.; Rutledge, C.E.; Wu, Z.S. Prospects for importation biological control of the soybean aphid: Anticipating potential costs and benefits. *Ann. Entomol. Soc. Am.* **2004**, *97*, 249–258. [CrossRef]

11. Raspor, P.; Mikličmilek, D.; Avbelj, M.; Čadež, N. Biocontrol of grey mould disease on grape caused by *Botrytis cinerea* with autochthonous wine yeasts. *Food Technol. Biotechnol.* **2010**, *48*, 336–343.

12. Elad, Y.; Kohl, J.; Fokkema, N.J. Control of infection and sporulation of *Botrytis cinerea* in bean and tomato by saprophytic bacteria and fungi. *Eur. J. Plant Pathol.* **1994**, *100*, 315–336. [CrossRef]

13. Stein, T. *Bacillus subtilis* antibiotics: Structures, syntheses and specific functions. *Mol. Microbiol.* **2005**, *56*, 845–857. [CrossRef] [PubMed]

14. Sonenshein, A.L.; Hoch, J.A.; Losick, R. *Bacillus Subtilis and Its Closest Relatives: From Genes to Cells*; American Society of Microbiology: Washington, DC, USA, 2002; pp. 263–264.

15. Dubnau, D.; Losick, R. Bistability in bacteria. *Mol. Microbiol.* **2010**, *61*, 564–572. [CrossRef] [PubMed]

16. Yavuztürk, B.G.; Imer, D.Y.; Park, P.K.; Koyuncu, I. Evaluation of a novel anti-biofouling microorganism (*Bacillus* sp. T5) for control of membrane biofouling and its effect on bacterial community structure in membrane bioreactors. *Water Sci. Technol.* **2018**, *77*, 971–978. [CrossRef] [PubMed]

17. Davey, M.E.; O'Toole, G.A. Microbial Biofilms: From Ecology to Molecular Genetics. *Microbiol. Mol. Biol. Rev.* **2000**, *64*, 847–867. [CrossRef] [PubMed]

18. Idris, E.S.E.; Iglesias, D.J.; Talon, M.; Borriss, R. Tryptophan-Dependent Production of Indole-3-Acetic Acid (IAA) Affects Level of Plant Growth Promotion by Bacillus amyloliquefaciens FZB42. *Mol. Plant-Microbe Interact.* **2007**, *20*, 619–626. [CrossRef] [PubMed]

19. Kakar, K.U.; Duan, Y.P.; Nawaz, Z.; Sun, G.; Almoneafy, A.A.; Hassan, M.A.; Elshakh, A.; Li, B.; Xie, G.L. A novel rhizobacterium Bk7 for biological control of brown sheath rot of rice caused by *Pseudomonas fuscovaginae* and its mode of action. *Eur. J. Plant Pathol.* **2014**, *138*, 819–834. [CrossRef]

20. Bais, H.P.; Fall, R.; Vivanco, J.M. Biocontrol of *Bacillus subtilis* against infection of Arabidopsis roots by *Pseudomonas syringae* is facilitated by biofilm formation and surfactin production. *Plant Physiol.* **2004**, *134*, 307–319. [CrossRef] [PubMed]

21. Kakar, K.U.; Nawaz, Z.; Cui, Z.; Almoneafy, A.A.; Zhu, B.; Xie, G.L. Characterizing the mode of action of *Brevibacillus laterosporus* B4 for control of bacterial brown strip of rice caused by *A. avenae* subsp. avenae RS-1. *World J. Microbiol. Biotechnol.* **2014**, *30*, 469–478. [CrossRef] [PubMed]

22. Höfte, M.; Bakker, P.A.H.M. *Competition for Iron and Induced Systemic Resistance by Siderophores of Plant Growth Promoting Rhizobacteria*; Springer: Berlin/Heidelberg, Germany, 2007; pp. 121–133.

23. Rashad, Y.M.; Al-Askar, A.A.; Ghoneem, K.M.; Saber, W.I.A.; Hafez, E.E. Chitinolytic *Streptomyces griseorubens* E44G enhances the biocontrol efficacy against *Fusarium* wilt disease of tomato. *Phytoparasitica* **2017**, *45*, 227–237. [CrossRef]

24. Abdalla, O.A.; Bibi, S.; Zhang, S. Application of plant growth-promoting rhizobacteria to control *Papaya ringspot* virus and Tomato chlorotic spot virus. *Arch. Phytopathol. Plant Protect.* **2017**, *54*, 584–597. [CrossRef]

25. Anton, D.; Bender, I.; Kaart, T.; Roasto, M.; Heinonen, M.; Luik, A.; Püssa, T. Changes in Polyphenols Contents and Antioxidant Capacities of Organically and Conventionally Cultivated Tomato (*Solanum lycopersicum* L.) Fruits during Ripening. *Int. J. Anal. Chem.* **2017**, *2017*, 2367453. [CrossRef] [PubMed]

26. Boukaew, S.; Prasertsan, P.; Troulet, C.; Bardin, M. Biological control of tomato gray mold caused by *Botrytis cinerea* by using *Streptomyces* spp. *Biocontrol* **2017**, *62*, 793–803. [CrossRef]

27. Asaka, O.; Shoda, M. Biocontrol of *Rhizoctonia solani* Damping-Off of Tomato with *Bacillus subtilis* RB14. *Appl. Environ. Microbiol.* **1996**, *62*, 4081–4085. [PubMed]

28.   Shali, A.; Rigi, G.; Pornour, M.; Ahmadian, G. Expression and Secretion of Cyan Fluorescent Protein (CFP) in *B. subtilis* using the Chitinase Promoter from *Bacillus pumilus* SG2. *Iran. Biomed. J.* **2017**, *21*, 240. [CrossRef] [PubMed]

29.   Al-Masri, M.I. Biological Control of Gray Mold Disease (*Botrytis cinerea*) on Tomato and Bean Plants by Using Local Isolates of Trichoderma harzianum. *Dirasat Agric. Sci.* **2005**, *32*, 145–156.

30.   Higgins, D.; Thompson, J.; Gibson, T.W. Improving the sensitivity of progressive multiple sequence alignment through sequence weighting, position-specific gap penalties and weight matrix. *Nucleic Acids Res.* **1994**, *22*, 4673–4680.

31.   Saitou, N.; Nei, M. The neighbor-joining method: A new method for reconstructing phylogenetic trees. *Mol. Biol. Evol.* **1987**, *4*, 406. [PubMed]

32.   Kimura, M. A simple method for estimating evolutionary rates of base substitutions through comparative studies of nucleotide sequences. *J. Mol. Evol.* **1980**, *16*, 111–120. [CrossRef] [PubMed]

33.   Boshart, M.; Weber, F.; Jahn, G.; Dorsch-Häsler, K.; Fleckenstein, B.; Schaffner, W. A very strong enhancer is located upstream of an immediate early gene of human cytomegalovirus. *Cell* **1985**, *41*, 521. [CrossRef]

International Journal of
*Molecular Sciences*

MDPI

*Article*

# Detection and Management of Mango Dieback Disease in the United Arab Emirates

Esam Eldin Saeed [1], Arjun Sham [1], Ayah AbuZarqa [1], Khawla A. Al Shurafa [2],
Tahra S. Al Naqbi [2], Rabah Iratni [1], Khaled El-Tarabily [1,*] and Synan F. AbuQamar [1,*]

[1]  Department of Biology, United Arab Emirates University, Al-Ain 15551, UAE;
    esameldin_saeed@uaeu.ac.ae (E.E.S.); arjunsham@uaeu.ac.ae (A.S.); 201250620@uaeu.ac.ae (A.A.);
    R_Iratni@uaeu.ac.ae (R.I.)
[2]  Ministry of Climate Change and Environment, Sharjah 1509, UAE; kaalshurafa@moccae.gov.ae (K.A.A.S.);
    tsalnaqbi@moccae.gov.ae (T.S.A.N.)
*   Correspondence: ktarabily@uaeu.ac.ae (K.E.-T.); sabuqamar@uaeu.ac.ae (S.F.A.); Tel.: +971-3-713-6518
    (K.E.-T.); +971-3-713-6733 (S.F.A.)

Received: 14 September 2017; Accepted: 28 September 2017; Published: 20 October 2017

**Abstract:** Mango is affected by different decline disorders causing significant losses to mango growers. In the United Arab Emirates (UAE), the pathogen was isolated from all tissues sampled from diseased trees affected by *Lasiodiplodia theobromae*. Symptoms at early stages of the disease included general wilting appearance of mango trees, and dieback of twigs. In advanced stages, the disease symptoms were also characterized by the curling and drying of leaves, leading to complete defoliation of the tree and discolouration of vascular regions of the stems and branches. To substantially reduce the devastating impact of dieback disease on mango, the fungus was first identified based on its morphological and cultural characteristics. Target regions of 5.8S rRNA (*ITS*) and elongation factor 1-α (*EF1-α*) genes of the pathogen were amplified and sequenced. We also found that the systemic chemical fungicides, Score®, Cidely® Top, and Penthiopyrad®, significantly inhibited the mycelial growth of *L. theobromae* both in vitro and in the greenhouse. Cidely® Top proved to be a highly effective fungicide against *L. theobromae* dieback disease also under field conditions. Altogether, the morphology of the fruiting structures, molecular identification and pathogenicity tests confirm that the causal agent of the mango dieback disease in the UAE is *L. theobromae*.

**Keywords:** dieback; disease management; *Lasiodiplodia theobromae*; mango; pathogenicity

## 1. Introduction

Mango (*Mangifera indica* L.) is an evergreen fruit tree that is adapted to tropical and subtropical conditions. Mango cultivars vary considerably in fruit size, colour, shape, flavor, texture, and taste [1], and is cultivated in many regions of the world, including India, China, Pakistan, Mexico, Brazil, Egypt, and Nigeria [2]. In addition, mango production has increased in non-traditional mango producing areas including the UAE. According to the FAO (2014), UAE has significantly increased the cultivated area and the number of trees of mango (FAOSTAT; Available online http://faostat.fao.org/site/339/default.aspx), and growers have widely cultivated this crop due to its nutritional and economical values, and their delicacy in flavour and taste. Recently, mango has become an increasingly popular fruit in the UAE markets, after dates and citrus. Mango suffers from diseases worldwide caused by a variety of pathogens that affect all parts of the tree and, therefore, reduce yield and quality of the fruit [3–5].

Mango decline or dieback is a serious disease of mango. The causal agent of this disease remained uncertain for many years due to different fungi associated with it [4]. Fungal pathogens, such as *Neofusicoccum ribis*, *Botryosphaeria dothidea*, *Diplodia* sp., *Pseudofusicoccum* sp., and *Ceratocystis* sp. may infect mango trees individually, or in combinations, to cause mango dieback in different parts of the world [5–10]. Botryosphaeriaceae species, such as *Lasiodiplodia hormozganensis*, *L. iraniensis*, and *L. egyptiacae* have also been associated with mango dieback in Iran, Australia, and Egypt [10–12]. *Lasiodiplodia theobromae* (Syn: *Botryodiplodia theobromae*) [13,14], however, it has been reported as the causal agent in destroying mango orchards within days or a few weeks of infection in India, USA, Pakistan, Brazil, Oman, and Korea [15–20]. *L. theobromae* is a soil-borne wound pathogen that can affect all parts of the mango tree at all ages. Consequently, mango dieback is considered to be an important problem confronting the mango industry and marketing [21]. To date, the mango dieback disease nor its causal organism has been reported from the UAE.

The fungus, *L. theobromae*, often invades twigs and branches from their tips of mango trees causing them to dry and the plant to wilt [22]. Under favourable conditions, infections are characterized by dying back of twigs from the top, downwards, followed by discolouration and the death of leaves, particularly in older trees, which gives an appearance of fire scorch. Symptoms can also be observed on reproductive structures [23]. In severe situations, branches start drying one after another in a sequence resulting in death of the trees of the mango plantation. Commonly, once the symptoms of decline or widespread dieback are evident, it is difficult to stop or reverse the progress of disease. The disease has also been observed on different mango varieties associated with the variation in their susceptibility towards the fungus. Reports have shown that certain varieties are highly susceptible [24,25]. In vivo studies demonstrated that *L. theobromae* becomes aggressive in colonizing host tissues when plants are under abiotic stress, such as heat, water stress, or drought stresses [26,27]. In general, dieback is one of the deadly diseases of mango, which causes a serious damage to the tree and its productivity.

To manage dieback disease, traditional horticultural practices have been applied to confront the fungal attack. In general, avoidance of wounding of trees can limit disease incidence [28]. Infected parts should be pruned from 7–10 cm below the infection site, removed, and burnt [29]. Attempts to arrest early infections have been made by treating with copper oxychloride or pasting with cow dung on pruned ends [30]. Biological control (e.g., *Trichoderma* spp.) have also been tried to reduce disease incidence of *L. theobromae* under in vitro and in field conditions [31,32]. Implementation of integrated disease management (IDM) programs which combine cultural, chemical, and biological approaches are highly recommended to control mango dieback, reduce cost, and improve production efficiency. Despite its negative impact on the environment and human health, the use of chemicals continues to be the major strategy to lessen the menace of crop diseases. In this study, we report fungicide treatments against *L. theobromae* as an effective and reliable approach to reduce the economic losses associated with mango dieback disease. Growers in the UAE and other mango producing countries experiencing this damaging disease are expected to directly benefit from the outcome of this study. Future physiological and molecular analyses will shed more light on dieback disease and its causal agent, which will ultimately lead to the development of effective IDM strategies to manage this disease. Here, we aimed not only to determine the etiology of this disease on mango trees in the UAE, but also to evaluate some of the available fungicides for their effect on the pathogen under in vitro and in vivo conditions.

## 2. Results

### 2.1. Symptoms of Dieback Disease on Mango

Trees manifested with disease symptoms from Kuwaitat, Al Ain—in the eastern region of Abu Dhabi Emirate, UAE—were reported. The pathogen was observed to attack different parts of the mango trees. First, we noticed the disease symptoms in all plant tissues, including leaves, twigs, and apical tips. When the fungus attacks the leaves, their margins roll upwards (Figure S1) turning them a brownish colour (Figure 1A). Later, a scorch-like appearance developed, followed by the dropping of the infected leaves. Moreover, twigs died from the tips back inwards (toward the vascular tissues) (Figure 1A), giving a scorched appearance to the branches (Figure S1). We observed browning in the vascular tissues when longitudinal cross-sections were made in diseased mango twigs

(Figure 1B). We also determined the disease symptoms associated with dieback on whole trees in the field.

**Figure 1.** Naturally-infested mango trees showing symptoms of dieback disease and morphological phenotypes of *Lasiodiplodia theobromae* conidia. Symptoms on (**A**) leaves; (**B**) twigs; (**C**) whole tree; and (**D**) *L. theobromae* hyaline, aseptate immature (red arrow) and brown, 1-septate, thick-walled mature conidia (black arrow) from a 10-day old potato dextrose agar (PDA) culture.

At later stages of invasion, disease symptoms such as wilting, complete drying of leaves and death of the apical region of plants, may also appear (Figure 1C) and at different ages of mango trees (Figure S1). In general, branches dry out one after another in a sequence resulting in the eventual death of the whole tree. These symptoms on mango are typical of the dieback disease.

*2.2. Morphological and Phylogenetic Identification of L. theobromae Associated with Dieback Disease*

The isolate obtained on potato dextrose agar (PDA) and sporulation from naturally-affected tissues associated with dieback disease on mango trees (Figure 1A–C) were microscopically examined. On PDA, colonies of *L. theobromae* (Pat.) Griffon and Maubl. [13,14] had initial white aerial mycelia that turned greenish-gray mycelium with age (Figure S1). The mycelium produced dark brown to black conidia. We also observed mycelial growth and production of immature and mature conidia (Figure 1D). Immature conidia were subovoid or ellipsoid, thick-walled, hyaline and one-celled, turning dark brown, two-celled and with irregular longitudinal striations when at maturity. The size of mature conidia averaged $26.6 \pm 0.51$ µm long and $12.9 \pm 0.28$ µm wide. This suggests that *L. theobromae* is most likely the causal organism of dieback in mango.

We also established a phylogenic analysis of the isolate. PCR amplification of internally-transcribed spacers (*ITS*) of the rDNA gene from mycelium of infected tissues subcultured on PDA was carried (Figure 1). Our results detected the *ITS* gene of all infected tissues (Figure 2A), confirming that *L. theobromae* is frequently associated with all dieback disease symptoms on mango trees in the UAE. To check if the DNA sequences of this species collected in the UAE belongs to any isolated *Lasiodiplodia* isolate, we compared the identified strain with those available in GenBank based on a phylogeny tree. For that purpose, the *ITS* rDNA and the translational elongation factor 1-α (*TEF1-α*) gene [33] were used as a single gene set. The concatenated two-gene set (*ITS* and *TEF1-α*) were sequenced and deposited in GenBank (accession number: MF114110 and MF097964, respectively).

We also determined the relationship among this obtained and other closely related *ITS/TEF1-α* sequences [12,30]. All sequences were aligned and maximum likelihood analyses were performed for estimation of the phylogenetic tree. The adaptation to different plant hosts has led to the evolution of at least 13 cryptic species within the *L. theobromae* species complex [12]. The generated *ITS/TEF1-α* sequence belonging to our strain clustered in one clade corresponding to *L. theobromae* from different sources, confirming its identity with this species (Figure 2B). Among the studied *Lasiodiplodia* species, our analysis revealed that this pathogen is placed adjacent to *L. theobromae* CBS130989, distinguishing the obtained isolate from those belonging to other species of *Lasiodiplodia*, *Diplodia*, or *Phyllosticta*.

Our phylogenetic analysis supports that the species *L. theobromae* (collection number DSM 105134) dominates in the UAE causing dieback disease on mango trees.

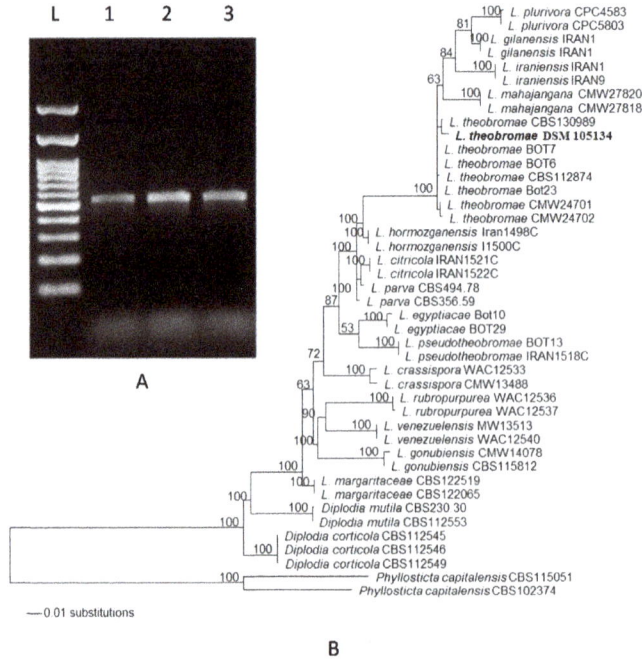

**Figure 2.** Molecular identification of *L. theobromae*. (**A**) PCR amplification of the *ITS* rDNA region in infected leaves, twigs and apical tips (lanes 1–3, respectively); and (**B**) dendrogram showing phylogenetic relationships of the fungal sequence of the specimen used in this study (DSM 105134) with the most related *ITS* and *TEF1-α* sequences in GenBank (accession number, MF114110), prepared by the neighbour-joining method. The maximum likelihood tree is obtained from combined *ITS*/*TEF1-α* sequence data. Numbers at the nodes are ML bootstrap values after 100 replicates are expressed as percentages (LnL = −3497.793130). The scale bar on the rooted tree indicates a 0.01 substitution per nucleotide position. The strain from this report is indicated in bold. *ITS*, internal transcribed spacer; *TEF1-α*, translational elongation factor *1-α*; L, DNA ladder.

### 2.3. Pathogenicity Tests of L. theobromae on Mango Leaves, Fruits, and Seedlings

To confirm our results, detached leaves were spray-inoculated with the isolated pathogen. Following inoculation, a black rot developed on the leaves after five days post-inoculation (dpi) (Figure 3A). No disease symptoms appeared on control leaves sprayed with sterilized distilled water. Similarly, we inoculated mango fruits with the same pathogen. On fruits, dark brown to black lesions averaged 26.4 mm in diameter, beneath the PDA plugs containing the pathogen were observed at 5 dpi (Figure 3B). No disease symptoms were evident under the control plug without the pathogen. The symptoms of the disease were evident on the inoculated leaves (Figure 3A) and fruits (Figure 3B), but not from the control tissues, fulfilling the Koch's postulates relating to the pathogenicity of *L. theobromae* (Figure 3C,D). Our data suggest that *L. theobromae* causes the disease on different tested tissues of mango.

In addition, we performed pathogenicity tests on healthy mango (cv Badami) seedlings, and monitored the disease progress. Plants were inoculated with 8 mm mycelial discs from 10-day old pure *L. theobromae* cultures grown in PDA, while control seedlings were inoculated with PDA without the pathogen. The seedlings were maintained under greenhouse conditions. Following inoculation, seedlings developed typical dieback symptoms showing a dark brown to black, necrotic tissues at the tip of the stem (point of inoculation). At the first week, black colour appeared on the stem at the site of inoculation (Figure S2). The disease progressed rapidly along the stem in the following weeks. At three weeks post-inoculation (wpi), symptoms often expressed as defoliated leaves and characterized by conidiomata development and tissue necrosis in inoculated plants (Figure 4A). At 5 wpi, seedlings showed complete black discolouration and necrosis of internal tissues of stems and branches (Figure 4B,C), forcing the leaves to fall (Figure 4D). Control leaf tissues remained symptomless. The pathogen was consistently re-isolated from the disease affected tissues; thus fulfilling Koch's postulates that these detected symptoms were associated with the inoculation with the pathogen *L. theobromae* (Figure 4E).

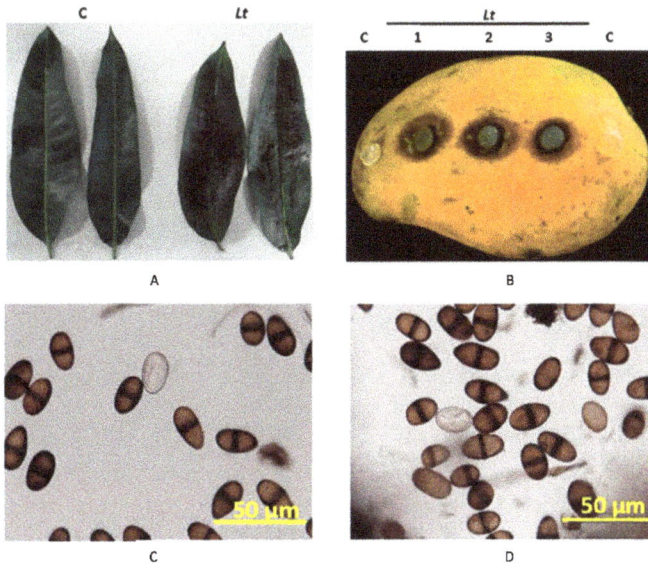

**Figure 3.** Pathogenicity assays of *L. theobromae* on mango leaves and fruits. Pathogenicity tests on inoculated (right) and non-inoculated (left) of (**A**) detached mango leaves; and (**B**) mango fruits, at 5 dpi. Conidia of the pathogen from the inoculated mango (**C**) leaves; and (**D**) fruits. C, control (no *L. theobromae*); Lt, *L. theobromae*.

**Figure 4.** Pathogenicity and Koch's postulate testing with *L. theobromae*. Pathogenicity test on inoculated (right) and non-inoculated (left) of seedlings at (**A**) 3 wpi; and (**B**) 5 wpi. Close-up views of symptomatic (right) and non-symptomatic (left) apical tip tissues at 3 and 5 wpi, respectively; (**C**) Longitudinal section of young stems showing browning of vascular tissues; (**D**) Defoliated leaves of inoculated (right) and control (left) seedlings; and (**E**) the number of defoliated leaves of inoculated (dark column) and control (clear column) seedlings, at 5 wpi. Asterisks are significantly different from the corresponding control at $p < 0.05$. (**F**) Conidia after re-isolation of the pathogen from colonized tissues.

## 2.4. In Vitro Evaluation of Systemic Fungicides Against L. theobromae

To evaluate the effect of fungicides, Score® (difenoconazole), Cidely® Top (difenoconazole and cyflufenamid), and Penthiopyrad® (Carboxamide), on the mycelial growth of *L. theobromae*, six concentrations, ranging between 25 and 1000 ppm of selected fungicides were applied in vitro (Figure S3). With the exception of Cidely® Top, there was significant difference among the concentrations of the other two tested fungicides below 250 ppm, in inhibiting the mycelial growth of the causal agent of dieback disease, *L. theobromae* (Figure S3). On the other hand, Cidely® Top fungicide increased fungal inhibition zone even at low concentration i.e., 25 ppm, and showed no, or slightly, significant deference when compared with other concentrations, ranging between 76–98% mycelial growth inhibition (Figure 5A). We also compared mycelial growth inhibition of *L. theobromae* in vitro at 250 ppm, which was considered as the most efficient concentration in the three fungicides. The results indicated that Score®, Cidely® Top, and Penthiopyrad® inhibited the mycelial growth and sporulation of *L. theobromae* by 77%, 92%, and 50%, respectively (Figure 5A,B). This suggests that the systemic fungicide, Cidely® Top, was the most effective fungicide at 250 ppm concentration among all tested fungicides; and that the fungal inhibition zones were observed, even when a low dosage of this fungicide was applied.

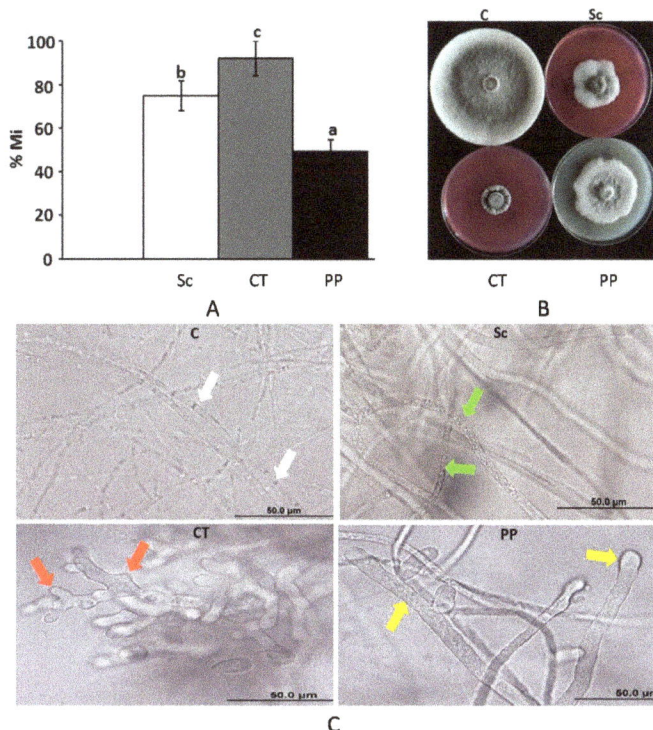

**Figure 5.** Efficacy of fungicides against *L. theobromae*. (**A**) Growth inhibition rate (% Mi) of *L. theobromae* using 250 ppm of the fungicides after 10 days. Values with different letters are significantly different from each other at $p < 0.05$; (**B**) the effect of fungicides (250 ppm) on in vitro mycelial growth; and (**C**) abnormalities in hyphal morphology, septum formation, and cytoplasmic contents of *L. theobromae*, following fungicide treatments, compared to control. White arrows indicate normal septate hyphal growth; green arrows indicate formation of non-septate hyphal formation and cytoplasmic coagulation; red arrows indicate hyphal swellings and branch deformation; yellow arrows indicate hyphal swellings and cytoplasmic coagulation. C, control (no fungicide); Sc, Score®; CT, Cidely® Top; PP, Penthiopyrad®.

In addition, microscopic examination was performed to find out the mode of action of the fungicides in inhibiting the growth of this fungal pathogen. The observations revealed that the fungicides caused diverse morphological alternations on *L. theobromae*. In comparison to the hyphal growth of *L. theorbromae* without any treatment, Score® was capable of causing septal malformations in the hyphal cells (Figure 5C). Striking morphological abnormalities were observed in cell cultures of *L. theobromae* treated with Cidely® Top. The fungicide affected the growth of the pathogen causing significant cytoplasmic coagulation, shrivelled or misshaped mycelia. On the other hand, Penthiopyrad® caused considerable thickening of hyphal tips and incomplete septal formation (Figure 5C). Altogether, the selected systemic fungicides inhibited the mycelial growth of *L. theobromae* by inducing morphological abnormalities of *L. theobromae*. Although many reports in literature have noted pronounced fungal growth inhibition with fungicides under in vitro conditions, many have failed to repeat these performances under greenhouse or field conditions [34].

## 2.5. Effect of Fungicides on Mango Plants Infected with L. theobromae

To confirm our results, we sprayed Score®, Cidely® Top, or Penthiopyrad® fungicides on diseased seedlings artificially inoculated with *L. theobromae* at 2 wpi, and measured the efficacy of the fungicide again after another four weeks (four weeks post treatment; wpt). Before the treatment with the fungicides (corresponding to 0 wpt), plants showed obvious dieback disease symptoms (Figure 6A–C). At 2 wpt with the fungicides, plants started to recover (Figure 6B) and prevented further disease progression at the assessment at 4 wpt (Figure 6A–C), which was in contrast to the plants sprayed with sterilized distilled water (*L. theobromae* control). We also observed the emergence of new leaves from the apical or auxiliary buds of seedlings treated with Score® or Penthiopyrad®, that were comparable to untreated control samples (Figure S4). Cidely® Top-treated disease affected seedlings not only recovered after 4 wpt (Figure 6B), but also showed vigorous vegetative growth (Figure S4). Since all fungicide-treated plants showed very limited disease symptoms with lesser leaf defoliation at 4 wpt, we decided to determine the effects of fungicides on conidia numbers and morphology. Although we did not notice any morphological malformation of the conidia obtained from plants treated with the fungicides, we expected a drop in the number of conidia produced (Figure 6D). Therefore, we counted the number of mature and immature conidia recovered from the tip of the stems of treated-mango plants. In general, there was a significant reduction in the number of mature conidia in all fungicide treatments (Figure 6E). The Cidely® Top caused a greater reduction in the number of mature conidia, followed by Score®- and Penthiopyrad®-treated plants (Figure 6E). Although Score® and Penthiopyrad® had a similar reducing effect on the number of mature conidia, Score® showed at least a three-fold reduction in immature conidia numbers of *L. theorbromae* when compared with Penthiopyrad® fungicide. Application of Cidely® Top resulted in the absence of the immature conidia. Our data suggest that *L. theobromae* appeared to lose some of its aggressiveness as a pathogen when the tested fungicides were applied; while a strong suppression was evident in the severity of the dieback disease in mango plants treated with Cidely® Top.

In the field trials, the promising fungicide Cidely® Top was applied to a mango orchard affected by dieback in order to confirm the results obtained from the in vitro and greenhouse experiments. Mango trees (cv Sindhri), were sprayed with 250 ppm of Cidely® Top fungicide. The disease severity in the fungicide treated mango plants was gradually reduced already at 4 wpt of spraying with the fungicide (Table 1). It was also noted in the trees treated with Cidely® Top that new vegetative growth comprising of fresh shoots increased after 12 wpt (Figure 7). As expected, the fungicide Cidely® Top did not elicit any phytotoxic response on the cultivar under the field conditions. In untreated control plants, disease severity indices (DSI) increased with time in contrast to sprayed plants with the fungicide (Table 1). This suggests that the application with Cidely® Top results in the complete disappearance of symptoms of the disease and the full recovery of the disease-affected trees.

**Table 1.** Disease severity index (DSI) after the application of Cidely® Top on naturally-infested mango trees cv. Sindhri in the field (*n* = 12).

| Treatment | DSI [1] | |
|---|---|---|
| | 4 Wpt | 12 Wpt |
| *Lt* | 3.42 (b) | 4.42 (b) |
| CT | 1.58 (a) | 0.42 (a) |

[1] DSI is on a scale of 5: 0 = no infection, 1 = 1–10%, 2 = 11–25%, 3 = 26–50%, 4 = 51–75%, and 5 = 76–100% damage necrotic, dark brown area or defoliation in leaves. Values with different letters are significantly different from each other at $p < 0.05$. *Lt*, naturally-infested trees with *L. theobromae* only; CT, naturally-infested trees with *L. theobromae* sprayed with Cidely® Top; wpt, weeks post-treatment.

**Figure 6.** Effect of fungicide treatments on artificially inoculated mango seedlings with *L. theobromae* in the greenhouse. Fungicidal suppression of dieback disease on mango seedlings using (**A**) Score; (**B**) Cidely® Top; (**C**) Penthiopyrad® at 0 (top panel), 2 (middle panel), and 4 (bottom panel) wpt; (**D**) conidia of the pathogen reisolated from affected tissues of fungicide-treated plants; and (**E**) the number of conidia/mL at 4 wpt (6 wpi with *L. theobromae*). Seedlings inoculated with *L. theobromae* at two weeks before the fungicide treatment. Values with different letters are significantly different from each other at *p* < 0.05. C, control (non-inoculated seedling); *Lt*, *L. theobromae*-inoculated seedling; Sc, Score®; CT, Cidely® Top; PP, Penthiopyrad®.

**Figure 7.** The effect of Cidely® Top treatments on mango trees (cv Sindhri) naturally infected with *L. theobromae* in the field. Fungicidal suppression of dieback disease symptoms on mango trees (*n* = 12) non-treated (**A**) and treated with the fungicide Cidely® Top (**B**) at 12 wpt. The photo on the left shows the condition of a diseased, affected tree; however, on the right, it shows another tree which was previously affected and has already recovered from severe disease symptoms.

## 3. Discussion

Mango (*Mangifera indica* L.) is known as "the king of fruits" because it is one of the most popular fruit of tropical regions [35]. The UAE has been motivated to widely grow mango in recent years [36]. *L. theobromae* [13,14] is a geographically widespread species of Botryosphaeriaceae [17,37], causing dieback disease in various mango growing areas in the world [9,10]. This fungal pathogen could be found alone or in combination with other fungal pathogens to cause dieback disease. Symptoms associated with this disease are expressed as twig tip dieback that advances into the old wood with branches that dry and die, and leaves scorch and fall, eventually causing death of plants. In the UAE, typical symptoms of dieback disease has been observed (Figure 1) and is yet to develop to an epidemic phase, causing fast-spreading death in mango orchards in a short period of time (i.e., about two months after the initial infection). Therefore, urgent need for appropriate and cost-effective research to properly manage this important disease. In this report, we aimed to determine the causal agent(s) of dieback on mango trees, and to find an effective solution for the potential threat associated with this disease in the UAE.

The pathogen was isolated and identified morphologically and phylogenetically. Microscopy demonstrated that the pathogen is a prolific producer of immature and mature conidia on PDA (Figure 1D). Consistent with Punithalingam [37], immature conidia were initially hyaline, unicellular, ellipsoid to oblong, thick walled with granular contents. We also observed that with age, mature conidia became two-celled, dark brown, with longitudinally striated appearance and an average size of 26.6 μm × 12.9 μm. On maturity, the size of conidia is about 20–30 μm × 10–15 μm [22,37]. In addition, the morphological characteristics of conidia were similar to those previously described [14]. The assessments of spore biology are important to distinguish the fungal survival, dispersal and pathogenicity among closely related species within *Botryosphaeriaceae* spp. [8,33], though we argue about the difficulty in identifying the species of the pathogen based merely on its conidial characteristics. To prove the microbial aetiology of the disease by verifying the existence of the pathogen and its progression in tissues, leaves, fruits and whole plants of mango were inoculated with the isolated pathogen (Figures 3 and 4). The results of inoculation on tissues were similar to the disease symptoms in the field and the re-isolation of the pathogen from the inoculated plants confirmed Koch's postulates. Our data match those in previous pathogenicity tests, which have been done on baobab (*Adansonia* sp.) [38], grapes (*Vitis vinifera*) [39], cocoa (*Theobroma cacao*) [40], yam (*Dioscorea alata*) [41], banana (*Musa* sp.) [42], and mango [12,22]. As in previous artificial inoculation trials on mango seedlings in Peru [43], our study, too, found that symptoms, such as blackening of shoot tips, partial death of crown areas, and defoliation of leaves, developed rapidly and were clearly evident after five weeks of inoculation. Thus, it would be virtually impossible to distinguish between *Lasiodiplodia* species based on their morphology only.

Specific genomic regions of *L. theobromae* that correspond to the two widely used loci *ITS* and *TEF1-α* were amplified and sequenced. Phylogenetic analysis of DNA sequences combining, *ITS* and *TEF1-α* [12,33,38], was also performed to discriminate between *Lasiodiplodia* species, and to identify the causal agent of the dieback disease on mango in the UAE. The adaptation to different plant hosts and environments has led to the evolution of at least 13 cryptic species within the *L. theobromae* species complex [12]. The identified *L. theobromae* DSM 105134 from the UAE fits into one clade with several *L. theobromae* strains from different sources. The most closely related *ITS/TEF1-α* was *L. theobromae* CBS 130989 (=BOT4), an isolate from mango in Egypt [12], which demonstrated an identity of 100%. Our data also showed that the *ITS/TEF1-α* identified in this study clustered together with *L. theobromae* isolates BOT 6, BOT 7, and BOT 23 from mango in Egypt [12]. The isolate CBS 112874 of *L. theobromae* was reported to infect grapes in South Africa [44]. Similarly, the *ITS/TEF1-α* which belongs to *L. theobromae* collected from the UAE showed 99% identity with that of both CMW 24701 and CMW 24702 strains isolated from *Eucalyptus* sp. in China [45]. None of the *ITS/TEF1-α* sequences that belong to *L. theobromae* including the pathogen from this study, clustered with other *Lasiodiplodia* spp. reported worldwide. This provides strong evidence that the isolate DSM 105134 in the current study belongs to *L. theobromae* sp. complex and is the main causal agent of dieback on mango in the UAE. Yet, it is probable this destructive strain of the fungal pathogen may have been introduced from Egypt.

This study was further extended to evaluate systemic fungicides to potentially control the pathogen under greenhouse and field conditions. There is now strong evidence that the indiscriminate use of chemicals does pose a potential risk to humans and other organisms, and unwelcomed side effects to the environment [46]. Yet, many studies urge to combine a number of antagonistic strategies, such as fungicides, biocontrol agents (BCA), and plant extracts, to prevent or reduce the activity of the pathogen growth and manage diseases in crops which, as a concept, is known as IDM [47,48]. IDM does not necessarily seek to eliminate the use of chemicals, but aims to minimize in a way that becomes least destructive to non-target life [49]. Several reports have focused on BCA against *L. theobromae*. In vitro studies showed that the antagonism of *Trichoderma* spp. (*T. harzianum* and *T. viride*) or *Aspergillus niger*, can be effective against *L. theobromae* [50,51]. Under laboratory and field conditions to protect bottle gourd (*Lagenaria siceraria*) against seedling and root rot diseases, plants treated with *Bacillus subtilis*, *T. harzianum* or *T. viride* were reported to reduce the pathogenic effect of *L. theobromae* by more than 90% [52]. Some researchers, on the other hand, have indicated that the efficacy of the BCA is dependent on many factors including the host age, the disease severity and the field environment. *Trichoderma* and other biological products, however, could serve as potential BCA against diseases associated with *L. theobromae*; indicating the potential for the integrated management of this disease. Until now, limited research has targeted the potential of the applicability of fungicides for the effective management of the dieback disease in mango.

With aim of searching for the successful fungicides to inhibit *L. theobromae*, we selected three systemic fungicides, Score®, Cidely® Top, and Penthiopyrad®, and tested their efficacy under in vitro, greenhouse, and field conditions. All fungicides used in this study, in general, inhibited the fungus at the tested concentration (250 ppm), evidenced by the altered hyphal morphology, septum formation and the integrity of the cytoplasmic contents. Among all the fungicides tested, Cidely® Top (difenoconazole and cyflufenamid) showed the strongest inhibition of mycelial growth with minor tolerance by the organism after 10 days of the in vitro experiment, and a significant reduction in disease symptoms in relation to the conidia counts in Cidely® Top-treated seedlings at 2–4 wpt. This suggests that this fungicide may serve as a candidate fungicide for the management of *L. theobromae* affected mango trees. To a lesser extent, the difenoconazole-based fungicide, Score®, too was significantly effective in the reduction of the pathogenic activities of *L. theobromae* in both the laboratory and greenhouse trials. This result is in agreement with previous findings that this chemical does inhibit the growth of *L. theobromae* in vitro and in vivo [53,54], although higher concentrations of Score® were used in their studies than was applied in our study. Although difenoconazole (Score® and Cidely® Top) was ineffective against *Fusarium magniferae* [55], this active ingredient was significantly effective and promising for managing other plant diseases [56–58], including dieback on mango (Figure 6). This could be attributed to the different growth conditions, fungicide application methods and/or the nature of responses to the chemicals by different fungal pathogens. It is noteworthy to mention that we found that the superior efficiency of the fungicide Cidely® Top over Score®, may possibly be due to the additional presence of the active ingredient cyflufenamid, which may have contributed to the increased levels of inhibition of *L. theobromae*. Penthiopyrad showed high inhibitory activity against a wide range of plant pathogens, including *Rhizoctonia solani*, *Botrytis cinerea*, *Fusarium oxysporum*, and *Leptosphaeria* spp. [58,59]. Our data showed that the application of the carboxamide-based fungicide, Penthiopyrad®, was significantly effective in the reduction the pathogen hyphal growth and the production of mature conidia, in addition to causing hyphal swellings and cytoplasmic coagulation of *L. theobromae* compared to the PDA and seedling controls. The result obtained for carboxamide (Penthiopyrad®) seems to be in disagreement with a previous finding reporting that this chemical showed strong inhibitory activity of spore germination in various plant pathogens compared with that of mycelial growth of *B. cinerea* [58]. This discordance could be due to the preventive control effect of Penthiopyrad® and the dosage of the fungicide treatment. To date, no reports exist relating to the evaluation of Cidely® Top on mango trees infected with *L. theobromae*; while the same fungicide, however, was found to be highly effective against the pathogenic fungus *Thielaviopsis punctulata* on date palm [48]. Therefore, a field experiment was conducted to test the efficacy of Cidely® Top, in infested mango orchards. Mango trees showed almost complete recovery, evident in the reduction of DSI by 54–91% in mango-sprayed trees with Cidely® Top after 4 and 12 wpt compared to the untreated control. In conclusion, Cidely® Top was useful in managing this destructive disease of mango in field, and could potentially be used as an effective component of IDM of dieback disease on mango.

"Omics" are useful approaches to identify molecular changes that occur during disease or even prior to it, when prospective data are available [60]. Such data assume that the differences between healthy and diseased groups are directly related to disease [61,62]. This report focusing on the phenotype, i.e., symptoms associated with dieback disease, could be considered as a starting point for future comparative "omic" analyses including genomes and responses to environmental variation. Ultimately, we aim to reach towards full protection, which could ideally be achieved by the employment of IDM programs as well as "omic" approaches. In this research, we identified *L. theobromae*, for the first time, as the causal agent of dieback disease on mango in the UAE. We were also successful in finding a chemical means (viz. Cidely® Top) to inhibit *L. theobromae* growth on mango trees. Investigation searching for other practices including IDM treatments to manage dieback in mango is in progress, ideally to promote the crop productivity and sustainability.

## 4. Materials and Methods

### 4.1. Fungal Isolation and Purification

Diseased trees in the Kuwaitat area in Al Ain City (Eastern region of Abu Dhabi Emirate; latitude/longitude: 24.21/55.74) with drying leaves on branches and twigs (Figure 1A–C) were studied in this investigation. A symptomatic tree (approximately five years old) was lifted and transferred to the Plant Microbiology Laboratory, Department of Biology, United Arab Emirates University in Al Ain City, for investigation. Longitudinal cross-sections were made of the diseased tree twigs and the pathogen was isolated from affected tissues (Figure S1). Tissues were cut into small pieces (2–5 mm long), washed, and surface-sterilized with mercuric chloride 0.1% for 1 min, followed by three consecutive washings in sterile distilled water. They were then transferred onto PDA (Lab M Limited, Lancashire, UK) plates, pH 6.0; supplemented with penicillin-streptomycin (Sigma-Aldrich Chemie GmbH, Taufkirchen, Germany) used at a rate of 25 mg/L of the growth medium in order to inhibit the bacterial contaminants. Petri dishes were incubated in an incubator at 28 ± 2 °C for five days. After this period of incubation, the mycelia growing out of the plated tissue was aseptically sub-cultured on fresh PDA and lastly purified by using hyphal-tip isolation technique [63]. The mycelium and conidia were observed using Nikon-Eclipse 50i light microscope (Nikon Instruments Inc., NY, USA) to characterize different fungal structures. A culture of the identified fungus, *L. theobromae* (Pat.) Griffon and Maubl. [13,14], has been deposited in Leibniz-Institute DSMZ-Deutsche Sammlung von Mikroorganismen und Zellkulturen GmbH (Braunschweig, Germany) under the collection number DSM 105134.

### 4.2. DNA Isolation, PCR, and Sequencing

The pathogen's DNA from infected tissues of leaf, twig and apical shoot tips was extracted from mycelium cultured for 10–14 days at 28 °C on PDA plates, using the plant/fungi DNA isolation kit (Norgen Biotek Corp., Thorold, ON, Canada) with some modifications. Target regions of internal transcribed spacer (*ITS*) of the nuclear rDNA for *L. theobromae* using ITS1 and ITS4 primers [33], and partial *TEF1-α* using EF1-728F and EF1-986R [64] were amplified using the PCR. All primer sequence sets can be found in Table S1. All protocols for amplification and sequencing were as described [33].

## 4.3. Phylogenetic Analysis

For the analysis of the phylogenetic placement of the fungal isolate the sequences of *ITS* rDNA and *TEF1-α* genes were used as a single gene set and a concatenated two-gene set, ITS/TEF1-α. The obtained *ITS* and *TEF1-α* sequences were deposited in GenBank (accession numbers MF114110 and MF097964, respectively) and were further combined for constructing the phylogenetic tree against the *Lasiodiplodia* species database managed by the National Centre for Biotechnology Information (NCBI; www.ncbi.nlm.nih.gov). The *ITS/TEF1-a* sequence of the isolate from the UAE was aligned with sequences retrieved from GenBank, representing isolates that belong to about 18 species of the genus *Lasiodiplodia* [12,33]. All sequences were compared and aligned and maximum likelihood analyses were performed for estimation of the phylogenetic tree [65]. Phylogenetic trees were constructed and validated with a statistical support of the branches with 100 bootstrap resamples. These belong to isolates are: *L. plurivora, L. gilanensis, L. iraniensis, L. mahajangana, L. theobromae, L. hormozganensis, L. citricola, L. parva, L. egyptiacae, L. pseudotheobromae, L. crassispora, L. rubropurpurea, L. venezuelensis, L. gonubiensis, L. margaritaceae, Diplodia mutila, D. corticola,* and *Phyllosticta capitalensis.*

## 4.4. Disease Assays and Pathogenicity Tests

Inoculated detached mango (cv Badami) leaves ($n = 12$) were surface-sterilized with 70% ethanol before spray-inoculation with $5 \times 10^4$ spores/mL of 10-day old culture of *L. theobromae* spore suspension. Control leaves were sprayed with sterilized distilled water without pathogen using a Preval sprayer (Valve Corp., Yonkers, NY, USA). Inoculated leaves were kept in a growth chamber at 28 °C and 80% relative humidity (RH). Inoculated detached leaves were examined for disease symptoms after five days.

Detached mango fruits ($n = 12$) were also tested to determine the effect of *L. theobromae*. Healthy mango fruits (cv Badami) were purchased from local fresh markets in the UAE. Fruits were stored at 4 °C and used within two days of purchase. Fruits were washed with sterile distilled water to remove dust and then the fruits were surface-sterilized with 70% ethanol. After air-drying in a flow cabinet, the mangoes were wounded using a sterilized scalpel (2 mm diameter, five wounds per mango), as described previously [66]. On each fruit, three agar plugs (11 mm in diameter) containing mycelium of *L. theobromae* (placed colonized surface down) and two agar control plugs containing no pathogen were applied. Inoculated fruits were further kept in a humid growth chamber at 28 °C and 80% RH, and were examined for disease symptoms after five days.

Disease was also assayed on whole mango seedlings (cv Badami). Twelve-month-old mango seedlings were inoculated with agar plugs (8 mm in diameter) containing mycelium of *L. theobromae* at the growing tip region of the stem, where the area of inoculation was wrapped with parafilm, as previously described [43]. Before inoculation, we surface-sterilized apical tips with 70% ethanol, and introduced mechanical wounding with sterilized scalpels. Control seedlings were treated/inoculated with PDA discs without pathogen. All inoculated seedlings were further maintained in a greenhouse with a photoperiod extended to 15 h under fluorescent lights (160 W/mol·m$^2$·s) at 28 °C, and were examined for disease symptoms at 1, 3, and 5 wpi.

To satisfy Koch's postulates, pieces of inoculated leaf and fruit tissues were removed from sites showing disease symptoms at 5 dpi, surface sterilized as mentioned above and plated on PDA. Similarly, pieces of infected stems showing disease symptoms at 5 wpi were surface sterilized as mentioned above, plated and incubated at 28 °C and the subsequent growth was recorded.

## 4.5. Evaluation of Fungicides Against L. theobromae

The fungicide experiment was carried out as previously described [48,57]. These fungicides selected were Score 250 EC® (Difenoconazole; Syngenta International AG, Basel, Switzerland), Cidely® Top 125/15 DC (Difenoconazole and Cyflufenamid; Syngenta), and Penthiopyrad 20SC® (Carboxamide; Mitsui Chemicals Agro Inc, Tokyo, Japan). Each fungicide was dissolved in water with final concentrations of 0, 25, 75, 125, 250, 500, and 1000 ppm, and was then introduced aseptically into sterilized PDA at room temperature (RT). Penicillin-streptomycin antibiotics were added to inhibit the bacterial growth. The homogenized mixtures were aseptically poured into sterile Petri dishes. To introduce the tested pathogen on the control (without fungicide) and treatment (with fungicide) medium, a sterile cork-borer (8 mm diameter) was used. Cultures were incubated at 28 °C for 10 days, and the percentage of the mycelial growth inhibition was measured according to:

$$\% \ Mi = (Mc - Mt)/Mc \times 100\% \tag{1}$$

where $Mi$ = the inhibition of the mycelial growth; $Mc$ = the colony diameter (in mm) of the control set; and $Mt$ = the colony diameter (in mm) of the target fungus on the medium with fungicide.

An in vivo evaluation of the fungicides was also carried out on one-year-old mango seedlings (cv Badami) under greenhouse conditions, as described above. Seedlings were previously inoculated with agar culture discs containing mycelium of *L. theobromae* at the apical tip as described above. Inoculated seedlings were further kept in the greenhouse at 28 °C for two weeks (until disease symptoms were evident). Plants were then either sprayed with the fungicide (250 ppm; treatment) or with sterilized distilled water (control). Fungal conidia counts and the number of falling leaves were recorded at 2 and 4 wpt, as previously described [48]. Basically, the method of conidia counts involves homogenizing of known weight of affected tissues in 5 mL of water and assessing the suspended material to estimate the number of conidia using haemocytometer (Agar Scientific Limited, Essex, UK). It should be noted that leaf drop symptoms and fungal conidia counts were used for monitoring disease progression in the greenhouse experiment.

The field trials were performed in an orchard located in Abu Al-Abyad Island (Northern region of Abu Dhabi city, UAE; Latitude/Longitude: 24.20/53.80). Cidely® Top was the only systemic fungicide tested on twelve mango trees cv Sindhri (four years old). Each tree was chosen so as to be surrounded by untreated trees (*L. theobromae* naturally-infested control) that can serve as a reservoir for recontamination. Treated mango trees were completely sprayed with the fungicide at the recommended dose used previously (250 ppm). The DSI corresponding to disease symptoms or recovery were recorded for disease assessment for fully grown trees at 4 and 12 wpt, using a scale of 0–5: 0 = no apparent symptoms, 1 = 1–10% necrotic, dark brown area on leaves or defoliating leaves, 2 = 11–25%, 3 = 26–50%, 4 = 51–75%, and 5 = 76–100% [67]. Experiments were repeated twice in March 2015 and March 2016 with similar results.

### 4.6. Statistical Analysis

For pathogenicity tests on leaves, fruits and seedlings, 12 tissues or seedlings for each treatment were used. For the in vitro evaluation of fungicides against *L. theobromae*, six plates for each treatment were used. For the fungal conidia counts and falling leaves of in vivo evaluation of fungicides under greenhouse conditions, a minimum of six plants for each treatment was used. Data represent the mean ± SD. Analysis of variance (ANOVA) and Duncan's multiple range test were performed to determine the statistical significance at $p < 0.05$. All experiments were independently repeated three times with similar results.

For the DSI of the fungicide treatment in the field trials against *L. theobromae*, two replicates were tested. Data (mean ± SD) from a minimum of 12 plants per replicate were performed. Statistical significance at $p < 0.05$ was determined by ANOVA and Duncan's multiple range test. Similar results were obtained in each replicate.

All statistical analyses were performed by using SAS Software version 9 (SAS Institute Inc., Cary, NC, USA).

**Supplementary Materials:** Supplementary materials can be found at www.mdpi.com/1422-0067/18/10/2086/s1.

**Acknowledgments:** We are grateful to Krishnapillai Sivasithamparam, University of Western Australia, for his critical reading of the manuscript. We also extend our thanks to Marwan Jaddou for his technical support.

This project was funded by the Khalifa Centre for Biotechnology and Genetic Engineering-UAEU (grant no. 31R081); and the UAEU Program for Advanced Research (grant no. 31S255) to SAQ.

**Author Contributions:** Khaled El-Tarabily and Synan F. AbuQamar conceived and designed the experiments; Esam Eldin Saeed, Arjun Sham, Ayah AbuZarqa, Khawla A. Al Shurafa, and Tahra S. Al Naqbi performed the experiments; Khaled El-Tarabily and Synan F. AbuQamar analysed the data; Rabah Iratni, Khaled El-Tarabily, and Synan F. AbuQamar contributed reagents/materials/analysis tools; and Khaled El-Tarabily and Synan F. AbuQamar wrote the paper. All authors have read and approved the manuscript.

**Conflicts of Interest:** The authors declare no conflict of interest.

## References

1. Berardini, N.; Fezer, R.; Conrad, J.; Beifuss, U.; Carle, R.; Schieber, A. Screening of mango (*Mangifera indica* L.) cultivars for their contents of flavonol O- and xanthone C-glycosides, anthocyanins, and pectin. *J. Agric. Food Chem.* **2005**, *53*, 1563–1570. [CrossRef] [PubMed]

2. Nelson, S. Mango anthracnose (*Colletotrichum gloeosporioides*). *Plant Dis.* **2008**, *48*, 1–9.

3. Prakash, O. *Compendium of Mango Diseases and Disorders*; Vedams eBooks (P) Ltd.: New Delhi, India, 2003; p. 84.

4. Ploetz, R.C. Diseases of mango. In *Diseases of Tropical Fruit Crops*; Ploetz, R.C., Ed.; APS Press: St. Paul, MN, USA, 2003; pp. 327–363.

5. Ploetz, R.C. The major diseases of mango: Strategies and potential for sustainable management. *Acta Hortic.* **2004**, *645*, 137–150. [CrossRef]

6. Smith, P.F.; Scudder, G.K. Some studies of mineral deficiency symptoms in mango. *Proc. Florida State Hort. Soc.* **1951**, *64*, 243–248.

7. Ramos, L.J.; Lara, S.P.; McMillan, R.T.; Narayanan, K.R. Tip die back of mango (*Mangifera indica*) caused by *Botryosphaeria ribis.. Plant Dis.* **1991**, *75*, 315–318. [CrossRef]

8. Slippers, B.; Johnson, G.I.; Crous, P.W.; Coutinho, T.A.; Wingfield, B.; Wingfield, M.J. Phylogenetic and morphological re-evolution of the *Botryosphaeria* species causing diseases of *Mangifera indica*. *Mycologia* **2005**, *97*, 99–110. [CrossRef] [PubMed]

9. Van Wyk, M.; Al-Adawi, A.O.; Wingfield, B.D.; Al-Subhi, A.M.; Deadman, M.L.; Wingfield, M.J. DNA based characterization of *Ceratocystis fimbriata* isolates associated with mango decline in Oman. *Australas. Plant Pathol.* **2005**, *34*, 587–590. [CrossRef]

10. Sakalidis, M.L.; Ray, J.D.; Lanoiselet, V.; Hardy, G.E.S.T.; Burgess, T.I. Pathogenic Botryosphaeriacea associated with *Mangifera indica* in the Kimberley region of Western Australia. *Eur. J. Plant Pathol.* **2011**, *130*, 379–391. [CrossRef]

11. Abdollahzadeh, J.; Javadi, A.; Mohammadi, G.E.; Zare, R.; Phillips, A.J.L. Phylogeny and morphology of four new species of *Lasiodiplodia* from Iran. *Persoonia* **2010**, *25*, 1–10. [CrossRef] [PubMed]

12. Ismail, A.M.; Cirvilleri, G.; Polizzi, G.; Crous, P.W.; Groenewald, J.Z.; Lombard, L. *Lasiodiplodia* species associated with dieback disease of mango (*Mangifera indica*) in Egypt. *Australas. Plant Pathol.* **2012**, *41*, 649–660. [CrossRef]

13. Zambettakis, E.C. Recherches sur la systematique des "Sphaeropsidales-Phaeodidymae". *Bull. Trimest. Soc. Mycol. Fr.* **1954**, *70*, 219–349.

14. Sutton, B.C. *The Coelomycetes, Fungi Imperfecti with Pycnidia, Acervuli and Stromata*; Commonwealth Mycological Institute: Surrey, UK, 1980.

15. Sharma, I.M.; Raj, H.; Kaul, J.L. Studies on postharvest diseases of mango and chemical control of stem end rot and anthracnose. *Indian Phytopathol.* **1994**, *47*, 197–200.

16. Ploetz, R.C.; Benscher, D.; Vázquez, A.; Colls, A.; Nagel, J.; Schaffer, B. A re-examination of mango decline in Florida. *Plant Dis.* **1996**, *80*, 664–668. [CrossRef]

17. Al Adawi, A.O.; Deadman, M.L.; Al Rawahi, A.K.; Khan, A.J.; Al Maqbali, Y.M. *Diplodia theobromae* associated with sudden decline of mango in the Sultanate of Oman. *Plant Pathol.* **2003**, *52*, 419. [CrossRef]

18. Khanzada, M.A.; Lodhi, A.M.; Shahzad, S. Chemical control of *Lasiodiplodia theobromae*, the causal agent of mango decline in Sindh. *Pak. J. Bot.* **2005**, *37*, 1023–1030.

19. De Oliveira Costa, V.S.; Michereff, S.J.; Martins, R.B.; Gava, C.A.T.; Mizubuti, E.S.G.; Câmara, M.P.S. Species of Botryosphaeriaceae associated on mango in Brazil. *Eur. J. Plant Pathol.* **2010**, *127*, 509–519. [CrossRef]

20. Hong, S.K.; Lee, S.Y.; Choi, H.W.; Lee, Y.K.; Joa, J.H.; Shim, H. Occurrence of stem-end rot on mango fruits caused by *Lasiodiplodia theobromae* in Korea. *Plant Pathol. J.* **2012**, *28*, 455. [CrossRef]
21. Haggag, W.M. Mango diseases in Egypt. *Agric. Biol. J. N. Am.* **2010**, *1*, 285–289. [CrossRef]
22. Khanzada, M.A.; Lodhi, A.M.; Shahzad, S. Mango dieback and gummosis in Sindh, Pakistan caused by *Lasiodiplodia theobromae*. *Plant Health Prog.* **2004**. [CrossRef]
23. Naqvi, S.A.H.; Perveen, R.; Malik, M.T.; Malik, O.; Umer, U.D.; Wazeer, M.S.; Rehman, A.; Majid, T.; Abbas, Z. Characterization of symptoms severity on various mango cultivars to quick decline of mango in district Multan. *Int. J. Biosci.* **2014**, *4*, 157–163.
24. Khanzada, M.A.; Lodhi, A.M.; Rajput, A.Q.; Syed, R.N.; Shahzad, S. Response of different mango cultivars to mango decline pathogen, *Lasiodiplodia theobromae* Pat. *Int. J. Biol. Biotechnol.* **2015**, *12*, 643–647.
25. Naqvi, S.A.H.; Perveen, R. Mango quick decline manifestation on various cultivars at plants of particular age in the vicinity of district Multan. *Pak. J. Phytopathol.* **2015**, *27*, 31–39.
26. Kazmi, M.; Fateh, F.; Majeed, K.; Kashkhely, A.M.; Hussain, I.; Ahmad, I.; Jabeen, A. Incidence and etiology of mango sudden death phenomenon in Pakistan. *Pak. J. Phytopathol.* **2005**, *17*, 154–158.
27. Paolinelli-Alfonso, M.; Villalobos-Escobedo, J.M.; Rolshausen, P.; Herrera-Estrella, A.; Galindo-Sánchez, C.; López-Hernández, J.F.; Hernandez-Martinez, R. Global transcriptional analysis suggests *Lasiodiplodia theobromae* pathogenicity factors involved in modulation of grapevine defensive response. *BMC Genom.* **2016**, *17*, 615. [CrossRef] [PubMed]
28. Alemu, K. Dynamics and management of major postharvest fungal diseases of mango fruits. *J. Biol. Agric. Healthc.* **2014**, *4*, 13–21.
29. Asrey, R.; Patel, V.B.; Barman, K.; Pal, R.K. Pruning affects fruit yield and postharvest quality in mango (*Mangifera indica* L.) cv. Amrapali. *Fruits* **2013**, *68*, 367–380. [CrossRef]
30. Garg, N.; Pathak, O.; Pathak, R.K. Use of Cow Dung Paste for Controlling Gummosis and Die Back Diseases of Mango. In Proceedings of the 43rd Annual Conference of Association of Microbiologists of India, Hisar (Haryana), India, 11–13 December 2002; p. 171.
31. Gupta, V.P.; Tewar, S.K.; Govidaiah; Bajpai, A.K. Ultrastructure of mycoparasitisms of *Trichoderma*, *Gliocladium* and *Laetisaria* species on *Botryodiplodia theobromae*. *J. Phytopathol.* **1999**, *147*, 19–24. [CrossRef]
32. Bhuvaneswari, V.; Rao, M.S. Evaluation of *Trichoderma viride* antagonistic to post harvest pathogens on mango. *Indian Phytopathol.* **2001**, *54*, 493–494.
33. Alves, A.; Crous, P.W.; Correia, A.; Phillips, A.J.L. Morphological and molecular data reveal cryptic speciation in *Lasiodiplodia theobromae*. *Fungal Divers.* **2008**, *28*, 1–13.
34. O'Callaghan, M. Microbial inoculation of seed for improved crop performance: Issues and opportunities. *Appl. Microbiol. Biotechnol.* **2016**, *100*, 5729–5746. [CrossRef] [PubMed]
35. Usman, M.; Fatima, B.; Muhammad, M.J. Breeding in Mango. *Int. J. Agric. Biol.* **2001**, *3*, 522–526.
36. Lauricella, M.; Emanuele, S.; Calvaruso, G.; Giuliano, M.; D'Anneo, A. Multifaceted health benefits of *Mangifera indica* L. (Mango): The inestimable value of orchards recently planted in sicilian rural. *Nutrients* **2017**, *9*, 525. [CrossRef] [PubMed]
37. Punithalingam, E. *Plant Diseases Attributed to Botryodiplodia theobromae Pat*; J. Cramer: Vaduz, Liechtenstein, 1980; p. 121.
38. Cruywagen, E.M.; Slippers, B.; Roux, J.; Wingfield, M.J. Phylogenetic species recognition and hybridization in *Lasiodiplodia*: A case study on species from baobabs. *Fungal Biol.* **2017**, *121*, 420–436. [CrossRef] [PubMed]
39. Rodríguez-Gálvez, E.; Alves, A. Identification and pathogenicity of *Lasiodiplodia theobromae* causing dieback of table grapes in Peru. *Eur. J. Plant Pathol.* **2015**, *141*, 477–489. [CrossRef]
40. Mbenoun, M.; Momo Zeutsa, E.H.; Samuels, G.; Nsouga Amougou, F.; Nyasse, S. Dieback due to *Lasiodiplodia theobromae*, a new constraint to cocoa production in Cameroon. *New Dis. Rep.* **2007**, *15*, 59. [CrossRef]
41. Amusa, N.A.; Adegbite, A.A.; Muhammed, S.; Baiyewu, R.A. Yam disease and its management in Nigeria. *Afr. J. Biotechnol.* **2003**, *2*, 497–502. [CrossRef]
42. Twumasi, P.; Ohene-Mensah, G.; Moses, E. The rot fungus *Botryodiplodia theobromae* strains cross infect cocoa, mango, banana and yam with significant tissue damage and economic losses. *Afr. J. Agric. Res.* **2014**, *9*, 613–619.
43. Rodríguez-Gálvez, E.; Guerrero, P.; Barradas, C.; Crous, P.W.; Alves, A. Phylogeny and pathogenicity of *Lasiodiplodia* species associated with dieback of mango in Peru. *Fungal Biol.* **2017**, *121*, 452–465. [CrossRef] [PubMed]

44. Pavlic, D.; Slippers, B.; Coutinho, T.A.; Gryzenhout, M.; Wingfield, M.J. *Lasiodiplodia gonubiensis* sp. nov., a new *Botryosphaeria* anamorph from native *Syzygium cordatum* in South Africa. *Stud. Mycol.* **2004**, *50*, 313–322.

45. Chen, S.F.; Pavlic, D.; Roux, J.; Slippers, B.; Xie, Y.J.; Wingfield, M.J.; Zhou, X.D. Characterization of Botryosphaeriaceae from plantation-grown *Eucalyptus* species in South China. *Plant Pathol.* **2011**, *60*, 739–751. [CrossRef]

46. Aktar, M.W.; Sengupta, D.; Chowdhury, A. Impact of pesticides use in agriculture: Their benefits and hazards. *Interdiscip. Toxicol.* **2009**, *2*, 1–12. [CrossRef] [PubMed]

47. AbuQamar, S.F.; Moustafa, K.; Tran, L.S. Mechanisms and strategies of plant defense against *Botrytis cinerea*. *Crit. Rev. Biotechnol.* **2017**, *37*, 263–275. [CrossRef] [PubMed]

48. Saeed, E.E.; Sham, A.; Salmin, Z.; Abdelmowla, Y.; Iratni, R.; El-Tarabily, K.A.; AbuQamar, S.F. *Streptomyces globosus* UAE1, a potential effective biocontrol agent for black scorch disease in date palm plantations. *Front. Microbiol.* **2017**, *8*, 1455. [CrossRef] [PubMed]

49. Razdan, V.; Sabitha, M. Integrated disease management: Concepts and practices. In *Integrated Pest Management: Innovation-Development Process*; Peshin, R., Dhawan, A.K., Eds.; Springer: Dordrecht, The Netherlands, 2009.

50. Golam Mortuza, M.; Ilag, L.L. Potential for biocontrol of *Lasiodiplodia theobromae (Pat.)* Griff. & Maubl. in banana fruits by *Trichoderma* species. *Biol. Control* **1999**, *15*, 235–240.

51. Adeniyi, D.O.; Adedeji, A.R.; Oduwaye, O.F.; Kolawole, O.O. Evaluation of Biocontrol agents against *Lasiodiplodia theobromae* causing inflorescence blight of cashew in Nigeria. *IOSR J. Agric. Vet. Sci.* **2013**, *5*, 46–48.

52. Sultana, N.; Ghaffar, A. Effect of fungicides and microbial antagonists in the control of *Lasiodiplodia theobromae*, the cause of seed rot, seedling and root infection of bottle gourd. *Pak. J. Agric. Res.* **2010**, *23*, 46–52.

53. Syed, R.N.; Mansha, N.; Khaskheli, M.A.; Khanzada, M.A.; Lodhi, A.M. Chemical control of stem end rot of mango caused by *Lasiodiplodia theobromae*. *Pak. J. Phytopathol.* **2014**, *26*, 201–206.

54. Rehman, A.U.; Naqvi, S.; Latif, M.; Khan, S.; Malik, M.; Freed, S. Emerging resistance against different fungicides in *Lasiodiplodia theobromae*, the cause of mango dieback in Pakistan. *Arch. Biol. Sci.* **2015**, *67*, 241–249. [CrossRef]

55. Iqbal, Z.; Pervez, M.A.; Ahmad, S.; Iftikhar, Y.; Yasin, M.; Nawaz, A.; Ghazanfar, M.U.; Dasti, A.A.; Saleem, A. Determination of minimum inhibitory concentrations of fungicides against fungus *Fusarium mangiferae*. *Pak. J. Bot.* **2010**, *42*, 3525–3532.

56. Khan, S.H.; Idrees, M.; Muhammad, F.; Mahmood, A.; Zaidi, S.H. Incidence of shisham (*Dalbergia sissoo* Roxb.) decline and in vitro response of isolated fungus spp. to various fungicides. *Int. J. Agric. Biol.* **2004**, *6*, 611–614.

57. Saeed, E.E.; Sham, A.; El-Tarabily, K.A.; Abu Elsamen, F.; Iratni, R.; AbuQamar, S.F. Chemical control of dieback disease on date palm caused by the fungal pathogen, *Thielaviopsis punctulata*, in United Arab Emirates. *Plant Dis.* **2016**, *100*, 2370–2376. [CrossRef]

58. Yanase, Y.; Katsuta, H.; Tomiya, K.; Enomoto, M.; Sakamoto, O. Development of a novel fungicide, penthiopyrad. *J. Pestic. Sci.* **2013**, *38*, 167–168. [CrossRef]

59. Sewell, T.R.; Moloney, S.; Ashworth, M.; Ritchie, F.; Mashanova, A.; Huang, Y.J.; Stotz, H.U.; Fitt, B.D.L. Effects of a penthiopyrad and picoxystrobin fungicide mixture on phoma stem canker (*Leptosphaeria* spp.) on UK winter oilseed rape. *Eur. J. Plant Pathol.* **2016**, *145*, 675–685. [CrossRef]

60. AbuQamar, S.F.; Moustafa, K.; Tran, L.S. 'Omics' and plant responses to *Botrytis cinerea*. *Front. Plant Sci.* **2016**, *7*, 1658. [CrossRef] [PubMed]

61. Sham, A.; Moustafa, K.; Al-Shamisi, S.; Alyan, S.; Iratni, R.; AbuQamar, S. Microarray analysis of Arabidopsis *WRKY33* mutants in response to the necrotrophic fungus *Botrytis cinerea*. *PLoS ONE* **2017**, *12*, e0172343. [CrossRef] [PubMed]

62. Sham, A.; Al-Azzawi, A.; Al-Ameri, S.; Al-Mahmoud, B.; Awwad, F.; Al-Rawashdeh, A.; Iratni, R.; AbuQamar, S.F. Transcriptome analysis reveals genes commonly induced by *Botrytis cinerea* infection, cold, drought and oxidative stresses in *Arabidopsis*. *PLoS ONE* **2014**, *9*, e113718. [CrossRef] [PubMed]

63. Kirsop, B.E.; Doyle, A. *Maintenance of Microorganisms and Cultured Cells, a Manual of Laboratory Methods*, 2nd ed.; Academic Press: London, UK, 1991.

64. Carbone, I.; Kohn, L.M. A method for designing primer sets for speciation studies in filamentous ascomycetes. *Mycologia* **1999**, *91*, 553–555. [CrossRef]

65. Tamura, K.; Stecher, G.; Peterson, D.; Filipski, A.; Kumar, S. MEGA6: Molecular evolutionary genetics analysis version 6.0. *Mol. Biol. Evol.* **2013**, *30*, 2725–2729. [CrossRef] [PubMed]

66. Rungjindamai, N. Isolation and evaluation of biocontrol agents in controlling anthracnose disease of mango in Thailand. *J. Plant Prot. Res.* **2016**, *56*, 306–311. [CrossRef]

67. Amponsah, N.T.; Jones, E.; Ridgway, H.J.; Jaspers, M.V. Evaluation of fungicides for the management of *Botryosphaeria* dieback diseases of grapevines. *Pest. Manag. Sci.* **2012**, *68*, 676–683. [CrossRef] [PubMed]

MDPI

St. Alban-Anlage 66

4052 Basel

Switzerland

Tel. +41 61 683 77 34

Fax +41 61 302 89 18

www.mdpi.com

*International Journal of Molecular Sciences* Editorial Office

E-mail: ijms@mdpi.com

www.mdpi.com/journal/ijms